T0358304

A SUPERGRAVITY PRIMER

From Geometrical Principles to
the Final Lagrangian

A SUPERGRAVITY PRIMER

From Geometrical Principles to the Final Lagrangian

Michel Rausch de Traubenberg
Strasbourg University, CNRS Institut Pluridisciplinaire Hubert Curien, France

Mauricio Valenzuela
Centro de Estudios Científicos (CECs), Chile

NEW JERSEY • LONDON • SINGAPORE • BEIJING • SHANGHAI • HONG KONG • TAIPEI • CHENNAI • TOKYO

Published by

World Scientific Publishing Co. Pte. Ltd.
5 Toh Tuck Link, Singapore 596224
USA office: 27 Warren Street, Suite 401-402, Hackensack, NJ 07601
UK office: 57 Shelton Street, Covent Garden, London WC2H 9HE

British Library Cataloguing-in-Publication Data
A catalogue record for this book is available from the British Library.

A SUPERGRAVITY PRIMER
From Geometrical Principles to the Final Lagrangian

Copyright © 2020 by World Scientific Publishing Co. Pte. Ltd.

All rights reserved. This book, or parts thereof, may not be reproduced in any form or by any means, electronic or mechanical, including photocopying, recording or any information storage and retrieval system now known or to be invented, without written permission from the publisher.

For photocopying of material in this volume, please pay a copying fee through the Copyright Clearance Center, Inc., 222 Rosewood Drive, Danvers, MA 01923, USA. In this case permission to photocopy is not required from the publisher.

ISBN 978-981-121-051-8

For any available supplementary material, please visit
https://www.worldscientific.com/worldscibooks/10.1142/11557#t=suppl

Printed in Singapore

Preface

Since their discovery in the 70's supersymmetry and supergravity rapidly turn out to be recognised as major achievements in theoretical physics. Supersymmetry and supergravity have a deep group theory flavor since they can be seen to be the more general symmetries compatible with the principles of Quantum Field Theory. Indeed, the two types of particles, bosons of integer spin and fermions of half-integer spin, by virtue of the Nœther and of the Spin Statistics Theorems, can have symmetry generators with half-integer spin which exchange their statistics. Consequently, supersymmetry and supergravity can be seen as an extension of the spacetime symmetries introducting new types of groups, namely the Lie supergroups. It did not take long before supersymmetry and supergravity were applied in Particles Physics and in the 80's supersymmetric extensions of the Standard Model were obtained.

Constructed under these new symmetry building principles, supergravity is regarded now as the most general theory of particles physics. Indeed, it is based on symmetry principles describing the fundamental interactions (gauge symmetries) of particles, the spacetime symmetry and supersymmetry (fermions-bosons exchange symmetry). While in supersymmetry the parameters of the transformations are constant, supergravity is a local version where now the parameters depend on the spacetime coordinates. This simply means that supergravity is automatically a theory of gravity. Stated differently this entails that gravity is naturally incorporated to supergravity together with the fundamental (gauge) interactions. Even though supersymmetry and supergravity have not been discovered yet (up to 2019) in Colliders, since they are based on strong arguments and powerful techniques, supersymmetry and supergravity will certainly remain in theoretical physics.

There are many ways to construct invariant Lagrangians in supergravity. In this book we have decided to focus on the superspace approach. In this approach the usual spacetime is extended to a superspace, where a fermionic counterpart described by Grassmann variables is introduced together with the usual spacetime coordinates.

Supergravity has the reputation of being complicated, of requiring long and convoluted computations to reproduce known and new results. Unfortunately, there are not many books on supergravity available and most of them don't contain the details of the calculations, which makes it all the more difficult to learn supergravity. This book is devoted to a pedagogical introduction to supergravity from a practical perspective. As a particular feature, explicit details are provided, which make the computations easier to follow for the interested reader. Each chapter has summary tables, which contain the main results and, in addition, we have collected important or additional material in the appendix. This may also help students to understand the existing literature. This book can be seen as a natural development of the book (in French) of one of us *Supersymétrie : exercices avec solutions* and is written in the same spirit. This book is also based on several lectures given at the Doctoral School of Strasbourg.

As stated previously there are not so many books of supergravity. Since we have chosen to construct supergravity within the superspace approach, we are closely following one of the major references, the book of Wess and Bagger. By doing so the $N = 1$ supergravity Lagrangian in four spacetime dimensions is re-constructed in a very detailed manner. All the steps are carefully expounded, from the geometric principles of curved superspace to the field redefinition necessary to obtain a correctly normalised Lagrangian. The second part of this book is devoted to more phenomenological aspects of supergravity. Thus, supergravity breaking, no-scale supergravity, super-Brout-Englert-Higgs mechanism, *etc.*, are also investigated with a special attention to detail. Finally, the relationship between supergravity and particles physics or cosmology is discussed. In particular, gravity mediated supersymmetry breaking is investigated in detail to show how the so-called soft-supersymmetric breaking terms are generated.

Even though this book was written mainly with Ph.D students in mind, it is also addressed at physicists or mathematicians interested in supergravity and supersymmetry and who would like to have a better understanding of these concepts. Moreover, the notions are introduced in such a way that the reader can use them as a springboard to study other notions which are barely or even not at all covered in this book. Finally, since this book is a book on supergravity and not a book on supersymmetry, the main notions of the latter are introduced cursorily in the first chapter. To gain maximum benefit from this book, the reader should already have some notions of supersymmetry; otherwise, we recommend that the reader explore first the available literature on the subject. This book is particularly well-suited for those who already know supersymmetry (theoreticians, phenomenologists, *etc.*) and who want to study (parts of) supergravity. For a reader who wants to study some specific part of supergravity, it is fully possible to read the corresponding chapter(s), since all important notions introduced in previous chapters are collected in the summary tables from which all information can be easily accessed.

As mentioned previously, this book is primarily geared towards young graduate students and researchers, but not exclusively, and can serve as a first contact with the subject. We further hope that it will be useful for all who want to gain a better understanding of supergravity and that it can be used as a text for complementary reading.

Strasbourg and Valdivia
July, 2019

Michel Rausch de Traubenberg
Mauricio Valenzuella

well [illegible] by those who [illegible] ages [illegible] molecular [illegible] and who want to study [illegible] be [illegible] have [illegible] corresponding [illegible] special [illegible] ions are collected in the summary tables [illegible] all [illegible] [illegible]ed.

[illegible] towards young [illegible] [illegible] and [illegible] [illegible]

[illegible]

Acknowledgments

We are indebted to Richard Grimm, Ian Jack and Stefan Theisen for their input and their help in different stages in the preparation of this manuscript. We would like to express our gratitude to Gilbert Moultaka, Noureddine Mohammedi, Christian Schubert, and Marcus Julius Slupinski, long time collaborators, for helpful criticisms and encouragements. Many thanks to Philippe Brax, Émilian Dudas, Christophe Grojean, Jerzy Lukierski, Jean Orloff and Vincent Vennin for their support and wise advice. We would also like to thank Benjamin Fuks for his participation in a preliminary stage of this project. We are grateful to Robin Ducrocq who reproduced part of the computations and found some typos. Alex Boeglin is kindly acknowledged for his careful reading of the manuscript and Catherine Berger is cordially thanked for the design of the cover. We are also beholden to Steven Duplij who helped us to concretise our project. M. V. thanks Institut Pluridisciplinaire Hubert CURIEN (IPHC) for support during extended visits.

This project was a long and arduous task, it took us several years to finalise it, and it would certainly never have been written, had we not benefited from the support and encouragements of many colleagues, who are too numerous to mention. The credit for the motivation of this book goes to students that constitute the guiding spirit for this book.

Finally we would like to thank warmly our families, Almeira, Aurélien, Teresita and Véronique for their continuous support and encouragements.

Acknowledgements

Contents

List of Tables

Chapter 1

Introduction and outlook of the book

Because of the homogeneity and the isotropy of spacetime, elementary particles are classified according to irreducible representations of the Poincaré algebra, and are therefore characterised by the eigenvalue of their corresponding Casimir operators, *i.e.*, mass and spin (or helicity in the massless case) [Wigner (1939); Bargmann and Wigner (1948); Weinberg (1995)]. Internal interactions are associated to a compact Lie groups describing gauge interactions [Yang and Mills (1954)]. These two concepts have allowed the construction of many particle physics models. In particular, the so-called *Standard Model* [Glashow (1961); Salam and Ward (1964); Weinberg (1967); Glashow et al. (1970); Weinberg (1972b); Gross and Wilczek (1973, 1974); Politzer (1974)] seems to be able to describe almost all the currently available experimental high-energy physics data [Tanabashi et al. (2018)]. Nevertheless, it has been realised that it is possible to circumvent a strong no-go theorem [Coleman and Mandula (1967); Haag et al. (1975)], and to extend the Poincaré symmetry in a non-trivial way [Golfand and Likhtman (1971)]. Introducing anticommutators, *i.e.*, symmetric counterparts of commutators, for a *fermionic* class of generators, the fermionic symmetries can be incorporated in an extended version of Lie algebras, referred as Lie *superalgebras*, which feature an associated *supersymmetry*. This superalgebra has the $\mathbb{Z}_2$–graded structure,

$$[B,B] \subset B\,, \qquad [B,F] \subset F\,, \qquad \{F,F\} \subset B\,,$$

where B denotes the *bosonic* subset of generators and F denotes the *fermionic* subset of generators. The bosonic sector $B = I\mathfrak{so}(1,3) \oplus \mathfrak{g}$ is formed by the direct sum of the set of Poincaré generators $I\mathfrak{so}(1,3)$, and internal symmetry set of generators $\mathfrak{g} = \mathfrak{g}_1 \oplus \mathfrak{g}_2$, with $\mathfrak{g}_1$ a compact Lie algebra and $\mathfrak{g}_2$ an abelian compact Lie algebra. Hence, the above (anti)commutators can be fleshed out to read

$$[I\mathfrak{so}(1,3), I\mathfrak{so}(1,3)] \subset I\mathfrak{so}(1,3)\,, \quad [\mathfrak{g},\mathfrak{g}] \subset \mathfrak{g}, \quad [I\mathfrak{so}(1,3),\mathfrak{g}] = 0\,,$$
$$[I\mathfrak{so}(1,3), F] \subset F\,, \quad [\mathfrak{g}, F] \subset F, \quad \{F,F\} \subset \mathbb{R}^{1,3} \oplus \mathfrak{g}_2\,,$$

where $\mathbb{R}^{1,3}$ are the spacetime translations and $\mathfrak{g}_2$ is the abelian part of $\mathfrak{g}$ that commutes with all generators and consequently its elements are called central charges [Wess and Bagger (1992)].

This discovery has subsequently led to the introduction of (global) supersymmetry in particle physics, where in contrast to non-supersymmetric theories, the (constant) parameters of a supersymmetric transformation are anticommuting Grassmann variables, so that supersymmetric multiplets contain both fermionic and bosonic states. The implementation of supersymmetry in quantum field theories was done first in a component formalism, *i.e.*, where the component fields of each multiplet are explicitly separated. Later, it was noticed that the supermultiplet components can be treated as a single *superfield*, *i.e.*, where bosons and fermions can be re-arranged as Taylor expansion coefficients in the so called *superspace*, *i.e.*, a space with fermionic directions [Volkov and Akulov (1973); Wess and Zumino (1974a,c,b); Salam and Strathdee (1974b, 1975); Ferrara et al. (1974); Ferrara and Zumino (1974)].

Local versions of supersymmetric theories can also be built. Since the anticommutator of two local supersymmetric transformations is a spacetime translation, local supersymmetric theories contain gravity. Therefore, they are termed *supergravity*. They include the description of the dynamics of a spin-two massless field, the *graviton* which is associated to general reparametrisations invariance, as well as a massless spin-$3/2$ field, the *gravitino* which is associated with local supersymmetric transformations. However, the construction of theories involving a matter sector, coupled *via* gauge and supergravity interactions, is not obvious and often necessitates lengthy and painstaking computations. We present in this book an extensive description of such computations in the context of four-dimensional supergravity with one single fermionic supercharge.

The first theory of (pure) supergravity was proposed in the nineteen seventies in the independent pioneering works of Freedman et al. (1976); Deser and Zumino (1976), starting from globally invariant Lagrangians and using the Nœther procedure to render them local. This procedure leads to an invariant action involving both the Einstein-Hilbert action for the graviton and the Rarita and Schwinger (1941) action for the gravitino. However, it can be shown that the supergravity algebra only closes on-shell [Freedman and van Nieuwenhuizen (1976)], *i.e.*, when the equations of motion of the different fields are satisfied. Therefore, auxiliary fields must be added in order to preserve local supersymmetry off-shell, *i.e.*, without invoking the equations of motion. These considerations have led to the design of the minimal *supergravity* multiplet containing, in addition to the graviton and the gravitino fields, a set of auxiliary fields constituted of a complex scalar and a real vector field [Stelle and West (1978c); Ferrara and van Nieuwenhuizen (1978a)]. Non-minimal options [Breitenlohner (1977); Siegel and Gates (1979); Akulov et al. (1977); Gates et al. (1982); Galperin et al. (1985); Sohnius and West (1981a); Girardi et al. (1984)] and anti-de-Sitter supergravity [Freedman and Das (1977); Townsend (1977); Deser and Zumino (1977)] have also been considered, although they go beyond the scope of this work. Here we shall focus on the minimal option.

Supergravity-gauge interactions have been later considered in the context of both Maxwell and, more generally, Yang-Mills theories [Ferrara et al. (1976b,a); Freedman and Schwarz (1977)]. Concerning the matter sector, the supergravity couplings of a single massless chiral superfield were first derived in [Ferrara et al. (1977a)], followed by partial results in the framework of non-linear σ-model [Cremmer and Scherk (1978)], and in turn by the most general results [Cremmer et al. (1978b, 1979, 1982, 1983b); Bagger and Witten (1982); Bagger (1983)] where the supergravity couplings of arbitrary interacting matter fields have been derived. In particular, the approach that has been followed in [Cremmer et al. (1978b, 1979, 1982, 1983b)] employs the component formalism, or appropriate tensor calculus techniques for the derivation of the results [Stelle and West (1978c); Ferrara and van Nieuwenhuizen (1978a); Stelle and West (1978b,a); Ferrara and van Nieuwenhuizen (1978b)]. It has however been found that a superspace adapted to supergravity can be defined and used for the calculations [Wess and Zumino (1977, 1978b,a); Grimm et al. (1978); Akulov et al. (1975); Zumino (1978)]. The cornerstone of this last approach is to define a tangent space at each superspace point that behaves like the traditional flat superspace of global supersymmetry. The corresponding structure group is hence the Lorentz group, although other choices are possible [Arnowitt et al. (1975); Nath and Arnowitt (1975, 1976); Brink et al. (1978); Ne'eman and Regge (1978)]. The key advantage of the (supergravity) superspace formalism lies in the fact that the superspace is invariant under supergravity transformations by construction, and naturally contains all the necessary auxiliary fields.

Both in the component and in the superspace approaches, the obtainment of the Lagrangian is complicated and the calculations are lengthy. In the component formalism, the action is obtained by means of the appropriate tensor calculus techniques, as stated above. In contrast, the superspace formalism relies on the expansion of the Lagrangian superfields in the various fields once a series of geometrical objects has been defined such as, *e.g.* the superspace covariant derivative. In both approaches the derived Lagrangians require further manipulations as the kinetic terms are not properly normalised. The canonical normalisations are obtained by performing a Weyl rescaling of all the fields, together with a shift of the gravitino [Wess and Bagger (1992)]. Due to the related technical difficulties, alternative techniques to construct supergravity Lagrangians have been developed. While a first series of methods is based on a geometric formulation of supergravity [Binetruy et al. (2001)], a second series of so-called conformal methods has been widely employed. They rely on an enlargement of the Poincaré structure group of supergravity to the superconformal group with the introduction of compensating fields [Ferrara et al. (1977b); Kaku et al. (1978); Kaku and Townsend (1978); Townsend and van Nieuwenhuizen (1979); Ferrara et al. (1978); Kugo and Uehara (1983b,a, 1985); Fradkin and Tseytlin (1985); Gates et al. (1983); Buchbinder and Kuzenko (1998); Freedman and Van Proeyen (2012)], as extensively discussed in the component formalism in the book of Freedman and Van Proeyen (2012).

Since its origin, supergravity has rapidly became a central research topic and many different but related aspects have been intensively investigated. In particular, supergravity has been considered in the context of particle physics [Fayet (1984); Nilles (1984); Haber and Kane (1985); Weinberg (2000); Binetruy (2006)] as it conceptually allows one, for the reasons aforementioned, to naturally include gravity together with the other fundamental interactions. Moreover, supergravity exhibits a series of interesting features shared with all other supersymmetric models. First, it leads a better quantum behaviour and provides a solution to the infamous hierarchy problem associated with the presence of massive scalar fields (such as the recently discovered Higgs boson) [Witten (1981a)]. The ultraviolet divergences related to the quantum corrections to the scalar masses are indeed reduced to being logarithmic, in contrast to the non-supersymmetric case where they are quadratic [Dimopoulos and Raby (1981); Dine et al. (1981); Witten (1981a); Dimopoulos and Georgi (1981); Sakai (1981); Kaul (1982); Kaul and Majumdar (1982)]. Second, a series of non-renormalisation theorems holds. Aside from the particle content, a supersymmetric theory is uniquely defined by three fundamental functions that govern the interactions between the particles, namely the *Kähler potential*, the *superpotential* and the *gauge kinetic function*. The supersymmetric non-renormalisation theorems state that the Kähler potential has to be renormalised order by order in perturbation theory, the gauge kinetic function only at the one-loop level and the superpotential does not need to be renormalised at all [Grisaru et al. (1979); Seiberg (1993); Weinberg (2000)]. Finally, at the two-loop level, the supersymmetric renormalisation group equations [Derendinger and Savoy (1984); Falck (1986); Martin and Vaughn (1994); Yamada (1993, 1994b,a); Jack et al. (1994)], predict that the three gauge coupling constants of the Standard Model meet at high energies [Ibanez and Ross (1981); Dimopoulos et al. (1981);

Ellis et al. (1991); Amaldi et al. (1991); Langacker and Luo (1991); Giunti et al. (1991); Carena et al. (1993); Martin (1997)].

However, both in supersymmetry and supergravity[1], each of the degrees of freedom of the theory is associated with a superpartner of opposite statistics and of the same mass. This observation becomes critical when building supersymmetric models which includes the fields of the Standard Model. Since not a single supersymmetric partner state with the mass of a Standard Model particle has been observed, supersymmetry and supergravity cannot be exact symmetries of nature and have to be spontaneously broken. The most simple mechanisms leading to supersymmetry breaking [O'Raifeartaigh (1975); Fayet and Iliopoulos (1974)] turn out not to be phenomenologically viable as they predict states lighter than some of the Standard Model particles, which contradicts experimental data. The situation is improved in supergravity and realistic and simple breaking mechanisms can be designed [Chamseddine et al. (1982); Barbieri et al. (1982b); Ibañez (1982); Ohta (1983); Ellis et al. (1983); Alvarez-Gaumé et al. (1983); Polonyi (1977); Nilles (1984); Dimopoulos and Raby (1981); Dine et al. (1981); Derendinger and Savoy (1982); Fayet (1978); Dine and Fischler (1982); Nappi and Ovrut (1982); Alvarez-Gaumé et al. (1982); Dine and Nelson (1993); Dine et al. (1995, 1996); Giudice and Rattazzi (1999); Arkani-Hamed et al. (1998); Randall and Sundrum (1999); Bagger et al. (2000)].

Supersymmetry and supergravity also play important roles in cosmology [Binetruy (2006); Bailin and Love (2004)]. For instance, in models with unbroken R-parity [Farrar and Fayet (1978)], the lightest supersymmetric particle is stable. If it is, in addition, colourless and electrically neutral, it can then be seen as a natural candidate for dark matter in the Universe [Goldberg (1983); Ellis et al. (1984a)]. As another example of interesting features from the cosmological viewpoint, supergravity predicts, in contrast to supersymmetry, the existence of a spin-$3/2$ particle, the gravitino, that becomes massive *via* the so-called super-Higgs mechanism after supergravity breaking [Deser and Zumino (1977); Cremmer et al. (1979)]. As a consequence, gravitino production in the early Universe implies strong constraints on the building of viable cosmological scenarios. Moreover, the gravitino can possibly be the lightest supersymmetric particle and thus a viable dark matter candidate [Binetruy (2006)]. Finally, supergravity allows one to construct appealing models with a vanishing (or small) cosmological constant [Cremmer et al. (1983c); Ellis et al. (1984b); Lahanas and Nanopoulos (1987); Binetruy (2006)].

Irrespective of the relations with particle physics and cosmology, supergravity has interesting formal aspects. First, supergravity can be extended to include N

[1] From now on, we denote by *supersymmetry* global supersymmetry and by *supergravity* local supersymmetry.

spinor supercharges. In four dimensions, in order to avoid the appearance of any state with a spin higher than two, the number of supercharges is only restricted to be smaller than or equal to eight, $N \leq 8$ [Nahm (1978)]. In particular, $N = 4$ super-Yang-Mills theories have received special attention in the last decades as the associated β-function vanishes to all orders in perturbation theory [Avdeev et al. (1980); Caswell and Zanon (1981); Grisaru et al. (1980); Jones (1977); Poggio and Pendleton (1977); Mandelstam (1983); Howe et al. (1984); Brink et al. (1983); Sohnius and West (1981b); White (1992)], a feature yielding exceptional finiteness properties. Second, one can also extend the number of spacetime dimensions and consider supergravity in higher-dimensions, as shown, *e.g.*, in [Strathdee (1987)]. The number of dimensions D can indeed be different from four as it is only bounded to be smaller than or equal to eleven, $D \leq 11$. This again avoids the appearance of particles of spin greater than two, whereas the $D = 12$ case is viable when considering a $(2, 10)$ signature for the metric. Among all higher-dimensional theories, eleven-dimensional supergravity is uniquely defined and plays an important role in the framework of M-theory [Cremmer et al. (1978a)]. Ten-dimensional supergravity can be classified using the correspondence with the string theory whose low energy limit are: type I [Brink et al. (1977); Gliozzi et al. (1977); Chamseddine (1981); Bergshoeff et al. (1982, 1983); Chapline and Manton (1983)], type IIA [Schwarz and West (1983); Howe and West (1984); Schwarz (1983)] or type IIB [Campbell and West (1984); Huq and Namazie (1985); Giani and Pernici (1984)]. Furthermore, the compactification of the extra dimensions allows one to relate higher-dimensional supergravity to four-dimensional supergravity with $N > 1$ supercharges. In this way, four-dimensional $N = 8$ supergravity can be linked to either the eleven-dimensional or the ten-dimensional (of type IIA or IIB) cases, while four-dimensional $N = 4$ supergravity can be associated with $N = 1$ ten-dimensional supergravity of type I [Cremmer and Julia (1979)].

On another level, it has also been shown that supergravity establishes a link between two apparently different classes of theories. The so-called Anti-de-Sitter/Conformal Field Theory (AdS/CFT) correspondence is based on a duality between a specific class of ten-dimensional $N = 1$ supergravity theories and four-dimensional $N = 4$ super-Yang-Mills theories [Maldacena (1998); Gubser et al. (1998); Witten (1998)]. This duality has been further extended to link gravity in $D + 1$ dimensions to quantum field theories (without gravity) in D dimensions. This last property has practical applications for computations in the context of classical supergravity and has led to a new vision of the structure of the spacetime structure in the context of supergravity or string theory [D'Hoker and Freedman (2002); Maldacena (2003); Terning (2006); Becker et al. (2007); Freedman and Van Proeyen (2012)].

Moving away from the historical perspective, we now get back to the main goal of this book, four-dimensional $N = 1$ supergravity. Although there are many books [Wess and Bagger (1992); West (1986); Buchbinder and Kuzenko (1998); Freedman and Van Proeyen (2012); Weinberg (2000); Gates et al. (1983); Binetruy (2006); Terning (2006); Becker et al. (2007); Bailin and Love (2004); Freund (1986); Muller (1989); Bailin and Love (1994); Dine (2007); Nath (2016)] and reviews [van Nieuwenhuizen (1981); Fradkin and Tseytlin (1985); Lahanas and Nanopoulos (1987); Nilles (1984); Dragon et al. (1987); Binetruy et al. (2001); Duff et al. (1986); Cvetic and Soleng (1997)] on the topic, there are few pedagogical introductions to supergravity. Moreover, most of the existing literature is either highly technical and avoids detailed calculations, or is devoted to a very specific aspect of supergravity. The purpose of this book is to fill this gap. We mainly focus on the conceptual and technical construction of invariant Lagrangians in supergravity. All computations are presented explicitly and the results are accompanied by extensive technical details. In the second part of this book, we address more phenomenological aspects of supergravity in the context of particle physics and cosmology. In this way, this book is the logical continuation of the book of Fuks and Rausch de Traubenberg (2011) where, through a series of solved exercises, one of us has constructed the most general supersymmetric Lagrangian coupling matter fields *via* gauge interactions and studied a set of specific phenomenological implications of supersymmetry in the framework of the Minimal Supersymmetric Standard Model and several of its extensions.

One of the major references on supergravity is the book of Wess and Bagger (1992), which we have based our work on their approach for the building of a supergravity Lagrangian. The comparison of our results to theirs is immediate after accounting for the different choice of sign conventions for the metric in spacetime, for the metric of spinors and for the torsion tensors. As mentioned, all the steps leading to the final Lagrangian are very technical and involve lengthy computations. For a pedagogical purpose, this book consequently contains many practical details allowing anyone to derive the results easily. Moreover, the reading of the document has been facilitated by the introduction of summary tables at the end of each chapter. In this way, the reader not interested in a specific aspect can use the associated summary table for guidance to skip the related chapter. Finally, we have collected important material in the appendices of the book.

In Chapter 2, we summarise a selection of results in (global) supersymmetry that are necessary prior to any foray into supergravity. We introduce chiral and vector superfields and next define the basic functions allowing one to build any supersymmetric Lagrangian where interacting matter fields (described by chiral superfields) are coupled *via* gauge interactions (mediated by vector superfields), namely the Kähler potential, the superpotential and the gauge kinetic function. We omit most of the details and refer to existing references instead [Salam and Strathdee (1974b, 1975); Ferrara et al. (1974); Salam and Strathdee (1978); Wess and Zumino (1974b); Ferrara and Zumino (1974); Derendinger (1989); Fuks and Rausch de Traubenberg (2011)], as the main topic of the present book is supergravity and not supersymmetry.

Chapter 3 is devoted to the basic geometrical objects that are needed to investigate supergravity within the superspace formalism and that are thus central to our construction of a supergravity Lagrangian. In Sec. 3.1 we first define the curved superspace itself, for which the structure group is chosen to be the Lorentz group. In this way, any point of the curved superspace can be associated to a tangent flat superspace where we have a local Lorentz invariance. Consequently, two dynamical superfields are required for the construction of covariant derivatives, the supervierbein (also called the frame superfield) and the superconnection, from which we can derive curvature and torsion tensor superfields (in Sec. 3.1.1). Nonetheless, all these superfields have too many component fields (*e.g.*, the supervierbein has $8\times8\times16 = 1024$ components) so that it becomes impossible to extend the superfield formalism of supersymmetry to supergravity while fulfilling at the same time the principle of equivalence of general relativity. This may however be achieved after imposing constraints on the torsion superfield *via* the Bianchi identities. The "reduction" of these identities is performed in Sec. 3.1.2 and we give a proof that the elements of the curvature and torsion tensor superfields can be all expressed in terms of only three superfields: a chiral scalar superfield $\mathcal{R}$, a chiral symmetric spinor superfield $W_{(\alpha\beta\gamma)}$ and a real vector superfield G_μ. Furthermore, we also show that in contrast to the case of general relativity, all the components of the curvature tensor can be expressed in terms of those of the torsion tensor in the framework of supergravity. This naturally brings us to study the symmetry properties of the constraints imposed on the torsion tensor superfield, and therefore we investigate, in Sec. 3.2, the Weyl supergroup and the related super-Weyl transformations, the supergravity analog of the Weyl group of general relativity.

The components of the supergravity multiplet, namely the graviton $e_{\tilde{\mu}}{}^{\mu}$, the gravitino $\psi_{\tilde{\mu}}$ and the two above-mentioned auxiliary fields M and b_μ, are defined in Sec. 3.3 by choosing an appropriate gauge. This leads to the introduction of the so-called supergravity transformations, which are specific combinations of local Lorentz transformations and local superspace translations that preserve the gauge choice. In Sec. 3.4, we perform the explicit computation of the components of the three superfields $\mathcal{R}$, $W_{(\alpha\beta\gamma)}$ and G_μ, as well as that of the elements of the superconnection $\omega_{\tilde{\mu}\tilde{\nu}\tilde{\rho}}$. To this end, we rely on both the constraints that we have imposed upon the torsion tensor and several relations following from our gauge choice.

Finally, the transformation laws of the supergravity multiplet under a supergravity transformation are presented in Sec. 3.5.

On the basis of the constraints that have been imposed on the components of the torsion tensor superfield, the chiral and vector superfields of supersymmetry can be extended to the curved superspace. As shown in Chapter 4, this relies on the fact that these constraints can be seen as specific integrability conditions in the supergravity geometrical framework. The components of a chiral superfield and its derivatives are calculated in Sec. 4.1. It turns out that it is possible to define a *projector* associating an antichiral superfield to any chiral superfield. In Sec. 4.2, we turn to vector superfields in both the abelian and non-abelian cases. In the context of an appropriate Wess-Zumino gauge, we derive the components of these superfields and their derivatives and, as in supersymmetry, link them to their spinorial (anti-)chiral superfield strength tensors. While the gauge interactions of chiral and vector superfields are investigated in Sec. 4.3, we show in Sec. 4.4 that supergravity transformations combined with (field-depending) gauge transformations allow one to preserve the Wess-Zumino gauge. The transformation properties of chiral and vector superfields are finally derived.

Chapter 5 aims at setting the general principles necessary for the building of any supergravity action. In Sec. 5.1 we begin with showing that the superdeterminant of the supervierbein leads to an invariant measure for the full eight-dimensional curved superspace. However, this formulation is not suitable for practical computations. The situation is made more manageable in Sec. 5.2 with the introduction of hybrid Grassmann variables and the associated invariant measure. This procedure relies on two key observations. First, the projectors introduced in Sec. 4.1 allows one to rewrite actions in a chiral form, *i.e.*, involving only integration upon the left-handed (or right-handed) part of the Grassmann variables. Second, the hybrid variables are defined in such a way that the computation of the components of the various superfields is facilitated. In Sec. 5.3 we finally study the behaviour of the action under a super-Weyl transformation and we establish a link with Kähler transformations.

The core of our book is presented in Chapter 6, where we explicitly compute the component field expansion of the most general supergravity action. In Sec. 6.1 we start with the simple case of pure supergravity, while in Sec. 6.2 we focus on the calculation of the most general action with supergravity and gauge interactions of matter fields. The first technical part of this last calculation consists of the field expansion of the Kähler, superpotential and gauge sector Lagrangians and the elimination of the auxiliary fields. It turns out that the action obtained in this way is not canonically normalised. Field redefinitions *via* a Weyl rescaling and a shift of the gravitino are in order to achieve a correctly normalised Lagrangian. The analysis of all the resulting terms is performed in Sec. 6.3, while Sec. 6.6 addresses the way to introduce Fayet-Iliopoulos terms. In Sec. 6.5 we derive the limit of the obtained Lagrangian for renormalisable (global) supersymmetric theories, and in Sec. 6.8 supergravity in a Jordan frame is presented. As stated above, geometrical and superconformal methods have been developed to simplify the calculation of actions in supergravity. A comparison of the two is briefly detailed in Sec. 6.7.

Supergravity breaking is discussed in Chapter 7. We start by investigating the property of the goldstino field in Sec. 7.1 and emphasise the main differences between supersymmetry and supergravity. In particular, we show that in broken supergravity, the goldstino can be eliminated by conferring a non-vanishing mass to the gravitino *via* the so-called *super-Higgs mechanism*. Then in Sec. 7.2 we analyse gravity-induced supersymmetry breaking, a popular mechanism for the breaking of supersymmetry, and, in Sec. 7.3, we briefly study no-scale supergravity models predicting a vanishing cosmological constant. We close this chapter by presenting a generic tree-level mass matrices and trace formulæ in Sec. 7.4.

The relationship between supergravity and particles physics or between supergravity and cosmology is investigated in Chapter 8. In Sec. 8.1 we focus on the way the so-called soft-supersymmetric breaking terms are generated. We also briefly mention how phenomenological analysis could be performed and how specific supersymmetric models could be tested in colliders. In Sec. 8.2.4 we introduce a class of supergravity models called superconformal supergravity models and study their relationship with an inflationary phase in the early Universe.

This work comes accompanied by a series of appendices containing useful results and additional material. In App. A, we detail our conventions. In App. B, we present basic properties of the Lorentz group, focusing in particular on the relations between its vector and spinor representations. App. C summarises the main identities that have been used throughout this book, and the key results (including the definition of several covariant derivatives and the transformation laws under a supergravity transformation) concerning matter, gauge and gravitation multiplets in supergravity are collected in App. D. In App. E, we present a flow diagram summarising the main steps leading to an invariant action in supergravity. Since Kähler manifolds are central in supergravity (in particular for the chiral content of the theory), some results concerning Kähler structures, including the associated Killing vectors, are given in App. F. Two-loop renormalisation group equations linking the model parameters at different scales and widely used in the study of phenomenologically viable models are presented in App. G. We dedicate App. H to the cornerstone of supergravity and supersymmetry, namely Lie superalgebras. In this appendix we present a selection of results for the superconformal, super-anti-de-Sitter and super-Poincaré cases, together with their relation to the orthosymplectic and unitary superalgebras.

Chapter 2

A brief summary of $N = 1$ supersymmetry

Supersymmetric quantum field theories can be built naturally using the superspace formalism. In this chapter, we recall the basic features of the $N = 1$ superspace that are relevant for this book. Since details can be found in many textbooks and reviews, we omit them from the present document and refer instead to the existing literature [Salam and Strathdee (1974b, 1975); Ferrara et al. (1974); Salam and Strathdee (1978); Wess and Zumino (1974b); Ferrara and Zumino (1974); Derendinger (1989); Fuks and Rausch de Traubenberg (2011)].

2.1 Superspace, supercharges and superderivatives

For $N = 1$ supersymmetric theories, the superspace is defined by adjoining a Majorana spinor, *i.e.*, the Grassmann variables $(\theta^\alpha, \overline{\theta}_{\dot\alpha})$, to the usual spacetime coordinates x^μ: $z^M = (x^\mu, \theta^\alpha, \overline{\theta}_{\dot\alpha})$. The supercharges and the superderivatives (or equivalently the supersymmetric covariant derivatives) are given by

$$\begin{aligned} Q_\alpha &= \frac{\partial}{\partial\theta^\alpha} - i\sigma^\mu{}_{\alpha\dot\alpha}\overline{\theta}^{\dot\alpha}\partial_\mu \,, & \overline{Q}_{\dot\alpha} &= \frac{\partial}{\partial\overline{\theta}^{\dot\alpha}} + i\theta^\alpha\sigma^\mu{}_{\alpha\dot\alpha}\partial_\mu \,, \\ D_\alpha &= \frac{\partial}{\partial\theta^\alpha} + i\sigma^\mu{}_{\alpha\dot\alpha}\overline{\theta}^{\dot\alpha}\partial_\mu \,, & \overline{D}_{\dot\alpha} &= \frac{\partial}{\partial\overline{\theta}^{\dot\alpha}} - i\theta^\alpha\sigma^\mu{}_{\alpha\dot\alpha}\partial_\mu \,, \end{aligned} \tag{2.1}$$

respectively, where the Pauli matrices are given in App. A. Using the conventions of App. A for the spinor derivatives we get the algebra

$$\begin{aligned} &\{Q_\alpha, \overline{Q}_{\dot\alpha}\} = 2i\sigma^\mu{}_{\alpha\dot\alpha}\partial_\mu \,, \quad \{D_\alpha, \overline{D}_{\dot\alpha}\} = -2i\sigma^\mu{}_{\alpha\dot\alpha}\partial_\mu \,, \\ &\{Q_\alpha, D_\beta\} = \{\overline{Q}_{\dot\alpha}, D_\beta\} = \{Q_\alpha, \overline{D}_{\dot\beta}\} = \{\overline{Q}_{\dot\alpha}, \overline{D}_{\dot\beta}\} = 0 \,. \end{aligned}$$

All these operators are anti-hermitian, *i.e.*,

$$\begin{aligned} (D_\alpha)^\dagger &= -\overline{D}_{\dot\alpha} \,, \\ (Q_\alpha)^\dagger &= -\overline{Q}_{\dot\alpha} \,. \end{aligned}$$

We recall that our conventions regarding spin and Lorentz indices, as well as those related to the definition of the superspace and the derivatives with respect to the superspace coordinates, are summarised in App. A. They slightly differ from those adopted in the previous work of one of us [Fuks and Rausch de Traubenberg (2011)], but are more suitable for supergravity. Historically, a similar change in the conventions can be noticed by comparing, for instance, those of the pioneering works [Ferrara et al. (1974)] where hermitian supercharges and superderivatives have been used, to those of the book of Wess and Bagger (1992).

2.2 Chiral superfields

The degrees of freedom included in an $N = 1$ supersymmetric matter multiplet consist of one complex scalar field ϕ, one left-handed Weyl fermion ψ and one auxiliary complex scalar field F. In the superspace formalism, this supermultiplet is described by a chiral superfield Φ, whose component fields are ϕ, ψ and F. In a similar fashion, the component fields of the conjugate antichiral superfield $\Phi^\dagger$ are the conjugate fields $\phi^\dagger$, $\overline{\psi}$ and $F^\dagger$. Chiral and antichiral superfields are defined through the constraints,

$$\overline{D}_{\dot\alpha}\Phi = 0, \qquad D_\alpha\Phi^\dagger = 0\,, \tag{2.2}$$

respectively, and can be expressed in terms of their component fields as

$$\begin{aligned}
\Phi(y,\theta) &= \phi(y) + \sqrt{2}\theta\cdot\psi(y) - \theta\cdot\theta F(y)\\
&= \phi(x) + \sqrt{2}\theta\cdot\psi(x) - \theta\cdot\theta F(x) + i\theta\sigma^\mu\overline{\theta}\,\partial_\mu\phi(x)\\
&\quad - \frac{i}{\sqrt{2}}\theta\cdot\theta\,\partial_\mu\psi(x)\sigma^\mu\overline{\theta} - \frac{1}{4}\theta\cdot\theta\overline{\theta}\cdot\overline{\theta}\,\Box\phi(x)\,,\\
\Phi^\dagger(y^\dagger,\overline{\theta}) &= \phi^\dagger(y^\dagger) + \sqrt{2}\overline{\theta}\cdot\overline{\psi}(y^\dagger) - \overline{\theta}\cdot\overline{\theta}F^\dagger(y^\dagger)\\
&= \phi^\dagger(x) + \sqrt{2}\overline{\theta}\cdot\overline{\psi}(x) - \overline{\theta}\cdot\overline{\theta}F^\dagger(x) - i\theta\sigma^\mu\overline{\theta}\,\partial_\mu\phi^\dagger(x)\\
&\quad + \frac{i}{\sqrt{2}}\overline{\theta}\cdot\overline{\theta}\theta\sigma^\mu\partial_\mu\overline{\psi}(x) - \frac{1}{4}\theta\cdot\theta\overline{\theta}\cdot\overline{\theta}\,\Box\phi^\dagger(x)\,,
\end{aligned}$$

where we have introduced the variable

$$y^\mu = x^\mu + i\theta\sigma^\mu\overline{\theta}\,,$$

satisfying $\overline{D}_{\dot\alpha}y^\mu = 0$, and such that $\overline{y}^{\mu\dagger} = x^\mu - i\theta\sigma^\mu\overline{\theta}$.

The associated free Lagrangian is obtained from $\Phi^\dagger\Phi$,

$$
\begin{aligned}
\Phi^\dagger\Phi = &\ \phi^\dagger\phi + \sqrt{2}\theta\cdot\phi^\dagger\psi + \sqrt{2}\bar\theta\cdot\bar\psi\phi - \theta\cdot\theta\,\phi^\dagger F - \bar\theta\cdot\bar\theta\, F^\dagger\phi \\
&+ \theta\sigma^\mu\bar\theta\Big[i\phi^\dagger\partial_\mu\phi - i\partial_\mu\phi^\dagger\phi - \bar\psi\bar\sigma_\mu\psi\Big] \\
&+ \theta\cdot\theta\bar\theta\cdot\Big[-\sqrt{2}\bar\psi F - \frac{i}{\sqrt{2}}\partial_\mu\phi^\dagger\bar\sigma^\mu\psi + \frac{i}{\sqrt{2}}\phi^\dagger\bar\sigma^\mu\partial_\mu\psi\Big] \\
&+ \bar\theta\cdot\bar\theta\theta\cdot\Big[-\sqrt{2}F^\dagger\psi - \frac{i}{\sqrt{2}}\sigma^\mu\bar\psi\partial_\mu\phi + \frac{i}{\sqrt{2}}\sigma^\mu\partial_\mu\bar\psi\phi\Big] \\
&+ \theta\cdot\theta\bar\theta\cdot\bar\theta\Big[-\frac{1}{4}\phi^\dagger\Box\phi + \frac{1}{2}\partial_\mu\phi^\dagger\partial^\mu\phi - \frac{1}{4}\Box\phi^\dagger\phi \\
&\qquad - \frac{i}{2}\bar\psi\bar\sigma^\mu\partial_\mu\psi + \frac{i}{2}\partial_\mu\bar\psi\bar\sigma^\mu\psi + F^\dagger F\Big]\,,
\end{aligned}
\tag{2.3}
$$

where we have used standard calculation techniques in the superspace. While only the term with the highest power in the Grassmann variables, *i.e.*, the term in $\theta^2\bar\theta^2$, is identified with the Lagrangian[1], we provide, for future references, all the components of the superfield product $\Phi^\dagger\Phi$. In addition, we have not performed any integration by parts, in contrast to most of the available literature. This point will be addressed at the end of the procedure leading to the construction of the complete action. By convention, we have also chosen to write all anti-holomorphic fields to the left of the holomorphic ones.

Interactions among chiral superfields can be built after computing products of chiral superfields. The product of two chiral superfields Φ^1 and Φ^2 is given, in terms of the y^μ variable, by,

$$
\begin{aligned}
\Phi^1\Phi^2 &= (\phi^1 + \sqrt{2}\theta\cdot\psi^1 - \theta\cdot\theta F^1)\,(\phi^2 + \sqrt{2}\theta\cdot\psi^2 - \theta\cdot\theta F^2) \\
&= \phi^1\phi^2 + \sqrt{2}\theta\cdot(\phi^1\psi^2 + \phi^2\psi^1) - \theta\cdot\theta(\phi^1F^2 + \phi^2F^1 + \psi^1\cdot\psi^2)\,,
\end{aligned}
\tag{2.4}
$$

where ϕ^i, ψ^i and F^i are the component fields of the superfield Φ^i. This relation allows one first to iteratively calculate the product of any number of chiral superfields and next to define the inverse of a chiral superfield, solving the equation $\Phi\Phi^{-1} = 1$:

$$
\Phi^{-1} = \phi^{-1} + \sqrt{2}\theta\cdot\Big[-\phi^{-2}\psi\Big] - \theta\cdot\theta\Big[-\phi^{-2}F + \phi^{-3}\psi\cdot\psi\Big]\,.
$$

A very practical way to define the components of a chiral superfield Φ, in particular in the framework of supergravity, relies on the extraction of the lowest-order component of the superfield resulting from the action of a single or of two

[1] The kinetic terms for the fermionic fields ψ and $\bar\psi$ agree with the results of Wess and Bagger (1992). They however seem to be in contradiction with the canonical normalisation expected for kinetic terms of two-component Weyl fermions by an overall sign. This discrepancy finds its explanation in the two equivalent choices for the square root of -1, *i.e.*, i and $-i$.

superderivatives on Φ. This procedure leads to

$$\begin{aligned} \phi(x) &= \Phi\big| , & \psi_\alpha(x) &= \frac{1}{\sqrt{2}} D_\alpha \Phi\big| , & F(x) &= \frac{1}{4} D\cdot D\Phi\big| , \\ \phi^\dagger(x) &= \Phi^\dagger\big| , & \bar\psi_{\dot\alpha}(x) &= \frac{1}{\sqrt{2}} \bar D_{\dot\alpha} \Phi^\dagger\big| , & F^\dagger(x) &= \frac{1}{4} \bar D\cdot \bar D\Phi^\dagger\big| , \end{aligned} \tag{2.5}$$

where we have introduced the notation $X\big| \equiv X\big|_{\theta=\bar\theta=0}$.

2.3 Vector superfields

Gauge supermultiplets are representations of the $N = 1$ supersymmetric algebra whose degrees of freedom consist of a massless vector boson v_μ, a massless Majorana fermion $(\lambda_\alpha, \bar\lambda^{\dot\alpha})$ and a real auxiliary field D. Within the superspace formalism, they are naturally embedded into vector superfields V defined by the reality condition $V^\dagger = V$. After an expansion in terms of the Grassmann variables θ and $\bar\theta$, one obtains the component field expression

$$V = C + \theta\cdot\chi + \bar\theta\cdot\bar\chi + \frac{1}{2}\theta\cdot\theta F + \frac{1}{2}\bar\theta\cdot\bar\theta F^\dagger + \theta\sigma^\mu\bar\theta v_\mu + \theta\cdot\theta\bar\theta\cdot\bar\lambda + \bar\theta\cdot\bar\theta\theta\cdot\lambda + \frac{1}{2}\theta\cdot\theta\bar\theta\cdot\bar\theta D\,. \tag{2.6}$$

Several fields, namely the real scalar field C, the complex scalar field F and the Majorana fermion $(\chi_\alpha, \bar\chi^{\dot\alpha})$ are unphysical since they can be eliminated through a gauge transformation as we will see below.

Using again standard techniques, we can compute the product of two vector superfields V_1 and V_2. Denoting the component fields of the superfield V_i by C_i, χ_i, F_i, v_i, λ_i and D_i, one obtains,

$$\begin{aligned} V_1 V_2 = \Big[& C_1 + \theta\cdot\chi_1 + \bar\theta\cdot\bar\chi_1 + \frac{1}{2}\theta\cdot\theta F_1 + \frac{1}{2}\bar\theta\cdot\bar\theta F_1^\dagger + \theta\sigma^\mu\bar\theta v_{1\mu} \\ & + \theta\cdot\theta\bar\theta\cdot\bar\lambda_1 + \bar\theta\cdot\bar\theta\theta\cdot\lambda_1 + \frac{1}{2}\theta\cdot\theta\bar\theta\cdot\bar\theta D_1\Big] \\ \Big[& C_2 + \theta\cdot\chi_2 + \bar\theta\cdot\bar\chi_2 + \frac{1}{2}\theta\cdot\theta F_2 + \frac{1}{2}\bar\theta\cdot\bar\theta F_2^\dagger + \theta\sigma^\mu\bar\theta v_{2\mu} \\ & + \theta\cdot\theta\bar\theta\cdot\bar\lambda_2 + \bar\theta\cdot\bar\theta\theta\cdot\lambda_2 + \frac{1}{2}\theta\cdot\theta\bar\theta\cdot\bar\theta D_2\Big] \\ = \; & C_1C_2 + \theta\cdot\Big[C_1\chi_2 + C_2\chi_1\Big] + \bar\theta\cdot\Big[C_1\bar\chi_2 + C_2\bar\chi_1\Big] \\ & + \frac{1}{2}\theta\cdot\theta\Big[C_1F_2 + C_2F_1 - \chi_1\cdot\chi_2\Big] + \frac{1}{2}\bar\theta\cdot\bar\theta\Big[C_1F_2^\dagger + C_2F_1^\dagger - \bar\chi_1\cdot\bar\chi_2\Big] \\ & + \theta\sigma^\mu\bar\theta\Big[C_1v_{2\mu} + C_2v_{1\mu} + \frac{1}{2}\chi_1\sigma_\mu\bar\chi_2 + \frac{1}{2}\chi_2\sigma_\mu\bar\chi_1\Big] \\ & + \theta\cdot\theta\bar\theta\cdot\Big[C_1\bar\lambda_2 + C_2\bar\lambda_1 + \frac{1}{2}F_1\bar\chi_2 + \frac{1}{2}F_2\bar\chi_1 + \frac{1}{2}\bar\sigma^\mu\chi_1v_{2\mu} + \frac{1}{2}\bar\sigma^\mu\chi_2v_{1\mu}\Big] \end{aligned} \tag{2.7}$$

$$+\bar\theta\cdot\bar\theta\theta\cdot\Big[C_1\lambda_2+C_2\lambda_1+\frac{1}{2}F_1^\dagger\chi_2+\frac{1}{2}F_2^\dagger\chi_1-\frac{1}{2}\sigma^\mu\bar\chi_1 v_{2\mu}-\frac{1}{2}\sigma^\mu\bar\chi_2 v_{1\mu}\Big]$$
$$+\frac{1}{2}\theta\cdot\theta\bar\theta\cdot\bar\theta\Big[C_1D_2+C_2D_1-\chi_1\cdot\lambda_2-\chi_2\cdot\lambda_1-\bar\lambda_1\cdot\bar\chi_2-\bar\lambda_2\cdot\bar\chi_1$$
$$+\frac{1}{2}F_1F_2^\dagger+\frac{1}{2}F_2F_1^\dagger+v_{1\mu}v_2^\mu\Big]\,.$$

This expression allows one to derive the inverse of a vector superfield V^{-1} by solving the equation $VV^{-1}=1$:

$$\begin{aligned}V^{-1}&=C^{-1}+\theta\cdot\Big[-C_1^{-2}\chi_1\Big]+\bar\theta\cdot\Big[-C_1^{-2}\bar\chi_1\Big]+\frac{1}{2}\theta\cdot\theta\Big[-C_1^{-2}F_1-C_1^{-3}\chi_1\cdot\chi_1\Big]\\
&+\frac{1}{2}\bar\theta\cdot\bar\theta\Big[-C_1^{-2}F_1^\dagger-C_1^{-3}\bar\chi_1\cdot\bar\chi_1\Big]+\theta\sigma^\mu\bar\theta\Big[-C_1^{-2}v_{1\mu}+C_1^{-3}\chi_1\sigma_\mu\bar\chi_1\Big]\\
&+\theta\cdot\theta\bar\theta\cdot\Big[-C_1^{-2}\bar\lambda_1+C_1^{-3}F_1\bar\chi_1+C_1^{-3}\bar\sigma^\mu\chi_1 v_{1\mu}+\frac{3}{2}C_1^{-4}\chi_1\cdot\chi_1\bar\chi_1\Big]\\
&+\bar\theta\cdot\bar\theta\theta\cdot\Big[-C_1^{-2}\lambda_1+C_1^{-3}F_1^\dagger\chi_1-C_1^{-3}\sigma^\mu\bar\chi_1 v_{1\mu}+\frac{3}{2}C_1^{-4}\bar\chi_1\cdot\bar\chi_1\chi_1\Big]\\
&+\frac{1}{2}\theta\cdot\theta\bar\theta\cdot\bar\theta\Big[-C_1^{-2}D_1-2C_1^{-3}\chi_1\cdot\lambda_1-2C_1^{-3}\bar\chi_1\cdot\bar\lambda_1+C_1^{-3}F_1F_1^\dagger+C_1^{-3}v_{1\mu}v_1^\mu\\
&+\frac{3}{2}C_1^{-4}F_1^\dagger\chi_1\cdot\chi_1+\frac{3}{2}C_1^{-4}F_1\bar\chi_1\cdot\bar\chi_1+3C_1^{-5}\chi_1\cdot\chi_1\bar\chi_1\cdot\bar\chi_1-3C_1^{-4}\chi_1\sigma^\mu\bar\chi_1 v_{1\mu}\Big]\,.\end{aligned}$$

As for chiral superfields, first- and second-order superderivatives can be used to define the lower-order components of a vector superfield,

$$\begin{aligned}C&=V\big|\,,\quad \chi_\alpha=D_\alpha V\big|\,,\quad \bar\chi_{\dot\alpha}=\bar D_{\dot\alpha}V\big|\,,\\
F&=-\tfrac{1}{2}D\cdot DV\big|\,,\quad F^\dagger=-\tfrac{1}{2}\bar D\cdot\bar DV\big|\,,\quad v_\mu\sigma^\mu{}_{\alpha\dot\alpha}=\tfrac{1}{2}\big[D_\alpha,\bar D_{\dot\alpha}\big]V\big|\,.\end{aligned}\tag{2.8}$$

The higher-order components are obtained by making use of third-order and fourth-order derivatives. As shown below, this is equivalent to considering the superfield strength tensors W_α and $\bar W_{\dot\alpha}$ and their derivatives.

The vector supermultiplets are the fundamental objects allowing one to embed gauge interactions into supersymmetry. We introduce a generic compact non-abelian Lie group G, with its associated Lie algebra $\mathfrak{g}$, and a unitary representation $\mathcal{R}$ specified by the generators T_a that fulfil the relations[2]

$$\begin{aligned}\big[T_a,T_b\big]&=if_{ab}{}^cT_c\,,\\
\mathrm{Tr}(T_aT_b)&=\tau_\mathcal{R}\delta_{ab}\,,\end{aligned}\tag{2.9}$$

[2]The abelian case can be easily recovered from the non-abelian one by replacing the structure constants by zeros and the generators T_a by the corresponding $\mathfrak{u}(1)$ charge.

where $f_{ab}{}^c$ are the real structure constants of the algebra and $\tau_{\mathcal{R}}$ is the Dynkin index of the representation $\mathcal{R}$. Associating to each generator T_a a vector superfield V^a, we define the vector superfield $V = V^a T_a$. At the superfield level the gauge transformation laws for vector fields become [Grisaru et al. (1979)]

$$V \to V + i\,\text{ad}(gV)\cdot(\Lambda + \Lambda^\dagger) - i\,\text{ad}(gV)\coth[\text{ad}(gV)]\cdot(\Lambda - \Lambda^\dagger)\,,$$

where g stands for the coupling constant of the gauge interactions, where Λ is a chiral superfield standing for the transformation parameters and where the operator ad is defined by $\text{ad}(X)\cdot Y \equiv [X, Y]$, $\text{ad}^2(X)\cdot Y \equiv [X, [X, Y]]$, *etc.* Gauge invariance allows us to rewrite V in a form more suitable for computations after choosing the so-called Wess-Zumino gauge. In this case, all unphysical components of the vector superfield (2.6) vanish,

$$V_{\text{W.Z.}} = \theta\sigma^\mu\bar\theta v_\mu - i\theta\cdot\theta\bar\theta\cdot\bar\lambda + i\bar\theta\cdot\bar\theta\theta\cdot\lambda + \frac{1}{2}\theta\cdot\theta\bar\theta\cdot\bar\theta D\,, \qquad (2.10)$$

and the component fields of this superfield are defined by $v_\mu = v_\mu^a T_a$, $\lambda = \lambda^a T_a$, $\bar\lambda = \bar\lambda^a T_a$ and $D = D^a T_a$ (which depend on the variables x^μ). Moreover, we have explicitly introduced factors of $\pm i$ for the gaugino components λ and $\bar\lambda$ in contrast to Eq. (2.6). Starting from (2.10), we can build the superfield strength tensors W_α and $\overline{W}_{\dot\alpha}$, *i.e.*, quantities corresponding to the supersymmetrised versions of the field strength tensor. These are chiral and antichiral spinor superfields defined by

$$\begin{aligned} W_\alpha &= -\frac{1}{4}\bar D\cdot\bar D\, e^{2gV} D_\alpha e^{-2gV}\,, \\ \overline{W}_{\dot\alpha} &= +\frac{1}{4}D\cdot D\, e^{-2gV}\bar D_{\dot\alpha} e^{2gV}\,. \end{aligned} \qquad (2.11)$$

Note that

$$e^{2gV} D_\alpha e^{-2gV} = -2g(D_\alpha V + g[V, D_\alpha V])\,,$$

and that the computation of the action of the operator $\bar D\cdot\bar D$ on this term can be further simplified if instead of x we use the variables y since we know that $\bar D_{\dot\alpha} y^\mu = 0$. If we proceed along these lines, we should also make use of the expansion

$$f(x) = f(y) - i\theta\sigma^\mu\bar\theta\partial_{y^\mu} f(y) - \frac{1}{2}\theta\sigma^\mu\bar\theta\theta\sigma^\nu\bar\theta\partial_{y^\mu}\partial_{y^\nu} f(y),$$

for each field. An analogous procedure may be applied to $\overline{W}_{\dot\alpha}$ but now with the variable $y^\dagger$.

After expanding W_α and $\overline{W}_{\dot\alpha}$ in terms of the Grassmann variables θ and $\bar\theta$, one obtains,

$$\begin{aligned} W_\alpha &= -2g\Big(+i\lambda_\alpha + \Big[\frac{i}{2}(\sigma^\mu\bar\sigma^\nu\theta)_\alpha F_{\mu\nu} + \theta_\alpha D\Big] - \theta\cdot\theta(\sigma^\mu D_\mu\bar\lambda)_\alpha\Big)\,, \\ \overline{W}_{\dot\alpha} &= -2g\Big(-i\bar\lambda_{\dot\alpha} + \Big[\frac{i}{2}(\bar\theta\bar\sigma^\mu\sigma^\nu)_{\dot\alpha} F_{\mu\nu} + \bar\theta_{\dot\alpha} D\Big] - \bar\theta\cdot\bar\theta(D_\mu\lambda\sigma^\mu)_{\dot\alpha}\Big)\,, \end{aligned}$$

where all the (anti)chiral components depend on y^μ $((y^\mu)^\dagger)$, and not on the usual spacetime coordinates x^μ. Furthermore, we have made those expressions compact by introducing the non-abelian field strength tensor $F_{\mu\nu}$ and the covariant derivative of λ and $\bar\lambda$

$$\begin{aligned}F_{\mu\nu} &= \partial_{y^\mu} v_\nu - \partial_{y^\nu} v_\mu + ig[v_\mu, v_\nu]\ ,\\ D_\mu\lambda &= \partial_{y^\mu}\lambda + ig[v_\mu, \lambda]\ ,\\ D_\mu\bar\lambda &= \partial_{y^\mu}\bar\lambda + ig[v_\mu, \bar\lambda]\ .\end{aligned}$$

Inspecting the component fields of the left-handed and right-handed spinor superfields W_α and $\overline{W}_{\dot\alpha}$, it is manifest that these objects are hermitian-conjugates of each other. Moreover, the higher-order component fields of the vector superfield (2.10) can be obtained by making use of the relations

$$\begin{aligned}\lambda_\alpha &= \frac{i}{2g} W_\alpha\Big|\ ,\\ \bar\lambda_{\dot\alpha} &= -\frac{i}{2g}\overline{W}_{\dot\alpha}\Big|\ ,\\ D &= \frac{1}{4g} D^\alpha W_\alpha\Big| = \frac{1}{4g}\overline{D}_{\dot\alpha}\overline{W}^{\dot\alpha}\Big|\ ,\end{aligned} \tag{2.12}$$

the vector component v_μ being expressed as already shown in Eq. (2.8).

Similar to the case of non-supersymmetric quantum field theories, the super-Yang-Mills Lagrangian can be obtained by squaring the superfield strength tensors and tracing over the gauge indices,

$$\begin{aligned}\frac{\mathrm{Tr}(W^\alpha W_\alpha)}{16\tau_{\mathcal{R}} g^2} &= \frac{1}{4}\Big[-\lambda^a\cdot\lambda_a + \theta\cdot\big[\sigma^\mu\bar\sigma^\nu\lambda_a F^a_{\mu\nu} + 2i\lambda_a D^a\big]\\ &\quad + \theta\cdot\theta\Big[D^a D_a - \frac{1}{2}F^a_{\mu\nu}F^{\mu\nu}_a - \frac{i}{2}F^a_{\mu\nu}\,{}^*F^{\mu\nu}_a - 2i\lambda^a\sigma^\mu D_\mu\bar\lambda_a\Big]\Big]\ ,\\ \frac{\mathrm{Tr}(\overline{W}_{\dot\alpha}\overline{W}^{\dot\alpha})}{16\tau_{\mathcal{R}} g^2} &= \frac{1}{4}\Big[-\bar\lambda^a\cdot\bar\lambda_a + \bar\theta\cdot\big[\bar\sigma^\mu\sigma^\nu\bar\lambda_a F^a_{\mu\nu} - 2i\bar\lambda_a D^a\big]\\ &\quad + \bar\theta\cdot\bar\theta\Big[D^a D_a - \frac{1}{2}F^a_{\mu\nu}F^{\mu\nu}_a + \frac{i}{2}F^a_{\mu\nu}\,{}^*F^{\mu\nu}_a + 2iD_\mu\lambda^a\sigma^\mu\bar\lambda_a\Big]\Big]\ ,\end{aligned} \tag{2.13}$$

where we have introduced the dual field strength tensor of the gauge field,

$${}^*F^{\mu\nu}_a = \frac{1}{2}\varepsilon^{\mu\nu\rho\sigma}F_{\rho\sigma a}\ ,$$

and

$$\begin{aligned}\lambda_a &= \delta_{ab}\lambda^b\ ,\\ \bar\lambda_a &= \delta_{ab}\bar\lambda^b\ ,\\ F_{\mu\nu a} &= \delta_{ab}F^b_{\mu\nu}\ ,\\ D_a &= \delta_{ab}D^b\ .\end{aligned}$$

As before, these expressions imply a dependence of the fields on the variables y^μ and $(y^\mu)^\dagger$. Although the Lagrangian is identified as the sum of the θ^2-coefficient of the first quantity and of the $\bar{\theta}^2$-coefficient of the second one, we have given all the components of the squared superfield strength tensors for future references.

2.4 Superpotential, Kähler potential and gauge kinetic function

Supersymmetric Lagrangians can be entirely written in terms of three fundamental functions of the chiral content of the theory, namely the superpotential, the Kähler potential and the gauge kinetic function. Computations with these three functions, including among others their supergravity couplings, can be found in many works [Cremmer et al. (1979, 1982, 1983b); Bagger (1983)], whereas a detailed derivation of the associated Lagrangians is also available in the framework of supersymmetry [Fuks and Rausch de Traubenberg (2011)].

The kinetic terms (and gauge interaction terms, if relevant) describing the dynamics of the chiral superfield content of the theory are embedded within the Kähler potential, its simplest form being given by Eq. (2.3). The Kähler potential has been first introduced (in the context of supersymmetry and supergravity) more than 30 years ago [Zumino (1979); Alvarez-Gaumé and Freedman (1981)] and then reconsidered somewhat later in the context of gauge theories [Bagger (1983); Hull et al. (1986)]. Supersymmetry also allows chiral superfields to couple via non-gauge interactions, the latter being included in the superpotential. Finally, the gauge kinetic function is, as indicated by its name, bound to the gauge sector of the theory and drives kinetic and gauge interaction terms for the associated vector superfields.

We begin by fixing the superfield content of a generic theory, together with a gauge symmetry leaving the Lagrangian invariant. As in Sec. 2.3, we introduce a gauge group G and a representation $\mathcal{R}$ of the corresponding algebra spanned by the hermitian matrices T_a. To this gauge group we associate a vector superfield $V = V^a T_a$ and define the chiral sector of the theory by a collection of chiral (antichiral) superfields Φ^i ($\Phi^\dagger_{i^*}$) lying in the representation $\mathcal{R}$ (the complex conjugate representation $\bar{\mathcal{R}}$) of G. These notations for the (anti)chiral superfields allows one to distinguish between the indices of the two representations $\mathcal{R}$ and $\bar{\mathcal{R}}$ that we denote by i and i^*, respectively.

The superpotential is an arbitrary holomorphic function W depending only on the chiral superfields Φ^i. Since the product of chiral superfields is itself a chiral superfield, the superpotential is a chiral superfield too. It can be shown that its most general form reads

$$W(\Phi) = W(\phi) + \sqrt{2}\theta\cdot W_i\psi^i - \theta\cdot\theta\Big(F^i W_i(\phi) + \frac{1}{2}W_{ij}(\phi)\psi^i\cdot\psi^j\Big) \tag{2.14}$$

with

$$\begin{aligned} W_i(\phi) &= \frac{\partial W(\phi)}{\partial \phi^i}\,, \\ W_{ij}(\phi) &= \frac{\partial^2 W(\phi)}{\partial \phi^i \partial \phi^j}\,. \end{aligned}$$

In what follows, any time we perform an expansion as above, in the RHS the explicit dependence in the scalar component ϕ of the superfields Φ will be omitted.

The Kähler potential is an arbitrary real superfield depending on Φ^i and $\Phi^\dagger_{i^*}$. Under its most general form, it can be written as a finite (or infinite) sum[3]

$$K(\Phi, \Phi^\dagger) = W_I(\Phi^\dagger)W^I(\Phi)\,. \tag{2.15}$$

In the expansion above, $W^I(\Phi)$ are holomorphic functions depending on the chiral superfields Φ^i whereas $W_I(\Phi^\dagger)$ are anti-holomorphic functions depending on the anti-chiral superfields $\Phi^\dagger_{i^*}$.

It should not be assumed that $W_I(\Phi^\dagger)$ and $W^I(\Phi)$ are related by conjugation as they may be different functions. Alternatively the Kähler potential can be expanded as

$$K(\Phi, \Phi^\dagger) = \sum_{nm} c^{i_1^* i_2^* \dots i_m^*}{}_{i_1\dots i_n} \Phi^\dagger_{i_1^*} \cdots \Phi^\dagger_{i_m^*} \Phi^{i_1} \cdots \Phi^{i_n}\,, \tag{2.16}$$

of which (2.15) is a more compact notation.

Since W^I and W_I are respectively holomorphic and anti-holomorphic, the form of these functions is given by (2.14), together with the conjugate relation, and thus

$$\begin{aligned} K(\Phi, \Phi^\dagger) = &\left[W_I + \sqrt{2}\bar\theta\cdot\bar\psi_{i^*}\frac{\partial W_I}{\partial \phi^\dagger_{i^*}} - \bar\theta\cdot\bar\theta\left(F^\dagger_{i^*}\frac{\partial W_I}{\partial \phi^\dagger_{i^*}} + \frac{1}{2}\bar\psi_{i^*}\cdot\bar\psi_{j^*}\frac{\partial^2 W_I}{\partial\phi^\dagger_{i^*}\partial\phi^\dagger_{j^*}}\right)\right] \\ &\left[W^I + \sqrt{2}\theta\cdot\psi^i\frac{\partial W^I}{\partial\phi^i} - \theta\cdot\theta\left(F^i\frac{\partial W^I}{\partial\phi^i} + \frac{1}{2}\psi^i\cdot\psi^j\frac{\partial^2 W^I}{\partial\phi^i\partial\phi^j}\right)\right], \end{aligned}$$

where the components of the chiral and antichiral superfields Φ^i and $\Phi^\dagger_{i^*}$ depend here on the y and $y^\dagger$ variables. Using (2.3), one can calculate the component fields

[3]Of course there exist functions which do not admit such an expansion; however, since the final result does not depend on this expansion, we always assume that (2.14) holds, at least formally.

of the Kähler potential,

$$
\begin{aligned}
K(\Phi, \Phi^\dagger) = {} & K(\phi, \phi^\dagger) + \sqrt{2}\theta\cdot\Big[K_i \psi^i\Big] + \sqrt{2}\bar{\theta}\cdot\Big[K^{i^*}\bar{\psi}_{i^*}\Big] - \theta\cdot\theta\Big[K_i F^i + \frac{1}{2}K_{ij}\psi^i\cdot\psi^j\Big] \\
& -\bar{\theta}\cdot\bar{\theta}\Big[K^{i^*}F^\dagger_{i^*} + \frac{1}{2}K^{i^*j^*}\bar{\psi}_{i^*}\cdot\bar{\psi}_{j^*}\Big] + \theta\sigma^\mu\bar{\theta}\Big[-iK^{i^*}\partial_\mu\phi^\dagger_{i^*} + iK_i\partial_\mu\phi^i + K^{i^*}{}_i\psi^i\sigma_\mu\bar{\psi}_{i^*}\Big] \\
& +\theta\cdot\theta\bar{\theta}\cdot\Big[\frac{i}{\sqrt{2}}\bar{\sigma}^\mu(K_i\mathcal{D}_\mu + \mathcal{D}_i K_j \partial_\mu\phi^j)\psi^i - \sqrt{2}K^{i^*}{}_i F^i\bar{\psi}_{i^*} \\
& \qquad -\frac{1}{\sqrt{2}}K^{i^*}{}_k\Gamma_i{}^k{}_j\psi^i\cdot\psi^j\bar{\psi}_{i^*} - \frac{i}{\sqrt{2}}K^{i^*}{}_i\bar{\sigma}^\mu\psi^i\partial_\mu\phi^\dagger_{i^*}\Big] \\
& +\bar{\theta}\cdot\bar{\theta}\theta\cdot\Big[\frac{i}{\sqrt{2}}\sigma^\mu(K^{i^*}\mathcal{D}_\mu + \mathcal{D}^{i^*}K^{j^*}\partial_\mu\phi^\dagger_{j^*})\bar{\psi}_{i^*} - \sqrt{2}K^{i^*}{}_i F^\dagger_{i^*}\psi^i \qquad\qquad (2.17)\\
& \qquad -\frac{1}{\sqrt{2}}K^{k^*}{}_i\Gamma^{i^*}{}_{k^*}{}^{j^*}\bar{\psi}_{i^*}\cdot\bar{\psi}_{j^*}\psi^i - \frac{i}{\sqrt{2}}K^{i^*}{}_i\sigma^\mu\bar{\psi}_{i^*}\partial_\mu\phi^i\Big] \\
& +\theta\cdot\theta\bar{\theta}\cdot\bar{\theta}\Big[-\frac{1}{4}\partial_\mu(K_i\partial^\mu\phi^i + K^{i^*}\partial^\mu\phi^\dagger_{i^*}) + K^{j^*}{}_i\partial_\mu\phi^i\partial^\mu\phi^\dagger_{j^*} \\
& \qquad +K^{j^*}{}_i F^i F^\dagger_{j^*} + \frac{1}{4}K^{k^*\ell^*}{}_{ij}\psi^i\cdot\psi^j\bar{\psi}_{k^*}\cdot\bar{\psi}_{\ell^*} \\
& \qquad -\frac{i}{2}(K^{j^*}{}_i\psi^i\sigma^\mu\mathcal{D}_\mu\bar{\psi}_{j^*} - K^{j^*}{}_i\mathcal{D}_\mu\psi^i\sigma^\mu\bar{\psi}_{j^*}) \\
& \qquad +\frac{1}{2}K^{j^*}{}_i\Gamma^{k^*}{}_{j^*}{}^{\ell^*}F^i\bar{\psi}_{k^*}\cdot\bar{\psi}_{\ell^*} + \frac{1}{2}K^{i^*}{}_\ell\Gamma_i{}^\ell{}_j F^\dagger_{i^*}\psi^i\cdot\psi^j\Big] .
\end{aligned}
$$

In the equation above, we have introduced the following derivatives of the Kähler potential up to the fourth order,

$$
\begin{aligned}
K_i &= \frac{\partial K(\phi,\phi^\dagger)}{\partial\phi^i}\ , \qquad K_{ij} = \frac{\partial^2 K(\phi,\phi^\dagger)}{\partial\phi^i\partial\phi^j}\ , \\
K^{i^*} &= \frac{\partial K(\phi,\phi^\dagger)}{\partial\phi^\dagger_{i^*}}\ , \qquad K^{i^*j^*} = \frac{\partial^2 K(\phi,\phi^\dagger)}{\partial\phi^\dagger_{i^*}\partial\phi^\dagger_{j^*}}\ , \\
K^{j^*}{}_i &= \frac{\partial^2 K(\phi,\phi^\dagger)}{\partial\phi^i\partial\phi^\dagger_{j^*}} \quad (\textit{i.e.}, \text{the Kähler metric})\ , \qquad\qquad (2.18)\\
K_i{}^{k^*}{}_j &= \frac{\partial^3 K(\phi,\phi^\dagger)}{\partial\phi^i\partial\phi^j\partial\phi^\dagger_{k^*}} = K^{k^*}{}_\ell\Gamma_i{}^\ell{}_j\ , \quad K^{i^*}{}_k{}^{j^*} = \frac{\partial^3 K(\phi,\phi^\dagger)}{\partial\phi^\dagger_{i^*}\partial\phi^\dagger_{j^*}\partial\phi^k} = K^{\ell^*}{}_k\Gamma^{i^*}{}_{\ell^*}{}^{j^*}\ , \\
K^{i^*j^*}{}_{k\ell} &= \frac{\partial^4 K(\phi,\phi^\dagger)}{\partial\phi^\dagger_{i^*}\partial\phi^\dagger_{j^*}\partial\phi^k\partial\phi^\ell}\ .
\end{aligned}
$$

Moreover, the quantities $\Gamma^{i^*}{}_{j^*}{}^{k^*}$ and $\Gamma_i{}^j{}_k$ stand for the Christoffel symbols allowing one to define derivatives which are covariant with respect to the Kähler manifold,

$$\begin{aligned}
\mathcal{D}_\mu \bar{\psi}_{i^*} &= \partial_\mu \bar{\psi}_{i^*} + \Gamma^{j^*}{}_{i^*}{}^{k^*} \partial_\mu \phi^\dagger_{j^*} \bar{\psi}_{k^*} \, , \\
\mathcal{D}_\mu \psi^i &= \partial_\mu \psi^i + \Gamma_j{}^i{}_k \partial_\mu \phi^j \psi^k \, , \\
\mathcal{D}^{i^*} K^{j^*} &= K^{i^* j^*} - \Gamma^{i^*}{}_{k^*}{}^{j^*} K^{k^*} \, , \\
\mathcal{D}_i K_j &= K_{ij} - \Gamma_i{}^k{}_j K_k \, ,
\end{aligned} \tag{2.19}$$

and we recall that the fourth-order derivative of the Kähler potential can be related to the curvature tensor $R_i{}^{j^*}{}_k{}^{\ell^*}$ by

$$R_i{}^{j^*}{}_k{}^{\ell^*} = K^{j^*\ell^*}{}_{ik} - K^{m^*}{}_n \Gamma^{j^*}{}_{m^*}{}^{\ell^*} \Gamma_i{}^n{}_k \, . \tag{2.20}$$

Two remarks are in order after inspecting Eq. (2.17). First, several of the component fields, *e.g.*, $K_i F^i + \frac{1}{2} K_{ij} \psi^i \cdot \psi^j$, do not seem covariant with respect to the Kähler manifold. However, by solving the equations of motion for the auxiliary fields and by inserting the solutions back into the Kähler potential one recovers a manifest invariance. In the case of the example above, this is achieved by making use of $F^i = (K^{-1})^i{}_{j^*} W^{\star j^*} - \frac{1}{2} \Gamma_j{}^i{}_k \psi^j \cdot \psi^k$, so that we can rewrite the $\theta \cdot \theta$ component of $K(\Phi, \Phi^\dagger)$ as the Kähler-covariant object $(K^{-1})^i{}_{i^*} K_i W^{\star i^*} + \frac{1}{2} \mathcal{D}_i W_j \psi^i \cdot \psi^j$, where $W^\star$ is the conjugate superpotential. The same procedure leads to four-fermion interactions involving the curvature tensor and removes all terms involving three fermionic fields. Second, the bilinear scalar terms of the higher-order coefficient have been simplified using the relation

$$\begin{aligned}
& -\frac{1}{4}(K_i \Box \phi^i + K_{ij} \partial_\mu \phi^i \partial^\mu \phi^j) - \frac{1}{4}(K^{i^*} \Box \phi^\dagger_{i^*} + K^{i^* j^*} \partial_\mu \phi^\dagger_{i^*} \partial^\mu \phi^\dagger_{j^*}) \\
= \; & \frac{1}{2} K^{j^*}{}_i \partial_\mu \phi^i \partial^\mu \phi^\dagger_{j^*} - \frac{1}{4} \partial_\mu [K_i \partial^\mu \phi^i + K^{i^*} \partial^\mu \phi^\dagger_{i^*}] \, .
\end{aligned}$$

The expression obtained in (2.17) does not however include gauge interactions of the matter fields. Under a gauge transformation, chiral and antichiral superfields transform as

$$\begin{aligned}
\Phi &\to e^{2ig\Lambda} \Phi \, , \\
\Phi^\dagger &\to \Phi^\dagger e^{-2ig\Lambda^\dagger} \, ,
\end{aligned}$$

respectively, where the transformation parameter Λ is a chiral superfield and we recall that g stands for the coupling constant associated with the considered gauge group. As a consequence, the Kähler potential $K(\Phi, \Phi^\dagger)$ is not a gauge-invariant quantity, since $\Lambda^\dagger \neq \Lambda$. Gauge invariance can be recovered after following a supersymmetric version of the Nœther procedure. Since a vector superfield V transforms as

$$e^{-2gV} \to e^{2ig\Lambda^\dagger} e^{-2gV} e^{-2ig\Lambda} \, ,$$

the modification of the Kähler potential

$$K(\Phi, \Phi^\dagger) \to \mathcal{K} = \frac{1}{2}\Big[K(\Phi, \Phi^\dagger e^{-2gV}) + K(e^{-2gV}\Phi, \Phi^\dagger)\Big] \tag{2.21}$$

is gauge invariant. To compute the component fields of this last quantity, we expand the Kähler potential in terms of the chiral and antichiral superfields Φ and $\Phi^\dagger$, as in (2.16), and use the property of the Wess-Zumino gauge,

$$\begin{aligned} e^{-2gV} &= 1 - 2gV + 2g^2V^2 \\ &= 1 - 2g\theta\sigma^\nu\bar\theta v_\nu + 2gi\theta\cdot\theta\,\bar\theta\cdot\bar\lambda - 2gi\bar\theta\cdot\bar\theta\,\theta\cdot\lambda + \frac{1}{2}\theta\cdot\theta\,\bar\theta\cdot\bar\theta\Big(-2gD + 2g^2 v^\mu v_\mu\Big)\,, \end{aligned}$$

in order to derive the components of $\mathcal{K}$. The first term on the right-hand side of (2.21) is deduced from the calculation of one of the monomial term of (2.16),

$$(\Phi^\dagger e^{-2gV})_{j_1^*}\cdots(\Phi^\dagger e^{-2gV})_{j_m^*}\Phi^{i_1}\cdots\Phi^{i_n} = \Lambda_J\Lambda^I\,.$$

The computation of the superfield Λ^I is straightforward,

$$\begin{aligned} \Lambda^I =& \Big[\phi^{i_1} + i\theta\sigma^{\rho_1}\bar\theta\partial_{\rho_1}\phi^{i_1} - \frac{1}{4}\theta\cdot\theta\,\bar\theta\cdot\bar\theta\,\Box\phi^{i_1} + \sqrt{2}\theta\cdot\psi^{i_1} \\ &\qquad - \frac{i}{\sqrt{2}}\theta\cdot\theta\,\partial_{\mu_1}\psi^{i_1}\sigma^{\mu_1}\bar\theta - \theta\cdot\theta F^{i_1}\Big] \\ &\vdots \\ &\Big[\phi^{i_n} + i\theta\sigma^{\rho_n}\bar\theta\partial_{\rho_n}\phi^{i_n} - \frac{1}{4}\theta\cdot\theta\,\bar\theta\cdot\bar\theta\,\Box\phi^{i_n} + \sqrt{2}\theta\cdot\psi^{i_n} \\ &\qquad - \frac{i}{\sqrt{2}}\theta\cdot\theta\,\partial_{\mu_n}\psi^{i_n}\sigma^{\mu_n}\bar\theta - \theta\cdot\theta F^{i_n}\Big] \\ =& \Big[X^I + \sqrt{2}\theta\cdot\psi^i\frac{\partial X^I}{\partial\phi^i} - \theta\cdot\theta\Big(F^i\frac{\partial X^I}{\partial\phi^i} + \frac{1}{2}\psi^i\cdot\psi^j\frac{\partial^2 X^I}{\partial\phi^i\partial\phi^j}\Big) + i\theta\sigma^\mu\bar\theta\partial_\mu\phi^i\frac{\partial X^I}{\partial\phi^i} \\ &+ \frac{i}{\sqrt{2}}\theta\cdot\theta\,\bar\theta\cdot\Big(\frac{\partial X^I}{\partial\phi^i}\bar\sigma^\mu\partial_\mu\psi^i + \frac{\partial^2 X^I}{\partial\phi^i\partial\phi^j}\bar\sigma^\mu\psi^i\partial_\mu\phi^j\Big) \\ &- \frac{1}{4}\theta\cdot\theta\,\bar\theta\cdot\bar\theta\Big(\frac{\partial^2 X^I}{\partial\phi^i\partial\phi^j}\partial_\mu\phi^i\partial^\mu\phi^j + \frac{\partial X^I}{\partial\phi^i}\Box\phi^i\Big)\Big]\,, \end{aligned}$$

where the results are made compact by introducing the notation $X^I = \phi^{i_1}\cdots\phi^{i_n}$. The components of the superfield Λ_J are more complicated to extract as one has to simplify the expression

$$
\begin{aligned}
\Lambda_J = \Big[&\phi^\dagger_{k_1^*} - i\theta\sigma^{\mu_1}\bar\theta\partial_{\mu_1}\phi^\dagger_{k_1^*} - \frac{1}{4}\theta\cdot\theta\,\bar\theta\cdot\bar\theta\,\Box\phi^\dagger_{k_1^*} + \sqrt{2}\,\bar\theta\cdot\bar\psi_{k_1^*} \\
&\qquad + \frac{i}{\sqrt{2}}\bar\theta\cdot\bar\theta\,\theta\sigma^{\mu_1}\partial_{\mu_1}\bar\psi_{k_1^*} - \bar\theta\cdot\bar\theta\,F^\dagger_{k_1^*}\Big] \\
\Big[&\delta^{k_1^*}{}_{j_1^*} - 2g\,\theta\sigma^{\nu_1}\bar\theta\,v^{a_1}_{\nu_1}T_{a_1}{}^{k_1^*}{}_{j_1^*} + 2ig\,\theta\cdot\theta\,\bar\theta\cdot\bar\lambda^{a_1}T_{a_1}{}^{k_1^*}{}_{j_1^*} - 2ig\,\bar\theta\cdot\bar\theta\,\theta\cdot\lambda^{a_1}T_{a_1}{}^{k_1^*}{}_{j_1^*} \\
&+\frac{1}{2}\theta\cdot\theta\,\bar\theta\cdot\bar\theta\Big(-2gD^{a_1}T_{a_1}{}^{k_1^*}{}_{j_1^*} + 2g^2 v^{a_1}_{\nu_1}v^{b_1\nu_1}(T_{a_1}T_{b_1})^{k_1^*}{}_{j_1^*}\Big)\Big] \\
&\vdots \\
\Big[&\phi^\dagger_{k_m^*} - i\theta\sigma^{\mu_m}\bar\theta\partial_{\mu_m}\phi^\dagger_{k_m^*} - \frac{1}{4}\theta\cdot\theta\,\bar\theta\cdot\bar\theta\,\Box\phi^\dagger_{k_m^*} + \sqrt{2}\,\bar\theta\cdot\bar\psi_{k_m^*} \\
&\qquad + \frac{i}{\sqrt{2}}\bar\theta\cdot\bar\theta\,\theta\sigma^{\mu_m}\partial_{\mu_m}\bar\psi_{k_m^*} - \bar\theta\cdot\bar\theta\,F^\dagger_{k_m^*}\Big] \\
\Big[&\delta^{k_m^*}{}_{j_m^*} - 2g\,\theta\sigma^{\nu_m}\bar\theta\,v^{a_m}_{\nu_m}T_{a_m}{}^{k_m^*}{}_{j_m^*} + 2ig\,\theta\cdot\theta\,\bar\theta\cdot\bar\lambda^{a_m}T_{a_m}{}^{k_m^*}{}_{j_m^*} - 2ig\,\bar\theta\cdot\bar\theta\,\theta\cdot\lambda^{a_m}T_{a_m}{}^{k_m^*}{}_{j_m^*} \\
&+\frac{1}{2}\theta\cdot\theta\,\bar\theta\cdot\bar\theta\Big(-2gD^{a_m}T_{a_m}{}^{k_m^*}{}_{j_m^*} + 2g^2 v^{a_m}_{\nu_m}v^{b_m\nu_m}(T_{a_m}T_{b_m})^{k_m^*}{}_{j_m^*}\Big)\Big]\,.
\end{aligned}
$$

As for Λ^I, the results can be further rewritten in a compact manner by introducing the quantity $X_J = \phi^\dagger_{j_1^*}\cdots\phi^\dagger_{j_m^*}$,

$$
\begin{aligned}
\Lambda_J = X_J &+ \sqrt{2}\bar\theta\cdot\bar\psi_{i^*}\frac{\partial X_J}{\partial\phi^\dagger_{i^*}} - \bar\theta\cdot\bar\theta\Big[\frac{1}{2}\frac{\partial^2 X_J}{\partial\phi^\dagger_{i^*}\partial\phi^\dagger_{j^*}}\bar\psi_{i^*}\cdot\bar\psi_{j^*} + \frac{\partial X_J}{\partial\phi^\dagger_{i^*}}F^\dagger_{i^*}\Big] \\
&-\theta\sigma^\mu\bar\theta\frac{\partial X_J}{\partial\phi^\dagger_{i^*}}\Big[i\partial_\mu\phi^\dagger_{i^*} + 2g\,(\phi^\dagger T_a)_{i^*}v^a_\mu\Big] + 2ig\frac{\partial X_J}{\partial\phi^\dagger_{i^*}}\theta\cdot\theta\,\bar\theta\cdot\bar\lambda^a(\phi^\dagger T_a)_{i^*} \\
&+\bar\theta\cdot\bar\theta\,\theta\cdot\Big[\frac{\partial X_J}{\partial\phi^\dagger_{i^*}}\Big(\frac{i}{\sqrt{2}}\sigma^\mu\partial_\mu\bar\psi_{i^*} + \sqrt{2}g\sigma^\mu(\bar\psi T_a)_{i^*}v^a_\mu - 2ig(\phi^\dagger T_a)_{i^*}\lambda^a\Big) \\
&\qquad + \frac{\partial^2 X_J}{\partial\phi^\dagger_{i^*}\partial\phi^\dagger_{j^*}}\Big(\frac{i}{\sqrt{2}}\sigma^\mu\bar\psi_{i^*}\partial_\mu\phi^\dagger_{j^*} + \sqrt{2}g(\phi^\dagger T_a)_{i^*}v^a_\mu\sigma^\mu\bar\psi_{j^*}\Big)\Big] \\
&+\bar\theta\cdot\bar\theta\,\theta\cdot\theta\Big[\frac{\partial X_J}{\partial\phi^\dagger_{i^*}}\Big(-\frac{1}{4}\Box\phi^\dagger_{i^*} + igv^{\mu a}(\partial_\mu\phi^\dagger T_a)_{i^*} - gD^a(\phi^\dagger T_a)_{i^*} \\
&\qquad -i\sqrt{2}g(\bar\psi T_a)_{i^*}\cdot\bar\lambda^a + g^2 v^a_\mu v^{\mu b}(\phi^\dagger T_aT_b)_{i^*}\Big) \\
&\quad + \frac{\partial^2 X_J}{\partial\phi^\dagger_{i^*}\partial\phi^\dagger_{j^*}}\Big(-\frac{1}{4}\partial_\mu\phi^\dagger_{i^*}\partial^\mu\phi^\dagger_{j^*} + ig\partial_\mu\phi^\dagger_{i^*}(\phi^\dagger T_a)_{j^*}v^{\mu a} \\
&\qquad + g^2 v^a_\mu v^{\mu b}(\phi^\dagger T_a)_{i^*}(\phi^\dagger T_b)_{j^*} - i\sqrt{2}g\bar\psi_{i^*}\cdot\bar\lambda^a(\phi^\dagger T_a)_{j^*}\Big)\Big]\,.
\end{aligned}
$$

With the help of (2.7) to simplify the superfield product $\Lambda_J\Lambda^I$, this leads, to

$$
\begin{aligned}
\Lambda_J\Lambda^I = {} & X_J X^I + \sqrt{2}\theta\cdot\psi^i\, X_J \frac{\partial X^I}{\partial\phi^i} + \sqrt{2}\bar\theta\cdot\bar\psi_{i^*}\, \frac{\partial X_J}{\partial\phi^\dagger_{i^*}} X^I \\
& -\theta\cdot\theta \left[\frac{1}{2} X_J \frac{\partial^2 X^I}{\partial\phi^i\partial\phi^j}\, \psi^i\cdot\psi^j + X_J \frac{\partial X^I}{\partial\phi^i}\, F^i\right] \\
& -\bar\theta\cdot\bar\theta \left[\frac{1}{2} \frac{\partial^2 X_J}{\partial\phi^\dagger_{i^*}\partial\phi^\dagger_{j^*}} X^I\, \bar\psi_{i^*}\cdot\bar\psi_{j^*} + \frac{\partial X_J}{\partial\phi^\dagger_{i^*}} X^I\, F^\dagger_{i^*}\right] \\
& +\theta\sigma^\mu\bar\theta\Big[- i\partial_\mu\phi^\dagger_{i^*} \frac{\partial X_J}{\partial\phi^\dagger_{i^*}} X^I + i\partial_\mu\phi^i X_J \frac{\partial X^I}{\partial\phi^i} \\
& \qquad -2g\,(\phi^\dagger T_a)_{i^*} v^a_\mu \frac{\partial X_J}{\partial\phi^\dagger_{i^*}} X^I + \frac{\partial X_J}{\partial\phi^\dagger_{i^*}} \frac{\partial X^I}{\partial\phi^i} \psi^i\sigma_\mu\bar\psi_{i^*}\Big] \\
& +\theta\cdot\theta\,\bar\theta\cdot\Big[- \sqrt{2}\frac{\partial X_J}{\partial\phi^\dagger_{i^*}} \frac{\partial X^I}{\partial\phi^i} F^i\bar\psi_{i^*} - \frac{1}{\sqrt{2}} \frac{\partial X_J}{\partial\phi^\dagger_{i^*}} \frac{\partial^2 X^I}{\partial\phi^i\partial\phi^j} \psi^i\cdot\psi^j\bar\psi_{i^*} \\
& \qquad -\frac{i}{\sqrt{2}} \frac{\partial X_J}{\partial\phi^\dagger_{i^*}} \frac{\partial X^I}{\partial\phi^i} \bar\sigma^\mu\psi^i\partial_\mu\phi^\dagger_{i^*} + \frac{i}{\sqrt{2}} X_J \frac{\partial X^I}{\partial\phi^i} \bar\sigma^\mu\partial_\mu\psi^i \\
& \qquad +\frac{i}{\sqrt{2}} X_J \frac{\partial^2 X^I}{\partial\phi^i\partial\phi^j} \bar\sigma^\mu\psi^i\partial_\mu\phi^j - \sqrt{2}g \frac{\partial X_J}{\partial\phi^\dagger_{i^*}} \frac{\partial X^I}{\partial\phi^i} \bar\sigma^\mu\psi^i v^a_\mu(\phi^\dagger T_a)_{i^*} \\
& \qquad +2ig \frac{\partial X_J}{\partial\phi^\dagger_{i^*}} X^I(\phi^\dagger T_a)_{i^*}\bar\lambda^a\Big] \\
& +\bar\theta\cdot\bar\theta\,\theta\cdot\Big[\frac{i}{\sqrt{2}} \frac{\partial X_J}{\partial\phi^\dagger_{i^*}} X^I\sigma^\mu\partial_\mu\bar\psi_{i^*} - \sqrt{2}\frac{\partial X_J}{\partial\phi^\dagger_{i^*}} \frac{\partial X^I}{\partial\phi^i}\psi^i F^\dagger_{i^*} \\
& \qquad -\frac{1}{\sqrt{2}} \frac{\partial^2 X_J}{\partial\phi^\dagger_{i^*}\partial\phi^\dagger_{j^*}} \frac{\partial X^I}{\partial\phi^i} \bar\psi_{i^*}\cdot\bar\psi_{j^*}\psi^i - \frac{i}{\sqrt{2}} \frac{\partial X_J}{\partial\phi^\dagger_{i^*}} \frac{\partial X^I}{\partial\phi^i} \sigma^\mu\bar\psi_{i^*}\partial_\mu\phi^i \\
& \qquad +\frac{i}{\sqrt{2}} \frac{\partial^2 X_J}{\partial\phi^\dagger_{i^*}\partial\phi^\dagger_{j^*}} X^I\sigma^\mu\bar\psi_{i^*}\partial_\mu\phi^\dagger_{j^*} + \sqrt{2}g \frac{\partial^2 X_J}{\partial\phi^\dagger_{i^*}\partial\phi^\dagger_{j^*}} X^I(\phi^\dagger T_a)_{i^*} v^a_\mu\sigma^\mu\bar\psi_{j^*} \\
& \qquad + \sqrt{2}g \frac{\partial X_J}{\partial\phi^\dagger_{i^*}} X^I\sigma^\mu(\bar\psi T_a)_{i^*} v^a_\mu - 2ig \frac{\partial X_J}{\partial\phi^\dagger_{i^*}} X^I(\phi^\dagger T_a)_{i^*}\lambda^a\Big] \\
& +\bar\theta\cdot\bar\theta\,\theta\cdot\theta\Big[- \frac{1}{4} X_J \frac{\partial X^I}{\partial\phi^i}\, \Box\phi^i - \frac{1}{4} X_J \frac{\partial^2 X^I}{\partial\phi^i\partial\phi^j} \partial_\mu\phi^i\partial^\mu\phi^j \\
& \qquad -\frac{1}{4}\frac{\partial X_J}{\partial\phi^\dagger_{i^*}} X^I\Box\phi^\dagger_{i^*} - \frac{1}{4}\frac{\partial^2 X_J}{\partial\phi^\dagger_{i^*}\partial\phi^\dagger_{j^*}} X^I\partial_\mu\phi^\dagger_{i^*}\partial^\mu\phi^\dagger_{j^*} + \frac{1}{2}\frac{\partial X_J}{\partial\phi^\dagger_{i^*}} \frac{\partial X^I}{\partial\phi^i}\partial_\mu\phi^i\partial^\mu\phi^\dagger_{i^*} \\
& \qquad +\frac{i}{2}\frac{\partial X_J}{\partial\phi^\dagger_{i^*}} \frac{\partial X^I}{\partial\phi^i}\partial_\mu\psi^i\sigma^\mu\bar\psi_{i^*} - \frac{i}{2}\frac{\partial X_J}{\partial\phi^\dagger_{i^*}} \frac{\partial X^I}{\partial\phi^i}\psi^i\sigma^\mu\partial_\mu\bar\psi_{i^*}
\end{aligned}
$$

$$
\begin{aligned}
&+\frac{1}{4}\frac{\partial^2 X_J}{\partial\phi^\dagger_{i^*}\partial\phi^\dagger_{j^*}}\frac{\partial^2 X^I}{\partial\phi^i\partial\phi^j}\bar\psi_{i^*}\cdot\bar\psi_{j^*}\psi^i\cdot\psi^j-\frac{i}{2}\frac{\partial^2 X_J}{\partial\phi^\dagger_{i^*}\partial\phi^\dagger_{j^*}}\frac{\partial X^I}{\partial\phi^i}\psi^i\sigma^\mu\bar\psi_{i^*}\partial_\mu\phi^\dagger_{j^*}\\
&+\frac{i}{2}\frac{\partial X_J}{\partial\phi^\dagger_{i^*}}\frac{\partial^2 X^I}{\partial\phi^i\partial\phi^j}\psi^i\sigma^\mu\bar\psi_{i^*}\partial_\mu\phi^j+\frac{\partial X_J}{\partial\phi^\dagger_{i^*}}\frac{\partial X^I}{\partial\phi^i}F^iF^\dagger_{i^*}\\
&+\frac{1}{2}\frac{\partial X_J}{\partial\phi^\dagger_{i^*}}\frac{\partial^2 X^I}{\partial\phi^i\partial\phi^j}F^\dagger_{i^*}\psi^i\cdot\psi^j+\frac{1}{2}\frac{\partial^2 X_J}{\partial\phi^\dagger_{i^*}\partial\phi^\dagger_{j^*}}\frac{\partial X^I}{\partial\phi^i}\bar\psi_{i^*}\cdot\bar\psi_{j^*}F^i\\
&+ig\frac{\partial X_J}{\partial\phi^\dagger_{i^*}}X^I v^{\mu a}(\partial_\mu\phi^\dagger T_a)_{i^*}-ig\frac{\partial X_J}{\partial\phi^\dagger_{i^*}}\frac{\partial X^I}{\partial\phi^i}\partial_\mu\phi^i(\phi^\dagger T_a)_{i^*}v^{\mu a}\\
&+ig\frac{\partial^2 X_J}{\partial\phi^\dagger_{i^*}\partial\phi^\dagger_{j^*}}X^I\partial_\mu\phi^\dagger_{i^*}(\phi^\dagger T_a)_{j^*}v^{\mu a}\\
&+g^2\frac{\partial^2 X_J}{\partial\phi^\dagger_{i^*}\partial\phi^\dagger_{j^*}}X^I v^a_\mu v^{\mu b}(\phi^\dagger T_a)_{i^*}(\phi^\dagger T_b)_{j^*}\\
&+g^2\frac{\partial X_J}{\partial\phi^\dagger_{i^*}}X^I v^a_\mu v^{\mu b}(\phi^\dagger T_aT_b)_{i^*}-g\frac{\partial^2 X_J}{\partial\phi^\dagger_{i^*}\partial\phi^\dagger_{j^*}}\frac{\partial X^I}{\partial\phi^i}\psi^i\sigma^\mu\bar\psi_{i^*}(\phi^\dagger T_a)_{j^*}v^a_\mu\\
&-g\frac{\partial X_J}{\partial\phi^\dagger_{i^*}}\frac{\partial X^I}{\partial\phi^i}\psi^i\sigma^\mu(\bar\psi T_a)_{i^*}v^a_\mu-g\frac{\partial X_J}{\partial\phi^\dagger_{i^*}}X^I D^a(\phi^\dagger T_a)_{i^*}\\
&+i\sqrt{2}g\frac{\partial X_J}{\partial\phi^\dagger_{i^*}}\frac{\partial X^I}{\partial\phi^i}\psi^i\cdot\lambda^a(\phi^\dagger T_a)_{i^*}-i\sqrt{2}g\frac{\partial X_J}{\partial\phi^\dagger_{i^*}}X^I(\bar\psi T_a)_{i^*}\cdot\bar\lambda^a\\
&-i\sqrt{2}g\frac{\partial^2 X_J}{\partial\phi^\dagger_{i^*}\partial\phi^\dagger_{j^*}}X^I\bar\psi_{i^*}\cdot\bar\lambda^a(\phi^\dagger T_a)_{j^*}\Big]\,.
\end{aligned}
\tag{2.22}
$$

Summing over all the terms arising from (2.16) and adding the hermitian conjugate piece, *i.e.*, the second contribution appearing on the right-hand side of (2.21), one obtains

$$
\begin{aligned}
\mathcal{K}=&\;K(\phi,\phi^\dagger)+\sqrt{2}\theta\cdot\big[K_i\psi^i\big]+\sqrt{2}\bar\theta\cdot\big[K^{i^*}\bar\psi_{i^*}\big]\\
&-\theta\cdot\theta\Big[K_iF^i+\frac{1}{2}K_{ij}\psi^i\cdot\psi^j\Big]-\bar\theta\cdot\bar\theta\Big[K^{i^*}F^\dagger_{i^*}+\frac{1}{2}K^{i^*j^*}\bar\psi_{i^*}\cdot\bar\psi_{j^*}\Big]\\
&+\theta\sigma^\mu\bar\theta\Big[-iK^{i^*}D_\mu\phi^\dagger_{i^*}+iK_iD_\mu\phi^i+K^{i^*}{}_i\psi^i\sigma_\mu\bar\psi_{i^*}\Big]\\
&+\theta\cdot\theta\,\bar\theta\cdot\Big[-\sqrt{2}K^{i^*}{}_iF^i\bar\psi_{i^*}-\frac{1}{\sqrt{2}}K^{i^*}{}_k\Gamma_i{}^k{}_j\psi^i\cdot\psi^j\bar\psi_{i^*}\\
&\qquad-\frac{i}{\sqrt{2}}K^{i^*}{}_i\bar\sigma^\mu\psi^iD_\mu\phi^\dagger_{i^*}+\frac{i}{\sqrt{2}}\bar\sigma^\mu(K_i\mathcal{D}_\mu+\mathcal{D}_iK_jD_\mu\phi^j)\psi^i\\
&\qquad+igK^{i^*}\bar\lambda^a(\phi^\dagger T_a)_{i^*}+igK_i\bar\lambda^a(T_a\phi)^i\Big]
\end{aligned}
\tag{2.23}
$$

$$
\begin{aligned}
&+\bar\theta\cdot\bar\theta\,\theta\cdot\Big[-\sqrt{2}K^{i^*}{}_i F^\dagger_{i^*}\psi^i - \frac{1}{\sqrt{2}}K^{k^*}{}_i\Gamma^{i^*}{}_{k^*}{}^{j^*}\bar\psi_{i^*}\cdot\bar\psi_{j^*}\psi^i \\
&\qquad -\frac{i}{\sqrt{2}}K^{i^*}{}_i\sigma^\mu\bar\psi_{i^*}D_\mu\phi^i + \frac{i}{\sqrt{2}}\sigma^\mu(K^{i^*}\mathcal{D}_\mu + \mathcal{D}^{i^*}K^{j^*}D_\mu\phi^\dagger_{j^*})\bar\psi_{i^*} \\
&\qquad -igK^{i^*}\lambda^a(\phi^\dagger T_a)_{i^*} - igK_i\lambda^a(T_a\phi)^i\Big] \\
&+\theta\cdot\theta\,\bar\theta\cdot\bar\theta\Big[-\frac{1}{4}\partial_\mu(K_iD^\mu\phi^i + K^{i^*}D^\mu\phi^\dagger_{i^*}) + K^{j^*}{}_iD_\mu\phi^iD^\mu\phi^\dagger_{j^*} + K^{i^*}{}_iF^iF^\dagger_{i^*} \\
&\qquad +\frac{1}{4}K^{i^*j^*}{}_{ij}\psi^i\cdot\psi^j\bar\psi_{i^*}\cdot\bar\psi_{j^*} - \frac{i}{2}K^{i^*}{}_i(\psi^i\sigma^\mu\mathcal{D}_\mu\bar\psi_{i^*} - \mathcal{D}_\mu\psi^i\sigma^\mu\bar\psi_{i^*}) \\
&\qquad +\frac{1}{2}K^{i^*}{}_i\Gamma^{j^*}{}_{i^*}{}^{k^*}F^i\bar\psi_{j^*}\cdot\bar\psi_{k^*} + \frac{1}{2}K^{i^*}{}_i\Gamma_j{}^i{}_kF^\dagger_{i^*}\psi^j\cdot\psi^k \\
&\qquad -\frac{g}{2}D^a((\phi^\dagger T_a)_{i^*}K^{i^*} + K^i(T_a\phi)^i) \\
&\qquad +ig\sqrt{2}(\phi^\dagger T_a)_{i^*}K^{i^*}{}_i\psi^i\cdot\lambda^a - ig\sqrt{2}\bar\lambda^a\cdot\bar\psi_{i^*}K^{i^*}{}_i(T_a\phi)^i\Big]\,,
\end{aligned}
$$

where the derivatives are now covariant both with respect to the gauge group and to the Kähler manifold. They can be expressed as

$$
\begin{aligned}
\mathcal{D}_\mu\psi^i &= D_\mu\psi^i + \Gamma_j{}^i{}_kD_\mu\phi^j\psi^k\,, & D_\mu\psi^i &= \partial_\mu\psi^i + igv^a_\mu(T_a\psi)^i\,, \\
\mathcal{D}_\mu\bar\psi_{i^*} &= D_\mu\bar\psi_{i^*} + \Gamma^{j^*}{}_{i^*}{}^{k^*}D_\mu\phi^\dagger_{j^*}\bar\psi_{k^*}\,, & D_\mu\bar\psi_{i^*} &= \partial_\mu\bar\psi_{i^*} - igv^a_\mu(\bar\psi T_a)_{i^*}\,, \\
D_\mu\phi^i &= \partial_\mu\phi^i + igv^a_\mu(T_a\phi)^i\,, & D_\mu\phi^\dagger_{i^*} &= \partial_\mu\phi^\dagger_{i^*} - igv^a_\mu(\phi^\dagger T_a)_{i^*}\,, \\
\mathcal{D}_iK_j &= K_{ij} - \Gamma_i{}^k{}_jK_k\,, & \mathcal{D}^{i^*}K^{j^*} &= K^{i^*j^*} - \Gamma^{i^*}{}_{k^*}{}^{i^*}K^{k^*}\,.
\end{aligned}
\tag{2.24}
$$

The final result has been simplified by making use of identities such as

$$
\begin{aligned}
0 = \delta_aK &= K_i(T_a\Phi)^i - K^{i^*}(\Phi^\dagger T_a)_{i^*}\,, \\
0 = \delta_a\delta_bK &= \Big[K_{ij}(T_a\Phi)^i(T_b\Phi)^j + K_i(T_aT_b\Phi)^i - K^{i^*}{}_i(T_a\Phi)^i(\Phi^\dagger T_b)_{i^*}\Big] \\
&\quad + \Big[K^{i^*j^*}(\Phi^\dagger T_a)_{i^*}(\Phi^\dagger T_a)_{j^*} + K^{i^*}(\Phi^\dagger T_aT_b)_{i^*} - K^{i^*}{}_i(T_a\Phi)^i(\Phi^\dagger T_b)_{i^*}\Big]\,,
\end{aligned}
$$

which express the gauge invariance of the Kähler potential. In this respect, $K_iD_\mu\phi^i$ and $K^{i^*}D_\mu\phi^\dagger_{i^*}$ are both gauge singlet quantities so that the term involving a total (spacetime) derivative in the $\theta\cdot\theta\bar\theta\cdot\bar\theta$ component can be rewritten using a total covariant derivative. Moreover, gauge invariance also ensures that all the terms of (2.22) that are proportional to $\bar\lambda\bar\psi\phi^\dagger$ and $\lambda\psi\phi$ cancel.

The gauge kinetic function is the last of the three functions necessary to build the most general supersymmetric Lagrangian. It allows one to address kinetic and gauge interaction terms for the gauge fields and is a chiral superfield $h_{ab}(\Phi)$, where a and b are indices related to the adjoint representation of the gauge group, coupled in a gauge-invariant fashion to the squared superfield strength tensor. The holomorphic piece of the gauge sector Lagrangian depends on the function h as

$$\begin{aligned}
\frac{1}{4g^2}h_{ab}(\Phi)W^{a\alpha}W^b_\alpha = &\Big[h_{ab}(\phi) + \sqrt{2}\theta\cdot\psi^i h_{abi} - \theta\cdot\theta\Big(F^i h_{abi} + \frac{1}{2}\psi^i\cdot\psi^j h_{abij}\Big)\Big] \\
&\Big[- \lambda^a\cdot\lambda^b + \theta\sigma^\mu\bar\sigma^\nu\lambda^a F^b_{\mu\nu} + 2i\theta\cdot\lambda^a D^b \\
&+\theta\cdot\theta\Big(- \frac{1}{2}F^a_{\mu\nu}F^{b\mu\nu} - \frac{i}{2}F^a_{\mu\nu}{}^*F^{b\mu\nu} + D^aD^b - 2i\lambda^a\sigma^\mu D_\mu\bar\lambda^b\Big)\Big] \\
= &-h_{ab}(\phi)\lambda^a\cdot\lambda^b \\
&+\theta\cdot\Big[h_{ab}(\phi)(\sigma^\mu\bar\sigma^\nu\lambda^a F^b_{\mu\nu}+2i\lambda^a D^b) - \sqrt{2}h_{abi}\lambda^a\cdot\lambda^b\psi^i\Big] \\
&+\theta\cdot\theta\Big[(F^i h_{abi}+\frac{1}{2}h_{abij}\psi^i\cdot\psi^j)\lambda^a\cdot\lambda^b - \frac{\sqrt{2}}{2}h_{abi}\psi^i\sigma^\mu\bar\sigma^\nu\lambda^a F^b_{\mu\nu} \\
&-\frac{1}{2}h_{ab}F^a_{\mu\nu}F^{b\mu\nu} - \frac{i}{2}h_{ab}F^a_{\mu\nu}{}^*F^{b\mu\nu} - 2ih_{ab}\lambda^a\sigma^\mu D_\mu\bar\lambda^b \\
&- \sqrt{2}ih_{abi}\psi^i\cdot\lambda^a D^b + h_{ab}D^aD^b\Big]\ ,
\end{aligned}$$

while the anti-holomorphic piece is similar and can be deduced from the second relation of (2.13), although the results depend instead on the complex conjugate function $h^\star(\Phi^\dagger)$. In the expression above, we have kept the dependence of the fields on the y-variable and simplifications have been performed on the basis of the symmetry properties of the gauge kinetic function under the exchange of its indices, $h_{ab} = h_{ba}$. Moreover, we have introduced our notations for the derivatives of the gauge kinetic function, namely

$$\begin{aligned}
h_{abi} &= \frac{\partial h_{ab}(\phi)}{\partial\phi^i}\ , \\
h_{abij} &= \frac{\partial^2 h_{ab}(\phi)}{\partial\phi^i\partial\phi^j}\ .
\end{aligned} \tag{2.25}$$

Collecting the results of Eq. (2.14), Eq. (2.23) and Eq. (2.25), the action of the most general supersymmetric model where matter superfields are coupled to gauge superfields is given by[4]

[4]The conventions adopted in this book are different from the ones in Fuks and Rausch de Traubenberg (2011) (see, *e.g.*, the covariant derivatives and the normalisation of the vector superfields). The comparison is however straightforward as it is sufficient to replace each factor of i by $-i$, except for the ${}^*F^{\mu\nu}F_{\mu\nu}$ term (due to the conventions for the rank-four Levi-Civita tensor). Additional differences occur in the case where the explicit expansion of the field strength tensor $F_{\mu\nu}$ and the covariant derivative $D_\mu\lambda$ is performed, since the commutation relation of Eq. (2.9) holds in both works.

$$
\begin{aligned}
S = & \int \mathrm{d}^4x\, \mathrm{d}^2\theta\, \mathrm{d}^2\bar{\theta}\, \frac{1}{2}\Big[K(\Phi, \Phi^\dagger e^{-2gV}) + K(e^{-2gV}\Phi, \Phi^\dagger)\Big] \\
& + \int \mathrm{d}^4x\, \mathrm{d}^2\theta\, W(\Phi) \\
& + \int \mathrm{d}^4x\, \mathrm{d}^2\bar{\theta}\, W^\star(\Phi^\dagger) \\
& + \frac{1}{16g^2} \int \mathrm{d}^4x\, \mathrm{d}^2\theta\, h_{ab}(\Phi) W^{a\alpha} W^b_\alpha \\
& + \frac{1}{16g^2} \int \mathrm{d}^4x\, \mathrm{d}^2\bar{\theta}\, h^\star_{ab}(\Phi^\dagger) \overline{W}^a_{\dot{\alpha}} \overline{W}^{b\dot{\alpha}} \,.
\end{aligned}
\tag{2.26}
$$

This action, invariant under both Yang-Mills and global supersymmetric transformations, describes a class of theories named *gauged supersymmetric* theories, in contrast to *ungauged supersymmetric* theories in which the gauge symmetry group is abelian and whose corresponding action can be trivially deduced from the results presented in this section.

In the following chapters, we detail the procedure leading to the generalisation of the action (2.26) when gravitational interactions are accounted for. As the result is invariant under both local super-Poincaré and Yang-Mills transformations, one often dubs the related theories *gauged supergravity* theories. This again contrasts with *ungauged supergravity* where the gauge group is reduced to be abelian.

Chapter 3

Curved superspace – Geometrical principles

This chapter is devoted to the introduction of an *ad hoc* curved superspace which extends the flat superspace of the previous chapter. In particular, the supergravity algebra, including local Lorentz transformations and superdiffeomorphims, will be defined. Dynamical superfields, *i.e.*, supervierbein and spin superconnection will be introduced together with their associated curvature and tensor superfields. Since these superfields have too many degrees of freedom, some constraints upon the torsion tensors need to be imposed. The reduction of the Bianchi identities will enable us to express all curvature and torsion superfields in terms of three superfields only.

It turns out that the structure of the algebra is larger than the supergravity algebra, and in particular, that the torsion constraints are invariant under superconformal transformations called Howe–Tucker transformations. These transformations will be studied in some details.

Selecting a specific gauge, supergravity transformations will be defined as a composition of a specific local Lorentz transformation with a specific superdiffeomorphim, which preserves the gauge fixing condition. Finally, all the lowest order components of the fundamental superfields will be explicitly computed.

3.1 Curved superspace

Curved superspace has been originally introduced in [Arnowitt et al. (1975); Nath and Arnowitt (1975, 1976); Brink et al. (1978); Ne'eman and Regge (1978)] and defined such that the symmetry group of the tangent superspace at each curved superspace point is the orthosymplectic group $OSP(1|4)$. As this tangent space does not coincide with the superspace of supersymmetry described in Chapter 2, the equivalence principle turns out to be difficult to accommodate. The solution to this issue, or equivalently the construction of a superspace adapted to supergravity computations, has been brought about by [Wess and Zumino (1977, 1978b,a); Grimm et al. (1978)] and it relies on two key features. First, a local Lorentz invariance is imposed at each point of the curved superspace, instead of an invariance under the orthosymplectic group. Second, the torsion tensor has been constrained in such a way that the equivalence principle remains valid. This allows one to extend the formalism of chiral and vector superfields introduced in supersymmetry to the curved superspace, so that these objects can be further used for Lagrangian building in supergravity.

While this chapter is entirely based on the Wess and Zumino approach, there exists an alternative way to construct a superspace dedicated to supergravity [Akulov et al. (1977)], which we choose not to address in the present work.

3.1.1 *Supervierbein, superconnection, torsion and curvature*

In this section, we introduce and define the dynamical superfields associated with a curved superspace, namely the supervierbein and the spin connection, together with the corresponding curvature and torsion tensors. Our construction is based on the approach of Wess and Zumino (1977) and begins with assuming that at each point of the superspace, the structure group is the Lorentz group. To account for local supersymmetry, we consider supersymmetric transformations with parameters $\epsilon(x)$ depending on the spacetime coordinates, and subsequently follow a standard path based on the Nœther procedure to build invariant actions. Among the different possible ways to proceed, we choose to use the differential form language. In this case, the calculations are very similar to those arising in general relativity problems, and only contain extra technical difficulties [Fuks and Rausch de Traubenberg (2011); Wess and Bagger (1992); West (1986); van Nieuwenhuizen (1981); Weinberg (1972a); Muller (1989); Gates et al. (1983); Freund (1986)].

As in general relativity, the main idea is based on the principle of equivalence. Locally, we choose a reference frame where gravitation is eliminated and the metric is Minkowskian, and define a (local) flat tangent space; whereas in contrast, gravitation holds in general reference frames of the curved superspace. This forces us to consider two types of transformations, local Lorentz transformations acting on the tangent space and general transformations acting on the curved space. Generically denoting the coordinates in the flat tangent superspace by $z^M = (x^\mu, \theta^\alpha, \bar\theta_{\dot\alpha})$, we define local Lorentz transformations by

$$\begin{aligned} x'^\mu &= x^\nu \Lambda_\nu{}^\mu(z) , \\ \theta'^\alpha &= \theta^\beta \Lambda_\beta{}^\alpha(z) , \\ \bar\theta'_{\dot\alpha} &= \bar\theta_{\dot\beta} \Lambda^{\dot\beta}{}_{\dot\alpha}(z) . \end{aligned} \tag{3.1}$$

In these relations, the matrices $\Lambda_\nu{}^\mu(z)$ ($\Lambda_\beta{}^\alpha(z)$ and $\Lambda^{\dot\beta}{}_{\dot\alpha}(z)$) correspond to Lorentz transformations in the vector (spinor) representation and depend on the coordinates. Since these local Lorentz transformations take the form above, they do not mix indices of different nature (vector, left-handed and right-handed spinor indices) and it is possible to define a superfield with a definite Lorentz nature. Introducing similar notations for the curved superspace, we generically denote the associated coordinates by $z^{\tilde M} = (x^{\tilde\mu}, \theta^{\tilde\alpha}, \bar\theta_{\tilde{\dot\alpha}})$. A general transformation is then defined by

$$z'^{\tilde M} = z^{\tilde M} + \xi^{\tilde M}(z)\,, \tag{3.2}$$

where the transformation parameters $\xi^{\tilde M}(z)$ are superfields depending on the coordinates. Following the conventions described in App. A, non-tilded indices stand for flat space indices while tilded indices are Einstein indices in the curved superspace. Moreover, Greek letters of the beginning of the alphabet are dedicated to spin indices while those of the middle of the alphabet to vector indices[1]. The variables conjugate to z^M are denoted by $\partial_M = (\partial_\mu, \partial_\alpha, \bar\partial^{\dot\alpha})$ and satisfy, by definition,

$$[\partial_M, z^N]_{|M||N|} = \delta_M{}^N\,,$$

where the quantity $|M|$ is the grading of the index M and is equal to zero for a bosonic index and one for a fermionic index. Additionally $[A_M, B_N]_{|M||N|} = A_M B_N - (-1)^{|M||N|} B_N A_N$ is dubbed *graded commutator.*

Flat and curved superspace indices are connected by the first of the dynamical variables, the *supervierbein* also sometimes called *vielbein*, while the *superconnection* leads to the definition of covariant derivatives[2]. The supervierbein $E_{\tilde M}{}^M$ and its inverse $E_M{}^{\tilde M}$ allow one to express a quantity V in flat (curved) superspace in terms of the corresponding object in curved (flat) superspace,

$$\begin{aligned} V_{\tilde M} &= E_{\tilde M}{}^M V_M\,,\\ V_M &= E_M{}^{\tilde M} V_{\tilde M} \end{aligned} \tag{3.3}$$

with

$$\begin{aligned} E_{\tilde M}{}^M E_M{}^{\tilde N} &= \delta_{\tilde M}{}^{\tilde N}\,,\\ E_M{}^{\tilde M} E_{\tilde M}{}^N &= \delta_M{}^N\,. \end{aligned}$$

This implies that for any curved superspace point, one can (locally) eliminate gravitation and construct a theory, in the flat tangent superspace, invariant under

[1] These notations which are not widely used follow those introduced in [Fuks and Rausch de Traubenberg (2011)].

[2] Strictly speaking, it is necessary to introduce Christoffel symbols for constructing a fully covariant derivative for tensors containing both Lorentzian and Einsteinian indices. They are however irrelevant for this work. As examples, the torsion tensor (see Eq. (3.17)) or the gravitino field strength tensor (see Eq. (3.116)) have symmetry properties such that the contributions of the Christoffel symbols vanish.

the Lorentz transformations (3.1). In contrast, in curved superspace, the theory will be invariant under the superdiffeomorphisms (Eq. (3.2)).

The second of the dynamical variables, the superconnection $\Omega_{\tilde{M}M}{}^{N}$, satisfies the symmetry property,

$$\Omega_{\tilde{M}MN} = -(-)^{|M||N|}\Omega_{\tilde{M}NM} \,. \tag{3.4}$$

Since $\Omega_{\tilde{M}MN}$ is Lie-algebra valued (see App. B), the only non-vanishing components of the superconnection are $\Omega_{\tilde{M}\mu\nu}$, $\Omega_{\tilde{M}\alpha\beta}$ and $\Omega_{\tilde{M}}{}^{\dot{\alpha}\dot{\beta}}$. As stated above, this quantity is introduced with the aim of defining covariant derivatives with respect to Lorentz transformations[3],

$$\begin{aligned} &\mathcal{D}_{\tilde{M}}X^M = \partial_{\tilde{M}}X^M + (-)^{|\tilde{M}||N|}X^N\Omega_{\tilde{M}N}{}^M \,, && \mathcal{D}_{\tilde{M}}X_M = \partial_{\tilde{M}}X_M - \Omega_{\tilde{M}M}{}^N X_N \,, \\ &\mathcal{D}_N X^M = E_N{}^{\tilde{M}}\mathcal{D}_{\tilde{M}}X^M \,, && \mathcal{D}_N X_M = E_N{}^{\tilde{M}}\mathcal{D}_{\tilde{M}}X_M \,. \end{aligned} \tag{3.5}$$

This relies on the assumption that under a Lorentz transformation of parameter Λ, the superconnection varies as

$$\Omega_{\tilde{M}M}{}^N \to \Omega'_{\tilde{M}M}{}^N = (\Lambda^{-1})_M{}^P\Omega_{\tilde{M}P}{}^Q\Lambda_Q{}^N - (\Lambda^{-1})_M{}^P\partial_{\tilde{M}}\Lambda_P{}^N \,. \tag{3.6}$$

Consequently, we can show that the derivatives (3.5) have the required behaviour under Lorentz transformations[4],

$$\begin{aligned} &\mathcal{D}_{\tilde{M}}X^M \to (\mathcal{D}_{\tilde{M}}X^N)\Lambda_N{}^M \,, && \mathcal{D}_{\tilde{M}}X_M \to (\Lambda^{-1})_M{}^N(\mathcal{D}_{\tilde{M}}X_N) \,, \\ &\mathcal{D}_N X^M \to (\Lambda^{-1})_N{}^P(\mathcal{D}_P X^Q)\Lambda_Q{}^M \,, && \mathcal{D}_N X_M \to (\Lambda^{-1})_M{}^P(\Lambda^{-1})_N{}^Q(\mathcal{D}_Q X_P) \,, \end{aligned} \tag{3.7}$$

and that

$$\mathcal{D}_{\tilde{M}}(X^M Y_M) = \partial_{\tilde{M}}(X^M Y_M) \,.$$

Notice that the dynamical variables introduced above are superfields which can be of a fermionic or bosonic nature, according to the gradings of their indices. The elements of the supervierbein $E_{\tilde{M}}{}^M$ are hence fermionic (bosonic) quantities if $|\tilde{M}| + |M| = 1$ ($|\tilde{M}| + |M| = 0$). Along the same lines, those of the superconnection $\Omega_{\tilde{M}M}{}^N$ are fermionic (bosonic) if $|\tilde{M}| + |M| + |N| = 1$ ($|\tilde{M}| + |M| + |N| = 0$), or equivalently if $|\tilde{M}| = 1$ ($|\tilde{M}| = 0$) since the indices M and N are of the same nature due to the Lie-algebra valuedness of $\Omega_{\tilde{M}}{}^{MN}$.

Considering a generic superfield V^P, the torsion and the curvature tensors are defined by

$$\begin{aligned} \Big[\mathcal{D}_M, \mathcal{D}_N\Big]_{|M||N|} V^P &= \Big(\mathcal{D}_M\mathcal{D}_N - (-)^{|M||N|}\mathcal{D}_N\mathcal{D}_M\Big)V^P \\ &= T_{MN}{}^Q\mathcal{D}_Q V^P + (-)^{|Q|(|M|+|N|)}V^Q R_{MNQ}{}^P \,. \end{aligned} \tag{3.8}$$

[3]The sign below is obtained because $\Omega_{\tilde{N}N}{}^M$ is pushed through X^N leading to the factor $(-1)^{|N|(|\tilde{M}|+|M|+|N|)} = (-1)^{|N||\tilde{M}|}$.

[4]Pay attention of the positions of the indices P and Q.

Since the tensor $R_{MNP}{}^{Q}$ is Lie-algebra valued in its last two indices P and Q, these are of the same fermionic or bosonic nature so that we could have replaced $|Q|$ by $|P|$ in Eq. (3.8). Using Eqs. (B.4) and (B.11), one derives the graded commutator operator

$$\left[\mathcal{D}_M, \mathcal{D}_N\right]_{|M||N|} = T_{MN}{}^{P}\mathcal{D}_P - \frac{1}{2}R_{MN\underline{\alpha}\underline{\beta}}J^{\underline{\beta}\underline{\alpha}} , \tag{3.9}$$

where $\underline{\alpha}$ stands for a generic spin index being either α or $\dot{\alpha}$. On the right-hand side of this expression, the operator $J^{\mu\nu}$ is absent since the algebra must be expressed in terms of independent operators, namely $J^{\beta\alpha}$ and $J^{\dot{\beta}\dot{\alpha}}$ (see App. B). This identity shows that the algebra of the operators $\mathcal{D}_M$, the generators of the supergravity transformations (see Sec. 3.3), closes only after accounting for Lorentz transformations in the flat tangent space. On the basis of the relation $[\mathcal{D}_M, \mathcal{D}_N]_{|M||N|} = -(-)^{|M||N|}[\mathcal{D}_N, \mathcal{D}_M]_{|M||N|}$ and of the Lie-algebra valuedness of the curvature tensor, we can derive the symmetry properties of the torsion and curvature tensors under the exchange of their indices,

$$\begin{aligned} T_{MNP} &= -(-)^{|M||N|}T_{NMP} ,\\ R_{MNPQ} &= -(-)^{|M||N|}R_{NMPQ} ,\\ R_{MNPQ} &= -(-)^{|P||Q|}R_{MNQP} . \end{aligned} \tag{3.10}$$

Moreover, since $\mathcal{D}_\mu^\dagger = -\mathcal{D}_\mu$ and $\mathcal{D}_\alpha^\dagger = -\overline{\mathcal{D}}_{\dot{\alpha}}$, we can derive the properties of the various torsion and curvature tensor elements under hermitian conjugation. For instance, considering the conjugation of the commutator $[\mathcal{D}_\mu, \mathcal{D}_\nu]$, one obtains the relation $(T_{\mu\nu}{}^{\alpha})^\dagger = T_{\mu\nu}{}^{\dot{\alpha}}$.

In order to explicitly compute the elements of the torsion and curvature tensors, we start by expressing the graded commutator (3.9) as

$$\begin{aligned} \left[\mathcal{D}_M, \mathcal{D}_N\right]_{|M||N|} &= \mathcal{D}_M\mathcal{D}_N - (-)^{|M||N|}\mathcal{D}_N\mathcal{D}_M\\ &= \left[E_M{}^{\tilde{M}}(\mathcal{D}_{\tilde{M}}E_N{}^{\tilde{N}}) - (-)^{|M||N|}E_N{}^{\tilde{M}}(\mathcal{D}_{\tilde{M}}E_M{}^{\tilde{N}})\right]\mathcal{D}_{\tilde{N}}\\ &\quad +(-)^{|\tilde{M}|(|\tilde{N}|+|N|)}E_M{}^{\tilde{M}}E_N{}^{\tilde{N}}\mathcal{D}_{\tilde{M}}\mathcal{D}_{\tilde{N}}\\ &\quad -(-)^{|M||N|}(-)^{|\tilde{N}|(|\tilde{M}|+|M|)}E_N{}^{\tilde{N}}E_M{}^{\tilde{M}}\mathcal{D}_{\tilde{N}}\mathcal{D}_{\tilde{M}}\\ &= \left[E_M{}^{\tilde{M}}\mathcal{D}_{\tilde{M}}E_N{}^{\tilde{N}} - (-)^{|M||N|}E_N{}^{\tilde{M}}\mathcal{D}_{\tilde{M}}E_M{}^{\tilde{N}}\right]\mathcal{D}_{\tilde{N}}\\ &\quad +(-)^{|\tilde{M}|(|\tilde{N}|+|N|)}E_M{}^{\tilde{M}}E_N{}^{\tilde{N}}\left[\mathcal{D}_{\tilde{M}}\mathcal{D}_{\tilde{N}} - (-)^{|\tilde{M}||\tilde{N}|}\mathcal{D}_{\tilde{N}}\mathcal{D}_{\tilde{M}}\right], \end{aligned} \tag{3.11}$$

and next apply this operator to the superfield V^Q. The covariant derivative of $V_N{}^{\tilde{N}}$ and $V_{\tilde{N}}{}^N$ being given by

$$\begin{aligned}\mathcal{D}_{\tilde{M}} V_N{}^{\tilde{N}} &= \partial_{\tilde{M}} V_N{}^{\tilde{N}} - \Omega_{\tilde{M}N}{}^P V_P{}^{\tilde{N}} ,\\ \mathcal{D}_{\tilde{M}} V_{\tilde{N}}{}^N &= \partial_{\tilde{M}} V_{\tilde{N}}{}^N + (-)^{|\tilde{M}|(|\tilde{N}|+|M|)} V_{\tilde{N}}{}^M \Omega_{\tilde{M}M}{}^N ,\end{aligned} \tag{3.12}$$

we can simplify the first term of (3.11) as

$$\begin{aligned}&\Big[E_M{}^{\tilde{M}}(\partial_{\tilde{M}} E_N{}^{\tilde{N}})E_{\tilde{N}}{}^P - (-)^{|M||N|}E_N{}^{\tilde{M}}(\partial_{\tilde{M}} E_M{}^{\tilde{N}})E_{\tilde{N}}{}^P - E_M{}^{\tilde{M}}\Omega_{\tilde{M}N}{}^P\\ &\qquad +(-)^{|M||N|}E_N{}^{\tilde{M}}\Omega_{\tilde{M}M}{}^P\Big]\mathcal{D}_P V^Q ,\end{aligned} \tag{3.13}$$

and the second term as

$$\begin{aligned}\Big[\mathcal{D}_{\tilde{M}}\mathcal{D}_{\tilde{N}} - (-)^{|\tilde{M}||\tilde{N}|}\mathcal{D}_{\tilde{N}}\mathcal{D}_{\tilde{M}}\Big]V^Q &= (-)^{|S|(|\tilde{M}|+|\tilde{N}|)}V^S\Big[\partial_{\tilde{M}}\Omega_{\tilde{N}S}{}^Q - (-)^{|\tilde{M}||\tilde{N}|}\partial_{\tilde{N}}\Omega_{\tilde{M}S}{}^Q\\ &\qquad - \Omega_{\tilde{M}S}{}^R\Omega_{\tilde{N}R}{}^Q + (-)^{|\tilde{M}|(|\tilde{N}|)}\Omega_{\tilde{N}S}{}^R\Omega_{\tilde{M}R}{}^Q\Big] .\end{aligned} \tag{3.14}$$

These two relations allow us to express the torsion and curvature tensors in terms of the supervierbein and the superconnection,

$$\begin{aligned}T_{MN}{}^Q =&\ E_M{}^{\tilde{M}}(\partial_{\tilde{M}} E_N{}^{\tilde{N}})E_{\tilde{N}}{}^Q - (-)^{|M||N|}E_N{}^{\tilde{M}}(\partial_{\tilde{M}} E_M{}^{\tilde{N}})E_{\tilde{N}}{}^Q\\ &-E_M{}^{\tilde{M}}\Omega_{\tilde{M}N}{}^Q + (-)^{|M||N|}E_N{}^{\tilde{M}}\Omega_{\tilde{M}M}{}^Q ,\\ R_{MNS}{}^Q =&\ (-)^{|\tilde{M}|(|\tilde{N}|+|N|)}E_M{}^{\tilde{M}}E_N{}^{\tilde{N}}\Big[\partial_{\tilde{M}}\Omega_{\tilde{N}S}{}^Q - \Omega_{\tilde{M}S}{}^R\Omega_{\tilde{N}R}{}^Q\\ &-(-)^{|\tilde{M}||\tilde{N}|}\Big(\partial_{\tilde{N}}\Omega_{\tilde{M}S}{}^Q - \Omega_{\tilde{N}S}{}^R\Omega_{\tilde{M}R}{}^Q\Big)\Big] ,\end{aligned} \tag{3.15}$$

where the derivation of the second expression entails the commutation of V^S and $E_{\tilde{M}}{}^M E_{\tilde{N}}{}^N$ which makes additional minus signs appear. Using the identity

$$0 = \partial_{\tilde{M}}\Big(E_N{}^{\tilde{N}}E_{\tilde{N}}{}^Q\Big) = \partial_{\tilde{M}}E_N{}^{\tilde{N}}E_{\tilde{N}}{}^Q + (-1)^{|\tilde{M}|(|N|+|\tilde{N}|)}E_N{}^{\tilde{N}}\partial_{\tilde{M}}E_{\tilde{N}}{}^Q ,$$

we can rewrite the torsion tensor as

$$\begin{aligned}T_{MN}{}^Q =& - (-)^{|\tilde{M}|(|N|+|\tilde{N}|)}E_M{}^{\tilde{M}}E_N{}^{\tilde{N}}\partial_{\tilde{M}}E_{\tilde{N}}{}^Q + (-)^{|M||N|+|\tilde{M}|(|M|+|\tilde{N}|)}E_N{}^{\tilde{M}}E_M{}^{\tilde{N}}\partial_{\tilde{M}}E_{\tilde{N}}{}^Q\\ &- E_M{}^{\tilde{M}}\Omega_{\tilde{M}N}{}^Q + (-)^{|M||N|}E_N{}^{\tilde{M}}\Omega_{\tilde{M}M}{}^Q ,\end{aligned} \tag{3.16}$$

which, after introducing Einsteinian indices and using Eq. (3.12), yields

$$\begin{aligned}T_{\tilde{P}\tilde{Q}}{}^Q &= (-)^{|\tilde{P}|(|N|+|\tilde{Q}|)}E_{\tilde{Q}}{}^N E_{\tilde{P}}{}^M T_{MN}{}^Q\\ &= -\partial_{\tilde{P}}E_{\tilde{Q}}{}^Q + (-1)^{|\tilde{P}||\tilde{Q}|}\partial_{\tilde{Q}}E_{\tilde{P}}{}^Q - (-1)^{|\tilde{P}|(|\tilde{Q}|+|N|)}E_{\tilde{Q}}{}^N\Omega_{\tilde{P}N}{}^Q\\ &\quad + (-1)^{|\tilde{Q}||M|}E_{\tilde{P}}{}^M\Omega_{\tilde{Q}M}{}^Q\\ &= -\mathcal{D}_{\tilde{P}}E_{\tilde{Q}}{}^Q + (-)^{|\tilde{P}||\tilde{Q}|}\mathcal{D}_{\tilde{Q}}E_{\tilde{P}}{}^Q .\end{aligned} \tag{3.17}$$

Table 3.1 Torsion and curvature tensors with flat and curved indices.

The torsion and curvature tensors are defined from the graded commutators of two superderivatives,

$$\Big[\mathcal{D}_M, \mathcal{D}_N\Big]_{|M||N|} = T_{MN}{}^P \mathcal{D}_P - \frac{1}{2} R_{MN\underline{\alpha}\underline{\beta}} J^{\underline{\beta}\underline{\alpha}} ,$$

with

$$\begin{aligned}
T_{MN}{}^Q &= E_M{}^{\tilde{M}}(\partial_{\tilde{M}} E_N{}^{\tilde{N}}) E_{\tilde{N}}{}^Q - E_M{}^{\tilde{M}} \Omega_{\tilde{M}N}{}^Q - (-)^{|M||N|}\Big(E_N{}^{\tilde{M}}(\partial_{\tilde{M}} E_M{}^{\tilde{N}}) E_{\tilde{N}}{}^Q - E_N{}^{\tilde{M}} \Omega_{\tilde{M}M}{}^Q\Big) , \\
R_{MNS}{}^Q &= (-)^{|\tilde{M}|(|\tilde{N}|+|N|)} E_M{}^{\tilde{M}} E_N{}^{\tilde{N}}\Big[\partial_{\tilde{M}} \Omega_{\tilde{N}S}{}^Q - \Omega_{\tilde{M}S}{}^R \Omega_{\tilde{N}R}{}^Q - (-)^{|\tilde{M}||\tilde{N}|}\Big(\partial_{\tilde{N}} \Omega_{\tilde{M}S}{}^Q - \Omega_{\tilde{N}S}{}^R \Omega_{\tilde{M}R}{}^Q\Big)\Big] , \\
T_{\tilde{P}\tilde{Q}}{}^Q &= -\mathcal{D}_{\tilde{P}} E_{\tilde{Q}}{}^Q + (-)^{|\tilde{P}||\tilde{Q}|} \mathcal{D}_{\tilde{Q}} E_{\tilde{P}}{}^Q , \\
R_{\tilde{M}\tilde{N}S}{}^Q &= \partial_{\tilde{M}} \Omega_{\tilde{N}S}{}^Q - (-)^{|\tilde{M}||\tilde{N}|} \partial_{\tilde{N}} \Omega_{\tilde{M}S}{}^Q - \Omega_{\tilde{M}S}{}^R \Omega_{\tilde{N}R}{}^Q + (-)^{|\tilde{M}||\tilde{N}|} \Omega_{\tilde{N}S}{}^R \Omega_{\tilde{M}R}{}^Q .
\end{aligned}$$

Converting the first two indices of the curvature tensor of Eq. (3.15) to Einsteinian indices, one obtains,

$$\begin{aligned}
R_{\tilde{M}\tilde{N}S}{}^Q &= (-)^{|\tilde{M}|(|N|+|\tilde{N}|)} E_{\tilde{N}}{}^N E_{\tilde{M}}{}^M R_{MNS}{}^Q \\
&= \partial_{\tilde{M}} \Omega_{\tilde{N}S}{}^Q - (-)^{|\tilde{M}||\tilde{N}|} \partial_{\tilde{N}} \Omega_{\tilde{M}S}{}^Q - \Omega_{\tilde{M}S}{}^R \Omega_{\tilde{N}R}{}^Q + (-)^{|\tilde{M}||\tilde{N}|} \Omega_{\tilde{N}S}{}^R \Omega_{\tilde{M}R}{}^Q .
\end{aligned} \tag{3.18}$$

These results are summarised in Table 3.1.

In (global) supersymmetry, the formalism above leads to drastic simplifications. First, the connection $\Omega_{\tilde{M}M}{}^P$ vanishes so that $\mathcal{D}_{\tilde{M}} = \partial_{\tilde{M}}$. Next, the superderivatives with flat indices are defined by

$$\begin{aligned}
D_\mu &= \partial_\mu = E_\mu{}^{\tilde{M}} \partial_{\tilde{M}} , \\
D_\alpha &= E_\alpha{}^{\tilde{M}} \partial_{\tilde{M}} , \\
\overline{D}^{\dot{\alpha}} &= E^{\dot{\alpha}\tilde{M}} \partial_{\tilde{M}} .
\end{aligned}$$

As these relations are equivalent to those of Eq. (2.1), the supervierbein and its inverse are given by

$$\begin{aligned}
E_M{}^{\tilde{M}} &= \begin{pmatrix} \delta_\mu{}^{\tilde{\mu}} & 0 & 0 \\ i\delta_\mu{}^{\tilde{\mu}}(\sigma^\mu\overline{\theta})_\alpha & \delta_\alpha{}^{\tilde{\alpha}} & 0 \\ -i\delta_\mu{}^{\tilde{\mu}}(\theta\sigma^\mu)_{\dot{\beta}}\,\varepsilon^{\dot{\alpha}\dot{\beta}} & 0 & \delta^{\dot{\alpha}}{}_{\tilde{\dot{\alpha}}} \end{pmatrix} , \\
E_{\tilde{M}}{}^M &= \begin{pmatrix} \delta_{\tilde{\mu}}{}^\mu & 0 & 0 \\ -i\delta_{\tilde{\mu}}{}^\mu(\sigma^{\tilde{\mu}}\overline{\theta})_{\tilde{\alpha}} & \delta_{\tilde{\alpha}}{}^\alpha & 0 \\ i\delta_{\tilde{\mu}}{}^\mu(\theta\sigma^{\tilde{\mu}})_{\tilde{\dot{\beta}}}\,\varepsilon^{\tilde{\dot{\alpha}}\tilde{\dot{\beta}}} & 0 & \delta^{\tilde{\dot{\alpha}}}{}_{\dot{\alpha}} \end{pmatrix} .
\end{aligned} \tag{3.19}$$

Assuming a real curved superspace, the coordinates $z^{\tilde{M}} = (x^{\tilde{\mu}}, \theta^{\tilde{\alpha}}, \overline{\theta}_{\tilde{\dot{\alpha}}})$ satisfy the conjugation relations

$$\begin{aligned}
(x^{\tilde{\mu}})^\dagger &= x^{\tilde{\mu}} , \\
(\theta^{\tilde{\alpha}})^\dagger &= \overline{\theta}^{\tilde{\dot{\alpha}}} .
\end{aligned} \tag{3.20}$$

First, in the context of supersymmetry, with the definitions (2.1) for the superderivatives and the vielbein $E_{\tilde{M}}{}^{M}$ given in (3.19), one obtains similar relations for a point $z^M = (x^\mu, \theta^\alpha, \bar{\theta}_{\dot{\alpha}})$ of the flat superspace,

$$\begin{aligned}(x^\mu)^\dagger &= x^\mu\,,\\ (\theta^\alpha)^\dagger &= \bar{\theta}^{\dot{\alpha}}\,.\end{aligned}\tag{3.21}$$

This can be shown by using, among others, the conjugation properties of the non-vanishing elements of $E_{\tilde{M}}{}^{M}$,

$$\begin{aligned}(E_{\tilde{\mu}}{}^{\mu})^\dagger &= E_{\tilde{\mu}}{}^{\mu}\,,\\ (E_{\tilde{\alpha}}{}^{\mu})^\dagger &= E_{\bar{\tilde{\alpha}}}{}^{\mu}\,, \qquad (E^{\bar{\tilde{\alpha}}\mu})^\dagger = E^{\tilde{\alpha}\mu}\,,\\ (E_{\tilde{\alpha}}{}^{\alpha})^\dagger &= -E_{\bar{\tilde{\alpha}}}{}^{\dot{\alpha}}\,, \quad (E^{\bar{\tilde{\alpha}}}{}_{\dot{\alpha}})^\dagger = -E^{\tilde{\alpha}}{}_{\alpha}\,,\end{aligned}\tag{3.22}$$

that are obtained from the identities

$$\begin{aligned}(\delta_{\tilde{\alpha}}{}^{\alpha})^\dagger &= \delta_{\bar{\tilde{\alpha}}}{}^{\dot{\alpha}}\,,\\ \varepsilon^{\bar{\tilde{\alpha}}\bar{\tilde{\beta}}}\,\delta_{\bar{\tilde{\beta}}}{}^{\dot{\beta}}\,\varepsilon_{\dot{\alpha}\dot{\beta}} &= -\delta^{\bar{\tilde{\alpha}}}{}_{\dot{\alpha}}\,.\end{aligned}$$

Second, in the context of supergravity, all the elements of the vielbein could be non-zero. The relations (3.22) are then supplemented by the identities

$$\begin{aligned}(E_{\tilde{\mu}}{}^{\alpha})^\dagger &= E_{\tilde{\mu}}{}^{\dot{\alpha}}\,, \quad (E_{\tilde{\mu}\dot{\alpha}})^\dagger = E_{\tilde{\mu}\alpha}\,,\\ (E_{\tilde{\alpha}\dot{\alpha}})^\dagger &= -E_{\bar{\tilde{\alpha}}\alpha} \quad (E^{\bar{\tilde{\alpha}}\alpha})^\dagger = -E^{\tilde{\alpha}\dot{\alpha}}\,,\end{aligned}\tag{3.23}$$

which ensures that Eq. (3.21) is fulfilled. By making use of Eq. (3.20) together with Eq. (3.22), Eq. (3.23) and the graded commutation relation

$$X^M E_M{}^{\tilde{M}} = (-)^{|M||\tilde{M}|} E^{M\tilde{M}} X_M\,,$$

one can indeed derive

$$\begin{aligned}(\mathrm{d}x^\mu)^\dagger &= (\mathrm{d}z^{\tilde{M}} E_{\tilde{M}}{}^{\mu})^\dagger = \mathrm{d}x^{\tilde{\mu}} E_{\tilde{\mu}}{}^{\mu} + \mathrm{d}\bar{\theta}_{\bar{\tilde{\alpha}}} E^{\bar{\tilde{\alpha}}\mu} + \mathrm{d}\theta^{\tilde{\alpha}} E_{\tilde{\alpha}}{}^{\mu} = \mathrm{d}x^\mu\,,\\ (\mathrm{d}\theta^\alpha)^\dagger &= (\mathrm{d}z^{\tilde{M}} E_{\tilde{M}}{}^{\alpha})^\dagger = \mathrm{d}x^{\tilde{\mu}} E_{\tilde{\mu}}{}^{\dot{\alpha}} + \mathrm{d}\bar{\theta}^{\bar{\tilde{\alpha}}}(-E_{\bar{\tilde{\alpha}}}{}^{\dot{\alpha}}) + \mathrm{d}\theta_{\tilde{\alpha}}(-E^{\tilde{\alpha}\dot{\alpha}})\\ &= \mathrm{d}x^{\tilde{\mu}} E_{\tilde{\mu}}{}^{\dot{\alpha}} + \mathrm{d}\theta^{\tilde{\alpha}} E_{\tilde{\alpha}}{}^{\dot{\alpha}} + \mathrm{d}\bar{\theta}_{\bar{\tilde{\alpha}}} E^{\bar{\tilde{\alpha}}\dot{\alpha}} = \mathrm{d}\bar{\theta}^{\dot{\alpha}}\,.\end{aligned}$$

In other words, this means that both $(\mathrm{d}\theta^\alpha, \mathrm{d}\bar{\theta}_{\dot{\alpha}})$ (or $(\theta^\alpha, \bar{\theta}_{\dot{\alpha}})$) and $(\mathrm{d}\theta^{\tilde{\alpha}}, \mathrm{d}\bar{\theta}_{\bar{\tilde{\alpha}}})$ (or $(\theta^{\tilde{\alpha}}, \bar{\theta}_{\bar{\tilde{\alpha}}})$) are Majorana fermions, while other choices for the definitions of the vielbein would have led to additional minus signs. For instance, adopting the conventions of [Fuks and Rausch de Traubenberg (2011)] for the superderivatives would have implied that $(\theta^\alpha, -\bar{\theta}_{\dot{\alpha}})$ is a Majorana fermion in the flat superspace, with a non-vanishing Majorana phase.

Finally, the algebra of the superderivatives in supersymmetry, $\{D_\alpha, \bar{D}_{\dot{\alpha}}\} = -2i\sigma^\mu{}_{\alpha\dot{\alpha}}\partial_\mu$, warrants that the $T_{\alpha\dot{\alpha}}{}^M$ components of the torsion tensor satisfy

$$\begin{aligned} T_{\alpha\dot{\alpha}}{}^\beta &= T_{\alpha\dot{\alpha}}{}^{\dot{\beta}} = 0 \,, \\ T_{\alpha\dot{\alpha}}{}^\mu &= -2i\sigma^\mu{}_{\alpha\dot{\alpha}} \,. \end{aligned}$$

These considerations naturally lead to the constraints that are imposed in local supersymmetry to reduce the Bianchi identities (see the next section).

3.1.2 *The Bianchi identities*

All the components of the torsion and curvature superfields are not independent since they satisfy the Bianchi identities

$$\begin{aligned} 0 = (-)^{|M_1||M_3|}&\Big[\mathcal{D}_{M_1}, [\mathcal{D}_{M_2}, \mathcal{D}_{M_3}]_{|M_2||M_3|}\Big]_{|M_1|(|M_2|+|M_3|)} \\ +(-)^{|M_2||M_1|}&\Big[\mathcal{D}_{M_2}, [\mathcal{D}_{M_3}, \mathcal{D}_{M_1}]_{|M_3||M_1|}\Big]_{|M_2|(|M_3|+|M_1|)} \\ +(-)^{|M_3||M_2|}&\Big[\mathcal{D}_{M_3}, [\mathcal{D}_{M_1}, \mathcal{D}_{M_2}]_{|M_1||M_2|}\Big]_{|M_3|(|M_1|+|M_2|)} \,. \end{aligned} \tag{3.24}$$

Using Eq. (3.9), the Lie-algebra valuedness of the curvature tensor (see Eq. (B.6)) and the property

$$[X, YZ]_{|X|(|Y|+|Z|)} = [X, Y]_{|X||Y|}Z + (-)^{|X||Y|}Y[X, Z]_{|X||Z|} \,,$$

we can calculate the double-graded commutator

$$\begin{aligned} &\Big[\mathcal{D}_{M_1}, [\mathcal{D}_{M_2}, \mathcal{D}_{M_3}]_{|M_2||M_3|}\Big]_{|M_1|(|M_2|+|M_3|)} \\ &= \Big[\mathcal{D}_{M_1}, T_{M_2M_3}{}^R\mathcal{D}_R\Big]_{|M_1|(|M_2|+|M_3|)} - \frac{1}{2}\Big[\mathcal{D}_{M_1}, R_{M_2M_3\underline{\alpha}\underline{\beta}}J^{\underline{\beta}\underline{\alpha}}\Big]_{|M_1|(|M_2|+|M_3|)} \\ &= \Big[\mathcal{D}_{M_1}, T_{M_2M_3}{}^R\Big]_{|M_1|(|M_2|+|M_3|+|R|)}\mathcal{D}_R + (-)^{|M_1|(|M_2|+|M_3|+|R|)}T_{M_2M_3}{}^R\Big[\mathcal{D}_{M_1}, \mathcal{D}_R\Big]_{|M_1||R|} \\ &\quad -\frac{1}{2}\Big[\mathcal{D}_{M_1}, R_{M_2M_3\underline{\alpha}\underline{\beta}}\Big]_{|M_1|(|M_2|+|M_3|)}J^{\underline{\beta}\underline{\alpha}} \\ &\quad -(-)^{|M_1|(|M_2|+|M_3|)}\frac{1}{2}R_{M_2M_3\underline{\alpha}\underline{\beta}}\Big[\mathcal{D}_{M_1}, J^{\underline{\beta}\underline{\alpha}}\Big] \\ &= \Big(\mathcal{D}_{M_1}T_{M_2M_3}{}^S + (-)^{|M_1|(|M_2|+|M_3|+|R|)}T_{M_2M_3}{}^RT_{M_1R}{}^S \\ &\qquad +(-)^{|M_1|(|M_2|+|M_3|)}R_{M_2M_3M_1}{}^S\Big)\mathcal{D}_S \\ &\quad +\frac{1}{2}\Big(-(-)^{|M_1|(|M_2|+|M_3|+|R|)}T_{M_2M_3}{}^RR_{M_1R\underline{\alpha}\underline{\beta}} - \mathcal{D}_{M_1}R_{M_2M_3\underline{\alpha}\underline{\beta}}\Big)J^{\underline{\beta}\underline{\alpha}} \,. \end{aligned} \tag{3.25}$$

For the last commutator we have used (B.6). Strictly speaking, the equation above is only correct in the case where the graded commutator acts on a given vector V^N (see Eq. (3.8), as well as Eq. (B.11) for N being a Lorentz index). However, the dependence on V^N has been omitted for clarity. Injecting the result of Eq. (3.25) into Eq. (3.24) and making use of the symmetry properties of the curvature and torsion tensors expressed by Eq. (3.10), the Bianchi identities are reduced to two sets of identities,

$$\begin{aligned}
I_{M_1M_2M_3S} : 0 = (-)^{|M_1||M_3|}\Big[&\mathcal{D}_{M_1}T_{M_2M_3S} - T_{M_1M_2}{}^R T_{RM_3S} + R_{M_1M_2M_3S}\Big] \\
+ (-)^{|M_2||M_1|}\Big[&\mathcal{D}_{M_2}T_{M_3M_1S} - T_{M_2M_3}{}^R T_{RM_1S} + R_{M_2M_3M_1S}\Big] \\
+ (-)^{|M_3||M_2|}\Big[&\mathcal{D}_{M_3}T_{M_1M_2S} - T_{M_3M_1}{}^R T_{RM_2S} + R_{M_3M_1M_2S}\Big] ,
\end{aligned} \tag{3.26}$$

$$\begin{aligned}
I_{M_1M_2M_3;PQ} : 0 = (-)^{|M_1||M_3|}\Big[&T_{M_1M_2}{}^R R_{RM_3PQ} - \mathcal{D}_{M_1}R_{M_2M_3PQ}\Big] \\
+ (-)^{|M_2||M_1|}\Big[&T_{M_2M_3}{}^R R_{RM_1PQ} - \mathcal{D}_{M_2}R_{M_3M_1PQ}\Big] \\
+ (-)^{|M_3||M_2|}\Big[&T_{M_3M_1}{}^R R_{RM_2PQ} - \mathcal{D}_{M_3}R_{M_1M_2PQ}\Big] .
\end{aligned}$$

In contrast to general relativity, in supergravity we can express the components of the curvature tensor in terms of those of the torsion tensor. This property is known as the Dragon (1979) theorem and is proven by rewriting of the Bianchi identity $I_{M_1M_2M_3S}$ of Eq. (3.26) as

$$\begin{aligned}
(-1)^{|M_1||M_3|}R_{M_1M_2M_3S} + (-1)^{|M_2||M_1|}R_{M_2M_3M_1S} + (-1)^{|M_3||M_2|}R_{M_3M_1M_2S} \\
= \Delta_{M_1M_2M_3S} .
\end{aligned} \tag{3.27}$$

In this expression, we have introduced a quantity $\Delta_{M_1M_2M_3S}$ that depends only on the components of the torsion tensor. The Lie-algebra valuedness of the curvature tensor ensures that $R_{M_1M_2M_3S}$ is non-vanishing only if M_3 and S are indices of the same nature. Moreover, we have the properties

$$\begin{aligned}
R_{M_1M_2\alpha}{}^\beta &= \frac{1}{4}(\sigma^\mu\bar\sigma^\nu)_\alpha{}^\beta R_{M_1M_2\mu\nu} , \\
R_{M_1M_2}{}^{\dot\alpha}{}_{\dot\beta} &= \frac{1}{4}(\bar\sigma^\mu\sigma^\nu)^{\dot\alpha}{}_{\dot\beta} R_{M_1M_2\mu\nu} \\
R_{\mu\nu\rho\sigma} &= \frac{1}{2}\bar\sigma_\mu{}^{\dot\alpha\alpha}\bar\sigma_\nu{}^{\dot\beta\beta}(\varepsilon_{\alpha\beta}R_{\dot\alpha\dot\beta\rho\sigma} - \varepsilon_{\dot\alpha\dot\beta}R_{\alpha\beta\rho\sigma}) ,
\end{aligned}$$

as shown in App. B. Making additional use of the symmetry properties of the curvature tensor in Eq. (3.10), one observes that it is sufficient to prove the Dragon theorem for $R_{M_1M_2\mu\nu}$ with $(M_1, M_2) = (\alpha,\beta)$, $(\alpha,\dot\alpha)$, (σ,ρ) and for the complex conjugate cases, since it then holds automatically for M_3 and S being spin indices. For $(M_1, M_2) = (\alpha, \beta)$ and $(M_1, M_2) = (\alpha, \dot{\alpha})$, the identity (3.27) reduces to

$$\begin{aligned} R_{\alpha\beta\mu\nu} &= \Delta_{\alpha\beta\mu\nu} \,, \\ R_{\alpha\dot{\alpha}\mu\nu} &= \Delta_{\alpha\dot{\alpha}\mu\nu} \,, \end{aligned}$$

which shows that $R_{\alpha\beta\mu\nu}$ and $R_{\alpha\dot{\alpha}\mu\nu}$ can be expressed entirely in terms of the elements of the torsion tensor. For $(M_1, M_2) = (\sigma, \rho)$, we employ the relation

$$\frac{1}{2}(R_{\alpha\rho\mu\nu} + R_{\mu\alpha\rho\nu}) - \frac{1}{2}(R_{\alpha\mu\nu\rho} + R_{\nu\alpha\mu\rho}) + \frac{1}{2}(R_{\alpha\nu\rho\mu} + R_{\rho\alpha\nu\mu}) = R_{\alpha\rho\mu\nu} \,,$$

together with Eq. (3.27), which leads to

$$R_{\alpha\rho\mu\nu} = \frac{1}{2}(\Delta_{\alpha\rho\mu\nu} - \Delta_{\alpha\mu\nu\rho} + \Delta_{\alpha\nu\rho\mu}) \,. \tag{3.28}$$

Consequently, all the components of the curvature tensor can be rewritten in terms of one or several of the Δ quantities, or equivalently in terms of the components of the torsion tensor. Moreover, the Dragon theorem further implies that the Bianchi identities $I_{M_1 M_2 M_2;PQ}$ are satisfied as soon as the identities $I_{M_1 M_2 M_3 S}$ hold [Dragon (1979); Grimm et al. (1979)].

The torsion and curvature tensors are two superfields with a large number of components. Therefore, the flat limit of Eq. (3.9) does not, in general, reproduce the supersymmetric algebra. Moreover, the decomposition of the involved superfields in terms of their component fields may introduce states with a spin higher than two. These issues can be solved by imposing well-chosen constraints on the curvature and torsion tensors. Although there is no general recipe to derive these constraints, the idea is based on three features:

- the number of degrees of freedom has to be reduced as much as possible;
- the supergravity algebra must be equivalent to the supersymmetric algebra in the flat limit;
- all imposed constraints must be compatible with the Bianchi identities.

One complete set of constraints to be imposed has been derived in the works of [Grimm et al. (1978); Wess and Zumino (1978b, 1977); Wess and Bagger (1992)] and reads

$$\begin{aligned} T_{\underline{\alpha}\underline{\beta}}{}^{\underline{\gamma}} = 0 \,, \quad T_{\alpha\beta}{}^{\mu} = T_{\dot{\alpha}\dot{\beta}}{}^{\mu} = 0 \,, \\ T_{\alpha\dot{\alpha}}{}^{\mu} = T_{\dot{\alpha}\alpha}{}^{\mu} = -2i\sigma^{\mu}{}_{\alpha\dot{\alpha}} \,, \\ T_{\underline{\alpha}\mu}{}^{\nu} = T_{\mu\underline{\alpha}}{}^{\nu} = 0 \,, \quad T_{\mu\nu}{}^{\rho} = 0 \,. \end{aligned} \tag{3.29}$$

These constraints have an algebraic origin [Gates et al. (1980); Siegel and Gates (1979); West (1986)] that we illustrate by two examples. First, we consider a chiral superfield Φ. By definition, this superfield satisfies $\overline{\mathcal{D}}_{\dot\alpha}\Phi = 0$ (see Sec. 4.1) so that the anticommutator $\{\overline{\mathcal{D}}_{\dot\alpha}, \overline{\mathcal{D}}_{\dot\beta}\}\Phi = T_{\dot\alpha\dot\beta}{}^{M}\mathcal{D}_M\Phi$ is vanishing. Consequently, one must require that $T_{\dot\alpha\dot\beta}{}^{\alpha} = T_{\dot\alpha\dot\beta}{}^{\mu} = 0$. Second, as mentioned at the end of Sec. 3.1.1, the supersymmetric algebra can only be recovered after imposing the pair of identities

$$T_{\alpha\dot\alpha}{}^{\mu} = T_{\dot\alpha\alpha}{}^{\mu} = -2i\sigma^{\mu}{}_{\alpha\dot\alpha} ,$$

and

$$T_{\alpha\dot\alpha}{}^{\beta} = T_{\alpha\dot\alpha}{}^{\dot\beta} = 0 .$$

The other constraints are further derived in a similar fashion.

Since the identities $I_{M_1M_2M_2;PQ}$ of Eq. (3.26) are automatically satisfied in the case where the identities $I_{M_1M_2M_3S}$ are fulfilled, we restrict our analysis to the set of identities $I_{M_1M_2M_3S}$. If we assume that the constraints of Eq. (3.29) are compatible with the Bianchi identities, the latter can be *reduced* to a smaller number of (simpler) equations. Decomposing the $I_{M_1M_2M_3S}$ relations with respect to the nature of the four indices, one obtains 30 different identities after accounting for the symmetry stemming from the exchange of the first three indices. Moreover, some pairs of identities are complex conjugate of each other, such as $I_{\mu\nu\rho\dot\alpha}$ and $I_{\mu\nu\rho\alpha}$, and several identities are trivially satisfied, such as $I_{\alpha\beta\gamma\mu}$ or $I_{\alpha\beta\gamma\dot\delta}$. We subsequently get thirteen independent identities

$$\begin{aligned}
&(1)\ \ I_{\dot\alpha\dot\beta\mu\nu} &&: R_{\dot\alpha\dot\beta\mu\nu} = 2i\sigma_{\nu\gamma\dot\alpha}T_{\dot\beta\mu}{}^{\gamma} + 2i\sigma_{\nu\gamma\dot\beta}T_{\dot\alpha\mu}{}^{\gamma} ,\\
&(2)\ \ I_{\alpha\beta\dot\gamma\dot\delta} &&: R_{\alpha\beta\dot\gamma\dot\delta} = 2i\sigma^{\mu}{}_{\beta\dot\gamma}T_{\alpha\mu\dot\delta} + 2i\sigma^{\mu}{}_{\alpha\dot\gamma}T_{\beta\mu\dot\delta} ,\\
&(3)\ \ I_{\dot\alpha\dot\beta\dot\gamma\dot\delta} &&: R_{\dot\alpha\dot\beta\dot\gamma\dot\delta} + R_{\dot\beta\dot\gamma\dot\alpha\dot\delta} + R_{\dot\gamma\dot\alpha\dot\beta\dot\delta} = 0 ,\\
&(4)\ \ I_{\alpha\dot\beta\gamma\delta} &&: R_{\alpha\dot\beta\gamma\delta} + R_{\dot\beta\gamma\alpha\delta} = 2i\sigma^{\mu}{}_{\alpha\dot\beta}T_{\gamma\mu\delta} + 2i\sigma^{\mu}{}_{\gamma\dot\beta}T_{\alpha\mu\delta} ,\\
&(5)\ \ I_{\alpha\dot\beta\mu\nu} &&: R_{\alpha\dot\beta\mu\nu} = -2i\sigma_{\nu\alpha\dot\gamma}T_{\dot\beta\mu}{}^{\dot\gamma} + 2i\sigma_{\nu\gamma\dot\beta}T_{\alpha\mu}{}^{\gamma} ,\\
&(6)\ \ I_{\dot\alpha\dot\beta\mu\gamma} &&: \overline{\mathcal{D}}_{\dot\alpha}T_{\dot\beta\mu\gamma} + \overline{\mathcal{D}}_{\dot\beta}T_{\dot\alpha\mu\gamma} = 0 ,\\
&(7)\ \ I_{\alpha\mu\nu\rho} &&: R_{\alpha\mu\nu\rho} + R_{\nu\alpha\mu\rho} = 2i\sigma_{\rho\alpha\dot\beta}T_{\mu\nu}{}^{\dot\beta} ,\\
&(8)\ \ I_{\alpha\mu\beta\gamma} &&: R_{\alpha\mu\beta\gamma} + R_{\beta\mu\alpha\gamma} = \mathcal{D}_{\alpha}T_{\beta\mu\gamma} + \mathcal{D}_{\beta}T_{\alpha\mu\gamma} ,\\
&(9)\ \ I_{\mu\alpha\dot\beta\dot\gamma} &&: R_{\mu\alpha\dot\beta\dot\gamma} = -\mathcal{D}_{\alpha}T_{\dot\beta\mu\dot\gamma} - \overline{\mathcal{D}}_{\dot\beta}T_{\alpha\mu\dot\gamma} - 2i\sigma^{\nu}{}_{\alpha\dot\beta}T_{\nu\mu\dot\gamma} ,\\
&(10)\ I_{\mu\nu\dot\alpha\dot\beta} &&: R_{\mu\nu\dot\alpha\dot\beta} = -\mathcal{D}_{\mu}T_{\nu\dot\alpha\dot\beta} - \mathcal{D}_{\nu}T_{\dot\alpha\mu\dot\beta} - \overline{\mathcal{D}}_{\dot\alpha}T_{\mu\nu\dot\beta} + T_{\nu\dot\alpha}{}^{\underline{\gamma}}T_{\underline{\gamma}\mu\dot\beta} + T_{\dot\alpha\mu}{}^{\underline{\gamma}}T_{\underline{\gamma}\nu\dot\beta} ,\\
&(11)\ I_{\mu\nu\rho\sigma} &&: R_{\mu\nu\rho\sigma} + R_{\nu\rho\mu\sigma} + R_{\rho\mu\nu\sigma} = 0 ,\\
&(12)\ I_{\mu\nu\dot\alpha\beta} &&: -\mathcal{D}_{\mu}T_{\nu\dot\alpha\beta} - \mathcal{D}_{\nu}T_{\dot\alpha\mu\beta} - \overline{\mathcal{D}}_{\dot\alpha}T_{\mu\nu\beta} + T_{\nu\dot\alpha}{}^{\underline{\gamma}}T_{\underline{\gamma}\mu\beta} + T_{\dot\alpha\mu}{}^{\underline{\gamma}}T_{\underline{\gamma}\nu\beta} = 0 ,\\
&(13)\ I_{\mu\nu\rho\alpha} &&: -\mathcal{D}_{\mu}T_{\nu\rho\alpha} - \mathcal{D}_{\nu}T_{\rho\mu\alpha} - \mathcal{D}_{\rho}T_{\mu\nu\alpha} + T_{\mu\nu}{}^{\underline{\gamma}}T_{\underline{\gamma}\rho\alpha} + T_{\nu\rho}{}^{\underline{\gamma}}T_{\underline{\gamma}\mu\alpha} + T_{\rho\mu}{}^{\underline{\gamma}}T_{\underline{\gamma}\nu\alpha}\\
& && \quad = 0 ,
\end{aligned} \tag{3.30}$$

after extensive use of the symmetry properties of Eq. (3.10) and of the fact that R_{MNPQ} vanishes if the indices P and Q are not of the same nature. We recall that the underlined summed index $\underline{\gamma}$ implies a summation over both the dotted and undotted spin indices γ and $\dot{\gamma}$. The related contributions originate from terms quadratic in the torsion tensor in Eq. (3.26),

$$\begin{aligned} T_{M_1M_2}{}^R T_{RM_3S} &= T_{M_1M_2}{}^\mu T_{\mu M_3S} + T_{M_1M_2}{}^{\underline{\gamma}} T_{\underline{\gamma} M_3S} \\ &= T_{M_1M_2}{}^\mu T_{\mu M_3S} + T_{M_1M_2}{}^\gamma T_{\gamma M_3S} + T_{M_1M_2\dot{\gamma}} T^{\dot{\gamma}}{}_{M_3S} \, . \end{aligned} \quad (3.31)$$

The thirteen identities above allow one to show that all the elements of the torsion and curvature superfields can be rewritten as functions of three independent superfields and their complex conjugates,

$$\mathcal{R}, \qquad G_{\alpha\dot{\beta}}, \qquad \text{and} \qquad W_{(\alpha\beta\gamma)} \, ,$$

where $W_{(\alpha\beta\gamma)}$ is symmetric with respect to the exchange of its indices [Grimm et al. (1979); Siegel (1978); Wess and Bagger (1992)]. This can be demonstrated by following standard techniques used in general relativity [Penrose (1960)] where all vector indices are converted to spinor indices using Eq. (B.7), and where each tensor is further decomposed in terms of irreducible representations of $SL(2,\mathbb{C})$ [Grimm et al. (1978)]. For instance, the torsion tensor $T_{\dot{\alpha}\mu\beta}$ is first converted to $T_{\dot{\alpha}\gamma\dot{\gamma}\beta} = \sigma^\mu{}_{\gamma\dot{\gamma}} T_{\dot{\alpha}\mu\beta}$ and then expanded as

$$T_{\dot{\alpha}\gamma\dot{\gamma}\beta} = \varepsilon_{\dot{\alpha}\dot{\gamma}}\varepsilon_{\gamma\beta} T + \varepsilon_{\gamma\beta} T_{(\dot{\alpha}\dot{\gamma})} + \varepsilon_{\dot{\alpha}\dot{\gamma}} T_{(\gamma\beta)} + T_{(\dot{\alpha}\dot{\gamma})(\gamma\beta)} \, , \quad (3.32)$$

where the new T-tensors are symmetric with respect to the exchange of the indices in the parentheses.

Identities (1), (2) and (3): $I_{\dot{\alpha}\dot{\beta}\mu\nu}$, $I_{\alpha\dot{\beta}\dot{\gamma}\dot{\delta}}$ and $I_{\dot{\alpha}\dot{\beta}\dot{\gamma}\dot{\delta}}$.

We start by considering the first three identities of Eq. (3.30). As mentioned above, the first step in the derivation of their solutions consists in converting all vector indices to spinor indices and expanding the resulting tensors in terms of their irreducible $SL(2,\mathbb{C})$ components. In this way, the curvature tensor $R_{\dot{\alpha}\dot{\beta}\mu\nu}$ becomes

$$\sigma^\mu{}_{\gamma\dot{\gamma}}\sigma^\nu{}_{\delta\dot{\delta}} R_{\dot{\alpha}\dot{\beta}\mu\nu} = R_{\dot{\alpha}\dot{\beta}\gamma\dot{\gamma}\delta\dot{\delta}} = 2\varepsilon_{\gamma\delta} R_{\dot{\alpha}\dot{\beta}\dot{\gamma}\dot{\delta}} - 2\varepsilon_{\dot{\gamma}\dot{\delta}} R_{\dot{\alpha}\dot{\beta}\gamma\delta} \, .$$

Making use, in addition, of Eq. (3.32), the first identity $I_{\dot{\alpha}\dot{\beta}\mu\nu}$ reduces to

$$\begin{aligned} \varepsilon_{\gamma\delta} R_{\dot{\alpha}\dot{\beta}\dot{\gamma}\dot{\delta}} - \varepsilon_{\dot{\gamma}\dot{\delta}} R_{\dot{\alpha}\dot{\beta}\gamma\delta} &= 2i\varepsilon_{\dot{\delta}\dot{\alpha}}\big(\varepsilon_{\dot{\beta}\dot{\gamma}}\varepsilon_{\gamma\delta} T + \varepsilon_{\gamma\delta} T_{(\dot{\beta}\dot{\gamma})} + \varepsilon_{\dot{\beta}\dot{\gamma}} T_{(\gamma\delta)} + T_{(\dot{\beta}\dot{\gamma})(\gamma\delta)}\big) \\ &\quad + 2i\varepsilon_{\dot{\delta}\dot{\beta}}\big(\varepsilon_{\dot{\alpha}\dot{\gamma}}\varepsilon_{\gamma\delta} T + \varepsilon_{\gamma\delta} T_{(\dot{\alpha}\dot{\gamma})} + \varepsilon_{\dot{\alpha}\dot{\gamma}} T_{(\gamma\delta)} + T_{(\dot{\alpha}\dot{\gamma})(\gamma\delta)}\big) \, . \end{aligned} \quad (3.33)$$

Contracting this equation with $\varepsilon^{\dot\gamma\dot\delta}\varepsilon^{\gamma\delta}$ and using the symmetry properties of $R_{\dot\alpha\dot\beta\gamma\delta}$ under the exchange of its last two indices (see Eq. (3.10)), one derives $T_{(\dot\alpha\dot\beta)} = 0$.

We next focus on the pieces of Eq. (3.33) that are symmetric under the exchange of the indices γ and δ,

$$-\varepsilon_{\dot\gamma\dot\delta}R_{\dot\alpha\dot\beta\gamma\delta} = 2i\varepsilon_{\dot\delta\dot\alpha}\big(\varepsilon_{\dot\beta\dot\gamma}T_{(\gamma\delta)} + T_{(\dot\beta\dot\gamma)(\gamma\delta)}\big) + 2i\varepsilon_{\dot\delta\dot\beta}\big(\varepsilon_{\dot\alpha\dot\gamma}T_{(\gamma\delta)} + T_{(\dot\alpha\dot\gamma)(\gamma\delta)}\big)\,, \qquad (3.34)$$

and contract the resulting expression with $\varepsilon^{\dot\gamma\dot\beta}\varepsilon^{\dot\delta\dot\alpha}$, $\varepsilon^{\dot\gamma\dot\delta}$ and $\varepsilon^{\dot\delta\dot\alpha}$, respectively. Employing again the symmetry properties of the curvature tensor $R_{\dot\alpha\dot\beta\gamma\delta}$, one obtains $T_{(\gamma\delta)} = R_{\dot\alpha\dot\beta\gamma\delta} = T_{(\dot\alpha\dot\beta)(\gamma\delta)} = 0$.

Defining $T = 2i\mathcal{R}$, the solution of the first of the thirteen Bianchi identities of Eq. (3.30) is given by [5]

$$R_{\dot\alpha\dot\beta\gamma\delta} = 0\,,\quad R_{\dot\alpha\dot\beta\dot\gamma\dot\delta} = 4(\varepsilon_{\dot\alpha\dot\delta}\varepsilon_{\dot\beta\dot\gamma} + \varepsilon_{\dot\beta\dot\delta}\varepsilon_{\dot\alpha\dot\gamma})\mathcal{R}\,,\quad T_{\dot\alpha\beta\dot\gamma\delta} = 2i\varepsilon_{\dot\alpha\dot\gamma}\varepsilon_{\beta\delta}\mathcal{R}\,, \qquad (3.35)$$

from which we derive the conjugate relations,

$$R_{\alpha\beta\dot\gamma\dot\delta} = 0\,,\quad R_{\alpha\beta\gamma\delta} = 4(\varepsilon_{\alpha\delta}\varepsilon_{\beta\gamma} + \varepsilon_{\beta\delta}\varepsilon_{\alpha\gamma})\mathcal{R}^\dagger\,,\quad T_{\alpha\beta\dot\gamma\dot\delta} = 2i\varepsilon_{\alpha\beta}\varepsilon_{\dot\gamma\dot\delta}\mathcal{R}^\dagger\,. \qquad (3.36)$$

It can be shown that Eq. (3.35) and Eq. (3.36) are also solutions for the second and third identities of Eq. (3.30).

Identities (4) and (5): $I_{\alpha\beta\gamma\delta}$ and $I_{\alpha\dot\beta\mu\nu}$.

In spinor notations and using the Lie-algebra valuedness of the curvature tensor, the two identities $I_{\alpha\dot\beta\gamma\delta}$ and $I_{\alpha\dot\beta\mu\nu}$ are rewritten as

$$\begin{aligned} R_{\alpha\dot\beta\gamma\delta} + R_{\dot\beta\gamma\alpha\delta} &= 2i(T_{\gamma\alpha\dot\beta\delta} + T_{\alpha\gamma\dot\beta\delta})\,,\\ 2\varepsilon_{\delta\epsilon}R_{\alpha\dot\beta\dot\delta\dot\epsilon} - 2\varepsilon_{\dot\delta\dot\epsilon}R_{\alpha\dot\beta\delta\epsilon} &= 4i(\varepsilon_{\dot\epsilon\dot\beta}T_{\alpha\delta\dot\delta\epsilon} - \varepsilon_{\epsilon\alpha}T_{\dot\beta\delta\dot\delta\dot\epsilon})\,. \end{aligned} \qquad (3.37)$$

The tensor $T_{\alpha\beta\dot\gamma\delta}$ including two spin-one $SL(2,\mathbb{C})$ representations, its decomposition in terms of irreducible components is not unique. We choose $T_{\alpha\beta\dot\gamma\delta} = \varepsilon_{\beta\delta}X_{\alpha\dot\gamma} + T_{\alpha\dot\gamma(\beta\delta)}$, which further leads to

$$T_{\alpha\beta\dot\gamma\delta} = \varepsilon_{\beta\delta}X_{\alpha\dot\gamma} + \varepsilon_{\alpha\beta}Y_{\delta\dot\gamma} + \varepsilon_{\alpha\delta}Y_{\beta\dot\gamma} + Z_{(\alpha\beta\delta)\dot\gamma}\,,$$

$$T_{\dot\alpha\gamma\dot\beta\dot\delta} = \varepsilon_{\dot\beta\dot\delta}\overline{X}_{\gamma\dot\alpha} + \varepsilon_{\dot\alpha\dot\beta}\overline{Y}_{\gamma\dot\delta} + \varepsilon_{\dot\alpha\dot\delta}\overline{Y}_{\gamma\dot\beta} + \overline{Z}_{(\dot\alpha\dot\beta\dot\delta)\gamma}\,,$$

---

[5] The element $R_{\dot\alpha\beta\gamma\delta}$ of the curvature tensor vanishes although $R_{\dot\alpha\dot\beta\dot\gamma\dot\delta}$ is non-zero. This implies that the complexified algebra of $\mathfrak{so}(1,3)$, *i.e.*, $\mathfrak{so}(1,3)\otimes\mathbb{C} \cong \mathfrak{so}(1,3,\mathbb{C}) \cong \mathfrak{so}(4,\mathbb{C})$, is being considered. As a consequence, the transformation parameters $\omega_{\dot\alpha\dot\beta\mu\nu} = \frac{1}{2}\bar\sigma_\mu{}^{\dot\gamma\gamma}\bar\sigma_\nu{}^{\dot\delta\delta}\varepsilon_{\gamma\delta}R_{\dot\alpha\dot\beta\dot\gamma\dot\delta}$ in Eq. (B.2) are complex quantities, which justifies their definition in terms of the complex superfields $\mathcal{R}$, $G_{\alpha\dot\beta}$ and $W_{(\alpha\beta\gamma)}$. Handling complex Lorentz transformation parameters is *a priori* surprising. However, the relevant transformation parameters induced by the action of the commutator $[\xi^M\mathcal{D}_M, \xi^N\mathcal{D}_N]$ on the vector component of a superfield V^P are given by $\xi^\beta\xi^\alpha R_{\alpha\beta\mu\nu} + \xi^{\dot\beta}\xi^{\dot\alpha}R_{\dot\alpha\dot\beta\mu\nu}$, which is a real quantity. The Lorentz transformations are therefore described by the real form $\mathfrak{so}(1,3)$ of $\mathfrak{so}(4,\mathbb{C})$, as expected.

where we have introduced the new tensors X, Y and Z with definite symmetry properties.

Inserting those results in Eq. (3.37) and contracting the second equation with $\varepsilon^{\dot\delta\dot\epsilon}\varepsilon^{\delta\epsilon}$, we obtain $X_{\alpha\dot\beta} = \overline{X}_{\alpha\dot\beta}$. Similarly, contracting the first relation of Eq. (3.37) with $\varepsilon^{\gamma\delta}$ and the second one with $\varepsilon^{\dot\delta\dot\epsilon}\varepsilon^{\alpha\epsilon}$, we derive

$$R_{\dot\beta\gamma\alpha}{}^{\gamma} = -6i(X_{\alpha\dot\beta} + Y_{\alpha\dot\beta}) = 3i(X_{\alpha\dot\beta} - Y_{\alpha\dot\beta}) ,$$

or equivalently $Y_{\alpha\dot\beta} = -3X_{\alpha\dot\beta}$. Finally, we symmetrise the first expression of Eq. (3.37) with respect to the three undotted indices. The result is then compared with the second relation of Eq. (3.37) after having first contracted it with $\varepsilon^{\dot\delta\dot\epsilon}$ and then symmetrised it with respect to the indices δ and ϵ. This allows one to get

$$Z_{(\alpha\gamma\delta)\dot\beta} = \varepsilon_{\delta\gamma}X_{\alpha\dot\beta} + \varepsilon_{\gamma\alpha}X_{\delta\dot\beta} + \varepsilon_{\alpha\delta}X_{\gamma\dot\beta} = 0 .$$

Defining $X_{\alpha\dot\beta} = {}^{i}\!/_{4}G_{\alpha\dot\beta}$, one obtains the solution to the fourth and fifth Bianchi identities (and their complex conjugates),

$$\begin{aligned} T_{\alpha\gamma\dot\beta\delta} &= \tfrac{i}{4}\Big(\varepsilon_{\gamma\delta}G_{\alpha\dot\beta} - 3\varepsilon_{\alpha\delta}G_{\gamma\dot\beta} - 3\varepsilon_{\alpha\gamma}G_{\delta\dot\beta}\Big) , \\ R_{\alpha\dot\beta\gamma\delta} &= \varepsilon_{\alpha\delta}G_{\gamma\dot\beta} + \varepsilon_{\alpha\gamma}G_{\delta\dot\beta} , \\ T_{\dot\alpha\beta\dot\gamma\dot\delta} &= \tfrac{i}{4}\Big(\varepsilon_{\dot\gamma\dot\delta}G_{\beta\dot\alpha} - 3\varepsilon_{\dot\alpha\dot\delta}G_{\beta\dot\gamma} - 3\varepsilon_{\dot\alpha\dot\gamma}G_{\beta\dot\delta}\Big) , \\ R_{\dot\alpha\beta\dot\gamma\dot\delta} &= \varepsilon_{\dot\alpha\dot\delta}G_{\beta\dot\gamma} + \varepsilon_{\dot\alpha\dot\gamma}G_{\beta\dot\delta} , \end{aligned} \tag{3.38}$$

where we have used $(G_{\alpha\dot\beta})^\dagger = G_{\beta\dot\alpha}$.

Identity (6): $I_{\dot\alpha\dot\beta\mu\gamma}$.

The sixth identity of Eq. (3.30) can be rewritten, after introducing spinor notations and using the results of Eq. (3.35), as

$$\varepsilon_{\dot\beta\dot\gamma}\varepsilon_{\gamma\beta}\overline{\mathcal{D}}_{\dot\alpha}\mathcal{R} + \varepsilon_{\dot\alpha\dot\gamma}\varepsilon_{\gamma\beta}\overline{\mathcal{D}}_{\dot\beta}\mathcal{R} = 0 .$$

This gives, after a contraction with $\varepsilon^{\dot\beta\dot\gamma}$,

$$\overline{\mathcal{D}}_{\dot\alpha}\mathcal{R} = 0 . \tag{3.39}$$

The superfield $\mathcal{R}$ is thus chiral, and by complex conjugation $\mathcal{R}^\dagger$ is antichiral.

Identities (7) and (8): $I_{\alpha\mu\nu\rho}$ and $I_{\alpha\mu\beta\gamma}$.

Symmetrising the seventh identity of Eq. (3.30) with respect to the three vector indices and using the symmetry properties of the curvature tensor, we start by rewriting the identity $I_{\alpha\mu\nu\rho}$ as

$$\begin{aligned} R_{\alpha\mu\nu\rho} &= \frac{1}{2}(R_{\alpha\mu\nu\rho} + R_{\nu\alpha\mu\rho}) - \frac{1}{2}(R_{\alpha\nu\rho\mu} + R_{\rho\alpha\nu\mu}) + \frac{1}{2}(R_{\alpha\rho\mu\nu} + R_{\mu\alpha\rho\nu}) \\ &= -i(\sigma_{\mu\alpha\dot\beta} T_{\nu\rho}{}^{\dot\beta} - \sigma_{\nu\alpha\dot\beta} T_{\rho\mu}{}^{\dot\beta} - \sigma_{\rho\alpha\dot\beta} T_{\mu\nu}{}^{\dot\beta}) \,. \end{aligned}$$

Following the technique described in the beginning of this section, we introduce spinor notations,

$$\varepsilon_{\beta\gamma} R_{\alpha\delta\dot\delta\dot\beta\dot\gamma} - \varepsilon_{\dot\beta\dot\gamma} R_{\alpha\delta\dot\delta\beta\gamma} = -i(\varepsilon_{\delta\alpha} T_{\beta\dot\beta\gamma\dot\gamma\dot\delta} - \varepsilon_{\beta\alpha} T_{\gamma\dot\gamma\delta\dot\delta\dot\beta} - \varepsilon_{\gamma\alpha} T_{\delta\dot\delta\beta\dot\beta\dot\gamma}) \,, \tag{3.40}$$

and further expand the rank-five tensors in terms of irreducible components,

$$T_{\alpha\dot\alpha\beta\dot\beta\dot\gamma} = -2\varepsilon_{\alpha\beta}\Big(\overline{W}_{(\dot\alpha\dot\beta\dot\gamma)} + \varepsilon_{\dot\beta\dot\gamma}\overline{W}_{\dot\alpha} + \varepsilon_{\dot\alpha\dot\gamma}\overline{W}_{\dot\beta}\Big) + 2\varepsilon_{\dot\alpha\dot\beta} W_{(\alpha\beta)\dot\gamma} \,. \tag{3.41}$$

In this expression, we have introduced several W-tensors with definite symmetry properties, using in addition the relation $T_{\alpha\dot\alpha\beta\dot\beta\gamma} = -T_{\beta\dot\beta\alpha\dot\alpha\gamma}$ to reduce the number of independent tensors. Inserting Eq. (3.41) back into Eq. (3.40) and contracting the result with $\varepsilon^{\dot\beta\dot\gamma}$, one gets

$$R_{\alpha\delta\dot\delta\beta\gamma} = i\Big[2\varepsilon_{\delta\alpha} W_{(\beta\gamma)\dot\delta} + \varepsilon_{\beta\alpha} W_{(\gamma\delta)\dot\delta} + \varepsilon_{\gamma\alpha} W_{(\delta\beta)\dot\delta} + 3(\varepsilon_{\alpha\beta}\varepsilon_{\gamma\delta} + \varepsilon_{\delta\beta}\varepsilon_{\gamma\alpha})\overline{W}_{\dot\delta}\Big] \,. \tag{3.42}$$

We now turn to the eighth Bianchi identity and make use of both Eq. (3.38) and Eq. (3.42) to rewrite $I_{\alpha\mu\beta\gamma}$ in spinor notations,

$$R_{\alpha\delta\dot\delta\beta\gamma} + R_{\beta\delta\dot\delta\alpha\gamma} = \mathcal{D}_\alpha T_{\beta\delta\dot\delta\gamma} + \mathcal{D}_\beta T_{\alpha\delta\dot\delta\gamma} \,.$$

Contracting with $\varepsilon^{\beta\gamma}\varepsilon^{\delta\alpha}$, $\varepsilon^{\gamma\alpha}$ and $\varepsilon^{\beta\gamma}$, respectively, one finds $\overline{W}_{\dot\alpha} = -{}^1\!/_4 \mathcal{D}^\epsilon G_{\epsilon\dot\alpha}$, $W_{(\delta\beta)\dot\alpha} = {}^1\!/_4(\mathcal{D}_\delta G_{\beta\dot\alpha} + \mathcal{D}_\beta G_{\delta\dot\alpha})$ and the relation

$$\varepsilon_{\delta\alpha}\mathcal{D}^\epsilon G_{\epsilon\dot\alpha} = \mathcal{D}_\delta G_{\alpha\dot\alpha} - \mathcal{D}_\alpha G_{\delta\dot\alpha} \,. \tag{3.43}$$

This allows us to derive

$$\begin{aligned} R_{\alpha\delta\dot\delta\beta\gamma} &= i(\varepsilon_{\alpha\beta}\varepsilon_{\delta\gamma} + \varepsilon_{\alpha\gamma}\varepsilon_{\delta\beta})\mathcal{D}^\epsilon G_{\epsilon\dot\delta} - \frac{i}{2}(\varepsilon_{\alpha\delta}\mathcal{D}_\beta + \varepsilon_{\alpha\beta}\mathcal{D}_\delta)G_{\gamma\dot\delta} \\ &\quad - \frac{i}{2}(\varepsilon_{\alpha\delta}\mathcal{D}_\gamma + \varepsilon_{\alpha\gamma}\mathcal{D}_\delta)G_{\beta\dot\delta} \,, \\ R_{\dot\alpha\delta\dot\delta\dot\beta\dot\gamma} &= i(\varepsilon_{\dot\alpha\dot\beta}\varepsilon_{\dot\delta\dot\gamma} + \varepsilon_{\dot\alpha\dot\gamma}\varepsilon_{\dot\delta\dot\beta})\overline{\mathcal{D}}^{\dot\epsilon} G_{\delta\dot\epsilon} - \frac{i}{2}(\varepsilon_{\dot\alpha\dot\delta}\overline{\mathcal{D}}_{\dot\beta} + \varepsilon_{\dot\alpha\dot\beta}\overline{\mathcal{D}}_{\dot\delta})G_{\delta\dot\gamma} \\ &\quad - \frac{i}{2}(\varepsilon_{\dot\alpha\dot\delta}\overline{\mathcal{D}}_{\dot\gamma} + \varepsilon_{\dot\alpha\dot\gamma}\overline{\mathcal{D}}_{\dot\delta})G_{\delta\dot\beta} \,, \end{aligned} \tag{3.44}$$

$$
\begin{aligned}
T_{\alpha\dot\alpha\beta\dot\beta\dot\gamma} &= -2\varepsilon_{\alpha\beta}\overline{W}_{(\dot\alpha\dot\beta\dot\gamma)} + \frac{1}{2}\varepsilon_{\beta\alpha}(\varepsilon_{\dot\gamma\dot\beta}\mathcal{D}^\epsilon G_{\epsilon\dot\alpha} + \varepsilon_{\dot\gamma\dot\alpha}\mathcal{D}^\epsilon G_{\epsilon\dot\beta}) \\
&\quad + \frac{1}{2}\varepsilon_{\dot\alpha\dot\beta}(\mathcal{D}_\alpha G_{\beta\dot\gamma} + \mathcal{D}_\beta G_{\alpha\dot\gamma})\,, \\
T_{\alpha\dot\alpha\beta\dot\beta\gamma} &= -2\varepsilon_{\dot\alpha\dot\beta}W_{(\alpha\beta\gamma)} + \frac{1}{2}\varepsilon_{\dot\beta\dot\alpha}(\varepsilon_{\gamma\beta}\overline{\mathcal{D}}^{\dot\epsilon} G_{\alpha\dot\epsilon} + \varepsilon_{\gamma\alpha}\overline{\mathcal{D}}^{\dot\epsilon} G_{\beta\dot\epsilon}) \\
&\quad + \frac{1}{2}\varepsilon_{\alpha\beta}(\overline{\mathcal{D}}_{\dot\alpha}G_{\gamma\dot\beta} + \overline{\mathcal{D}}_{\dot\beta}G_{\gamma\dot\alpha})\,.
\end{aligned}
$$

Contracting Eq. (3.40) instead with $\varepsilon^{\beta\gamma}$, we obtain the $R_{\alpha\delta\dot\delta\dot\beta\dot\gamma}$ and $R_{\dot\alpha\delta\dot\delta\beta\gamma}$ elements of the curvature tensor

$$
\begin{aligned}
R_{\alpha\delta\dot\delta\dot\beta\dot\gamma} &= -4i\varepsilon_{\alpha\delta}\overline{W}_{(\dot\beta\dot\gamma\dot\delta)} - \frac{i}{2}(\varepsilon_{\dot\delta\dot\beta}\mathcal{D}_\alpha G_{\delta\dot\gamma} + \varepsilon_{\dot\delta\dot\gamma}\mathcal{D}_\alpha G_{\delta\dot\beta})\,, \\
R_{\dot\alpha\delta\dot\delta\beta\gamma} &= -4i\varepsilon_{\dot\alpha\dot\delta}W_{(\beta\gamma\delta)} - \frac{i}{2}(\varepsilon_{\delta\beta}\overline{\mathcal{D}}_{\dot\alpha}G_{\gamma\dot\delta} + \varepsilon_{\delta\gamma}\overline{\mathcal{D}}_{\dot\alpha}G_{\beta\dot\delta})\,.
\end{aligned}
\tag{3.45}
$$

Identity (9): $I_{\mu\alpha\dot\beta\dot\gamma}$.

Converting first the identity $I_{\mu\alpha\dot\beta\dot\gamma}$ into spinor notations,

$$
R_{\alpha\delta\dot\delta\dot\beta\dot\gamma} = \mathcal{D}_\alpha T_{\dot\beta\delta\dot\delta\dot\gamma} + \overline{\mathcal{D}}_{\dot\beta}T_{\alpha\delta\dot\delta\dot\gamma} + 2iT_{\alpha\dot\beta\delta\dot\delta\dot\gamma}\,, \tag{3.46}
$$

the results of Eq. (3.36), Eq. (3.38), Eq. (3.44) and Eq. (3.45) constrain the $\mathcal{R}$ and $G_{\alpha\dot\alpha}$ superfields to satisfy

$$
\begin{aligned}
\mathcal{D}^\beta G_{\beta\dot\alpha} &= -\overline{\mathcal{D}}_{\dot\alpha}\mathcal{R}^\dagger\,, \\
\overline{\mathcal{D}}^{\dot\beta} G_{\alpha\dot\beta} &= -\mathcal{D}_\alpha\mathcal{R}\,.
\end{aligned}
\tag{3.47}
$$

Identities (10) and (11): $I_{\mu\nu\dot\alpha\dot\beta}$ and $I_{\mu\nu\rho\sigma}$.

We now turn to the computation of the $R_{\mu\nu\dot\alpha\dot\beta}$ and $R_{\mu\nu\alpha\beta}$ elements of the curvature tensor. Decomposing them into irreducible components X and Ψ with given symmetry properties, we have

$$
\begin{aligned}
R_{\mu\nu\dot\alpha\dot\beta} &\to R_{\gamma\dot\gamma\delta\dot\delta\dot\alpha\dot\beta} = 2\varepsilon_{\gamma\delta}\overline{X}_{(\dot\gamma\dot\delta)(\dot\alpha\dot\beta)} - 2\varepsilon_{\dot\gamma\dot\delta}\Psi_{(\gamma\delta)(\dot\alpha\dot\beta)}\,, \\
R_{\mu\nu\alpha\beta} &\to R_{\gamma\dot\gamma\delta\dot\delta\alpha\beta} = -2\varepsilon_{\dot\gamma\dot\delta}X_{(\gamma\delta)(\alpha\beta)} + 2\varepsilon_{\gamma\delta}\overline{\Psi}_{(\dot\gamma\dot\delta)(\alpha\beta)}\,.
\end{aligned}
\tag{3.48}
$$

The tensors X and Ψ can be entirely expressed in terms of the $\mathcal{R}$, $G_{\alpha\dot\alpha}$ and $\overline{W}_{(\alpha\beta\gamma)}$ superfields by analysing the tenth Bianchi identity $I_{\mu\nu\dot\alpha\dot\beta}$. After switching to spinor notations and identifying terms with given symmetry properties, this indeed leads to

$$
\overline{X}_{(\dot\gamma\dot\delta)(\dot\alpha\dot\beta)} = \frac{1}{4}\Big[-\mathcal{D}_{\gamma\dot\gamma}T_{\dot\alpha}{}^{\gamma}{}_{\dot\delta\dot\beta} - \mathcal{D}_{\gamma\dot\delta}T_{\dot\alpha}{}^{\gamma}{}_{\dot\gamma\dot\beta} + \overline{\mathcal{D}}_{\dot\alpha}T_{\gamma\dot\gamma}{}^{\gamma}{}_{\dot\delta\dot\beta} - T_{\dot\alpha\gamma\dot\delta}{}^{\underline{\epsilon}}T_{\underline{\epsilon}}{}^{\gamma}{}_{\dot\gamma\dot\beta} - T_{\dot\alpha\gamma\dot\gamma}{}^{\underline{\epsilon}}T_{\underline{\epsilon}}{}^{\gamma}{}_{\dot\delta\dot\beta}\Big]\,,
\tag{3.49}
$$

$$
\Psi_{(\gamma\delta)(\dot\alpha\dot\beta)} = \frac{1}{4}\Big[\mathcal{D}_{\gamma\dot\gamma}T_{\dot\alpha\delta}{}^{\dot\gamma}{}_{\dot\beta} + \mathcal{D}_{\delta\dot\gamma}T_{\dot\alpha\gamma}{}^{\dot\gamma}{}_{\dot\beta} - \overline{\mathcal{D}}_{\dot\alpha}T_{\gamma\dot\gamma\delta}{}^{\dot\gamma}{}_{\dot\beta} + T_{\dot\alpha\delta\dot\gamma}{}^{\underline{\epsilon}}T_{\underline{\epsilon}\gamma}{}^{\dot\gamma}{}_{\dot\beta} + T_{\dot\alpha\gamma\dot\gamma}{}^{\underline{\epsilon}}T_{\underline{\epsilon}\delta}{}^{\dot\gamma}{}_{\dot\beta}\Big]\,,
$$

where we have defined $\mathcal{D}_{\alpha\dot\alpha} = \sigma^\mu{}_{\alpha\dot\alpha}\mathcal{D}_\mu$. The different contributions to the $\overline{X}_{(\dot\gamma\dot\delta)(\dot\alpha\dot\beta)}$ tensor can be further rewritten, by employing the results of Eq. (3.35), Eq. (3.36), Eq. (3.38) and Eq. (3.44), as

$$
\begin{aligned}
\overline{\mathcal{D}}_{\dot\alpha} T_{\gamma\dot\gamma}{}^\gamma{}_{\dot\delta\dot\beta} &= 4\overline{\mathcal{D}}_{\dot\alpha}\overline{W}_{(\dot\beta\dot\gamma\dot\delta)} + \frac{1}{2}(\varepsilon_{\dot\alpha\dot\gamma}\varepsilon_{\dot\beta\dot\delta} + \varepsilon_{\dot\alpha\dot\delta}\varepsilon_{\dot\beta\dot\gamma})\overline{\mathcal{D}}\cdot\overline{\mathcal{D}}\mathcal{R}^\dagger \,, \\
\mathcal{D}_{\gamma\dot\gamma} T_{\dot\alpha}{}^\gamma{}_{\dot\delta\dot\beta} + \mathcal{D}_{\gamma\dot\delta} T_{\dot\alpha}{}^\gamma{}_{\dot\gamma\dot\beta} &= \frac{i}{4}\Big[(\varepsilon_{\dot\gamma\dot\beta}\mathcal{D}_{\gamma\dot\delta} + \varepsilon_{\dot\delta\dot\beta}\mathcal{D}_{\gamma\dot\gamma})G^\gamma{}_{\dot\alpha} - 3\varepsilon_{\dot\alpha\dot\beta}(\mathcal{D}_{\gamma\dot\gamma}G^\gamma{}_{\dot\delta} + \mathcal{D}_{\gamma\dot\delta}G^\gamma{}_{\dot\gamma}) \\
&\quad -3(\varepsilon_{\dot\alpha\dot\delta}\mathcal{D}_{\gamma\dot\gamma} + \varepsilon_{\dot\alpha\dot\gamma}\mathcal{D}_{\gamma\dot\delta})G^\gamma{}_{\dot\beta}\Big] \,, \\
T_{\dot\alpha\gamma\dot\delta}{}^\epsilon T_\epsilon{}^\gamma{}_{\dot\gamma\dot\beta} + T_{\dot\alpha\gamma\dot\gamma}{}^\epsilon T_\epsilon{}^\gamma{}_{\dot\delta\dot\beta} &= -8(\varepsilon_{\dot\alpha\dot\gamma}\varepsilon_{\dot\delta\dot\beta} + \varepsilon_{\dot\alpha\dot\delta}\varepsilon_{\dot\gamma\dot\beta})\mathcal{R}\mathcal{R}^\dagger \,, \qquad (3.50) \\
T_{\dot\alpha\gamma\dot\delta\dot\epsilon} T^{\dot\epsilon\gamma}{}_{\dot\gamma\dot\beta} + T_{\dot\alpha\gamma\dot\gamma\dot\epsilon} T^{\dot\epsilon\gamma}{}_{\dot\delta\dot\beta} &= \frac{1}{16}\Big[(\varepsilon_{\dot\gamma\dot\beta}G_{\gamma\dot\delta} + \varepsilon_{\dot\delta\dot\beta}G_{\gamma\dot\gamma})G^\gamma{}_{\dot\alpha} - 9(\varepsilon_{\dot\alpha\dot\gamma}G_{\gamma\dot\delta} + \varepsilon_{\dot\alpha\dot\delta}G_{\gamma\dot\gamma})G^\gamma{}_{\dot\beta} \\
&\quad +3(\varepsilon_{\dot\alpha\dot\gamma}\varepsilon_{\dot\delta\dot\beta} + \varepsilon_{\dot\alpha\dot\delta}\varepsilon_{\dot\gamma\dot\beta})G_{\epsilon\dot\epsilon}G^{\epsilon\dot\epsilon}\Big] \,.
\end{aligned}
$$

By definition, the $\overline{X}_{(\dot\gamma\dot\delta)(\dot\alpha\dot\beta)}$ tensor is symmetric under the exchange of the pairs of indices $(\dot\alpha, \dot\beta)$ and $(\dot\gamma, \dot\delta)$. In the four equations above, the symmetry under the exchange of the $\dot\gamma$ and $\dot\delta$ indices is manifest, while the one under the exchange of $\dot\alpha$ and $\dot\beta$ is not. Contracting all these equations with $\varepsilon^{\dot\alpha\dot\beta}$, one obtains the constraint

$$
\varepsilon^{\dot\alpha\dot\beta}\overline{X}_{(\dot\gamma\dot\delta)(\dot\alpha\dot\beta)} = -\overline{\mathcal{D}}^{\dot\alpha}\,\overline{W}_{(\dot\alpha\dot\gamma\dot\delta)} - \frac{i}{2}(\mathcal{D}_{\gamma\dot\gamma}G^\gamma{}_{\dot\delta} + \mathcal{D}_{\gamma\dot\delta}G^\gamma{}_{\dot\gamma}) = 0 \,,
$$

so that, on the basis of the property $\overline{\mathcal{D}}_{\dot\alpha}\overline{W}_{(\dot\beta\dot\gamma\dot\delta)} - \overline{\mathcal{D}}_{\dot\beta}\overline{W}_{(\dot\alpha\dot\gamma\dot\delta)} = \varepsilon_{\dot\alpha\dot\beta}\overline{\mathcal{D}}^{\dot\epsilon}\,\overline{W}_{(\dot\epsilon\dot\gamma\dot\delta)} = -i/2\varepsilon_{\dot\alpha\dot\beta}(\mathcal{D}_{\gamma\dot\gamma}G^\gamma{}_{\dot\delta} + D_{\gamma\dot\delta}G^\gamma{}_{\dot\gamma})$, we may derive

$$
\begin{aligned}
\overline{\mathcal{D}}_{\dot\alpha}\,\overline{W}_{(\dot\beta\dot\gamma\dot\delta)} &= \frac{1}{4}\Big[\overline{\mathcal{D}}_{\dot\alpha}\overline{W}_{(\dot\beta\dot\gamma\dot\delta)} + \overline{\mathcal{D}}_{\dot\beta}\overline{W}_{(\dot\gamma\dot\delta\dot\alpha)} + \overline{\mathcal{D}}_{\dot\gamma}\overline{W}_{(\dot\delta\dot\alpha\dot\beta)} + \overline{\mathcal{D}}_{\dot\delta}\overline{W}_{(\dot\alpha\dot\beta\dot\gamma)}\Big] \\
&\quad + \frac{1}{4}\Big[\overline{\mathcal{D}}_{\dot\alpha}\overline{W}_{(\dot\beta\dot\gamma\dot\delta)} - \overline{\mathcal{D}}_{\dot\beta}\overline{W}_{(\dot\gamma\dot\delta\dot\alpha)}\Big] \\
&\quad + \frac{1}{4}\Big[\overline{\mathcal{D}}_{\dot\alpha}\overline{W}_{(\dot\beta\dot\gamma\dot\delta)} - \overline{\mathcal{D}}_{\dot\gamma}\overline{W}_{(\dot\delta\dot\alpha\dot\beta)}\Big] + \frac{1}{4}\Big[\overline{\mathcal{D}}_{\dot\alpha}\overline{W}_{(\dot\beta\dot\gamma\dot\delta)} - \overline{\mathcal{D}}_{\dot\delta}\overline{W}_{(\dot\alpha\dot\beta\dot\gamma)}\Big] \qquad (3.51) \\
&= \frac{1}{4}\Big[\overline{\mathcal{D}}_{\dot\alpha}\overline{W}_{(\dot\beta\dot\gamma\dot\delta)} + \overline{\mathcal{D}}_{\dot\beta}\overline{W}_{(\dot\gamma\dot\delta\dot\alpha)} + \overline{\mathcal{D}}_{\dot\gamma}\overline{W}_{(\dot\delta\dot\alpha\dot\beta)} + \overline{\mathcal{D}}_{\dot\delta}\overline{W}_{(\dot\alpha\dot\beta\dot\gamma)}\Big] \\
&\quad - \frac{i}{8}\Big[\varepsilon_{\dot\alpha\dot\beta}(\mathcal{D}_{\gamma\dot\gamma}G^\gamma{}_{\dot\delta} + \mathcal{D}_{\gamma\dot\delta}G^\gamma{}_{\dot\gamma}) + \varepsilon_{\dot\alpha\dot\gamma}(\mathcal{D}_{\gamma\dot\beta}G^\gamma{}_{\dot\delta} + \mathcal{D}_{\gamma\dot\delta}G^\gamma{}_{\dot\beta}) \\
&\quad + \varepsilon_{\dot\alpha\dot\delta}(\mathcal{D}_{\gamma\dot\beta}G^\gamma{}_{\dot\gamma} + \mathcal{D}_{\gamma\dot\gamma}G^\gamma{}_{\dot\beta})\Big] \,.
\end{aligned}
$$

Identifying and collecting the terms containing first derivatives of G in this last equation as well as the four equations of Eq. (3.50), we obtain, after symmetrising with respect to $\dot\alpha$ and $\dot\beta$,

$$
\begin{aligned}
-\tfrac{i}{16}\Big[&(\varepsilon_{\dot\gamma\dot\beta}\mathcal{D}_{\gamma\dot\delta} + \varepsilon_{\dot\delta\dot\beta}\mathcal{D}_{\gamma\dot\gamma})G^\gamma{}_{\dot\alpha} + (\varepsilon_{\dot\gamma\dot\alpha}\mathcal{D}_{\gamma\dot\delta} + \varepsilon_{\dot\delta\dot\alpha}\mathcal{D}_{\gamma\dot\gamma})G^\gamma{}_{\dot\beta} + (\varepsilon_{\dot\beta\dot\gamma}\mathcal{D}_{\gamma\dot\alpha} + \varepsilon_{\dot\alpha\dot\gamma}\mathcal{D}_{\gamma\dot\beta})G^\gamma{}_{\dot\delta} \\
&+ (\varepsilon_{\dot\beta\dot\delta}\mathcal{D}_{\gamma\dot\alpha} + \varepsilon_{\dot\alpha\dot\delta}\mathcal{D}_{\gamma\dot\beta})G^\gamma{}_{\dot\gamma}\Big] = \tfrac{i}{8}(\varepsilon_{\dot\alpha\dot\gamma}\varepsilon_{\dot\beta\dot\delta} + \varepsilon_{\dot\alpha\dot\delta}\varepsilon_{\dot\beta\dot\gamma})\mathcal{D}_{\epsilon\dot\epsilon}G^{\epsilon\dot\epsilon} \,. \qquad (3.52)
\end{aligned}
$$

This expression can be further simplified on the basis of Eq. (3.8), Eq. (3.29) and Eq. (B.2), to read

$$\begin{aligned}
\bar{\sigma}_\mu{}^{\epsilon\dot{\epsilon}}\{\mathcal{D}_\epsilon, \overline{\mathcal{D}}_{\dot{\epsilon}}\}G^\mu &= \bar{\sigma}_\mu{}^{\epsilon\dot{\epsilon}}[T_{\epsilon\dot{\epsilon}}{}^\nu \mathcal{D}_\nu G^\mu + G^\nu R_{\epsilon\dot{\epsilon}\nu}{}^\mu] \\
&= -4i\mathcal{D}_\mu G^\mu \\
&\quad + \frac{1}{2}\bar{\sigma}_\mu{}^{\epsilon\dot{\epsilon}} G^\nu R_{\epsilon\dot{\epsilon}\sigma\rho}[\eta^{\mu\rho}\delta^\sigma{}_\nu - \eta^{\mu\sigma}\delta^\rho{}_\nu] \\
&= -4i\mathcal{D}_\mu G^\mu - \frac{1}{2}\varepsilon^{\dot{\epsilon}\dot{\delta}}\varepsilon^{\epsilon\delta}\bar{\sigma}_\rho{}^{\gamma\dot{\gamma}} G^\rho R_{\epsilon\dot{\epsilon}\delta\dot{\delta}\gamma\dot{\gamma}} \\
&= -2i\mathcal{D}_{\epsilon\dot{\epsilon}} G^{\epsilon\dot{\epsilon}} \, ,
\end{aligned}$$

where the term containing the curvature tensor vanishes because of its Lie-algebra valuedness and of Eq. (3.38). Making additional use of Eq. (3.47), Eq. (3.53), one gets

$$\begin{aligned}
\frac{i}{8}\mathcal{D}_{\epsilon\dot{\epsilon}} G^{\epsilon\dot{\epsilon}} &= -\frac{1}{16}\bar{\sigma}_\mu{}^{\epsilon\dot{\epsilon}}[\mathcal{D}_\epsilon \overline{\mathcal{D}}_{\dot{\epsilon}} + \overline{\mathcal{D}}_{\dot{\epsilon}}\mathcal{D}_\epsilon]G^\mu \\
&= -\frac{1}{16}[\mathcal{D}_\epsilon \overline{\mathcal{D}}_{\dot{\epsilon}} + \overline{\mathcal{D}}_{\dot{\epsilon}}\mathcal{D}_\epsilon]G^{\epsilon\dot{\epsilon}} \\
&= \frac{1}{16}[\mathcal{D}\cdot\mathcal{D}\mathcal{R} - \overline{\mathcal{D}}\cdot\overline{\mathcal{D}}\mathcal{R}^\dagger] \, .
\end{aligned} \tag{3.53}$$

Collecting all the contributions to the $\overline{X}$-tensor and using the identity

$$\begin{aligned}
\varepsilon_{\dot{\gamma}\dot{\beta}} G_{\gamma\dot{\delta}} G^\gamma{}_{\dot{\alpha}} &= \frac{1}{2}\varepsilon_{\dot{\gamma}\dot{\beta}}\Big(G_{\gamma\dot{\delta}}G^\gamma{}_{\dot{\alpha}} - G_{\gamma\dot{\alpha}}G^\gamma{}_{\dot{\delta}}\Big) \\
&= \frac{1}{2}\varepsilon_{\dot{\gamma}\dot{\beta}}\varepsilon_{\dot{\alpha}\dot{\delta}} G_{\epsilon\dot{\epsilon}} G^{\epsilon\dot{\epsilon}} \, ,
\end{aligned}$$

we finally derive

$$\begin{aligned}
\overline{X}_{(\dot{\gamma}\dot{\delta})(\dot{\alpha}\dot{\beta})} &= \frac{1}{4}\Big[\overline{\mathcal{D}}_{\dot{\alpha}} W_{(\dot{\beta}\dot{\gamma}\dot{\delta})} + \overline{\mathcal{D}}_{\dot{\beta}} W_{(\dot{\gamma}\dot{\delta}\dot{\alpha})} + \overline{\mathcal{D}}_{\dot{\gamma}} W_{(\dot{\delta}\dot{\alpha}\dot{\beta})} + \overline{\mathcal{D}}_{\dot{\delta}} W_{(\dot{\alpha}\dot{\beta}\dot{\gamma})}\Big] \\
&\quad + (\varepsilon_{\dot{\alpha}\dot{\delta}}\varepsilon_{\dot{\beta}\dot{\gamma}} + \varepsilon_{\dot{\beta}\dot{\delta}}\varepsilon_{\dot{\alpha}\dot{\gamma}})\Big[-2\mathcal{R}\mathcal{R}^\dagger + \frac{1}{8}G_{\epsilon\dot{\epsilon}}G^{\epsilon\dot{\epsilon}} + \frac{1}{16}\Big(\overline{\mathcal{D}}\cdot\overline{\mathcal{D}}\mathcal{R}^\dagger + \mathcal{D}\cdot\mathcal{D}\mathcal{R}\Big)\Big] \, .
\end{aligned} \tag{3.54}$$

In a similar fashion, the different terms contributing to the Ψ-tensor of Eq. (3.50) are rewritten as

$$\begin{aligned}
-\overline{\mathcal{D}}_{\dot{\alpha}} T_{\gamma\dot{\gamma}\delta}{}^{\dot{\gamma}}{}_{\dot{\beta}} &= \overline{\mathcal{D}}_{\dot{\alpha}}\mathcal{D}_\gamma G_{\delta\dot{\beta}} + \overline{\mathcal{D}}_{\dot{\alpha}}\mathcal{D}_\delta G_{\gamma\dot{\beta}} \, , \\
\mathcal{D}_{\gamma\dot{\gamma}} T_{\dot{\alpha}\delta}{}^{\dot{\gamma}}{}_{\dot{\beta}} + \mathcal{D}_{\delta\dot{\gamma}} T_{\dot{\alpha}\gamma}{}^{\dot{\gamma}}{}_{\dot{\beta}} &= \frac{i}{4}\Big[\mathcal{D}_{\gamma\dot{\beta}} G_{\delta\dot{\alpha}} + 3\mathcal{D}_{\gamma\dot{\alpha}} G_{\delta\dot{\beta}} - 3\varepsilon_{\dot{\alpha}\dot{\beta}}\mathcal{D}_{\gamma\dot{\gamma}} G_\delta{}^{\dot{\gamma}} + \mathcal{D}_{\delta\dot{\beta}} G_{\gamma\dot{\alpha}} \\
&\quad + 3\mathcal{D}_{\delta\dot{\alpha}} G_{\gamma\dot{\beta}} - 3\varepsilon_{\dot{\alpha}\dot{\beta}}\mathcal{D}_{\delta\dot{\gamma}} G_\gamma{}^{\dot{\gamma}}\Big] \, , \\
T_{\dot{\alpha}\delta\dot{\gamma}}{}^\epsilon T_{\epsilon\gamma}{}^{\dot{\gamma}}{}_{\dot{\beta}} + T_{\dot{\alpha}\gamma\dot{\gamma}}{}^\epsilon T_{\epsilon\delta}{}^{\dot{\gamma}}{}_{\dot{\beta}} &= 0 \, , \\
T_{\dot{\alpha}\delta\dot{\gamma}\dot{\epsilon}} T^{\dot{\epsilon}}{}_\gamma{}^{\dot{\gamma}}{}_{\dot{\beta}} + T_{\dot{\alpha}\gamma\dot{\gamma}\dot{\epsilon}} T^{\dot{\epsilon}}{}_\delta{}^{\dot{\gamma}}{}_{\dot{\beta}} &= -(G_{\delta\dot{\alpha}}G_{\gamma\dot{\beta}} + G_{\gamma\dot{\alpha}}G_{\delta\dot{\beta}}) \, .
\end{aligned} \tag{3.55}$$

Contracting with $\varepsilon^{\dot\alpha\dot\beta}$, one derives the constraint

$$\overline{\mathcal{D}}_{\dot\gamma}\mathcal{D}_\gamma G_\delta{}^{\dot\gamma} + \overline{\mathcal{D}}_{\dot\gamma}\mathcal{D}_\delta G_\gamma{}^{\dot\gamma} = -2i\Big[\mathcal{D}_{\gamma\dot\gamma}G_\delta{}^{\dot\gamma} + \mathcal{D}_{\delta\dot\gamma}G_\gamma{}^{\dot\gamma}\Big]\,,$$

which may then be used to get

$$\begin{aligned}\Psi_{(\delta\gamma)(\dot\alpha\dot\beta)} &= -\frac{1}{4}\Big(G_{\delta\dot\alpha}G_{\gamma\dot\beta} + G_{\delta\dot\beta}G_{\gamma\dot\alpha}\Big) + \frac{i}{8}\Big(\mathcal{D}_{\gamma\dot\beta}G_{\delta\dot\alpha} + \mathcal{D}_{\gamma\dot\alpha}G_{\delta\dot\beta} + \mathcal{D}_{\delta\dot\beta}G_{\gamma\dot\alpha} + \mathcal{D}_{\delta\dot\alpha}G_{\gamma\dot\beta}\Big)\\ &\quad+\frac{1}{8}\Big(\overline{\mathcal{D}}_{\dot\alpha}\mathcal{D}_\gamma G_{\delta\dot\beta} + \overline{\mathcal{D}}_{\dot\alpha}\mathcal{D}_\delta G_{\gamma\dot\beta} + \overline{\mathcal{D}}_{\dot\beta}\mathcal{D}_\gamma G_{\delta\dot\alpha} + \overline{\mathcal{D}}_{\dot\beta}\mathcal{D}_\delta G_{\gamma\dot\alpha}\Big)\,. \end{aligned}\tag{3.56}$$

It can be shown that the solution to the tenth Bianchi identity is also a solution for the eleventh identity $I_{\mu\nu\rho\sigma}$, the Bianchi identity of general relativity. Introducing spinor notations and decomposing the $R_{\mu\nu\rho\sigma}$ tensor into its irreducible components, we get

$$\begin{aligned}R_{\alpha\dot\alpha\beta\dot\beta\gamma\dot\gamma\delta\dot\delta} &= 4\varepsilon_{\alpha\beta}\varepsilon_{\gamma\delta}\overline{X}_{(\dot\alpha\dot\beta)(\dot\gamma\dot\delta)} + 4\varepsilon_{\dot\alpha\dot\beta}\varepsilon_{\dot\gamma\dot\delta}X_{(\alpha\beta)(\gamma\delta)}\\ &\quad-4\varepsilon_{\alpha\beta}\varepsilon_{\dot\gamma\dot\delta}\overline{\Psi}_{(\dot\alpha\dot\beta)(\gamma\delta)} - 4\varepsilon_{\dot\alpha\dot\beta}\varepsilon_{\gamma\delta}\Psi_{(\alpha\beta)(\dot\gamma\dot\delta)}\,. \end{aligned}\tag{3.57}$$

Inserting this expression back into the identity $I_{\mu\nu\rho\sigma}$, contracting it with $\varepsilon^{\alpha\beta}\varepsilon^{\dot\gamma\dot\delta}$ and identifying all the terms with a given symmetry property under the exchange of the indices γ and δ, we obtain

$$\begin{aligned}\Psi_{(\gamma\delta)(\dot\alpha\dot\beta)} &= \overline{\Psi}_{(\dot\alpha\dot\beta)(\gamma\delta)}\,,\\ \overline{X}_{(\dot\alpha\dot\gamma)(\dot\beta}{}^{\dot\gamma}{}_{)} &= \varepsilon_{\dot\alpha\dot\beta}\Lambda\,,\\ X_{(\alpha\gamma)(\beta}{}^{\gamma}{}_{)} &= \varepsilon_{\alpha\beta}\Lambda\,,\end{aligned}\tag{3.58}$$

where the superfield Λ is real. It is straightforward to prove that these constraints are fulfilled by the results of Eq. (3.54) and Eq. (3.56).

Identities (12) and (13): $I_{\mu\nu\dot\alpha\beta}$ and $I_{\mu\nu\rho\alpha}$.

Converting the identity $I_{\mu\nu\dot\alpha\beta}$ into spinor notations and using Eq. (3.35), Eq. (3.38) and Eq. (3.44), we obtain, after having symmetrised the identity with respect to all undotted indices,

$$\overline{\mathcal{D}}_{\dot\alpha}W_{(\alpha\beta\gamma)} = 0\,.$$

The superfield $W_{(\alpha\beta\gamma)}$ is thus a chiral superfield. Similarly, it can be shown that $\overline{W}_{(\dot\alpha\dot\beta\dot\gamma)}$ is antichiral. Under these additional conditions, the thirteenth Bianchi identity is automatically satisfied, as can be demonstrated after using, among others, Eq. (3.8) and Eq. (3.29).

As a summary, in this subsection, we have shown that all the non-vanishing components of the torsion and curvature tensors can be expressed as functions of three superfields $\mathcal{R}$, $G_{\alpha\dot{\alpha}}$ and $W_{(\alpha\beta\gamma)}$ which satisfy the constraints summarised in Table 3.2. Expanding the graded commutator of Eq. (3.9) with respect to all possible choices for the nature of the indices, we have, retaining only the non-zero contributions,

$$\begin{aligned}
\left\{\mathcal{D}_\alpha, \mathcal{D}_\beta\right\} &= -\frac{1}{2} R_{\alpha\beta\gamma\delta} J^{\delta\gamma} , \\
\left\{\overline{\mathcal{D}}_{\dot{\alpha}}, \overline{\mathcal{D}}_{\dot{\beta}}\right\} &= -\frac{1}{2} R_{\dot{\alpha}\dot{\beta}\dot{\gamma}\dot{\delta}} J^{\dot{\delta}\dot{\gamma}} , \\
\left\{\mathcal{D}_\alpha, \overline{\mathcal{D}}_{\dot{\alpha}}\right\} &= T_{\alpha\dot{\alpha}}{}^{\mu} \mathcal{D}_\mu - \frac{1}{2} R_{\alpha\dot{\alpha}\beta\gamma} J^{\gamma\beta} - \frac{1}{2} R_{\alpha\dot{\alpha}\dot{\beta}\dot{\gamma}} J^{\dot{\gamma}\dot{\beta}} , \\
\left[\mathcal{D}_\alpha, \mathcal{D}_\mu\right] &= T_{\alpha\mu}{}^{\beta} \mathcal{D}_\beta - T_{\alpha\mu}{}^{\dot{\beta}} \overline{\mathcal{D}}_{\dot{\beta}} - \frac{1}{2} R_{\alpha\mu\beta\gamma} J^{\gamma\beta} - \frac{1}{2} R_{\alpha\mu\dot{\beta}\dot{\gamma}} J^{\dot{\gamma}\dot{\beta}} , \\
\left[\overline{\mathcal{D}}_{\dot{\alpha}}, \mathcal{D}_\mu\right] &= T_{\dot{\alpha}\mu}{}^{\beta} \mathcal{D}_\beta - T_{\dot{\alpha}\mu}{}^{\dot{\beta}} \overline{\mathcal{D}}_{\dot{\beta}} - \frac{1}{2} R_{\dot{\alpha}\mu\beta\gamma} J^{\gamma\beta} - \frac{1}{2} R_{\dot{\alpha}\mu\dot{\beta}\dot{\gamma}} J^{\dot{\gamma}\dot{\beta}} , \\
\left[\mathcal{D}_\mu, \mathcal{D}_\nu\right] &= T_{\mu\nu}{}^{\beta} \mathcal{D}_\beta - T_{\mu\nu}{}^{\dot{\beta}} \overline{\mathcal{D}}_{\dot{\beta}} - \frac{1}{2} R_{\mu\nu\beta\gamma} J^{\gamma\beta} - \frac{1}{2} R_{\mu\nu\dot{\beta}\dot{\gamma}} J^{\dot{\gamma}\dot{\beta}} .
\end{aligned} \tag{3.59}$$

The definitions of the elements of the torsion and curvature tensors appearing in these expressions are gathered into Table 3.3 and Table 3.4, respectively, and we recall that R_{MNPQ} is Lie-algebra valued in its last two indices so that $R_{MN\mu\nu}$ is related to $R_{MN\alpha\beta}$ and $R_{MN\dot{\alpha}\dot{\beta}}$ by

$$R_{MN\mu\nu} = \frac{1}{2} \bar{\sigma}_\mu{}^{\dot{\gamma}\gamma} \bar{\sigma}_\nu{}^{\dot{\delta}\delta} (\varepsilon_{\gamma\delta} R_{MN\dot{\gamma}\dot{\delta}} - \varepsilon_{\dot{\gamma}\dot{\delta}} R_{MN\gamma\delta}) . \tag{3.60}$$

Moreover, the structure of the algebra obeying Eq. (3.59) is only compatible with chiral (antichiral) superfields carrying undotted (dotted) spinor indices, as induced in particular by the second (first) relation of the algebra. Conversely, it is not possible to define chiral (antichiral) superfields carrying vector or right-handed (left-handed) spinor indices. This feature is closely related to the non-vanishing $R_{\alpha\beta\gamma\delta}$ and $R_{\dot{\alpha}\dot{\beta}\dot{\gamma}\dot{\delta}}$ elements of the curvature tensor, a property specific to supergravity theories.

3.2 The Weyl supergroup

The supergravity algebra (3.59) is obviously invariant under both the local Lorentz transformations (3.1) and the superdiffeomorphisms (3.2). However, the structure of this algebra is preserved by transformations induced by a larger group, the so-called Weyl supergroup. This is analogous to the Weyl group extending the symmetry group of general relativity, and which entails the choice of rescaling

Table 3.2 Constraints on the superfields $\mathcal{R}$, $G_{\alpha\dot\alpha}$ and $W_{(\alpha\beta\gamma)}$.

$$\begin{gathered}
\overline{\mathcal{D}}_{\dot\alpha}\mathcal{R} = 0\,, \qquad \mathcal{D}_\alpha\mathcal{R}^\dagger = 0\,,\\
\mathcal{D}^\beta G_{\beta\dot\alpha} = -\overline{\mathcal{D}}_{\dot\alpha}\mathcal{R}^\dagger\,, \qquad \overline{\mathcal{D}}^{\dot\beta} G_{\alpha\dot\beta} = -\mathcal{D}_\alpha\mathcal{R}\,,\\
\overline{\mathcal{D}}_{\dot\alpha} W_{(\alpha\beta\gamma)} = 0\,, \qquad \mathcal{D}_\alpha \overline{W}_{(\dot\alpha\dot\beta\dot\gamma)} = 0\,,\\
\overline{\mathcal{D}}^{\dot\alpha}\, \overline{W}_{(\dot\alpha\dot\gamma\dot\delta)} + \tfrac{i}{2}(\mathcal{D}_{\gamma\dot\gamma}G^\gamma{}_{\dot\delta} + \mathcal{D}_{\gamma\dot\delta}G^\gamma{}_{\dot\gamma}) = 0\,,\\
\mathcal{D}^\alpha\, W_{(\alpha\gamma\delta)} + \tfrac{i}{2}(\mathcal{D}_{\gamma\dot\gamma}G_\delta{}^{\dot\gamma} + \mathcal{D}_{\delta\dot\gamma}G_\gamma{}^{\dot\gamma}) = 0\,,\\
\overline{\mathcal{D}}_{\dot\gamma}\mathcal{D}_\gamma G_\delta{}^{\dot\gamma} + \overline{\mathcal{D}}_{\dot\gamma}\mathcal{D}_\delta G_\gamma{}^{\dot\gamma} = -2i(\mathcal{D}_{\gamma\dot\gamma}G_\delta{}^{\dot\gamma} + \mathcal{D}_{\delta\dot\gamma}G_\gamma{}^{\dot\gamma})\,.
\end{gathered}$$

Table 3.3 Non-zero components of the torsion tensor.

$$\begin{gathered}
T_{\alpha\dot\alpha}{}^\mu = T_{\dot\alpha\alpha}{}^\mu = -2i\sigma^\mu{}_{\alpha\dot\alpha}\,,\\
T_{\alpha\mu}{}^\beta = -T_{\mu\alpha}{}^\beta = \tfrac{1}{2}\bar\sigma_\mu{}^{\dot\gamma\gamma}T_{\alpha\gamma\dot\gamma}{}^\beta\,, \qquad T_{\alpha\mu}{}^{\dot\beta} = -T_{\mu\alpha}{}^{\dot\beta} = \tfrac{1}{2}\bar\sigma_\mu{}^{\dot\gamma\gamma}T_{\alpha\gamma\dot\gamma}{}^{\dot\beta}\,,\\
T_{\dot\alpha\mu}{}^\beta = -T_{\mu\dot\alpha}{}^\beta = \tfrac{1}{2}\bar\sigma_\mu{}^{\dot\gamma\gamma}T_{\dot\alpha\gamma\dot\gamma}{}^\beta\,, \qquad T_{\dot\alpha\mu}{}^{\dot\beta} = -T_{\mu\dot\alpha}{}^{\dot\beta} = \tfrac{1}{2}\bar\sigma_\mu{}^{\dot\gamma\gamma}T_{\dot\alpha\gamma\dot\gamma}{}^{\dot\beta}\,,\\
T_{\mu\nu}{}^\alpha = -T_{\nu\mu}{}^\alpha = \tfrac{1}{4}\bar\sigma_\mu{}^{\dot\delta\delta}\bar\sigma_\nu{}^{\dot\gamma\gamma}T_{\delta\dot\delta\gamma\dot\gamma}{}^\alpha\,, \qquad T_{\mu\nu}{}^{\dot\alpha} = -T_{\nu\mu}{}^{\dot\alpha} = \tfrac{1}{4}\bar\sigma_\mu{}^{\dot\delta\delta}\bar\sigma_\nu{}^{\dot\gamma\gamma}T_{\delta\dot\delta\gamma\dot\gamma}{}^{\dot\alpha}\,,
\end{gathered}$$

with

$$\begin{gathered}
T_{\alpha\gamma\dot\gamma\dot\beta} = 2i\varepsilon_{\alpha\gamma}\varepsilon_{\dot\gamma\dot\beta}\mathcal{R}^\dagger\,, \qquad T_{\dot\alpha\gamma\dot\gamma\beta} = 2i\varepsilon_{\dot\alpha\dot\gamma}\varepsilon_{\gamma\beta}\mathcal{R}\,,\\
T_{\alpha\gamma\dot\gamma\beta} = \tfrac{i}{4}\Big(\varepsilon_{\gamma\beta}G_{\alpha\dot\gamma} - 3\varepsilon_{\alpha\beta}G_{\gamma\dot\gamma} - 3\varepsilon_{\alpha\gamma}G_{\beta\dot\gamma}\Big)\,, \qquad T_{\dot\alpha\gamma\dot\gamma\dot\beta} = \tfrac{i}{4}\Big(\varepsilon_{\dot\gamma\dot\beta}G_{\gamma\dot\alpha} - 3\varepsilon_{\dot\alpha\dot\beta}G_{\gamma\dot\gamma} - 3\varepsilon_{\dot\alpha\dot\gamma}G_{\gamma\dot\beta}\Big)\,,\\
T_{\delta\dot\delta\gamma\dot\gamma\alpha} = -2\varepsilon_{\dot\delta\dot\gamma}W_{(\delta\gamma\alpha)} + \tfrac{1}{2}\varepsilon_{\dot\gamma\dot\delta}\Big(\varepsilon_{\alpha\gamma}\overline{\mathcal{D}}^{\dot\epsilon}G_{\delta\dot\epsilon} + \varepsilon_{\alpha\delta}\overline{\mathcal{D}}^{\dot\epsilon}G_{\gamma\dot\epsilon}\Big) + \tfrac{1}{2}\varepsilon_{\delta\gamma}\Big(\overline{\mathcal{D}}_{\dot\delta}G_{\alpha\dot\gamma} + \overline{\mathcal{D}}_{\dot\gamma}G_{\alpha\dot\delta}\Big)\,,\\
T_{\delta\dot\delta\gamma\dot\gamma\dot\alpha} = -2\varepsilon_{\delta\gamma}\overline{W}_{(\dot\delta\dot\gamma\dot\alpha)} + \tfrac{1}{2}\varepsilon_{\gamma\delta}\Big(\varepsilon_{\dot\alpha\dot\gamma}\mathcal{D}^\epsilon G_{\epsilon\dot\delta} + \varepsilon_{\dot\alpha\dot\delta}\mathcal{D}^\epsilon G_{\epsilon\dot\gamma}\Big) + \tfrac{1}{2}\varepsilon_{\dot\delta\dot\gamma}\Big(\mathcal{D}_\delta G_{\gamma\dot\alpha} + \mathcal{D}_\gamma G_{\delta\dot\alpha}\Big)\,.
\end{gathered}$$

the vierbein. The existence of the Weyl supergroup has been shown for the first time by Howe and Tucker (1978); Siegel (1979a), using calculation techniques developed in [Wess and Zumino (1978b); Siegel (1979b)] to prove the invariance of the supergravity algebra under the so-called Howe–Tucker–Weyl rescaling of the supervierbein. More specifically, this invariance has an algebraic origin [Gates et al. (1980); Gates and Siegel (1980)] that can be understood as an automorphism of the supergravity algebra parameterised by a chiral superfield [Dragon et al. (1987)], as shown in App. H.

3.2.1 *Howe–Tucker transformations*

In order to derive the automorphism of the algebra (3.59), we start by using Eq. (B.4) to show that the Lorentz generators $J_{\alpha\beta}$ have to be invariant under the associated symmetry transformations. Next, we turn to the superderivative operator

Table 3.4 Non-zero components of the curvature tensor.

$$R_{\alpha\beta\gamma\delta} = 4(\varepsilon_{\alpha\gamma}\varepsilon_{\beta\delta} + \varepsilon_{\beta\gamma}\varepsilon_{\alpha\delta})\mathcal{R}^\dagger\,, \qquad R_{\dot\alpha\dot\beta\dot\gamma\dot\delta} = 4(\varepsilon_{\dot\alpha\dot\gamma}\varepsilon_{\dot\beta\dot\delta} + \varepsilon_{\dot\beta\dot\gamma}\varepsilon_{\dot\alpha\dot\delta})\mathcal{R}\,,$$

$$R_{\alpha\dot\alpha\beta\gamma} = R_{\dot\alpha\alpha\beta\gamma} = \varepsilon_{\alpha\gamma}G_{\beta\dot\alpha} + \varepsilon_{\alpha\beta}G_{\gamma\dot\alpha}\,, \qquad R_{\dot\alpha\alpha\dot\beta\dot\gamma} = R_{\alpha\dot\alpha\dot\beta\dot\gamma} = \varepsilon_{\dot\alpha\dot\gamma}G_{\alpha\dot\beta} + \varepsilon_{\dot\alpha\dot\beta}G_{\alpha\dot\gamma}\,,$$

$$R_{\alpha\mu\beta\gamma} = -R_{\mu\alpha\beta\gamma} = \tfrac{1}{2}\bar\sigma_\mu{}^{\dot\delta\delta}R_{\alpha\delta\dot\delta\beta\gamma} \qquad R_{\alpha\mu\dot\beta\dot\gamma} = -R_{\mu\alpha\dot\beta\dot\gamma} = \tfrac{1}{2}\bar\sigma_\mu{}^{\dot\delta\delta}R_{\alpha\delta\dot\delta\dot\beta\dot\gamma}\,,$$

$$R_{\dot\alpha\mu\beta\gamma} = -R_{\mu\dot\alpha\beta\gamma} = \tfrac{1}{2}\bar\sigma_\mu{}^{\dot\delta\delta}R_{\dot\alpha\delta\dot\delta\beta\gamma}\,, \qquad R_{\dot\alpha\mu\dot\beta\dot\gamma} = -R_{\mu\dot\alpha\dot\beta\dot\gamma} = \tfrac{1}{2}\bar\sigma_\mu{}^{\dot\delta\delta}R_{\dot\alpha\delta\dot\delta\dot\beta\dot\gamma}\,,$$

$$R_{\mu\nu\alpha\beta} = \tfrac{1}{4}\bar\sigma_\mu{}^{\dot\gamma\gamma}\bar\sigma_\nu{}^{\dot\delta\delta}R_{\gamma\dot\gamma\delta\dot\delta\alpha\beta}\,, \qquad R_{\mu\nu\dot\alpha\dot\beta} = \tfrac{1}{4}\bar\sigma_\mu{}^{\dot\gamma\gamma}\bar\sigma_\nu{}^{\dot\delta\delta}R_{\gamma\dot\gamma\delta\dot\delta\dot\alpha\dot\beta}\,,$$

with

$$R_{\alpha\delta\dot\delta\beta\gamma} = i(\varepsilon_{\alpha\beta}\varepsilon_{\delta\gamma} + \varepsilon_{\alpha\gamma}\varepsilon_{\delta\beta})\mathcal{D}^\epsilon G_{\epsilon\dot\delta} - \tfrac{i}{2}(\varepsilon_{\alpha\delta}\mathcal{D}_\beta + \varepsilon_{\alpha\beta}\mathcal{D}_\delta)G_{\gamma\dot\delta} - \tfrac{i}{2}(\varepsilon_{\alpha\delta}\mathcal{D}_\gamma + \varepsilon_{\alpha\gamma}\mathcal{D}_\delta)G_{\beta\dot\delta}\,,$$

$$R_{\dot\alpha\delta\dot\delta\dot\beta\dot\gamma} = i(\varepsilon_{\dot\alpha\dot\beta}\varepsilon_{\dot\delta\dot\gamma} + \varepsilon_{\dot\alpha\dot\gamma}\varepsilon_{\dot\delta\dot\beta})\overline{\mathcal{D}}^{\dot\epsilon}G_{\delta\dot\epsilon} - \tfrac{i}{2}(\varepsilon_{\dot\alpha\dot\delta}\overline{\mathcal{D}}_{\dot\beta} + \varepsilon_{\dot\alpha\dot\beta}\overline{\mathcal{D}}_{\dot\delta})G_{\delta\dot\gamma} - \tfrac{i}{2}(\varepsilon_{\dot\alpha\dot\delta}\overline{\mathcal{D}}_{\dot\gamma} + \varepsilon_{\dot\alpha\dot\gamma}\overline{\mathcal{D}}_{\dot\delta})G_{\delta\dot\beta}\,,$$

$$R_{\alpha\delta\dot\delta\dot\beta\dot\gamma} = -4i\varepsilon_{\alpha\delta}\overline{W}_{(\dot\delta\dot\beta\dot\gamma)} - \tfrac{i}{2}(\varepsilon_{\dot\delta\dot\beta}\mathcal{D}_\alpha G_{\delta\dot\gamma} + \varepsilon_{\dot\delta\dot\gamma}\mathcal{D}_\alpha G_{\delta\dot\beta})\,,$$

$$R_{\dot\alpha\delta\dot\delta\beta\gamma} = -4i\varepsilon_{\dot\alpha\dot\delta}W_{(\delta\beta\gamma)} - \tfrac{i}{2}(\varepsilon_{\delta\beta}\overline{\mathcal{D}}_{\dot\alpha}G_{\gamma\dot\delta} + \varepsilon_{\delta\gamma}\overline{\mathcal{D}}_{\dot\alpha}G_{\beta\dot\delta})\,,$$

$$R_{\gamma\dot\gamma\delta\dot\delta\alpha\beta} = -2\varepsilon_{\dot\gamma\dot\delta}X_{(\gamma\delta)(\alpha\beta)} + 2\varepsilon_{\gamma\delta}\overline{\Psi}_{(\dot\gamma\dot\delta)(\alpha\beta)}\,, \qquad R_{\gamma\dot\gamma\delta\dot\delta\dot\alpha\dot\beta} = 2\varepsilon_{\gamma\delta}\overline{X}_{(\dot\gamma\dot\delta)(\dot\alpha\dot\beta)} - 2\varepsilon_{\dot\gamma\dot\delta}\Psi_{(\gamma\delta)(\dot\alpha\dot\beta)}\,,$$

where the X and Ψ tensors are given by

$$\begin{aligned}
X_{(\gamma\delta)(\alpha\beta)} &= -\frac{1}{4}\Big[\mathcal{D}_\alpha W_{(\beta\gamma\delta)} + \mathcal{D}_\beta W_{(\gamma\delta\alpha)} + \mathcal{D}_\gamma W_{(\delta\alpha\beta)} + \mathcal{D}_\delta W_{(\alpha\beta\gamma)}\Big] \\
&\quad +(\varepsilon_{\alpha\delta}\varepsilon_{\beta\gamma} + \varepsilon_{\beta\delta}\varepsilon_{\alpha\gamma})\Big[-2\mathcal{R}\mathcal{R}^\dagger + \frac{1}{8}G_{\epsilon\dot\epsilon}G^{\epsilon\dot\epsilon} + \frac{1}{16}\big(\overline{\mathcal{D}}\cdot\overline{\mathcal{D}}\mathcal{R}^\dagger + \mathcal{D}\cdot\mathcal{D}\mathcal{R}\big)\Big]\,, \\
\overline{X}_{(\dot\gamma\dot\delta)(\dot\alpha\dot\beta)} &= \frac{1}{4}\Big[\overline{\mathcal{D}}_{\dot\alpha}\overline{W}_{(\dot\beta\dot\gamma\dot\delta)} + \overline{\mathcal{D}}_{\dot\beta}\overline{W}_{(\dot\gamma\dot\delta\dot\alpha)} + \overline{\mathcal{D}}_{\dot\gamma}\overline{W}_{(\dot\delta\dot\alpha\dot\beta)} + \overline{\mathcal{D}}_{\dot\delta}\overline{W}_{(\dot\alpha\dot\beta\dot\gamma)}\Big] \\
&\quad +(\varepsilon_{\dot\alpha\dot\delta}\varepsilon_{\dot\beta\dot\gamma} + \varepsilon_{\dot\beta\dot\delta}\varepsilon_{\dot\alpha\dot\gamma})\Big[-2\mathcal{R}\mathcal{R}^\dagger + \frac{1}{8}G_{\epsilon\dot\epsilon}G^{\epsilon\dot\epsilon} + \frac{1}{16}\big(\overline{\mathcal{D}}\cdot\overline{\mathcal{D}}\mathcal{R}^\dagger + \mathcal{D}\cdot\mathcal{D}\mathcal{R}\big)\Big]\,, \\
\Psi_{(\delta\gamma)(\dot\alpha\dot\beta)} &= \overline{\Psi}_{(\dot\alpha\dot\beta)(\delta\gamma)} \\
&= -\frac{1}{4}\big(G_{\delta\dot\alpha}G_{\gamma\dot\beta} + G_{\delta\dot\beta}G_{\gamma\dot\beta}\big) + \frac{i}{8}\big(\mathcal{D}_{\gamma\dot\beta}G_{\delta\dot\alpha} + \mathcal{D}_{\gamma\dot\alpha}G_{\delta\dot\beta} + \mathcal{D}_{\delta\dot\beta}G_{\gamma\dot\alpha} + \mathcal{D}_{\delta\dot\alpha}G_{\gamma\dot\beta}\big) \\
&\quad +\frac{1}{8}\big(\overline{\mathcal{D}}_{\dot\alpha}\mathcal{D}_\gamma G_{\delta\dot\beta} + \overline{\mathcal{D}}_{\dot\alpha}\mathcal{D}_\delta G_{\gamma\dot\beta} + \overline{\mathcal{D}}_{\dot\beta}\mathcal{D}_\gamma G_{\delta\dot\alpha} + \overline{\mathcal{D}}_{\dot\beta}\mathcal{D}_\delta G_{\gamma\dot\alpha}\big)\,.
\end{aligned}$$

The tensors $R_{\alpha\beta\mu\nu}$, $R_{\dot\alpha\dot\beta\mu\nu}$, $R_{\alpha\dot\alpha\mu\nu} = R_{\dot\alpha\alpha\mu\nu}$, $R_{\alpha\mu\nu\rho} = -R_{\mu\alpha\nu\rho}$, $R_{\dot\alpha\mu\nu\rho} = -R_{\mu\dot\alpha\nu\rho}$ and $R_{\mu\nu\rho\sigma}$ may be deduced using Eq. (3.60).

$\mathcal{D}_\alpha$ whose most general infinitesimal deformation reads

$$\delta\mathcal{D}_\alpha = -\Phi\mathcal{D}_\alpha + (\Phi_\alpha)^{\beta\gamma}J_{\beta\gamma}\,,$$

where Φ and $\Phi_\alpha^{\beta\gamma}$ are arbitrary superfields, $\Phi_\alpha^{\beta\gamma}$ being symmetric under the exchange of the indices β and γ. In order to preserve the algebra (3.59), the variation of the anticommutator $\{\mathcal{D}_\alpha, \mathcal{D}_\beta\}$ must be proportional to the Lorentz generators in the left-handed spinor representation. Recalling that the action of the Lorentz generators on the superderivative $\mathcal{D}_\alpha$ is given by

$$[\mathcal{D}_\alpha, J_{\gamma\delta}] = \varepsilon_{\delta\alpha}\mathcal{D}_\gamma + \varepsilon_{\gamma\alpha}\mathcal{D}_\delta\,, \qquad (3.61)$$

as shown in App. B, the variation $\delta\{\mathcal{D}_\alpha, \mathcal{D}_\beta\}$ can be written as

$$\delta\{\mathcal{D}_\alpha, \mathcal{D}_\beta\} = -2\Phi\{\mathcal{D}_\alpha, \mathcal{D}_\beta\} - \mathcal{D}_\beta\Phi\mathcal{D}_\alpha - \mathcal{D}_\alpha\Phi\mathcal{D}_\beta - (\Phi_\alpha)^{\gamma\delta}\Big[\varepsilon_{\delta\beta}\mathcal{D}_\gamma + \varepsilon_{\gamma\beta}\mathcal{D}_\delta\Big] \\ -(\Phi_\beta)^{\gamma\delta}\Big[\varepsilon_{\delta\alpha}\mathcal{D}_\gamma + \varepsilon_{\gamma\alpha}\mathcal{D}_\delta\Big] + \mathcal{D}_\beta(\Phi_\alpha)^{\gamma\delta}J_{\delta\gamma} + \mathcal{D}_\alpha(\Phi_\beta)^{\gamma\delta}J_{\delta\gamma}\,. \tag{3.62}$$

Imposing that the terms not proportional to $J^{\gamma\delta}$ vanish, one obtains the condition

$$\mathcal{D}_\beta\Phi\mathcal{D}_\alpha + \mathcal{D}_\alpha\Phi\mathcal{D}_\beta = (\Phi_\alpha)^{\gamma\delta}\Big[\varepsilon_{\beta\delta}\mathcal{D}_\gamma + \varepsilon_{\beta\gamma}\mathcal{D}_\delta\Big] + (\Phi_\beta)^{\gamma\delta}\Big[\varepsilon_{\alpha\delta}\mathcal{D}_\gamma + \varepsilon_{\alpha\gamma}\mathcal{D}_\delta\Big]\,.$$

Investigating explicitly all the possible values for α and β in this relation, one is led to

$$(\Phi_\alpha)^{\beta\gamma} = \frac{1}{2}\Big[\delta^\gamma{}_\alpha\mathcal{D}^\beta\Phi + \delta^\beta{}_\alpha\mathcal{D}^\gamma\Phi\Big]\,, \tag{3.63}$$

which further yields

$$\delta\mathcal{D}_\alpha = -\Phi\mathcal{D}_\alpha + \mathcal{D}^\gamma\Phi J_{\alpha\gamma}\,. \tag{3.64}$$

Similarly, the variation of the superderivatives $\overline{\mathcal{D}}_{\dot\alpha}$ under an infinitesimal deformation reads

$$\delta\overline{\mathcal{D}}_{\dot\alpha} = -\Phi^\dagger\overline{\mathcal{D}}_{\dot\alpha} - \overline{\mathcal{D}}^{\dot\gamma}\Phi^\dagger J_{\dot\alpha\dot\gamma}\,. \tag{3.65}$$

We now make use of the results of Eq. (3.64) and Eq. (3.65) to calculate the variation of the anticommutator $\{\mathcal{D}_\alpha, \overline{\mathcal{D}}_{\dot\alpha}\}$,

$$\delta\{\mathcal{D}_\alpha, \overline{\mathcal{D}}_{\dot\alpha}\} = -[\Phi + \Phi^\dagger]\{\mathcal{D}_\alpha, \overline{\mathcal{D}}_{\dot\alpha}\} - \overline{\mathcal{D}}_{\dot\alpha}\Phi\mathcal{D}_\alpha - \mathcal{D}_\alpha\Phi^\dagger\overline{\mathcal{D}}_{\dot\alpha} \\ +\overline{\mathcal{D}}_{\dot\alpha}\mathcal{D}^\gamma\Phi J_{\gamma\alpha} - \mathcal{D}_\alpha\overline{\mathcal{D}}^{\dot\gamma}\Phi^\dagger J_{\dot\gamma\dot\alpha}\,. \tag{3.66}$$

Starting from the right-hand side of the third equation in (3.59) and accounting for the constraints listed in (3.29), the variation of this anticommutator may also be expressed as

$$\delta\{\mathcal{D}_\alpha, \overline{\mathcal{D}}_{\dot\alpha}\} = -2i\sigma^\mu{}_{\alpha\dot\alpha}\delta\mathcal{D}_\mu - \frac{1}{2}\delta R_{\alpha\dot\alpha\beta\gamma}J^{\gamma\beta} - \frac{1}{2}\delta R_{\alpha\dot\alpha\dot\beta\dot\gamma}J^{\dot\gamma\dot\beta}\,, \tag{3.67}$$

so that by comparing the two results, one derives

$$\delta\mathcal{D}_\mu = -[\Phi + \Phi^\dagger]\mathcal{D}_\mu - \frac{i}{4}\bar\sigma_\mu{}^{\dot\alpha\alpha}[\overline{\mathcal{D}}_{\dot\alpha}\Phi\mathcal{D}_\alpha + \mathcal{D}_\alpha\Phi^\dagger\overline{\mathcal{D}}_{\dot\alpha}] + (\Phi_\mu)^{\rho\sigma}J_{\rho\sigma}\,, \tag{3.68}$$

where we have introduced a contribution depending on the unknown superfield Φ_μ, which is proportional to the Lorentz generators in the vector representation $J^{\rho\sigma}$. Although this operator is not strictly speaking an independent operator, its use allows one to simplify the notations.

Constraints on the superfield Φ can be obtained by calculating the variation of the commutator $[\mathcal{D}_\alpha, \mathcal{D}_\mu]$ and by preventing the initially vanishing contributions of the $T_{\alpha\mu}{}^{\nu}$ element of the torsion tensor (see Eq. (3.29)) from being generated by a Howe–Tucker transformation. We first compute $\delta[\mathcal{D}_\alpha, \mathcal{D}_\mu]$,

$$\begin{aligned}\delta[\mathcal{D}_\alpha, \mathcal{D}_\mu] = &-\Phi[\mathcal{D}_\alpha, \mathcal{D}_\mu] + \mathcal{D}_\mu \Phi \mathcal{D}_\alpha - [\Phi + \Phi^\dagger][\mathcal{D}_\alpha, \mathcal{D}_\mu] - \mathcal{D}_\alpha(\Phi + \Phi^\dagger)\mathcal{D}_\mu \\ &+ 2\mathcal{D}^\gamma \Phi \varepsilon_{\alpha\beta}(\sigma_{\mu\nu})_\gamma{}^\beta \mathcal{D}^\nu - \mathcal{D}_\mu \mathcal{D}^\gamma \Phi J_{\gamma\alpha} \\ &- \frac{i}{4}\bar{\sigma}_\mu{}^{\dot\beta\beta}\Big[\mathcal{D}_\alpha \overline{\mathcal{D}}_{\dot\beta} \Phi \mathcal{D}_\beta - \overline{\mathcal{D}}_{\dot\beta}\Phi\{\mathcal{D}_\alpha, \mathcal{D}_\beta\} + \mathcal{D}_\alpha \mathcal{D}_\beta \Phi^\dagger \overline{\mathcal{D}}_{\dot\beta} \\ &+ 2i\sigma^\nu{}_{\alpha\dot\beta}\mathcal{D}_\beta \Phi^\dagger\, \mathcal{D}_\nu + \frac{1}{2}\mathcal{D}_\beta \Phi^\dagger R_{\alpha\dot\beta\gamma\delta} J^{\delta\gamma} + \frac{1}{2}\mathcal{D}_\beta \Phi^\dagger R_{\alpha\dot\beta\dot\gamma\dot\delta} J^{\dot\delta\dot\gamma}\Big] \\ &+ \mathcal{D}_\alpha(\Phi_\mu)^{\rho\sigma} J_{\rho\sigma} + (\Phi_\mu)^{\rho\sigma}(\sigma_{\rho\sigma})_\alpha{}^\beta \mathcal{D}_\beta\,,\end{aligned}$$

which results from Eq. (3.64), Eq. (3.68) and the properties

$$\begin{aligned}[J_{\alpha\beta}, \mathcal{D}_\mu] &= 2\varepsilon_{\alpha\gamma}(\sigma_{\mu\nu})_\beta{}^\gamma\, \mathcal{D}^\nu\,, \\ [\mathcal{D}_\alpha, J_{\mu\nu}] &= (\sigma_{\mu\nu})_\alpha{}^\beta \mathcal{D}_\beta\end{aligned}$$

which have been derived from the relations between the spinor and vector representations of the Lorentz algebra presented in App. B. Identifying all the terms proportional to $\mathcal{D}_\mu$ and forcing them to cancel, the variation of $T_{\alpha\mu}{}^\nu$ vanishes. One hence arrives at

$$-\frac{1}{2}\mathcal{D}_\alpha[2\Phi + \Phi^\dagger]\eta_{\mu\nu} + (\sigma_{\nu\mu})_\alpha{}^\beta \mathcal{D}_\beta[2\Phi + \Phi^\dagger] = 0\,.$$

Since the superfield $3\Sigma = 2\Phi + \Phi^\dagger$ must be chiral, we may reexpress the superfield Φ and its conjugate $\Phi^\dagger$ as functions of Σ and $\Sigma^\dagger$ by

$$\begin{aligned}\Phi &= 2\Sigma^\dagger - \Sigma\,, \\ \Phi^\dagger &= 2\Sigma - \Sigma^\dagger\,.\end{aligned} \tag{3.69}$$

The superfield $(\Phi_\mu)^{\rho\sigma}$ appearing in the last term of the variation $\delta\mathcal{D}_\mu$ in Eq. (3.68) is derived from the variation of the commutator $[\mathcal{D}_\mu, \mathcal{D}_\nu]$, which reads

$$\begin{aligned}\delta[\mathcal{D}_\mu, \mathcal{D}_\nu] = &-2[\Sigma + \Sigma^\dagger][\mathcal{D}_\mu, \mathcal{D}_\nu] + \mathcal{D}_\nu[\Sigma + \Sigma^\dagger]\mathcal{D}_\mu - \mathcal{D}_\mu[\Sigma + \Sigma^\dagger]\mathcal{D}_\nu \\ &+2(\Phi_\mu)_{\rho\nu}\mathcal{D}^\rho - \mathcal{D}_\nu(\Phi_\mu)^{\rho\sigma} J_{\rho\sigma} - 2(\Phi_\nu)_{\rho\mu}\mathcal{D}^\rho + \mathcal{D}_\mu(\Phi_\nu)^{\rho\sigma} J_{\rho\sigma} \\ &-\frac{i}{2}\bar\sigma_\mu{}^{\dot\alpha\alpha}\Big[\overline{\mathcal{D}}_{\dot\alpha}\Sigma^\dagger[\mathcal{D}_\alpha, \mathcal{D}_\nu] - \mathcal{D}_\nu \overline{\mathcal{D}}_{\dot\alpha}\Sigma^\dagger \mathcal{D}_\alpha \\ &\qquad +\mathcal{D}_\alpha \Sigma[\overline{\mathcal{D}}_{\dot\alpha}, \mathcal{D}_\nu] - \mathcal{D}_\nu \mathcal{D}_\alpha \Sigma \overline{\mathcal{D}}_{\dot\alpha}\Big] \\ &+\frac{i}{2}\bar\sigma_\nu{}^{\dot\alpha\alpha}\Big[\overline{\mathcal{D}}_{\dot\alpha}\Sigma^\dagger[\mathcal{D}_\alpha, \mathcal{D}_\mu] - \mathcal{D}_\mu \overline{\mathcal{D}}_{\dot\alpha}\Sigma^\dagger \mathcal{D}_\alpha \\ &\qquad +\mathcal{D}_\alpha \Sigma[\overline{\mathcal{D}}_{\dot\alpha}, \mathcal{D}_\mu] - \mathcal{D}_\mu \mathcal{D}_\alpha \Sigma \overline{\mathcal{D}}_{\dot\alpha}\Big]\,,\end{aligned} \tag{3.70}$$

after employing Eq. (3.68), Eq. (3.69) and the relation

$$[J_{\mu\nu}, \mathcal{D}_\rho] = \eta_{\nu\rho}\mathcal{D}_\mu - \eta_{\mu\rho}\mathcal{D}_\nu$$

which has been derived from App. B. Since the variation of $T_{\mu\nu}{}^\rho$ has to vanish to preserve the structure of the superalgebra defined by Eq. (3.59), the terms proportional to $\mathcal{D}_\mu$ must compensate, so that

$$\mathcal{D}_\nu[\Sigma + \Sigma^\dagger]\mathcal{D}_\mu - \mathcal{D}_\mu[\Sigma + \Sigma^\dagger]\mathcal{D}_\nu = 2(\Phi_\mu)_{\nu\rho}\mathcal{D}^\rho - 2(\Phi_\nu)_{\mu\rho}\mathcal{D}^\rho .$$

Symmetrising with respect to all indices, we further derive

$$(\Phi_\mu)_{\nu\rho} + (\Phi_\nu)_{\rho\mu} + (\Phi_\rho)_{\mu\nu} = 0 ,$$

and, since $(\Phi_\mu)_{\nu\rho}$ is antisymmetric in ν and ρ, we also have

$$(\Phi_\mu)_{\nu\rho} = \frac{1}{2}\Big[\eta_{\mu\rho}\mathcal{D}_\nu[\Sigma + \Sigma^\dagger] - \eta_{\mu\nu}\mathcal{D}_\rho[\Sigma + \Sigma^\dagger]\Big] . \tag{3.71}$$

Combining the results of Eq. (3.64), Eq. (3.65) and Eq. (3.68) with the definitions of Eq. (3.69) and Eq. (3.71), we can write a Howe–Tucker rescaling of all operators relevant in the context of the supergravity algebra, in the infinitesimal form

$$\begin{aligned}
\delta_\Sigma\mathcal{D}_\alpha &= [\Sigma - 2\Sigma^\dagger]\mathcal{D}_\alpha - \mathcal{D}^\gamma\Sigma J_{\alpha\gamma} ,\\
\delta_\Sigma\overline{\mathcal{D}}_{\dot\alpha} &= [\Sigma^\dagger - 2\Sigma]\overline{\mathcal{D}}_{\dot\alpha} + \overline{\mathcal{D}}^{\dot\gamma}\Sigma^\dagger J_{\dot\alpha\dot\gamma} ,\\
\delta_\Sigma\mathcal{D}_\mu &= -(\Sigma + \Sigma^\dagger)\mathcal{D}_\mu - \frac{i}{2}[\overline{\mathcal{D}}\Sigma^\dagger\bar\sigma_\mu\mathcal{D} + \mathcal{D}\Sigma\sigma_\mu\overline{\mathcal{D}}] - \mathcal{D}^\nu(\Sigma + \Sigma^\dagger)J_{\mu\nu} ,\\
\delta_\Sigma J_{\alpha\beta} &= 0 ,\\
\delta_\Sigma J_{\dot\alpha\dot\beta} &= 0 ,
\end{aligned} \tag{3.72}$$

the superfield Σ being the transformation parameter and referred to in the variation by the symbol δ_Σ.

3.2.2 *Howe–Tucker transformations of the superfields $\mathcal{R}$, $G_{\alpha\dot\alpha}$ and $W_{(\alpha\beta\gamma)}$*

In this subsection, we study the variations under a Howe–Tucker transformation of the three key superfields that are necessary for deriving all the components of the curvature and torsion tensors, namely the three superfields $\mathcal{R}$, $G_{\alpha\dot\alpha}$ and $W_{(\alpha\beta\gamma)}$ introduced in Sec. 3.1.2.

The variation of the superfield $\mathcal{R}$ is computed from the one of the anticommutator $\{\mathcal{D}_\alpha, \mathcal{D}_\beta\}$. Using Eq. (3.59), Eq. (3.62), Eq. (3.63) and Eq. (3.69), one gets

$$\delta_\Sigma\{\mathcal{D}_\alpha, \mathcal{D}_\beta\} = \Big[(2\Sigma^\dagger - \Sigma)R_{\alpha\beta\gamma\delta} + \frac{1}{2}\varepsilon_{\alpha\delta}\varepsilon_{\beta\gamma}\mathcal{D}\cdot\mathcal{D}\Sigma + \frac{1}{2}\varepsilon_{\beta\delta}\varepsilon_{\alpha\gamma}\mathcal{D}\cdot\mathcal{D}\Sigma\Big]J^{\delta\gamma} . \tag{3.73}$$

In this expression, we have simplified the result using the property $\mathcal{D}_\alpha \mathcal{D}_\beta \Sigma = 1/2\varepsilon_{\alpha\beta}\mathcal{D}\cdot\mathcal{D}\Sigma$, which is derived from the two relations

$$\{\mathcal{D}_\alpha, \mathcal{D}_\beta\}\Sigma = 0\,,$$

$$[\mathcal{D}_\alpha, \mathcal{D}_\beta]\Sigma = \varepsilon_{\alpha\beta}\mathcal{D}\cdot\mathcal{D}\Sigma\,.$$

The first equation above is due to the algebra (3.59) and the second one follows from (anti)symmetry properties. Computing the variation of the right-hand side of the first equation of the supergravity algebra given by Eq. (3.59), the anticommutator of Eq. (3.73) can also be written as

$$\delta_\Sigma\{\mathcal{D}_\alpha, \mathcal{D}_\beta\} = -\frac{1}{2}\delta_\Sigma R_{\alpha\beta\gamma\delta}J^{\delta\gamma}\,,$$

which, when comparing the two equations and replacing the curvature tensor $R_{\alpha\beta\gamma\delta}$ by its value given in Table 3.4, leads to

$$\delta_\Sigma\mathcal{R}^\dagger = -4\Sigma^\dagger\mathcal{R}^\dagger - \frac{1}{4}(\mathcal{D}\cdot\mathcal{D} - 8\mathcal{R}^\dagger)\Sigma\,. \tag{3.74}$$

Similarly, but starting from the anticommutator $\{\overline{\mathcal{D}}_{\dot\alpha}, \overline{\mathcal{D}}_{\dot\beta}\}$, one derives

$$\delta_\Sigma\mathcal{R} = -4\Sigma\mathcal{R} - \frac{1}{4}(\overline{\mathcal{D}}\cdot\overline{\mathcal{D}} - 8\mathcal{R})\Sigma^\dagger\,. \tag{3.75}$$

The second term on the right-hand side of these two identities being antichiral and chiral (see Sec. 4.1), respectively, the chirality properties of the superfields $\mathcal{R}$ and $\mathcal{R}^\dagger$ are preserved by Howe–Tucker transformations.

The variation of the superfield $G_{\alpha\dot\alpha}$ is derived from the one of the anticommutator $\{\mathcal{D}_\alpha, \overline{\mathcal{D}}_{\dot\alpha}\}$. Making use of Eq. (3.69) together with Eq. (3.66), one finds,

$$\begin{aligned}
\delta_\Sigma\{\mathcal{D}_\alpha, \overline{\mathcal{D}}_{\dot\alpha}\} &= -[\Sigma + \Sigma^\dagger]\{\mathcal{D}_\alpha, \overline{\mathcal{D}}_{\dot\alpha}\} - 2\overline{\mathcal{D}}_{\dot\alpha}\Sigma^\dagger\mathcal{D}_\alpha - 2\mathcal{D}_\alpha\Sigma\overline{\mathcal{D}}_{\dot\alpha}\\
&\quad - \overline{\mathcal{D}}_{\dot\alpha}\mathcal{D}^\gamma\Sigma J_{\gamma\alpha} + \mathcal{D}_\alpha\overline{\mathcal{D}}^{\dot\gamma}\Sigma^\dagger J_{\dot\gamma\dot\alpha}\\
&= 2i\sigma^\mu{}_{\alpha\dot\alpha}[\Sigma + \Sigma^\dagger]\mathcal{D}_\mu + \frac{1}{2}[\Sigma + \Sigma^\dagger]R_{\alpha\dot\alpha\beta\gamma}J^{\beta\gamma} + \frac{1}{2}[\Sigma + \Sigma^\dagger]R_{\alpha\dot\alpha\dot\beta\dot\gamma}J^{\dot\beta\dot\gamma}\\
&\quad - 2\overline{\mathcal{D}}_{\dot\alpha}\Sigma^\dagger\mathcal{D}_\alpha - 2\mathcal{D}_\alpha\Sigma\overline{\mathcal{D}}_{\dot\alpha} - 2i\sigma^\mu{}_{\gamma\dot\alpha}\mathcal{D}_\mu\Sigma J_\alpha{}^\gamma + 2i\sigma^\mu{}_{\alpha\dot\gamma}\mathcal{D}_\mu\Sigma^\dagger J^{\dot\gamma}{}_{\dot\alpha}\\
&= 2i\sigma^\mu{}_{\alpha\dot\alpha}[\Sigma + \Sigma^\dagger]\mathcal{D}_\mu + [\Sigma + \Sigma^\dagger]G_{\gamma\dot\alpha}J_\alpha{}^\gamma + [\Sigma + \Sigma^\dagger]G_{\alpha\dot\gamma}J^{\dot\gamma}{}_{\dot\alpha}\\
&\quad - 2\overline{\mathcal{D}}_{\dot\alpha}\Sigma^\dagger\mathcal{D}_\alpha - 2\mathcal{D}_\alpha\Sigma\overline{\mathcal{D}}_{\dot\alpha} - 2i\sigma^\mu{}_{\gamma\dot\alpha}\mathcal{D}_\mu\Sigma J_\alpha{}^\gamma + 2i\sigma^\mu{}_{\alpha\dot\gamma}\mathcal{D}_\mu\Sigma^\dagger J^{\dot\gamma}{}_{\dot\alpha}\,.
\end{aligned} \tag{3.76}$$

For the second equality, we have used (3.59), the constraints (3.29) and the relations

$$\begin{aligned}
\overline{\mathcal{D}}_{\dot\alpha}\mathcal{D}_\gamma\Sigma &= \{\overline{\mathcal{D}}_{\dot\alpha}, \mathcal{D}_\gamma\}\Sigma &&= -2i\sigma^\mu{}_{\gamma\dot\alpha}\mathcal{D}_\mu\Sigma\,,\\
\mathcal{D}_\alpha\overline{\mathcal{D}}_{\dot\gamma}\Sigma^\dagger &= \{\mathcal{D}_\alpha, \overline{\mathcal{D}}_{\dot\gamma}\}\Sigma^\dagger &&= -2i\sigma^\mu{}_{\alpha\dot\gamma}\mathcal{D}_\mu\Sigma^\dagger\,,
\end{aligned} \tag{3.77}$$

implied by the chirality properties of the superfields Σ and $\Sigma^\dagger$. For the third equality of Eq. (3.76), we have replaced the elements of the curvature tensor by the expressions of Table 3.4. On a different footing, the variation $\delta_\Sigma\{\mathcal{D}_\alpha, \overline{\mathcal{D}}_{\dot\alpha}\}$ can also be computed from Eq. (3.67),

$$\delta\{\mathcal{D}_\alpha, \overline{\mathcal{D}}_{\dot\alpha}\} = -2i\sigma^\mu{}_{\alpha\dot\alpha}\delta\mathcal{D}_\mu - \frac{1}{2}\delta R_{\alpha\dot\alpha\beta\gamma}J^{\gamma\beta} - \frac{1}{2}\delta R_{\alpha\dot\alpha\dot\beta\dot\gamma}J^{\dot\gamma\dot\beta}\,.$$

The first term on the right-hand side of this equation is calculated from the results in Eq. (3.72), together with the relations between the vector and spinor representations of the generators of the Lorentz algebra in Eq. (B.10) and the identities among the Pauli matrices of App. C,

$$\begin{aligned}-2i\sigma^\mu{}_{\alpha\dot\alpha}\delta_\Sigma\mathcal{D}_\mu &= 2i\sigma^\mu{}_{\alpha\dot\alpha}[\Sigma+\Sigma^\dagger]\mathcal{D}_\mu - 2\overline{\mathcal{D}}_{\dot\alpha}\Sigma^\dagger\mathcal{D}_\alpha - 2\mathcal{D}_\alpha\Sigma\overline{\mathcal{D}}_{\dot\alpha}\\ &\quad + i\mathcal{D}^\nu[\Sigma+\Sigma^\dagger][\sigma_{\nu\alpha\dot\gamma}J^{\dot\gamma}{}_{\dot\alpha} - \sigma_{\nu\gamma\dot\alpha}J_\alpha{}^\gamma]\,.\end{aligned} \tag{3.78}$$

In contrast, the second and third terms of this equation are obtained from Table 3.4 and read

$$\begin{aligned}-\frac{1}{2}\delta_\Sigma R_{\alpha\dot\alpha\beta\gamma}J^{\gamma\beta} &= -\delta_\Sigma G_{\gamma\dot\alpha}J_\alpha{}^\gamma\\ -\frac{1}{2}\delta_\Sigma R_{\alpha\dot\alpha\dot\beta\dot\gamma}J^{\dot\gamma\dot\beta} &= -\delta_\Sigma G_{\alpha\dot\gamma}J^{\dot\gamma}{}_{\dot\alpha}\,.\end{aligned} \tag{3.79}$$

Collecting the results of Eq. (3.78) and Eq. (3.79), one obtains

$$\begin{aligned}\delta_\Sigma\{\mathcal{D}_\alpha, \overline{\mathcal{D}}_{\dot\alpha}\} = 2i\sigma^\mu{}_{\alpha\dot\alpha}&[\Sigma+\Sigma^\dagger]\mathcal{D}_\mu - 2\overline{\mathcal{D}}_{\dot\alpha}\Sigma^\dagger\mathcal{D}_\alpha - 2\mathcal{D}_\alpha\Sigma\overline{\mathcal{D}}_{\dot\alpha}\\ &- \Big[i\sigma_{\nu\gamma\dot\alpha}\mathcal{D}^\nu[\Sigma+\Sigma^\dagger] + \delta_\Sigma G_{\gamma\dot\alpha}\Big]J_\alpha{}^\gamma\\ &+ \Big[i\sigma_{\nu\alpha\dot\gamma}\mathcal{D}^\nu[\Sigma+\Sigma^\dagger] - \delta_\Sigma G_{\alpha\dot\gamma}\Big]J^{\dot\gamma}{}_{\dot\alpha}\,,\end{aligned}$$

which, by comparison with Eq. (3.76) allows one to derive the variation of G ,

$$\delta_\Sigma G_{\alpha\dot\alpha} = -[\Sigma+\Sigma^\dagger]G_{\alpha\dot\alpha} + i\sigma^\mu{}_{\alpha\dot\alpha}\mathcal{D}_\mu[\Sigma-\Sigma^\dagger]\,. \tag{3.80}$$

After restoring vector indices according to Eq. (B.7), this last relation can be equivalently rewritten, as

$$\delta_\Sigma G_\mu = -[\Sigma+\Sigma^\dagger]G_\mu + i\mathcal{D}_\mu[\Sigma-\Sigma^\dagger]\,. \tag{3.81}$$

The variation of the $W_{(\alpha\beta\gamma)}$ superfield is derived by evaluating $\delta_\Sigma[\mathcal{D}_\mu, \mathcal{D}_\nu]$. More precisely, this requires the variation of the $T_{\mu\nu\beta}$ element of the torsion tensor to be computed. Using both Eq. (3.70) and Eq. (3.71), one derives

$$\begin{aligned}\delta_\Sigma[\mathcal{D}_\mu, \mathcal{D}_\nu] = -2[\Sigma + \Sigma^\dagger]\,[\mathcal{D}_\mu, \mathcal{D}_\nu] + \mathcal{D}_\nu \mathcal{D}^\rho[\Sigma + \Sigma^\dagger] J_{\mu\rho} - \mathcal{D}_\mu \mathcal{D}^\rho[\Sigma + \Sigma^\dagger] J_{\nu\rho} \\ -\frac{i}{2}\bar{\sigma}_\mu{}^{\dot\alpha\alpha}\Big[\overline{\mathcal{D}}_{\dot\alpha}\Sigma^\dagger[\mathcal{D}_\alpha, \mathcal{D}_\nu] - \mathcal{D}_\nu\overline{\mathcal{D}}_{\dot\alpha}\Sigma^\dagger\mathcal{D}_\alpha \\ +\mathcal{D}_\alpha\Sigma[\overline{\mathcal{D}}_{\dot\alpha}, \mathcal{D}_\nu] - \mathcal{D}_\nu\mathcal{D}_\alpha\Sigma\overline{\mathcal{D}}_{\dot\alpha}\Big] \\ +\frac{i}{2}\bar{\sigma}_\nu{}^{\dot\alpha\alpha}\Big[\overline{\mathcal{D}}_{\dot\alpha}\Sigma^\dagger[\mathcal{D}_\alpha, \mathcal{D}_\mu] - \mathcal{D}_\mu\overline{\mathcal{D}}_{\dot\alpha}\Sigma^\dagger\mathcal{D}_\alpha \\ +\mathcal{D}_\alpha\Sigma[\overline{\mathcal{D}}_{\dot\alpha}, \mathcal{D}_\mu] - \mathcal{D}_\mu\mathcal{D}_\alpha\Sigma\overline{\mathcal{D}}_{\dot\alpha}\Big]\,.\end{aligned} \tag{3.82}$$

The variation of $T_{\mu\nu\beta}$ is obtained by extracting all the terms proportional to the $\mathcal{D}^\beta$ operator. Starting with the first term of the first line, we make use of the expression $T_{\mu\nu\beta}$ given in Table 3.3 and convert the spinor indices of the $G_{\alpha\dot\alpha}$ quantity into vector indices as indicated in Eq. (B.7). The results are then simplified by means of Eq. (C.5),

$$\begin{aligned}-2[\Sigma{+}\Sigma^\dagger][\mathcal{D}_\mu, \mathcal{D}_\nu] &\to 2[\Sigma{+}\Sigma^\dagger] T_{\mu\nu\beta}\mathcal{D}^\beta \\ &= [\Sigma{+}\Sigma^\dagger]\Big[2\varepsilon^{\delta\alpha}(\sigma_{\mu\nu})_\alpha{}^\gamma W_{(\beta\gamma\delta)} + \sigma_{\mu\beta\dot\alpha}\overline{\mathcal{D}}^{\dot\alpha} G_\nu - \sigma_{\nu\beta\dot\alpha}\overline{\mathcal{D}}^{\dot\alpha} G_\mu\Big]\mathcal{D}^\beta\,.\end{aligned} \tag{3.83}$$

The other terms of the first line of the right-hand side of Eq. (3.82) do not depend on the $\mathcal{D}^\beta$ superderivatives so that none of them contributes to the variation of $T_{\mu\nu\beta}$. We therefore move on to the first term of both the second and third lines of Eq. (3.82). Using the supergravity algebra (Eq. (3.59)) to calculate the two commutators, we get

$$\begin{aligned}&-\frac{i}{2}\bar{\sigma}_\mu{}^{\dot\alpha\alpha}\overline{\mathcal{D}}_{\dot\alpha}\Sigma^\dagger[\mathcal{D}_\alpha, \mathcal{D}_\nu] + \frac{i}{2}\bar{\sigma}_\nu{}^{\dot\alpha\alpha}\overline{\mathcal{D}}_{\dot\alpha}\Sigma^\dagger[\mathcal{D}_\alpha, \mathcal{D}_\mu] \\ &\to \frac{i}{2}\overline{\mathcal{D}}_{\dot\alpha}\Sigma^\dagger\Big[\bar{\sigma}_\mu{}^{\dot\alpha\alpha}T_{\alpha\nu\beta} - \bar{\sigma}_\nu{}^{\dot\alpha\alpha}T_{\alpha\mu\beta}\Big]\mathcal{D}^\beta \\ &= \overline{\mathcal{D}}^{\dot\alpha}\Sigma^\dagger\Big[\frac{3}{4}G_\nu\sigma_{\mu\beta\dot\alpha} - \frac{3}{4}G_\mu\sigma_{\nu\beta\dot\alpha} + \frac{i}{2}\varepsilon_{\mu\nu\rho\sigma}G^\rho\sigma^\sigma{}_{\beta\dot\alpha}\Big]\mathcal{D}^\beta\,,\end{aligned} \tag{3.84}$$

where the equality arises from the definition of $T_{\alpha\nu\beta}$ shown in Table 3.3, from the relations among the Pauli matrices given in App. C and from Eq. (B.7). In a similar fashion, the sum of the third term of both the second and third lines of Eq. (3.82) adds up to

$$\begin{aligned}&-\frac{i}{2}\bar{\sigma}_\mu{}^{\dot\alpha\alpha}\mathcal{D}_\alpha\Sigma[\overline{\mathcal{D}}_{\dot\alpha}, \mathcal{D}_\nu] + \frac{i}{2}\bar{\sigma}_\nu{}^{\dot\alpha\alpha}\mathcal{D}_\alpha\Sigma[\overline{\mathcal{D}}_{\dot\alpha}, \mathcal{D}_\mu] \\ &\to \frac{i}{2}\mathcal{D}_\alpha\Sigma\Big[\bar{\sigma}_\mu{}^{\dot\alpha\alpha}T_{\dot\alpha\nu\beta} - \bar{\sigma}_\nu{}^{\dot\alpha\alpha}T_{\dot\alpha\mu\beta}\Big]\mathcal{D}^\beta \\ &= 2(\sigma_{\nu\mu})_\beta{}^\alpha\mathcal{D}_\alpha\Sigma\,\mathcal{R}\,\mathcal{D}^\beta\,.\end{aligned} \tag{3.85}$$

Finally, the second terms of both the second and third lines of Eq. (3.82) directly contribute to the variation of $T_{\mu\nu\beta}$

$$\frac{i}{2}\Big[\sigma_{\mu\beta\dot\alpha}\mathcal{D}_\nu\overline{\mathcal{D}}^{\dot\alpha}\Sigma^\dagger - \sigma_{\nu\beta\dot\alpha}\mathcal{D}_\mu\overline{\mathcal{D}}^{\dot\alpha}\Sigma^\dagger\Big]\mathcal{D}^\beta\,, \tag{3.86}$$

whereas all the remaining terms are proportional to $\overline{\mathcal{D}}_{\dot\alpha}$ and can thus be ignored. Collecting the results of Eq. (3.83), Eq. (3.84), Eq. (3.85) and Eq. (3.86), one gets

$$\begin{aligned}-\delta_\Sigma[T_{\mu\nu\beta}\mathcal{D}^\beta]=\Big\{&2[\Sigma+\Sigma^\dagger]\varepsilon^{\delta\alpha}(\sigma_{\mu\nu})_\alpha{}^\gamma W_{(\beta\gamma\delta)}+[\Sigma+\Sigma^\dagger][\sigma_{\mu\beta\dot\alpha}\overline{\mathcal{D}}^{\dot\alpha}G_\nu-\sigma_{\nu\beta\dot\alpha}\overline{\mathcal{D}}^{\dot\alpha}G_\mu]\\&+\overline{\mathcal{D}}^{\dot\alpha}\Sigma^\dagger\Big[\frac{3}{4}G_\nu\sigma_{\mu\beta\dot\alpha}-\frac{3}{4}G_\mu\sigma_{\nu\beta\dot\alpha}+\frac{i}{2}\varepsilon_{\mu\nu\rho\sigma}G^\rho\sigma^\sigma{}_{\beta\dot\alpha}\Big]+2(\sigma_{\nu\mu})_\beta{}^\alpha\mathcal{D}_\alpha\Sigma\,\mathcal{R}\\&+\frac{i}{2}\Big[\sigma_{\mu\beta\dot\alpha}\mathcal{D}_\nu-\sigma_{\nu\beta\dot\alpha}\mathcal{D}_\mu\Big]\overline{\mathcal{D}}^{\dot\alpha}\Sigma^\dagger\Big\}\mathcal{D}^\beta+\ldots\,,\end{aligned}\tag{3.87}$$

where the dots stand for all contributions not proportional to the $\mathcal{D}^\beta$ operator. The same variation can also be calculated directly from the last relation in (3.59),

$$\delta_\Sigma[\mathcal{D}_\mu,\mathcal{D}_\nu]=-T_{\mu\nu\beta}\delta_\Sigma\mathcal{D}^\beta-\delta_\Sigma T_{\mu\nu\beta}D^\beta+\ldots\,,\tag{3.88}$$

where the dots stand again for terms not proportional to the $\mathcal{D}^\beta$ superderivative. The first term of the equation above, $-T_{\mu\nu\beta}\delta_\Sigma\mathcal{D}^\beta$, is calculated by means of Eq. (3.72). Selecting only the contributions proportional to the spinor derivative and converting spinor to vector indices, one gets

$$\begin{aligned}-T_{\mu\nu\beta}\delta_\Sigma\mathcal{D}^\beta=\Big[&-[\Sigma-2\Sigma^\dagger]\varepsilon^{\delta\alpha}(\sigma_{\mu\nu})_\alpha{}^\gamma W_{(\beta\gamma\delta)}\\&-\frac{1}{2}[\Sigma-2\Sigma^\dagger][\sigma_{\mu\beta\dot\alpha}\overline{\mathcal{D}}^{\dot\alpha}G_\nu-\sigma_{\nu\beta\dot\alpha}\overline{\mathcal{D}}^{\dot\alpha}G_\mu]\Big]\mathcal{D}^\beta+\ldots\,.\end{aligned}\tag{3.89}$$

After using the results of Table 3.3 for the torsion tensor, Eq. (3.80) for the variation of G as well as the relations of App. C and Eq. (B.7) to simplify spinor indices, the second term of Eq. (3.88) reads

$$-\delta_\Sigma T_{\mu\nu\beta}\mathcal{D}^\beta=-\Big[\varepsilon^{\delta\alpha}(\sigma_{\mu\nu})_\alpha{}^\gamma\delta_\Sigma W_{(\beta\gamma\delta)}+\frac{1}{2}\delta_\Sigma\Big[\sigma_{\mu\beta\dot\alpha}\overline{\mathcal{D}}^{\dot\alpha}G_\nu-\sigma_{\nu\beta\dot\alpha}\overline{\mathcal{D}}^{\dot\alpha}G_\mu\Big]\Big]\mathcal{D}^\beta\,.\tag{3.90}$$

Finally, the variation of the quantity $\overline{\mathcal{D}}^{\dot\alpha}G_\nu$ is obtained from Eq. (3.72) and Eq. (3.81),

$$\begin{aligned}\delta_\Sigma[\overline{\mathcal{D}}^{\dot\alpha}G_\nu]&=[\delta_\Sigma\overline{\mathcal{D}}^{\dot\alpha}]G_\nu+\overline{\mathcal{D}}^{\dot\alpha}[\delta_\Sigma G_\nu]\\&=-3\Sigma\overline{\mathcal{D}}^{\dot\alpha}G_\nu+(\bar\sigma_{\nu\rho})^{\dot\alpha}{}_{\dot\gamma}\overline{\mathcal{D}}^{\dot\gamma}\Sigma^\dagger G^\rho-2\overline{\mathcal{D}}^{\dot\alpha}\Sigma^\dagger G_\nu\\&\quad-i\mathcal{D}_\nu\overline{\mathcal{D}}^{\dot\alpha}\Sigma^\dagger+\bar\sigma_\nu{}^{\dot\alpha\gamma}\mathcal{D}_\gamma\Sigma\,\mathcal{R}\,.\end{aligned}$$

Inserting this result into Eq. (3.90), simplifying the products of Pauli matrices with the help of the relations of App. C and also using Eq. (3.89), one gets,

$$\delta_\Sigma[T_{\mu\nu\beta}\mathcal{D}^\beta] = \Big\{ -\varepsilon^{\delta\alpha}(\sigma_{\mu\nu})_\alpha{}^\gamma \delta_\Sigma W_{(\beta\gamma\delta)} + [\Sigma+\Sigma^\dagger][\sigma_{\mu\beta\dot\alpha}\overline{\mathcal{D}}^{\dot\alpha} G_\nu - \sigma_{\nu\beta\dot\alpha}\overline{\mathcal{D}}^{\dot\alpha} G_\mu]$$
$$+\overline{\mathcal{D}}^{\dot\alpha}\Sigma^\dagger\Big[\frac{3}{4}G_\nu\sigma_{\mu\beta\dot\alpha} - \frac{3}{4}G_\mu\sigma_{\nu\beta\dot\alpha} + \frac{i}{2}\varepsilon_{\mu\nu\rho\sigma}G^\rho\sigma^\sigma{}_{\beta\dot\alpha}\Big]$$
$$+\frac{i}{2}\Big[\sigma_{\mu\beta\dot\alpha}\mathcal{D}_\nu - \sigma_{\nu\beta\dot\alpha}\mathcal{D}_\mu\Big]\overline{\mathcal{D}}^{\dot\alpha}\Sigma^\dagger + 2(\sigma_{\nu\mu})_\beta{}^\alpha\mathcal{D}_\alpha\Sigma\,\mathcal{R}$$
$$-[\Sigma - 2\Sigma^\dagger]\varepsilon^{\delta\alpha}(\sigma_{\mu\nu})_\alpha{}^\gamma W_{(\beta\gamma\delta)}\Big\}\mathcal{D}^\beta + \dots .$$

Comparing with (3.87), one deduces

$$\delta_\Sigma W_{(\alpha\beta\gamma)} = -3\Sigma W_{(\alpha\beta\gamma)} ,$$

and in a similar way, one can establish the conjugate relation

$$\delta_\Sigma \overline{W}_{(\dot\alpha\dot\beta\dot\gamma)} = -3\Sigma^\dagger \overline{W}_{(\dot\alpha\dot\beta\dot\gamma)} .$$

3.2.3 *Howe–Tucker transformations of the supervierbein and the superconnection*

Before moving on, we dedicate this subsection to the calculation of the transformation laws of the vielbein and the superconnection under a Howe–Tucker transformation. They can be derived from the property

$$\mathcal{D}_M = E_M{}^{\tilde M}(\partial_{\tilde M} - \frac{1}{2}\Omega_{\tilde M\underline{\alpha\beta}}J^{\underline{\beta\alpha}})$$

which implies

$$\delta_\Sigma\mathcal{D}_M = \delta_\Sigma E_M{}^{\tilde M}\mathcal{D}_{\tilde M} - \frac{1}{2}E_M{}^{\tilde M}\delta_\Sigma\Omega_{\tilde M\underline{\alpha\beta}}J^{\underline{\beta\alpha}} ,$$

which is obtained after making the superderivative operator act on a generic vector V^N. Considering the three cases $M = \alpha$, $M = \dot\alpha$ and $M = \mu$ separately and comparing with Eq. (3.72), one deduces,

$$\begin{aligned}
\delta_\Sigma\mathcal{D}_\alpha &= [\Sigma - 2\Sigma^\dagger]\mathcal{D}_\alpha - \mathcal{D}^\gamma\Sigma J_{\alpha\gamma} \\
&= \delta_\Sigma E_\alpha{}^{\tilde M}\mathcal{D}_{\tilde M} - \frac{1}{2}E_\alpha{}^{\tilde M}\delta_\Sigma\Omega_{\tilde M\underline{\alpha\beta}}J^{\underline{\beta\alpha}} , \\
\delta_\Sigma\overline{\mathcal{D}}^{\dot\alpha} &= [\Sigma^\dagger - 2\Sigma]\overline{\mathcal{D}}^{\dot\alpha} - \overline{\mathcal{D}}_{\dot\gamma}\Sigma^\dagger J^{\dot\alpha\dot\gamma} \\
&= \delta_\Sigma E^{\dot\alpha\tilde M}\mathcal{D}_{\tilde M} - \frac{1}{2}E^{\dot\alpha\tilde M}\delta_\Sigma\Omega_{\tilde M\underline{\alpha\beta}}J^{\underline{\beta\alpha}} , \\
\delta_\Sigma\mathcal{D}_\mu &= -(\Sigma+\Sigma^\dagger)\mathcal{D}_\mu - \frac{i}{2}[\overline{\mathcal{D}}\Sigma^\dagger\bar\sigma_\mu\mathcal{D} + \mathcal{D}\Sigma\sigma_\mu\overline{\mathcal{D}}] - \mathcal{D}^\nu(\Sigma+\Sigma^\dagger)J_{\mu\nu} \\
&= \delta_\Sigma E_\mu{}^{\tilde M}\mathcal{D}_{\tilde M} - \frac{1}{2}E_\mu{}^{\tilde M}\delta_\Sigma\Omega_{\tilde M\underline{\alpha\beta}}J^{\underline{\beta\alpha}} .
\end{aligned} \tag{3.91}$$

Identifying the torsion pieces, one obtains the variations of the vielbein,

$$\begin{aligned}
\delta_\Sigma E_\alpha{}^{\tilde M} &= [\Sigma - 2\Sigma^\dagger] E_\alpha{}^{\tilde M}\,, \\
\delta_\Sigma E^{\dot\alpha \tilde M} &= [\Sigma^\dagger - 2\Sigma] E^{\dot\alpha \tilde M}\,, \\
\delta_\Sigma E_\mu{}^{\tilde M} &= -\,[\Sigma + \Sigma^\dagger] E_\mu{}^{\tilde M} - \frac{i}{2} (\overline{\mathcal{D}}\Sigma^\dagger \bar\sigma_\mu)^\alpha E_\alpha{}^{\tilde M} - \frac{i}{2} (\mathcal{D}\Sigma \sigma_\mu)_{\dot\alpha} E^{\dot\alpha \tilde M}\,.
\end{aligned} \tag{3.92}$$

Defining $\overline{\Omega}_{MPQ} = E_M{}^{\tilde M} \delta_\Sigma \Omega_{\tilde M PQ}$ and identifying the contributions in Eq. (3.91) proportional to the generators of the Lorentz algebra, one finds

$$\begin{aligned}
\overline{\Omega}_{\alpha\beta\gamma} &= -\,\varepsilon_{\alpha\beta} \mathcal{D}_\gamma \Sigma - \varepsilon_{\alpha\gamma} \mathcal{D}_\beta \Sigma\,, \\
\overline{\Omega}_{\dot\alpha\dot\beta\dot\gamma} &= \varepsilon_{\dot\alpha\dot\beta} \overline{\mathcal{D}}_{\dot\gamma} \Sigma^\dagger + \varepsilon_{\dot\alpha\dot\gamma} \overline{\mathcal{D}}_{\dot\beta} \Sigma^\dagger\,, \\
\overline{\Omega}_{\mu\nu\rho} &= \mathcal{D}_\nu [\Sigma + \Sigma^\dagger] \eta_{\mu\rho} - \mathcal{D}_\rho [\Sigma + \Sigma^\dagger] \eta_{\mu\nu}\,.
\end{aligned} \tag{3.93}$$

In order to calculate the variation of the elements of the superconnection, we invert these relations with the help of the properties

$$\begin{aligned}
\delta_\Sigma \Omega_{\tilde M \beta\gamma} &= E_{\tilde M}{}^\mu E_\mu{}^{\tilde N} \delta_\Sigma \Omega_{\tilde N \beta\gamma} + E_{\tilde M}{}^\alpha E_\alpha{}^{\tilde N} \delta_\Sigma \Omega_{\tilde N \beta\gamma} + E_{\tilde M \dot\alpha} E^{\dot\alpha \tilde N} \delta_\Sigma \Omega_{\tilde N \beta\gamma}\,, \\
\delta_\Sigma \Omega_{\tilde M \dot\beta\dot\gamma} &= E_{\tilde M}{}^\mu E_\mu{}^{\tilde N} \delta_\Sigma \Omega_{\tilde N \dot\beta\dot\gamma} + E_{\tilde M}{}^\alpha E_\alpha{}^{\tilde N} \delta_\Sigma \Omega_{\tilde N \dot\beta\dot\gamma} + E_{\tilde M \dot\alpha} E^{\dot\alpha \tilde N} \delta_\Sigma \Omega_{\tilde N \dot\beta\dot\gamma}\,, \\
\delta_\Sigma \Omega_{\tilde M \nu\rho} &= E_{\tilde M}{}^\mu E_\mu{}^{\tilde N} \delta_\Sigma \Omega_{\tilde N \nu\rho} + E_{\tilde M}{}^\alpha E_\alpha{}^{\tilde N} \delta_\Sigma \Omega_{\tilde N \nu\rho} + E_{\tilde M \dot\alpha} E^{\dot\alpha \tilde N} \delta_\Sigma \Omega_{\tilde N \nu\rho}\,,
\end{aligned} \tag{3.94}$$

so that, after converting all spinor to vector indices (using Eq. (B.1)) for the two first relations of Eq. (3.93), and vector to spinor indices (using Eq. (B.8)) for the last relation of Eq. (3.93), we obtain

$$\begin{aligned}
\delta_\Sigma \Omega_{\tilde M \beta\gamma} &= E_{\tilde M \beta} \mathcal{D}_\gamma \Sigma + E_{\tilde M \gamma} \mathcal{D}_\beta \Sigma - E_{\tilde M \nu} \varepsilon_{\gamma\alpha} (\sigma^{\nu\rho})_\beta{}^\alpha \mathcal{D}_\rho [\Sigma + \Sigma^\dagger]\,, \\
\delta_\Sigma \Omega_{\tilde M \dot\beta\dot\gamma} &= E_{\tilde M \dot\beta} \overline{\mathcal{D}}_{\dot\gamma} \Sigma^\dagger + E_{\tilde M \dot\gamma} \overline{\mathcal{D}}_{\dot\beta} \Sigma^\dagger - E_{\tilde M \nu} \varepsilon_{\dot\beta\dot\alpha} (\bar\sigma^{\nu\rho})^{\dot\alpha}{}_{\dot\gamma} \mathcal{D}_\rho [\Sigma + \Sigma^\dagger]\,, \\
\delta_\Sigma \Omega_{\tilde M \nu\rho} &= -E_{\tilde M \nu} \mathcal{D}_\rho [\Sigma + \Sigma^\dagger] + E_{\tilde M \rho} \mathcal{D}_\nu [\Sigma + \Sigma^\dagger] \\
&\quad -2 E_{\tilde M}{}^\alpha (\sigma_{\nu\rho})_\alpha{}^\beta \mathcal{D}_\beta \Sigma - 2 E_{\tilde M \dot\alpha} (\bar\sigma_{\nu\rho})^{\dot\alpha}{}_{\dot\beta} \overline{\mathcal{D}}^{\dot\beta} \Sigma^\dagger\,.
\end{aligned} \tag{3.95}$$

As will be shown in Sec. 3.4, the constraints of Eq. (3.29) allow us to express the superconnection in terms of the elements of the vielbein. Therefore, the variations of the elements of the superconnection under a Howe–Tucker transformation could have been obtained from those of the vielbein.

Finally, since $\delta_\Sigma(E_{\tilde M}{}^M E_M{}^{\tilde N}) = 0$, we arrive at the transformation laws of the vielbein $\delta_\Sigma E_{\tilde M}{}^M = -E_{\tilde M}{}^N \delta_\Sigma E_N{}^{\tilde N} E_{\tilde N}{}^M$ from the results of Eq. (3.92),

$$\begin{aligned}
\delta_\Sigma E_{\tilde M}{}^\alpha &= [2\Sigma^\dagger - \Sigma] E_{\tilde M}{}^\alpha + \frac{i}{2} E_{\tilde M}{}^\mu (\overline{\mathcal{D}}\Sigma^\dagger \bar\sigma_\mu)^\alpha\,, \\
\delta_\Sigma E_{\tilde M \dot\alpha} &= [2\Sigma - \Sigma^\dagger] E_{\tilde M \dot\alpha} + \frac{i}{2} E_{\tilde M}{}^\mu (\mathcal{D}\Sigma \sigma_\mu)_{\dot\alpha}\,, \\
\delta_\Sigma E_{\tilde M}{}^\mu &= (\Sigma + \Sigma^\dagger) E_{\tilde M}{}^\mu\,.
\end{aligned} \tag{3.96}$$

To conclude this section, some comments are in order. In supergravity, different types of Lagrangians can be constructed. While they are all invariant, by definition, under supergravity transformations, superconformal supergravity models encompass, in addition, invariance under super-Weyl rescaling. This symmetry however also plays a key role in the context of Poincaré supergravity since the transformations (3.72) preserve the structure of the supergravity algebra (3.59) (and thus of the constraints (3.29)). While we collect all the results derived in this section in Table 3.5, more details can be found in Sec. 5.3.

3.3 Supergravity transformations and gauge fixing conditions

3.3.1 *Transformation properties*

Under a transformation corresponding to both a general coordinate change (3.2) and a local Lorentz transformation (3.1), a vector superfield V^M, the supervierbein $E_{\tilde{M}}{}^M$ and the superconnection $\Omega_{\tilde{M}P}{}^Q$ transform according to

$$\begin{aligned} V^M(z) &\rightarrow V'^M(z') = V^N(z)\Lambda_N{}^M(z)\,, \\ E_{\tilde{M}}{}^M(z) &\rightarrow E'_{\tilde{M}}{}^M(z') = \frac{\partial z^{\tilde{N}}}{\partial z'^{\tilde{M}}} E_{\tilde{N}}{}^N(z)\Lambda_N{}^M(z)\,, \\ \Omega_{\tilde{M}P}{}^Q(z) &\rightarrow \Omega'_{\tilde{M}P}{}^Q(z') = \frac{\partial z^{\tilde{N}}}{\partial z'^{\tilde{M}}}(\Lambda^{-1})_P{}^R(z)\Big[\Omega_{\tilde{N}R}{}^S(z)\Lambda_S{}^Q(z) - \partial_{\tilde{N}}\Lambda_R{}^Q(z)\Big]\,, \end{aligned} \tag{3.97}$$

where the position of the Lorentz transformation parameters $\Lambda_M{}^N$ is irrelevant since their grading is zero (M and N are indices of the same nature). By expanding $\Lambda_M{}^N$ around the identity, $\Lambda_M{}^N = \delta_M{}^N + L_M{}^N$ where $L_M{}^N$ are infinitesimal quantities, and by introducing the covariant derivatives given in (3.5), we derive the variation of V^M from the first relation,

$$\delta V^M = V'^M(z) - V^M(z) = -\xi^{\tilde{M}}\mathcal{D}_{\tilde{M}}V^M + V^N(\xi^{\tilde{M}}\Omega_{\tilde{M}N}{}^M + L_N{}^M)\,. \tag{3.98}$$

Using, in addition, the property

$$\partial_{\tilde{M}}\xi^N = \partial_{\tilde{M}}(\xi^{\tilde{N}}E_{\tilde{N}}{}^N) = \partial_{\tilde{M}}\xi^{\tilde{N}}\;E_{\tilde{N}}{}^N + (-)^{|\tilde{M}||\tilde{N}|}\xi^{\tilde{N}}\partial_{\tilde{M}}E_{\tilde{N}}{}^N\,,$$

together with the definitions, the torsion (3.17) and curvature tensors (3.18), the last two equalities of Eq. (3.97) allow us to compute the variations of the vielbein and of the superconnection,

Table 3.5 Howe–Tucker transformations.

The variations of the superderivatives and the generators of the Lorentz algebra under a Howe–Tucker transformation parameterised by a chiral superfield Σ read

$$\begin{aligned}
\delta_\Sigma \mathcal{D}_\alpha &= [\Sigma - 2\Sigma^\dagger]\mathcal{D}_\alpha - \mathcal{D}^\gamma \Sigma J_{\alpha\gamma}\,,\\
\delta_\Sigma \overline{\mathcal{D}}_{\dot\alpha} &= [\Sigma^\dagger - 2\Sigma]\overline{\mathcal{D}}_{\dot\alpha} + \overline{\mathcal{D}}^{\dot\gamma}\Sigma^\dagger J_{\dot\alpha\dot\gamma}\\
\delta_\Sigma \mathcal{D}_\mu &= -(\Sigma + \Sigma^\dagger)\mathcal{D}_\mu - \frac{i}{2}[\overline{\mathcal{D}}\Sigma^\dagger \bar\sigma_\mu \mathcal{D} + \mathcal{D}\Sigma\sigma_\mu\overline{\mathcal{D}}] - \mathcal{D}^\nu(\Sigma + \Sigma^\dagger)J_{\mu\nu}\,,\\
\delta_\Sigma J_{\alpha\beta} &= 0\,,\\
\delta_\Sigma J_{\dot\alpha\dot\beta} &= 0\,.
\end{aligned}$$

The ones of the superfields $\mathcal{R}$, $G_{\alpha\dot\alpha}$ and $W_{(\alpha\beta\gamma)}$, necessary for the computation of all the elements of the torsion and curvature tensors, are given by

$$\begin{aligned}
\delta_\Sigma \mathcal{R} &= -4\Sigma\mathcal{R} - \frac{1}{4}(\overline{\mathcal{D}}\cdot\overline{\mathcal{D}} - 8\mathcal{R})\Sigma^\dagger\,,\\
\delta_\Sigma \mathcal{R}^\dagger &= -4\Sigma^\dagger\mathcal{R}^\dagger - \frac{1}{4}(\mathcal{D}\cdot\mathcal{D} - 8\mathcal{R}^\dagger)\Sigma\,,\\
\delta_\Sigma G_{\alpha\dot\alpha} &= -[\Sigma + \Sigma^\dagger]G_{\alpha\dot\alpha} + i\sigma^\mu{}_{\gamma\dot\alpha}\mathcal{D}_\mu[\Sigma - \Sigma^\dagger]\,,\\
\delta_\Sigma G_\mu &= -[\Sigma + \Sigma^\dagger]G_\mu + i\mathcal{D}_\mu[\Sigma - \Sigma^\dagger]\,,\\
\delta_\Sigma W_{(\alpha\beta\gamma)} &= -3\Sigma W_{(\alpha\beta\gamma)}\,,\\
\delta_\Sigma \overline{W}_{(\dot\alpha\dot\beta\dot\gamma)} &= -3\Sigma^\dagger W_{(\dot\alpha\dot\beta\dot\gamma)}\,.
\end{aligned}$$

The transformation laws of the supervierbein and its inverse are

$$\begin{aligned}
\delta_\Sigma E_\alpha{}^{\tilde M} &= [\Sigma - 2\Sigma^\dagger]E_\alpha{}^{\tilde M}\,,\\
\delta_\Sigma E^{\dot\alpha\tilde M} &= [\Sigma^\dagger - 2\Sigma]E^{\dot\alpha\tilde M}\,,\\
\delta_\Sigma E_\mu{}^{\tilde M} &= -[\Sigma + \Sigma^\dagger]E_\mu{}^{\tilde M} - \frac{i}{2}(\overline{\mathcal{D}}\Sigma^\dagger\bar\sigma_\mu)^\alpha E_\alpha{}^{\tilde M} - \frac{i}{2}(\mathcal{D}\Sigma\sigma_\mu)_{\dot\alpha}E^{\dot\alpha\tilde M}\,,\\
\delta_\Sigma E_{\tilde M}{}^\alpha &= [2\Sigma^\dagger - \Sigma]E_{\tilde M}{}^\alpha + \frac{i}{2}E_{\tilde M}{}^\mu(\overline{\mathcal{D}}\Sigma^\dagger\bar\sigma_\mu)^\alpha\,,\\
\delta_\Sigma E_{\tilde M\dot\alpha} &= [2\Sigma - \Sigma^\dagger]E_{\tilde M\dot\alpha} + \frac{i}{2}E_{\tilde M}{}^\mu(\mathcal{D}\Sigma\sigma_\mu)_{\dot\alpha}\,,\\
\delta_\Sigma E_{\tilde M}{}^\mu &= (\Sigma + \Sigma^\dagger)E_{\tilde M}{}^\mu\,.
\end{aligned}$$

and the ones of the superconnection are

$$\begin{aligned}
\delta_\Sigma \Omega_{\tilde M\beta\gamma} &= E_{\tilde M\beta}\mathcal{D}_\gamma\Sigma + E_{\tilde M\gamma}\mathcal{D}_\beta\Sigma - E_{\tilde M\nu}\varepsilon_{\gamma\alpha}(\sigma^{\nu\rho})_\beta{}^\alpha\mathcal{D}_\rho[\Sigma + \Sigma^\dagger]\,,\\
\delta_\Sigma \Omega_{\tilde M\dot\beta\dot\gamma} &= E_{\tilde M\dot\beta}\overline{\mathcal{D}}_{\dot\gamma}\Sigma^\dagger + E_{\tilde M\dot\gamma}\overline{\mathcal{D}}_{\dot\beta}\Sigma^\dagger - E_{\tilde M\nu}\varepsilon_{\dot\beta\dot\alpha}(\bar\sigma^{\nu\rho})^{\dot\alpha}{}_{\dot\gamma}\mathcal{D}_\rho[\Sigma + \Sigma^\dagger]\,,\\
\delta_\Sigma \Omega_{\tilde M\nu\rho} &= E_{\tilde M\rho}\mathcal{D}_\nu[\Sigma + \Sigma^\dagger] - E_{\tilde M\nu}\mathcal{D}_\rho[\Sigma + \Sigma^\dagger] - 2E_{\tilde M}{}^\alpha(\sigma_{\nu\rho})_\alpha{}^\beta\mathcal{D}_\beta\Sigma - 2E_{\tilde M\dot\alpha}(\bar\sigma_{\nu\rho})^{\dot\alpha}{}_{\dot\beta}\overline{\mathcal{D}}^{\dot\beta}\Sigma^\dagger\,.
\end{aligned}$$

$$\begin{aligned}
\delta E_{\tilde{M}}{}^{M} &= E'_{\tilde{M}}{}^{M}(z) - E_{\tilde{M}}{}^{M}(z) = -\xi^{\tilde{N}}\partial_{\tilde{N}}E_{\tilde{M}}{}^{M} - \partial_{\tilde{M}}\xi^{\tilde{N}}E_{\tilde{N}}{}^{M} + E_{\tilde{M}}{}^{N}L_{N}{}^{M} \\
&= -\xi^{\tilde{N}}\Big[-T_{\tilde{N}\tilde{M}}{}^{M} - (-)^{|\tilde{N}|(|P|+|\tilde{M}|)}E_{\tilde{M}}{}^{P}\Omega_{\tilde{N}P}{}^{M} + (-)^{|\tilde{M}||P|}E_{\tilde{N}}{}^{P}\Omega_{\tilde{M}P}{}^{M}\Big] \\
&\quad -\partial_{\tilde{M}}\xi^{M} + E_{\tilde{M}}{}^{N}L_{N}{}^{M} \\
&= -\mathcal{D}_{\tilde{M}}\xi^{N} + \xi^{\tilde{N}}T_{\tilde{N}\tilde{M}}{}^{M} + E_{\tilde{M}}{}^{P}(\xi^{\tilde{N}}\Omega_{\tilde{N}P}{}^{M} + L_{P}{}^{M}), \qquad (3.99) \\
\delta\Omega_{\tilde{M}P}{}^{Q} &= -\xi^{\tilde{N}}\partial_{\tilde{N}}\Omega_{\tilde{M}P}{}^{Q} - \partial_{\tilde{M}}\xi^{\tilde{N}}\Omega_{\tilde{N}P}{}^{Q} + \Omega_{\tilde{M}P}{}^{R}L_{R}{}^{Q} - L_{P}{}^{R}\Omega_{\tilde{M}R}{}^{Q} - \partial_{\tilde{M}}L_{P}{}^{Q} \\
&= -\xi^{\tilde{N}}R_{\tilde{N}\tilde{M}P}{}^{Q} + \Omega_{\tilde{M}P}{}^{R}(L_{R}{}^{Q} + \xi^{\tilde{N}}\Omega_{\tilde{N}R}{}^{Q}) - (L_{P}{}^{R} + \xi^{\tilde{N}}\Omega_{\tilde{N}P}{}^{R})\Omega_{\tilde{M}R}{}^{Q} \\
&\quad -\partial_{\tilde{M}}(L_{P}{}^{Q} + \xi^{\tilde{N}}\Omega_{\tilde{N}P}{}^{Q}).
\end{aligned}$$

The expressions of δV^{M}, $\delta E_{\tilde{\mu}}{}^{M}$ and $\delta\Omega_{\tilde{M}P}{}^{Q}$ can be further simplified by redefining the coordinate-depending Lorentz transformation parameters as [Wess and Zumino (1978a); Wess and Bagger (1992)]

$$L_{P}{}^{Q} \to L_{P}{}^{Q} - \xi^{\tilde{M}}\Omega_{\tilde{M}P}{}^{Q},$$

so that, since $\xi^{N}\mathcal{D}_{N} = \xi^{\tilde{N}}\mathcal{D}_{\tilde{N}}$, we get

$$\begin{aligned}
\delta V^{M} &= -\xi^{N}\mathcal{D}_{N}V^{M} + V^{N}L_{N}{}^{M}, \\
\delta E_{\tilde{M}}{}^{M} &= -\mathcal{D}_{\tilde{M}}\xi^{M} + \xi^{\tilde{N}}T_{\tilde{N}\tilde{M}}{}^{M} + E_{\tilde{M}}{}^{P}L_{P}{}^{M}, \qquad (3.100) \\
\delta\Omega_{\tilde{M}P}{}^{Q} &= -\xi^{\tilde{N}}R_{\tilde{N}\tilde{M}P}{}^{Q} + \Omega_{\tilde{M}P}{}^{R}L_{R}{}^{Q} - L_{P}{}^{R}\Omega_{\tilde{M}R}{}^{Q} - \partial_{\tilde{M}}L_{P}{}^{Q}.
\end{aligned}$$

Finally, the variation of a scalar superfield U is trivially given by

$$\delta U = -\xi^{\tilde{M}}\partial_{\tilde{M}}U = -\xi^{\tilde{M}}\mathcal{D}_{\tilde{M}}U = -\xi^{M}\mathcal{D}_{M}U. \qquad (3.101)$$

Making use of (3.100) and (3.101), one can derive the variations of the superfields $\mathcal{R}$ and $G^{\mu} = 1/2\bar{\sigma}^{\mu\dot{\alpha}\alpha}G_{\alpha\dot{\alpha}}$ to obtain

$$\begin{aligned}
\delta\mathcal{R} &= -\xi^{\tilde{M}}\mathcal{D}_{\tilde{M}}\mathcal{R}, \\
\delta G^{\mu} &= -\xi^{\tilde{M}}\mathcal{D}_{\tilde{M}}G^{\mu} + G^{\nu}L_{\nu}{}^{\mu}.
\end{aligned}$$

3.3.2 *Gauge fixing conditions*

In the transformation laws (3.100), the parameters are all superfields which can be expanded in terms of their component fields. For instance, the parameters of the general transformation can be expanded in terms of the Einstein-valued Grassmann variables as

$$\begin{aligned}
\xi^{M}(z) &= \xi^{(0,0)M}(x) + \theta^{\tilde{\alpha}}\xi^{(1,0)M}_{\tilde{\alpha}}(x) + \bar{\theta}_{\tilde{\dot{\alpha}}}\xi^{(0,1)M\tilde{\dot{\alpha}}}(x) + \theta\sigma_{\tilde{\mu}}\bar{\theta}\xi^{(1,1)M\tilde{\mu}}(x) \\
&\quad +\theta\cdot\theta\xi^{(2,0)M}(x) + \bar{\theta}\cdot\bar{\theta}\xi^{(0,2)M}(x) \\
&\quad +\bar{\theta}\cdot\bar{\theta}\,\theta^{\tilde{\alpha}}\xi^{(1,2)M}_{\tilde{\alpha}}(x) + \theta\cdot\theta\,\bar{\theta}_{\tilde{\dot{\alpha}}}\xi^{(2,1)M\tilde{\dot{\alpha}}}(x) + \theta\cdot\theta\,\bar{\theta}\cdot\bar{\theta}\xi^{(2,2)M}(x). \qquad (3.102)
\end{aligned}$$

Similar expressions hold for the parameters of the local Lorentz transformation. As a consequence, there is a huge arbitrariness from which we can benefit in adopting a gauge where the $\theta/\bar{\theta}$-independent components of the vielbein and the superconnection take a simple form. This restriction to the lowest-order components is further motivated by the fact that only the lowest-order component fields (or equivalently the $\theta = \bar{\theta} = 0$ components) of $E_M{}^{\tilde{M}}$, $E_{\tilde{M}}{}^M$, $\Omega_{\tilde{M}P}{}^Q$, $\mathcal{R}$ and $G_{\alpha\dot{\alpha}}$ are relevant for the calculation of all the elements of the curvature and torsion tensors. Subsequently, most of the higher-order components of the transformation parameters appearing in (3.100) can be left unspecified [Wess and Bagger (1992)].

We start by analysing the variation of the superconnection given in (3.100) and calculate the pieces independent of the θ and $\bar{\theta}$ variables

$$\begin{aligned}
\delta\Omega_{\tilde{\alpha}P}{}^Q\Big| &= -\xi^{(0,0)\tilde{N}}\, R_{\tilde{N}\tilde{\alpha}P}{}^Q\Big| + \Omega_{\tilde{\alpha}P}{}^R\Big|\, (L^{(0,0)})_R{}^Q - (L^{(0,0)})_P{}^R\, \Omega_{\tilde{\alpha}R}{}^Q\Big| - (L^{(1,0)}_{\tilde{\alpha}})_P{}^Q\,, \\
\delta\Omega^{\tilde{\dot{\alpha}}}{}_P{}^Q\Big| &= -\xi^{(0,0)\tilde{N}}\, R_{\tilde{N}}{}^{\tilde{\dot{\alpha}}}{}_P{}^Q\Big| + \Omega^{\tilde{\dot{\alpha}}}{}_P{}^R\Big|\, (L^{(0,0)})_R{}^Q - (L^{(0,0)})_P{}^R\, \Omega^{\tilde{\dot{\alpha}}}{}_R{}^Q\Big| - (L^{(0,1)\tilde{\dot{\alpha}}})_P{}^Q\,,
\end{aligned} \tag{3.103}$$

where we have used the shortland notation $X\big| \equiv X\big|_{\theta=\bar{\theta}=0}$. This shows that a suitable choice of the Lorentz transformation parameters $L^{(1,0)}_{\tilde{\alpha}}$ and $L^{(0,1)\tilde{\dot{\alpha}}}$ allows us to eliminate the superconnection components $\Omega_{\tilde{\alpha}M}{}^N\big|$ and $\Omega^{\tilde{\dot{\alpha}}}{}_M{}^N\big|$. Adopting such a gauge, the lowest-order components of the superconnection are given by [Wess and Zumino (1977); Wess and Bagger (1992)]:

$$\begin{aligned}
\Omega_{\tilde{\mu}M}{}^N\Big| &= \omega_{\tilde{\mu}M}{}^N(x)\,, \\
\Omega_{\tilde{\alpha}M}{}^N\Big| &= 0\,, \\
\Omega^{\tilde{\dot{\alpha}}}{}_M{}^N\Big| &= 0\,,
\end{aligned} \tag{3.104}$$

where the ω-tensor, called the spin-connection, does not depend on the Grassmann variables and will be specified below.

Making use of (3.102) and (3.104), we observe that

$$\begin{aligned}
\mathcal{D}_{\tilde{\mu}}\xi^M\Big| &= \partial_{\tilde{\mu}}\xi^{(0,0)M} + \xi^{(0,0)N}\Omega_{\tilde{\mu}N}{}^M\Big| = \partial_{\tilde{\mu}}\xi^{(0,0)M} + \xi^{(0,0)N}\omega_{\tilde{\mu}N}{}^M\,, \\
\mathcal{D}_{\tilde{\alpha}}\xi^M\Big| &= \xi^{(1,0)M}_{\tilde{\alpha}} + (-)^{|N|}\xi^{(0,0)N}\,\Omega_{\tilde{\alpha}N}{}^M\Big| = \xi^{(1,0)M}_{\tilde{\alpha}}\,, \\
\overline{\mathcal{D}}^{\tilde{\dot{\alpha}}}\xi^M\Big| &= \xi^{(0,1)M\tilde{\dot{\alpha}}} + (-)^{|N|}\,\xi^{(0,0)N}\Omega^{\tilde{\dot{\alpha}}}{}_N{}^M\Big| = \xi^{(0,1)M\tilde{\dot{\alpha}}}\,.
\end{aligned} \tag{3.105}$$

This allows us to calculate the variation of six of the components of the vielbein,

$$
\begin{aligned}
\delta E_{\tilde{\alpha}}{}^{\mu}\Big| &= -\xi_{\tilde{\alpha}}^{(1,0)\mu} + \xi^{(0,0)M} T_{M\tilde{\alpha}}{}^{\mu}\Big| + E_{\tilde{\alpha}}{}^{\nu}\Big|(L^{(0,0)})_{\nu}{}^{\mu}\,,\\
\delta E^{\tilde{\dot{\alpha}}\mu}\Big| &= -\xi^{(0,1)\mu\tilde{\dot{\alpha}}} + \xi^{(0,0)M} T_{M}{}^{\tilde{\dot{\alpha}}\mu}\Big| + E^{\tilde{\dot{\alpha}}\nu}\Big|(L^{(0,0)})_{\nu}{}^{\mu}\,,\\
\delta E_{\tilde{\alpha}\dot{\beta}}\Big| &= -\xi_{\dot{\beta}\tilde{\alpha}}^{(1,0)} + \xi^{(0,0)M} T_{M\tilde{\alpha}\dot{\beta}} + E_{\tilde{\alpha}\dot{\alpha}}\Big|(L^{(0,0)})^{\dot{\alpha}}{}_{\dot{\beta}}\,,\\
\delta E^{\tilde{\dot{\alpha}}\beta}\Big| &= -\xi^{(0,1)\beta\tilde{\dot{\alpha}}} + \xi^{(0,0)M} T_{M}{}^{\tilde{\dot{\alpha}}\beta}\Big| + E^{\tilde{\dot{\alpha}}\gamma}\Big|\,(L^{(0,0)})_{\gamma}{}^{\beta}\,,\\
\delta E_{\tilde{\alpha}}{}^{\beta}\Big| &= -\xi_{\tilde{\alpha}}^{(1,0)\beta} + \xi^{(0,0)\mu}(E_{\tilde{\alpha}}{}^{\gamma} T_{\mu\gamma}{}^{\beta})\Big| + E_{\tilde{\alpha}}{}^{\gamma}\Big|\,(L^{(0,0)})_{\gamma}{}^{\beta}\,,\\
\delta E^{\tilde{\dot{\alpha}}}{}_{\dot{\beta}}\Big| &= -\xi^{(0,1)\tilde{\dot{\alpha}}}{}_{\dot{\beta}} + \xi^{(0,0)\mu}(E^{\tilde{\dot{\alpha}}}{}_{\dot{\gamma}} T_{\mu}{}^{\dot{\gamma}}{}_{\dot{\beta}})\Big| + E^{\tilde{\dot{\alpha}}}{}_{\dot{\gamma}}\Big|\,(L^{(0,0)})^{\dot{\gamma}}{}_{\dot{\beta}}\,,
\end{aligned}
$$

where we have employed (3.104) to get rid of the terms depending on the superconnection. The last two relations have also been further simplified by means of the constraints on the torsion tensor (3.29). Consequently, a proper choice of the parameters $\xi_{\tilde{\alpha}}^{(1,0)\mu}$, $\xi^{(0,1)\mu\tilde{\dot{\alpha}}}$, $\xi_{\dot{\beta}\tilde{\alpha}}^{(1,0)}$ and $\xi^{(0,1)\beta\tilde{\dot{\alpha}}}$ implies that the lowest-order components of the superfields $E_{\tilde{\alpha}}{}^{\mu}$, $E^{\tilde{\dot{\alpha}}\mu}$, $E_{\tilde{\alpha}\dot{\beta}}$ and $E^{\tilde{\dot{\alpha}}\beta}$ can be eliminated. Since the vielbein must be invertible, the diagonal elements of $E_{\tilde{\alpha}}{}^{\beta}|$ and $E^{\tilde{\dot{\alpha}}}{}_{\dot{\beta}}|$ must be prevented from vanishing. The transformation parameters $\xi_{\tilde{\alpha}}^{(1,0)\beta}$ and $\xi^{(0,1)\tilde{\dot{\alpha}}}{}_{\dot{\beta}}$ are hence conveniently chosen such that $E_{\tilde{\alpha}}{}^{\beta}| = \delta_{\tilde{\alpha}}{}^{\beta}$ and $E^{\tilde{\dot{\alpha}}}{}_{\dot{\beta}}| = \delta^{\tilde{\dot{\alpha}}}{}_{\dot{\beta}}$. Under these gauge fixing conditions, the $\theta = \bar{\theta} = 0$ pieces of the supervierbein and its inverse take the form [Wess and Zumino (1977); Wess and Bagger (1992)],

$$
\begin{aligned}
E_{\tilde{M}}{}^{M}(z)\Big| &= \begin{pmatrix} e_{\tilde{\mu}}{}^{\mu}(x) & \frac{1}{2}\psi_{\tilde{\mu}}{}^{\alpha}(x) & \frac{1}{2}\bar{\psi}_{\tilde{\mu}\dot{\alpha}}(x) \\ 0 & \delta_{\tilde{\alpha}}{}^{\alpha} & 0 \\ 0 & 0 & \delta^{\tilde{\dot{\alpha}}}{}_{\dot{\alpha}} \end{pmatrix},\\
E_{M}{}^{\tilde{M}}(z)\Big| &= \begin{pmatrix} e_{\mu}{}^{\tilde{\mu}}(x) & -\frac{1}{2}\psi_{\mu}{}^{\tilde{\alpha}}(x) & -\frac{1}{2}\bar{\psi}_{\mu\tilde{\dot{\alpha}}}(x) \\ 0 & \delta_{\alpha}{}^{\tilde{\alpha}} & 0 \\ 0 & 0 & \delta^{\dot{\alpha}}{}_{\tilde{\dot{\alpha}}}\,, \end{pmatrix}
\end{aligned}
\tag{3.106}
$$

where the remaining, unconstrained, degrees of freedom correspond to the spin-two graviton $e_{\tilde{\mu}}{}^{\mu}$ and to the two Weyl components of the spin-3/2 gravitino field $(\psi_{\tilde{\mu}}{}^{\alpha}, \bar{\psi}_{\tilde{\mu}\dot{\alpha}})$. Moreover, the expressions (3.106) are compatible with the relations (3.22) and (3.23). Since $E_{\tilde{M}}{}^{M} E_{M}{}^{\tilde{N}} = \delta_{\tilde{M}}{}^{\tilde{N}}$ and $E_{M}{}^{\tilde{M}} E_{\tilde{M}}{}^{N} = \delta_{M}{}^{N}$, we have also

$$
\begin{aligned}
e_{\tilde{\mu}}{}^{\mu} e_{\mu}{}^{\tilde{\nu}} &= \delta_{\tilde{\mu}}{}^{\tilde{\nu}}\,, & e_{\mu}{}^{\tilde{\mu}} e_{\tilde{\mu}}{}^{\nu} &= \delta_{\mu}{}^{\nu}\,,\\
\psi_{\mu}{}^{\tilde{\alpha}} &= e_{\mu}{}^{\tilde{\mu}} \psi_{\tilde{\mu}}{}^{\alpha} \delta_{\alpha}{}^{\tilde{\alpha}}\,, & \bar{\psi}_{\mu\tilde{\dot{\alpha}}} &= e_{\mu}{}^{\tilde{\mu}} \bar{\psi}_{\tilde{\mu}\dot{\alpha}} \delta^{\dot{\alpha}}{}_{\tilde{\dot{\alpha}}}\,.
\end{aligned}
\tag{3.107}
$$

Gauge choices different from the one taken in (3.104) and in (3.106) are also possible. We refer to page 444 of [Zumino (1978)] for an explicit example.

The computation of the lowest-order component of the metric tensor in curved superspace $\eta_{\tilde{M}\tilde{N}}$ can be derived from Eq. (3.106),

$$\begin{aligned}\eta_{\tilde{M}\tilde{N}}\Big| &= (-)^{|\tilde{M}|(|N|+|\tilde{N}|)} E_{\tilde{N}}{}^N E_{\tilde{M}}{}^M \eta_{MN}\Big| \\ &= \begin{pmatrix} g_{\tilde{\mu}\tilde{\nu}} - \frac{1}{4}(\psi_{\tilde{\mu}}\cdot\psi_{\tilde{\nu}} + \bar{\psi}_{\tilde{\mu}}\cdot\bar{\psi}_{\tilde{\nu}}) & -\frac{1}{2}\psi_{\tilde{\mu}\tilde{\beta}} & -\frac{1}{2}\bar{\psi}_{\tilde{\mu}}{}^{\tilde{\bar{\beta}}} \\ -\frac{1}{2}\psi_{\tilde{\nu}\tilde{\alpha}} & \varepsilon_{\tilde{\alpha}\tilde{\beta}} & 0 \\ -\frac{1}{2}\bar{\psi}_{\tilde{\nu}}{}^{\tilde{\bar{\alpha}}} & 0 & \varepsilon^{\tilde{\bar{\alpha}}\tilde{\bar{\beta}}} \end{pmatrix},\end{aligned} \tag{3.108}$$

with $g_{\tilde{\mu}\tilde{\nu}} = e_{\tilde{\mu}}{}^{\mu} e_{\tilde{\nu}}{}^{\nu}\eta_{\mu\nu}$, $\psi_{\tilde{\mu}\tilde{\alpha}} = \psi_{\tilde{\mu}\alpha}\delta_{\tilde{\alpha}}{}^{\alpha}$, $\bar{\psi}_{\tilde{\mu}}{}^{\tilde{\bar{\alpha}}} = \bar{\psi}_{\tilde{\mu}}{}^{\dot{\alpha}}\delta^{\tilde{\bar{\alpha}}}{}_{\dot{\alpha}}$ and η_{MN} defined in App. B. This metric allows one to lower Einsteinian indices as in

$$X_{\tilde{M}}\Big| = X^{\tilde{N}}\eta_{\tilde{M}\tilde{N}}\Big| \,.$$

In the particular case of the graviton field, one can further show that

$$e_{\mu\tilde{\rho}} = E_{\mu\tilde{\rho}}\Big| = E_{\mu}{}^{\tilde{N}}\eta_{\tilde{\rho}\tilde{N}}\Big| = e_{\mu}{}^{\tilde{\mu}} g_{\tilde{\rho}\tilde{\mu}} = e_{\mu}{}^{\tilde{\mu}} e_{\tilde{\rho}}{}^{\nu} e_{\tilde{\mu}}{}^{\rho}\eta_{\nu\rho} = e_{\tilde{\rho}}{}^{\nu}\eta_{\nu\mu} = e_{\tilde{\rho}\mu} \,. \tag{3.109}$$

The results of (3.102) mean that we are not allowed to eliminate the lowest-order components of the $\mathcal{R}$ and $G_{\alpha\dot{\alpha}}$ superfields, so that these new degrees of freedom must be kept. We therefore define a complex scalar field M and a real vector field $b_{\alpha\dot{\alpha}}$ [Wess and Zumino (1978a); Wess and Bagger (1992)],

$$\begin{aligned}\mathcal{R}(z)\Big| &= -\frac{1}{6}M(x)\,, \\ G_{\alpha\dot{\alpha}}(z)\Big| &= -\frac{1}{3}b_{\alpha\dot{\alpha}}(x)\,.\end{aligned} \tag{3.110}$$

More precisely, this implies that the supergravity multiplet is defined by both Eq. (3.106) and Eq. (3.110) and that it contains a helicity-two massless particle $e_{\tilde{\mu}}{}^{\mu}$ called the graviton, a helicity-3/2 massless particle $(\psi_{\tilde{\mu}}{}^{\alpha}, \bar{\psi}_{\tilde{\mu}\dot{\alpha}})$ called the gravitino and two auxiliary fields M and $b_{\alpha\dot{\alpha}}$; M being a complex scalar field and b a real vector field. In the supergravity framework considered in this work and called minimal supergravity, the supergravity multiplet is taken minimal and contains twelve bosonic and twelve fermionic degrees of freedom [Stelle and West (1978c); Ferrara and van Nieuwenhuizen (1978a)]. Other consistent sets of constraints on the elements of the torsion tensor could have been postulated, which would have led to different supergravity multiplets [Girardi et al. (1984)]. Furthermore, supergravity multiplets with different field content could have been defined, either with twelve bosonic and twelve fermionic degrees of freedom (the so-called new minimal models) [Akulov et al. (1977); Sohnius and West (1981a); Gates et al. (1982)] or in non-minimal setups with 20 bosonic and 20 fermionic degrees of freedom [Akulov et al. (1977); Sohnius and West (1981a); Gates et al. (1982)].

Table 3.6 Gauge fixing conditions on the vielbein, the superconnection, and the superfields $\mathcal{R}$ and $G_{\alpha\dot{\alpha}}$.

$$E_{\tilde{M}}{}^{M}(z)\Big| = \begin{pmatrix} e_{\tilde{\mu}}{}^{\mu}(x) & \frac{1}{2}\psi_{\tilde{\mu}}{}^{\alpha}(x) & \frac{1}{2}\bar{\psi}_{\tilde{\mu}\dot{\alpha}}(x) \\ 0 & \delta_{\tilde{\alpha}}{}^{\alpha} & 0 \\ 0 & 0 & \delta^{\tilde{\bar{\alpha}}}{}_{\dot{\alpha}} \end{pmatrix}, \qquad E_{M}{}^{\tilde{M}}(z)\Big| = \begin{pmatrix} e_{\mu}{}^{\tilde{\mu}}(x) & -\frac{1}{2}\psi_{\mu}{}^{\tilde{\alpha}}(x) & -\frac{1}{2}\bar{\psi}_{\mu\tilde{\bar{\alpha}}}(x) \\ 0 & \delta_{\alpha}{}^{\tilde{\alpha}} & 0 \\ 0 & 0 & \delta^{\dot{\alpha}}{}_{\tilde{\bar{\alpha}}} \end{pmatrix}$$

$$\Omega_{\tilde{\mu}M}{}^{N}\Big| = \omega_{\tilde{\mu}M}{}^{N}(x)\,, \qquad \Omega_{\tilde{\alpha}M}{}^{N}\Big| = 0\,, \qquad \Omega^{\tilde{\bar{\alpha}}}{}_{M}{}^{N}\Big| = 0$$

$$\mathcal{R}(z)\Big| = -\tfrac{1}{6}M(x)$$

$$G_{\alpha\dot{\alpha}}(z)\Big| = -\tfrac{1}{3}b_{\alpha\dot{\alpha}}(x)\,.$$

The gauge fixing conditions that are taken in this book, and that we have imposed on the lowest-order components of the supervierbein, the superconnection and the superfields $\mathcal{R}$ and $G_{\alpha\dot{\alpha}}$, are summarised in Table 3.6. In Sec. 3.4, we will demonstrate that all the components of the superfields $\mathcal{R}$, $G_{\alpha\dot{\alpha}}$ and $W_{(\alpha\beta\gamma)}$ can be rewritten in terms of the components of the supergravity multiplet alone.

3.3.3 *Supergravity transformations*

The gauge fixing constraints (3.104) and (3.106) are not preserved under arbitrary Lorentz and general transformations. The set of all possible transformations must be restricted to keep the theory invariant with respect to our gauge choice [Wess and Zumino (1977); Wess and Bagger (1992)]. To this end, we first assume that

$$\xi^{\mu}(z)\Big| = 0\,, \qquad \xi^{\alpha}(z)\Big| = \varepsilon^{\alpha}(x)\,, \qquad \bar{\xi}_{\dot{\alpha}}(z)\Big| = \bar{\varepsilon}_{\dot{\alpha}}(x)\,, \qquad L_{MN}(z)\Big| = 0\,, \quad (3.111)$$

and investigate, under these conditions, the variations of the vielbein and the superconnection. Since in our gauge, the variations $\delta E_{\tilde{\alpha}}{}^{\mu}\big|$ and $\delta E^{\tilde{\bar{\alpha}}\mu}\big|$ have to vanish, we derive

$$\begin{aligned} \delta E_{\tilde{\alpha}}{}^{\mu}\Big| &= -\mathcal{D}_{\tilde{\alpha}}\xi^{\mu}\Big| + \xi^{N}T_{N\tilde{\alpha}}{}^{\mu}\Big| + E_{\tilde{\alpha}}{}^{\nu}L_{\nu}{}^{\mu}\Big| = -\partial_{\tilde{\alpha}}\xi^{\mu}\Big| + \xi^{N}E_{\tilde{\alpha}}{}^{\alpha}T_{N\alpha}{}^{\mu}\Big| \\ &= -\,\partial_{\tilde{\alpha}}\xi^{\mu}\Big| + 2i\delta_{\tilde{\alpha}}{}^{\alpha}\sigma^{\mu}{}_{\alpha\dot{\alpha}}\bar{\varepsilon}^{\dot{\alpha}} = 0\,, \\ \delta E^{\tilde{\bar{\alpha}}\mu}\Big| &= -\bar{\mathcal{D}}^{\tilde{\bar{\alpha}}}\xi^{\mu}\Big| + \xi^{N}T_{N}{}^{\tilde{\bar{\alpha}}\mu}\Big| + E^{\tilde{\bar{\alpha}}\nu}L_{\nu}{}^{\mu}\Big| = -\bar{\partial}^{\tilde{\bar{\alpha}}}\xi^{\mu}\Big| + \xi^{N}E^{\tilde{\bar{\alpha}}}{}_{\dot{\alpha}}\varepsilon^{\dot{\alpha}\dot{\beta}}T_{N\dot{\beta}}{}^{\mu}\Big| \\ &= -\,\bar{\partial}^{\tilde{\bar{\alpha}}}\xi^{\mu}\Big| - 2i\delta^{\tilde{\bar{\alpha}}}{}_{\dot{\alpha}}\varepsilon^{\dot{\alpha}\dot{\beta}}\varepsilon^{\alpha}\sigma^{\mu}{}_{\alpha\dot{\beta}} = 0\,. \end{aligned} \quad (3.112)$$

These results have been obtained by relying on the gauge fixing conditions (3.104) and (3.106), the assumption (3.111) and the constraints on the torsion tensor (3.29) that imply that $\xi^{N}T_{N\alpha}{}^{\mu} = \bar{\varepsilon}_{\dot{\alpha}}T^{\dot{\alpha}}{}_{\alpha}{}^{\mu} = -T_{\dot{\alpha}\alpha}{}^{\mu}\bar{\varepsilon}^{\dot{\alpha}}$ and $\xi^{N}T_{N\dot{\beta}}{}^{\mu} = \varepsilon^{\alpha}T_{\alpha\dot{\beta}}{}^{\mu}$. One possible solution for these equations is given by

$$\xi^{\mu}(z) = 2i(\theta^{\tilde{\alpha}}\delta_{\tilde{\alpha}}{}^{\alpha}\sigma^{\mu}{}_{\alpha\dot{\alpha}}\bar{\varepsilon}^{\dot{\alpha}} + \varepsilon^{\alpha}\sigma^{\mu}{}_{\alpha\dot{\beta}}\varepsilon^{\dot{\alpha}\dot{\beta}}\bar{\theta}_{\tilde{\bar{\alpha}}}\delta^{\tilde{\bar{\alpha}}}{}_{\dot{\alpha}}) + \cdots\,,$$

where the dots stand for higher order component fields. In order to promote this result to a quantity invariant under the general transformations (3.2), we substitute $(\delta_{\tilde{\alpha}}{}^{\alpha}, \delta^{\tilde{\bar{\alpha}}}{}_{\dot{\alpha}})$ by $(E_{\tilde{\alpha}}{}^{\alpha}, E^{\tilde{\bar{\alpha}}}{}_{\dot{\alpha}})$ and embed in this way all higher-order components into the vielbein. After converting all indices to flat indices the simplest solution for ξ^{μ} is then given by

$$\xi^{\mu}(z) = 2i(\tilde{\theta}\sigma^{\mu}\bar{\varepsilon} - \varepsilon\sigma^{\mu}\bar{\tilde{\theta}}) , \tag{3.113}$$

with $\tilde{\theta}_{\alpha} = \theta^{\tilde{\alpha}} E_{\tilde{\alpha}}{}^{\beta}\varepsilon_{\alpha\beta} \neq \theta_{\alpha}$ and a similar relation for $\bar{\theta}_{\tilde{\dot{\alpha}}}$. Similarly, we choose the simplest solution for the fermionic transformation parameters by setting all higher-order components to zero,

$$\begin{aligned} \xi^{\alpha}(z) &= \varepsilon^{\alpha}(x) , \\ \bar{\xi}_{\dot{\alpha}}(z) &= \bar{\varepsilon}_{\dot{\alpha}}(x) . \end{aligned}$$

Since the investigation of the other components of the vielbein does not lead to additional conditions on the transformation parameters, we move on with the analysis of the superconnection for which the variations $\delta\Omega_{\tilde{\alpha}M}{}^{N}\big|$ and $\delta\Omega^{\tilde{\bar{\alpha}}}{}_{M}{}^{N}\big|$ must vanish. Since in the adopted gauge (see Table 3.4),

$$\begin{aligned} \delta\Omega_{\tilde{\alpha}\alpha\beta}\Big| &= -\xi^{M} R_{M\tilde{\alpha}\alpha\beta}\Big| - \partial_{\tilde{\alpha}} L_{\alpha\beta}\Big| = -\varepsilon^{\gamma}\ E_{\tilde{\alpha}}{}^{\delta} R_{\gamma\delta\alpha\beta}\Big| + \bar{\varepsilon}^{\dot{\gamma}}\ E_{\tilde{\alpha}}{}^{\delta} R_{\dot{\gamma}\delta\alpha\beta}\Big| - \partial_{\tilde{\alpha}} L_{\alpha\beta}\Big| \\ &= -\frac{2}{3}\delta_{\tilde{\alpha}}{}^{\delta}\Big(\varepsilon_{\beta}\varepsilon_{\delta\alpha} + \varepsilon_{\delta\beta}\varepsilon_{\alpha}\Big)M^{*} - \frac{1}{3}\delta_{\tilde{\alpha}}{}^{\delta}\Big(\varepsilon_{\delta\beta} b_{\alpha\dot{\gamma}}\bar{\varepsilon}^{\dot{\gamma}} + \varepsilon_{\delta\alpha} b_{\beta\dot{\gamma}}\bar{\varepsilon}^{\dot{\gamma}}\Big) - \partial_{\tilde{\alpha}} L_{\alpha\beta}\Big| = 0 , \\ \delta\Omega^{\tilde{\bar{\alpha}}}{}_{\dot{\alpha}\dot{\beta}}\Big| &= -\xi^{M} R_{M}{}^{\tilde{\bar{\alpha}}}{}_{\dot{\alpha}\dot{\beta}}\Big| - \partial^{\tilde{\bar{\alpha}}} L_{\dot{\alpha}\dot{\beta}}\Big| = -\varepsilon^{\gamma}\varepsilon^{\dot{\delta}\dot{\epsilon}}\ E^{\tilde{\bar{\alpha}}}{}_{\dot{\delta}} R_{\gamma\dot{\epsilon}\dot{\alpha}\dot{\beta}}\Big| + \bar{\varepsilon}^{\dot{\gamma}}\varepsilon^{\dot{\delta}\dot{\epsilon}}\ E^{\tilde{\bar{\alpha}}}{}_{\dot{\delta}} R_{\dot{\gamma}\dot{\epsilon}\dot{\alpha}\dot{\beta}}\Big| - \partial^{\tilde{\bar{\alpha}}} L_{\dot{\alpha}\dot{\beta}}\Big| \\ &= \frac{1}{3}\delta^{\tilde{\bar{\alpha}}}{}_{\dot{\delta}}\Big(\delta^{\dot{\delta}}{}_{\dot{\beta}}\ \varepsilon^{\gamma} b_{\gamma\dot{\alpha}} + \delta^{\dot{\delta}}{}_{\dot{\alpha}}\ \varepsilon^{\gamma} b_{\gamma\dot{\beta}}\Big) + \frac{2}{3}\delta^{\tilde{\bar{\alpha}}}{}_{\dot{\delta}}\Big(\bar{\varepsilon}_{\dot{\alpha}}\delta^{\dot{\delta}}{}_{\dot{\beta}} + \bar{\varepsilon}_{\dot{\beta}}\delta^{\dot{\delta}}{}_{\dot{\alpha}}\Big)M - \partial^{\tilde{\bar{\alpha}}} L_{\dot{\alpha}\dot{\beta}}\Big| = 0 , \end{aligned} \tag{3.114}$$

we compute the transformation parameters $L_{\alpha\beta}$ and $L_{\dot{\alpha}\dot{\beta}}$ by following the strategy used for the computation of the general transformation parameter ξ^{μ}. Replacing $(\delta_{\tilde{\alpha}}{}^{\alpha}, \delta^{\tilde{\bar{\alpha}}}{}_{\dot{\alpha}})$ by $(E_{\tilde{\alpha}}{}^{\alpha}, E^{\tilde{\bar{\alpha}}}{}_{\dot{\alpha}})$ to account for higher-order terms, we derive (with $\tilde{\theta}_{\alpha}$ = and $\bar{\theta}_{\dot{\alpha}}$ defined before)

$$\begin{aligned} L_{\alpha\beta} &= \frac{1}{3}\Big[\tilde{\theta}_{\alpha}(2\varepsilon_{\beta}M^{*} + b_{\beta\dot{\gamma}}\bar{\varepsilon}^{\dot{\gamma}}) + \tilde{\theta}_{\beta}(2\varepsilon_{\alpha}M^{*} + b_{\alpha\dot{\gamma}}\bar{\varepsilon}^{\dot{\gamma}})\Big] , \\ L_{\dot{\alpha}\dot{\beta}} &= \frac{1}{3}\Big[\bar{\theta}_{\dot{\alpha}}(2\bar{\varepsilon}_{\dot{\beta}}M + \varepsilon^{\gamma} b_{\gamma\dot{\beta}}) + \bar{\theta}_{\dot{\beta}}(2\bar{\varepsilon}_{\dot{\alpha}}M + \varepsilon^{\gamma} b_{\gamma\dot{\alpha}})\Big] . \end{aligned} \tag{3.115}$$

The $L_{\mu\nu}$ parameters are subsequently obtained from $L_{\alpha\beta}$ and $L_{\dot{\alpha}\dot{\beta}}$ and the relations of App. B. Moreover, since $L^{\dagger}_{\alpha\beta} = -L_{\dot{\alpha}\dot{\beta}}$, the parameters $L_{\mu\nu}$ are real.

Since no extra constraint can be obtained from the other components of the superconnection, we have completed our derivation of the subset of general coordinate and Lorentz transformations preserving the gauge choice of the previous

Table 3.7 Supergravity transformations.

A supergravity transformation contains a general transformation ξ and a Lorentz transformation L defined by

$$\begin{aligned}
\xi^\mu(z) &= 2i\big[\tilde{\theta}\sigma^\mu\bar{\varepsilon} - \varepsilon\sigma^\mu\bar{\theta}\big]\,, \\
\xi^\alpha(z) &= \varepsilon^\alpha\,, \\
\bar{\xi}_{\dot\alpha}(z) &= \bar{\varepsilon}_{\dot\alpha}\,, \\
L_{\alpha\beta}(z) &= \frac{1}{3}\big[\tilde{\theta}_\alpha(2\varepsilon_\beta M^* + b_{\beta\dot\gamma}\bar{\varepsilon}^{\dot\gamma}) + \tilde{\theta}_\beta(2\varepsilon_\alpha M^* + b_{\alpha\dot\gamma}\bar{\varepsilon}^{\dot\gamma})\big]\,, \\
L_{\dot\alpha\dot\beta}(z) &= \frac{1}{3}\big[\bar{\theta}_{\dot\alpha}(2\bar{\varepsilon}_{\dot\beta} M + \varepsilon^\gamma b_{\gamma\dot\beta}) + \bar{\theta}_{\dot\beta}(2\bar{\varepsilon}_{\dot\alpha} M + \varepsilon^\gamma b_{\gamma\dot\alpha})\big]\,, \\
L_{\mu\nu}(z) &= -\frac{1}{2}\big[(\bar{\sigma}_\mu\sigma_\nu)^{\dot\alpha}{}_{\dot\beta} L_{\dot\alpha}{}^{\dot\beta} + (\sigma_\mu\bar{\sigma}_\nu)_\alpha{}^\beta L^\alpha{}_\beta\big]\,,
\end{aligned}$$

where the fields $(\varepsilon, \bar{\varepsilon})$, b and M depend only on the spacetime coordinates x and all indices are flat superspace indices.

section. The results (3.111), (3.113) and (3.115), summarised in Table 3.7, form the so-called supergravity transformations.

In the flat limit, $\mathcal{D}_{\tilde M} = \partial_{\tilde M}$, the supervierbein and its inverse are given by Eq. (3.19) and the superderivatives by

$$\begin{aligned}
D_\mu &= E_\mu{}^{\tilde M}\partial_{\tilde M} = \partial_\mu\,, \\
D_\alpha &= E_\alpha{}^{\tilde M}\partial_{\tilde M} = \partial_\alpha + i\sigma^\mu{}_{\alpha\dot\alpha}\bar{\theta}^{\dot\alpha}\partial_\mu\,, \\
\bar{D}^{\dot\alpha} &= E^{\dot\alpha\tilde M}\partial_{\tilde M} = \bar{\partial}^{\dot\alpha} - i\theta^\alpha\sigma^\mu{}_{\alpha\dot\beta}\,\varepsilon^{\dot\alpha\dot\beta}\partial_\mu\,.
\end{aligned}$$

The flat limit form of the supergravity transformations thus reads

$$\xi^M D_M = \varepsilon\cdot Q + \bar{Q}\cdot\bar{\varepsilon}\,,$$

and is equivalent to standard supersymmetric transformations with supercharges defined as in Eq. (2.1). Starting from transformation parameters defined by

$$\begin{aligned}
\xi^{\tilde\alpha} &= \xi^M E_M{}^{\tilde\alpha} = \varepsilon^{\tilde\alpha}\,, \\
\bar{\xi}_{\tilde{\dot\alpha}} &= \bar{\xi}^M E_{M\tilde{\dot\alpha}} = \bar{\varepsilon}_{\tilde{\dot\alpha}}\,, \\
\xi^{\tilde\mu} &= \xi^M E_M{}^{\tilde\mu} = -i(\varepsilon\sigma^{\tilde\mu}\bar{\theta} - \theta\sigma^{\tilde\mu}\bar{\varepsilon})\,,
\end{aligned}$$

we could have derived the same conclusion by using quantities with Einsteinian indices since $\xi^{\tilde M}\partial_{\tilde M} = \xi^M D_M$.

3.4 Explicit computation of the lowest order component fields

Using the relations of Table 3.1 and the gauge defined by Table 3.6, we can rewrite the lowest order components of the torsion and curvature tensors with Einsteinian indices as functions of the vierbein, the connection and the gravitino,

$$
\begin{aligned}
T_{\tilde{\mu}\tilde{\nu}}{}^{\mu}\Big| &= -\partial_{\tilde{\mu}} e_{\tilde{\nu}}{}^{\mu} + \partial_{\tilde{\nu}} e_{\tilde{\mu}}{}^{\mu} - \omega_{\tilde{\mu}\tilde{\nu}}{}^{\mu} + \omega_{\tilde{\nu}\tilde{\mu}}{}^{\mu} \,, \\
T_{\tilde{\mu}\tilde{\nu}}{}^{\alpha}\Big| &= -\frac{1}{2}\Big(\partial_{\tilde{\mu}}\psi_{\tilde{\nu}}{}^{\alpha} - \partial_{\tilde{\nu}}\psi_{\tilde{\mu}}{}^{\alpha} - \psi_{\tilde{\mu}}{}^{\beta}\omega_{\tilde{\nu}\beta}{}^{\alpha} + \psi_{\tilde{\nu}}{}^{\beta}\omega_{\tilde{\mu}\beta}{}^{\alpha}\Big) \\
&= -\frac{1}{2}\Big(\mathcal{D}_{\tilde{\mu}}\psi_{\tilde{\nu}}{}^{\alpha} - \mathcal{D}_{\tilde{\nu}}\psi_{\tilde{\mu}}{}^{\alpha}\Big) = -\frac{1}{2}\psi_{\tilde{\mu}\tilde{\nu}}{}^{\alpha} \,, \\
T_{\tilde{\mu}\tilde{\nu}\dot{\alpha}}\Big| &= -\frac{1}{2}\Big(\partial_{\tilde{\mu}}\bar{\psi}_{\tilde{\nu}\dot{\alpha}} - \partial_{\tilde{\nu}}\bar{\psi}_{\tilde{\mu}\dot{\alpha}} - \bar{\psi}_{\tilde{\mu}\dot{\beta}}\omega_{\tilde{\nu}}{}^{\dot{\beta}}{}_{\dot{\alpha}} + \bar{\psi}_{\tilde{\nu}\dot{\beta}}\omega_{\tilde{\mu}}{}^{\dot{\beta}}{}_{\dot{\alpha}}\Big) \\
&= -\frac{1}{2}\Big(\mathcal{D}_{\tilde{\mu}}\bar{\psi}_{\tilde{\nu}\dot{\alpha}} - \mathcal{D}_{\tilde{\nu}}\bar{\psi}_{\tilde{\mu}\dot{\alpha}}\Big) = -\frac{1}{2}\bar{\psi}_{\tilde{\mu}\tilde{\nu}\dot{\alpha}} \,, \\
R_{\tilde{\mu}\tilde{\nu}\mu}{}^{\nu}\Big| &= \partial_{\tilde{\mu}}\omega_{\tilde{\nu}\mu}{}^{\nu} - \partial_{\tilde{\nu}}\omega_{\tilde{\mu}\mu}{}^{\nu} + \omega_{\tilde{\nu}\mu}{}^{\rho}\omega_{\tilde{\mu}\rho}{}^{\nu} - \omega_{\tilde{\mu}\mu}{}^{\rho}\omega_{\tilde{\nu}\rho}{}^{\nu} \,.
\end{aligned}
\tag{3.116}
$$

This relies on the Lie-algebra valuedness of the connection and on the introduction of the reduced quantities

$$
\begin{aligned}
\omega_{\tilde{\mu}\tilde{\nu}}{}^{\mu} &= e_{\tilde{\nu}}{}^{\nu}\omega_{\tilde{\mu}\nu}{}^{\mu} \,, \\
\mathcal{D}_{\tilde{\mu}}\psi_{\tilde{\nu}}{}^{\alpha} &= \partial_{\tilde{\mu}}\psi_{\tilde{\nu}}{}^{\alpha} + \psi_{\tilde{\nu}}{}^{\beta}\omega_{\tilde{\mu}\beta}{}^{\alpha} \,, \\
\mathcal{D}_{\tilde{\mu}}\bar{\psi}_{\tilde{\nu}\dot{\alpha}} &= \partial_{\tilde{\mu}}\bar{\psi}_{\tilde{\nu}\dot{\alpha}} + \bar{\psi}_{\tilde{\nu}\dot{\beta}}\omega_{\tilde{\mu}}{}^{\dot{\beta}}{}_{\dot{\alpha}} \,, \\
\psi_{\tilde{\mu}\tilde{\nu}}{}^{\alpha} &= \mathcal{D}_{\tilde{\mu}}\psi_{\tilde{\nu}}{}^{\alpha} - \mathcal{D}_{\tilde{\nu}}\psi_{\tilde{\mu}}{}^{\alpha} \,, \\
\bar{\psi}_{\tilde{\mu}\tilde{\nu}\dot{\alpha}} &= \mathcal{D}_{\tilde{\mu}}\bar{\psi}_{\tilde{\nu}\dot{\alpha}} - \mathcal{D}_{\tilde{\nu}}\bar{\psi}_{\tilde{\mu}\dot{\alpha}} \,.
\end{aligned}
\tag{3.117}
$$

It is possible to convert an Einstein index to a Lorentz or group structure index:

$$
\begin{aligned}
\psi_{\mu\nu} = e_{\mu}{}^{\tilde{\mu}} e_{\nu}{}^{\tilde{\nu}}\psi_{\tilde{\mu}\tilde{\nu}} &= e_{\mu}{}^{\tilde{\mu}} e_{\nu}{}^{\tilde{\nu}}\Big(\mathcal{D}_{\tilde{\mu}}\psi_{\tilde{\nu}} - \mathcal{D}_{\tilde{\nu}}\psi_{\tilde{\mu}}\Big) \,, \\
&= e_{\mu}{}^{\tilde{\mu}} e_{\nu}{}^{\tilde{\nu}}\Big(\mathcal{D}_{\tilde{\mu}}(e_{\tilde{\nu}}{}^{\rho}\psi_{\rho}) - \mathcal{D}_{\tilde{\nu}}(e_{\tilde{\mu}}{}^{\sigma}\psi_{\sigma})\Big) \,,
\end{aligned}
$$

meaning that $\psi_{\mu\nu}$

$$
\psi_{\mu\nu} \neq \mathcal{D}_{\mu}\psi_{\nu} - \mathcal{D}_{\nu}\psi_{\mu} \,,
$$

as we could have naively expected. The set of relations in (3.116) constitutes the cornerstone for the derivation of the components of the connection, as well as of the $W_{(\alpha\beta\gamma)}$, $G_{\alpha\dot{\alpha}}$ and R superfields. It is based firstly on the conversion of all Einsteinian indices to Lorentzian indices and secondly on the constraints obtained from the Bianchi identities. This is detailed in the next subsections and the results will be summarised in Table 3.8. Extra information, in particular on the physical meaning of the different tensors, can be found *e.g.* in [Ferrara and Zumino (1978)].

3.4.1 Computation of the connection $\omega_{\tilde\mu\tilde\nu\tilde\rho}$

Following the technique sketched above, we express the torsion tensor $T_{\tilde\mu\tilde\nu}{}^{\mu}\big|$ in terms of the vielbein and quantities carrying exclusively flat indices,

$$\begin{aligned} T_{\tilde\mu\tilde\nu}{}^{\mu}\Big| &= E_{\tilde\nu}{}^{N}E_{\tilde\mu}{}^{M}T_{MN}{}^{\mu}\Big| = \Big(E_{\tilde\nu}{}^{\beta}E_{\tilde\mu\dot\gamma} + E_{\tilde\nu\dot\gamma}E_{\tilde\mu}{}^{\beta}\Big)T_{\beta}{}^{\dot\gamma\mu}\Big| \\ &= -\frac{i}{2}\Big(\psi_{\tilde\mu}\sigma^{\mu}\bar\psi_{\tilde\nu} - \psi_{\tilde\nu}\sigma^{\mu}\bar\psi_{\tilde\mu}\Big)\,. \end{aligned} \tag{3.118}$$

This derivation relies both on the constraints (3.29) and on the lowest-order coefficients of the supervierbein presented in Table 3.6. Comparing the first relation of Eq. (3.116) to Eq. (3.118), using the quantity $e_{\mu\tilde\rho}$ introduced in Eq. (3.109) and the definitions of Eq. (3.117), one gets

$$\omega_{\tilde\mu\tilde\nu\tilde\rho} - \omega_{\tilde\nu\tilde\mu\tilde\rho} = -e_{\mu\tilde\rho}(\partial_{\tilde\mu}e_{\tilde\nu}{}^{\mu} - \partial_{\tilde\nu}e_{\tilde\mu}{}^{\mu}) + \frac{i}{2}e_{\mu\tilde\rho}(\psi_{\tilde\mu}\sigma^{\mu}\bar\psi_{\tilde\nu} - \psi_{\tilde\nu}\sigma^{\mu}\bar\psi_{\tilde\mu})\,.$$

However, the connection $\omega_{\tilde\mu\tilde\nu\tilde\rho}$ is antisymmetric in its last two indices. Therefore, $\omega_{\tilde\mu\tilde\nu\tilde\rho} = -\omega_{\tilde\mu\tilde\rho\tilde\nu}$ and we get

$$\begin{aligned} \omega_{\tilde\mu\tilde\nu\tilde\rho} &= \frac{1}{2}(\omega_{\tilde\mu\tilde\nu\tilde\rho} - \omega_{\tilde\nu\tilde\mu\tilde\rho}) - \frac{1}{2}(\omega_{\tilde\nu\tilde\rho\tilde\mu} - \omega_{\tilde\rho\tilde\nu\tilde\mu}) + \frac{1}{2}(\omega_{\tilde\rho\tilde\mu\tilde\nu} - \omega_{\tilde\mu\tilde\rho\tilde\nu}) \\ &= -\frac{1}{2}e_{\mu\tilde\rho}(\partial_{\tilde\mu}e_{\tilde\nu}{}^{\mu} - \partial_{\tilde\nu}e_{\tilde\mu}{}^{\mu}) + \frac{1}{2}e_{\mu\tilde\mu}(\partial_{\tilde\nu}e_{\tilde\rho}{}^{\mu} - \partial_{\tilde\rho}e_{\tilde\nu}{}^{\mu}) \\ &\quad - \frac{1}{2}e_{\mu\tilde\nu}\Big(\partial_{\tilde\rho}e_{\tilde\mu}{}^{\mu} - \partial_{\tilde\mu}e_{\tilde\rho}{}^{\mu}\Big) + \frac{i}{4}e_{\mu\tilde\rho}(\psi_{\tilde\mu}\sigma^{\mu}\bar\psi_{\tilde\nu} - \psi_{\tilde\nu}\sigma^{\mu}\bar\psi_{\tilde\mu}) \\ &\quad - \frac{i}{4}e_{\mu\tilde\mu}(\psi_{\tilde\nu}\sigma^{\mu}\bar\psi_{\tilde\rho} - \psi_{\tilde\rho}\sigma^{\mu}\bar\psi_{\tilde\nu}) + \frac{i}{4}e_{\mu\tilde\nu}(\psi_{\tilde\rho}\sigma^{\mu}\bar\psi_{\tilde\mu} - \psi_{\tilde\mu}\sigma^{\mu}\bar\psi_{\tilde\rho})\,. \end{aligned} \tag{3.119}$$

This expression, first obtained in [Deser and Zumino (1976); Freedman et al. (1976)], is analogous to one valid in general relativity, where the spin connection is related to the vierbein (and eventually to the torsion tensor) [Weinberg (1972a); van Nieuwenhuizen (1981); Fuks and Rausch de Traubenberg (2011)]. Our results are also similar to those obtained in the Palatini formalism for pure supergravity systems, in particular due to the constraint $T_{\mu\nu}{}^{\rho} = 0$ [van Nieuwenhuizen (1981); Fuks and Rausch de Traubenberg (2011)].

3.4.2 Computation of the superfields $W_{(\alpha\beta\gamma)}$ and $\overline{W}_{(\dot\alpha\dot\beta\dot\gamma)}$

In order to express the W and $\overline{W}$ superfields in terms of the elements of the supergravity multiplet, we reexpress the torsion tensors $T_{\tilde\mu\tilde\nu}{}^{\alpha}$ and $T_{\tilde\mu\tilde\nu\dot\alpha}$ in terms of quantities carrying Lorentz indices, then use the constraints of Eq. (3.29) and the results for the vielbein of Table 3.6. In this way, we may write

$$
\begin{aligned}
T_{\tilde{\mu}\tilde{\nu}}{}^{\alpha}\Big| &= E_{\tilde{\nu}}{}^{\nu}E_{\tilde{\mu}}{}^{\mu}T_{\mu\nu}{}^{\alpha}\Big| + E_{\tilde{\nu}}{}^{\nu}E_{\tilde{\mu}}{}^{\beta}T_{\beta\nu}{}^{\alpha}\Big| + E_{\tilde{\nu}}{}^{\beta}E_{\tilde{\mu}}{}^{\mu}T_{\mu\beta}{}^{\alpha}\Big| \\
&\quad + E_{\tilde{\nu}}{}^{\nu}E_{\tilde{\mu}\dot{\beta}}T^{\dot{\beta}}{}_{\nu}{}^{\alpha}\Big| + E_{\tilde{\nu}\dot{\beta}}E_{\tilde{\mu}}{}^{\mu}T_{\mu}{}^{\dot{\beta}\alpha}\Big| \\
&= e_{\tilde{\nu}}{}^{\nu}e_{\tilde{\mu}}{}^{\mu}T_{\mu\nu}{}^{\alpha}\Big| + \frac{1}{2}e_{\tilde{\nu}}{}^{\nu}\psi_{\tilde{\mu}}{}^{\beta}T_{\beta\nu}{}^{\alpha}\Big| + \frac{1}{2}\psi_{\tilde{\nu}}{}^{\beta}e_{\tilde{\mu}}{}^{\mu}T_{\mu\beta}{}^{\alpha}\Big| \\
&\quad + \frac{1}{2}e_{\tilde{\nu}}{}^{\nu}\bar{\psi}_{\tilde{\mu}\dot{\beta}}T^{\dot{\beta}}{}_{\nu}{}^{\alpha}\Big| + \frac{1}{2}\bar{\psi}_{\tilde{\nu}\dot{\beta}}e_{\tilde{\mu}}{}^{\mu}T_{\mu}{}^{\dot{\beta}\alpha}\Big| ,
\end{aligned}
$$

$$
\begin{aligned}
T_{\tilde{\mu}\tilde{\nu}\dot{\alpha}}\Big| &= E_{\tilde{\nu}}{}^{\nu}E_{\tilde{\mu}}{}^{\mu}T_{\mu\nu\dot{\alpha}}\Big| + E_{\tilde{\nu}}{}^{\nu}E_{\tilde{\mu}}{}^{\beta}T_{\beta\nu\dot{\alpha}}\Big| + E_{\tilde{\nu}}{}^{\beta}E_{\tilde{\mu}}{}^{\mu}T_{\mu\beta\dot{\alpha}}\Big| \\
&\quad + E_{\tilde{\nu}}{}^{\nu}E_{\tilde{\mu}\dot{\beta}}T^{\dot{\beta}}{}_{\nu\dot{\alpha}}\Big| + E_{\tilde{\nu}\dot{\beta}}E_{\tilde{\mu}}{}^{\mu}T_{\mu}{}^{\dot{\beta}}{}_{\dot{\alpha}}\Big| \\
&= e_{\tilde{\nu}}{}^{\nu}e_{\tilde{\mu}}{}^{\mu}T_{\mu\nu\dot{\alpha}}\Big| + \frac{1}{2}e_{\tilde{\nu}}{}^{\nu}\psi_{\tilde{\mu}}{}^{\beta}T_{\beta\nu\dot{\alpha}}\Big| + \frac{1}{2}\psi_{\tilde{\nu}}{}^{\beta}e_{\tilde{\mu}}{}^{\mu}T_{\mu\beta\dot{\alpha}}\Big| \\
&\quad + \frac{1}{2}e_{\tilde{\nu}}{}^{\nu}\bar{\psi}_{\tilde{\mu}\dot{\beta}}T^{\dot{\beta}}{}_{\nu\dot{\alpha}}\Big| + \frac{1}{2}\bar{\psi}_{\tilde{\nu}\dot{\beta}}e_{\tilde{\mu}}{}^{\mu}T_{\mu}{}^{\dot{\beta}}{}_{\dot{\alpha}}\Big| .
\end{aligned}
$$

Comparing with the second and third equations in (3.116), respectively, and converting all vector indices to spinor indices, one can express the lowest order coefficient of the torsion tensors $T_{\delta\dot{\delta}\gamma\dot{\gamma}}{}^{\alpha} = \sigma^{\mu}{}_{\delta\dot{\delta}}\sigma^{\nu}{}_{\gamma\dot{\gamma}}T_{\mu\nu}{}^{\alpha}$ and $T_{\delta\dot{\delta}\gamma\dot{\gamma}\dot{\alpha}} = \sigma^{\mu}{}_{\delta\dot{\delta}}\sigma^{\nu}{}_{\gamma\dot{\gamma}}T_{\mu\nu\dot{\alpha}}$ as

$$
\begin{aligned}
T_{\delta\dot{\delta}\gamma\dot{\gamma}}{}^{\alpha}\Big| &= -\frac{1}{2}\psi_{\delta\dot{\delta}\gamma\dot{\gamma}}{}^{\alpha} - \frac{1}{2}\left(\psi_{\delta\dot{\delta}}{}^{\beta}T_{\beta\gamma\dot{\gamma}}{}^{\alpha}\big| - \psi_{\gamma\dot{\gamma}}{}^{\beta}T_{\beta\delta\dot{\delta}}{}^{\alpha}\big|\right) \\
&\quad + \frac{1}{2}\left(\bar{\psi}_{\delta\dot{\delta}}{}^{\dot{\beta}}T_{\dot{\beta}\gamma\dot{\gamma}}{}^{\alpha}\big| - \bar{\psi}_{\gamma\dot{\gamma}}{}^{\dot{\beta}}T_{\dot{\beta}\delta\dot{\delta}}{}^{\alpha}\big|\right), \\
T_{\delta\dot{\delta}\gamma\dot{\gamma}\dot{\alpha}}\Big| &= -\frac{1}{2}\bar{\psi}_{\delta\dot{\delta}\gamma\dot{\gamma}\dot{\alpha}} - \frac{1}{2}\left(\psi_{\delta\dot{\delta}}{}^{\beta}T_{\beta\gamma\dot{\gamma}\dot{\alpha}}\big| - \psi_{\gamma\dot{\gamma}}{}^{\beta}T_{\beta\delta\dot{\delta}\dot{\alpha}}\big|\right) \\
&\quad + \frac{1}{2}\left(\bar{\psi}_{\delta\dot{\delta}}{}^{\dot{\beta}}T_{\dot{\beta}\gamma\dot{\gamma}\dot{\alpha}}\big| - \bar{\psi}_{\gamma\dot{\gamma}}{}^{\dot{\beta}}T_{\dot{\beta}\delta\dot{\delta}\dot{\alpha}}\big|\right),
\end{aligned}
\tag{3.120}
$$

where we have used the symmetry properties of the torsion tensor (see (3.10)) and introduced the quantities

$$
\begin{aligned}
\psi_{\beta\dot{\beta}}{}^{\alpha} &= \sigma^{\mu}{}_{\beta\dot{\beta}}e_{\mu}{}^{\tilde{\mu}}\psi_{\tilde{\mu}}{}^{\alpha} , \\
\bar{\psi}_{\beta\dot{\beta}\dot{\alpha}} &= \sigma^{\mu}{}_{\beta\dot{\beta}}e_{\mu}{}^{\tilde{\mu}}\bar{\psi}_{\tilde{\mu}\dot{\alpha}} , \\
\psi_{\beta\dot{\beta}\gamma\dot{\gamma}}{}^{\alpha} &= \sigma^{\mu}{}_{\beta\dot{\beta}}\sigma^{\nu}{}_{\gamma\dot{\gamma}}e_{\mu}{}^{\tilde{\mu}}e_{\nu}{}^{\tilde{\nu}}\psi_{\tilde{\mu}\tilde{\nu}}{}^{\alpha} , \\
\bar{\psi}_{\beta\dot{\beta}\gamma\dot{\gamma}\dot{\alpha}} &= \sigma^{\mu}{}_{\beta\dot{\beta}}\sigma^{\nu}{}_{\gamma\dot{\gamma}}e_{\mu}{}^{\tilde{\mu}}e_{\nu}{}^{\tilde{\nu}}\bar{\psi}_{\tilde{\mu}\tilde{\nu}\dot{\alpha}} .
\end{aligned}
$$

Using the expressions for the $T_{\delta\dot{\delta}\gamma\dot{\gamma}\alpha}$ and $T_{\delta\dot{\delta}\gamma\dot{\gamma}\dot{\alpha}}$ tensors derived from the Bianchi identities (see Table 3.3) and recalling that $W_{(\alpha\beta\gamma)}$ and $\overline{W}_{(\dot{\alpha}\dot{\beta}\dot{\gamma})}$ are invariant under the exchange of any two indices, one can rewrite these superfields as

$$\begin{aligned}
W_{(\alpha_1\alpha_2\alpha_3)} &= -\frac{1}{4}\sum_{\tau\in S_3}\frac{1}{3!}\varepsilon^{\dot\gamma\dot\delta}T_{\alpha_{\tau(1)}\dot\delta\alpha_{\tau(2)}\dot\gamma\alpha_{\tau(3)}}\,,\\
\overline{W}_{(\dot\alpha_1\dot\alpha_2\dot\alpha_3)} &= -\frac{1}{4}\sum_{\tau\in S_3}\frac{1}{3!}\varepsilon^{\gamma\delta}T_{\delta\dot\alpha_{\tau(1)}\gamma\dot\alpha_{\tau(2)}\dot\alpha_{\tau(3)}}\,,
\end{aligned}$$

where S_3 is the permutation group of three elements. Inserting Eq. (3.120) into the relations above, making use of the expressions of the various components of the torsion tensors presented in Table 3.3, and applying the constraints of Table 3.6, we find

$$\begin{aligned}
W_{(\alpha_1\alpha_2\alpha_3)}\Big| &= -\frac{1}{48}\sum_{\tau\in S_3}\Big(\psi_{\alpha_{\tau(1)}\dot\delta\alpha_{\tau(2)}}{}^{\dot\delta}{}_{\alpha_{\tau(3)}} - i\psi_{\alpha_{\tau(1)}\dot\delta\alpha_{\tau(2)}}b_{\alpha_{\tau(3)}}{}^{\dot\delta}\Big)\,,\\
\overline{W}_{(\dot\alpha_1\dot\alpha_2\dot\alpha_3)}\Big| &= -\frac{1}{48}\sum_{\tau\in S_3}\Big(\bar\psi_{\delta\dot\alpha_{\tau(1)}}{}^{\delta}{}_{\dot\alpha_{\tau(2)}\dot\alpha_{\tau(3)}} + i\bar\psi_{\delta\dot\alpha_{\tau(1)}\dot\alpha_{\tau(2)}}b^{\delta}{}_{\dot\alpha_{\tau(3)}}\Big)\,,
\end{aligned}$$

in agreement with the results of Wess and Bagger (1992).

3.4.3 *Computation of $\mathcal{D}_\delta G_{\gamma\dot\alpha}$ and $\overline{\mathcal{D}}_{\dot\delta}G_{\alpha\dot\gamma}$*

The lowest-order component of the $G_{\gamma\dot\alpha}$ superfield has been defined in (3.110), so that we can directly move on with the calculation of the first order derivatives of this superfield. To this end, we start with the expressions of the $T_{\delta\dot\delta\gamma\dot\gamma\alpha}$ and $T_{\delta\dot\delta\gamma\dot\gamma\dot\alpha}$ torsion tensors (see Table 3.3) that we respectively contract with $\varepsilon^{\gamma\delta}$ and $\varepsilon^{\dot\gamma\dot\delta}$. This leads to

$$\begin{aligned}
\mathcal{D}_\delta G_{\gamma\dot\alpha} + \mathcal{D}_\gamma G_{\delta\dot\alpha} &= \varepsilon^{\dot\gamma\dot\delta}T_{\delta\dot\delta\gamma\dot\gamma\dot\alpha}\,,\\
\overline{\mathcal{D}}_{\dot\delta}G_{\alpha\dot\gamma} + \overline{\mathcal{D}}_{\dot\gamma}G_{\alpha\dot\delta} &= \varepsilon^{\gamma\delta}T_{\delta\dot\delta\gamma\dot\gamma\alpha}\,.
\end{aligned}\tag{3.121}$$

Moreover, on the basis of the symmetry properties of (3.43) (and the associated conjugate relations), we deduce

$$\begin{aligned}
\varepsilon_{\delta\gamma}\Big[\varepsilon^{\beta\eta}\varepsilon^{\dot\gamma\dot\delta}T_{\eta\dot\delta\beta\dot\alpha\dot\gamma}\Big] &= 3\varepsilon_{\delta\gamma}\big[\mathcal{D}^{\epsilon}G_{\epsilon\dot\alpha}\big] = 3(\mathcal{D}_\delta G_{\gamma\dot\alpha} - \mathcal{D}_\gamma G_{\delta\dot\alpha})\,,\\
\varepsilon_{\dot\delta\dot\gamma}\Big[\varepsilon^{\dot\beta\dot\eta}\varepsilon^{\gamma\delta}T_{\delta\dot\eta\alpha\dot\beta\gamma}\Big] &= 3\varepsilon_{\dot\delta\dot\gamma}\big[\overline{\mathcal{D}}^{\dot\epsilon}G_{\gamma\dot\epsilon}\big] = 3(\overline{\mathcal{D}}_{\dot\delta}G_{\alpha\dot\gamma} - \overline{\mathcal{D}}_{\dot\gamma}G_{\alpha\dot\delta})\,.
\end{aligned}\tag{3.122}$$

The first-order derivatives of the G-superfield are then obtained from (3.121) and (3.122),

$$\begin{aligned}
\mathcal{D}_\delta G_{\gamma\dot\alpha} &= \frac{1}{2}\varepsilon^{\dot\gamma\dot\delta}T_{\delta\dot\delta\gamma\dot\gamma\dot\alpha} + \frac{1}{6}\varepsilon_{\delta\gamma}\varepsilon^{\epsilon\eta}\varepsilon^{\dot\eta\dot\delta}T_{\eta\dot\delta\epsilon\dot\alpha\dot\eta}\,,\\
\overline{\mathcal{D}}_{\dot\delta}G_{\alpha\dot\gamma} &= \frac{1}{2}\varepsilon^{\gamma\delta}T_{\delta\dot\delta\gamma\dot\gamma\alpha} + \frac{1}{6}\varepsilon_{\dot\delta\dot\gamma}\varepsilon^{\dot\epsilon\dot\eta}\varepsilon^{\eta\delta}T_{\delta\dot\eta\alpha\dot\epsilon\eta}\,.
\end{aligned}\tag{3.123}$$

Using (3.120) and Table 3.3, we can calculate the different terms included in the right-hand side of these equations. Focusing on the computation of $\mathcal{D}_\delta G_{\gamma\dot\alpha}$, for the first contribution, one gets

$$\begin{aligned}\frac{1}{2}\varepsilon^{\dot\gamma\dot\delta}T_{\delta\dot\delta\gamma\dot\gamma\dot\alpha}\Big| &= -\frac{1}{4}\bar\psi_\delta{}^{\dot\epsilon}{}_{\gamma\dot\epsilon\dot\alpha}+\frac{i}{12}(\psi_{\gamma\dot\alpha\delta}+\psi_{\delta\dot\alpha\gamma})M^*\\ &\quad+\frac{i}{48}[\bar\psi_{\gamma\dot\alpha}{}^{\dot\epsilon}b_{\delta\dot\epsilon}+\bar\psi_{\delta\dot\alpha}{}^{\dot\epsilon}b_{\gamma\dot\epsilon}+3\bar\psi_{\gamma\dot\epsilon\dot\alpha}b_\delta{}^{\dot\epsilon}+3\bar\psi_{\delta\dot\epsilon\dot\alpha}b_\gamma{}^{\dot\epsilon}+3\bar\psi_{\gamma\dot\epsilon}{}^{\dot\epsilon}b_{\delta\dot\alpha}+3\bar\psi_{\delta\dot\epsilon}{}^{\dot\epsilon}b_{\gamma\dot\alpha}]\\ &= -\frac{1}{4}\bar\psi_\delta{}^{\dot\epsilon}{}_{\gamma\dot\epsilon\dot\alpha}+\frac{i}{12}(\psi_{\gamma\dot\alpha\delta}+\psi_{\delta\dot\alpha\gamma})M^*\\ &\quad+\frac{i}{24}[\bar\psi_{\gamma\dot\epsilon\dot\alpha}b_\delta{}^{\dot\epsilon}+\bar\psi_{\delta\dot\epsilon\dot\alpha}b_\gamma{}^{\dot\epsilon}+2\bar\psi_{\gamma\dot\epsilon}{}^{\dot\epsilon}b_{\delta\dot\alpha}+2\bar\psi_{\delta\dot\epsilon}{}^{\dot\epsilon}b_{\gamma\dot\alpha}]\,.\end{aligned}$$

The second equality is based on the symmetry property

$$\begin{gathered}\varepsilon^{\dot\gamma\dot\delta}(\bar\psi_{\gamma\dot\gamma\dot\alpha}b_{\delta\dot\delta}+\bar\psi_{\gamma\dot\alpha\dot\delta}b_{\delta\dot\gamma}+\bar\psi_{\gamma\dot\delta\dot\gamma}b_{\delta\dot\alpha})=0\\ \Longleftrightarrow\\ \bar\psi_{\gamma\dot\epsilon\dot\alpha}b_\delta{}^{\dot\epsilon}+\bar\psi_{\gamma\dot\alpha}{}^{\dot\epsilon}b_{\delta\dot\epsilon}-\bar\psi_{\gamma\dot\epsilon}{}^{\dot\epsilon}b_{\delta\dot\alpha}=0\,.\end{gathered}\tag{3.124}$$

which is in fact more general and holds for any rank-three tensor. This can be proven by injecting explicit values for the independent spin indices. The second contribution to $\mathcal{D}_\delta G_{\gamma\dot\alpha}$ is given by

$$\begin{aligned}\frac{1}{6}\varepsilon_{\delta\gamma}\varepsilon^{\epsilon\eta}\varepsilon^{\dot\eta\dot\delta}T_{\eta\dot\delta\epsilon\dot\alpha\dot\eta}\Big| &= -\frac{1}{12}\varepsilon_{\delta\gamma}\bar\psi^{\epsilon\dot\epsilon}{}_{\epsilon\dot\alpha\dot\epsilon}+\frac{i}{12}\varepsilon_{\delta\gamma}\psi_{\epsilon\dot\alpha}{}^{\epsilon}M^*\\ &\quad-\frac{i}{48}\varepsilon_{\delta\gamma}\Big(\bar\psi_{\epsilon\dot\alpha\dot\epsilon}b^{\epsilon\dot\epsilon}+\bar\psi_{\epsilon\dot\epsilon}{}^{\dot\epsilon}b^{\epsilon}{}_{\dot\alpha}+\bar\psi_{\epsilon\dot\epsilon\dot\alpha}b^{\epsilon\dot\epsilon}\Big)\\ &= -\frac{1}{12}\varepsilon_{\delta\gamma}\bar\psi^{\epsilon\dot\epsilon}{}_{\epsilon\dot\alpha\dot\epsilon}+\frac{i}{12}(\psi_{\gamma\dot\alpha\delta}-\psi_{\delta\dot\alpha\gamma})M^*\\ &\quad+\frac{i}{24}\Big(\bar\psi_{\delta\dot\epsilon\dot\alpha}b_\gamma{}^{\dot\epsilon}-\bar\psi_{\gamma\dot\epsilon\dot\alpha}b_\delta{}^{\dot\epsilon}\Big)\,,\end{aligned}\tag{3.125}$$

where, for the last equality, we have used again the relation in Eq. (3.124), together with the symmetry properties

$$\begin{aligned}\varepsilon_{\delta\gamma}\psi_{\varepsilon\dot\alpha}{}^{\varepsilon} &= \psi_{\gamma\dot\alpha\delta}-\psi_{\delta\dot\alpha\gamma}\,,\\ \varepsilon_{\delta\gamma}\bar\psi_{\varepsilon\dot\epsilon\dot\gamma}b^{\varepsilon}{}_{\dot\delta} &= \bar\psi_{\gamma\dot\epsilon\dot\gamma}b_{\delta\dot\delta}-\bar\psi_{\delta\dot\epsilon\dot\gamma}b_{\gamma\dot\delta}\,.\end{aligned}$$

Combining (3.124) and (3.125), we finally get

$$\begin{aligned}\mathcal{D}_\delta G_{\gamma\dot\alpha}\Big| &= -\frac{1}{4}\bar\psi_\delta{}^{\dot\epsilon}{}_{\gamma\dot\epsilon\dot\alpha}-\frac{1}{12}\varepsilon_{\delta\gamma}\bar\psi^{\epsilon\dot\epsilon}{}_{\epsilon\dot\alpha\dot\epsilon}+\frac{i}{6}\psi_{\gamma\dot\alpha\delta}M^*\\ &\quad+\frac{i}{12}\Big(\bar\psi_{\delta\dot\epsilon\dot\alpha}b_\gamma{}^{\dot\epsilon}+\bar\psi_{\gamma\dot\epsilon}{}^{\dot\epsilon}b_{\delta\dot\alpha}+\bar\psi_{\delta\dot\epsilon}{}^{\dot\epsilon}b_{\gamma\dot\alpha}\Big)\,,\end{aligned}\tag{3.126}$$

which agrees with the results of Wess and Bagger (1992). The lowest-order coefficient of the other first-order superderivative of the G-superfield, $\bar{\mathcal{D}}_{\dot\delta}G_{\alpha\dot\gamma}\big|$, can be either calculated along the same lines, or directly deduced by hermitian conjugation,

$$\overline{\mathcal{D}}_{\dot\delta} G_{\alpha\dot\gamma}\Big| = -\frac{1}{4}\psi^{\epsilon}{}_{\dot\delta\epsilon\dot\gamma\alpha} - \frac{1}{12}\varepsilon_{\dot\delta\dot\gamma}\psi_{\alpha}{}^{\dot\epsilon\epsilon}{}_{\dot\epsilon\epsilon} - \frac{i}{6}\bar\psi_{\alpha\dot\gamma\dot\delta}M$$
$$-\frac{i}{12}\Big(\psi_{\epsilon\dot\delta\alpha}b^{\epsilon}{}_{\dot\gamma} + \psi_{\epsilon\dot\gamma}{}^{\epsilon}b_{\alpha\dot\delta} + \psi_{\epsilon\dot\delta}{}^{\epsilon}b_{\alpha\dot\gamma}\Big), \tag{3.127}$$

where we have used the symmetry property $\psi_{\gamma\dot\gamma\delta\dot\delta\varepsilon} = -\psi_{\delta\dot\delta\gamma\dot\gamma\varepsilon}$. Converting all spinor indices to vector indices using (B.7), (3.126) and (3.127) are rewritten as

$$\begin{aligned}
\mathcal{D}_\delta G^\mu\Big| &= \frac{1}{3}(\sigma^\nu\bar\psi_\nu{}^\mu)_\delta - \frac{i}{12}\varepsilon^{\mu\nu\rho\sigma}(\sigma_\sigma\bar\psi_{\nu\rho})_\delta + \frac{i}{6}\psi^\mu{}_\delta M^* + \frac{i}{6}(\sigma^\nu\bar\psi_\nu)_\delta b^\mu \\
&\quad + \frac{1}{12}\varepsilon^{\mu\nu\rho\sigma}(\sigma_\sigma\bar\psi_\nu)_\delta\, b_\rho\,, \\
\overline{\mathcal{D}}^{\dot\delta} G^\mu\Big| &= -\frac{1}{3}(\bar\sigma^\nu\psi_\nu{}^\mu)^{\dot\delta} - \frac{i}{12}\varepsilon^{\mu\nu\rho\sigma}(\bar\sigma_\sigma\psi_{\nu\rho})^{\dot\delta} - \frac{i}{6}\bar\psi^{\mu\dot\delta}M + \frac{i}{6}(\bar\sigma^\nu\psi_\nu)^{\dot\delta}b^\mu \\
&\quad - \frac{1}{12}\varepsilon^{\mu\nu\rho\sigma}(\bar\sigma_\sigma\psi_\nu)^{\dot\delta}\, b_\rho\,,
\end{aligned} \tag{3.128}$$

where we have used the identities (C.5).

3.4.4 *Computation of $\mathcal{D}_\alpha\mathcal{R}$ and $\overline{\mathcal{D}}_{\dot\alpha}\mathcal{R}^\dagger$*

It is straightforward to calculate the derivatives $\mathcal{D}_\alpha\mathcal{R}$ and $\overline{\mathcal{D}}_{\dot\alpha}\mathcal{R}^\dagger$ from (3.126) and (3.127), since the first derivatives of the superfields $\mathcal{R}$ and $G_{\alpha\dot\alpha}$ are related by Eq. (3.47). Reintroducing vector indices, one gets [Wess and Bagger (1992)]

$$\begin{aligned}
\mathcal{D}_\alpha\mathcal{R}\Big| &= -\overline{\mathcal{D}}^{\dot\gamma}G_{\alpha\dot\gamma}\Big| = \varepsilon^{\dot\delta\dot\gamma}\overline{\mathcal{D}}_{\dot\delta}G_{\alpha\dot\gamma}\Big| \\
&= \frac{1}{4}\mathrm{Tr}(\sigma^\mu\bar\sigma^\nu)\psi_{\mu\nu\alpha} - \frac{1}{6}(\sigma^\mu\bar\sigma^\nu\psi_{\mu\nu})_\alpha + \frac{i}{6}(\sigma^\mu\bar\psi_\mu)_\alpha M - \frac{i}{12}\mathrm{Tr}(\sigma^\mu\bar\sigma^\nu)\psi_{\mu\alpha}b_\nu \\
&= -\frac{1}{3}(\sigma^{\mu\nu}\psi_{\mu\nu})_\alpha + \frac{i}{6}(\sigma^\mu\bar\psi_\mu)_\alpha M - \frac{i}{6}\psi_{\mu\alpha}b^\mu\,,
\end{aligned} \tag{3.129}$$

where we have used (C.8) and we recall that $\psi_{\mu\nu\alpha}$ is antisymmetric under the exchange of its vector indices. Hermitian conjugation leads to

$$\overline{\mathcal{D}}^{\dot\alpha}\mathcal{R}^\dagger\Big| = -\frac{1}{3}(\bar\sigma^{\mu\nu}\bar\psi_{\mu\nu})^{\dot\alpha} + \frac{i}{6}(\bar\sigma^\mu\psi_\mu)^{\dot\alpha}M^* + \frac{i}{6}\bar\psi_\mu{}^{\dot\alpha}b^\mu\,, \tag{3.130}$$

after having raised the free spin index. This relation can also be directly recovered from $\overline{\mathcal{D}}_{\dot\alpha}\mathcal{R}^\dagger = -\mathcal{D}^\gamma G_{\gamma\dot\alpha}$ and using (3.126).

3.4.5 Computation of $\mathcal{D}\cdot\mathcal{D}\mathcal{R}$ and $\overline{\mathcal{D}}\cdot\overline{\mathcal{D}}\mathcal{R}^\dagger$

To compute $\mathcal{D}\cdot\mathcal{D}\mathcal{R}$, the second derivative of the superfield $\mathcal{R}$, we compare the last relation given in (3.116) to the expression of $R_{\tilde\mu\tilde\nu}{}^{\mu\nu}|$ derived from the curvature tensors carrying flat indices,

$$\begin{aligned} R_{\tilde\mu\tilde\nu}{}^{\mu\nu}\Big| &= E_{\tilde\nu}{}^N E_{\tilde\mu}{}^M R_{MN}{}^{\mu\nu}\Big| \\ &= e_{\tilde\nu}{}^\sigma e_{\tilde\mu}{}^\rho R_{\rho\sigma}{}^{\mu\nu}\Big| + \frac{1}{2}(e_{\tilde\nu}{}^\rho \psi_{\tilde\mu}{}^\alpha - e_{\tilde\mu}{}^\rho \psi_{\tilde\nu}{}^\alpha) R_{\alpha\rho}{}^{\mu\nu}\Big| + \frac{1}{2}\Big(e_{\tilde\nu}{}^\rho \overline{\psi}_{\tilde\mu\dot\alpha} - e_{\tilde\mu}{}^\rho \overline{\psi}_{\tilde\nu\dot\alpha}\Big) R^{\dot\alpha}{}_\rho{}^{\mu\nu}\Big| \\ &\quad + \frac{1}{4}\psi_{\tilde\nu}{}^\beta \psi_{\tilde\mu}{}^\alpha R_{\alpha\beta}{}^{\mu\nu}\Big| + \frac{1}{4}\overline{\psi}_{\tilde\nu\dot\beta}\overline{\psi}_{\tilde\mu\dot\alpha} R^{\dot\alpha\dot\beta\mu\nu}\Big| + \frac{1}{4}(\psi_{\tilde\nu}{}^\alpha \overline{\psi}_{\tilde\mu\dot\alpha} + \overline{\psi}_{\tilde\nu\dot\alpha}\psi_{\tilde\mu}{}^\alpha) R_\alpha{}^{\dot\alpha\mu\nu}\Big| , \end{aligned}$$

where we have used the symmetry properties of the curvature tensor given in (3.10) and the definitions (3.107). Contracting with $e_\mu{}^{\tilde\mu} e_\nu{}^{\tilde\nu}$ and using again the symmetry properties of the curvature tensor, one obtains

$$\begin{aligned} e_\mu{}^{\tilde\mu} e_\nu{}^{\tilde\nu} R_{\tilde\mu\tilde\nu}{}^{\mu\nu}\Big| &= R_{\mu\nu}{}^{\mu\nu}\Big| + \psi_\mu{}^\alpha R_{\alpha\nu}{}^{\mu\nu}\Big| + \overline{\psi}_{\mu\dot\alpha} R^{\dot\alpha}{}_\nu{}^{\mu\nu}\Big| \\ &\quad + \frac{1}{4}\psi_\nu{}^\beta \psi_\mu{}^\alpha R_{\alpha\beta}{}^{\mu\nu}\Big| + \frac{1}{4}\overline{\psi}_{\nu\dot\beta}\overline{\psi}_{\mu\dot\alpha} R^{\dot\alpha\dot\beta\mu\nu}\Big| - \frac{1}{2}\psi_\mu{}^\alpha \overline{\psi}_{\nu\dot\alpha} R_\alpha{}^{\dot\alpha\mu\nu}\Big| . \end{aligned} \tag{3.131}$$

Each of these terms can be re-expressed as functions of the lowest order components of the superfields $\mathcal{R}$ and $G_{\alpha\dot\alpha}$, their derivatives and the gravitino field $(\psi_\mu{}^\alpha, \overline{\psi}_{\mu\dot\alpha})$, using the relations presented in Table 3.4. Converting vector to spinor indices as in (B.7), we can rewrite the first term of the right-hand side of (3.131) as

$$R^{\mu\nu}{}_{\mu\nu}\Big| = \frac{1}{4}R^{\alpha\dot\alpha\beta\dot\beta}{}_{\alpha\dot\alpha\beta\dot\beta}\Big| = 2(X^{\alpha\beta}{}_{\alpha\beta} + \overline{X}^{\dot\alpha\dot\beta}{}_{\dot\alpha\dot\beta})\Big| ,$$

where the contributions of the tensors $\Psi_{(\alpha\beta)(\dot\alpha\dot\beta)}$ and $\overline{\Psi}_{(\dot\alpha\dot\beta)(\alpha\beta)}$ vanish after being contracted with a rank-two antisymmetric tensor, since they are symmetric in both their pairs of indices. Inserting the definitions of the tensors $X_{(\alpha\beta)(\gamma\delta)}$ and $\overline{X}_{(\dot\alpha\dot\beta)(\dot\gamma\dot\delta)}$ and observing that the terms involving the derivatives of the superfields W and $\overline{W}$ cancel due to their symmetry properties, we get

$$R^{\mu\nu}{}_{\mu\nu}\Big| = -48\mathcal{R}\mathcal{R}^\dagger\Big| + 6G_\mu G^\mu\Big| + \frac{3}{2}\Big(\mathcal{D}\cdot\mathcal{D}\mathcal{R} + \overline{\mathcal{D}}\cdot\overline{\mathcal{D}}\mathcal{R}^\dagger\Big)\Big| .$$

Along the same lines, we can calculate the second term of the right-hand side of (3.131), which, after converting vector indices to spinor indices, reads

$$\begin{aligned} \psi_\mu{}^\alpha R_{\alpha\nu}{}^{\mu\nu}\Big| &= \frac{1}{2}\varepsilon^{\dot\gamma\dot\epsilon}\varepsilon^{\gamma\epsilon}\varepsilon^{\dot\beta\dot\delta}\varepsilon^{\beta\delta}\psi_{\beta\dot\beta}{}^\alpha\Big(\varepsilon_{\delta\epsilon}R_{\alpha\gamma\dot\gamma\dot\delta\dot\epsilon}\Big| - \varepsilon_{\dot\delta\dot\epsilon}R_{\alpha\gamma\dot\gamma\delta\epsilon}\Big|\Big) \\ &= i\psi^{\alpha\dot\beta\gamma}\Big[-\mathcal{D}_\gamma G_{\alpha\dot\beta}\Big| + \varepsilon_{\gamma\alpha}\overline{\mathcal{D}}_{\dot\beta}\mathcal{R}^\dagger\Big|\Big] , \end{aligned}$$

using the definitions of Table 3.4, the constraints of Table 3.2 as well as (3.43). Reconverting spinor to vector indices and adding the third term of the right-hand side of (3.131), which is simply obtained by hermitian conjugation, we then have

$$\psi_\mu{}^\alpha R_{\alpha\nu}{}^{\mu\nu}\Big| + \bar\psi_{\mu\dot\alpha} R^{\dot\alpha}{}_\nu{}^{\mu\nu}\Big| = -2i\Big(\psi_\mu{}^\alpha \mathcal{D}_\alpha - \bar\psi_{\mu\dot\alpha}\overline{\mathcal{D}}^{\dot\alpha}\Big)G^\mu\Big| \\ -i\Bigg[(\psi_\mu\sigma^\mu)_{\dot\alpha}\overline{\mathcal{D}}^{\dot\alpha}\mathcal{R}^\dagger\Big| + (\bar\psi_\mu\bar\sigma^\mu)^\alpha\mathcal{D}_\alpha\mathcal{R}\Big|\Bigg].$$

It is immediate to calculate the fourth term of the right-hand side of (3.131), which reads

$$\frac{1}{4}\psi_\nu{}^\beta\psi_\mu{}^\alpha R_{\alpha\beta}{}^{\mu\nu}\Big| = \frac{1}{8}\psi_\nu{}^\beta\psi_\mu{}^\alpha\bar\sigma^{\mu\dot\gamma\gamma}\bar\sigma^{\nu\dot\delta\delta}\Bigg[\varepsilon_{\gamma\delta}R_{\alpha\beta\dot\gamma\dot\delta}\Big| - \varepsilon_{\dot\gamma\dot\delta}R_{\alpha\beta\gamma\delta}\Big|\Bigg] = 2\psi_\mu\sigma^{\mu\nu}\psi_\nu\mathcal{R}^\dagger\Big|,$$

where we have again used the relations of Table 3.4. Summing with the (hermitian conjugate) fifth term, we obtain,

$$\frac{1}{4}\psi_\nu{}^\beta\psi_\mu{}^\alpha R_{\alpha\beta}{}^{\mu\nu}\Big| + \frac{1}{4}\bar\psi_{\nu\dot\beta}\bar\psi_{\mu\dot\alpha}R^{\dot\alpha\dot\beta\mu\nu}\Big| = 2\Bigg(\bar\psi_\mu\bar\sigma^{\mu\nu}\bar\psi_\nu\mathcal{R}\Big| + \psi_\mu\sigma^{\mu\nu}\psi_\nu\mathcal{R}^\dagger\Big|\Bigg).$$

Finally, the last term of the right-hand side of (3.131) is computed in the same way, and gives

$$-\frac{1}{2}\psi_\mu{}^\alpha\bar\psi_{\nu\dot\alpha}R_\alpha{}^{\dot\alpha\mu\nu}\Big| = \frac{1}{4}\psi_\mu^\alpha\bar\psi_\nu^{\dot\alpha}\bar\sigma^{\mu\dot\gamma\gamma}\bar\sigma^{\nu\dot\delta\delta}\Bigg[\varepsilon_{\gamma\delta}R_{\alpha\dot\alpha\dot\gamma\dot\delta}\Big| - \varepsilon_{\dot\gamma\dot\delta}R_{\alpha\dot\alpha\gamma\delta}\Big|\Bigg] \\ = \Big[\psi_\mu{}^\alpha(\bar\sigma^{\mu\nu}\bar\psi_\nu)^{\dot\alpha} + (\psi_\mu\sigma^{\mu\nu})^\alpha\bar\psi_\nu{}^{\dot\alpha}\Big]G_{\alpha\dot\alpha}\Big|.$$

Collecting all the terms, we establish the relation

$$\begin{aligned} e_\mu{}^{\tilde\mu}e_\nu{}^{\tilde\nu}R_{\tilde\mu\tilde\nu}{}^{\mu\nu}\Big| = &-48\mathcal{R}\mathcal{R}^\dagger\Big| + 6G_\mu G^\mu\Big| + \frac{3}{2}\Big[\mathcal{D}\cdot\mathcal{D}\mathcal{R} + \overline{\mathcal{D}}\cdot\overline{\mathcal{D}}\mathcal{R}^\dagger\Big]\Big| \\ &- 2i\Big[\psi_\mu{}^\alpha\mathcal{D}_\alpha - \bar\psi_{\mu\dot\alpha}\overline{\mathcal{D}}^{\dot\alpha}\Big]G^\mu\Big| \\ &- i\Big[(\psi_\mu\sigma^\mu)_{\dot\alpha}\overline{\mathcal{D}}^{\dot\alpha}\mathcal{R}^\dagger + (\bar\psi_\mu\bar\sigma^\mu)^\alpha\mathcal{D}_\alpha\mathcal{R}\Big]\Big| \\ &+ 2\Big[\bar\psi_\mu\bar\sigma^{\mu\nu}\bar\psi_\nu\mathcal{R} + \psi_\mu\sigma^{\mu\nu}\psi_\nu\mathcal{R}^\dagger\Big]\Big| \\ &+ \Big[\psi_\mu{}^\alpha(\bar\sigma^{\mu\nu}\bar\psi_\nu)^{\dot\alpha} + (\psi_\mu\sigma^{\mu\nu})^\alpha\bar\psi_\nu{}^{\dot\alpha}\Big]G_{\alpha\dot\alpha}\Big|, \end{aligned} \tag{3.132}$$

which allows for the computation of the sum of the second order derivatives of the superfield $\mathcal{R}$. From (3.47), we can calculate the difference between these derivatives,

$$\mathcal{D}\cdot\mathcal{D}\mathcal{R} - \overline{\mathcal{D}}\cdot\overline{\mathcal{D}}\mathcal{R}^\dagger = -\{\mathcal{D}^\alpha, \overline{\mathcal{D}}^{\dot\alpha}\}G_{\alpha\dot\alpha} \\ = -\bar\sigma^{\mu\dot\alpha\alpha}\Big[T_{\alpha\dot\alpha}{}^\nu\mathcal{D}_\nu G_\mu - G^\nu R_{\alpha\dot\alpha\nu\mu}\Big] = 4i\mathcal{D}_\mu G^\mu,$$

where we have used the results (3.53). This, after combining this last equation with (3.132), allows us to derive $\mathcal{D}\cdot\mathcal{D}\mathcal{R}|$ [Wess and Bagger (1992)],

$$\begin{aligned}\mathcal{D}\cdot\mathcal{D}\mathcal{R}\Big| &= \frac{1}{3}e_\mu{}^{\tilde\mu}e_\nu{}^{\tilde\nu}R_{\tilde\mu\tilde\nu}{}^{\mu\nu}\Big| + 16\mathcal{R}\mathcal{R}^\dagger\Big| - 2G^\mu G_\mu\Big| - \frac{2}{3}\Big[\psi_\mu\sigma^{\mu\nu}\psi_\nu\mathcal{R}^\dagger\Big| + \bar\psi_\mu\bar\sigma^{\mu\nu}\bar\psi_\nu\mathcal{R}\Big|\Big]\\ &-\frac{1}{3}\Big[\psi_\mu{}^\alpha(\bar\sigma^{\mu\nu}\bar\psi_\nu)^{\dot\alpha} + (\psi_\mu\sigma^{\mu\nu})^\alpha\bar\psi_\nu{}^{\dot\alpha}\Big]G_{\alpha\dot\alpha}\Big| + \frac{i}{3}\Big[(\psi_\mu\sigma^\mu)_{\dot\alpha}\overline{\mathcal{D}}^{\dot\alpha}\mathcal{R}^\dagger\Big|\\ &+(\bar\psi_\mu\bar\sigma^\mu)^\alpha\mathcal{D}_\alpha\mathcal{R}\Big|\Big] + \frac{2}{3}i\Big[\psi_\mu{}^\alpha\mathcal{D}_\alpha - \bar\psi_{\mu\dot\alpha}\overline{\mathcal{D}}^{\dot\alpha}\Big]G^\mu\Big| + 2i\mathcal{D}_\mu G^\mu\Big|\,. \end{aligned}\tag{3.133}$$

Replacing the lowest order coefficients of the superfields $\mathcal{R}$ and G, as well as those of their derivatives, by the expressions previously calculated, (3.133) can be further simplified. The first line of (3.133) can, adopting the gauge choice presented in Table 3.6, be rewritten as

$$\frac{1}{3}e_\mu{}^{\tilde\mu}e_\nu{}^{\tilde\nu}R_{\tilde\mu\tilde\nu}{}^{\mu\nu}\Big| + \frac{4}{9}MM^* - \frac{2}{9}b_\mu b^\mu + \frac{1}{9}\Big[\psi^\mu\sigma_{\mu\nu}\psi^\nu M^* + \bar\psi^\mu\bar\sigma_{\mu\nu}\bar\psi^\nu M\Big]\,. \tag{3.134}$$

Using the identities (C.5) as well as the relations of Table 3.6, one can rewrite the first term of the second line of Eq. (3.133) as

$$-\frac{i}{9}\varepsilon^{\mu\nu\rho\sigma}\psi_\mu\sigma_\sigma\bar\psi_\nu b_\rho\,.$$

The identities (C.5), together with the expressions of the lowest order components of the first derivatives of the superfield $\mathcal{R}$ given in (3.129) and (3.130), allow us to reduce the second term of the second line of (3.133) to

$$\begin{aligned}&\frac{1}{18}\varepsilon^{\mu\nu\rho\sigma}\Big[\bar\psi_\mu\bar\sigma_\sigma\psi_{\nu\rho} - \psi_\mu\sigma_\sigma\bar\psi_{\nu\rho}\Big] - \frac{i}{9}\Big[\psi^\mu\sigma^\nu\bar\psi_{\mu\nu} + \bar\psi^\mu\bar\sigma^\nu\psi_{\mu\nu}\Big]\\ &-\frac{1}{18}\Big[\psi_\mu\cdot\psi^\mu M^* + \bar\psi_\mu\cdot\bar\psi^\mu M\Big] - \frac{1}{9}\Big[\psi_\mu\sigma^{\mu\nu}\psi_\nu M^* + \bar\psi_\mu\bar\sigma^{\mu\nu}\bar\psi_\nu M\Big]\\ &-\frac{1}{18}\Big[\psi_\nu\sigma^\nu\bar\psi^\mu - \bar\psi_\nu\bar\sigma^\nu\psi^\mu\Big]b_\mu\,.\end{aligned}\tag{3.135}$$

In order to compute the last two terms of (3.133), we recall that[6]

$$\begin{aligned}2i\mathcal{D}_\mu G^\mu\Big| &= 2iE_\mu{}^{\tilde\mu}\mathcal{D}_{\tilde\mu}G^\mu\Big| + 2i\Big[E_\mu{}^{\tilde\alpha}\mathcal{D}_{\tilde\alpha}G^\mu\Big| + E_{\mu\tilde{\dot\alpha}}\overline{\mathcal{D}}^{\tilde{\dot\alpha}}G^\mu\Big|\Big]\\ &= -\frac{2i}{3}e_\mu{}^{\tilde\mu}\mathcal{D}_{\tilde\mu}b^\mu - i\Big[\psi_\mu{}^\alpha\mathcal{D}_\alpha G^\mu\Big| + \bar\psi_{\mu\dot\alpha}\overline{\mathcal{D}}^{\dot\alpha}G^\mu\Big|\Big]\,.\end{aligned}\tag{3.136}$$

The last two terms of the equation above can be combined and calculated together with the first term of the last line of (3.133). This is performed with the help of

[6]These manipulations regarding the covariant derivatives in curved and flat spaces are central in supergravity and are investigated in Chapter 4.

(3.128), as well as again with the identities (C.5). Converting spinor indices to vector indices as in (B.7), we obtain

$$\begin{aligned}\frac{i}{3}\Big[5\bar{\psi}_\mu{}^{\dot\delta}\,\bar{\mathcal{D}}_{\dot\delta}G^\mu\Big| - \psi_\mu{}^\delta\,\mathcal{D}_\delta G^\mu\Big|\Big] = &-\frac{1}{36}\varepsilon^{\mu\nu\rho\sigma}\Big[\psi_\mu\sigma_\sigma\bar{\psi}_{\nu\rho} + 5\bar{\psi}_\mu\bar{\sigma}_\sigma\psi_{\nu\rho}\Big] \\ &-\frac{i}{9}\Big[\psi^\nu\sigma^\mu\bar{\psi}_{\mu\nu} - 5\bar{\psi}^\nu\bar{\sigma}^\mu\psi_{\mu\nu}\Big] \\ &+\frac{1}{18}\Big[\psi_\mu\cdot\psi^\mu M^* - 5\bar{\psi}_\mu\cdot\bar{\psi}^\mu M\Big] \\ &+\frac{i}{9}\varepsilon^{\mu\nu\rho\sigma}\psi_\mu\sigma_\sigma\bar{\psi}_\nu b_\rho \\ &+\frac{1}{18}\Big[\psi_\mu\sigma^\nu\bar{\psi}_\nu - 5\psi_\nu\sigma^\nu\bar{\psi}_\mu\Big]b^\mu\,. \end{aligned} \tag{3.137}$$

Collecting the results (3.134), (3.135), (3.136) and (3.137), we derive the lowest order component of the second order derivative of the superfield $\mathcal{R}$,

$$\begin{aligned}\mathcal{D}\cdot\mathcal{D}\mathcal{R}\Big| = &\frac{1}{3}e_\mu{}^{\tilde\mu}e_\nu{}^{\tilde\nu}R_{\tilde\mu\tilde\nu}{}^{\mu\nu}\Big| + \frac{4}{9}MM^* - \frac{2}{9}b_\mu b^\mu - \frac{2i}{3}e_\mu{}^{\tilde\mu}\mathcal{D}_{\tilde\mu}b^\mu \\ &-\frac{1}{3}\bar{\psi}_\mu\cdot\bar{\psi}^\mu M - \frac{1}{3}\psi_\nu\sigma^\nu\bar{\psi}_\mu b^\mu - \frac{2i}{3}\bar{\psi}^\mu\bar{\sigma}^\nu\psi_{\mu\nu} \\ &-\frac{1}{12}\varepsilon^{\mu\nu\rho\sigma}\Big[\psi_\mu\sigma_\sigma\bar{\psi}_{\nu\rho} + \bar{\psi}_\mu\bar{\sigma}_\sigma\psi_{\nu\rho}\Big]\,. \end{aligned} \tag{3.138}$$

Hermitian conjugation (or equivalently, a calculation similar to the one above), yields the second-order derivative of the superfield $\mathcal{R}^\dagger$,

$$\begin{aligned}\bar{\mathcal{D}}\cdot\bar{\mathcal{D}}\mathcal{R}^\dagger\Big| = &\frac{1}{3}e_\mu{}^{\tilde\mu}e_\nu{}^{\tilde\nu}R_{\tilde\mu\tilde\nu}{}^{\mu\nu}\Big| + \frac{4}{9}MM^* - \frac{2}{9}b_\mu b^\mu + \frac{2i}{3}e_\mu{}^{\tilde\mu}\mathcal{D}_{\tilde\mu}b^\mu \\ &-\frac{1}{3}\psi_\mu\cdot\psi^\mu M^* - \frac{1}{3}\psi_\mu\sigma^\nu\bar{\psi}_\nu b^\mu - \frac{2i}{3}\psi^\mu\sigma^\nu\bar{\psi}_{\mu\nu} \\ &+\frac{1}{12}\varepsilon^{\mu\nu\rho\sigma}\Big[\psi_\mu\sigma_\sigma\bar{\psi}_{\nu\rho} + \bar{\psi}_\mu\bar{\sigma}_\sigma\psi_{\nu\rho}\Big]\,. \end{aligned} \tag{3.139}$$

3.5 Transformation laws for the components of the supergravity multiplet

In this section, we derive the variations of the component fields of the supergravity multiplet under the supergravity transformations described in Sec. 3.3. Adopting the gauge choice of Table 3.6, the transformations are parameterised as in Table 3.7. In this case, the variations of the fields can be written as functions of the zeroth order components of the vielbein as well as of the superfields $\mathcal{R}$ and $G_{\alpha\dot\alpha}$ and their derivatives.

Table 3.8 Lowest order components of the superfields $\mathcal{R}, G_\mu, W_{(\alpha\beta\gamma)}$, and their derivatives.

All the torsion and curvature tensors can be expressed as functions of the superfields $\mathcal{R}$, G_μ and $W_{(\alpha\beta\gamma)}$, as well as of their derivatives. The latter can be written using the component fields of the gravitation supermultiplet, *i.e.*, the graviton field $e_{\tilde{\mu}}{}^\mu$, the gravitino field $(\psi_{\tilde{\mu}}{}^\alpha, \bar{\psi}_{\tilde{\mu}\dot{\alpha}})$ and the two auxiliary fields M and b^μ,

$$
\begin{aligned}
\mathcal{R}\Big| &= -\frac{1}{6}M\,,\\
\mathcal{R}^\dagger\Big| &= -\frac{1}{6}M^*\,,\\
\mathcal{D}_\alpha\mathcal{R}\Big| &= -\frac{1}{3}(\sigma^{\mu\nu}\psi_{\mu\nu})_\alpha + \frac{i}{6}(\sigma^\mu\bar{\psi}_\mu)_\alpha M - \frac{i}{6}\psi_{\mu\alpha}b^\mu\,,\\
\overline{\mathcal{D}}^{\dot{\alpha}}\mathcal{R}^\dagger\Big| &= -\frac{1}{3}(\bar{\sigma}^{\mu\nu}\bar{\psi}_{\mu\nu})^{\dot{\alpha}} + \frac{i}{6}(\bar{\sigma}^\mu\psi_\mu)^{\dot{\alpha}}M^* + \frac{i}{6}\bar{\psi}_\mu{}^{\dot{\alpha}}b^\mu\,,\\
\mathcal{D}\cdot\mathcal{D}\mathcal{R}\Big| &= \frac{1}{3}e_\mu{}^{\tilde{\mu}}e_\nu{}^{\tilde{\nu}}R_{\tilde{\mu}\tilde{\nu}}{}^{\mu\nu}\Big| + \frac{4}{9}MM^* - \frac{2}{9}b_\mu b^\mu - \frac{2i}{3}e_\mu{}^{\tilde{\mu}}\mathcal{D}_{\tilde{\mu}}b^\mu - \frac{1}{3}\bar{\psi}_\mu\cdot\bar{\psi}^\mu M - \frac{1}{3}\psi_\nu\sigma^\nu\bar{\psi}_\mu b^\mu\\
&\quad -\frac{2i}{3}\bar{\psi}^\mu\bar{\sigma}^\nu\psi_{\mu\nu} - \frac{1}{12}\varepsilon^{\mu\nu\rho\sigma}\big[\psi_\mu\sigma_\sigma\bar{\psi}_{\nu\rho} + \bar{\psi}_\mu\bar{\sigma}_\sigma\psi_{\nu\rho}\big]\,,\\
\overline{\mathcal{D}}\cdot\overline{\mathcal{D}}\mathcal{R}^\dagger\Big| &= \frac{1}{3}e_\mu{}^{\tilde{\mu}}e_\nu{}^{\tilde{\nu}}R_{\tilde{\mu}\tilde{\nu}}{}^{\mu\nu}\Big| + \frac{4}{9}MM^* - \frac{2}{9}b_\mu b^\mu + \frac{2i}{3}e_\mu{}^{\tilde{\mu}}\mathcal{D}_{\tilde{\mu}}b^\mu - \frac{1}{3}\psi_\mu\cdot\psi^\mu M^* - \frac{1}{3}\psi_\mu\sigma^\nu\bar{\psi}_\nu b^\mu\\
&\quad -\frac{2i}{3}\psi^\mu\sigma^\nu\bar{\psi}_{\mu\nu} + \frac{1}{12}\varepsilon^{\mu\nu\rho\sigma}\big[\psi_\mu\sigma_\sigma\bar{\psi}_{\nu\rho} + \bar{\psi}_\mu\bar{\sigma}_\sigma\psi_{\nu\rho}\big]\,,\\
G_{\alpha\dot{\alpha}}\Big| &= -\frac{1}{3}b_{\alpha\dot{\alpha}}\,,\\
\mathcal{D}_\delta G_{\gamma\dot{\alpha}}\Big| &= -\frac{1}{4}\bar{\psi}_\delta{}^{\dot{\epsilon}}{}_{\gamma\dot{\epsilon}\dot{\alpha}} - \frac{1}{12}\varepsilon_{\delta\gamma}\bar{\psi}^{\epsilon\dot{\epsilon}}{}_{\epsilon\dot{\alpha}\dot{\epsilon}} + \frac{i}{6}\psi_{\gamma\dot{\alpha}\delta}M^* + \frac{i}{12}\Big(\bar{\psi}_{\delta\dot{\epsilon}\dot{\alpha}}b_\gamma{}^{\dot{\epsilon}} + \bar{\psi}_{\gamma\dot{\epsilon}}{}^{\dot{\epsilon}}b_{\delta\dot{\alpha}} + \bar{\psi}_{\delta\dot{\epsilon}}{}^{\dot{\epsilon}}b_{\gamma\dot{\alpha}}\Big)\,,\\
\overline{\mathcal{D}}_{\dot{\delta}}G_{\alpha\dot{\gamma}}\Big| &= -\frac{1}{4}\psi^\epsilon{}_{\dot{\delta}\epsilon\dot{\gamma}\alpha} - \frac{1}{12}\varepsilon_{\dot{\delta}\dot{\gamma}}\psi_\alpha{}^{\dot{\epsilon}\epsilon}{}_{\dot{\epsilon}\epsilon} - \frac{i}{6}\bar{\psi}_{\alpha\dot{\gamma}\dot{\delta}}M - \frac{i}{12}\Big(\psi_{\epsilon\dot{\delta}\alpha}b^\epsilon{}_{\dot{\gamma}} + \psi_{\epsilon\dot{\gamma}}{}^\epsilon b_{\alpha\dot{\delta}} + \psi_{\epsilon\dot{\delta}}{}^\epsilon b_{\alpha\dot{\gamma}}\Big)\,,\\
\mathcal{D}_\delta G^\mu\Big| &= \frac{1}{3}(\sigma^\nu\bar{\psi}_\nu{}^\mu)_\delta - \frac{i}{12}\varepsilon^{\mu\nu\rho\sigma}(\sigma_\sigma\bar{\psi}_{\nu\rho})_\delta + \frac{i}{6}\psi^\mu{}_\delta M^* + \frac{i}{6}(\sigma^\nu\bar{\psi}_\nu)_\delta b^\mu + \frac{1}{12}\varepsilon^{\mu\nu\rho\sigma}(\sigma_\sigma\bar{\psi}_\nu)_\delta\, b_\rho\,,\\
\overline{\mathcal{D}}^{\dot{\delta}}G^\mu\Big| &= -\frac{1}{3}(\bar{\sigma}^\nu\psi_\nu{}^\mu)^{\dot{\delta}} - \frac{i}{12}\varepsilon^{\mu\nu\rho\sigma}(\bar{\sigma}_\sigma\psi_{\nu\rho})^{\dot{\delta}} - \frac{i}{6}\bar{\psi}^{\mu\dot{\delta}}M + \frac{i}{6}(\bar{\sigma}^\nu\psi_\nu)^{\dot{\delta}}b^\mu - \frac{1}{12}\varepsilon^{\mu\nu\rho\sigma}(\bar{\sigma}_\sigma\psi_\nu)^{\dot{\delta}}\, b_\rho\,,\\
W_{(\alpha_1\alpha_2\alpha_3)}\Big| &= -\frac{1}{48}\sum_{\tau\in S_3}\Big(\psi_{\alpha_{\tau(1)}\dot{\delta}\alpha_{\tau(2)}}{}^{\dot{\delta}}{}_{\alpha_{\tau(3)}} - i\psi_{\alpha_{\tau(1)}\dot{\delta}\alpha_{\tau(2)}}b_{\alpha_{\tau(3)}}{}^{\dot{\delta}}\Big)\,,\\
\overline{W}_{(\dot{\alpha}_1\dot{\alpha}_2\dot{\alpha}_3)}\Big| &= -\frac{1}{48}\sum_{\tau\in S_3}\Big(\bar{\psi}_{\delta\dot{\alpha}_{\tau(1)}}{}^\delta{}_{\dot{\alpha}_{\tau(2)}\dot{\alpha}_{\tau(3)}} + i\bar{\psi}_{\delta\dot{\alpha}_{\tau(1)}\dot{\alpha}_{\tau(2)}}b^\delta{}_{\dot{\alpha}_{\tau(3)}}\Big)\,.
\end{aligned}
$$

The second relation in (3.100) reduces, for the case of the graviton field, to

$$
\delta e_{\tilde{\mu}}{}^\mu = \delta E_{\tilde{\mu}}{}^\mu\Big| = -\mathcal{D}_{\tilde{\mu}}\xi^\mu\Big| + \xi^M T_{M\tilde{\mu}}{}^\mu\Big| + E_{\tilde{\mu}}{}^\nu L_\nu{}^\mu\Big| = \xi^\beta T_{\beta\tilde{\mu}}{}^\mu\Big| + \bar{\xi}_{\dot{\beta}}T^{\dot{\beta}}{}_{\tilde{\mu}}{}^\mu\Big|\,, \tag{3.140}
$$

where we have used the transformation parameters given in Table 3.7. The torsion tensors can be calculated as

$$
\begin{aligned}
T_{\alpha\tilde{\mu}}{}^{\mu}\Big| &= (-)^{|\alpha|(|\tilde{\mu}|+|M|)} E_{\tilde{\mu}}{}^{M} T_{\alpha M}{}^{\mu}\Big| = -E_{\tilde{\mu}\dot{\alpha}} T_{\alpha}{}^{\dot{\alpha}\mu}\Big| \\
&= (\frac{1}{2}\psi_{\tilde{\mu}}{}^{\dot{\alpha}})(-2i\sigma^{\mu}{}_{\alpha\dot{\alpha}}) = -i(\sigma^{\mu}\bar{\psi}_{\tilde{\mu}})_{\alpha}\,, \\
T^{\dot{\alpha}}{}_{\tilde{\mu}}{}^{\mu}\Big| &= (-)^{|\dot{\alpha}|(|\tilde{\mu}|+|M|)} E_{\tilde{\mu}}{}^{M} T^{\dot{\alpha}}{}_{M}{}^{\mu}\Big| = -E_{\tilde{\mu}}{}^{\alpha} T^{\dot{\alpha}}{}_{\alpha}{}^{\mu}\Big| \\
&= -(\frac{1}{2}\psi_{\tilde{\mu}}{}^{\alpha})\varepsilon^{\dot{\alpha}\dot{\beta}}(-2i\sigma^{\mu}{}_{\alpha\dot{\beta}}) = -i(\bar{\sigma}^{\mu}\psi_{\tilde{\mu}})^{\dot{\alpha}}\,,
\end{aligned}
$$

under the constraints (3.29) and the results of Table 3.6. Therefore, the variation of the graviton reads [Wess and Zumino (1978a); Wess and Bagger (1992)]

$$
\delta e_{\tilde{\mu}}{}^{\mu} = -i\Big[\varepsilon(x)\sigma^{\mu}\bar{\psi}_{\tilde{\mu}} - \psi_{\tilde{\mu}}\sigma^{\mu}\bar{\varepsilon}(x)\Big]\,. \tag{3.141}
$$

In addition, since $e_{\tilde{\mu}}{}^{\mu}e_{\mu}{}^{\tilde{\nu}} = \delta_{\tilde{\mu}}{}^{\tilde{\nu}}$, we get

$$
\delta e_{\mu}{}^{\tilde{\mu}} = -\delta(e_{\tilde{\rho}}{}^{\nu})e_{\mu}{}^{\tilde{\rho}}e_{\nu}{}^{\tilde{\mu}} = ie_{\nu}{}^{\tilde{\mu}}e_{\mu}{}^{\tilde{\rho}}\Big[\varepsilon(x)\sigma^{\nu}\bar{\psi}_{\tilde{\rho}} - \psi_{\tilde{\rho}}\sigma^{\nu}\bar{\varepsilon}(x)\Big]\,. \tag{3.142}
$$

We now turn to the gravitino field. Similarly to (3.140), we have

$$
\begin{aligned}
\delta\psi_{\tilde{\mu}}{}^{\alpha} = 2\delta E_{\tilde{\mu}}{}^{\alpha}\Big| &= -2\mathcal{D}_{\tilde{\mu}}\xi^{\alpha}\Big| + 2\xi^{M}T_{M\tilde{\mu}}{}^{\alpha}\Big| + 2E_{\tilde{\mu}}{}^{\beta}L_{\beta}{}^{\alpha}\Big| \\
&= -2\mathcal{D}_{\tilde{\mu}}\xi^{\alpha}\Big| + 2\xi^{\beta}T_{\beta\tilde{\mu}}{}^{\alpha}\Big| + 2\bar{\xi}_{\dot{\beta}}T^{\dot{\beta}}{}_{\tilde{\mu}}{}^{\alpha}\Big|\,.
\end{aligned}
$$

The calculation of the torsion tensors, using Table 3.3, leads to

$$
\begin{aligned}
T_{\beta\tilde{\mu}}{}^{\alpha}\Big| &= (-)^{|\beta|(|\tilde{\mu}|+|M|)}E_{\tilde{\mu}}{}^{M}T_{\beta M}{}^{\alpha}\Big| = \frac{1}{2}e_{\tilde{\mu}}{}^{\nu}\bar{\sigma}_{\nu}{}^{\dot{\gamma}\gamma}\varepsilon^{\alpha\delta}\,T_{\beta\gamma\dot{\gamma}\delta}\Big| \\
&= \frac{i}{6}e_{\tilde{\mu}}{}^{\nu}\Big[(\sigma_{\rho}\bar{\sigma}_{\nu})_{\beta}{}^{\alpha}\,b^{\rho} - 3\delta_{\beta}{}^{\alpha}b_{\nu}\Big]\,, \\
T^{\dot{\beta}}{}_{\tilde{\mu}}{}^{\alpha}\Big| &= (-)^{|\dot{\beta}|(|\tilde{\mu}|+|M|)}E_{\tilde{\mu}}{}^{M}T^{\dot{\beta}}{}_{M}{}^{\alpha}\Big| = \frac{1}{2}e_{\tilde{\mu}}{}^{\nu}\bar{\sigma}_{\nu}^{\dot{\gamma}\gamma}\,\varepsilon^{\dot{\beta}\dot{\delta}}\varepsilon^{\alpha\delta}T_{\dot{\delta}\gamma\dot{\gamma}\delta}\Big| \\
&= \frac{i}{6}e_{\tilde{\mu}}{}^{\nu}\bar{\sigma}_{\nu}{}^{\dot{\beta}\alpha}M\,,
\end{aligned}
\tag{3.143}
$$

where we have used (B.7) to convert spinor to vector indices and *vice versa*. Altogether, this allows us to write the variation of the gravitino field under a supergravity transformation, as [Wess and Zumino (1978a); Wess and Bagger (1992)]

$$
\begin{aligned}
\delta\psi_{\tilde{\mu}}{}^{\alpha} &= -2\mathcal{D}_{\tilde{\mu}}\varepsilon^{\alpha}(x) + \frac{i}{3}e_{\tilde{\mu}}{}^{\nu}(\bar{\varepsilon}(x)\bar{\sigma}_{\nu})^{\alpha}M + \frac{i}{3}e_{\tilde{\mu}}{}^{\nu}\Big[(\varepsilon(x)\sigma_{\rho}\bar{\sigma}_{\nu})^{\alpha}b^{\rho} - 3\varepsilon^{\alpha}(x)\,b_{\nu}\Big]\,, \\
\delta\bar{\psi}_{\tilde{\mu}\dot{\alpha}} &= -2\mathcal{D}_{\tilde{\mu}}\bar{\varepsilon}_{\dot{\alpha}}(x) + \frac{i}{3}e_{\tilde{\mu}}{}^{\nu}(\varepsilon(x)\sigma_{\nu})_{\dot{\alpha}}M^{*} - \frac{i}{3}e_{\tilde{\mu}}{}^{\nu}\Big[(\bar{\varepsilon}(x)\bar{\sigma}_{\rho}\sigma_{\nu})_{\dot{\alpha}}b^{\rho} - 3\bar{\varepsilon}_{\dot{\alpha}}(x)\,b_{\nu}\Big]\,,
\end{aligned}
\tag{3.144}
$$

where the second relation has been derived along the same lines (or has equivalently been obtained by hermitian conjugation of the variation $\delta\psi_{\tilde{\mu}}{}^{\alpha}$). The variations of the auxiliary fields are given by (3.102), and using (3.128) and (3.129), we get [Wess and Zumino (1978a); Wess and Bagger (1992)]

$$
\begin{aligned}
\delta M &= -6\delta\mathcal{R}\big| = 6\xi^M\mathcal{D}_M\mathcal{R}\big| = 6\xi^\alpha\mathcal{D}_\alpha\mathcal{R}\big| \\
&= -2(\varepsilon\sigma^{\mu\nu}\psi_{\mu\nu}) + i(\varepsilon\sigma^\mu\bar{\psi}_\mu)M - i(\varepsilon\cdot\psi_\mu)b^\mu\,, \\
\delta M^* &= -6\delta\mathcal{R}^\dagger\big| = 6\xi^M\mathcal{D}_M\mathcal{R}^\dagger\big| = 6\bar{\xi}_{\dot\alpha}\bar{\mathcal{D}}^{\dot\alpha}\mathcal{R}^\dagger\big| \\
&= -2(\bar{\varepsilon}\bar{\sigma}^{\mu\nu}\bar{\psi}_{\mu\nu}) + i(\bar{\varepsilon}\bar{\sigma}^\mu\psi_\mu)M^* + i(\bar{\varepsilon}\cdot\bar{\psi}_\mu)b^\mu\,, \\
\delta b^\mu &= -3\delta G^\mu\big| = 3\xi^M\mathcal{D}_M G^\mu\big| - 3G^\nu L_\nu{}^\mu\big| = 3\xi^\alpha\mathcal{D}_\alpha G^\mu\big| + 3\bar{\xi}_{\dot\alpha}\bar{\mathcal{D}}^{\dot\alpha}G^\mu\big| \\
&= \varepsilon\sigma^\nu\bar{\psi}_\nu{}^\mu - \bar{\varepsilon}\bar{\sigma}^\nu\psi_\nu{}^\mu - \frac{i}{4}\varepsilon^{\mu\nu\rho\sigma}\Big[\varepsilon\sigma_\sigma\bar{\psi}_{\nu\rho} + \bar{\varepsilon}\bar{\sigma}_\sigma\psi_{\nu\rho}\Big] \\
&\quad + \frac{i}{2}\Big[\varepsilon\cdot\psi^\mu M^* - \bar{\varepsilon}\cdot\bar{\psi}^\mu M\Big] + \frac{i}{2}\Big[\varepsilon\sigma^\nu\bar{\psi}_\nu + \bar{\varepsilon}\bar{\sigma}^\nu\psi_\nu\Big]b^\mu \\
&\quad + \frac{1}{4}\varepsilon^{\mu\nu\rho\sigma}\Big[\varepsilon\sigma_\sigma\bar{\psi}_\nu - \bar{\varepsilon}\bar{\sigma}_\sigma\psi_\nu\Big]b_\rho\,.
\end{aligned}
\tag{3.145}
$$

Note that these transformation properties have already been obtained in [Stelle and West (1978c); Ferrara and van Nieuwenhuizen (1978a); Ferrara et al. (1977a)] and partially, *i.e.*, without the auxiliary fields in [Deser and Zumino (1976); Freedman et al. (1976)].

The connection transforms under a supergravity transformation as given by the last relation of Eq. (3.100), and thus reads as

$$
\delta\omega_{\tilde\mu\mu}{}^\nu = \delta\Omega_{\tilde\mu\mu}{}^\nu\big| = -\xi^M R_{M\tilde\mu\mu}{}^\nu\big| + \Omega_{\tilde\mu\mu}{}^\rho L_\rho{}^\nu\big| - L_\mu{}^\rho\Omega_{\tilde\mu\rho}{}^\nu\big| - \partial_{\tilde\mu}L_\mu{}^\nu\big| = -\xi^\alpha R_{\alpha\tilde\mu\mu}{}^\nu\big| - \bar{\xi}_{\dot\alpha}R^{\dot\alpha}{}_{\tilde\mu\mu}{}^\nu\big|\,.
$$

Using the various curvature tensors summarised in Table 3.4 and converting vector to spinor indices, the variation of the connection can be calculated explicitly and expressed in terms of the component fields of the gravitation multiplet, the superfields $\mathcal{R}$ and G and their derivatives. Note that one could proceed differently, starting from Eq. (3.119) and using the transformations derived in Eqs. (3.141) and (3.144). Since this result is irrelevant for the sequel of our work, it is not given here. Note that the variation of the connection was first given in [Deser and Zumino (1976); Ferrara et al. (1977a)] and then in [Ferrara and van Nieuwenhuizen (1978a)], see also [van Nieuwenhuizen (1981)].

Chapter 4

Superfields in curved superspace

Similarly to the case of global supersymmetry, the degrees of freedom included in an $N = 1$ chiral and vector supermultiplet can be embedded into chiral and vector superfields, respectively. In this chapter, we present a detailed calculation of the component fields associated to chiral and vector superfields and their superderivatives, together with their transformation properties under supergravity transformations. Projectors mapping antichiral (chiral) scalar superfields to chiral (antichiral) superfields are introduced. Gauge interactions of chiral superfields, *i.e.*, the way to couple vector and chiral superfields in a gauge and supergravity invariant fashion is also addressed.

4.1 Chiral superfields

Chiral and vector multiplets were independently introduced in supergravity within the tensor calculus approach [Stelle and West (1978b,a); Ferrara and van Nieuwenhuizen (1978b)], the superspace approach [Wess and Zumino (1978a)] or when constructing the first invariant Lagrangians [Ferrara et al. (1976b,a, 1977a); Cremmer et al. (1979, 1982, 1978b, 1983b)]. In this chapter, we introduce the superfields within the superspace approach, and follow [Wess and Zumino (1978a); Wess and Bagger (1992)]. In the context of supergravity, the superfields must be defined on the curved superspace. Due to the torsion constraints (3.29) it is possible to safely extend to supergravity the chiral and vector superfields introduced in supersymmetry. Actions, or equivalently Lagrangians, invariant under supergravity transformations can then be constructed with the help of the two above-mentioned superfields, together with the components of the gravitation supermultiplet, as will be shown in Chapter 6. It must be mention that in [Ogievetsky and Sokatchev (1980c,b,a,d)] an alternative superspace was introduce. However, we will not follow this alternative here.

Within the superspace formalism, chiral and antichiral supermultiplets are described by chiral and antichiral superfields, respectively. As seen in Sec. 2.2, a chiral superfield Φ and its associated antichiral superfield $\Phi^\dagger$ are defined, in the framework of global supersymmetry, by the constraints $\overline{D}_{\dot\alpha}\Phi = 0$ and $D_\alpha\Phi^\dagger = 0$, respectively, where the superderivatives are given by Eq. (2.1). In curved superspace, these constraints must be made covariant, and read

$$\begin{aligned} \overline{\mathcal{D}}_{\dot\alpha}\Phi &= 0\,, \\ \mathcal{D}_\alpha\Phi^\dagger &= 0\,, \end{aligned} \tag{4.1}$$

where regular superderivatives have been replaced by their covariant versions which however have flat indices. Due to the vanishing elements of the torsion tensor $T_{\dot\alpha\dot\beta}{}^\mu = T_{\dot\alpha\dot\beta}{}^\alpha = 0$, this constraint is compatible with supergravity transformations. Moreover, the torsion constraint ensure that (4.1) is compatible with the supergravity algebra [Gates et al. (1980)].

4.1.1 *Properties of chiral superfields and their superderivatives*

In this section, we focus on the analysis of the components of a chiral superfield Φ, an antichiral superfield $\Phi^\dagger$ and those of their derivatives up to the fourth order. This study will in turn be essential in order to define the components of the basic superfields necessary to construct the final action.

Definition of the component fields.

A chiral superfield Φ depends on the curved coordinates $z^{\tilde M} = (x^{\tilde\mu}, \theta^{\tilde\alpha}, \overline{\theta}_{\tilde{\dot\alpha}})$. Similarly to the calculations performed in Chapter 2, we could expand the superfield as a series in terms of the Grassmann variables. However, in this case, manipulations of superfield expressions become very unpractical, since it is mandatory to rewrite the series expansion in terms of variables with flat indices, which is highly cumbersome due to the unknown (complete) expansion of various quantities, such as the supervierbein where only the lowest order component has been derived, as shown in Table 3.6. A better approach starts from the fact that the three components of Φ, *i.e.*, the scalar, fermionic and auxiliary fields ϕ, χ_α and F, respectively, can be defined in analogy to (2.5), as [Wess and Zumino (1978a); Wess and Bagger (1992)][1]

$$\phi = \Phi\Big|\,, \qquad \chi_\alpha = \frac{1}{\sqrt{2}}\mathcal{D}_\alpha\Phi\Big|\,, \qquad F = \frac{1}{4}\mathcal{D}\cdot\mathcal{D}\Phi\Big|\,, \tag{4.2}$$

[1]It turns out that in Sec. 5.2 pertinent Grassmann variables will be introduced such that the field expansion of chiral superfields drastically simplifies.

By hermitian conjugation, the component fields of the antichiral superfield $\Phi^\dagger$ are given by

$$\phi^\dagger = \Phi^\dagger\Big| , \qquad \overline{\chi}^{\dot\alpha} = \frac{1}{\sqrt{2}}\overline{\mathcal{D}}^{\dot\alpha}\Phi^\dagger\Big| , \qquad F^\dagger = \frac{1}{4}\overline{\mathcal{D}}\cdot\overline{\mathcal{D}}\Phi^\dagger\Big| . \tag{4.3}$$

We emphasise again that the component fields always carry Lorentzian indices and not Einsteinian indices.

In many manipulations of chiral superfields defined on the curved superspace, the calculation of higher-order covariant derivatives with flat indices, such as the quantity $\overline{\mathcal{D}}_{\dot\alpha}\mathcal{D}_\alpha\mathcal{D}_\beta\Phi\big|$, is central. However, derivatives with Lorentz indices are only defined in terms of derivatives with Einstein indices, as shown in (3.5), and we therefore dedicate the rest of this section to the computation of the lowest order components of superderivatives (with flat indices) up to the fourth order of the chiral superfield Φ and the conjugate antichiral superfield $\Phi^\dagger$. The computations of the various derivatives are standard and follows [Wess and Bagger (1992)].

First-order superderivatives of chiral and antichiral superfields $\mathcal{D}_M\Phi\big|$ and $\mathcal{D}_M\Phi^\dagger\big|$.

Whilst the computation of $\mathcal{D}_\alpha\Phi\big|$ and $\overline{\mathcal{D}}^{\dot\alpha}\Phi\big|$ is trivial, due to the constraint (4.1) and to the definition of Eq. (4.2), the evaluation of $\mathcal{D}_\mu\Phi\big|$ must be deduced from

$$\mathcal{D}_\mu\Phi = E_\mu{}^{\tilde\mu}\mathcal{D}_{\tilde\mu}\Phi + E_\mu{}^{\tilde\alpha}\mathcal{D}_{\tilde\alpha}\Phi + E_{\mu\tilde{\dot\alpha}}\overline{\mathcal{D}}^{\tilde{\dot\alpha}}\Phi . \tag{4.4}$$

Adopting the gauge choice given in Table 3.6, for the lowest order component we obtain

$$\mathcal{D}_\mu\Phi\Big| = e_\mu{}^{\tilde\mu}\Big(\partial_{\tilde\mu}\phi - \frac{\sqrt{2}}{2}\psi_{\tilde\mu}\cdot\chi\Big) \equiv \hat{D}_\mu\phi , \tag{4.5}$$

where we have again used the relations (4.1) and (4.2), as well as Eq. (3.5) for the definition of the covariant superderivative $\mathcal{D}_\mu$. The different terms which appear in the equation above have been combined into a derivative of the scalar component of the considered chiral superfield, $\hat{D}_\mu\phi$, which is covariant with respect to the supergravity transformations through the additional term depending on the gravitino field.

To summarise, the first derivatives (with flat indices) of a chiral superfield Φ read

$$\begin{aligned} \mathcal{D}_\mu\Phi\Big| &= \hat{D}_\mu\phi , \\ \mathcal{D}_\alpha\Phi\Big| &= \sqrt{2}\chi_\alpha , \\ \overline{\mathcal{D}}^{\dot\alpha}\Phi\Big| &= 0 , \end{aligned} \tag{4.6}$$

whilst the ones of the antichiral superfield $\Phi^\dagger$, deduced by hermitian conjugation, are given by

$$\begin{aligned}
\mathcal{D}_\mu \Phi^\dagger\Big| &= \hat{D}_\mu \phi^\dagger = e_\mu{}^{\tilde{\mu}}\Big(\partial_{\tilde{\mu}}\phi^\dagger - \frac{\sqrt{2}}{2}\bar{\psi}_{\tilde{\mu}}\cdot\bar{\chi}\Big)\,, \\
\mathcal{D}_\alpha \Phi^\dagger\Big| &= 0\,, \\
\overline{\mathcal{D}}^{\dot{\alpha}}\Phi^\dagger\Big| &= \sqrt{2}\bar{\chi}^{\dot{\alpha}}
\end{aligned} \tag{4.7}$$

where we have introduced the covariant derivative $\hat{D}_\mu\phi^\dagger$ defined in analogy with $\hat{D}_\mu\phi$.

Second-order derivatives of chiral and antichiral superfields $\mathcal{D}_M\mathcal{D}_N\Phi\big|$ and $\mathcal{D}_M\mathcal{D}_N\Phi^\dagger\big|$.

Addressing the calculation of the lowest order component of the nine second-order derivatives $\mathcal{D}_M\mathcal{D}_N\Phi\big|$, we first note that the constraint (4.1) enforces three of these derivatives to vanish, $\mathcal{D}_M\overline{\mathcal{D}}_{\dot{\alpha}}\Phi\big| = 0$. We now turn to the computation of the three quantities $\mathcal{D}_M\mathcal{D}_\alpha\Phi\big|$, and start with $\mathcal{D}_\alpha\mathcal{D}_\beta\Phi\big|$, which can be directly deduced from the definition of the F-term in Eq. (4.2). Using, on the one hand, the antisymmetry property

$$X_\alpha Y_\beta - X_\beta Y_\alpha = \varepsilon_{\alpha\beta} X\cdot Y\,, \tag{4.8}$$

and on the other hand, Eq. (3.8) together with the constraints $T_{\alpha\beta}{}^\mu = T_{\alpha\beta}{}^{\underline{\gamma}} = 0$ given in Eq. (3.29), we get

$$\begin{aligned}
\frac{1}{2}(\mathcal{D}_\alpha\mathcal{D}_\beta - \mathcal{D}_\beta\mathcal{D}_\alpha)\Phi\Big| &= 2\varepsilon_{\alpha\beta}F\,, \\
\frac{1}{2}(\mathcal{D}_\alpha\mathcal{D}_\beta + \mathcal{D}_\beta\mathcal{D}_\alpha)\Phi\Big| &= \frac{1}{2}\Big\{\mathcal{D}_\alpha, \mathcal{D}_\beta\Big\}\Phi\Big| = 0\,,
\end{aligned} \tag{4.9}$$

which leads to

$$\mathcal{D}_\alpha\mathcal{D}_\beta\Phi\Big| = 2\varepsilon_{\alpha\beta}F\,. \tag{4.10}$$

The computation of $\mathcal{D}_\mu\mathcal{D}_\alpha\Phi\big|$ follows the same lines as the one of $\mathcal{D}_\mu\Phi$ in (4.5),

$$\mathcal{D}_\mu\mathcal{D}_\alpha\Phi\Big| = \Big(E_\mu{}^{\tilde{\mu}}\mathcal{D}_{\tilde{\mu}} + E_\mu{}^{\tilde{\alpha}}\mathcal{D}_{\tilde{\alpha}} + E_{\mu\tilde{\dot{\alpha}}}\widetilde{\overline{\mathcal{D}}}^{\dot{\alpha}}\Big)\mathcal{D}_\alpha\Phi\Big|\,.$$

Then, adopting the gauge introduced in Table 3.6 and recalling the definition of the lowest order component of the superfields Φ and $\mathcal{D}_M\Phi$ (see Eqs. (4.1) and (4.6)), we get,

$$
\begin{aligned}
E_\mu{}^{\tilde{\mu}}\mathcal{D}_{\tilde{\mu}}\mathcal{D}_\alpha\Phi\Big| &= e_\mu{}^{\tilde{\mu}}\Big[\partial_{\tilde{\mu}}\mathcal{D}_\alpha\Phi\Big| - \Omega_{\tilde{\mu}\alpha}{}^\beta\mathcal{D}_\beta\Phi\Big|\ \Big] = \sqrt{2}e_\mu{}^{\tilde{\mu}}\Big[\partial_{\tilde{\mu}}\delta_\alpha{}^\beta - \omega_{\tilde{\mu}\alpha}{}^\beta\Big]\chi_\beta\,,\\
E_\mu{}^{\tilde{\alpha}}\mathcal{D}_{\tilde{\alpha}}\mathcal{D}_\alpha\Phi\Big| &= -\frac{1}{2}\psi_\mu{}^\beta\mathcal{D}_\beta\mathcal{D}_\alpha\Phi\Big| = \psi_{\mu\alpha}F\,,\\
E_{\mu\tilde{\dot\alpha}}\overline{\mathcal{D}}^{\tilde{\dot\alpha}}\mathcal{D}_\alpha\Phi\Big| &= \frac{1}{2}\bar{\psi}_\mu{}^{\dot\alpha}\overline{\mathcal{D}}_{\dot\alpha}\mathcal{D}_\alpha\Phi\Big| = \frac{1}{2}\bar{\psi}_\mu{}^{\dot\alpha}\{\overline{\mathcal{D}}_{\dot\alpha},\mathcal{D}_\alpha\}\Phi\Big| = -i\sigma^\nu{}_{\alpha\dot\alpha}\bar{\psi}_\mu{}^{\dot\alpha}\,\hat{D}_\nu\phi\,,
\end{aligned}
\tag{4.11}
$$

where, we have used the definition of the covariant superderivative given in (3.5) for the first equation, the relation (4.10) for the second equation, and (3.8) together with the constraints (3.29) for the third equation. Let us also note that Eq. (3.3) allows us to replace summed spinorial Einsteinian indices by summed spinorial Lorentz indices in the second and third equations above, since in the gauge which we have adopted, $E_{\tilde{\alpha}}{}^\alpha\Big| = \delta_{\tilde{\alpha}}{}^\alpha$, $E_{\tilde{\alpha}\dot\alpha}\Big| = 0$, $E_{\tilde{\alpha}}{}^\mu\Big| = 0$ and $E^{\tilde{\dot\alpha}}{}_{\dot\alpha}\Big| = \delta^{\tilde{\dot\alpha}}{}_{\dot\alpha}$, $E^{\tilde{\dot\alpha}\alpha}\Big| = 0$, $E^{\tilde{\dot\alpha}\mu}\Big| = 0$. Such kind of manipulations are central and will be recurrent in the sequel. A typical instance of such a replacement is $\mathcal{D}_\alpha\Phi\Big| = \delta_\alpha{}^{\tilde{\alpha}}\mathcal{D}_{\tilde{\alpha}}\Phi\Big|$ (with a similar relation holding for the antichiral superfield $\Phi^\dagger$). Collecting all terms, we have then

$$
\begin{aligned}
\mathcal{D}_\mu\mathcal{D}_\alpha\Phi\Big| &\equiv \sqrt{2}\hat{D}_\mu\chi_\alpha\\
&= \sqrt{2}e_\mu{}^{\tilde{\mu}}\Big[\partial_{\tilde{\mu}}\chi_\alpha - \omega_{\tilde{\mu}\alpha}{}^\beta\chi_\beta + \frac{1}{\sqrt{2}}\psi_{\tilde{\mu}\alpha}F - \frac{i}{\sqrt{2}}(\sigma^\nu\bar{\psi}_\mu)_\alpha\hat{D}_\nu\phi\Big],
\end{aligned}
\tag{4.12}
$$

where we have introduced the covariant derivative $\hat{D}_\mu\chi_\alpha$. This last calculation, as well as the one leading to (4.5), illustrate that for a given component ξ_A (carrying indices or not), defined as $X_A\Big| = \xi_A$ with X_A being a superfield or any of its derivatives, the action of the flat superderivative $e_\mu{}^{\tilde{\mu}}\mathcal{D}_{\tilde{\mu}}$, covariant with respect to the supergravity transformations, reads[2]

$$
e_\mu{}^{\tilde{\mu}}\mathcal{D}_{\tilde{\mu}}X_A\Big| = e_\mu{}^{\tilde{\mu}}\mathcal{D}_{\tilde{\mu}}\xi_A\,. \tag{4.13}
$$

The last relation in (4.11) also shows that

$$
\overline{\mathcal{D}}_{\dot\alpha}\mathcal{D}_\alpha\Phi\Big| = \{\overline{\mathcal{D}}_{\dot\alpha},\mathcal{D}_\alpha\}\Phi\Big| = -2i\sigma^\mu{}_{\alpha\dot\alpha}\hat{D}_\mu\phi\,.
$$

The superderivatives $\mathcal{D}_\alpha\mathcal{D}_\mu\Phi\Big|$ and $\overline{\mathcal{D}}_{\dot\alpha}\mathcal{D}_\mu\Phi\Big|$ are obtained using again the definition of the covariant superderivative (3.5) and the expressions (4.2), together with the relations of Tables 3.3 and 3.8 and the previously calculated

[2]Since the covariant derivatives $\mathcal{D}_\mu\xi_A$ involves the superconnection $\Omega_{\tilde{\mu}}$, the second member of Eq. (4.13) must be read as $\mathcal{D}_\mu|\xi_A$. However, in order to simplify the notations, we employ the more compact form $\mathcal{D}_\mu\xi_A$, assuming the superconnection to be replaced by its lowest order component, $\Omega_{\tilde{\mu}} \to \omega_{\tilde{\mu}}$.

quantities $\mathcal{D}_\mu \mathcal{D}_M \Phi\big|$,

$$
\begin{aligned}
\mathcal{D}_\alpha \mathcal{D}_\mu \Phi\Big| &= \mathcal{D}_\mu \mathcal{D}_\alpha \Phi\Big| + [\mathcal{D}_\alpha, \mathcal{D}_\mu]\Phi\Big| = \sqrt{2}\hat{D}_\mu \chi_\alpha + T_{\alpha\mu}{}^\beta \mathcal{D}_\beta \Phi\Big| \\
&= \sqrt{2}\hat{D}_\mu \chi_\alpha - \frac{\sqrt{2}i}{3}\Big[\chi_\alpha b_\mu + b^\nu(\sigma_{\mu\nu}\chi)_\alpha\Big] , \qquad (4.14) \\
\overline{\mathcal{D}}_{\dot\alpha} \mathcal{D}_\mu \Phi\Big| &= [\overline{\mathcal{D}}_{\dot\alpha}, \mathcal{D}_\mu]\Phi\Big| = T_{\dot\alpha\mu}{}^\alpha \mathcal{D}_\alpha \Phi\Big| = -\frac{\sqrt{2}i}{6}(\chi\sigma_\mu)_{\dot\alpha}\, M .
\end{aligned}
$$

The commutators appearing in the relations above have been simplified with the help of (3.8). It is now straightforward to deduce the last of the nine second-order superderivatives, $\mathcal{D}_\mu \mathcal{D}_\nu \Phi\big|$, since

$$
\begin{aligned}
\mathcal{D}_\mu \mathcal{D}_\nu \Phi\Big| &= E_\mu{}^{\tilde\mu} \mathcal{D}_{\tilde\mu} \mathcal{D}_\nu \Phi\Big| + E_\mu{}^{\tilde\alpha} \mathcal{D}_{\tilde\alpha} \mathcal{D}_\nu \Phi\Big| + E_{\mu\tilde{\dot\alpha}} \overline{\mathcal{D}}^{\tilde{\dot\alpha}} \mathcal{D}_\nu \Phi\Big| \\
&= e_\mu{}^{\tilde\mu}\bigg[\partial_{\tilde\mu} \hat{D}_\nu \phi - \omega_{\tilde\mu\nu}{}^\rho \hat{D}_\rho \phi - \frac{\sqrt{2}}{2}\psi_{\tilde\mu}\cdot \hat{D}_\nu \chi \\
&\quad + \frac{\sqrt{2}i}{6}\Big(\psi_{\tilde\mu}\cdot\chi b_\nu + \psi_{\tilde\mu}\sigma_{\nu\rho}\chi b^\rho\Big) + \frac{\sqrt{2}i}{12}(\chi\sigma_\nu \bar\psi_{\tilde\mu})M\bigg] .
\end{aligned}
$$

Collecting all results, the second-order derivatives of a chiral superfield Φ read

$$
\begin{aligned}
\mathcal{D}_\mu \mathcal{D}_\nu \Phi\Big| &= e_\mu{}^{\tilde\mu}\bigg[\partial_{\tilde\mu} \hat{D}_\nu \phi - \omega_{\tilde\mu\nu}{}^\rho \hat{D}_\rho \phi - \frac{\sqrt{2}}{2}\psi_{\tilde\mu}\cdot \hat{D}_\nu \chi \\
&\quad + \frac{\sqrt{2}i}{6}\Big(\psi_{\tilde\mu}\cdot\chi b_\nu + \psi_{\tilde\mu}\sigma_{\nu\rho}\chi b^\rho\Big) + \frac{\sqrt{2}i}{12}(\chi\sigma_\nu \bar\psi_{\tilde\mu})M\bigg] , \\
\mathcal{D}_\mu \mathcal{D}_\alpha \Phi\Big| &= \sqrt{2}\hat{D}_\mu \chi_\alpha , \\
\mathcal{D}_\alpha \mathcal{D}_\mu \Phi\Big| &= \sqrt{2}\hat{D}_\mu \chi_\alpha - \frac{\sqrt{2}i}{3}\Big[\chi_\alpha b_\mu + b^\nu(\sigma_{\mu\nu}\chi)_\alpha\Big] , \qquad (4.15) \\
\overline{\mathcal{D}}_{\dot\alpha} \mathcal{D}_\mu \Phi\Big| &= -\frac{\sqrt{2}i}{6}(\chi\sigma_\mu)_{\dot\alpha}\, M , \\
\overline{\mathcal{D}}_{\dot\alpha} \mathcal{D}_\alpha \Phi\Big| &= -2i\sigma^\mu{}_{\alpha\dot\alpha} \hat{D}_\mu \phi , \\
\mathcal{D}_\alpha \mathcal{D}_\beta \Phi\Big| &= 2\varepsilon_{\alpha\beta} F , \\
\mathcal{D}_\mu \overline{\mathcal{D}}_{\dot\alpha} \Phi\Big| &= \mathcal{D}_\alpha \overline{\mathcal{D}}_{\dot\alpha} \Phi\Big| = \overline{\mathcal{D}}_{\dot\alpha} \overline{\mathcal{D}}_{\dot\beta} \Phi\Big| = 0 .
\end{aligned}
$$

Similarly, or equivalently by hermitian conjugation, we obtain the corresponding relations in the case of an antichiral superfield $\Phi^\dagger$,

$$
\begin{aligned}
\mathcal{D}_\mu \mathcal{D}_\nu \Phi^\dagger \Big| &= e_\mu{}^{\tilde{\mu}} \Big[\partial_{\tilde{\mu}} \hat{D}_\nu \phi^\dagger - \omega_{\tilde{\mu}\nu}{}^\rho \hat{D}_\rho \phi^\dagger - \frac{\sqrt{2}}{2} \bar{\psi}_{\tilde{\mu}} \cdot \hat{D}_\nu \bar{\chi} \\
&\quad - \frac{\sqrt{2}i}{6} \big(\bar{\psi}_{\tilde{\mu}} \cdot \bar{\chi} b_\nu + \bar{\chi} \bar{\sigma}_{\rho\nu} \bar{\psi}_{\tilde{\mu}} b^\rho \big) - \frac{\sqrt{2}i}{12} (\psi_{\tilde{\mu}} \sigma_\nu \bar{\chi}) M^* \Big] , \\
\mathcal{D}_\mu \overline{\mathcal{D}}_{\dot{\alpha}} \Phi^\dagger \Big| &= \sqrt{2} \hat{D}_\mu \bar{\chi}_{\dot{\alpha}} \\
&\equiv \sqrt{2} e_\mu{}^{\tilde{\mu}} \Big(\partial_{\tilde{\mu}} \bar{\chi}_{\dot{\alpha}} + \omega_{\tilde{\mu}\dot{\alpha}}{}^{\dot{\beta}} \bar{\chi}_{\dot{\beta}} + \frac{1}{\sqrt{2}} \bar{\psi}_{\tilde{\mu}\dot{\alpha}} F^\dagger + \frac{i}{\sqrt{2}} (\psi_{\tilde{\mu}} \sigma^\nu)_{\dot{\alpha}} \hat{D}_\nu \phi^\dagger \Big) , \\
\overline{\mathcal{D}}_{\dot{\alpha}} \mathcal{D}_\mu \Phi^\dagger \Big| &= \sqrt{2} \hat{D}_\mu \bar{\chi}_{\dot{\alpha}} + \frac{\sqrt{2}i}{3} \Big[\bar{\chi}_{\dot{\alpha}} b_\mu + b^\nu (\bar{\chi} \bar{\sigma}_{\nu\mu})_{\dot{\alpha}} \Big] , \qquad (4.16) \\
\mathcal{D}_\alpha \mathcal{D}_\mu \Phi^\dagger \Big| &= \frac{\sqrt{2}i}{6} (\sigma_\mu \bar{\chi})_\alpha \, M^* , \\
\mathcal{D}_\alpha \overline{\mathcal{D}}_{\dot{\alpha}} \Phi^\dagger \Big| &= -2i \sigma^\mu{}_{\alpha\dot{\alpha}} \hat{D}_\mu \phi^\dagger , \\
\overline{\mathcal{D}}_{\dot{\alpha}} \overline{\mathcal{D}}_{\dot{\beta}} \Phi^\dagger \Big| &= -2 \varepsilon_{\dot{\alpha}\dot{\beta}} F^\dagger , \\
\mathcal{D}_\mu \mathcal{D}_\alpha \Phi^\dagger \Big| &= \mathcal{D}_\beta \mathcal{D}_\alpha \Phi^\dagger \Big| = \overline{\mathcal{D}}_{\dot{\alpha}} \mathcal{D}_\alpha \Phi^\dagger \Big| = 0 .
\end{aligned}
$$

Third-order derivatives of chiral and antichiral superfields $\mathcal{D}_M \mathcal{D}_N \mathcal{D}_P \Phi\big|$ and $\mathcal{D}_M \mathcal{D}_N \mathcal{D}_P \Phi^\dagger\big|$.

In Sec. 4.1.2, we will derive the variation laws of the component fields of chiral and antichiral superfields under supergravity transformations. This requires the computation of the lowest order component of several third-order derivatives, $\mathcal{D}_\alpha \mathcal{D} \cdot \mathcal{D} \Phi$, $\overline{\mathcal{D}}_{\dot{\alpha}} \mathcal{D}_\alpha \mathcal{D}_\beta \Phi$ and the associated complex conjugate quantities. We first recall the property (3.124), which applied to the rank-three tensor $\mathcal{D}_\alpha \mathcal{D}_\beta \mathcal{D}_\gamma$, gives

$$
\mathcal{D}_\alpha \mathcal{D}_\beta \mathcal{D}_\gamma + \mathcal{D}_\beta \mathcal{D}_\gamma \mathcal{D}_\alpha + \mathcal{D}_\gamma \mathcal{D}_\alpha \mathcal{D}_\beta - \mathcal{D}_\alpha \mathcal{D}_\gamma \mathcal{D}_\beta - \mathcal{D}_\beta \mathcal{D}_\alpha \mathcal{D}_\gamma - \mathcal{D}_\gamma \mathcal{D}_\beta \mathcal{D}_\alpha = 0 .
$$

Since $\{\mathcal{D}_\alpha, \mathcal{D}_\beta\}\Phi = 0$, as shown in (4.9), we can derive from this last relation the property

$$
\Big(\mathcal{D}_\alpha \mathcal{D}_\beta \mathcal{D}_\gamma + \mathcal{D}_\beta \mathcal{D}_\gamma \mathcal{D}_\alpha + \mathcal{D}_\gamma \mathcal{D}_\alpha \mathcal{D}_\beta \Big) \Phi = 0
$$

which implies,

$$
\mathcal{D}_\alpha \mathcal{D}_\beta \mathcal{D}_\gamma \Phi = - \Big(\mathcal{D}_\beta \mathcal{D}_\gamma \mathcal{D}_\alpha + \mathcal{D}_\gamma \mathcal{D}_\alpha \mathcal{D}_\beta \Big) \Phi .
$$

Hence, (4.17), together with the fact that the anticommutator $\{\mathcal{D}_\alpha, \mathcal{D}_\beta\}\Phi$ vanishes (as shown in (3.8)), leads to

$$
\begin{aligned}
&\frac{1}{3}\Big[\{\mathcal{D}_\alpha,\mathcal{D}_\beta\}\mathcal{D}_\gamma-\{\mathcal{D}_\alpha,\mathcal{D}_\gamma\}\mathcal{D}_\beta\Big]\Phi\\
&=\frac{1}{3}\Big[\mathcal{D}_\alpha\mathcal{D}_\beta\mathcal{D}_\gamma+\mathcal{D}_\beta\mathcal{D}_\alpha\mathcal{D}_\gamma-\mathcal{D}_\alpha\mathcal{D}_\gamma\mathcal{D}_\beta-\mathcal{D}_\gamma\mathcal{D}_\alpha\mathcal{D}_\beta\Big]\Phi\\
&=\frac{1}{3}\Big[\mathcal{D}_\alpha\mathcal{D}_\beta\mathcal{D}_\gamma-\mathcal{D}_\beta\mathcal{D}_\gamma\mathcal{D}_\alpha+\mathcal{D}_\alpha\mathcal{D}_\beta\mathcal{D}_\gamma-\mathcal{D}_\gamma\mathcal{D}_\alpha\mathcal{D}_\beta\Big]\Phi=\mathcal{D}_\alpha\mathcal{D}_\beta\mathcal{D}_\gamma\Phi\,.
\end{aligned}
$$

Contracting the indices with the help of the $\varepsilon^{\gamma\beta}$ tensor, one gets

$$
\mathcal{D}_\alpha\mathcal{D}\cdot\mathcal{D}\Phi=-\frac{2}{3}\{\mathcal{D}_\alpha,\mathcal{D}_\beta\}\mathcal{D}^\beta\Phi=\frac{2}{3}R_{\alpha\gamma}{}^{\delta\gamma}\mathcal{D}_\delta\Phi=8\mathcal{R}^\dagger\mathcal{D}_\alpha\Phi\,, \tag{4.17}
$$

where we have used the constraints of (3.29) and the definition of the curvature tensor $R_{\alpha\beta\gamma\delta}$ given in Table 3.4. Recalling that $\mathcal{R}$ and $\mathcal{R}^\dagger$ are chiral and antichiral superfields, respectively, this last relation, together with its hermitian conjugate, leads to two essential properties in supergravity,

$$
\begin{aligned}
\mathcal{D}_\alpha\Big[\mathcal{D}\cdot\mathcal{D}-8\mathcal{R}^\dagger\Big]\Phi&=0\,,\\
\overline{\mathcal{D}}^{\dot\alpha}\Big[\overline{\mathcal{D}}\cdot\overline{\mathcal{D}}-8\mathcal{R}\Big]\Phi^\dagger&=0\,,
\end{aligned} \tag{4.18}
$$

which hold for any (anti)chiral superfield Φ ($\Phi^\dagger$). Hence, it is possible to associate to any chiral (antichiral) superfield Φ ($\Phi^\dagger$), an antichiral (chiral) superfield $\Xi^\dagger$ (Ξ),

$$
\begin{aligned}
\Xi&=\Big[\overline{\mathcal{D}}\cdot\overline{\mathcal{D}}-8\mathcal{R}\Big]\Phi^\dagger\,,\\
\Xi^\dagger&=\Big[\mathcal{D}\cdot\mathcal{D}-8\mathcal{R}^\dagger\Big]\Phi\,.
\end{aligned} \tag{4.19}
$$

The operators $\overline{\mathcal{D}}\cdot\overline{\mathcal{D}}-8\mathcal{R}$ and $\mathcal{D}\cdot\mathcal{D}-8\mathcal{R}^\dagger$ are two "projectors" mapping antichiral to chiral superfields and chiral to antichiral superfields. Strictly speaking these operators are not projector in the mathematical sense but in the supergravity literature they are conventionally called "projectors". They were introduced in [Wess and Zumino (1978a); Gates et al. (1980)]. In the sequel, we will show that these projectors are more general and allow to map any scalar superfield to chiral and antichiral superfields. They are also the cornerstones in the construction of invariant actions. Starting from Eq. (4.17) and the relations given in Table 3.6, we can finally calculate the lowest order component of the third-order superderivatives $\mathcal{D}_\alpha\mathcal{D}\cdot\mathcal{D}\Phi$ and $\overline{\mathcal{D}}^{\dot\alpha}\overline{\mathcal{D}}\cdot\overline{\mathcal{D}}\Phi^\dagger$,

$$
\begin{aligned}
\mathcal{D}_\alpha\mathcal{D}\cdot\mathcal{D}\Phi\Big|&=-\frac{4\sqrt{2}}{3}\chi_\alpha M^*\,,\\
\overline{\mathcal{D}}^{\dot\alpha}\overline{\mathcal{D}}\cdot\overline{\mathcal{D}}\Phi^\dagger\Big|&=-\frac{4\sqrt{2}}{3}\bar\chi^{\dot\alpha}M\,.
\end{aligned} \tag{4.20}
$$

We now turn to the computation of the lowest order component of the second third-order superderivatives relevant for the derivation of the variation laws of the components of chiral and antichiral superfields under supergravity transformation, $\overline{\mathcal{D}}_{\dot\alpha}\mathcal{D}_\alpha\mathcal{D}_\beta\Phi\big|$ and $\mathcal{D}_\alpha\overline{\mathcal{D}}_{\dot\alpha}\overline{\mathcal{D}}_{\dot\beta}\Phi^\dagger\big|$. Since the superfield Φ is chiral, the first of these derivatives can be rewritten as

$$\overline{\mathcal{D}}_{\dot\alpha}\mathcal{D}_\alpha\mathcal{D}_\beta\Phi = \{\overline{\mathcal{D}}_{\dot\alpha},\mathcal{D}_\alpha\}\mathcal{D}_\beta\Phi - \mathcal{D}_\alpha\{\overline{\mathcal{D}}_{\dot\alpha},\mathcal{D}_\beta\}\Phi\,.$$

From the relations of Tables 3.3 and 3.4, and using (3.8), the two above terms become

$$\begin{aligned}
\{\overline{\mathcal{D}}_{\dot\alpha},\mathcal{D}_\alpha\}\mathcal{D}_\beta\Phi &= T_{\alpha\dot\alpha}{}^M\mathcal{D}_M\mathcal{D}_\beta\Phi + R_{\dot\alpha\alpha\gamma\beta}\mathcal{D}^\gamma\Phi\\
&= -2i\sigma^\mu{}_{\alpha\dot\alpha}\mathcal{D}_\mu\mathcal{D}_\beta\Phi + \Big[\varepsilon_{\alpha\beta}G_{\gamma\dot\alpha}\mathcal{D}^\gamma\Phi + G_{\beta\dot\alpha}\mathcal{D}_\alpha\Phi\Big]\,,\\
\mathcal{D}_\alpha\{\overline{\mathcal{D}}_{\dot\alpha},\mathcal{D}_\beta\}\Phi &= \mathcal{D}_\alpha T_{\dot\alpha\beta}{}^M\mathcal{D}_M\Phi\\
&= -2i\sigma^\mu{}_{\beta\dot\alpha}\mathcal{D}_\alpha\mathcal{D}_\mu\Phi = -2i\sigma^\mu{}_{\beta\dot\alpha}\mathcal{D}_\mu\mathcal{D}_\alpha\Phi - 2i\sigma^\mu{}_{\beta\dot\alpha}[\mathcal{D}_\alpha,\mathcal{D}_\mu]\Phi\\
&= -2i\sigma^\mu{}_{\beta\dot\alpha}\mathcal{D}_\mu\mathcal{D}_\alpha\Phi - 2i\sigma^\mu{}_{\beta\dot\alpha}T_{\alpha\mu}{}^\gamma\mathcal{D}_\gamma\Phi\\
&= -2i\sigma^\mu{}_{\beta\dot\alpha}\mathcal{D}_\mu\mathcal{D}_\alpha\Phi - \frac{1}{2}\Big[G_{\alpha\dot\alpha}\mathcal{D}_\beta\Phi - 3G_{\beta\dot\alpha}\mathcal{D}_\alpha\Phi - 3\varepsilon_{\alpha\beta}G_{\gamma\dot\alpha}\mathcal{D}^\gamma\Phi\Big]\,,
\end{aligned}$$

where for the second equality, we have converted vector indices of the torsion tensor $T_{\alpha\mu}{}^\beta$ to spinor indices, using (B.7). Collecting the two terms, we get,

$$\overline{\mathcal{D}}_{\dot\alpha}\mathcal{D}_\alpha\mathcal{D}_\beta\Phi = \varepsilon_{\alpha\beta}\Big[2i\sigma^\mu{}_{\gamma\dot\alpha}\mathcal{D}_\mu\mathcal{D}^\gamma\Phi - G_{\gamma\dot\alpha}\mathcal{D}^\gamma\Phi\Big]\,, \tag{4.21}$$

where we have used the antisymmetry property (4.8). In a similar fashion, one can show that

$$\mathcal{D}_\alpha\overline{\mathcal{D}}_{\dot\alpha}\overline{\mathcal{D}}_{\dot\beta}\Phi^\dagger = \varepsilon_{\dot\alpha\dot\beta}\Big[2i\sigma^\mu{}_{\alpha\dot\gamma}\mathcal{D}_\mu\overline{\mathcal{D}}^{\dot\gamma}\Phi^\dagger + G_{\alpha\dot\gamma}\overline{\mathcal{D}}^{\dot\gamma}\Phi^\dagger\Big]\,. \tag{4.22}$$

It is now trivial to deduce the lowest order component of these derivatives, using the relations of Table 3.8 as well as (4.6), (4.7), (4.15) and (4.16)[3]

$$\begin{aligned}
\overline{\mathcal{D}}_{\dot\alpha}\mathcal{D}_\alpha\mathcal{D}_\beta\Phi\Big| &= \sqrt{2}\varepsilon_{\alpha\beta}\Big[2i(\hat{D}_\mu\chi\sigma^\mu)_{\dot\alpha} + \frac{1}{3}b_{\gamma\dot\alpha}\chi^\gamma\Big]\,,\\
\mathcal{D}_\alpha\overline{\mathcal{D}}_{\dot\alpha}\overline{\mathcal{D}}_{\dot\beta}\Phi^\dagger\Big| &= \sqrt{2}\varepsilon_{\dot\alpha\dot\beta}\Big[2i(\sigma^\mu\hat{D}_\mu\overline{\chi})_\alpha - \frac{1}{3}b_{\alpha\dot\gamma}\overline{\chi}^{\dot\gamma}\Big]\,.
\end{aligned} \tag{4.23}$$

Finally, the last third-order derivatives to be computed are $\overline{\mathcal{D}}_{\dot\alpha}\overline{\mathcal{D}}_{\dot\beta}\mathcal{D}_\alpha\Phi\big|$ and the hermitian conjugate quantity, $\mathcal{D}_\alpha\mathcal{D}_\beta\overline{\mathcal{D}}_{\dot\alpha}\Phi^\dagger\big|$. It is straightforward to derive them

[3]The results of Eq. (4.23) may seem surprising, since the two relations are not hermitian conjugate. However, they will lead, in Sec. 4.1.2, to conjugate variations for the component fields F and $F^\dagger$ (as in Eqs. (4.30)) under supergravity transformations.

using the graded commutation relations (3.8), since

$$\begin{aligned}\overline{\mathcal{D}}_{\dot\alpha}\overline{\mathcal{D}}_{\dot\beta}\mathcal{D}_\alpha\Phi &= \overline{\mathcal{D}}_{\dot\alpha}\{\overline{\mathcal{D}}_{\dot\beta},\mathcal{D}_\alpha\}\Phi\\ &= -2i\sigma^\mu{}_{\alpha\dot\beta}\overline{\mathcal{D}}_{\dot\alpha}\mathcal{D}_\mu\Phi\\ &= -2i\sigma^\mu{}_{\alpha\dot\beta}[\overline{\mathcal{D}}_{\dot\alpha},\mathcal{D}_\mu]\Phi\\ &= -2iT_{\dot\alpha\alpha\dot\beta}{}^\gamma\mathcal{D}_\gamma\Phi\\ &= -4\varepsilon_{\dot\alpha\dot\beta}\mathcal{R}\mathcal{D}_\alpha\Phi\,,\end{aligned} \tag{4.24}$$

where we have used the fact that Φ is a chiral superfield, the constraints (3.29) and the results of Table 3.3. The lowest-order component thus reads

$$\overline{\mathcal{D}}_{\dot\alpha}\overline{\mathcal{D}}_{\dot\beta}\mathcal{D}_\alpha\Phi\Big| = \frac{2}{3}\sqrt{2}\varepsilon_{\dot\alpha\dot\beta}M\chi_\alpha\,,$$

employing the results of Table 3.6 and Eq. (4.6). In a similar way, we obtain the conjugate relations

$$\mathcal{D}_\alpha\mathcal{D}_\beta\overline{\mathcal{D}}_{\dot\alpha}\Phi^\dagger = 4\varepsilon_{\alpha\beta}\mathcal{R}^\dagger\overline{\mathcal{D}}_{\dot\alpha}\Phi^\dagger$$

and

$$\mathcal{D}_\alpha\mathcal{D}_\beta\overline{\mathcal{D}}_{\dot\alpha}\Phi^\dagger\Big| = -\frac{2}{3}\sqrt{2}\varepsilon_{\alpha\beta}M^*\bar\chi_{\dot\alpha}\,.$$

Fourth-order derivatives of chiral and antichiral superfields $\overline{\mathcal{D}}\cdot\overline{\mathcal{D}}\mathcal{D}\cdot\mathcal{D}\Phi\big|$ and $\mathcal{D}\cdot\mathcal{D}\overline{\mathcal{D}}\cdot\overline{\mathcal{D}}\Phi^\dagger\big|$.

In order to express chiral actions in terms of components, it is necessary to compute all the component fields of the superfields Ξ and $\Xi^\dagger$ introduced in (4.19). In addition to all the previously calculated derivatives this requires, the computation of the lowest order components of two fourth-order derivatives, *i.e.*, $\overline{\mathcal{D}}\cdot\overline{\mathcal{D}}\mathcal{D}\cdot\mathcal{D}\Phi\big|$ and $\mathcal{D}\cdot\mathcal{D}\overline{\mathcal{D}}\cdot\overline{\mathcal{D}}\Phi^\dagger\big|$. Starting from (4.21) and (4.22), they read

$$\begin{aligned}\overline{\mathcal{D}}\cdot\overline{\mathcal{D}}\mathcal{D}\cdot\mathcal{D}\Phi &= -4i\bar\sigma^{\mu\dot\alpha\alpha}\overline{\mathcal{D}}_{\dot\alpha}\mathcal{D}_\mu\mathcal{D}_\alpha\Phi + 2\mathcal{D}^\alpha\Phi\mathcal{D}_\alpha\mathcal{R} - 8iG^\mu\mathcal{D}_\mu\Phi\,,\\ \mathcal{D}\cdot\mathcal{D}\overline{\mathcal{D}}\cdot\overline{\mathcal{D}}\Phi^\dagger &= -4i\bar\sigma^{\mu\dot\alpha\alpha}\mathcal{D}_\alpha\mathcal{D}_\mu\overline{\mathcal{D}}_{\dot\alpha}\Phi^\dagger + 2\overline{\mathcal{D}}_{\dot\alpha}\Phi^\dagger\overline{\mathcal{D}}^{\dot\alpha}\mathcal{R}^\dagger + 8iG^\mu\mathcal{D}_\mu\Phi^\dagger\,,\end{aligned} \tag{4.25}$$

where we have used the relations given in Table 3.2, as well as (3.8) together with the constraints (3.29). In addition, spinor indices have been converted to vector indices (see (B.7)). We also recall that since Φ and $\Phi^\dagger$ are chiral and antichiral superfields, respectively, we may write $\overline{\mathcal{D}}_{\dot\beta}\mathcal{D}_\gamma\Phi = \{\overline{\mathcal{D}}_{\dot\beta},\mathcal{D}_\gamma\}\Phi$ and $\mathcal{D}_\beta\overline{\mathcal{D}}_{\dot\gamma}\Phi^\dagger = \{\mathcal{D}_\beta,\overline{\mathcal{D}}_{\dot\gamma}\}\Phi^\dagger$. We now need to calculate the third-order derivatives $\bar\sigma^{\mu\dot\alpha\alpha}\overline{\mathcal{D}}_{\dot\alpha}\mathcal{D}_\mu\mathcal{D}_\alpha\Phi$

and $\bar{\sigma}^{\mu\dot{\alpha}\alpha}\mathcal{D}_\alpha\mathcal{D}_\mu\overline{\mathcal{D}}_{\dot{\alpha}}\Phi^\dagger$. Using (3.8), the constraints (3.29), the relations (4.6), (4.7), (4.15) and (4.16), as well as Tables 3.2, 3.3, 3.4 and 3.8, we obtain

$$\begin{aligned}
\overline{\mathcal{D}}_{\dot{\alpha}}\mathcal{D}_\mu\mathcal{D}_\alpha\Phi &= \Big[\overline{\mathcal{D}}_{\dot{\alpha}},\mathcal{D}_\mu\Big]\mathcal{D}_\alpha\Phi + \mathcal{D}_\mu\Big\{\overline{\mathcal{D}}_{\dot{\alpha}},\mathcal{D}_\alpha\Big\}\Phi\\
&= T_{\dot{\alpha}\mu}{}^{\beta}\mathcal{D}_\beta\mathcal{D}_\alpha\Phi + T_{\dot{\alpha}\mu\dot{\beta}}\overline{\mathcal{D}}^{\dot{\beta}}\mathcal{D}_\alpha\Phi + R_{\dot{\alpha}\mu\gamma\alpha}\mathcal{D}^\gamma\Phi - 2i\sigma^\nu{}_{\alpha\dot{\alpha}}\mathcal{D}_\mu\mathcal{D}_\nu\Phi\,,\\
\mathcal{D}_\alpha\mathcal{D}_\mu\overline{\mathcal{D}}_{\dot{\alpha}}\Phi^\dagger &= \Big[\mathcal{D}_\alpha,\mathcal{D}_\mu\Big]\overline{\mathcal{D}}_{\dot{\alpha}}\Phi^\dagger + \mathcal{D}_\mu\Big\{\mathcal{D}_\alpha,\overline{\mathcal{D}}_{\dot{\alpha}}\Big\}\Phi^\dagger\\
&= T_{\alpha\mu}{}^{\beta}\mathcal{D}_\beta\overline{\mathcal{D}}_{\dot{\alpha}}\Phi^\dagger + T_{\alpha\mu\dot{\beta}}\overline{\mathcal{D}}^{\dot{\beta}}\overline{\mathcal{D}}_{\dot{\alpha}}\Phi^\dagger - R_{\alpha\mu\dot{\gamma}\dot{\alpha}}\overline{\mathcal{D}}^{\dot{\gamma}}\Phi^\dagger - 2i\sigma^\nu{}_{\alpha\dot{\alpha}}\mathcal{D}_\mu\mathcal{D}_\nu\Phi^\dagger\,.
\end{aligned}\tag{4.26}$$

Contracting with $\bar{\sigma}^{\mu\dot{\alpha}\alpha}$, the lowest order components of the different terms above are given by

$$\begin{aligned}
\bar{\sigma}^{\mu\dot{\alpha}\alpha}T_{\dot{\alpha}\mu\dot{\beta}}\overline{\mathcal{D}}^{\dot{\beta}}\mathcal{D}_\alpha\Phi\Big| &= \frac{10}{3}b^\nu\hat{D}_\nu\phi\,,\\
\bar{\sigma}^{\mu\dot{\alpha}\alpha}T_{\alpha\mu}{}^{\beta}\,\mathcal{D}_\beta\overline{\mathcal{D}}_{\dot{\alpha}}\Phi^\dagger\Big| &= -\frac{10}{3}b^\nu\hat{D}_\nu\phi^\dagger\,,\\
\bar{\sigma}^{\mu\dot{\alpha}\alpha}T_{\dot{\alpha}\mu}{}^{\beta}\,\mathcal{D}_\beta\mathcal{D}_\alpha\Phi\Big| &= -\frac{8i}{3}MF\,,\\
\bar{\sigma}^{\mu\dot{\alpha}\alpha}T_{\alpha\mu\dot{\beta}}\,\overline{\mathcal{D}}^{\dot{\beta}}\overline{\mathcal{D}}_{\dot{\alpha}}\Phi^\dagger\Big| &= -\frac{8i}{3}M^*F^\dagger\,,\\
\bar{\sigma}^{\mu\dot{\alpha}\alpha}R_{\dot{\alpha}\mu\gamma\alpha}\,\mathcal{D}^\gamma\Phi\Big| &= \frac{3i}{2}\mathcal{D}^\alpha\Phi\mathcal{D}_\alpha\mathcal{R}\Big|\\
&= -\frac{\sqrt{2}i}{2}\chi\sigma^{\mu\nu}\psi_{\mu\nu} - \frac{\sqrt{2}}{4}\chi\sigma^\mu\bar{\psi}_\mu M + \frac{\sqrt{2}}{4}\chi\cdot\psi_\mu b^\mu\,,\\
\bar{\sigma}^{\mu\dot{\alpha}\alpha}R_{\alpha\mu\dot{\gamma}\dot{\alpha}}\,\overline{\mathcal{D}}^{\dot{\gamma}}\Phi^\dagger\Big| &= -\frac{3i}{2}\overline{\mathcal{D}}_{\dot{\alpha}}\Phi^\dagger\overline{\mathcal{D}}^{\dot{\alpha}}\mathcal{R}^\dagger\Big|\\
&= \frac{\sqrt{2}i}{2}\bar{\chi}\bar{\sigma}^{\mu\nu}\bar{\psi}_{\mu\nu} + \frac{\sqrt{2}}{4}\bar{\chi}\bar{\sigma}^\mu\psi_\mu M^* + \frac{\sqrt{2}}{4}\bar{\chi}\cdot\bar{\psi}_\mu b^\mu\,,
\end{aligned}$$

$$\begin{aligned}
\bar{\sigma}^{\mu\dot{\alpha}\alpha}\sigma^\nu{}_{\alpha\dot{\alpha}}\mathcal{D}_\mu\mathcal{D}_\nu\Phi\Big| &= 2e_\mu{}^{\tilde{\mu}}\mathcal{D}_{\tilde{\mu}}\hat{D}^\mu\phi + \frac{\sqrt{2}i}{6}\chi\sigma^\mu\bar{\psi}_\mu M - \sqrt{2}\psi^\mu\cdot\hat{D}_\mu\chi\\
&\quad+\frac{\sqrt{2}i}{3}\psi_\mu\sigma^{\mu\rho}\chi b_\rho + \frac{\sqrt{2}i}{3}b^\mu\chi\cdot\psi_\mu\,,\\
\bar{\sigma}^{\mu\dot{\alpha}\alpha}\sigma^\nu{}_{\alpha\dot{\alpha}}\mathcal{D}_\mu\mathcal{D}_\nu\Phi^\dagger\Big| &= 2e_\mu{}^{\tilde{\mu}}\mathcal{D}_{\tilde{\mu}}\hat{D}^\mu\phi^\dagger - \frac{\sqrt{2}i}{6}\psi_\mu\sigma^\mu\bar{\chi}M^* - \sqrt{2}\bar{\psi}^\mu\cdot\hat{D}_\mu\bar{\chi}\\
&\quad-\frac{\sqrt{2}i}{3}\bar{\chi}\bar{\sigma}^{\rho\mu}\bar{\psi}_\mu b_\rho - \frac{\sqrt{2}i}{3}b^\mu\bar{\chi}\cdot\bar{\psi}_\mu\,.
\end{aligned}$$

Collecting all contributions, we arrive at

$$\begin{aligned}
\bar{\sigma}^{\mu\dot{\alpha}\alpha}\overline{\mathcal{D}}_{\dot{\alpha}}\mathcal{D}_\mu\mathcal{D}_\alpha\Phi\Big| &= \frac{10}{3}b^\mu\hat{D}_\mu\phi - \frac{8i}{3}MF - \frac{\sqrt{2}i}{2}\chi\sigma^{\mu\nu}\psi_{\mu\nu} \\
&+ \frac{\sqrt{2}}{12}\chi\sigma^\mu\bar{\psi}_\mu M + \frac{11\sqrt{2}}{12}\chi\cdot\psi_\mu b^\mu \\
&- 4ie_\mu{}^{\tilde{\mu}}\mathcal{D}_{\tilde{\mu}}\hat{D}^\mu\phi + 2\sqrt{2}i\psi^\mu\cdot\hat{D}_\mu\chi + \frac{2\sqrt{2}}{3}\psi_\mu\sigma^{\mu\nu}\chi b_\nu\,, \\
\bar{\sigma}^{\mu\dot{\alpha}\alpha}\mathcal{D}_\alpha\mathcal{D}_\mu\overline{\mathcal{D}}_{\dot{\alpha}}\Phi^\dagger\Big| &= -\frac{10}{3}b^\mu\hat{D}_\mu\phi^\dagger - \frac{8i}{3}M^*F^\dagger - \frac{\sqrt{2}i}{2}\bar{\chi}\bar{\sigma}^{\mu\nu}\bar{\psi}_{\mu\nu} \\
&- \frac{\sqrt{2}}{12}\psi_\mu\sigma^\mu\bar{\chi}M^* - \frac{11\sqrt{2}}{12}\bar{\chi}\cdot\bar{\psi}_\mu b^\mu \\
&- 4ie_\mu{}^{\tilde{\mu}}\mathcal{D}_{\tilde{\mu}}\hat{D}^\mu\phi^\dagger + 2\sqrt{2}i\bar{\psi}^\mu\cdot\hat{D}_\mu\bar{\chi} - \frac{2\sqrt{2}}{3}\bar{\chi}\bar{\sigma}^{\nu\mu}\bar{\psi}_\mu b_\nu\,,
\end{aligned} \tag{4.27}$$

and consequently,

$$\begin{aligned}
\overline{\mathcal{D}}\cdot\overline{\mathcal{D}}\mathcal{D}\cdot\mathcal{D}\Phi\Big| &= -16e_\mu{}^{\tilde{\mu}}\mathcal{D}_{\tilde{\mu}}\hat{D}^\mu\phi - \frac{32i}{3}b^\mu\hat{D}_\mu\phi + 8\sqrt{2}\psi_\mu\cdot\hat{D}^\mu\chi \\
&- \frac{8\sqrt{2}}{3}\chi\sigma^{\mu\nu}\psi_{\mu\nu} - \frac{32}{3}MF - 4\sqrt{2}i\chi\cdot\psi_\mu b^\mu - \frac{8\sqrt{2}i}{3}\psi_\mu\sigma^{\mu\nu}\chi b_\nu\,, \\
\mathcal{D}\cdot\mathcal{D}\overline{\mathcal{D}}\cdot\overline{\mathcal{D}}\Phi^\dagger\Big| &= -16e_\mu{}^{\tilde{\mu}}\mathcal{D}_{\tilde{\mu}}\hat{D}^\mu\phi^\dagger + \frac{32i}{3}b^\mu\hat{D}_\mu\phi^\dagger + 8\sqrt{2}\bar{\psi}_\mu\cdot\hat{D}^\mu\bar{\chi} \\
&- \frac{8\sqrt{2}}{3}\bar{\chi}\bar{\sigma}^{\mu\nu}\bar{\psi}_{\mu\nu} - \frac{32}{3}M^*F^\dagger + 4\sqrt{2}i\bar{\chi}\cdot\bar{\psi}_\mu b^\mu + \frac{8\sqrt{2}i}{3}\bar{\chi}\bar{\sigma}^{\nu\mu}\bar{\psi}_\mu b_\nu\,.
\end{aligned} \tag{4.28}$$

4.1.2 *Transformation laws of chiral and antichiral superfields under supergravity transformations*

Considering a supergravity transformation of parameters ξ^M and $L_M{}^N$, we have now all the ingredients to easily derive the transformation laws of the component fields of the chiral and antichiral superfields Φ and $\Phi^\dagger$. Applying (3.101) to the (scalar) superfields themselves, as well as to the (scalar) quantities $\mathcal{D}\cdot\mathcal{D}\Phi$ and $\overline{\mathcal{D}}\cdot\overline{\mathcal{D}}\Phi^\dagger$ [Wess and Zumino (1978a); Wess and Bagger (1992)],

$$\begin{aligned}
\delta\Phi &= -\xi^M\mathcal{D}_M\Phi = -\xi^\alpha\mathcal{D}_\alpha\Phi - \xi^\mu\mathcal{D}_\mu\Phi\,, \\
\delta\Phi^\dagger &= -\xi^M\mathcal{D}_M\Phi^\dagger = -\bar{\xi}_{\dot{\alpha}}\overline{\mathcal{D}}^{\dot{\alpha}}\Phi^\dagger - \xi^\mu\mathcal{D}_\mu\Phi^\dagger\,, \\
\delta(\mathcal{D}\cdot\mathcal{D}\Phi) &= -\xi^M\mathcal{D}_M(\mathcal{D}\cdot\mathcal{D}\Phi) \\
&= -\xi^\alpha\mathcal{D}_\alpha\mathcal{D}\cdot\mathcal{D}\Phi - \bar{\xi}_{\dot{\alpha}}\overline{\mathcal{D}}^{\dot{\alpha}}\mathcal{D}\cdot\mathcal{D}\Phi - \xi^\mu\mathcal{D}_\mu\mathcal{D}\cdot D\Phi\,,
\end{aligned}$$

$$\begin{aligned}\delta(\overline{\mathcal{D}}\cdot\overline{\mathcal{D}}\Phi^\dagger) &= -\xi^M \mathcal{D}_M(\overline{\mathcal{D}}\cdot\overline{\mathcal{D}}\Phi^\dagger)\\ &= -\xi^\alpha \mathcal{D}_\alpha \overline{\mathcal{D}}\cdot\overline{\mathcal{D}}\Phi^\dagger - \bar{\xi}_{\dot\alpha}\overline{\mathcal{D}}^{\dot\alpha}\overline{\mathcal{D}}\cdot\overline{\mathcal{D}}\Phi^\dagger - \xi^\mu \mathcal{D}_\mu \overline{\mathcal{D}}\cdot\overline{\mathcal{D}}\Phi^\dagger\,,\end{aligned}$$

and using these relations together with the results of Table 3.7, we deduce the variations of the scalar components from (4.6) and (4.7) [Wess and Zumino (1978a); Wess and Bagger (1992)],

$$\begin{aligned}\delta\phi &= \delta\Phi\big| = -\varepsilon^\alpha \mathcal{D}_\alpha\Phi\big| = -\sqrt{2}\varepsilon\cdot\chi\,,\\ \delta\phi^\dagger &= \delta\Phi^\dagger\big| = -\bar\varepsilon_{\dot\alpha}\overline{\mathcal{D}}_{\dot\alpha}\Phi^\dagger\big| = -\sqrt{2}\bar\varepsilon\cdot\bar\chi\,,\end{aligned} \tag{4.29}$$

while those of the auxiliary components follow from (4.20) and (4.23) [Wess and Zumino (1978a); Wess and Bagger (1992)],

$$\begin{aligned}\delta F &= \frac{1}{4}\delta(\mathcal{D}\cdot\mathcal{D}\Phi)\big| = \frac{\sqrt{2}}{3}\varepsilon\cdot\chi M^* - i\sqrt{2}\hat{D}_\mu\chi\sigma^\mu\bar\varepsilon - \frac{\sqrt{2}}{6}b_\mu\chi\sigma^\mu\bar\varepsilon\,,\\ \delta F^\dagger &= \frac{1}{4}\delta(\overline{\mathcal{D}}\cdot\overline{\mathcal{D}}\Phi^\dagger)\big| = \frac{\sqrt{2}}{3}\bar\varepsilon\cdot\bar\chi M + i\sqrt{2}\varepsilon\sigma^\mu\hat{D}_\mu\bar\chi - \frac{\sqrt{2}}{6}b_\mu\varepsilon\sigma^\mu\bar\chi\,.\end{aligned} \tag{4.30}$$

Notice that as opposed to the flat case, the auxiliary term does not transform anymore as a total derivative. Applying the first relation of (3.100) to the spinorial superfields $\mathcal{D}_\alpha\Phi$ and $\overline{\mathcal{D}}_{\dot\alpha}\Phi^\dagger$, one gets the transformation laws [Wess and Zumino (1978a); Wess and Bagger (1992)]

$$\begin{aligned}\delta(\mathcal{D}_\alpha\Phi) &= -\xi^M\mathcal{D}_M\mathcal{D}_\alpha\Phi + \mathcal{D}^M\Phi L_{M\alpha}\\ &= -\xi^\beta\mathcal{D}_\beta\mathcal{D}_\alpha\Phi - \bar\xi_{\dot\beta}\overline{\mathcal{D}}^{\dot\beta}\mathcal{D}_\alpha\Phi - \xi^\mu\mathcal{D}_\mu\mathcal{D}_\alpha\Phi + \mathcal{D}^\beta\Phi L_{\beta\alpha}\,,\\ \delta(\overline{\mathcal{D}}_{\dot\alpha}\Phi^\dagger) &= -\xi^M\mathcal{D}_M\overline{\mathcal{D}}_{\dot\alpha}\Phi^\dagger + \mathcal{D}^M\Phi^\dagger L_{M\dot\alpha}\\ &= -\xi^\beta\mathcal{D}_\beta\overline{\mathcal{D}}_{\dot\alpha}\Phi^\dagger - \bar\xi_{\dot\beta}\overline{\mathcal{D}}^{\dot\beta}\overline{\mathcal{D}}_{\dot\alpha}\Phi^\dagger - \xi^\mu\mathcal{D}_\mu\overline{\mathcal{D}}_{\dot\alpha}\Phi^\dagger + \overline{\mathcal{D}}_{\dot\beta}\Phi^\dagger L^{\dot\beta}{}_{\dot\alpha}\,,\end{aligned}$$

which allow to deduce the variation laws of the fermionic components of the superfields Φ and $\Phi^\dagger$, using (4.15) and (4.16),

$$\begin{aligned}\delta\chi_\alpha &= \frac{1}{\sqrt{2}}\delta(\mathcal{D}_\alpha\Phi)\big| = \sqrt{2}\left[\varepsilon_\alpha F - i(\sigma^\mu\bar\varepsilon)_\alpha\hat{D}_\mu\phi\right],\\ \delta\bar\chi_{\dot\alpha} &= \frac{1}{\sqrt{2}}\delta(\overline{\mathcal{D}}_{\dot\alpha}\Phi^\dagger)\big| = \sqrt{2}\left[\bar\varepsilon_{\dot\alpha}F^\dagger + i(\varepsilon\sigma^\mu)_{\dot\alpha}\hat{D}_\mu\phi^\dagger\right].\end{aligned} \tag{4.31}$$

The transformation laws of the chiral multiplet was also given partially in [Ferrara et al. (1976a, 1977a)] (without the auxiliary fields) and in [Ferrara and van Nieuwenhuizen (1978b); Stelle and West (1978a)].

It might be interesting to obtain also the variations of $\hat{D}_\mu\phi$ or $\hat{D}_\mu\chi$ under supergravity transformations, *e.g.* to check that the supergravity action we will compute in Sec. 6.2 is invariant. These computations seems to be *a priori* complicated. Indeed, among many things $\hat{D}_\mu\chi$ involves the spin connection, meaning that one needs to calculate the variation of the spin-connection . However, the derivative $\hat{D}_\mu\phi$ or $\hat{D}_\mu\chi$ are by definition covariant with respect to supergravity transformations. Therefore their transformation should be obtained naturally, without having to compute any variation of the spin-connection. In particular, from $\hat{D}_\mu\phi = \mathcal{D}_\mu\Phi\big|$, we may write

$$\begin{aligned}
\delta(\hat{D}_\mu\phi) &= -\xi^M\mathcal{D}_M\mathcal{D}_\mu\Phi\Big| + \mathcal{D}^M\Phi L_{M\mu}\Big| \\
&= -\varepsilon^\alpha\mathcal{D}_\alpha\mathcal{D}_\mu\Phi\Big| + \varepsilon^{\dot\alpha}\overline{\mathcal{D}}_{\dot\alpha}\mathcal{D}_\mu\Phi\Big| \\
&= -\sqrt{2}\varepsilon\cdot\hat{D}_\mu\chi + \frac{\sqrt{2}}{3}i\Big(\varepsilon\cdot\chi b_\mu + \varepsilon\sigma_{\mu\nu}\chi b^\nu\Big) + \frac{\sqrt{2}}{6}i\bar{\epsilon}\bar{\sigma}_\mu\chi M\,,
\end{aligned}$$

where we have used (4.14). And in a similar way

$$\delta(\hat{D}_\mu\phi^\dagger) = -\sqrt{2}\bar{\varepsilon}\cdot\hat{D}_\mu\bar{\chi} - \frac{\sqrt{2}}{3}i\Big(\bar{\varepsilon}\cdot\bar{\chi}b_\mu + \bar{\varepsilon}\bar{\sigma}_{\mu\nu}\bar{\chi}b^\nu\Big) + \frac{\sqrt{2}}{6}i\epsilon\bar{\sigma}_\mu\bar{\chi}M^*\,.$$

The computation of the variation of the derivative of the fermionic field is more involved. We only give the steps for the variation of $\hat{D}_\mu\bar{\chi}$. We first need to define $\hat{D}_\mu F^\dagger$ by $\mathcal{D}_\mu\mathcal{D}_{\dot\alpha}\mathcal{D}_{\dot\beta}\Phi^\dagger\big| = -2\varepsilon_{\dot\alpha\dot\beta}\hat{D}_\mu F^\dagger$. Using (4.20) and (4.23) we obtain

$$\begin{aligned}
D_\mu\mathcal{D}_{\dot\alpha}\mathcal{D}_{\dot\beta}\Phi^\dagger &\equiv -2\varepsilon_{\dot\alpha\dot\beta}\hat{D}_\mu F^\dagger \\
&= E_\mu{}^{\tilde\mu}\mathcal{D}_{\tilde\mu}\mathcal{D}_{\dot\alpha}\mathcal{D}_{\dot\beta}\Phi^\dagger\Big| + E_\mu{}^{\tilde\alpha}\mathcal{D}_{\tilde\alpha}\mathcal{D}_{\dot\alpha}\mathcal{D}_{\dot\beta}\Phi^\dagger\Big| + E_{\mu\bar{\tilde\alpha}}\mathcal{D}^{\bar{\tilde\alpha}}\mathcal{D}_{\dot\alpha}\mathcal{D}_{\dot\beta}\Phi^\dagger\Big| \\
&= -2\varepsilon_{\dot\alpha\dot\beta}\Big(\partial_\mu F^\dagger + \frac{\sqrt{2}}{2}i\psi_\mu\sigma^\nu\hat{D}_\nu\bar{\chi} - \frac{\sqrt{2}}{12}\psi_\mu\sigma^\nu\bar{\chi}b_\nu + \frac{\sqrt{2}}{6}\bar{\psi}_\mu\cdot\bar{\chi}M\Big)\,.
\end{aligned}$$

The variation of $\hat{D}_\mu\bar{\chi}_{\dot\alpha}$ is given by

$$\begin{aligned}
\delta(\hat{D}_\mu\bar{\chi}_{\dot\alpha}) &= \frac{1}{\sqrt{2}}\delta(\mathcal{D}_\mu\overline{\mathcal{D}}_{\dot\alpha}\Phi^\dagger)\Big| \\
&= \frac{1}{\sqrt{2}}\Big[-\xi^M\mathcal{D}_M\mathcal{D}_\mu\overline{\mathcal{D}}_{\dot\alpha}\Phi^\dagger\Big| + \mathcal{D}^M\overline{\mathcal{D}}_{\dot\alpha}\Phi^\dagger L_{M\mu}\Big| + \mathcal{D}_\mu\mathcal{D}^M\Phi^\dagger L_{M\dot\alpha}\Big|\Big] \\
&= -\frac{1}{\sqrt{2}}\varepsilon^\alpha\mathcal{D}_\alpha\mathcal{D}_\mu\overline{\mathcal{D}}_{\dot\alpha}\Phi^\dagger\Big| - \frac{1}{\sqrt{2}}\bar{\varepsilon}_{\dot\gamma}\overline{\mathcal{D}}^{\dot\gamma}\mathcal{D}_\mu\overline{\mathcal{D}}_{\dot\alpha}\Phi^\dagger\Big|\,.
\end{aligned}$$

Using (3.8) we obtain

$$\begin{aligned}
\mathcal{D}_\alpha\mathcal{D}_\mu\overline{\mathcal{D}}_{\dot\alpha}\Phi^\dagger\Big| &= T_{\alpha\mu}{}^\beta\mathcal{D}_\beta\overline{\mathcal{D}}_{\dot\alpha}\Phi^\dagger\Big| - T_{\alpha\mu}{}^{\dot\beta}\overline{\mathcal{D}}_{\dot\beta}\overline{\mathcal{D}}_{\dot\alpha}\Phi^\dagger\Big| \\
&\quad -R_{\alpha\mu\dot\gamma\dot\alpha}\overline{\mathcal{D}}^{\dot\gamma}\Phi^\dagger\Big| - 2i\sigma^\nu{}_{\alpha\dot\alpha}\mathcal{D}_\mu\mathcal{D}_\nu\Phi^\dagger\Big|\,,
\end{aligned}$$

$$\overline{\mathcal{D}}_{\dot\beta}\mathcal{D}_\mu\overline{\mathcal{D}}_{\dot\alpha}\Phi^\dagger\Big| = T_{\dot\beta\mu}{}^\beta\mathcal{D}_\beta\overline{\mathcal{D}}_{\dot\alpha}\Phi^\dagger\Big| - T_{\dot\beta\mu}{}^{\dot\gamma}\overline{\mathcal{D}}_{\dot\gamma}\overline{\mathcal{D}}_{\dot\alpha}\Phi^\dagger\Big| - R_{\dot\beta\mu\dot\gamma\alpha}\overline{\mathcal{D}}^{\dot\gamma}\Phi^\dagger\Big| + \mathcal{D}_\mu\overline{\mathcal{D}}_{\dot\beta}\overline{\mathcal{D}}_{\dot\alpha}\Phi^\dagger\Big|\,.$$

To finish up the computations we just have to use Tables 3.3 and 3.4 for the zero components of the torsion and curvature tensors and the results of Sec. 4.1.1 for the various derivatives. However, the final result is not very enlightening. We will see in section 4.1.3 that it will be more convenient to obtain the variation of some linear combinations involving $\hat{D}\overline{\chi}$. Notice, that there is an alternative way to determine the transformation of covariant derivatives without invoking any reference to any superspace, using powerful principles to avoid the lengthy computations of $\delta\hat{D}_\mu\chi$. See [Freedman and Van Proeyen (2012)] for more details.

4.1.3 *The components of the chiral superfield Ξ associated to an antichiral superfield $\Phi^\dagger$*

In this section, we apply the results of Sec. 4.1.1 and Table 3.8 to the computation of the component fields of the chiral superfield Ξ, associated to the antichiral superfield $\Phi^\dagger$. Recall that the chiral superfield Ξ is defined as

$$\Xi = \Big(\overline{\mathcal{D}}\cdot\overline{\mathcal{D}} - 8\mathcal{R}\Big)\Phi^\dagger ,$$

which leads to[4]

$$\begin{aligned} \Xi\Big| &= \overline{\mathcal{D}}\cdot\overline{\mathcal{D}}\Phi^\dagger\Big| - 8\mathcal{R}\Phi^\dagger\Big| ,\\ \mathcal{D}_\alpha\Xi\Big| &= \mathcal{D}_\alpha\overline{\mathcal{D}}\cdot\overline{\mathcal{D}}\Phi^\dagger\Big| - 8(\mathcal{D}_\alpha\mathcal{R})\Phi^\dagger\Big| ,\\ \mathcal{D}\cdot\mathcal{D}\Xi\Big| &= \mathcal{D}\cdot\mathcal{D}\overline{\mathcal{D}}\cdot\overline{\mathcal{D}}\Phi^\dagger\Big| - 8(\mathcal{D}\cdot\mathcal{D}\mathcal{R})\Phi^\dagger\Big| , \end{aligned}$$

where we have used the fact that $\Phi^\dagger$ is antichiral, *i.e.*, $\mathcal{D}_\alpha\Phi^\dagger = 0$. Similarly, the components of the antichiral superfield $\Xi^\dagger$ associated to the chiral superfield Φ,

$$\Xi^\dagger = \Big(\mathcal{D}\cdot\mathcal{D} - 8\mathcal{R}^\dagger\Big)\Phi ,$$

are given by

$$\begin{aligned} \Xi^\dagger\Big| &= \mathcal{D}\cdot\mathcal{D}\Phi\Big| - 8\mathcal{R}^\dagger\Phi\Big| ,\\ \overline{\mathcal{D}}_{\dot\alpha}\Xi^\dagger\Big| &= \overline{\mathcal{D}}_{\dot\alpha}\mathcal{D}\cdot\mathcal{D}\Phi\Big| - 8(\overline{\mathcal{D}}_{\dot\alpha}\mathcal{R}^\dagger)\Phi\Big| ,\\ \overline{\mathcal{D}}\cdot\overline{\mathcal{D}}\Xi^\dagger\Big| &= \overline{\mathcal{D}}\cdot\overline{\mathcal{D}}\mathcal{D}\cdot\mathcal{D}\Phi\Big| - 8(\overline{\mathcal{D}}\cdot\overline{\mathcal{D}}\mathcal{R}^\dagger)\Phi\Big| . \end{aligned}$$

The complete analytical results are presented in Table 4.1, together with derivatives of the physical degrees of freedom covariant with respect to the supergravity transformations.

We would like to conclude this section by a remark. In the first papers on supergravity [Cremmer et al. (1979, 1982, 1983b)] (as in the first papers on supersymmetry), the computations were done without invoking superfields. Consider a

[4]The normalisation chosen is slightly different from that of the usual chiral superfields. This choice will be clarified in Sec. 6.2.

Table 4.1 Chiral superfields.

A chiral superfield is defined by $\overline{\mathcal{D}}_{\dot\alpha}\Phi = 0$. The component fields of Φ, *i.e.*, its scalar, fermionic and auxiliary components are given by $\phi = \Phi\big|$, $\chi_\alpha = \frac{1}{\sqrt{2}}\mathcal{D}_\alpha\Phi\big|$ and $F = \frac{1}{4}\mathcal{D}\cdot\mathcal{D}\Phi\big|$. Similarly, if $\Phi^\dagger$ is an antichiral superfield, it satisfies $\mathcal{D}_\alpha\Phi^\dagger = 0$ and its component fields are $\phi^\dagger = \Phi^\dagger\big|$, $\bar\chi^{\dot\alpha} = \frac{1}{\sqrt{2}}\overline{\mathcal{D}}^{\dot\alpha}\Phi^\dagger\big|$ and $F^\dagger = \frac{1}{4}\overline{\mathcal{D}}\cdot\overline{\mathcal{D}}\Phi^\dagger\big|$.
To the antichiral superfield $\phi^\dagger$ we can associate the $\Xi = (\overline{\mathcal{D}}\cdot\overline{\mathcal{D}} - 8\mathcal{R})\Phi^\dagger$, which is a chiral superfield, and to the chiral superfield Φ, the antichiral superfield $\Xi^\dagger = (\mathcal{D}\cdot\mathcal{D} - 8\mathcal{R}^\dagger)\Phi$. The components of Ξ read

$$
\begin{aligned}
\Xi\Big| &= 4F^\dagger + \frac{4}{3}M\phi^\dagger\,,\\
\mathcal{D}_\alpha\Xi\Big| &= -4i\sqrt{2}(\sigma^\mu\hat{D}_\mu\bar\chi)_\alpha + \frac{2\sqrt{2}}{3}b_\mu(\sigma^\mu\bar\chi)_\alpha + \frac{8}{3}(\sigma^{\mu\nu}\psi_{\mu\nu})_\alpha\phi^\dagger - \frac{4i}{3}(\sigma^\mu\bar\psi_\mu)_\alpha M\phi^\dagger + \frac{4i}{3}\psi^\mu_\alpha b_\mu\phi^\dagger\,,\\
\mathcal{D}\cdot\mathcal{D}\Xi\Big| &= -16e_\mu{}^{\tilde\mu}\mathcal{D}_{\tilde\mu}\hat{D}^\mu\phi^\dagger + \frac{32i}{3}b^\mu\hat{D}_\mu\phi^\dagger + 8\sqrt{2}\bar\psi_\mu\cdot\hat{D}^\mu\bar\chi - \frac{8\sqrt{2}}{3}\bar\chi\bar\sigma^{\mu\nu}\bar\psi_{\mu\nu} - \frac{32}{3}M^*F^\dagger + 4\sqrt{2}i\bar\chi\cdot\bar\psi_\mu b^\mu\\
&\quad + \frac{8\sqrt{2}i}{3}\bar\chi\bar\sigma^{\nu\mu}\bar\psi_\mu b_\nu + \phi^\dagger\Big[-\frac{8}{3}e_\mu{}^{\tilde\mu}e_\nu{}^{\tilde\nu}R_{\tilde\mu\tilde\nu}{}^{\mu\nu}\Big| - \frac{32}{9}MM^* + \frac{16}{9}b_\mu b^\mu + \frac{16i}{3}e_\mu{}^{\tilde\mu}\mathcal{D}_{\tilde\mu}b^\mu\\
&\quad + \frac{8}{3}\bar\psi_\mu\cdot\bar\psi^\mu M + \frac{8}{3}\psi_\nu\sigma^\nu\bar\psi_\mu b^\mu + \frac{16i}{3}\bar\psi^\mu\bar\sigma^\nu\psi_{\mu\nu} + \frac{2}{3}\varepsilon^{\mu\nu\rho\sigma}\big(\psi_\mu\sigma_\sigma\bar\psi_{\nu\rho} + \bar\psi_\mu\bar\sigma_\sigma\psi_{\nu\rho}\big)\Big]\,,
\end{aligned}
$$

while those of $\Xi^\dagger$ are

$$
\begin{aligned}
\Xi^\dagger\Big| &= 4F + \frac{4}{3}M^*\phi\,,\\
\overline{\mathcal{D}}_{\dot\alpha}\Xi^\dagger\Big| &= 4i\sqrt{2}(\hat{D}_\mu\chi\sigma^\mu)_{\dot\alpha} + \frac{2\sqrt{2}}{3}b_\mu(\chi\sigma^\mu)_{\dot\alpha} - \frac{8}{3}(\bar\psi_{\mu\nu}\bar\sigma^{\mu\nu})_{\dot\alpha}\phi + \frac{4i}{3}(\psi_\mu\sigma^\mu)_\alpha M^*\phi - \frac{4i}{3}\bar\psi^\mu_{\dot\alpha}b_\mu\phi\,,\\
\overline{\mathcal{D}}\cdot\overline{\mathcal{D}}\Xi^\dagger\Big| &= -16e_\mu{}^{\tilde\mu}\mathcal{D}_{\tilde\mu}\hat{D}^\mu\phi - \frac{32i}{3}b^\mu\hat{D}_\mu\phi + 8\sqrt{2}\psi_\mu\cdot\hat{D}^\mu\chi - \frac{8\sqrt{2}}{3}\chi\sigma^{\mu\nu}\psi_{\mu\nu} - \frac{32}{3}MF - 4\sqrt{2}i\chi\cdot\psi_\mu b^\mu\\
&\quad - \frac{8\sqrt{2}i}{3}\chi\sigma^{\mu\nu}\psi_\mu b_\nu + \phi\Big[-\frac{8}{3}e_\mu{}^{\tilde\mu}e_\nu{}^{\tilde\nu}R_{\tilde\mu\tilde\nu}{}^{\mu\nu}\Big| - \frac{32}{9}MM^* + \frac{16}{9}b_\mu b^\mu - \frac{16i}{3}e_\mu{}^{\tilde\mu}\mathcal{D}_{\tilde\mu}b^\mu\\
&\quad + \frac{8}{3}\psi_\mu\cdot\psi^\mu M^* + \frac{8}{3}\psi_\mu\sigma^\nu\bar\psi_\nu b^\mu + \frac{16i}{3}\psi^\mu\sigma^\nu\bar\psi_{\mu\nu} - \frac{2}{3}\varepsilon^{\mu\nu\rho\sigma}\big(\psi_\mu\sigma_\sigma\bar\psi_{\nu\rho} + \bar\psi_\mu\bar\sigma_\sigma\psi_{\nu\rho}\big)\Big]\,,
\end{aligned}
$$

and depend on the components of the gravitation supermultiplet, *i.e.*, the gravitino $(\psi_{\tilde\mu}{}^\alpha, \bar\psi_{\tilde\mu\dot\alpha})$, the graviton $e_{\tilde\mu}{}^\mu$ and the auxiliary fields M and b^μ. We have also introduced the covariant derivatives

$$
\begin{aligned}
\hat{D}_\mu\phi &= e_\mu{}^{\tilde\mu}\big(\partial_{\tilde\mu}\phi - \tfrac{\sqrt{2}}{2}\psi_{\tilde\mu}\cdot\chi\big)\,, &\quad \hat{D}_\mu\chi_\alpha &= e_\mu{}^{\tilde\mu}\big(\mathcal{D}_{\tilde\mu}\chi_\alpha + \tfrac{1}{\sqrt{2}}\psi_{\tilde\mu\alpha}F - \tfrac{i}{\sqrt{2}}(\sigma^\nu\bar\psi_{\tilde\mu})_\alpha\hat{D}_\nu\phi\big)\\
\hat{D}_\mu\phi^\dagger &= e_\mu{}^{\tilde\mu}\big(\partial_{\tilde\mu}\phi^\dagger - \tfrac{\sqrt{2}}{2}\bar\psi_{\tilde\mu}\cdot\bar\chi\big)\,, &\quad \hat{D}_\mu\bar\chi_{\dot\alpha} &= e_\mu{}^{\tilde\mu}\big(\mathcal{D}_{\tilde\mu}\bar\chi_{\dot\alpha} + \tfrac{1}{\sqrt{2}}\bar\psi_{\tilde\mu\dot\alpha}F^\dagger + \tfrac{i}{\sqrt{2}}(\psi_{\tilde\mu}\sigma^\nu)_{\dot\alpha}\hat{D}_\nu\phi^\dagger\big)
\end{aligned}
$$

composite multiplet, *i.e.*, such that its components are expressed in terms of the components of elementary multiplets. In this language the multiplet Φ is elementary whereas the multiplet Ξ is composite. Thus, knowing the transformations law of multiplets allows to compute all the components of a composite multiplet from its first component. Indeed, from this component and its transformation properties it is possible to obtain recursively the higher components of the composite multiplet. In practice, we just have to compare the transformation of the composite multiplet with the same transformation computed in terms of the transformation laws of the elementary multiplets appearing in the composite field. For the chiral superfield $\Xi = (\tilde{\phi}, \tilde{\chi}, \tilde{F})$, starting from $\tilde{\phi} = 4F^\dagger + \frac{4}{3}M\phi^\dagger$ and with the transformation of the chiral multiplet ((4.29), (4.30) and (4.31)) together with the transformation of the gravity multiplet ((3.142), (3.144) and (3.145)), we easily obtain $\tilde{\chi} = \mathcal{D}\,\Xi|$ from $\delta\tilde{\phi}$. However here, the computation of the higher order term is much more tedious since we have to use the variation of $\hat{D}_\mu\bar{\chi}$ or $\psi_{\mu\nu}$. For these computations, we can use the result of the previous section for the variation of $\hat{D}_\mu\bar{\chi}$. For the variation of $\psi_{\mu\nu}$, we can use the variation of $\mathcal{D}_\alpha\mathcal{R}$ which is not complicated to obtain. There is a way to avoid this computation. Indeed, we have

$$\begin{aligned}
\delta\tilde{\chi}_\alpha &= -\xi^M\mathcal{D}_M\mathcal{D}_\alpha(\overline{\mathcal{D}}\cdot\overline{\mathcal{D}} - 8\mathcal{R})\Phi^\dagger\Big| + \mathcal{D}^M(\overline{\mathcal{D}}\cdot\overline{\mathcal{D}} - 8\mathcal{R})\Phi^\dagger L_{M\alpha}\Big| \\
&= -\varepsilon^\beta\mathcal{D}_\beta\mathcal{D}_\alpha\overline{\mathcal{D}}\cdot\overline{\mathcal{D}}\Phi^\dagger\Big| + 8\varepsilon^\beta\mathcal{D}_\beta\mathcal{D}_\alpha R\Big|\phi^\dagger - \bar{\varepsilon}_{\dot\beta}\overline{\mathcal{D}}^{\dot\beta}\mathcal{D}_\alpha(\overline{\mathcal{D}}\cdot\overline{\mathcal{D}} - 8\mathcal{R})\Phi^\dagger\Big| \\
&\equiv \Big(\frac{1}{2}\varepsilon_\alpha\tilde{F} - 2i(\sigma^\mu\bar{\varepsilon})_\alpha\hat{D}_\mu\tilde{\phi}\Big)\,.
\end{aligned}$$

The first term readily gives $\tilde{F}$ and the second leads to $\hat{D}_\mu\tilde{\phi}$. In particular this give the variation,

$$\begin{aligned}
&\delta\Big[-4i\sqrt{2}\big(\sigma^\mu\hat{D}_\mu\bar{\chi}\big)_\alpha + \frac{2\sqrt{2}}{3}b_\mu(\sigma^\mu\bar{\chi})_\alpha + \frac{8}{3}(\sigma^{\mu\nu}\psi_{\mu\nu})_\alpha\phi^\dagger \\
&\quad -\frac{4i}{3}(\sigma^\mu\bar{\psi}_\mu)_\alpha M\phi^\dagger + \frac{4i}{3}\psi^\mu_\alpha b_\mu\phi^\dagger\Big] \\
&= \epsilon_\alpha\Big\{-8e_\mu{}^{\tilde\mu}\mathcal{D}_{\tilde\mu}\hat{D}^\mu\phi^\dagger + \frac{16i}{3}b^\mu\hat{D}_\mu\phi^\dagger + 4\sqrt{2}\bar{\psi}_\mu\cdot\hat{D}^\mu\bar{\chi} - \frac{4\sqrt{2}}{3}\bar{\chi}\bar{\sigma}^{\mu\nu}\bar{\psi}_{\mu\nu} \\
&\quad -\frac{16}{3}M^*F^\dagger + 2\sqrt{2}i\bar{\chi}\cdot\bar{\psi}_\mu b^\mu + \frac{4\sqrt{2}i}{3}\bar{\chi}\bar{\sigma}^{\nu\mu}\bar{\psi}_\mu b_\nu \\
&\quad +\phi^\dagger\Big[-\frac{4}{3}e_\mu{}^{\tilde\mu}e_\nu{}^{\tilde\nu}R_{\tilde\mu\tilde\nu}{}^{\mu\nu}\Big| - \frac{16}{9}MM^* + \frac{8}{9}b_\mu b^\mu + \frac{8i}{3}e_\mu{}^{\tilde\mu}\mathcal{D}_{\tilde\mu}b^\mu \\
&\quad +\frac{4}{3}\bar{\psi}_\mu\cdot\bar{\psi}^\mu M + \frac{4}{3}\psi_\nu\sigma^\nu\bar{\psi}_\mu b^\mu + \frac{8i}{3}\bar{\psi}^\mu\bar{\sigma}^\nu\psi_{\mu\nu} \\
&\quad +\frac{1}{3}\varepsilon^{\mu\nu\rho\sigma}\big(\psi_\mu\sigma_\sigma\bar{\psi}_{\nu\rho} + \bar{\psi}_\mu\bar{\sigma}_\sigma\psi_{\nu\rho}\big)\Big]\Big\} \\
&\quad -2i(\sigma^\mu\bar{\varepsilon})_\alpha\Big\{4\hat{D}_\mu F^\dagger + \frac{4}{3}M\hat{D}_\mu\phi^\dagger + \frac{4}{3}\phi^\dagger\hat{D}_\mu M\Big\}\,,
\end{aligned}$$

whence we easily obtain

$$
\begin{aligned}
\hat{D}_\mu M &\equiv -6\mathcal{D}_\mu \mathcal{R}\Big| \\
&= -6E_\mu{}^{\tilde{M}} \mathcal{D}_{\tilde{M}} \mathcal{R}\Big| \\
&= \partial_\mu M - \psi_\mu \sigma^{\nu\rho} \psi_{\nu\rho} + \frac{i}{2} \psi_\mu \sigma^\nu \overline{\psi}_\nu M - \frac{i}{2} \psi_\mu \cdot \psi_\nu b^\nu .
\end{aligned}
$$

4.2 Vector superfields

In Sec. 2.3, we have introduced vector superfields to describe gauge supermultiplets in $N = 1$ global supersymmetry. This hold over to local supersymmetry, so that both in flat and curved superspace, vector superfields V are defined by the reality condition

$$
V = V^\dagger . \tag{4.32}
$$

4.2.1 *Properties of vector superfields and their superderivatives*

In this section, we focus on a detailed calculation of the component fields of a vector superfield V associated to a given gauge group, abelian or not, as well as on the lowest-order components of its derivatives up to the third order which are necessary for the construction of the action. We also investigate the consequences of the choice of the Wess-Zumino gauge, where all the non-physical fields are vanishing, and which we adopt to undertake our computations. Vector supermultiplets/superfields were introduced in supergravity in superspace formalism [Wess and Zumino (1978a)] in the tensor calculus approach [Stelle and West (1978b)] or when considering the first invariant Lagrangians [Ferrara et al. (1976b); Freedman and Schwarz (1977); Cremmer et al. (1979, 1982, 1978b, 1983b)].

Definition of the component fields.

Like for the case of chiral superfields (see Sec. 4.1.1), defining the component fields after expanding superfield expressions in terms of Grassmann variables with flat indices is not the most convenient choice. It is better to similarly define the component fields in terms of the lowest-order components of the vector superfield itself and its superderivatives, as was done in (2.8) [Wess and Zumino (1978a); Wess and Bagger (1992)],

$$
\begin{aligned}
C &= V\Big| , \\
\chi_\alpha = \mathcal{D}_\alpha V\Big| , &\quad \overline{\chi}_{\dot\alpha} = \overline{\mathcal{D}}_{\dot\alpha} V\Big| , \\
F = -\frac{1}{2}\mathcal{D}\cdot\mathcal{D}V\Big| , &\quad F^\dagger = -\frac{1}{2}\overline{\mathcal{D}}\cdot\overline{\mathcal{D}}V\Big| , \\
v_\mu &= \frac{1}{4}\overline{\sigma}_\mu{}^{\dot\alpha\alpha}[\mathcal{D}_\alpha, \overline{\mathcal{D}}_{\dot\alpha}]V\Big| ,
\end{aligned}
\tag{4.33}
$$

where C is a real scalar field, where F is a complex scalar field, where $(\chi_\alpha, \overline{\chi}^{\dot\alpha})$ is a Majorana fermion and where v_μ is a real vector field. We recall that all these fields except the vector boson v_μ are non-physical (as initially explained in Sec. 2.3) and can be eliminated after a suitable gauge choice. The construction of supersymmetric and gauge-invariant Lagrangians associated to the superfield V requires to square the superfield strength tensors W_α and $\overline{W}_{\dot\alpha}$ introduced in (2.11), a chiral and an antichiral spinorial superfield related to the vector superfield V. Those quantities can be made covariant with respect to supergravity transformations with the use of the (covariant) projector operators introduced in (4.19), and, in the abelian case [Wess and Bagger (1992)], they read

$$\begin{aligned} W_\alpha &= -\frac{1}{4}\Big[\overline{\mathcal{D}}\cdot\overline{\mathcal{D}} - 8\mathcal{R}\Big]\mathcal{D}_\alpha V\,,\\ \overline{W}_{\dot\alpha} &= -\tfrac{1}{4}\Big[\mathcal{D}\cdot\mathcal{D} - 8\mathcal{R}^\dagger\Big]\overline{\mathcal{D}}_{\dot\alpha} V\,, \end{aligned} \tag{4.34}$$

whilst, in the non-abelian case, they are given by [Wess and Bagger (1992)]

$$\begin{aligned} W_\alpha &= -\tfrac{1}{4}\Big[\overline{\mathcal{D}}\cdot\overline{\mathcal{D}} - 8\mathcal{R}\Big]e^{2gV}\mathcal{D}_\alpha e^{-2gV}\,,\\ \overline{W}_{\dot\alpha} &= \tfrac{1}{4}\Big[\mathcal{D}\cdot\mathcal{D} - 8\mathcal{R}^\dagger\Big]e^{-2gV}\overline{\mathcal{D}}_{\dot\alpha} e^{2gV}\,, \end{aligned} \tag{4.35}$$

where g denotes the coupling constant associated to the gauge group related to V. Generalising (2.12) to the curved superspace case, the superfields W_α and $\overline{W}_{\dot\alpha}$ and their superderivatives allow to define the higher-order component fields of the vector superfield V, *i.e.*, the physical gaugino Majorana spinor $(\lambda_\alpha, \overline{\lambda}^{\dot\alpha})$ and the auxiliary D-term. In the abelian case they are hence defined by

$$\begin{aligned} \lambda_\alpha &= -iW_\alpha\big|\,,\\ \overline{\lambda}_{\dot\alpha} &= i\overline{W}_{\dot\alpha}\big|\,,\\ D &= -\frac{1}{2}\mathcal{D}^\alpha W_\alpha\big| = -\frac{1}{2}\overline{\mathcal{D}}_{\dot\alpha}\overline{W}^{\dot\alpha}\big|\,, \end{aligned} \tag{4.36}$$

whilst, in the non-abelian case, they read

$$\begin{aligned} \lambda_\alpha &= \frac{i}{2g}W_\alpha\big|\,,\\ \overline{\lambda}_{\dot\alpha} &= -\frac{i}{2g}\overline{W}_{\dot\alpha}\big|\,,\\ D &= \frac{1}{4g}\mathcal{D}^\alpha W_\alpha\big| = \frac{1}{4g}\overline{\mathcal{D}}_{\dot\alpha}\overline{W}^{\dot\alpha}\big|\,. \end{aligned} \tag{4.37}$$

The Wess-Zumino gauge.

In global supersymmetry, the superfields W_α and $\overline{W}_{\dot\alpha}$ associated to a vector superfield V are chiral and antichiral, respectively. These properties can be extended to the case of local supersymmetry. It can indeed be shown, in a fashion analogous to (4.17), that the operator $\overline{\mathcal{D}}\cdot\overline{\mathcal{D}} - 8\mathcal{R}$ is projecting any superfield X containing an arbitrary number of undotted indices onto a chiral superfield,

$$\Xi_{\alpha_1\ldots\alpha_n} = \Big[\overline{\mathcal{D}}\cdot\overline{\mathcal{D}} - 8\mathcal{R}\Big] X_{\alpha_1\ldots\alpha_n}$$

and

$$\overline{\mathcal{D}}_{\dot\alpha}\Xi_{\alpha_1\ldots\alpha_n} = 0\ ,$$

since $T_{\dot\beta\dot\gamma}{}^{\underline{\alpha}} = T_{\dot\beta\dot\gamma}{}^{\mu} = R_{\dot\beta\dot\gamma\alpha\beta} = 0$ as given in (3.29) and (3.35). In the case of superfields containing at least one dotted index, as, *e.g.*, a vector superfield, however this does not hold anymore due to additional terms depending on the curvature tensor. Subsequently, the superfield W_α is chiral, both in the abelian and non-abelian cases, and similarly, one can show that the $\overline{W}_{\dot\alpha}$ superfield is always antichiral.

For any chiral superfield Λ, the (abelian) gauge transformation $V \to V+\Lambda+\Lambda^\dagger$ (trivially) preserves the reality condition (4.32). Under such a transformation, the spinorial superfield W_α transforms as

$$W_\alpha \ \to\ W_\alpha - \frac{1}{4}\Big[\overline{\mathcal{D}}\cdot\overline{\mathcal{D}} - 8\mathcal{R}\Big]\mathcal{D}_\alpha\Lambda = W_\alpha\ , \tag{4.38}$$

due to (4.24). Generalising to non-abelian gauge groups, for which the associated gauge transformation laws for a vector superfield V are defined to be

$$e^{2gV} \to e^{-2ig\Lambda} e^{2gV} e^{2ig\Lambda^\dagger}\ ,$$

one gets

$$\begin{aligned} W_\alpha &\to e^{-2ig\Lambda} W_\alpha e^{2ig\Lambda} - \frac{1}{4} e^{-2ig\Lambda}\Big[\overline{\mathcal{D}}\cdot\overline{\mathcal{D}} - 8\mathcal{R}\Big]\mathcal{D}_\alpha e^{2ig\Lambda} \\ &= e^{-2ig\Lambda} W_\alpha e^{2ig\Lambda}\ , \end{aligned} \tag{4.39}$$

where we have once again used (4.24). Since this property can be easily extended to the antichiral spinorial superfield $\overline{W}_{\dot\alpha}$, the square of the superfield strength tensors are thus gauge invariant quantities both in the abelian and the non-abelian gauge theories and the spinorial superfields W_α and $\overline{W}_{\dot\alpha}$ can be safely used in supersymmetric and gauge-invariant Lagrangian building, as well as in global supersymmetry.

From these properties, we adopt, like in global supersymmetry, a specific gauge to perform our calculations, the so-called Wess-Zumino gauge, where the non-physical fields C, $(\chi_\alpha, \overline{\chi}^{\dot\alpha})$ and F vanish. Consequently, the Wess-Zumino gauge is defined by the following relations

$$V\big| = 0\,, \quad \mathcal{D}_\alpha V\big| = 0\,, \quad \overline{\mathcal{D}}_{\dot\alpha} V\big| = 0\,, \quad \mathcal{D}\cdot\mathcal{D}V\big| = 0\,, \quad \overline{\mathcal{D}}\cdot\overline{\mathcal{D}}V\big| = 0\,. \tag{4.40}$$

The construction of invariant actions related to supergravity, and containing chiral and vector supermultiplets requires, in addition to the calculation of the superderivatives of chiral superfields addressed in Sec. 4.1, the computation of several superderivatives of vector superfields, which are described in the rest of this section. This would allow, *e.g.*, to build superfield strength tensors both in the abelian and non-abelian cases with the help of the spinorial superfield W_α or to compute gauge interaction terms between vector and chiral superfields. Let us stress that all the calculations performed in the sequel are only valid in the Wess-Zumino gauge and that the definitions (4.40) are assumed to always hold in the following.

First-order derivatives of a vector superfield in the Wess-Zumino gauge.

In the Wess-Zumino gauge, the fermionic quantities $\mathcal{D}_\alpha V\big|$ and $\overline{\mathcal{D}}_{\dot\alpha} V\big|$ are vanishing, by definition, as seen in Eq. (4.40) and the only non-trivial first-order superderivative to calculate is $\mathcal{D}_\mu V\big|$. However, since $V\big| = 0$, as shown in (4.40), we also get $\mathcal{D}_{\tilde\mu} V\big| = 0$ because of (3.5), and furthermore

$$\mathcal{D}_\mu V\big| = E_\mu{}^{\tilde\mu}\mathcal{D}_{\tilde\mu} V\big| + E_\mu{}^{\tilde\alpha}\mathcal{D}_{\tilde\alpha} V\big| + E_{\mu\tilde{\dot\alpha}}\overline{\mathcal{D}}^{\tilde{\dot\alpha}} V\big| = 0\,. \tag{4.41}$$

The lowest-order component of all the first-order derivatives of a vector superfield V are then vanishing in the Wess-Zumino gauge,

$$\begin{aligned}
\mathcal{D}_\alpha V\big| &= 0\,,\\
\overline{\mathcal{D}}_{\dot\alpha} V\big| &= 0\,,\\
\mathcal{D}_\mu V\big| &= 0\,.
\end{aligned}$$

Second-order derivatives of a vector superfield in the Wess-Zumino gauge.

We now turn to the calculation of the lowest-order component of the second-order superderivatives of a vector superfield V. Firstly, using the constraints (3.29) together with (3.8) and (4.42) one obtains

$$\begin{aligned}
\{\mathcal{D}_\alpha, \overline{\mathcal{D}}_{\dot\alpha}\}V\big| &= -2i\sigma^\mu{}_{\alpha\dot\alpha}\mathcal{D}_\mu V\big| = 0\,,\\
\{\mathcal{D}_\alpha, \mathcal{D}_\beta\}V\big| &= 0\,,\\
\{\overline{\mathcal{D}}_{\dot\alpha}, \overline{\mathcal{D}}_{\dot\beta}\}V\big| &= 0\,.
\end{aligned} \tag{4.42}$$

Secondly, (4.33) together with (B.7) give

$$v_\mu = \frac{1}{4}\bar{\sigma}_\mu{}^{\dot{\alpha}\alpha}[\mathcal{D}_\alpha, \overline{\mathcal{D}}_{\dot{\alpha}}]V\big| \tag{4.43}$$

and hence

$$v_{\alpha\dot{\alpha}} = \frac{1}{2}[\mathcal{D}_\alpha, \overline{\mathcal{D}}_{\dot{\alpha}}]V\big|\,,$$

whilst the antisymmetry property (4.8), together with (4.40), lead to

$$\begin{aligned}
[\mathcal{D}_\alpha, \mathcal{D}_\beta]V\big| &= \varepsilon_{\alpha\beta}\mathcal{D}\cdot\mathcal{D}V\big| = 0\,,\\
[\overline{\mathcal{D}}_{\dot{\alpha}}, \overline{\mathcal{D}}_{\dot{\beta}}]V\big| &= \varepsilon_{\dot{\beta}\dot{\alpha}}\overline{\mathcal{D}}\cdot\overline{\mathcal{D}}V\big| = 0\,.
\end{aligned}$$

It follows immediately that

$$\begin{aligned}
&\mathcal{D}_\alpha\overline{\mathcal{D}}_{\dot{\alpha}}V\big| = -\overline{\mathcal{D}}_{\dot{\alpha}}\mathcal{D}_\alpha V\big| = v_{\alpha\dot{\alpha}}\,,\\
&\mathcal{D}_\alpha\mathcal{D}_\beta V\big| = 0\,,\\
&\overline{\mathcal{D}}_{\dot{\alpha}}\overline{\mathcal{D}}_{\dot{\beta}}V\big| = 0\,.
\end{aligned} \tag{4.44}$$

The computation of the two superderivatives $\mathcal{D}_\mu\mathcal{D}_\alpha V\big|$ and $\mathcal{D}_\mu\overline{\mathcal{D}}_{\dot{\alpha}}V\big|$ is immediate after converting the Lorentz index μ to an Einstein index,

$$\begin{aligned}
\mathcal{D}_\mu\mathcal{D}_\alpha V\big| &= E_\mu{}^{\tilde{\mu}}\mathcal{D}_{\tilde{\mu}}\mathcal{D}_\alpha V\big| + E_\mu{}^{\tilde{\alpha}}\mathcal{D}_{\tilde{\alpha}}\mathcal{D}_\alpha V\big| + E_{\mu\tilde{\dot{\alpha}}}\overline{\mathcal{D}}^{\tilde{\dot{\alpha}}}\mathcal{D}_\alpha V\big|\\
&= -\frac{1}{2}v_{\alpha\dot{\alpha}}\bar{\psi}_\mu{}^{\dot{\alpha}}\,,\\
\mathcal{D}_\mu\overline{\mathcal{D}}_{\dot{\alpha}}V\big| &= E_\mu{}^{\tilde{\mu}}\mathcal{D}_{\tilde{\mu}}\overline{\mathcal{D}}_{\dot{\alpha}}V\big| + E_\mu{}^{\tilde{\alpha}}\mathcal{D}_{\tilde{\alpha}}\overline{\mathcal{D}}_{\dot{\alpha}}V\big| + E_{\mu\tilde{\dot{\alpha}}}\overline{\mathcal{D}}^{\tilde{\dot{\alpha}}}\overline{\mathcal{D}}_{\dot{\alpha}}V\big|\\
&= -\frac{1}{2}\psi_\mu{}^\alpha v_{\alpha\dot{\alpha}}\,,
\end{aligned}$$

due to (4.44). These two relations allow to further deduce the two quantities

$$\begin{aligned}
\mathcal{D}_\alpha\mathcal{D}_\mu V\big| &= [\mathcal{D}_\alpha, \mathcal{D}_\mu]V\big| + \mathcal{D}_\mu\mathcal{D}_\alpha V\big|\\
&= T_{\alpha\mu}{}^M\mathcal{D}_M V\big| + \mathcal{D}_\mu\mathcal{D}_\alpha V\big|\\
&= -\frac{1}{2}v_{\alpha\dot{\alpha}}\bar{\psi}_\mu{}^{\dot{\alpha}}\,,\\
\overline{\mathcal{D}}_{\dot{\alpha}}\mathcal{D}_\mu V\big| &= [\overline{\mathcal{D}}_{\dot{\alpha}}, \mathcal{D}_\mu]V\big| + \mathcal{D}_\mu\mathcal{D}_{\dot{\alpha}}V\big|\\
&= T_{\dot{\alpha}\mu}{}^M\mathcal{D}_M V\big| + \mathcal{D}_\mu\overline{\mathcal{D}}_{\dot{\alpha}}V\big|\\
&= -\frac{1}{2}\psi_\mu{}^\alpha v_{\alpha\dot{\alpha}}\,,
\end{aligned}$$

where the results of (4.42) leads to the cancellation of the terms containing components of the torsion tensor. It is finally straightforward to derive the remaining second-order superderivative:

$$\begin{aligned}
\mathcal{D}_\mu \mathcal{D}_\nu V\big| &= E_\mu{}^{\tilde{\mu}} \mathcal{D}_{\tilde{\mu}} \mathcal{D}_\nu V\big| + E_\mu{}^{\tilde{\alpha}} \mathcal{D}_{\tilde{\alpha}} \mathcal{D}_\nu V\big| + E_{\mu\tilde{\dot{\alpha}}} \overline{\mathcal{D}}^{\tilde{\dot{\alpha}}} \mathcal{D}_\nu V\big| \\
&= \frac{1}{4}\Big(\psi_\mu{}^\alpha \overline{\psi}_\nu{}^{\dot{\alpha}} + \psi_\nu{}^\alpha \overline{\psi}_\mu{}^{\dot{\alpha}}\Big) v_{\alpha\dot{\alpha}} \\
&= \frac{1}{4}\Big(\psi_\mu \sigma^\rho \overline{\psi}_\nu + \psi_\nu \sigma^\rho \overline{\psi}_\mu\Big) v_\rho \,. \qquad (4.45)
\end{aligned}$$

Here we have used a property (4.13) which states that in the Wess-Zumino gauge, the term containing the graviton field $e_\mu{}^{\tilde{\mu}}$ vanishes. Moreover, in the second line of (4.45), we have converted spinor indices to vector indices using (B.7).

Collecting the results, in the Wess-Zumino gauge, the second-order superderivatives of a vector superfield V read

$$\begin{aligned}
\mathcal{D}_\alpha \overline{\mathcal{D}}_{\dot{\alpha}} V\big| &= -\overline{\mathcal{D}}_{\dot{\alpha}} \mathcal{D}_\alpha V\big| = v_{\alpha\dot{\alpha}} \,, \\
\mathcal{D}_\alpha \mathcal{D}_\beta V\big| &= \overline{\mathcal{D}}_{\dot{\alpha}} \overline{\mathcal{D}}_{\dot{\beta}} V\big| = 0 \,, \\
\mathcal{D}_\alpha \mathcal{D}_\mu V\big| &= \mathcal{D}_\mu \mathcal{D}_\alpha V\big| = -\frac{1}{2} v_{\alpha\dot{\alpha}} \overline{\psi}_\mu{}^{\dot{\alpha}} = -\frac{1}{2}(\sigma^\nu \overline{\psi}_\mu)_\alpha \, v_\nu \,, \qquad (4.46) \\
\overline{\mathcal{D}}_{\dot{\alpha}} \mathcal{D}_\mu V\big| &= \mathcal{D}_\mu \overline{\mathcal{D}}_{\dot{\alpha}} V\big| = -\frac{1}{2} \psi_\mu{}^\alpha v_{\alpha\dot{\alpha}} = -\frac{1}{2}(\psi_\mu \sigma^\nu)_{\dot{\alpha}} \, v_\nu \,, \\
\mathcal{D}_\mu \mathcal{D}_\nu V\big| &= \frac{1}{4}\Big(\psi_\mu \sigma^\rho \overline{\psi}_\nu + \psi_\nu \sigma^\rho \overline{\psi}_\mu\Big) v_\rho \,,
\end{aligned}$$

where we have converted the remaining spinor indices into vector indices.

Third-order derivatives of a vector superfield in the Wess-Zumino gauge.

In Sec. 4.2.2, we will calculate the component fields of the spinorial superfields W_α and $\overline{W}_{\dot{\alpha}}$ associated to a vector superfield V as defined in (4.34) and (4.35) in the abelian and non-abelian cases, respectively. This requires the computation of several third-order superderivatives of V which we now address.

In the abelian case, and in the Wess-Zumino gauge, (4.34) and (4.36) lead immediately, to

$$
\begin{aligned}
W_\alpha\big| &= i\lambda_\alpha \\
&= -\frac{1}{4}\left[\overline{\mathcal{D}}\cdot\overline{\mathcal{D}} - 8\mathcal{R}\right]\mathcal{D}_\alpha V\big| \\
&= -\frac{1}{4}\overline{\mathcal{D}}\cdot\overline{\mathcal{D}}\mathcal{D}_\alpha V\big|\,, \\
\overline{W}_{\dot\alpha}\big| &= -i\bar\lambda_{\dot\alpha} \\
&= -\frac{1}{4}\left[\mathcal{D}\cdot\mathcal{D} - 8\mathcal{R}^\dagger\right]\overline{\mathcal{D}}_{\dot\alpha} V\big| \\
&= -\frac{1}{4}\mathcal{D}\cdot\mathcal{D}\overline{\mathcal{D}}_{\dot\alpha} V\big|\,.
\end{aligned}
\tag{4.47}
$$

Furthermore, adopting the Wess-Zumino gauge ensures that those relations still hold for non-abelian vector superfields. Anticipating the results of Sec. 4.2.2, Eq. (4.35) can be rewritten as

$$
\begin{aligned}
W_\alpha &= \frac{g}{2}\left[\overline{\mathcal{D}}\cdot\overline{\mathcal{D}} - 8\mathcal{R}\right]\left[\mathcal{D}_\alpha V + g[V,\mathcal{D}_\alpha V]\right], \\
\overline{W}_{\dot\alpha} &= \frac{g}{2}\left[\mathcal{D}\cdot\mathcal{D} - 8\mathcal{R}^\dagger\right]\left[\overline{\mathcal{D}}_{\dot\alpha} V - g[V,\overline{\mathcal{D}}_{\dot\alpha} V]\right].
\end{aligned}
\tag{4.48}
$$

In the Wess-Zumino gauge, the lowest-order components of the terms containing the commutators trivially vanish, and hence $[V,\mathcal{D}_\alpha V]\big| = 0$, $[V,\overline{\mathcal{D}}_{\dot\alpha} V]\big| = 0$. Therefore, extracting the lowest-order component of the spinorial superfields W_α and $\overline{W}_{\dot\alpha}$ in (4.48) allows to recover (4.47) in the non-abelian case.

Next, from the antisymmetry property (4.8), one gets

$$
\begin{aligned}
\left(\mathcal{D}_\alpha\mathcal{D}_\beta\overline{\mathcal{D}}_{\dot\alpha} - \mathcal{D}_\beta\mathcal{D}_\alpha\overline{\mathcal{D}}_{\dot\alpha}\right)V\big| &= 4i\varepsilon_{\alpha\beta}\bar\lambda_{\dot\alpha}\,, \\
\left(\overline{\mathcal{D}}_{\dot\alpha}\overline{\mathcal{D}}_{\dot\beta}\mathcal{D}_\alpha - \overline{\mathcal{D}}_{\dot\beta}\overline{\mathcal{D}}_{\dot\alpha}\mathcal{D}_\alpha\right)V\big| &= 4i\varepsilon_{\dot\alpha\dot\beta}\lambda_\alpha\,,
\end{aligned}
$$

whilst the constraints (3.29), together with (3.8), lead to

$$
\left(\mathcal{D}_\alpha\mathcal{D}_\beta\overline{\mathcal{D}}_{\dot\alpha} + \mathcal{D}_\beta\mathcal{D}_\alpha\overline{\mathcal{D}}_{\dot\alpha}\right)V\big| = \left(\overline{\mathcal{D}}_{\dot\alpha}\overline{\mathcal{D}}_{\dot\beta}\mathcal{D}_\alpha + \overline{\mathcal{D}}_{\dot\beta}\overline{\mathcal{D}}_{\dot\alpha}\mathcal{D}_\alpha\right)V\big| = 0\,,
$$

where the terms containing the curvature tensors are proportional to first-order superderivatives of V and thus vanish due to (4.44). Consequently,

$$
\begin{aligned}
\mathcal{D}_\alpha\mathcal{D}_\beta\overline{\mathcal{D}}_{\dot\alpha} V\big| &= 2i\varepsilon_{\alpha\beta}\bar\lambda_{\dot\alpha}\,, \\
\overline{\mathcal{D}}_{\dot\alpha}\overline{\mathcal{D}}_{\dot\beta}\mathcal{D}_\alpha V\big| &= 2i\varepsilon_{\dot\alpha\dot\beta}\lambda_\alpha\,.
\end{aligned}
\tag{4.49}
$$

Starting from this last relation, we can compute additional third-order superderivatives of V,

$$\begin{aligned}
\mathcal{D}_\alpha\overline{\mathcal{D}}_{\dot\alpha}\mathcal{D}_\beta V\big| &= \mathcal{D}_\alpha\{\overline{\mathcal{D}}_{\dot\alpha},\mathcal{D}_\beta\}V\big| - \mathcal{D}_\alpha\mathcal{D}_\beta\overline{\mathcal{D}}_{\dot\alpha}V\big| \\
&= -2i\sigma^\mu{}_{\beta\dot\alpha}\mathcal{D}_\alpha\mathcal{D}_\mu V\big| - \mathcal{D}_\alpha\mathcal{D}_\beta\overline{\mathcal{D}}_{\dot\alpha}V\big| \\
&= i\sigma^\mu{}_{\beta\dot\alpha}(\sigma^\nu\bar\psi_\mu)_\alpha v_\nu - 2i\varepsilon_{\alpha\beta}\bar\lambda_{\dot\alpha}\ , \\
\overline{\mathcal{D}}_{\dot\alpha}\mathcal{D}_\alpha\overline{\mathcal{D}}_{\dot\beta}V\big| &= \overline{\mathcal{D}}_{\dot\alpha}\{\mathcal{D}_\alpha,\overline{\mathcal{D}}_{\dot\beta}\}V\big| - \overline{\mathcal{D}}_{\dot\alpha}\overline{\mathcal{D}}_{\dot\beta}\mathcal{D}_\alpha V\big| \\
&= -2i\sigma^\mu{}_{\alpha\dot\beta}\overline{\mathcal{D}}_{\dot\alpha}\mathcal{D}_\mu V\big| - \overline{\mathcal{D}}_{\dot\alpha}\overline{\mathcal{D}}_{\dot\beta}\mathcal{D}_\alpha V\big| \\
&= i\sigma^\mu{}_{\alpha\dot\beta}(\psi_\mu\sigma^\nu)_{\dot\alpha}v_\nu - 2i\varepsilon_{\dot\alpha\dot\beta}\lambda_\alpha\ , \qquad (4.50)\\
\overline{\mathcal{D}}_{\dot\alpha}\mathcal{D}_\alpha\mathcal{D}_\beta V\big| &= \{\overline{\mathcal{D}}_{\dot\alpha},\mathcal{D}_\alpha\}\mathcal{D}_\beta V\big| - \mathcal{D}_\alpha\overline{\mathcal{D}}_{\dot\alpha}\mathcal{D}_\beta V\big| \\
&= i\varepsilon_{\alpha\beta}\Big[2\bar\lambda_{\dot\alpha} + (\bar\psi_\mu\bar\sigma^\nu\sigma^\mu)_{\dot\alpha}v_\nu\Big]\ , \\
\mathcal{D}_\alpha\overline{\mathcal{D}}_{\dot\alpha}\overline{\mathcal{D}}_{\dot\beta}V\big| &= \{\mathcal{D}_\alpha,\overline{\mathcal{D}}_{\dot\alpha}\}\overline{\mathcal{D}}_{\dot\beta}V\big| - \overline{\mathcal{D}}_{\dot\alpha}\mathcal{D}_\alpha\overline{\mathcal{D}}_{\dot\beta}V\big| \\
&= i\varepsilon_{\dot\alpha\dot\beta}\Big[2\lambda_\alpha + (\sigma^\mu\bar\sigma^\nu\psi_\mu)_\alpha v_\nu\Big]\ ,
\end{aligned}$$

by recalling the expressions of the lowest-order coefficients of the second-order superderivatives of V in (4.46) as well as the graded commutators (3.8) and through simplifications made possible by the constraints (3.29) and our gauge choice (4.40). Please let us note that the last two relations are obtained using the antisymmetry property (4.8) and that the terms containing the curvature tensor vanish once again due to Eq. (4.44).

The next third-order derivatives which remain to be calculated are $\mathcal{D}_\alpha\mathcal{D}_\beta\mathcal{D}_\gamma V\big|$ and $\overline{\mathcal{D}}_{\dot\alpha}\overline{\mathcal{D}}_{\dot\beta}\overline{\mathcal{D}}_{\dot\gamma}V\big|$. They can be computed by applying the property (3.124) to the two rank-three tensors $\mathcal{D}_\alpha\mathcal{D}_\beta\mathcal{D}_\gamma V\big|$ and $\overline{\mathcal{D}}_{\dot\alpha}\overline{\mathcal{D}}_{\dot\beta}\overline{\mathcal{D}}_{\dot\gamma}V\big|$, to get

$$\begin{aligned}
0 &= \Big[\mathcal{D}_\alpha\mathcal{D}_\beta\mathcal{D}_\gamma + \mathcal{D}_\beta\mathcal{D}_\gamma\mathcal{D}_\alpha + \mathcal{D}_\gamma\mathcal{D}_\alpha\mathcal{D}_\beta - \mathcal{D}_\alpha\mathcal{D}_\gamma\mathcal{D}_\beta - \mathcal{D}_\beta\mathcal{D}_\alpha\mathcal{D}_\gamma - \mathcal{D}_\gamma\mathcal{D}_\beta\mathcal{D}_\alpha\Big]V\Big| \\
&= \mathcal{D}_\alpha\mathcal{D}_\beta\mathcal{D}_\gamma V\big|\ , \\
0 &= \Big[\overline{\mathcal{D}}_{\dot\alpha}\overline{\mathcal{D}}_{\dot\beta}\overline{\mathcal{D}}_{\dot\gamma} + \overline{\mathcal{D}}_{\dot\beta}\overline{\mathcal{D}}_{\dot\gamma}\overline{\mathcal{D}}_{\dot\alpha} + \overline{\mathcal{D}}_{\dot\gamma}\overline{\mathcal{D}}_{\dot\alpha}\overline{\mathcal{D}}_{\dot\beta} - \overline{\mathcal{D}}_{\dot\alpha}\overline{\mathcal{D}}_{\dot\gamma}\overline{\mathcal{D}}_{\dot\beta} - \overline{\mathcal{D}}_{\dot\beta}\overline{\mathcal{D}}_{\dot\alpha}\overline{\mathcal{D}}_{\dot\gamma} - \overline{\mathcal{D}}_{\dot\gamma}\overline{\mathcal{D}}_{\dot\beta}\overline{\mathcal{D}}_{\dot\alpha}\Big]V\Big| \\
&= \overline{\mathcal{D}}_{\dot\alpha}\overline{\mathcal{D}}_{\dot\beta}\overline{\mathcal{D}}_{\dot\gamma}V\big|\ .
\end{aligned} \qquad (4.51)$$

The second pair of equalities come from the fact that in the Wess-Zumino gauge,

$$
\begin{aligned}
\{\mathcal{D}_\alpha, \mathcal{D}_\beta\}V\big| &= \{\overline{\mathcal{D}}_{\dot\alpha}, \overline{\mathcal{D}}_{\dot\beta}\}V\big| = 0\,,\\
\{\mathcal{D}_\alpha, \mathcal{D}_\beta\}\mathcal{D}_\gamma V\big| &= R_{\alpha\beta\gamma\delta}\mathcal{D}^\gamma V\big| = 0\,,\\
\{\overline{\mathcal{D}}_{\dot\alpha}, \overline{\mathcal{D}}_{\dot\beta}\}\overline{\mathcal{D}}_{\dot\gamma} V\big| &= -R_{\dot\alpha\dot\beta\dot\gamma\dot\delta}\mathcal{D}^{\dot\gamma} V\big| = 0\,.
\end{aligned}
$$

We now turn to the computation of third-order superderivatives of the vector superfield V which contain one single derivative with a vector index $\mathcal{D}_\mu$. They are readily obtained from (4.44), (4.46) and (4.51),

$$
\begin{aligned}
\mathcal{D}_\mu\mathcal{D}_\alpha\mathcal{D}_\beta V\big| &= E_\mu{}^{\tilde\mu}\mathcal{D}_{\tilde\mu}\mathcal{D}_\alpha\mathcal{D}_\beta V\big| + E_\mu{}^{\tilde\alpha}\mathcal{D}_{\tilde\alpha}\mathcal{D}_\alpha\mathcal{D}_\beta V\big| + E_{\mu\tilde{\dot\alpha}}\overline{\mathcal{D}}^{\tilde{\dot\alpha}}\mathcal{D}_\alpha\mathcal{D}_\beta V\big|\\
&= -i\varepsilon_{\alpha\beta}\Big[\bar\lambda\cdot\bar\psi_\mu + \frac{1}{2}\bar\psi_\mu\bar\sigma^\nu\sigma^\rho\bar\psi_\nu\, v_\rho\Big]\,,
\end{aligned}
$$

where the $E_\mu{}^{\tilde\mu}\mathcal{D}_{\tilde\mu}\mathcal{D}_\alpha\mathcal{D}_\beta V\big|$ term vanishes due (4.13). Consequently, we are left with

$$
\begin{aligned}
\mathcal{D}_\alpha\mathcal{D}_\mu\mathcal{D}_\beta V\big| &= \mathcal{D}_\mu\mathcal{D}_\alpha\mathcal{D}_\beta V\big| + [\mathcal{D}_\alpha, \mathcal{D}_\mu]\mathcal{D}_\beta V\big|\\
&= -\, i\varepsilon_{\alpha\beta}\Big[\bar\lambda\cdot\bar\psi_\mu + \frac{1}{2}\bar\psi_\mu\bar\sigma^\nu\sigma^\rho\bar\psi_\nu\, v_\rho\Big] - \frac{i}{6}\sigma_{\mu\alpha\dot\gamma}v_\beta{}^{\dot\gamma}M^*\,,\\
\mathcal{D}_\alpha\mathcal{D}_\beta\mathcal{D}_\mu V\big| &= \mathcal{D}_\alpha\mathcal{D}_\mu\mathcal{D}_\beta V\big| + \mathcal{D}_\alpha[\mathcal{D}_\beta, \mathcal{D}_\mu]V\big|\\
&= -\, i\varepsilon_{\alpha\beta}\Big[\bar\lambda\cdot\bar\psi_\mu + \frac{1}{2}\bar\psi_\mu\bar\sigma^\nu\sigma^\rho\bar\psi_\nu\, v_\rho - \frac{1}{3}v_\mu M^*\Big]\,.
\end{aligned}
$$

To derive those last relations, we have used (3.8) together with the constraints (3.29) and the results of Tables 3.3 and 3.6, as well as the antisymmetry property (4.8). Once again, all the terms containing the components of the curvature tensor can be omitted since they contain first-order superderivatives of V (see Eq. (4.44)). Similarly, one obtains the following conjugate expressions

$$
\begin{aligned}
\mathcal{D}_\mu\overline{\mathcal{D}}_{\dot\alpha}\overline{\mathcal{D}}_{\dot\beta} V\big| &= -\, i\varepsilon_{\dot\alpha\dot\beta}\Big[\psi_\mu\cdot\lambda - \frac{1}{2}\psi_\mu\sigma^\nu\bar\sigma^\rho\psi_\nu\, v_\rho\Big]\,,\\
\overline{\mathcal{D}}_{\dot\alpha}\mathcal{D}_\mu\overline{\mathcal{D}}_{\dot\beta} V\big| &= -\, i\varepsilon_{\dot\alpha\dot\beta}\Big[\psi_\mu\cdot\lambda - \frac{1}{2}\psi_\mu\sigma^\nu\bar\sigma^\rho\psi_\nu\, v_\rho\Big] - \frac{i}{6}\sigma^\mu{}_{\gamma\dot\alpha}v^\gamma{}_{\dot\beta}M\,,\\
\overline{\mathcal{D}}_{\dot\alpha}\overline{\mathcal{D}}_{\dot\beta}\mathcal{D}_\mu V\big| &= -\, i\varepsilon_{\dot\alpha\dot\beta}\Big[\psi_\mu\cdot\lambda - \frac{1}{2}\psi_\mu\sigma^\nu\bar\sigma^\rho\psi_\nu\, v_\rho - \frac{1}{3}v_\mu M\Big]\,.
\end{aligned}
$$

We finally address the computation of the lowest-order component of the superderivative $\mathcal{D}_\mu\mathcal{D}_\alpha\overline{\mathcal{D}}_{\dot\alpha}V$ as well as the one of all the quantities obtained by permutation of the operators $\mathcal{D}_\alpha$, $\overline{\mathcal{D}}_{\dot\alpha}$ and $\mathcal{D}_\mu$. Converting flat indices to curved indices, one gets

$$\begin{aligned}\mathcal{D}_\mu\mathcal{D}_\alpha\overline{\mathcal{D}}_{\dot\alpha}V\big| &\equiv \hat{D}_\mu v_{\alpha\dot\alpha}\\ &= E_\mu{}^{\tilde\mu}\mathcal{D}_{\tilde\mu}\mathcal{D}_\alpha\overline{\mathcal{D}}_{\dot\alpha}V\big| + E_\mu{}^{\tilde\alpha}\mathcal{D}_{\tilde\alpha}\mathcal{D}_\alpha\overline{\mathcal{D}}_{\dot\alpha}V\big| + E_{\mu\tilde{\dot\alpha}}\overline{\mathcal{D}}^{\tilde{\dot\alpha}}\mathcal{D}_\alpha\overline{\mathcal{D}}_{\dot\alpha}V\big|\\ &= e_\mu{}^{\tilde\mu}\mathcal{D}_{\tilde\mu}v_{\alpha\dot\alpha} + i\big[\psi_{\mu\alpha}\bar\lambda_{\dot\alpha} + \bar\psi_{\mu\dot\alpha}\lambda_\alpha\big] - \frac{i}{2}\sigma^\nu{}_{\alpha\dot\alpha}(\psi_\nu\sigma^\rho\bar\psi_\mu)v_\rho\end{aligned}$$

where we have introduced a derivative of the vector component field of V, $\hat{D}_\mu v_{\alpha\dot\alpha}$, which is covariant with respect to the supergravity transformations. Equivalently, converting spinor indices to vector indices, we get

$$\hat{D}_\mu v_\nu = e_\mu{}^{\tilde\mu}\mathcal{D}_{\tilde\mu}v_\nu - \frac{i}{2}\big[\bar\lambda\bar\sigma_\nu\psi_\mu - \bar\psi_\mu\bar\sigma_\nu\lambda\big] - \frac{i}{2}(\psi_\nu\sigma^\rho\bar\psi_\mu)v_\rho\ . \tag{4.52}$$

Adopting the Wess-Zumino gauge, as above, we deduce from (3.8), together with the constraints (3.29) and the results of Tables 3.3 and 3.6, the relations

$$\begin{aligned}\mathcal{D}_\alpha\mathcal{D}_\mu\overline{\mathcal{D}}_{\dot\alpha}V\big| &= \mathcal{D}_\mu\mathcal{D}_\alpha\overline{\mathcal{D}}_{\dot\alpha}V\big| + [\mathcal{D}_\alpha,\mathcal{D}_\mu]\overline{\mathcal{D}}_{\dot\alpha}V\big|\\ &= \mathcal{D}_\mu\mathcal{D}_\alpha\overline{\mathcal{D}}_{\dot\alpha}V\big| + T_{\alpha\mu}{}^\beta\mathcal{D}_\beta\overline{\mathcal{D}}_{\dot\alpha}V\big|\\ &= \hat{D}_\mu v_{\alpha\dot\alpha} + \frac{i}{24}\bar\sigma_\mu{}^{\dot\gamma\gamma}\big[b_{\alpha\dot\gamma}v_{\gamma\dot\alpha} - 3b_{\gamma\dot\gamma}v_{\alpha\dot\alpha} - 3\varepsilon_{\alpha\gamma}\,b_{\beta\dot\gamma}v^\beta{}_{\dot\alpha}\big]\\ &= \hat{D}_\mu v_{\alpha\dot\alpha} + \frac{i}{6}\big[(v_\mu b_\nu - 2v_\nu b_\mu)\sigma^\nu{}_{\alpha\dot\alpha} - b_\nu v^\nu\sigma_{\mu\alpha\dot\alpha}\big] - \frac{1}{6}\varepsilon_{\mu\nu\rho\sigma}b^\nu v^\rho\sigma^\sigma{}_{\alpha\dot\alpha}\ ,\\ \mathcal{D}_\alpha\overline{\mathcal{D}}_{\dot\alpha}\mathcal{D}_\mu V\big| &= \mathcal{D}_\alpha\mathcal{D}_\mu\overline{\mathcal{D}}_{\dot\alpha}V\big| + \mathcal{D}_\alpha[\overline{\mathcal{D}}_{\dot\alpha},\mathcal{D}_\mu]V\big|\\ &= \mathcal{D}_\alpha\mathcal{D}_\mu\overline{\mathcal{D}}_{\dot\alpha}V\big| + T_{\dot\alpha\mu\dot\beta}\mathcal{D}_\alpha\overline{\mathcal{D}}^{\dot\beta}V\big|\\ &= \hat{D}_\mu v_{\alpha\dot\alpha} + \frac{i}{6}\bar\sigma_\mu{}^{\dot\gamma\gamma}\big[b_{\alpha\dot\gamma}v_{\gamma\dot\alpha} - b_{\gamma\dot\alpha}v_{\alpha\dot\gamma}\big]\\ &= \hat{D}_\mu v_{\alpha\dot\alpha} - \frac{1}{3}\varepsilon_{\mu\nu\rho\sigma}b^\nu v^\rho\sigma^\sigma{}_{\alpha\dot\alpha}\ ,\end{aligned} \tag{4.53}$$

where we have converted spinor indices to vector indices and used the property of the Pauli matrices given in (C.5). In the derivation of the last relation above, we have employed the property

$$\begin{aligned}\bar\sigma_\mu^{\dot\gamma\gamma}\big[b_{\alpha\dot\gamma}v_{\gamma\dot\alpha} - 3b_{\gamma\dot\gamma}v_{\alpha\dot\alpha} - 3\varepsilon_{\alpha\gamma}b_{\beta\dot\gamma}v^\beta{}_{\dot\alpha}\big] - \bar\sigma_\mu^{\dot\gamma\gamma}\big[b_{\gamma\dot\alpha}v_{\alpha\dot\gamma} - 3b_{\gamma\dot\gamma}v_{\alpha\dot\alpha} - 3\varepsilon_{\dot\alpha\dot\gamma}b_{\gamma\dot\beta}v_\alpha{}^{\dot\beta}\big]\\ = 4\bar\sigma_\mu^{\dot\gamma\gamma}\big[b_{\alpha\dot\gamma}v_{\gamma\dot\alpha} - b_{\gamma\dot\alpha}v_{\alpha\dot\gamma}\big]\ ,\end{aligned}$$

which can be directly proven by converting spinor to vector indices, in a similar fashion, we obtain, the conjugate relations

$$\begin{aligned}
\mathcal{D}_\mu \overline{\mathcal{D}}_{\dot\alpha} \mathcal{D}_\alpha V\big| &= -(\hat{D}_\mu v_{\alpha\dot\alpha})^\dagger \,,\\
\overline{\mathcal{D}}_{\dot\alpha} \mathcal{D}_\mu \mathcal{D}_\alpha V\big| &= -(\hat{D}_\mu v_{\alpha\dot\alpha})^\dagger + \frac{i}{24}\bar\sigma_\mu{}^{\dot\gamma\gamma}\Big[b_{\gamma\dot\alpha} v_{\alpha\dot\gamma} - 3 b_{\gamma\dot\gamma} v_{\alpha\dot\alpha} - 3\varepsilon_{\dot\alpha\dot\gamma} b_{\gamma\dot\beta} v_\alpha{}^{\dot\beta}\Big]\\
&= -(\hat{D}_\mu v_{\alpha\dot\alpha})^\dagger + \frac{i}{6}\Big[(v_\mu b_\nu - 2 v_\nu b_\mu)\sigma^\nu{}_{\alpha\dot\alpha} - b_\nu v^\nu \sigma_{\mu\alpha\dot\alpha}\Big]\\
&\quad + \frac{1}{6}\varepsilon_{\mu\nu\rho\sigma} b^\nu v^\rho \sigma^\sigma{}_{\alpha\dot\alpha} \,,\\
\overline{\mathcal{D}}_{\dot\alpha} \mathcal{D}_\alpha \mathcal{D}_\mu V\big| &= -(\hat{D}_\mu v_{\alpha\dot\alpha})^\dagger + \frac{i}{6}\bar\sigma_\mu{}^{\dot\gamma\gamma}\Big[b_{\gamma\dot\alpha} v_{\alpha\dot\gamma} - b_{\alpha\dot\gamma} v_{\gamma\dot\alpha}\Big]\\
&= -(\hat{D}_\mu v_{\alpha\dot\alpha})^\dagger + \frac{1}{3}\varepsilon_{\mu\nu\rho\sigma} b^\nu v^\rho \sigma^\sigma{}_{\alpha\dot\alpha} \,.
\end{aligned}$$

4.2.2 *The components of the spinorial superfields associated to vector superfields*

In this section using all the results of the previous section and adopting the Wess-Zumino gauge (4.40), we calculate the components of the spinorial superfields W_α and $\overline{W}_{\dot\alpha}$ associated to a vector superfield V defined in (4.34) and (4.35) in the abelian and non-abelian cases, respectively.

Abelian vector superfields.

We first compute the components of the superfield strength tensors W_α and $\overline{W}_{\dot\alpha}$ associated to a superfield V related to an abelian gauge group. These superfields are defined by (4.34) while $\theta = \bar\theta = 0$ coefficients are deduced from the definitions of (4.36),

$$\begin{aligned}
W_\alpha\big| &= i\lambda_\alpha \,,\\
\overline{W}_{\dot\alpha}\big| &= -i\bar\lambda_{\dot\alpha} \,.
\end{aligned}$$

The second component fields of the superfield strength tensors are obtained *via* the computation of the first-order derivatives $\mathcal{D}_\beta W_\alpha\big|$ and $\overline{\mathcal{D}}_{\dot\beta} \overline{W}_{\dot\alpha}\big|$. The first quantity reduces to

$$\begin{aligned}
\mathcal{D}_\beta W_\alpha\big| &= -\frac{1}{4}\mathcal{D}_\beta\Big[\overline{\mathcal{D}}\cdot\overline{\mathcal{D}} - 8\mathcal{R}\Big]\mathcal{D}_\alpha V\big|\\
&= -\frac{1}{4}\mathcal{D}_\beta \overline{\mathcal{D}}\cdot\overline{\mathcal{D}}\mathcal{D}_\alpha V\big| \,,
\end{aligned} \tag{4.54}$$

using (4.44) and (4.46), and can be calculated from the relation

$$\begin{aligned}
\mathcal{D}_\beta \overline{\mathcal{D}}_{\dot\alpha} \overline{\mathcal{D}}_{\dot\beta} \mathcal{D}_\alpha V &= \{\mathcal{D}_\beta, \overline{\mathcal{D}}_{\dot\alpha}\}\overline{\mathcal{D}}_{\dot\beta}\mathcal{D}_\alpha V - \overline{\mathcal{D}}_{\dot\alpha}\{\mathcal{D}_\beta, \overline{\mathcal{D}}_{\dot\beta}\}\mathcal{D}_\alpha V + \overline{\mathcal{D}}_{\dot\alpha}\overline{\mathcal{D}}_{\dot\beta}\{\mathcal{D}_\beta, \mathcal{D}_\alpha\}V\\
&\quad - \overline{\mathcal{D}}_{\dot\alpha}\{\overline{\mathcal{D}}_{\dot\beta}, \mathcal{D}_\alpha\}\mathcal{D}_\beta V + \{\overline{\mathcal{D}}_{\dot\alpha}, \mathcal{D}_\alpha\}\overline{\mathcal{D}}_{\dot\beta}\mathcal{D}_\beta V - \mathcal{D}_\alpha\overline{\mathcal{D}}_{\dot\alpha}\overline{\mathcal{D}}_{\dot\beta}\mathcal{D}_\beta V \,.
\end{aligned}$$

Simplifying the anticommutators by introducing the torsion and curvature tensors using (3.8) and applying the constraints (3.29) together with the results of Tables 3.3 and 3.4, we obtain,

$$\begin{aligned}
&\mathcal{D}_\beta\overline{\mathcal{D}}_{\dot\alpha}\overline{\mathcal{D}}_{\dot\beta}\mathcal{D}_\alpha V\big| + \mathcal{D}_\alpha\overline{\mathcal{D}}_{\dot\alpha}\overline{\mathcal{D}}_{\dot\beta}\mathcal{D}_\beta V\big| \\
&\quad = -2i\sigma^\mu{}_{\beta\dot\alpha}\mathcal{D}_\mu\overline{\mathcal{D}}_{\dot\beta}\mathcal{D}_\alpha V\big| + 2i\sigma^\mu{}_{\beta\dot\beta}\overline{\mathcal{D}}_{\dot\alpha}\mathcal{D}_\mu\mathcal{D}_\alpha V\big| + 2i\sigma^\mu{}_{\alpha\dot\beta}\overline{\mathcal{D}}_{\dot\alpha}\mathcal{D}_\mu\mathcal{D}_\beta V\big| \\
&\qquad -2i\sigma^\mu{}_{\alpha\dot\alpha}\mathcal{D}_\mu\overline{\mathcal{D}}_{\dot\beta}\mathcal{D}_\beta V\big| - R_{\beta\dot\alpha\dot\beta\dot\gamma}\overline{\mathcal{D}}^{\dot\gamma}\mathcal{D}_\alpha V\big| + R_{\beta\dot\alpha\alpha\gamma}\overline{\mathcal{D}}_{\dot\beta}\mathcal{D}^\gamma V\big| - R_{\beta\dot\beta\alpha\gamma}\overline{\mathcal{D}}_{\dot\alpha}\mathcal{D}^\gamma V\big| \\
&\qquad -R_{\alpha\dot\beta\beta\gamma}\overline{\mathcal{D}}_{\dot\alpha}\mathcal{D}^\gamma V\big| - R_{\alpha\dot\alpha\dot\beta\dot\gamma}\overline{\mathcal{D}}^{\dot\gamma}\mathcal{D}_\beta V\big| + R_{\alpha\dot\alpha\beta\gamma}\overline{\mathcal{D}}_{\dot\beta}\mathcal{D}^\gamma V\big| \,.
\end{aligned}$$

After contracting this last expression with $\varepsilon^{\dot\alpha\dot\beta}$, we first focus on the terms containing a $\mathcal{D}_\mu$ derivative and related to the piece of the anticommutators depending on the torsion tensor. They are calculated as

$$\begin{aligned}
&\Big[\mathcal{D}_\beta\overline{\mathcal{D}}\cdot\overline{\mathcal{D}}\mathcal{D}_\alpha V\big| + \mathcal{D}_\alpha\overline{\mathcal{D}}\cdot\overline{\mathcal{D}}\mathcal{D}_\beta V\big|\Big]^{\text{torsion}} \\
&\qquad = 16i\,(\sigma^{\mu\nu})_\alpha{}^\gamma\varepsilon_{\beta\gamma}(\hat D_\mu v_\nu)^\dagger - \frac{5}{3}(b_{\beta\dot\alpha}v_\alpha{}^{\dot\alpha} + b_{\alpha\dot\alpha}v_\beta{}^{\dot\alpha}) \\
&\qquad = 16i\,(\sigma^{\mu\nu})_\alpha{}^\gamma\varepsilon_{\beta\gamma}(\hat D_\mu v_\nu)^\dagger + \frac{20}{3}v_\mu b_\nu(\sigma^{\mu\nu})_\alpha{}^\gamma\varepsilon_{\beta\gamma}\,,
\end{aligned}$$

using either spinor or vector indices. Next, the terms related to the piece of the anticommutators depending on the curvature tensor lead to

$$\begin{aligned}
\Big[\mathcal{D}_\beta\overline{\mathcal{D}}\cdot\overline{\mathcal{D}}\mathcal{D}_\alpha V\big| + \mathcal{D}_\alpha\overline{\mathcal{D}}\cdot\overline{\mathcal{D}}\mathcal{D}_\beta V\big|\Big]^{\text{curv}} &= \frac{5}{3}(b_{\beta\dot\alpha}v_\alpha{}^{\dot\alpha} + b_{\alpha\dot\alpha}v_\beta{}^{\dot\alpha}) \\
&= -\frac{20}{3}v_\mu b_\nu(\sigma^{\mu\nu})_\alpha{}^\gamma\varepsilon_{\beta\gamma}\,,
\end{aligned}$$

where the second equality is proved with the help of the self-duality property,

$$\frac{1}{2}\varepsilon_{\mu\nu\rho\sigma}\sigma^{\rho\sigma} = -i\sigma_{\mu\nu}\,.$$

On the one hand, adding all the contributions leads to

$$\mathcal{D}_\alpha W_\beta\big| + \mathcal{D}_\beta W_\alpha\big| = -4i\,(\sigma^{\mu\nu})_\alpha{}^\gamma\varepsilon_{\beta\gamma}(\hat D_\mu v_\nu)^\dagger\,,$$

while on the other hand, the definition of the D-term in (4.36) and the antisymmetry property (4.8) give

$$\mathcal{D}_\alpha W_\beta\big| - \mathcal{D}_\beta W_\alpha\big| = \varepsilon_{\alpha\beta}\mathcal{D}^\gamma W_\gamma = -2\varepsilon_{\alpha\beta}D\,.$$

Consequently, one gets

$$\mathcal{D}_\alpha W_\beta\big| = -i\,(\sigma^{\mu\nu})_{\alpha\beta}(\hat F^0_{\mu\nu}) - \varepsilon_{\alpha\beta}D\,, \tag{4.55}$$

and in a fully analogous way,

$$\overline{\mathcal{D}}_{\dot\alpha}\overline{W}_{\dot\beta}\big| = i\,(\bar\sigma^{\mu\nu})_{\dot\alpha\dot\beta}(\hat F^0_{\mu\nu}) + \varepsilon_{\dot\alpha\dot\beta}D\,, \tag{4.56}$$

where we recall that these conjugation properties stem directly from the anti-hermiticity of the superderivatives $\mathcal{D}_\alpha$ and $\overline{\mathcal{D}}_{\dot\alpha}$. In these two last expressions, we have introduced the abelian field strength tensor of the vector boson v^μ,

$$\hat{F}^0_{\mu\nu} = \hat{D}_\mu v_\nu - \hat{D}_\nu v_\mu \ ,$$

the superscript 0 denoting its abelian nature. Note also that $(\hat{D}_\mu v_\nu)^\dagger - (\hat{D}_\nu v_\mu)^\dagger = (\hat{D}_\mu v_\nu) - (\hat{D}_\nu v_\mu)$ since it is real. Converting all Lorentz indices to Einstein indices and taking the derivative, the curved field strength tensor is defined as

$$\begin{aligned}
\hat{F}^0_{\tilde\mu\tilde\nu} &= e_{\tilde\mu}{}^\mu e_{\tilde\nu}{}^\nu F^0_{\mu\nu} \\
&= F^0_{\tilde\mu\tilde\nu} - \frac{i}{2}\Big[\bar\lambda\bar\sigma_{\tilde\nu}\psi_{\tilde\mu} + \lambda\sigma_{\tilde\nu}\bar\psi_{\tilde\mu} - \bar\lambda\bar\sigma_{\tilde\mu}\psi_{\tilde\nu} - \lambda\sigma_{\tilde\mu}\bar\psi_{\tilde\nu}\Big]\ , \qquad (4.57)\\
F^0_{\tilde\mu\tilde\nu} &= \partial_{\tilde\mu} v_{\tilde\nu} - \partial_{\tilde\nu} v_{\tilde\mu} = \mathcal{D}_{\tilde\mu} v_{\tilde\nu} - \mathcal{D}_{\tilde\nu} v_{\tilde\mu}\ .
\end{aligned}$$

The last equality is based on the definition of the hatted derivative of the vector field in (4.52), and the property

$$\begin{aligned}
e_{\tilde\nu}{}^\nu \mathcal{D}_{\tilde\mu} v_\nu &= e_{\tilde\nu}{}^\nu\Big[\partial_{\tilde\mu}(e_\nu{}^{\tilde\rho} v_{\tilde\rho}) - \omega_{\tilde\mu\nu}{}^\rho(e_\rho{}^{\tilde\rho} v_{\tilde\rho})\Big] \\
&= \partial_{\tilde\mu} v_{\tilde\nu} + e_{\tilde\nu}{}^\nu(\partial_{\tilde\mu} e_\nu{}^{\tilde\rho} - \omega_{\tilde\mu\nu}{}^\rho e_\rho{}^{\tilde\rho}) v_{\tilde\rho} \\
&= \partial_{\tilde\mu} v_{\tilde\nu} + e_{\tilde\nu}{}^\nu \mathcal{D}_{\tilde\mu}(e_\nu{}^{\tilde\rho}) v_{\tilde\rho} \\
&= \partial_{\tilde\mu} v_{\tilde\nu} - e_\nu{}^{\tilde\rho} \mathcal{D}_{\tilde\mu}(e_{\tilde\nu}{}^\nu) v_{\tilde\rho}\ ,
\end{aligned}$$

derived from (3.5) and (3.12), and which yield

$$\begin{aligned}
e_{\tilde\nu}{}^\nu \mathcal{D}_{\tilde\mu} v_\nu - e_{\tilde\mu}{}^\nu \mathcal{D}_{\tilde\nu} v_\nu &= \partial_{\tilde\mu} v_{\tilde\nu} - \partial_{\tilde\nu} v_{\tilde\mu} + e_\nu{}^{\tilde\rho} T_{\tilde\mu\tilde\nu}{}^\nu v_{\tilde\rho} \\
&= \partial_{\tilde\mu} v_{\tilde\nu} - \partial_{\tilde\nu} v_{\tilde\mu} - \frac{i}{2}[\psi_{\tilde\mu}\sigma^{\tilde\rho}\bar\psi_{\tilde\nu} - \psi_{\tilde\nu}\sigma^{\tilde\rho}\bar\psi_{\tilde\mu}] v_{\tilde\rho}\ ,
\end{aligned}$$

according to the results of Table 3.1 and Eq. (3.118). Let us note that (4.57) is covariant under $U(1)$ gauge transformations. It is now immediate to define a derivative of the gaugino field covariant with respect to the supergravity transformation, *i.e.*, to compute the quantities $\hat{D}^0_\mu\lambda_\alpha$ and $\hat{D}^0_\mu\bar\lambda_{\dot\alpha}$, since

$$\begin{aligned}
\hat{D}^0_\mu\lambda_\alpha \equiv \mathcal{D}_\mu(-iW_\alpha)\big| &= E_\mu{}^{\tilde\mu}\mathcal{D}_{\tilde\mu}(-iW_\alpha)\big| - iE_\mu{}^{\tilde\alpha}\mathcal{D}_{\tilde\alpha}W_\alpha\big| - iE_{\mu\tilde{\dot\alpha}}\overline{\mathcal{D}}^{\tilde{\dot\alpha}}W_\alpha\big| \\
&= e_\mu{}^{\tilde\mu}\mathcal{D}_{\tilde\mu}\lambda_\alpha - \frac{1}{2}(\sigma^{\nu\rho}\psi_\mu)_\alpha(\hat{F}^0_{\nu\rho}) + \frac{i}{2}D\psi_{\mu\alpha}\ , \\
\hat{D}^0_\mu\bar\lambda_{\dot\alpha} \equiv \mathcal{D}_\mu(i\overline{W}_{\dot\alpha})\big| &= E_\mu{}^{\tilde\mu}\mathcal{D}_{\tilde\mu}(i\overline{W}_{\dot\alpha})\big| + iE_\mu{}^{\tilde\alpha}\mathcal{D}_{\tilde\alpha}\overline{W}_{\dot\alpha}\big| + iE_{\mu\tilde{\dot\alpha}}\overline{\mathcal{D}}^{\tilde{\dot\alpha}}\overline{W}_{\dot\alpha}\big| \qquad (4.58)\\
&= e_\mu{}^{\tilde\mu}\mathcal{D}_{\tilde\mu}\bar\lambda_{\dot\alpha} + \frac{1}{2}(\bar\psi_\mu\bar\sigma^{\nu\rho})_{\dot\alpha}(\hat{F}^0_{\nu\rho}) - \frac{i}{2}D\bar\psi_{\mu\dot\alpha}\ ,
\end{aligned}$$

where we have used $(\sigma^{\mu\nu})_{\alpha\beta} = (\sigma^{\mu\nu})_{\beta\alpha}$ and $(\bar\sigma^{\mu\nu})_{\dot\alpha\dot\beta} = (\bar\sigma^{\mu\nu})_{\dot\beta\dot\alpha}$. Accordingly to the Majorana nature of the gaugino, the two derivatives above are found to be hermitian conjugates to each other.

The highest-order component of the superfields W_α and $\overline{W}_{\dot\alpha}$ are given by the second-order derivatives $\mathcal{D}{\cdot}\mathcal{D}\,W_\alpha$ and $\overline{\mathcal{D}}{\cdot}\overline{\mathcal{D}}\,\overline{W}_{\dot\alpha}$. The first of these quantities reads,

$$\mathcal{D}\cdot\mathcal{D}W_\alpha\Big| = -\frac{1}{4}\mathcal{D}\cdot\mathcal{D}\,\overline{\mathcal{D}}\cdot\overline{\mathcal{D}}\mathcal{D}_\alpha V\Big| ,$$

because $\mathcal{D}_\alpha V\big| = \mathcal{D}_\beta\mathcal{D}_\alpha V\big| = \mathcal{D}_\gamma\mathcal{D}_\beta\mathcal{D}_\alpha V\big| = 0$. Starting from the relation

$$\begin{aligned}\mathcal{D}_\beta\mathcal{D}_\gamma\overline{\mathcal{D}}_{\dot\alpha}\overline{\mathcal{D}}_{\dot\beta}\mathcal{D}_\alpha V\Big| = \;& \mathcal{D}_\beta\mathcal{D}_\gamma\overline{\mathcal{D}}_{\dot\alpha}\{\overline{\mathcal{D}}_{\dot\beta},\mathcal{D}_\alpha\}V\Big| \\ & -\mathcal{D}_\beta\mathcal{D}_\gamma\{\overline{\mathcal{D}}_{\dot\alpha},\mathcal{D}_\alpha\}\overline{\mathcal{D}}_{\dot\beta}V\Big| \\ & +\mathcal{D}_\beta\mathcal{D}_\gamma\mathcal{D}_\alpha\overline{\mathcal{D}}_{\dot\alpha}\overline{\mathcal{D}}_{\dot\beta}V\Big| ,\end{aligned}$$

contracting with $\varepsilon^{\dot\alpha\dot\beta}\varepsilon^{\gamma\beta}$, and after applying (iteratively) the definition of the graded commutators (3.8) together with the constraints (3.29), one obtains

$$\begin{aligned}\mathcal{D}\cdot\mathcal{D}\,\overline{\mathcal{D}}\cdot\overline{\mathcal{D}}\mathcal{D}_\alpha V\Big| = \;& 4i\sigma^\mu{}_{\alpha\dot\alpha}\mathcal{D}\cdot\mathcal{D}\,\mathcal{D}_\mu\overline{\mathcal{D}}^{\dot\alpha}V\Big| + \mathcal{D}\cdot\mathcal{D}\,\mathcal{D}_\alpha\overline{\mathcal{D}}\cdot\overline{\mathcal{D}}V\Big| + \mathcal{D}\cdot\mathcal{D}(R_{\dot\alpha\alpha\dot\gamma}{}^{\dot\alpha}\overline{\mathcal{D}}^{\dot\gamma}V)\Big| \\ & +2i\mathcal{D}\cdot\mathcal{D}(T^{\dot\alpha}{}_{\alpha\dot\alpha}{}^{\beta}\mathcal{D}_\beta V + T^{\dot\alpha}{}_{\alpha\dot\alpha\dot\beta}\overline{\mathcal{D}}^{\dot\beta}V)\Big| .\end{aligned} \tag{4.59}$$

Commuting the $\mathcal{D}_\mu$ derivative, again through (3.8) together with the constraints (3.29), one can evaluate the first term of this last equation,

$$\begin{aligned}4i\sigma^\mu{}_{\alpha\dot\alpha}\mathcal{D}\cdot\mathcal{D}\,\mathcal{D}_\mu\overline{\mathcal{D}}^{\dot\alpha}V\Big| = \;& 4i\sigma^\mu{}_{\alpha\dot\alpha}\mathcal{D}_\mu\mathcal{D}\cdot\mathcal{D}\,\overline{\mathcal{D}}^{\dot\alpha}V\Big| + 4i\mathcal{D}^\beta T_{\beta\alpha\dot\alpha}{}^{\gamma}\mathcal{D}_\gamma\overline{\mathcal{D}}^{\dot\alpha}V\Big| \\ & -4iT_{\beta\alpha\dot\alpha\gamma}\Big(\mathcal{D}^\beta\mathcal{D}^\gamma\overline{\mathcal{D}}^{\dot\alpha} - \mathcal{D}^\gamma\mathcal{D}^\beta\overline{\mathcal{D}}^{\dot\alpha}\Big)V\Big| \\ & +4iT_{\beta\alpha\dot\alpha\dot\gamma}\Big(\mathcal{D}^\beta\overline{\mathcal{D}}^{\dot\gamma}\overline{\mathcal{D}}^{\dot\alpha} - \overline{\mathcal{D}}^{\dot\gamma}\mathcal{D}^\beta\overline{\mathcal{D}}^{\dot\alpha}\Big)V\Big| \\ & +4iR^\beta{}_{\alpha\dot\alpha\gamma\beta}\mathcal{D}^\gamma\overline{\mathcal{D}}^{\dot\alpha}V\Big| - 8iR_{\beta\alpha\dot\alpha}{}^{\dot\gamma\dot\alpha}\mathcal{D}^\beta\overline{\mathcal{D}}_{\dot\gamma}V\Big| .\end{aligned} \tag{4.60}$$

The terms appearing in (4.60) can be computed through the quantities obtained in Sec. 4.2.1, as well as those derived in the beginning of this section. In more detail, the contribution of the first term in (4.60) reads

$$\begin{aligned}4i\sigma^\mu{}_{\alpha\dot\alpha}\mathcal{D}_\mu\mathcal{D}\cdot\mathcal{D}\,\overline{\mathcal{D}}^{\dot\alpha}V\Big| &= -\,16i\sigma^\mu{}_{\alpha\dot\alpha}\mathcal{D}_\mu\overline{W}^{\dot\alpha}\Big| + 32i\sigma^\mu{}_{\alpha\dot\alpha}\mathcal{D}_\mu(\mathcal{R}^\dagger\overline{\mathcal{D}}^{\dot\alpha}V)\Big| \\ &= -\,16\sigma^\mu{}_{\alpha\dot\alpha}\hat{D}^0_\mu\bar\lambda^{\dot\alpha} - \frac{8i}{3}M^* v_\nu(\sigma^\mu\bar\sigma^\nu\psi_\mu)_\alpha ,\end{aligned}$$

where we have used the definition of $\overline{W}_{\dot\alpha}$ in (4.34) as well as the results of Eqs. (4.46) and (4.58) together with those presented in Table 3.8. Similarly, (4.46) allows to simplify the second, fifth and sixth terms of Eq. (4.60),

$$4i\mathcal{D}^\beta T_{\beta\alpha\dot\alpha}{}^\gamma \mathcal{D}_\gamma \overline{\mathcal{D}}^{\dot\alpha} V\big| = 10v^\mu \mathcal{D}_\alpha G_\mu\big| + 4v_\nu(\sigma^{\mu\nu})_\alpha{}^\beta \mathcal{D}_\beta G_\mu\big| ,$$
$$4iR^\beta{}_{\alpha\dot\alpha\gamma\beta}\mathcal{D}^\gamma \overline{\mathcal{D}}^{\dot\alpha} V\big| = 6v^\mu \mathcal{D}_\alpha G_\mu\big| + 28v_\nu(\sigma^{\mu\nu})_\alpha{}^\beta \mathcal{D}_\beta G_\mu\big| ,$$
$$-8iR_{\beta\alpha\dot\alpha}{}^{\dot\gamma\dot\alpha}\mathcal{D}^\beta \overline{\mathcal{D}}_{\dot\gamma} V\big| = -12v^\mu \mathcal{D}_\alpha G_\mu\big| - 24v_\nu(\sigma^{\mu\nu})_\alpha{}^\beta \mathcal{D}_\beta G_\mu\big| ,$$

together with the expressions of the non-zero components of the torsion and curvature tensors given in Tables 3.3 and 3.4 and after conversion of spinor indices to vector indices using (B.7). Using finally the results of Tables 3.3 and 3.8 in association with Eqs. (4.49) and (4.50), one can compute the third and fourth terms of Eq. (4.60) to get

$$-4iT_{\beta\alpha\dot\alpha\gamma}\Big(\mathcal{D}^\beta \mathcal{D}^\gamma \overline{\mathcal{D}}^{\dot\alpha} - \mathcal{D}^\gamma \mathcal{D}^\beta \overline{\mathcal{D}}^{\dot\alpha}\Big)V\big| = \frac{32i}{3} b_\mu(\sigma^\mu \bar\lambda)_\alpha ,$$
$$4iT_{\beta\alpha\dot\alpha\dot\gamma}\Big(\mathcal{D}^\beta \overline{\mathcal{D}}^{\dot\gamma} \overline{\mathcal{D}}^{\dot\alpha} - \overline{\mathcal{D}}^{\dot\gamma} \mathcal{D}^\beta \overline{\mathcal{D}}^{\dot\alpha}\Big)V\big| = \frac{32i}{3} M^* \lambda_\alpha + 4iM^* v_\nu(\sigma^\mu \bar\sigma^\nu \psi_\mu)_\alpha .$$

The second term in (4.59), like its first term, is also further reduced with the help of (3.8) together with the constraints (3.29),

$$\begin{aligned}\mathcal{D}\cdot\mathcal{D}\, \mathcal{D}_\alpha \overline{\mathcal{D}}\cdot\overline{\mathcal{D}} V\big| &= \mathcal{D}^\beta\{\mathcal{D}_\beta, \mathcal{D}_\alpha\}\overline{\mathcal{D}}\cdot\overline{\mathcal{D}} V\big| - \{\mathcal{D}^\beta, \mathcal{D}_\alpha\}\mathcal{D}_\beta \overline{\mathcal{D}}\cdot\overline{\mathcal{D}} V\big| \\ &\quad + \mathcal{D}_\alpha \mathcal{D}\cdot\mathcal{D}\, \overline{\mathcal{D}}\cdot\overline{\mathcal{D}} V\big| \\ &= R_{\beta\alpha\gamma}{}^\beta \mathcal{D}^\gamma \overline{\mathcal{D}}\cdot\overline{\mathcal{D}} V\big| + \mathcal{D}_\alpha \mathcal{D}\cdot\mathcal{D}\, \overline{\mathcal{D}}\cdot\overline{\mathcal{D}} V\big| .\end{aligned}$$

It is immediate to evaluate the two contributions to the sum above,

$$\begin{aligned}R_{\beta\alpha\gamma}{}^\beta \mathcal{D}^\gamma \overline{\mathcal{D}}\cdot\overline{\mathcal{D}} V\big| &= -8iM^*\lambda_\alpha - 4iM^* v_\nu(\sigma^\mu\bar\sigma^\nu\psi_\mu)_\alpha , \\ \mathcal{D}_\alpha \mathcal{D}\cdot\mathcal{D}\, \overline{\mathcal{D}}\cdot\overline{\mathcal{D}} V\big| &= \mathcal{D}_\alpha(\mathcal{D}\cdot\mathcal{D} - 8\mathcal{R}^\dagger)\overline{\mathcal{D}}\cdot\overline{\mathcal{D}} V\big| + 8\mathcal{D}_\alpha(\mathcal{R}^\dagger \overline{\mathcal{D}}\cdot\overline{\mathcal{D}} V)\big| \\ &= \frac{16i}{3} M^*\lambda_\alpha + \frac{8i}{3} M^* v_\nu(\sigma^\mu\bar\sigma^\nu\psi_\mu)_\alpha ,\end{aligned}$$

where we have used the expressions of the components of the curvature tensor given in Table 3.4, the results of Table 3.8, Eqs. (4.49) and (4.50) and the chirality properties of the projector operator $(\mathcal{D}\cdot\mathcal{D} - 8\mathcal{R}^\dagger)$. The same identities, with in addition (4.46) and the expressions of the components of the torsion tensor shown in Table 3.3, allow to calculate the two remaining terms of Eq. (4.59),

$$\mathcal{D}\cdot\mathcal{D}(R_{\dot\alpha\alpha\dot\gamma}{}^{\dot\alpha}\overline{\mathcal{D}}^{\dot\gamma}V)\big| = \Big[-2\mathcal{D}_\beta R_{\dot\alpha\alpha\dot\gamma}{}^{\dot\alpha}\mathcal{D}^\beta\overline{\mathcal{D}}^{\dot\gamma}V\big|\Big] + \Big[R_{\dot\alpha\alpha\dot\gamma}{}^{\dot\alpha}\mathcal{D}\cdot\mathcal{D}\,\overline{\mathcal{D}}^{\dot\gamma}V\big|\Big]$$
$$= \Big[6v^\mu\mathcal{D}_\alpha G_\mu\big| + 12v_\nu(\sigma^{\mu\nu})_\alpha{}^\beta\mathcal{D}_\beta G_\mu\big|\Big] + \Big[4ib_\mu(\sigma^\mu\bar\lambda)_\alpha\Big]\,,$$

$$2i\mathcal{D}\cdot\mathcal{D}(T^{\dot\alpha}{}_{\alpha\dot\alpha}{}^\beta\mathcal{D}_\beta V + T^{\dot\alpha}{}_{\alpha\dot\alpha\dot\beta}\overline{\mathcal{D}}^{\dot\beta}V)\big|$$
$$= \Big[4i\mathcal{D}^\beta T^{\dot\alpha}{}_{\alpha\dot\alpha\dot\beta}\mathcal{D}_\beta\overline{\mathcal{D}}^{\dot\beta}V\big|\Big] + \Big[2iT^{\dot\alpha}{}_{\alpha\dot\alpha\dot\beta}\mathcal{D}\cdot\mathcal{D}\,\overline{\mathcal{D}}^{\dot\beta}V\big|\Big]$$
$$= \Big[-10v^\mu\mathcal{D}_\alpha G_\mu\big| - 20v_\nu(\sigma^{\mu\nu})_\alpha{}^\beta\mathcal{D}_\beta G_\mu\big|\Big] + \Big[-\frac{20i}{3}b_\mu(\sigma^\mu\bar\lambda)_\alpha\Big]\,.$$

Collecting all the various contributions, we obtain $\mathcal{D}\cdot\mathcal{D}\,W_\alpha\big|$,

$$\mathcal{D}\cdot\mathcal{D}\,W_\alpha\big| = -\frac{1}{4}\mathcal{D}\cdot\mathcal{D}\,\overline{\mathcal{D}}\cdot\overline{\mathcal{D}}\mathcal{D}_\alpha V\big|$$
$$= 4\sigma^\mu{}_{\alpha\dot\alpha}\hat{D}^0_\mu\bar\lambda^{\dot\alpha} - 2ib_\mu(\sigma^\mu\bar\lambda)_\alpha - 2iM^*\lambda_\alpha\,. \tag{4.61}$$

Similarly, one obtains the conjugate relation

$$\overline{\mathcal{D}}\cdot\overline{\mathcal{D}}\,\overline{W}_{\dot\alpha}\big| = 4\sigma^\mu{}_{\alpha\dot\alpha}\hat{D}^0_\mu\lambda^\alpha + 2ib_\mu(\lambda\sigma^\mu)_{\dot\alpha} + 2iM\bar\lambda_{\dot\alpha}\,. \tag{4.62}$$

We summarise the most important analytical results regarding abelian vector superfield V in Table 4.2, *i.e.*, the component fields of V, in the Wess-Zumino gauge, those of the associated spinorial superfields W_α and $\overline{W}_{\dot\alpha}$, as well as the derivatives of the physical degrees of freedom covariant with respect to the supergravity transformations.

Non-abelian vector superfields.

We now turn to the derivation of the component fields of the spinorial superfields W_α and $\overline{W}_{\dot\alpha}$ associated to a non-abelian vector superfield $V = V^aT_a$ where T_a are hermitian representation matrices of the gauge group considered. We first prove (4.48) which we have only assumed to hold in Sec. 4.2.1. Expanding the exponentials appearing in the definitions of the superfield strength tensors in (4.35), we get

$$\begin{aligned}
W_\alpha &= -\frac{1}{4}\Big[\overline{\mathcal{D}}\cdot\overline{\mathcal{D}} - 8\mathcal{R}\Big]e^{2gV}\mathcal{D}_\alpha e^{-2gV}\\
&= -\frac{1}{4}\Big[\overline{\mathcal{D}}\cdot\overline{\mathcal{D}} - 8\mathcal{R}\Big]\Big[-2g\mathcal{D}_\alpha V + 2g^2\mathcal{D}_\alpha V^2 - 4g^2V\mathcal{D}_\alpha V + \ldots\Big]\,,\\
\overline{W}_{\dot\alpha} &= \frac{1}{4}\Big[\mathcal{D}\cdot\mathcal{D} - 8\mathcal{R}^\dagger\Big]e^{-2gV}\overline{\mathcal{D}}_{\dot\alpha}e^{2gV}\\
&= \frac{1}{4}\Big[\mathcal{D}\cdot\mathcal{D} - 8\mathcal{R}^\dagger\Big]\Big[2g\overline{\mathcal{D}}_{\dot\alpha}V + 2g^2\overline{\mathcal{D}}_{\dot\alpha}V^2 - 4g^2V\overline{\mathcal{D}}_{\dot\alpha}V + \ldots\Big]\,,
\end{aligned} \tag{4.63}$$

Table 4.2 Abelian vector superfields.

An abelian vector superfield V is defined by the constraint $V = V^\dagger$. In the Wess-Zumino gauge,

$$V\Big| = 0\,,\qquad \mathcal{D}_\alpha V\Big| = 0\,,\qquad \overline{\mathcal{D}}_{\dot\alpha} V\Big| = 0\,,\qquad \mathcal{D}_\alpha\mathcal{D}_\beta V\Big| = 0\,,\qquad \overline{\mathcal{D}}_{\dot\alpha}\overline{\mathcal{D}}_{\dot\beta} V\Big| = 0\,,$$

so that all non-physical component fields are eliminated. The remaining physical degrees of freedom, *i.e.*, the gaugino $(\lambda_\alpha, \bar\lambda^{\dot\alpha})$ and the gauge boson v_μ, as well as the auxiliary D-term, are given by

$$v_\mu = \frac{1}{4}\bar\sigma_\mu{}^{\dot\alpha\alpha}[\mathcal{D}_\alpha, \overline{\mathcal{D}}_{\dot\alpha}]V\Big|\,,\quad \lambda_\alpha = -iW_\alpha\Big|\,,\quad \bar\lambda_{\dot\alpha} = i\overline{W}_{\dot\alpha}\Big|\,,\quad D = -\frac{1}{2}\mathcal{D}^\alpha W_\alpha\Big| = -\frac{1}{2}\overline{\mathcal{D}}_{\dot\alpha}\overline{W}^{\dot\alpha}\Big|\,,$$

where we have introduced the chiral and antichiral spinorial superfields W_α and $\overline{W}_{\dot\alpha}$ defined by

$$W_\alpha = -\frac{1}{4}\Big[\overline{\mathcal{D}}\cdot\overline{\mathcal{D}} - 8\mathcal{R}\Big]\mathcal{D}_\alpha V\,,\qquad \overline{W}_{\dot\alpha} = -\frac{1}{4}\Big[\mathcal{D}\cdot\mathcal{D} - 8\mathcal{R}^\dagger\Big]\overline{\mathcal{D}}_{\dot\alpha} V\,.$$

The component fields of these two last quantities are given by

$$\begin{aligned}
W_\alpha\Big| &= i\lambda_\alpha\,,\\
\overline{W}_{\dot\alpha}\Big| &= -i\bar\lambda_{\dot\alpha}\,,\\
\mathcal{D}_\alpha W_\beta\Big| &= -i\,(\sigma^{\mu\nu})_{\alpha\beta}(\hat F^0_{\mu\nu}) - \varepsilon_{\alpha\beta}D\,,\\
\overline{\mathcal{D}}_{\dot\alpha}\overline{W}_{\dot\beta}\Big| &= i\,(\bar\sigma^{\mu\nu})_{\dot\alpha\dot\beta}(\hat F^0_{\mu\nu}) + \varepsilon_{\dot\alpha\dot\beta}D\,,\\
\mathcal{D}\cdot\mathcal{D}\,W_\alpha\Big| &= 4\sigma^\mu{}_{\alpha\dot\alpha}\hat D^0_\mu\bar\lambda^{\dot\alpha} - 2ib_\mu(\sigma^\mu\bar\lambda)_\alpha - 2iM^*\lambda_\alpha\,,\\
\overline{\mathcal{D}}\cdot\overline{\mathcal{D}}\,\overline{W}_{\dot\alpha}\Big| &= 4\sigma^\mu{}_{\alpha\dot\alpha}\hat D^0_\mu\lambda^\alpha + 2ib_\mu(\lambda\sigma^\mu)_{\dot\alpha} + 2iM\bar\lambda_{\dot\alpha}\,,
\end{aligned}$$

and depend on the components of the gravitation supermultiplet, *i.e.*, the gravitino $(\psi_{\tilde\mu}{}^\alpha, \bar\psi_{\tilde\mu\dot\alpha})$, the graviton $e_{\tilde\mu}{}^\mu$ and the auxiliary fields M and b^μ. The field strength tensor of the vector field and the covariant derivatives are defined by

$$\begin{aligned}
\hat F^0_{\mu\nu} &= \hat D_\mu v_\nu - \hat D_\nu v_\mu = e_\mu{}^{\tilde\mu} e_\nu{}^{\tilde\nu}\hat F^0_{\tilde\mu\tilde\nu}\,,\\
\hat F^0_{\tilde\mu\tilde\nu} &= F^0_{\tilde\mu\tilde\nu} - \frac{i}{2}\Big[\lambda\bar\sigma_{\tilde\nu}\psi_{\tilde\mu} + \lambda\sigma_{\tilde\nu}\bar\psi_{\tilde\mu} - \bar\lambda\bar\sigma_{\tilde\mu}\psi_{\tilde\nu} - \lambda\sigma_{\tilde\mu}\bar\psi_{\tilde\nu}\Big]\,,\\
\hat D_\mu v_\nu &= e_\mu{}^{\tilde\mu}\mathcal{D}_{\tilde\mu} v_\nu - \frac{i}{2}\Big[\lambda\bar\sigma_\nu\psi_\mu - \bar\psi_\mu\bar\sigma_\nu\lambda\Big] - \frac{i}{2}(\psi_\nu\sigma^\rho\bar\psi_\mu)v_\rho\,,\\
\hat D^0_\mu\lambda_\alpha &= e_\mu{}^{\tilde\mu}\mathcal{D}_{\tilde\mu}\lambda_\alpha - \frac{1}{2}(\sigma^{\nu\rho}\psi_\mu)_\alpha(\hat F^0_{\nu\rho}) + \frac{i}{2}D\psi_{\mu\alpha}\,,\\
\hat D^0_\mu\bar\lambda_{\dot\alpha} &= e_\mu{}^{\tilde\mu}\mathcal{D}_{\tilde\mu}\bar\lambda_{\dot\alpha} + \frac{1}{2}(\bar\psi_\mu\bar\sigma^{\nu\rho})_{\dot\alpha}(\hat F^0_{\nu\rho}) - \frac{i}{2}D\bar\psi_{\mu\dot\alpha}\,.
\end{aligned}$$

where the dots stand for term of order three and higher in V. By definition, the component fields of the chiral superfield W_α are given by $W_\alpha\Big|$, $\mathcal{D}_\beta W_\alpha\Big|$ and $\mathcal{D}\mathcal{D}\,W_\alpha\Big|$ while those of the antichiral superfield $\overline{W}_{\dot\alpha}$ read $\overline{W}_{\dot\alpha}\Big|$, $\overline{\mathcal{D}}_{\dot\beta}\overline{W}_{\dot\alpha}\Big|$ and $\overline{\mathcal{D}}\cdot\overline{\mathcal{D}}\,\overline{W}_{\dot\alpha}\Big|$ and we must therefore compute derivatives of V up to order five. Consequently, the components associated to the terms of the series expansions in V which are at least trilinear in V, *i.e.*, those included in the dots in (4.63), vanish in the Wess-Zumino gauge since $V\Big| = \mathcal{D}_M V\Big| = 0$. Moreover, $\mathcal{D}_\alpha V^2 = \mathcal{D}_\alpha V V + V\mathcal{D}_\alpha V$, and we can therefore rewrite the two superfields W_α and $\overline{W}_{\dot\alpha}$ as

$$\begin{aligned} W_\alpha &= \frac{g}{2}\Big[\overline{\mathcal{D}}\cdot\overline{\mathcal{D}} - 8\mathcal{R}\Big]\Big[\mathcal{D}_\alpha V + g[V, \mathcal{D}_\alpha V]\Big] , \\ \overline{W}_{\dot\alpha} &= \frac{g}{2}\Big[\mathcal{D}\cdot\mathcal{D} - 8\mathcal{R}^\dagger\Big]\Big[\overline{\mathcal{D}}_{\dot\alpha} V - g[V, \overline{\mathcal{D}}_{\dot\alpha} V]\Big] , \end{aligned}$$

keeping in mind that these two expressions only hold in the Wess-Zumino gauge. We now calculate the component fields of the W_α and $\overline{W}_{\dot\alpha}$ superfields. The$\theta = \bar\theta = 0$ component, by definition, as in (4.37), read

$$\begin{aligned} W_\alpha\Big| &= -2ig\lambda_\alpha , \\ \overline{W}_{\dot\alpha}\Big| &= 2ig\bar\lambda_{\dot\alpha} . \end{aligned}$$

The first-order superderivatives of the superfield strength tensors are given (in the Wess-Zumino gauge) by

$$\begin{aligned} \mathcal{D}_\beta W_\alpha\Big| &= \frac{g}{2}\mathcal{D}_\beta\overline{\mathcal{D}}\cdot\overline{\mathcal{D}}\mathcal{D}_\alpha V\Big| + \frac{g^2}{2}\mathcal{D}_\beta\overline{\mathcal{D}}\cdot\overline{\mathcal{D}}[V, \mathcal{D}_\alpha V]\Big| , \\ \overline{\mathcal{D}}_{\dot\beta}\overline{W}_{\dot\alpha}\Big| &= \frac{g}{2}\overline{\mathcal{D}}_{\dot\beta}\mathcal{D}\cdot\mathcal{D}\overline{\mathcal{D}}_{\dot\alpha} V\Big| - \frac{g^2}{2}\overline{\mathcal{D}}_{\dot\beta}\mathcal{D}\cdot\mathcal{D}[V, \overline{\mathcal{D}}_{\dot\alpha} V]\Big| . \end{aligned}$$

The contributions linear in V can be deduced from the calculations in the abelian case, *i.e.*, from (4.55) and (4.56), whilst the other contributions read

$$\begin{aligned} \mathcal{D}_\beta\overline{\mathcal{D}}\cdot\overline{\mathcal{D}}[V, \mathcal{D}_\alpha V]\Big| &= 2[\mathcal{D}_\beta\overline{\mathcal{D}}_{\dot\alpha} V, \overline{\mathcal{D}}^{\dot\alpha}\mathcal{D}_\alpha V]\Big| \\ &= -4(\sigma^{\mu\nu})_{\alpha\beta}[v_\mu, v_\nu] , \\ \overline{\mathcal{D}}_{\dot\beta}\mathcal{D}\cdot\mathcal{D}[V, \overline{\mathcal{D}}_{\dot\alpha} V]\Big| &= 2[\overline{\mathcal{D}}_{\dot\beta}\mathcal{D}^\alpha V, \mathcal{D}_\alpha\overline{\mathcal{D}}_{\dot\alpha} V]\Big| \\ &= -4(\bar\sigma^{\mu\nu})_{\dot\alpha\dot\beta}[v_\mu, v_\nu] . \end{aligned} \tag{4.64}$$

We therefore deduce the θ-coefficients of the superfields W_α and $\overline{W}_{\dot\alpha}$ to be

$$\begin{aligned} \mathcal{D}_\alpha W_\beta\Big| &= 2gi\,(\sigma^{\mu\nu})_{\alpha\beta}\Big[(\hat F^0_{\mu\nu}) + ig[v_\mu, v_\nu]\Big] + 2g\varepsilon_{\alpha\beta} D \\ &= 2gi\,(\sigma^{\mu\nu})_{\alpha\beta}(\hat F_{\mu\nu}) + 2g\varepsilon_{\alpha\beta} D , \\ \overline{\mathcal{D}}_{\dot\alpha}\overline{W}_{\dot\beta}\Big| &= -2gi\,(\bar\sigma^{\mu\nu})_{\dot\alpha\dot\beta}\Big[(\hat F^0_{\mu\nu}) + ig[v_\mu, v_\nu]\Big] - 2g\varepsilon_{\dot\alpha\dot\beta} D \\ &= -2gi\,(\bar\sigma^{\mu\nu})_{\dot\alpha\dot\beta}(\hat F_{\mu\nu}) - 2g\varepsilon_{\dot\alpha\dot\beta} D , \end{aligned} \tag{4.65}$$

where we have introduced the Yang-Mills field strength tensor for the gauge boson $v_\mu = v_\mu^a T_a$, namely

$$\begin{aligned} \hat F_{\mu\nu} &= \hat F^0_{\mu\nu} + ig[v_\mu, v_\nu] \\ &= \hat D_\mu v_\nu - \hat D_\nu v_\mu + ig[v_\mu, v_\nu] . \end{aligned}$$

Converting Lorentzian to Einsteinian indices, (4.57) can be extended to the non-abelian case as

$$\begin{aligned}\hat{F}_{\tilde{\mu}\tilde{\nu}} &= e_{\tilde{\mu}}{}^{\mu} e_{\tilde{\nu}}{}^{\nu}\Big(\hat{D}_\mu v_\nu - \hat{D}_\nu v_\mu + ig[v_\mu, v_\nu]\Big)\\ &= \partial_{\tilde{\mu}} v_{\tilde{\nu}} - \partial_{\tilde{\mu}} v_{\tilde{\nu}} + ig[v_{\tilde{\mu}}, v_{\tilde{\nu}}] - \frac{i}{2}\Big[\bar{\lambda}\bar{\sigma}_{\tilde{\nu}}\psi_{\tilde{\mu}} + \lambda\sigma_{\tilde{\nu}}\bar{\psi}_{\tilde{\mu}} - \bar{\lambda}\bar{\sigma}_{\tilde{\mu}}\psi_{\tilde{\nu}} - \lambda\sigma_{\tilde{\mu}}\bar{\psi}_{\tilde{\nu}}\Big]\\ &\equiv F_{\tilde{\mu}\tilde{\nu}} - \frac{i}{2}\Big[\bar{\lambda}\bar{\sigma}_{\tilde{\nu}}\psi_{\tilde{\mu}} + \lambda\sigma_{\tilde{\nu}}\bar{\psi}_{\tilde{\mu}} - \bar{\lambda}\bar{\sigma}_{\tilde{\mu}}\psi_{\tilde{\nu}} - \lambda\sigma_{\tilde{\mu}}\bar{\psi}_{\tilde{\nu}}\Big]\,. \end{aligned} \tag{4.66}$$

We finally address the calculation of the second-order superderivatives of W_α and $\overline{W}_{\dot{\alpha}}$,

$$\begin{aligned}\mathcal{D}\cdot\mathcal{D}\, W_\alpha\Big| &= \frac{g}{2}\mathcal{D}\cdot\mathcal{D}\,\overline{\mathcal{D}}\cdot\overline{\mathcal{D}}\mathcal{D}_\alpha V\Big| + \frac{g^2}{2}\mathcal{D}\cdot\mathcal{D}\,\overline{\mathcal{D}}\cdot\overline{\mathcal{D}}[V, \mathcal{D}_\alpha V]\Big|\,,\\ \overline{\mathcal{D}}\cdot\overline{\mathcal{D}}\,\overline{W}_{\dot{\alpha}}\Big| &= \frac{g}{2}\overline{\mathcal{D}}\cdot\overline{\mathcal{D}}\,\mathcal{D}\cdot\mathcal{D}\overline{\mathcal{D}}_{\dot{\alpha}} V\Big| - \frac{g^2}{2}\overline{\mathcal{D}}\cdot\overline{\mathcal{D}}\mathcal{D}\cdot\mathcal{D}[V, \overline{\mathcal{D}}_{\dot{\alpha}} V]\Big|\,.\end{aligned}$$

Once again, the first contributions on the right hand side of the equations above are deduced from our calculations in the abelian case, as shown in (4.61) and (4.62), whilst the other terms are given by

$$\begin{aligned}\mathcal{D}\cdot\mathcal{D}\,\overline{\mathcal{D}}\cdot\overline{\mathcal{D}}[V, \mathcal{D}_\alpha V]\Big| &= -2[\mathcal{D}\cdot\mathcal{D}\overline{\mathcal{D}}^{\dot{\alpha}} V, \overline{\mathcal{D}}_{\dot{\alpha}}\mathcal{D}_\alpha V]\Big| + 4[\mathcal{D}^\beta\overline{\mathcal{D}}^{\dot{\alpha}} V, \mathcal{D}_\beta\overline{\mathcal{D}}_{\dot{\alpha}}\mathcal{D}_\alpha V]\Big|\\ &= 16i[(\sigma^\mu\bar{\lambda})_\alpha, v_\mu] + 8i(\sigma^\mu\bar{\sigma}^{\nu\rho}\bar{\psi}_\mu)_\alpha[v_\nu, v_\rho]\,,\\ \overline{\mathcal{D}}\cdot\overline{\mathcal{D}}\,\mathcal{D}\cdot\mathcal{D}[V, \overline{\mathcal{D}}_{\dot{\alpha}} V]\Big| &= 2[\overline{\mathcal{D}}\cdot\overline{\mathcal{D}}\mathcal{D}^\alpha V, \mathcal{D}_\alpha\overline{\mathcal{D}}_{\dot{\alpha}} V]\Big| + 4[\overline{\mathcal{D}}^{\dot{\beta}}\mathcal{D}^\alpha V, \overline{\mathcal{D}}_{\dot{\beta}}\mathcal{D}_\alpha\overline{\mathcal{D}}_{\dot{\alpha}} V]\Big|\\ &= -16i[(\lambda\sigma^\mu)_{\dot{\alpha}}, v_\mu] + 8i(\psi_\mu\sigma^{\nu\rho}\sigma^\mu)_{\dot{\alpha}}[v_\nu, v_\rho]\,.\end{aligned} \tag{4.67}$$

Subsequently, one obtains

$$\begin{aligned}\mathcal{D}\cdot\mathcal{D}\, W_\alpha\Big| &= -2g\Big[4\sigma^\mu{}_{\alpha\dot{\alpha}}\hat{D}_\mu\bar{\lambda}^{\dot{\alpha}} - 2ib_\mu(\sigma^\mu\bar{\lambda})_\alpha - 2iM^*\lambda_\alpha\Big]\,,\\ \overline{\mathcal{D}}\cdot\overline{\mathcal{D}}\,\overline{W}_{\dot{\alpha}}\Big| &= -2g\Big[4\sigma^\mu{}_{\alpha\dot{\alpha}}\hat{D}_\mu\lambda^\alpha + 2ib_\mu(\lambda\sigma^\mu)_{\dot{\alpha}} + 2iM\bar{\lambda}_{\dot{\alpha}}\Big]\,,\end{aligned} \tag{4.68}$$

where we have introduced a derivative of the gaugino field covariant with respect to both the supergravity and the gauge transformations,

$$\begin{aligned}\hat{D}_\mu\lambda_\alpha &= \hat{D}^0_\mu\lambda_\alpha + ig[v_\mu, \lambda_\alpha] - \frac{i}{2}g(\sigma^{\nu\rho}\psi_\mu)_\alpha[v_\nu, v_\rho]\\ &= e_\mu{}^{\tilde{\mu}}\Big(\mathcal{D}_{\tilde{\mu}}\lambda_\alpha + ig[v_{\tilde{\mu}}, \lambda_\alpha]\Big) - \frac{1}{2}(\sigma^{\nu\rho}\psi_\mu)_\alpha\Big((\hat{F}^0_{\nu\rho}) + ig[v_\nu, v_\rho]]\Big) + \frac{i}{2}D\psi_{\mu\alpha}\,,\\ \hat{D}_\mu\bar{\lambda}_{\dot{\alpha}} &= \hat{D}^0_\mu\bar{\lambda}_{\dot{\alpha}} + ig[v_\mu, \bar{\lambda}_{\dot{\alpha}}] + \frac{i}{2}g(\bar{\psi}\bar{\sigma}^{\nu\rho})_{\dot{\alpha}}[v_\nu, v_\rho]\\ &= e_\mu{}^{\tilde{\mu}}\Big(\mathcal{D}_{\tilde{\mu}}\bar{\lambda}_{\dot{\alpha}} + ig[v_{\tilde{\mu}}, \bar{\lambda}_{\dot{\alpha}}]\Big) + \frac{1}{2}(\bar{\psi}\bar{\sigma}^{\nu\rho})_{\dot{\alpha}}\Big((\hat{F}^0_{\nu\rho}) + ig[v_\nu, v_\rho]\Big) - \frac{i}{2}D\bar{\psi}_{\mu\dot{\alpha}}\,,\end{aligned} \tag{4.69}$$

Table 4.3 Non-abelian vector superfields.

A non-abelian vector superfield $V = V^a T_a$, T_a being hermitian representation matrices of the gauge group considered, is defined by the constraint $V = V^\dagger$. In the Wess-Zumino gauge,

$$V\big| = 0\,, \qquad \mathcal{D}_\alpha V\big| = 0\,, \qquad \overline{\mathcal{D}}_{\dot\alpha} V\big| = 0\,, \qquad \mathcal{D}_\alpha \mathcal{D}_\beta V\big| = 0\,, \qquad \overline{\mathcal{D}}_{\dot\alpha}\overline{\mathcal{D}}_{\dot\beta} V\big| = 0\,,$$

so that all non-physical component fields are eliminated. The remaining physical degrees of freedom, *i.e.*, the gaugino $(\lambda_\alpha, \bar\lambda^{\dot\alpha})$ and the gauge boson v_μ, as well as the auxiliary D-term, are given by

$$v_\mu = \frac{1}{4}\bar\sigma_\mu{}^{\dot\alpha\alpha}\big[\mathcal{D}_\alpha, \overline{\mathcal{D}}_{\dot\alpha}\big]V\big|\,, \quad \lambda_\alpha = \frac{i}{2g}W_\alpha\big|\,, \quad \bar\lambda_{\dot\alpha} = -\frac{i}{2g}\overline{W}_{\dot\alpha}\big|\,, \quad D = \frac{1}{4g}\mathcal{D}^\alpha W_\alpha\big| = \frac{1}{4g}\overline{\mathcal{D}}_{\dot\alpha}\overline{W}^{\dot\alpha}\big|\,,$$

where we have introduced the chiral and antichiral spinorial superfields W_α and $\overline{W}_{\dot\alpha}$ defined by

$$W_\alpha = -\frac{1}{4}\big[\overline{\mathcal{D}}\cdot\overline{\mathcal{D}} - 8\mathcal{R}\big]e^{2gV}\mathcal{D}_\alpha e^{-2gV}\,, \qquad \overline{W}_{\dot\alpha} = -\frac{1}{4}\big[\mathcal{D}\cdot\mathcal{D} - 8\mathcal{R}^\dagger\big]e^{-2gV}\overline{\mathcal{D}}_{\dot\alpha}e^{2gV}\,.$$

The component fields of these two last quantities are given by

$$\begin{aligned}
W_\alpha\big| &= -2ig\lambda_\alpha\,,\\
\overline{W}_{\dot\alpha}\big| &= 2ig\bar\lambda_{\dot\alpha}\,,\\
\mathcal{D}_\alpha W_\beta\big| &= 2g\Big[i(\sigma^{\mu\nu})_{\alpha\beta}(\hat F_{\mu\nu}) + \varepsilon_{\alpha\beta}D\Big]\,,\\
\overline{\mathcal{D}}_{\dot\alpha}\overline{W}_{\dot\beta}\big| &= -2g\Big[i\,(\bar\sigma^{\mu\nu})_{\dot\alpha\dot\beta}(\hat F_{\mu\nu}) + \varepsilon_{\dot\alpha\dot\beta}D\Big]\,,\\
\mathcal{D}\cdot\mathcal{D}\,W_\alpha\big| &= -2g\Big[4\sigma^\mu{}_{\alpha\dot\alpha}\hat D_\mu\bar\lambda^{\dot\alpha} - 2ib_\mu(\sigma^\mu\bar\lambda)_\alpha - 2iM^*\lambda_\alpha\Big]\,,\\
\overline{\mathcal{D}}\cdot\overline{\mathcal{D}}\,\overline{W}_{\dot\alpha}\big| &= -2g\Big[4\sigma^\mu{}_{\alpha\dot\alpha}\hat D_\mu\lambda^\alpha + 2ib_\mu(\lambda\sigma^\mu)_{\dot\alpha} + 2iM\bar\lambda_{\dot\alpha}\Big]\,,
\end{aligned}$$

and depend on the components of the gravitation supermultiplet, *i.e.*, the gravitino $(\psi_{\tilde\mu}{}^\alpha, \bar\psi_{\tilde\mu\dot\alpha})$, the graviton $e_{\tilde\mu}{}^\mu$ and the auxiliary fields M and b^μ. The field strength tensor of the vector field and the covariant derivatives are defined by

$$\begin{aligned}
\hat F_{\mu\nu} &= \hat D_\mu v_\nu - \hat D_\nu v_\mu + ig[v_\mu, v_\nu] = e_\mu{}^{\tilde\mu}e_\nu{}^{\tilde\nu}\hat F_{\tilde\mu\tilde\nu}\,,\\
\hat F_{\tilde\mu\tilde\nu} &= F_{\tilde\mu\tilde\nu} - \frac{i}{2}\Big[\bar\lambda\bar\sigma_{\tilde\nu}\psi_{\tilde\mu} + \lambda\sigma_{\tilde\nu}\bar\psi_{\tilde\mu} - \bar\lambda\bar\sigma_{\tilde\mu}\psi_{\tilde\nu} - \lambda\sigma_{\tilde\mu}\bar\psi_{\tilde\nu}\Big]\,,\\
\hat D_\mu v_\nu &= e_\mu{}^{\tilde\mu}\mathcal{D}_{\tilde\mu}v_\nu - \frac{i}{2}\Big[\bar\lambda\bar\sigma_\nu\psi_\mu - \bar\psi_\mu\bar\sigma_\nu\lambda\Big] - \frac{i}{2}(\psi_\nu\sigma^\rho\bar\psi_\mu)v_\rho\,,\\
\hat D_\mu\lambda_\alpha &= e_\mu{}^{\tilde\mu}\Big(\mathcal{D}_{\tilde\mu}\lambda_\alpha + ig[v_{\tilde\mu}, \lambda_\alpha]\Big) - \frac{1}{2}(\sigma^{\nu\rho}\psi_\mu)_\alpha(\hat F_{\nu\rho}) + \frac{i}{2}D\psi_{\mu\alpha}\,,\\
\hat D_\mu\bar\lambda_{\dot\alpha} &= e_\mu{}^{\tilde\mu}\Big(\mathcal{D}_{\tilde\mu}\bar\lambda_{\dot\alpha} + ig[v_{\tilde\mu}, \bar\lambda_{\dot\alpha}]\Big) + \frac{1}{2}(\bar\psi_\mu\bar\sigma^{\nu\rho})_{\dot\alpha}(\hat F_{\nu\rho}) - \frac{i}{2}D\bar\psi_{\mu\dot\alpha}\,.
\end{aligned}$$

the additional terms in $[v_\nu, v_\rho]$, $[v_\mu, \lambda]$ and $[v_\mu, \bar\lambda]$ making the derivative covariant with respect to the gauge transformations. Notice that the expressions in (4.69) can be directly obtained along the same lines as in (4.58). Indeed, starting from

$$\hat{D}_\mu \lambda_\alpha = \frac{i}{2g}\Big(\mathcal{D}_\mu W_\alpha + \frac{i}{2} g \bar{\sigma}_\mu{}^{\dot{\beta}\beta}[\mathcal{D}_\beta \overline{\mathcal{D}}_{\dot{\beta}} V, W_\alpha]\Big)\Big| ,$$

$$\hat{D}_\mu \bar{\lambda}_{\dot{\alpha}} = -\frac{i}{2g}\Big(\mathcal{D}_\mu \overline{W}_{\dot{\alpha}} - \frac{i}{2} g \bar{\sigma}_\mu{}^{\dot{\beta}\beta}[\overline{\mathcal{D}}_{\dot{\beta}} \mathcal{D}_\beta V, \overline{W}_{\dot{\alpha}}]\Big)\Big| ,$$

the first terms contain, with respect to the abelian case, additional contributions involving $[v_\nu, v_\rho]$, as shown in (4.65), leading naturally to the non-abelian field strength tensor $\hat{F}_{\nu\rho} = \hat{F}^0_{\nu\rho} + ig[v_{\underline{\nu}}, v_\rho]$ in (4.69), whilst the second terms contain contributions in $[v_\mu, \lambda_\alpha]$ and $[v_\mu, \bar{\lambda}_{\dot{\alpha}}]$ making the derivatives $\mathcal{D}_{\tilde{\mu}} \lambda_\alpha$ and $\mathcal{D}_{\tilde{\mu}} \bar{\lambda}_{\dot{\alpha}}$ covariant with respect to the gauge transformations.

We summarise the most important analytical results with respect to non-abelian vector superfields in Table 4.3, *i.e.*, its component fields, in the Wess-Zumino gauge, those of the associated spinorial superfields, and the derivatives of the physical degrees of freedom covariant with respect to both the supergravity and the gauge transformations.

4.3 Gauge interactions of chiral superfields

In global supersymmetry (see (2.21)), the construction of gauge invariant kinetic and interaction Lagrangians for chiral supermultiplets proceeds through the computation of the quantities $\Phi^\dagger e^{-2gV}$ and $e^{-2gV}\Phi$, where Φ is a chiral superfield and $V = V^a T_a$ is a vector superfield associated to a specific non-abelian gauge group[5]. As in the previous section, the gauge coupling constant associated to the gauge group considered will be denoted by g. It is of course assumed that the superfield Φ lies in a given (non-trivial) unitary representation of the gauge group specified by the matrices T_a. In supergravity, building gauge invariant actions proceeds analogously to the global case and requires to make the quantities $\Phi^\dagger e^{-2gV}$ and $e^{-2gV}\Phi$ covariant with respect to the supergravity transformations. Therefore, we dedicate this section to the computation of the component fields of the conjugate chiral and antichiral superfields

$$\begin{aligned} \mathcal{X} &= (\overline{\mathcal{D}}\cdot\overline{\mathcal{D}} - 8\mathcal{R})\Phi^\dagger e^{-2gV} , \\ \mathcal{X}^\dagger &= (\mathcal{D}\cdot\mathcal{D} - 8\mathcal{R}^\dagger) e^{-2gV}\Phi , \end{aligned} \tag{4.70}$$

after adopting the Wess-Zumino gauge (4.40). By definition, the component fields are obtained through the calculation of the quantities $\mathcal{X}|$, $\mathcal{D}_\alpha \mathcal{X}|$ and $\mathcal{D}\cdot\mathcal{D}\mathcal{X}|$ and their hermitian conjugate counterparts. Expanding the exponentials (4.70) as

[5]The abelian case can be recovered by replacing the representation matrices by the charge operator of the abelian group.

series in the vector superfield V and recalling that in the Wess-Zumino gauge, $V\big| = \mathcal{D}_M V\big| = 0$, all terms of the expansion which are at least trilinear in V vanish. Therefore, we are left with

$$\begin{aligned}\mathcal{X} &= \Xi - 2g(\overline{\mathcal{D}}\cdot\overline{\mathcal{D}} - 8\mathcal{R})\Phi^\dagger V + 2g^2(\overline{\mathcal{D}}\cdot\overline{\mathcal{D}} - 8\mathcal{R})\Phi^\dagger V^2\,,\\ \mathcal{X}^\dagger &= \Xi^\dagger - 2g(\mathcal{D}\cdot\mathcal{D} - 8\mathcal{R}^\dagger)V\Phi + 2g^2(\mathcal{D}\cdot\mathcal{D} - 8\mathcal{R}^\dagger)V^2\Phi\,,\end{aligned} \tag{4.71}$$

where the chiral and antichiral superfields Ξ and $\Xi^\dagger$ have been defined in Eq. (4.19) and their component fields are given in Table 4.1.

Components of the $(\overline{\mathcal{D}}\cdot\overline{\mathcal{D}} - 8\mathcal{R})\Phi^\dagger V$ and $(\mathcal{D}\cdot\mathcal{D} - 8\mathcal{R}^\dagger)V\Phi$ superfields.

In the Wess-Zumino gauge, the lowest order component fields $(\overline{\mathcal{D}}\overline{\mathcal{D}} - 8\mathcal{R})\Phi^\dagger V\big|$ and $(\mathcal{D}\cdot\mathcal{D} - 8\mathcal{R}^\dagger)\Phi V\big|$ vanish since $\overline{\mathcal{D}}_{\dot\alpha}\overline{\mathcal{D}}_{\dot\beta} V\big| = \mathcal{D}_\alpha\mathcal{D}_\beta V\big| = 0$. The θ-coefficient of the chiral superfield $(\overline{\mathcal{D}}\cdot\overline{\mathcal{D}} - 8\mathcal{R})\Phi^\dagger V$ reads

$$\begin{aligned}\mathcal{D}_\alpha(\overline{\mathcal{D}}\cdot\overline{\mathcal{D}} - 8\mathcal{R})\Phi^\dagger V\big| &= 2\overline{\mathcal{D}}^{\dot\alpha}\Phi^\dagger\mathcal{D}_\alpha\overline{\mathcal{D}}_{\dot\alpha}V\big| + \Phi^\dagger\mathcal{D}_\alpha\overline{\mathcal{D}}\cdot\overline{\mathcal{D}}V\big|\\ &= 2\sqrt{2}(\sigma^\mu\bar\chi)_\alpha v_\mu - 4i\phi^\dagger\lambda_\alpha - 2i\phi^\dagger v_\nu(\sigma^\mu\bar\sigma^\nu\psi_\mu)_\alpha\,,\end{aligned}$$

using (4.3) and (4.7) for the lowest-order component of the antichiral superfield $\Phi^\dagger$ and for its first order derivative $\overline{\mathcal{D}}^{\dot\alpha}\Phi^\dagger$ as well as (4.46) and (4.50) for those of the second-order and third-order derivatives of the vector superfield $\mathcal{D}_\alpha\overline{\mathcal{D}}_{\dot\alpha}V\big|$ and $\mathcal{D}_\alpha\overline{\mathcal{D}}\cdot\overline{\mathcal{D}}V$. Similarly, the $\bar\theta$-coefficient of the antichiral superfield $(\mathcal{D}\cdot\mathcal{D} - 8\mathcal{R}^\dagger)V\Phi$ is given by

$$\overline{\mathcal{D}}_{\dot\alpha}(\mathcal{D}\cdot\mathcal{D} - 8\mathcal{R}^\dagger)V\Phi\big| = 2\sqrt{2}(\chi\sigma^\mu)_{\dot\alpha}v_\mu + 4i\phi\bar\lambda_{\dot\alpha} + 2i\phi v_\nu(\bar\psi_\mu\bar\sigma^\nu\sigma^\mu)_{\dot\alpha}\,.$$

Finally, the highest-order components of $(\overline{\mathcal{D}}\cdot\overline{\mathcal{D}} - 8\mathcal{R})\Phi^\dagger V$ can be rewritten as

$$\mathcal{D}\cdot\mathcal{D}(\overline{\mathcal{D}}\overline{\mathcal{D}} - 8\mathcal{R})\Phi^\dagger V\big| = 4\mathcal{D}^\alpha\overline{\mathcal{D}}^{\dot\alpha}\Phi^\dagger\mathcal{D}_\alpha\overline{\mathcal{D}}_{\dot\alpha}V\big| - 2\overline{\mathcal{D}}^{\dot\alpha}\Phi^\dagger\mathcal{D}\cdot\mathcal{D}\overline{\mathcal{D}}_{\dot\alpha}V\big| + \Phi^\dagger\mathcal{D}\cdot\mathcal{D}\,\overline{\mathcal{D}}\cdot\overline{\mathcal{D}}V\big|\,.$$

The first two terms are directly obtained from (4.7), (4.16), (4.46) and (4.49), while the last one can be calculated from the relation

$$\begin{aligned}\mathcal{D}\cdot\mathcal{D}\,\overline{\mathcal{D}}\cdot\overline{\mathcal{D}}V\big| &= \{\mathcal{D}^\alpha,\mathcal{D}_\alpha\}\overline{\mathcal{D}}\cdot\overline{\mathcal{D}}V\big| - \mathcal{D}_\alpha\{\mathcal{D}^\alpha,\overline{\mathcal{D}}_{\dot\alpha}\}\overline{\mathcal{D}}^{\dot\alpha}V\big| \\ &\quad + \mathcal{D}_\alpha\overline{\mathcal{D}}_{\dot\alpha}\{\mathcal{D}^\alpha,\overline{\mathcal{D}}^{\dot\alpha}\}V\big| - \mathcal{D}_\alpha\overline{\mathcal{D}}\cdot\overline{\mathcal{D}}\mathcal{D}^\alpha V\big| \\ &= -2i\bar\sigma^{\mu\dot\alpha\alpha}\Big[\mathcal{D}_\alpha\mathcal{D}_\mu\overline{\mathcal{D}}_{\dot\alpha}V\big| + \mathcal{D}_\alpha\overline{\mathcal{D}}_{\dot\alpha}\mathcal{D}_\mu V\big|\Big] \\ &\quad - R_{\alpha\dot\alpha\dot\gamma}{}^{\dot\alpha}\mathcal{D}^\alpha\overline{\mathcal{D}}^{\dot\gamma}V\big| - \mathcal{D}_\alpha\overline{\mathcal{D}}\cdot\overline{\mathcal{D}}\mathcal{D}^\alpha V\big|\,,\end{aligned}$$

where we have used the definition of the anticommutators given in (3.8) together with the constraints (3.29). The different contributions to $\mathcal{D}\cdot\mathcal{D}\,\overline{\mathcal{D}}\cdot\overline{\mathcal{D}}V\big|$ can now been deduced from (4.53) for the first term, from Eq. (4.46) and Table 3.4 for the second term and from Eqs. (4.54) and (4.55) for the last term. Thus we get

$$\begin{aligned}-2i\bar\sigma^{\mu\dot\alpha\alpha}\Big[\mathcal{D}_\alpha\mathcal{D}_\mu\overline{\mathcal{D}}_{\dot\alpha}V\big| + \mathcal{D}_\alpha\overline{\mathcal{D}}_{\dot\alpha}\mathcal{D}_\mu V\big|\Big] &= -8i\hat{D}_\mu v^\mu - \frac{10}{3}v_\mu b^\mu\,, \\ -R_{\alpha\dot\alpha\dot\gamma}{}^{\dot\alpha}\mathcal{D}^\alpha\overline{\mathcal{D}}^{\dot\gamma}V\big| &= -2b^\mu v_\mu\,, \\ -\mathcal{D}_\alpha\overline{\mathcal{D}}\cdot\overline{\mathcal{D}}\mathcal{D}^\alpha V\big| &= 8D\,.\end{aligned}$$

Collecting all the contributions, one obtains

$$\begin{aligned}\mathcal{D}\cdot\mathcal{D}(\overline{\mathcal{D}}\cdot\overline{\mathcal{D}} - 8\mathcal{R})\Phi^\dagger V\big| &= -16iv_\mu\hat{D}^\mu\phi^\dagger + 8\sqrt{2}i\bar\chi\cdot\bar\lambda - 8i\phi^\dagger\hat{D}_\mu v^\mu \\ &\quad - \frac{16}{3}\phi^\dagger v_\mu b^\mu + 8D\phi^\dagger\,.\end{aligned}$$

In an identical way, one gets the conjugate quantity

$$\begin{aligned}\overline{\mathcal{D}}\cdot\overline{\mathcal{D}}(\mathcal{D}\cdot\mathcal{D} - 8\mathcal{R}^\dagger)V\Phi\big| &= 16iv_\mu\hat{D}^\mu\phi - 8\sqrt{2}i\chi\cdot\lambda + 8i\phi(\hat{D}_\mu v^\mu)^\dagger \\ &\quad - \frac{16}{3}\phi v_\mu b^\mu + 8D\phi\,.\end{aligned}$$

Components of the $(\overline{\mathcal{D}}\cdot\overline{\mathcal{D}} - 8\mathcal{R})\Phi^\dagger V^2$ and $(\mathcal{D}\cdot\mathcal{D} - 8\mathcal{R}^\dagger)V^2\Phi$ superfields.

In the Wess-Zumino gauge, only the highest-order components of the superfields $(\overline{\mathcal{D}}\cdot\overline{\mathcal{D}} - 8\mathcal{R})\Phi^\dagger V^2$ and $(\mathcal{D}\cdot\mathcal{D} - 8\mathcal{R}^\dagger)V^2\Phi$ are non vanishing. They read

$$\begin{aligned}\mathcal{D}\cdot\mathcal{D}(\overline{\mathcal{D}}\cdot\overline{\mathcal{D}} - 8\mathcal{R})\Phi^\dagger V^2\big| &= 4\Phi^\dagger\mathcal{D}_\alpha\overline{\mathcal{D}}_{\dot\alpha}V\mathcal{D}^\alpha\overline{\mathcal{D}}^{\dot\alpha}V\big| = 8\phi^\dagger v_\mu v^\mu\,, \\ \overline{\mathcal{D}}\cdot\overline{\mathcal{D}}(\mathcal{D}\cdot\mathcal{D} - 8\mathcal{R}^\dagger)V^2\Phi\big| &= 4\overline{\mathcal{D}}_{\dot\alpha}\mathcal{D}_\alpha V\overline{\mathcal{D}}^{\dot\alpha}\mathcal{D}^\alpha V\Phi\big| = 8\phi v_\mu v^\mu\,,\end{aligned}$$

where we have used (4.2), (4.3) and (4.46). We collect all the results related to the superfields $\mathcal{X}$ and $\mathcal{X}^\dagger$ in Table 4.4.

Table 4.4 Gauge interactions of chiral superfields.

The gauge interactions of chiral and antichiral superfields Φ and $\Phi^\dagger$ are related to the antichiral and chiral superfields $\mathcal{X}^\dagger$ and $\mathcal{X}$, respectively, defined in the Wess-Zumino gauge by

$$\mathcal{X} = \Xi - 2g(\overline{\mathcal{D}}\cdot\overline{\mathcal{D}} - 8\mathcal{R})\Phi^\dagger V + 2g^2(\overline{\mathcal{D}}\cdot\overline{\mathcal{D}} - 8\mathcal{R})\Phi^\dagger V^2 ,$$
$$\mathcal{X}^\dagger = \Xi^\dagger - 2g(\mathcal{D}\cdot\mathcal{D} - 8\mathcal{R}^\dagger)V\Phi + 2g^2(\mathcal{D}\cdot\mathcal{D} - 8\mathcal{R}^\dagger)V^2\Phi ,$$

where V is the vector superfield associated to the gauge group considered and where g denotes its coupling constant. The component fields of $\mathcal{X}$ and $\mathcal{X}^\dagger$ are given by

$$\begin{aligned}
\mathcal{X}\big| &= \Xi\big| ,\\
\mathcal{X}^\dagger\big| &= \Xi^\dagger\big| ,\\
\mathcal{D}_\alpha\mathcal{X}\big| &= \mathcal{D}_\alpha\Xi\big| - 4\sqrt{2}g(\sigma^\mu\bar{\chi})_\alpha v_\mu + 8ig\phi^\dagger\lambda_\alpha + 4ig\phi^\dagger v_\nu(\sigma^\mu\bar{\sigma}^\nu\psi_\mu)_\alpha ,\\
\overline{\mathcal{D}}_{\dot\alpha}\mathcal{X}^\dagger\big| &= \overline{\mathcal{D}}_{\dot\alpha}\Xi^\dagger\big| - 4\sqrt{2}g(\chi\sigma^\mu)_{\dot\alpha}v_\mu - 8ig\phi\bar{\lambda}_{\dot\alpha} - 4ig\phi v_\nu(\bar{\psi}_\mu\bar{\sigma}^\nu\sigma^\mu)_{\dot\alpha} ,\\
\mathcal{D}\cdot\mathcal{D}\mathcal{X}\big| &= \mathcal{D}\cdot\mathcal{D}\Xi\big| + 32igv_\mu\hat{D}^\mu\phi^\dagger - 16\sqrt{2}ig\bar{\chi}\cdot\bar{\lambda} + 16ig\phi^\dagger\hat{D}_\mu v^\mu + \frac{32}{3}g\phi^\dagger v_\mu b^\mu\\
&\quad - 16gD\phi^\dagger + 16g^2\phi^\dagger v_\mu v^\mu ,\\
\overline{\mathcal{D}}\cdot\overline{\mathcal{D}}\mathcal{X}^\dagger\big| &= \overline{\mathcal{D}}\cdot\overline{\mathcal{D}}\Xi^\dagger\big| - 32igv_\mu\hat{D}^\mu\phi + 16\sqrt{2}ig\chi\cdot\lambda - 16ig\phi(\hat{D}_\mu v^\mu)^\dagger + \frac{32}{3}g\phi v_\mu b^\mu - 16gD\phi + 16g^2\phi v_\mu v^\mu ,
\end{aligned}$$

and depend on the components of the superfield Φ ($\Phi^\dagger$), *i.e.*, the scalar field ϕ ($\phi^\dagger$), the fermionic field χ_α ($\bar{\chi}_{\dot\alpha}$) and the auxiliary F-term ($F^\dagger$), on those of the superfields Ξ and $\Xi^\dagger$ defined in Table 4.1, on those of the vector superfield V, *i.e.*, the gaugino ($\lambda_\alpha, \bar{\lambda}^{\dot\alpha}$), the vector boson v_μ and the auxiliary D-term, as well as on the components of the gravitation supermultiplet, *i.e.*, the gravitino ($\psi_{\tilde\mu}{}^\alpha, \bar{\psi}_{\tilde\mu\dot\alpha}$), the graviton $e_{\tilde\mu}{}^\mu$ and the auxiliary fields M and b^μ. The covariant derivatives appearing in the equations above are given in Tables 4.1, 4.2 and 4.3.

4.4 Transformation laws of vector multiplet under supergravity transformations

Now, considering a supergravity transformation specified by ξ^M and L_{MN}, we can compute the transformation laws of the components of a vector multiplet. Observing that $\xi^\mu\big| = 0, L_{MN}\big| = 0$ and $\xi^\alpha\big| = \varepsilon^\alpha, \bar{\xi}_{\dot\alpha}\big| = \bar{\varepsilon}_{\dot\alpha}$ (see Table 3.6), if a field is defined by $\phi_{\dots} = X_{\dots}\big|$ where $\cdots$ indicates any type of indices, we may write $\delta(\phi_{\dots}) = -\varepsilon^\alpha\mathcal{D}_\alpha X_{\dots}\big| + \bar{\varepsilon}^{\dot\alpha}\overline{\mathcal{D}}_{\dot\alpha}X_{\dots}\big|$.

The transformation of

$$C = V\big| = 0 ,$$
$$\chi_\alpha = \mathcal{D}_\alpha V\big| = 0 , \quad \bar{\chi}_{\dot\alpha} = \overline{\mathcal{D}}_{\dot\alpha}V\big| = 0 ,$$
$$f = -1/2\mathcal{D}\cdot\mathcal{D}V\big| = 0 , \quad f^\dagger = -1/2\overline{\mathcal{D}}\cdot\overline{\mathcal{D}}V\big| = 0 ,$$

are obviously given by

$$
\begin{aligned}
\delta C &= -\varepsilon^\alpha \mathcal{D}_\alpha V\Big| + \bar{\varepsilon}^{\dot\alpha} \overline{\mathcal{D}}_{\dot\alpha} V\Big| = 0\,, \\
\delta \chi_\alpha &= -\varepsilon^\beta \mathcal{D}_\beta \mathcal{D}_\alpha V\Big| + \bar{\varepsilon}^{\dot\beta} \overline{\mathcal{D}}_{\dot\beta} \mathcal{D}_\alpha V\Big| = -(\sigma^\mu \bar\varepsilon)_\alpha v_\mu\,, \\
\delta \bar\chi_{\dot\alpha} &= -\varepsilon^\beta \mathcal{D}_\beta \overline{\mathcal{D}}_{\dot\alpha} V\Big| + \bar{\varepsilon}^{\dot\beta} \overline{\mathcal{D}}_{\dot\beta} \overline{\mathcal{D}}_{\dot\alpha} V\Big| = -(\varepsilon \sigma^\mu)_{\dot\alpha} v_\mu\,, \qquad (4.72)\\
\delta f &= \frac{1}{2}\varepsilon^\alpha \mathcal{D}_\alpha \mathcal{D}\cdot\mathcal{D} V\Big| - \frac{1}{2}\bar\varepsilon^{\dot\alpha} \overline{\mathcal{D}}_{\dot\alpha} \mathcal{D}\cdot\mathcal{D} V\Big| = 2i\bar\varepsilon\cdot\bar\lambda + i\bar\psi_\mu \bar\sigma^\nu \sigma^\mu \bar\varepsilon v_\nu\,, \\
\delta f^\dagger &= \frac{1}{2}\varepsilon^\alpha \mathcal{D}_\alpha \overline{\mathcal{D}}\cdot\overline{\mathcal{D}} V\Big| - \frac{1}{2}\bar\varepsilon^{\dot\alpha} \overline{\mathcal{D}}_{\dot\alpha} \overline{\mathcal{D}}\cdot\overline{\mathcal{D}} V\Big| = -2i\varepsilon\cdot\lambda - i\varepsilon\sigma^\mu\bar\sigma^\nu \psi_\mu v_\nu\,,
\end{aligned}
$$

and follow from (4.40), (4.44),(4.42), and (4.51), (4.50). In particular this means that if we perform a supergravity transformation we do not remain in the Wess-Zumino gauge. We will come back to this point latter on. Now we turn to the transformation of the physical fields. For the gauge boson, since $v_{\alpha\dot\alpha} = 1/2[\mathcal{D}_\alpha, \overline{\mathcal{D}}_{\dot\alpha}]V\big|$ and $v_\mu = 1/2\bar\sigma_\mu{}^{\dot\alpha\alpha} v_{\alpha\dot\alpha}$, we have

$$
\begin{aligned}
\delta v_\mu &= \frac{1}{4}\bar\sigma_\mu{}^{\dot\alpha\alpha}\Big\{ -\varepsilon^\beta \mathcal{D}_\beta \mathcal{D}_\alpha \overline{\mathcal{D}}_{\dot\alpha} V\Big| + \bar\varepsilon^{\dot\beta} \overline{\mathcal{D}}_{\dot\beta} \mathcal{D}_\alpha \overline{\mathcal{D}}_{\dot\alpha} V\Big| \\
&\qquad + \varepsilon^\beta \mathcal{D}_\beta \overline{\mathcal{D}}_{\dot\alpha} \mathcal{D}_\alpha V\Big| - \bar\varepsilon^{\dot\beta} \overline{\mathcal{D}}_{\dot\beta} \overline{\mathcal{D}}_{\dot\alpha} \mathcal{D}_\alpha V\Big| \Big\} \\
&= i(\varepsilon\sigma_\mu\bar\lambda - \lambda\sigma_\mu\bar\varepsilon) - \frac{i}{2}\big(\psi_\mu\sigma^\nu\bar\varepsilon - \varepsilon\sigma^\nu\bar\psi_\mu\big)v_\nu\,, \qquad (4.73)
\end{aligned}
$$

according (4.49) and (4.50). For the transformations of the gauginos we use $\lambda_\alpha = i/(2g)W_\alpha\big|, \bar\lambda_{\dot\alpha} = -i/(2g)\overline{W}_{\dot\alpha}\big|$ to get

$$
\begin{aligned}
\delta\lambda_\alpha &= \frac{i}{2g}\Big(-\varepsilon^\beta \mathcal{D}_\beta W_\alpha + \bar\varepsilon^{\dot\beta} \mathcal{D}_{\dot\beta} W_\alpha\Big) = (\sigma^{\mu\nu}\varepsilon)_\alpha \hat F_{\mu\nu} + i\varepsilon_\alpha D\,, \\
\delta\bar\lambda_{\dot\alpha} &= -(\bar\varepsilon\bar\sigma^{\mu\nu})_{\dot\alpha}\hat F_{\mu\nu} - i\bar\varepsilon_{\dot\alpha} D\,,
\end{aligned}
\qquad (4.74)
$$

with the help of (4.65), and the fact that W_α is a chiral spinor superfield. Finally, for the auxiliary field, we start from $D = 1/(4g)\mathcal{D}^a W_\alpha\big| = 1/(4g)\overline{\mathcal{D}}_{\dot\alpha}\overline{W}^{\dot\alpha}\big|$ to arrive at

$$
\begin{aligned}
\delta D &= \frac{1}{4g}\Big(-\varepsilon^\beta \mathcal{D}_\beta \mathcal{D}^a W_\alpha\Big| - \bar\varepsilon_{\dot\beta}\overline{\mathcal{D}}^{\dot\beta}\overline{\mathcal{D}}_{\dot\alpha}\overline{W}^{\dot\alpha}\Big|\Big) \\
&= -\varepsilon\sigma^\mu \hat D^0_\mu \bar\lambda - \hat D^0_\mu \lambda\sigma^\mu\bar\varepsilon + \frac{i}{2}b_\mu\big(\varepsilon\sigma^\mu\bar\lambda - \lambda\sigma^\mu\bar\varepsilon\big) + \frac{i}{2}\big(M^*\varepsilon\cdot\lambda - M\bar\varepsilon\cdot\bar\lambda\big)\,,
\end{aligned}
\qquad (4.75)
$$

after using (4.68) and $\mathcal{D}_\alpha\mathcal{D}_\beta = 1/2\varepsilon_{\alpha\beta}\mathcal{D}\cdot\mathcal{D}$, $\overline{\mathcal{D}}_{\dot\alpha}\overline{\mathcal{D}}_{\dot\beta} = -1/2\varepsilon_{\dot\alpha\dot\beta}\overline{\mathcal{D}}\cdot\overline{\mathcal{D}}$. Note that the derivative $\hat D^0_\mu$ are only covariant with respect to supergravity and not with respect to gauge transformations (see Table 4.2).

At this point some important remarks are in order. Since by a supergravity transformation we are no more in the Wess-Zumino gauge, the relationship between the components of the multiplet and the various derivatives are no longer valid. Remember that the assumption that V is in the Wess-Zumino gauge was essential when computing the various derivatives of V. In order to maintain the validity of these relationships, we just have to combine the supergravity transformations with a gauge transformation with superfield Λ such that the gauge transformation compensates the supergravity variation of the C, χ, f fields. We recall that a vector field transforms like

$$e^{-2gV} \to e^{-2ig\Lambda} e^{-2gV} e^{2ig\Lambda^\dagger} ,$$

in a gauge transformation. Using the Baker-Cambell-Hausdorff formula in the first order in Λ, we obtain [Grisaru et al. (1979)]

$$\delta V = i\text{ad}(gV) \cdot (\Lambda + \Lambda^\dagger) + i[\text{ad}(gV) \coth(\text{ad(gV)})].(\Lambda - \Lambda^\dagger) , \qquad (4.76)$$

with $\text{ad}(X) \cdot Y = [X, Y]$, $\text{ad}^2(X) \cdot Y = [X, [X, Y]]$, *etc.* From the relation

$$x\coth(x) = 1 + \frac{x^3}{3} - \frac{x^5}{45} + o(x^5) ,$$

and the property that in the Wess-Zumino gauge all the terms trilinear in V vanish, (4.76) reduces to

$$\delta V_{W.Z} = i(\Lambda - \Lambda^\dagger) + ig[V_{W.Z}, \Lambda + \Lambda^\dagger] .$$

The second term will ensure that we have the correct relationship between the derivatives of the physical fields transformed by the gauge transformation. In particular it legitimates *a posteriori* the relations (4.73), (4.74) and (4.75). The first term is chosen such that it compensates exactly the variations of C, χ, f under a supergravity transformation leading to,

$$\begin{aligned}
&\Lambda\Big| = 0 , && \Lambda^\dagger\Big| = 0 , \\
&\mathcal{D}_\alpha \Lambda\Big| = -i(\sigma^\mu \bar{\varepsilon})_\alpha v_\mu , && \bar{\mathcal{D}}_{\dot\alpha} \Lambda^\dagger\Big| = i(\varepsilon \sigma^\mu)_{\dot\alpha} v_\mu , \\
&\mathcal{D} \cdot \mathcal{D}\Lambda\Big| = 4\bar{\varepsilon} \cdot \bar{\lambda} + 2\, \bar{\psi}_\mu \bar{\sigma}^\nu \sigma^\mu \bar{\varepsilon} v_\nu, && \bar{\mathcal{D}} \cdot \bar{\mathcal{D}}\Lambda^\dagger\Big| = 4\varepsilon \cdot \lambda + 2\varepsilon \sigma^\mu \bar{\sigma}^\nu \psi_\mu v_\nu .
\end{aligned} \qquad (4.77)$$

Thus combining a supergravity transformation and a gauge transformation we get, using $\delta_\Lambda W_\alpha = -2ig[\Lambda, W_\alpha]$) that only $\delta_\Lambda D$ is affected by the gauge transformation:

$$\delta_\Lambda D = \frac{1}{4g} D^\alpha \delta_\Lambda W_\alpha\Big| + \frac{1}{4g} D_{\dot\alpha} \delta_\Lambda \overline{W}^{\dot\alpha}\Big| .$$

Adding both contribution we obtain

$$\begin{aligned}
\delta v_\mu &= i(\varepsilon\sigma_\mu\bar{\lambda} - \lambda\sigma_\mu\bar{\varepsilon}) - \frac{i}{2}\left(\psi_\mu\sigma^\nu\bar{\varepsilon} - \varepsilon\sigma^\nu\bar{\psi}_\mu\right)v_\nu\,, \\
\delta\lambda_\alpha &= (\sigma^{\mu\nu}\varepsilon)_\alpha \hat{F}_{\mu\nu} + i\varepsilon_\alpha D\,, \\
\delta\bar{\lambda}_{\dot{\alpha}} &= -(\bar{\varepsilon}\bar{\sigma}^{\mu\nu})_{\dot{\alpha}}\hat{F}_{\mu\nu} - i\bar{\varepsilon}_{\dot{\alpha}} D\,, \\
\delta D &= -\varepsilon\sigma^\mu\hat{D}_\mu\bar{\lambda} - \hat{D}_\mu\lambda\sigma^\mu\bar{\varepsilon} + \frac{i}{2}b_\mu\left(\varepsilon\sigma^\mu\bar{\lambda} - \lambda\sigma^\mu\bar{\varepsilon}\right) + \frac{i}{2}\left(M^*\varepsilon\cdot\lambda - M\bar{\varepsilon}\cdot\bar{\lambda}\right),
\end{aligned} \tag{4.78}$$

and the covariant derivative $\hat{D}_\mu$ of the gaugino, is now covariant with respect to supergravity and gauge transformations (see Table 4.3).

Similarly we have to combine the gauge transformation (4.77) with the supergravity transformation (4.29), (4.31) and (4.30) for a chiral superfield. Since under a gauge transformation we have,

$$\Phi \to e^{2ig\Lambda}\Phi\,,$$

we easily obtain, the variation under the gauge transformation

$$\begin{aligned}
\delta_\Lambda\phi^i &= 2ig\Lambda\Phi^i\Big| = 0\,, \\
\delta_\Lambda\chi^i_\alpha &= i\frac{2}{\sqrt{2}}g\mathcal{D}_\alpha(\Lambda\Phi^i)\Big| = \sqrt{2}g(\sigma^\mu\bar{\varepsilon})_\alpha v^a_\mu(T_a\phi)^i\,, \\
\delta_\Lambda F^i &= i\frac{1}{2}g\mathcal{D}\cdot\mathcal{D}(\Lambda\Phi) = (2ig\bar{\varepsilon}\cdot\bar{\lambda}^a + ig\bar{\psi}_\mu\bar{\sigma}^\nu\sigma^\mu\bar{\varepsilon}v^a_\nu)(T_a\phi)^i + \sqrt{2}(T_a\chi)^i\sigma^\mu\bar{\varepsilon}v^a_\mu\,.
\end{aligned}$$

Thus combining the two transformations we get:

$$\begin{aligned}
\delta\phi^i &= -\sqrt{2}\varepsilon\cdot\chi^i\,, & \delta\phi^\dagger_{i^*} &= -\sqrt{2}\bar{\varepsilon}\cdot\bar{\chi}_{i^*}\,, \\
\delta\chi^i_\alpha &= \sqrt{2}\left[\varepsilon_\alpha F^i - i(\sigma^\mu\bar{\varepsilon})_\alpha\hat{\hat{D}}_\mu\phi^i\right], & \delta\bar{\chi}_{\dot{\alpha}i^*} &= \sqrt{2}\left[\bar{\varepsilon}_{\dot{\alpha}}F^\dagger_{i^*} + i(\varepsilon\sigma^\mu)_{\dot{\alpha}}\hat{\hat{D}}_\mu\phi^\dagger_{i^*}\right], \\
\delta F^i &= \frac{\sqrt{2}}{3}\varepsilon\cdot\chi^i M^* - i\sqrt{2}\hat{\hat{D}}_\mu\chi^i\sigma^\mu\bar{\varepsilon} & \delta F^\dagger_{i^*} &= \frac{\sqrt{2}}{3}\bar{\varepsilon}\cdot\bar{\chi}_{i^*}M + i\sqrt{2}\varepsilon\sigma^\mu\hat{\hat{D}}_\mu\bar{\chi}_{i^*} \\
&\quad -\frac{\sqrt{2}}{6}b_\mu\chi^i\sigma^\mu\bar{\varepsilon} + 2ig\bar{\varepsilon}\cdot\bar{\lambda}^a(T_a\phi)^i\,, & &\quad -\frac{\sqrt{2}}{6}b_\mu\varepsilon\sigma^\mu\bar{\chi}_{i^*} - 2ig\varepsilon\cdot\lambda^a(\phi^\dagger T_a)_{i^*}
\end{aligned} \tag{4.79}$$

where the derivatives

$$\begin{aligned}
\hat{\hat{D}}_\mu\phi &= \hat{D}_\mu\phi + igv_\mu\phi\,, \\
\hat{\hat{D}}_\mu\chi &= \hat{D}_\mu\chi + igv_\mu\chi\,,
\end{aligned}$$

are now covariant with respect to supergravity and gauge symmetries

From now on we call supergravity transformations the transformations given by (4.79) and (4.78). With the reference to Table 3.7 this means that supergravity transformations are the composition of a local translations in the curved superspace, together with local Lorentz transformations and gauge transformations.

It should be noted that the parameters of the gauge transformations and of the Lorentz transformations depend on the field. In particular this means that when one computes the (anti)commutator of two transformations, the structure constant depends now on the fields and should more precisely be called structure functions. This is a general feature in supergravity. These types of algebras are called soft algebras (see *e.g.* [Freedman and Van Proeyen (2012); Batalin and Vilkovisky (1983); Henneaux (1990); Gomis et al. (1995); Sohnius (1983)]). Note finally that the vector multiplets were introduced firstly in [Ferrara et al. (1976b); Freedman and Schwarz (1977)] then in [Stelle and West (1978b,a)]. In the two first papers only partial transformation law (without the auxiliary fields of supergravity) were given.

Chapter 5

General principles to construct invariant actions

The formalism introduced in the previous chapters allows for the construction of invariant actions. To this end, we adopt the standard procedure of simply substituting regular flat superderivatives with covariant derivatives acting on curved superspace. In Secs 4.1 and 4.2, we have defined such derivatives for the various components of chiral and vector superfields needed in our requisite substitution procedure. Our expressions will exhibit the strong correlation between curved superspace through their link to the zeroth order components of the vielbein, the connection and the $\mathcal{R}$ and G_μ superfields (see Secs.3.1 and 3.3). However, the construction of an invariant action in curved superspace requires one additional ingredient, an invariant measure of integration. Its definition is the object of Sec. 5.1. Later in the present chapter, we will see that this leads to (at least) two different ways of building actions, either performing calculations in the full eight-dimensional superspace or in the six-dimensional chiral superspace, after having integrated out the antichiral variables.

5.1 The superdeterminant and Lagrangians in curved superspace

Following the notations introduced in Sec. 3.1, we generically denote by $z = (x,\theta) \equiv z^M = (x^\mu, \theta^\alpha, \bar\theta_{\dot\alpha})$ and $\tilde z = (\tilde x, \tilde\theta) \equiv z^{\tilde M} = (x^{\tilde\mu}, \theta^{\tilde\alpha}, \bar\theta_{\tilde{\dot\alpha}})$ the coordinates in flat and curved superspace, respectively. The key observation in order to define an invariant integration measure lies in the relation between the flat and curved set of coordinates through the vielbein. Recalling the compact notations for the indices $\underline{\alpha} = (\alpha, \dot\alpha)$ and $\underline{\tilde\alpha} = (\tilde\alpha, \tilde{\dot\alpha})$, coordinate transformation may be expressed as

$$\begin{pmatrix} \mathrm{d}x^{\tilde\mu} & \mathrm{d}\theta^{\underline{\tilde\alpha}} \end{pmatrix} \longrightarrow \begin{pmatrix} \mathrm{d}x^{\mu} & \mathrm{d}\theta^{\underline{\alpha}} \end{pmatrix} = \begin{pmatrix} \mathrm{d}x^{\tilde\mu} & \mathrm{d}\theta^{\underline{\tilde\alpha}} \end{pmatrix} \begin{pmatrix} E_{\tilde\mu}{}^{\mu} & E_{\tilde\mu}{}^{\underline{\alpha}} \\ E_{\underline{\tilde\alpha}}{}^{\mu} & E_{\underline{\tilde\alpha}}{}^{\underline{\alpha}} \end{pmatrix} ,$$

where $\theta^{\underline{\alpha}} \equiv (\theta^\alpha, \bar{\theta}_{\dot{\alpha}})$ and $\theta^{\underline{\tilde{\alpha}}} \equiv (\theta^{\tilde{\alpha}}, \bar{\theta}_{\dot{\tilde{\alpha}}})$. This allows to adopt the following invariant measure in the curved superspace,

$$\mathrm{d}^4x\,\mathrm{d}^4\theta = \mathrm{sdet}\Big(E_{\tilde{M}}{}^M\Big)\,\mathrm{d}^4\tilde{x}\,\mathrm{d}^4\tilde{\theta}\,. \tag{5.1}$$

In order to compute the expression of the superdeterminant in the general case, we start from two limiting cases [van Nieuwenhuizen (1981)]. Firstly, for a pure spacetime transformation, acting only on the x-variables and not on the Grassmann variables, the superdeterminant is given by the standard Jacobian,

$$\mathrm{sdet}\Big(E_{\tilde{M}}{}^M\Big) = \det\Big(E_{\tilde{\mu}}{}^\mu\Big)\,. \tag{5.2}$$

Secondly, the properties of the Grassmann variables under integration and (5.1) lead to

$$\begin{aligned} 1 &= \int \mathrm{d}\theta^{\tilde{1}}\mathrm{d}\theta^{\tilde{2}}\mathrm{d}\bar{\theta}_{\dot{\tilde{1}}}\mathrm{d}\bar{\theta}_{\dot{\tilde{2}}}\left(\bar{\theta}_{\dot{\tilde{2}}}\bar{\theta}_{\dot{\tilde{1}}}\theta^{\tilde{2}}\theta^{\tilde{1}}\right) \\ &= \int \mathrm{d}\theta^{1}\mathrm{d}\theta^{2}\mathrm{d}\bar{\theta}_{\dot{1}}\mathrm{d}\bar{\theta}_{\dot{2}}\left(\bar{\theta}_{\dot{2}}\bar{\theta}_{\dot{1}}\theta^{2}\theta^{1}\right) \\ &= \int \mathrm{d}\theta^{\tilde{1}}\mathrm{d}\theta^{\tilde{2}}\mathrm{d}\bar{\theta}_{\dot{\tilde{1}}}\mathrm{d}\bar{\theta}_{\dot{\tilde{2}}}\left(\bar{\theta}_{\dot{2}}\bar{\theta}_{\dot{1}}\theta^{2}\theta^{1}\right)\mathrm{sdet}\Big(E_{\underline{\tilde{\alpha}}}{}^{\underline{\alpha}}\Big)\,, \end{aligned} \tag{5.3}$$

considering this time a linear transformation acting only on the Grassmann variables θ and not on the regular spacetime coordinates. Since, on the one hand, $\theta^{\underline{\alpha}} = \theta^{\underline{\tilde{\alpha}}} E_{\underline{\tilde{\alpha}}}{}^{\underline{\alpha}}$ and, on the other hand, the θ-variables anticommute, we obtain

$$\bar{\theta}_{\dot{2}}\bar{\theta}_{\dot{1}}\theta^{2}\theta^{1} = \bar{\theta}_{\dot{\tilde{2}}}\bar{\theta}_{\dot{\tilde{1}}}\theta^{\tilde{2}}\theta^{\tilde{1}}\,\det(E_{\underline{\tilde{\alpha}}}{}^{\underline{\alpha}})\,,$$

which, together with (5.3), leads to

$$\mathrm{sdet}\Big(E_{\underline{\tilde{\alpha}}}{}^{\underline{\alpha}}\Big) = \det{}^{-1}\Big(E_{\underline{\tilde{\alpha}}}{}^{\underline{\alpha}}\Big)\,. \tag{5.4}$$

Rewriting the vielbein as a 2×2 block matrix,

$$\begin{aligned} E_{\tilde{M}}{}^M &= \begin{pmatrix} A & B \\ C & D \end{pmatrix} = \begin{pmatrix} E_{\tilde{\mu}}{}^\mu & E_{\tilde{\mu}}{}^{\underline{\alpha}} \\ E_{\underline{\tilde{\alpha}}}{}^\mu & E_{\underline{\tilde{\alpha}}}{}^{\underline{\alpha}} \end{pmatrix}, \\ E_M{}^{\tilde{M}} &= \begin{pmatrix} \tilde{A} & \tilde{B} \\ \tilde{C} & \tilde{D} \end{pmatrix} = \begin{pmatrix} E_\mu{}^{\tilde{\mu}} & E_\mu{}^{\underline{\tilde{\alpha}}} \\ E_{\underline{\alpha}}{}^{\tilde{\mu}} & E_{\underline{\alpha}}{}^{\underline{\tilde{\alpha}}} \end{pmatrix}, \end{aligned}$$

a simple calculation gives

$$\begin{pmatrix} A & B \\ C & D \end{pmatrix} = \begin{pmatrix} 1 & BD^{-1} \\ 0 & 1 \end{pmatrix}\begin{pmatrix} A - BD^{-1}C & 0 \\ 0 & D \end{pmatrix}\begin{pmatrix} 1 & 0 \\ D^{-1}C & 1 \end{pmatrix}. \tag{5.5}$$

This last equation allows to derive the expression of the superdeterminant in the general case starting from the two limiting cases previously addressed. Indeed, the first matrix of the right-hand side consists in a translation upon the bosonic variables, whilst the third one is a translation upon the fermionic variables. Their determinant equals unity then. From the second matrix and (5.2) and (5.4), one obtains

$$E \equiv \text{sdet}\left(E_{\tilde{M}}{}^{M}\right) = \det\left(E_{\tilde{\mu}}{}^{\mu} - E_{\tilde{\mu}}{}^{\underline{\beta}}(E^{-1})_{\underline{\beta}}{}^{\underline{\tilde{\beta}}}E_{\underline{\tilde{\beta}}}{}^{\mu}\right)\,\det{}^{-1}\left(E_{\underline{\tilde{\alpha}}}{}^{\underline{\alpha}}\right). \tag{5.6}$$

Let us note that in the general case, the inverse of the vielbein,

$$(E^{-1})_{\underline{\beta}}{}^{\underline{\tilde{\beta}}} = (E_{\underline{\tilde{\beta}}}{}^{\underline{\beta}})^{-1}$$

does not equate $E_{\underline{\beta}}{}^{\underline{\tilde{\beta}}}$.

We now turn to the proof that the superdeterminant transforms as a total derivative under supergravity transformations. So far, the indices M and $\tilde{M}$ were considered as independent quantities. In the context of the present demonstration, *and only here*, we define a compact notation where we assume that the indices are related, the curved index corresponding to a specific flat index M being denoted as $\tilde{M}$. The supertrace of a graded matrix $\mathcal{M}_{\tilde{M}}{}^{N}$ is then defined by $\text{str}(\mathcal{M}) = (-)^{|\tilde{M}|}\mathcal{M}_{\tilde{M}}{}^{M}$, with an implicit summation over M. We can relate supertrace and superdeterminant through

$$\text{sdet}(\mathcal{M}) = e^{\text{str}\ln(\mathcal{M})}\,,$$

which yields

$$\begin{aligned}
\delta\text{sdet}\mathcal{M} &= \text{sdet}(\mathcal{M}+\delta\mathcal{M}) - \text{sdet}(\mathcal{M}) \\
&= e^{\text{str}\ln(\mathcal{M}+\delta\mathcal{M})} - e^{\text{str}\ln\mathcal{M}} \\
&= \left[e^{\text{str}\ln(1+\delta\mathcal{M}\mathcal{M}^{-1})} - 1\right]e^{\text{str}\ln\mathcal{M}} \\
&\sim \text{str}(\delta\mathcal{M}\mathcal{M}^{-1})e^{\text{str}\ln\mathcal{M}} \\
&= \text{sdet}(\mathcal{M})\,\text{str}(\delta\mathcal{M}\mathcal{M}^{-1})\,.
\end{aligned} \tag{5.7}$$

In particular, applying this last relation to the vielbein and using the definition of the supertrace and the transformation properties (3.99), one gets

$$\begin{aligned}
\delta E &= E(-)^{|\tilde{M}|}\delta E_{\tilde{M}}{}^{N}E_{N}{}^{\tilde{M}} \\
&= -(-)^{|\tilde{M}|}E\left(\xi^{\tilde{N}}\partial_{\tilde{N}}E_{\tilde{M}}{}^{N} + \partial_{\tilde{M}}\xi^{\tilde{N}}E_{\tilde{N}}{}^{N} - E_{\tilde{M}}{}^{P}L_{P}{}^{N}\right)E_{N}{}^{\tilde{M}} \\
&= -(-)^{|\tilde{M}|}\left[\xi^{\tilde{N}}E\partial_{\tilde{N}}E_{\tilde{M}}{}^{N}E_{N}{}^{\tilde{M}} + E\partial_{\tilde{M}}\xi^{\tilde{M}}\right] + (-)^{|N|}EL_{N}{}^{N} \\
&= -(-)^{|\tilde{M}|}\left[\xi^{\tilde{N}}E\partial_{\tilde{N}}E_{\tilde{M}}{}^{N}E_{N}{}^{\tilde{M}} + E\partial_{\tilde{M}}\xi^{\tilde{M}}\right],
\end{aligned} \tag{5.8}$$

the last equality being related to the antisymmetry/symmetric properties of Lorentz transformation parameters. Since, in addition,

$$\begin{aligned}\partial_{\tilde{M}} E &= \partial_{\tilde{M}}\Big[e^{\operatorname{str}\ln E_{\tilde{N}}{}^{N}}\Big] \\ &= (-)^{|\tilde{N}|} E(\partial_{\tilde{M}} E_{\tilde{N}}{}^{N}) E_{N}{}^{\tilde{N}} ,\end{aligned} \tag{5.9}$$

(5.8) can be further simplified to

$$\delta E = -(-)^{|\tilde{N}|}\partial_{\tilde{N}}(E\xi^{\tilde{N}}) , \tag{5.10}$$

which shows that the superdeterminant transforms as a total derivative under a general change of coordinates. Let us note that the superdeterminant, also named the Berezian, has first been introduced in [Berezin (1987)], followed by [van Nieuwenhuizen (1981); Arnowitt et al. (1975)] in the context of supergravity. In the second paper, it has been proved that the variation of the superdeterminant under a supergravity transformation is a total derivative. This quantity has consequently been called a superfield density and can thus be used in building an invariant action in a straightforward fashion. Indeed, let U be an arbitrary scalar superfield, which varies under a supergravity transformation as in (3.101), *i.e.*, $\delta U = -\xi^{\tilde{M}}\mathcal{D}_{\tilde{M}} U = -\xi^{\tilde{M}}\partial_{\tilde{M}} U$. The variation of the product EU is thus given by

$$\begin{aligned}\delta(EU) &= (\delta E)U + E\delta(U) \\ &= -(-)^{|\tilde{M}|}\partial_{\tilde{M}}(\xi^{\tilde{M}} EU) ,\end{aligned}$$

EU being therefore invariant under supergravity transformations.

In Chapter 2, we have computed the most general supersymmetric Lagrangian describing the dynamics of matter superfields coupled to gauge interactions, its expression being given in (2.26). We now have all the ingredients to couple this Lagrangian to supergravity. Firstly, as previously mentioned, we replace all flat space derivatives by derivatives covariant with respect to supergravity transformations. Secondly, we modify the integration volume by introducing the invariant measure (5.6). However, it is important to note that chiral and vector superfield terms in (2.26) must be treated in different ways. For a vector (or a general, but real) superfield V, we are integrating over the whole eight-dimensional superspace, and in order to obtain an invariant action in the curved superspace, we are lead to

$$\mathcal{S}_{\text{flat}} = \int \mathrm{d}^4x\, \mathrm{d}^4\theta\, V \quad\to\quad \mathcal{S}_{\text{curved}} = \int \mathrm{d}^4\tilde{x}\, \mathrm{d}^4\tilde{\theta}\, EV ,$$

where we recall that in our notations, $\mathrm{d}^4\theta \equiv \mathrm{d}^2\theta \mathrm{d}^2\bar{\theta}$ and $\mathrm{d}^4\tilde{\theta} \equiv \mathrm{d}^2\tilde{\theta}\mathrm{d}^2\bar{\tilde{\theta}}$. We now turn to the case of chiral superfields where we are integrating over the six-dimensional (chiral) part of the superspace. In Secs. 4.1 and 4.2, we have shown that for any chiral superfield Φ, there exists a general superfield U so that $\Phi = (\overline{\mathcal{D}\mathcal{D}}-8\mathcal{R})U$. Inverting this relation, one gets $U = -1/(8\mathcal{R})\Phi + 1/(8R)\overline{\mathcal{D}}\cdot\overline{\mathcal{D}}U$, which can be used to obtain an invariant action in curved superspace from the corresponding one in flat superspace,

$$\mathcal{S}_{\text{flat}} = \int d^4x\, d^2\theta\, \Phi \quad\to\quad \mathcal{S}_{\text{curved}} = \int d^4\tilde{x} d^4\tilde{\theta}\, E\Big(-\tfrac{1}{8\mathcal{R}}\Phi + \tfrac{1}{8\mathcal{R}}\overline{\mathcal{D}}\cdot\overline{\mathcal{D}}U\Big)$$
$$= -\tfrac{1}{8}\int d^4\tilde{x}\, d^4\tilde{\theta}\, \tfrac{E}{\mathcal{R}}\Phi.$$

where the last equality follows by partial integration because $\mathcal{R}$ is a chiral superfield. Hence, on curved superspace, the complete Lagrangian (2.26) can be cast as [Wess and Bagger (1992)]

$$\begin{aligned}\mathcal{L} = &-3\int d^2\tilde{\theta}\, d^2\overline{\tilde{\theta}}\,\frac{1}{2}E\Big(e^{-\frac{1}{3}K(\Phi,\Phi^\dagger e^{-2gV})} + e^{-\frac{1}{3}K(e^{-2gV}\Phi,\Phi^\dagger)}\Big)\\ &-\frac{1}{8}\int d^2\tilde{\theta}\, d^2\overline{\tilde{\theta}}\Big(\frac{E}{\mathcal{R}}W(\Phi) + \frac{E}{\mathcal{R}^\dagger}W^\star(\Phi^\dagger)\Big)\\ &-\frac{1}{8}\frac{1}{16g^2}\int d^2\tilde{\theta}\, d^2\overline{\tilde{\theta}}\Big(\frac{E}{\mathcal{R}}h_{ab}(\Phi)W^{a\alpha}W^b_\alpha + \frac{E}{\mathcal{R}^\dagger}h^\star_{ab}(\Phi^\dagger)\overline{W}^a_{\dot{\alpha}}\overline{W}^{b\dot{\alpha}}\Big).\end{aligned} \tag{5.11}$$

The two last integrals are obtained after a direct application of (5.11), noting that the spinor superfields W^a_α and $\overline{W}^{\dot{\alpha}a}$ are this time defined in curved superspace, as in Sec. 4.2, instead of in flat superspace as for the Lagrangian of Chapter 2. In contrast, the Lagrangian terms involving the Kähler potential do not lead themselves to a simple application of (5.11) which would lead to $\int EK$. The reason, as will be shown below, is that only this form of Lagrangian gives rise to correctly normalised terms, once the appropriate Weyl rescaling is performed. Accounting that the Kähler potential K has a [mass]^2 dimension and that the measure E is dimensionless, we can reintroduce mass units in terms of the Planck mass m_{P} in the form

$$\begin{aligned}&-3m_{\text{P}}^2\int d^2\tilde{\theta}\, d^2\overline{\tilde{\theta}}\,\frac{1}{2}E\Big(e^{-\frac{1}{3m_{\text{P}}^2}K(\Phi,\Phi^\dagger e^{-2gV})} + e^{-\frac{1}{3m_{\text{P}}^2}K(e^{-2gV}\Phi,\Phi^\dagger)}\Big)\\ &= -3m_{\text{P}}^2\int\Big\{d^2\tilde{\theta}\, d^2\overline{\tilde{\theta}}E + \frac{1}{2}E\Big[K(\Phi,\Phi^\dagger e^{-2gV}) + K(e^{-2gV}\Phi,\Phi^\dagger)\Big]\Big\} + o\Big(\frac{1}{m_{\text{P}}^2}\Big),\end{aligned} \tag{5.12}$$

where on the right-hand side of the equation, we have expanded the Lagrangian in the infinite Planck mass limit. In this limit, the first of the two terms decouples and consists in the pure supergravity action [Freedman et al. (1976); Deser and Zumino (1976)], whilst the second term gives rise to the correct Lagrangian terms induced by the Kähler potential in the flat limit (where $E \to 1$), as seen from (2.26). In the next subsections, we will start from the Lagrangian (5.11) and rewrite it in a form more suitable for practical computations, using hybrid variables that are different from the usual superspace coordinates.

Let us finally note that choosing the Lagrangian (5.11) as the basis of a supergravity-invariant action is not the only possible choice, and we could have started from a conformal supergravity Lagrangian, proportional to [Ferrara and Zumino (1978)]

$$\mathcal{L} \sim \int d\tilde{\theta}^2 \, d\bar{\tilde{\theta}}^2 \left(\frac{E}{\mathcal{R}} W^{(\alpha\beta\gamma)} W_{(\alpha\beta\gamma)} + \frac{E}{\mathcal{R}^\dagger} \overline{W}_{(\dot{\alpha}\dot{\beta}\dot{\gamma})} \overline{W}^{(\dot{\alpha}\dot{\beta}\dot{\gamma})} \right). \tag{5.13}$$

The expression of the superdeterminant E in terms of the θ-variables can in principle be computed [Muller (1982)]. This will not be considered here.

We choose to rewrite the Lagrangian (5.11) in another chiral form, more suitable for practical calculations. The trick is to introduce the so-called Θ−variables (see next section) related to a hybrid system of coordinates, containing both Lorentzian and Einsteinian indices. This idea of introducing different Grassmann variables can be considered as a real *tour de force* by Julius Wess [Grimm (2011)]. We would like to mention that chiral superspaces have also been introduced in [Ogievetsky and Sokatchev (1978)] and that coupling of non-minimal and new minimal models to supergravity has been done in [Grimm et al. (1984); Ferrara et al. (1983)].

Finally, before moving on, note for completeness, that chiral density was also introduced in the tensor calculus approach [Ferrara and van Nieuwenhuizen (1978b); Stelle and West (1978b)] and that correspondingly invariant actions were obtained [Cremmer et al. (1979, 1982, 1978b, 1983b)].

5.2 The introduction of hybrid variables

In curved superspace, the components of a chiral superfield $\Phi(x^{\tilde{\mu}}, \theta^{\tilde{\alpha}}, \bar{\theta}_{\tilde{\dot{\alpha}}})$ are given by its successive covariant superderivatives, as illustrated in (4.2), these derivatives carrying flat Lorentz indices. In contrast, the expansion of the superfield $\Phi(x^{\tilde{\mu}}, \theta^{\tilde{\alpha}}, \bar{\theta}_{\tilde{\dot{\alpha}}})$ as a series of the Grassmann variables is given as a series in the curved superspace. Therefore, the component fields have, *a priori*, a complicated expression. Consequently, in order to simplify any computation, we introduce a set of hybrid variables, $z^{\underline{M}} = (x^{\tilde{\mu}}, \Theta^\alpha, \overline{\Theta}_{\dot{\alpha}})$, where the bosonic variables $x^{\tilde{\mu}}$ carry Einstein indices and the fermionic variables, Θ^α and $\overline{\Theta}_{\dot{\alpha}}$ Lorentz indices [Wess and Zumino (1978a)]. These new Grassmann variables are defined in such as way that the expansion of a chiral superfield is now exactly given by the covariant derivatives, or the component fields themselves [Wess and Bagger (1992)],

$$\begin{aligned}\Phi(x,\Theta) &= \phi(x) + \sqrt{2}\Theta^\alpha\chi_\alpha(x) - \Theta^\alpha\Theta_\alpha F(x)\\ &= \phi(x) + \sqrt{2}\Theta\cdot\chi(x) - \Theta\cdot\Theta F(x)\,. \end{aligned} \tag{5.14}$$

The main advantage of these new hybrid Θ−variables lies in the fact that a supergravity-invariant action can now be written under a fully chiral form,

$$S_{\text{curved}} = \int d^4x\, d^2\Theta\ \Delta f(\Phi) + \text{h.c.}\,, \tag{5.15}$$

where we only integrate over a six-dimensional superspace. In the expression above, the superfield $f(\Phi)$ is a chiral function of the chiral superfield content of the theory whilst Δ is a so-called *chiral density*, a quantity equivalent to the invariant measure E, but in the context of the $(\Theta,\bar\Theta)$ variables. Analogously to E, the variation of the chiral density under a supergravity transformation is defined as in (5.10),

$$\delta\Delta = -(-)^{|\bar M|}\partial_{\bar M}(\eta^{\bar M}\Delta)\,, \tag{5.16}$$

where the derivative operators $\partial_{\bar M}$ act on the spacetime variables $x^{\tilde\mu}$ and on the $(\Theta,\bar\Theta)$ variables, whilst the quantities $\eta^{\tilde M}$ are the supergravity transformation parameters depending on the hybrid coordinates,

$$\eta^{\bar M}(x,\Theta) = \eta^{\bar M}_{(0)}(x) + \Theta^\alpha\eta^{\bar M}_{(1)\alpha}(x) + \Theta\cdot\Theta\eta^{\bar M}_{(2)}(x)\,. \tag{5.17}$$

The invariance of the action (5.15) is guaranteed by the definition of a chiral density through its transformation law (5.16), which implies that the product of a chiral density by a chiral superfield Φ is a chiral density,

$$\begin{aligned}\delta(\Delta\Phi) &= -(-)^{|\bar M|}\partial_{\bar M}(\eta^{\bar M}\Delta)\Phi - \Delta\eta^{\bar M}\partial_{\bar M}\Phi\\ &= -(-)^{|\bar M|}\partial_{\bar M}(\eta^{\bar M}\Delta\Phi)\,,\end{aligned}$$

where, similarly to (5.16), we have imposed the variation of the chiral superfield to be given by

$$\delta\Phi = -\eta^{\bar M}(x,\Theta)\partial_{\bar M}\Phi\,. \tag{5.18}$$

The components of the transformation parameters in the hybrid system, *i.e.*, those of the superfields $\eta^{\bar M}$, can be related to the transformation parameters ξ^M and $L_M{}^N$ defined in Table 3.7 by investigating the variations of a chiral superfield Φ.

Starting from (5.18) and using (5.14) and (5.17), the variation of Φ reads:

$$\begin{aligned}\delta\Phi &= \Big[-\eta^{\tilde\mu}_{(0)}\partial_{\tilde\mu}\phi - \sqrt{2}\eta^{\alpha}_{(0)}\chi_\alpha\Big]\\ &+ \sqrt{2}\Theta^\alpha\Big[\sqrt{2}\eta_{\alpha(0)}F - \eta^{\mu}_{(0)}\partial_{\tilde\mu}\chi_\alpha - \frac{1}{\sqrt{2}}\eta^{\tilde\mu}_{(1)\alpha}\partial_{\tilde\mu}\phi - \eta^{\beta}_{(1)\alpha}\chi_\beta\Big]\\ &- \Theta\cdot\Theta\Big[-\eta^{\tilde\mu}_{(0)}\partial_{\tilde\mu}F + \eta^{\tilde\mu}_{(2)}\partial_{\tilde\mu}\phi - \frac{1}{\sqrt{2}}\eta^{\tilde\mu}_{(1)}{}^{\alpha}\partial_{\tilde\mu}\chi_\alpha + \sqrt{2}\eta^{\alpha}_{(2)}\chi_\alpha - \eta^{\alpha}_{(1)\alpha}F\Big]\,,\end{aligned}$$

in the hybrid system of coordinates and omitting the x-dependence for clarity. Identifying this with the results obtained in Sec. 4.1.2 and replacing the covariant derivatives of the fields by their definitions (summarised in Table 4.1), *i.e.*,

$$\begin{aligned}\delta\phi &= -\sqrt{2}\varepsilon\cdot\chi\,,\\ \delta\chi_\alpha &= \sqrt{2}\Big[\varepsilon_\alpha F - ie_\mu{}^{\tilde\mu}(\sigma^\mu\bar\varepsilon)_\alpha\big(\partial_{\tilde\mu}\phi - \frac{\sqrt{2}}{2}\psi_{\tilde\mu}\cdot\chi\big)\Big]\,,\\ \delta F &= \frac{\sqrt{2}}{3}\varepsilon\cdot\chi M^* - i\sqrt{2}e_\mu{}^{\tilde\mu}\Big[\partial_{\tilde\mu}\chi^\alpha - \omega_{\tilde\mu}{}^{\alpha\beta}\chi_\beta + \tfrac{1}{\sqrt{2}}\psi_{\tilde\mu}{}^\alpha F\\ &\quad + \frac{i}{\sqrt{2}}(\bar\psi_{\tilde\mu}\bar\sigma^\nu)^\alpha e_\nu{}^{\tilde\nu}\big(\partial_{\tilde\nu}\phi - \tfrac{\sqrt{2}}{2}\psi_{\tilde\nu}\cdot\chi\big)\Big](\sigma^\mu\bar\varepsilon)_\alpha - \tfrac{\sqrt{2}}{6}b_\mu\chi\sigma^\mu\bar\varepsilon\,,\end{aligned}\tag{5.19}$$

one may then derive the expression of the η-superfields. In more details, the transformation laws of the zeroth order component of the superfield Φ, *i.e.*, those of the field ϕ, imply

$$\begin{aligned}\eta^{\tilde\mu}_{(0)} &= 0\,,\\ \eta^{\alpha}_{(0)} &= \varepsilon^\alpha\,,\end{aligned}$$

those of the fermionic component χ_α lead to

$$\begin{aligned}\eta^{\tilde\mu}_{(1)\alpha} &= 2ie_\mu{}^{\tilde\mu}(\sigma^\mu\bar\varepsilon)_\alpha\,,\\ \eta^{\beta}_{(1)\alpha} &= -ie_\mu{}^{\tilde\mu}(\sigma^\mu\bar\varepsilon)_\alpha\psi_{\tilde\mu}{}^\beta\,,\end{aligned}$$

whilst those of the auxiliary component F yield

$$\begin{aligned}\eta^{\alpha}_{(2)} &= \frac{1}{3}M^*\varepsilon^\alpha - ie_\mu{}^{\tilde\mu}\omega_{\tilde\mu}{}^{\alpha\beta}(\sigma^\mu\bar\varepsilon)_\beta + \frac{1}{6}b_\mu(\bar\varepsilon\bar\sigma^\mu)^\alpha - \frac{1}{2}e_\mu{}^{\tilde\mu}e_\nu{}^{\tilde\nu}(\bar\psi_{\tilde\mu}\bar\sigma^\nu\sigma^\mu\bar\varepsilon)\,\psi_{\tilde\nu}{}^\alpha\,,\\ \eta^{\tilde\nu}_{(2)} &= e_\mu{}^{\tilde\mu}e_\nu{}^{\tilde\nu}\bar\psi_{\tilde\mu}\bar\sigma^\nu\sigma^\mu\bar\varepsilon\,.\end{aligned}$$

Collecting the results and introducing the notation $\sigma^{\tilde\mu} = e_\mu{}^{\tilde\mu}\sigma^\mu$, we get

$$\begin{aligned}\eta^{\tilde\mu} &= 2i\Theta\sigma^{\tilde\mu}\bar\varepsilon + \Theta\cdot\Theta\,\bar\psi_{\tilde\nu}\bar\sigma^{\tilde\mu}\sigma^{\tilde\nu}\bar\varepsilon\,,\\ \eta^\alpha &= \varepsilon^\alpha - i\Theta\sigma^{\tilde\mu}\bar\varepsilon\psi_{\tilde\mu}{}^\alpha\\ &\quad + \Theta\cdot\Theta\Big[\frac{1}{3}M^*\varepsilon^\alpha - i\omega_{\tilde\mu}{}^{\alpha\beta}(\sigma^{\tilde\mu}\bar\varepsilon)_\beta + \frac{1}{6}b_\mu(\bar\varepsilon\bar\sigma^\mu)^\alpha - \frac{1}{2}\psi_{\tilde\nu}{}^\alpha\bar\psi_{\tilde\mu}\bar\sigma^{\tilde\nu}\sigma^{\tilde\mu}\bar\varepsilon\Big]\,.\end{aligned}\tag{5.20}$$

We now turn to the evaluation of the variation of a chiral density Δ, recalling that the latter can, in general, be written as

$$\Delta = a + \sqrt{2}\Theta \cdot \rho - \Theta\cdot\Theta f\ .$$

The variation of a chiral density is imposed by (5.16), therefore

$$\delta\Delta = -(-)^{|\bar{M}|}\partial_{\bar{M}}(\eta^{\bar{M}}\Delta) = -(-)^{|\bar{M}|}\partial_{\bar{M}}\eta^{\bar{M}}\Delta - \eta^{\bar{M}}\partial_{\bar{M}}\Delta\ .$$

The two terms appearing on the right-hand side above may be written as respectively

$$\begin{aligned}
-(-)^{|\bar{M}|}(\partial_{\bar{M}}\eta^{\bar{M}})\Delta &= a\ \eta^{\alpha}_{(1)\alpha} + \sqrt{2}\Theta^{\alpha}\Big[-\frac{1}{\sqrt{2}}a\partial_{\tilde{\mu}}\eta^{\tilde{\mu}}_{(1)\alpha} - \sqrt{2}a\ \eta_{\alpha(2)} + \eta^{\beta}_{(1)\beta}\rho_{\alpha}\Big] \\
&\quad -\Theta\cdot\Theta\Big[a\ \partial_{\tilde{\mu}}\eta^{\tilde{\mu}}_{(2)} + f\ \eta^{\alpha}_{(1)\alpha} - \frac{1}{\sqrt{2}}(\partial_{\tilde{\mu}}\eta^{\tilde{\mu}}_{(1)}{}^{\alpha} + 2\eta^{\alpha}_{(2)})\rho_{\alpha}\Big]\,, \\
-\eta^{\bar{M}}\partial_{\bar{M}}\Delta &= -\sqrt{2}\eta^{\alpha}_{(0)}\rho_{\alpha} + \sqrt{2}\Theta^{\alpha}\Big[-\frac{1}{\sqrt{2}}\eta^{\tilde{\mu}}_{(1)\alpha}\ \partial_{\tilde{\mu}}a - \eta^{\beta}_{(1)\alpha}\ \rho_{\beta} + \sqrt{2}\eta_{\alpha(0)}f\Big] \\
&\quad -\Theta\cdot\Theta\Big[\eta^{\tilde{\mu}}_{(2)}\ \partial_{\tilde{\mu}}a - \frac{1}{\sqrt{2}}\eta^{\tilde{\mu}}_{(1)}{}^{\alpha}\partial_{\tilde{\mu}}\rho_{\alpha} + \sqrt{2}\eta^{\alpha}_{(2)}\rho_{\alpha} - \eta^{\alpha}_{(1)\alpha}f\Big]\,.
\end{aligned}$$

Injecting the results of (5.20), we derive the variation of the component fields a, ρ and f to be

$$\begin{aligned}
\delta a &= i\psi_{\tilde{\mu}}\sigma^{\tilde{\mu}}\bar{\varepsilon}\ a - \sqrt{2}\varepsilon\cdot\rho\,, \\
\delta\rho_{\alpha} &= \sqrt{2}\varepsilon_{\alpha}f - i\sqrt{2}\mathcal{D}_{\tilde{\mu}}[(\sigma^{\tilde{\mu}}\bar{\varepsilon})_{\alpha}a] - \frac{\sqrt{2}}{3}M^{*}\varepsilon_{\alpha}a + \frac{\sqrt{2}}{6}b_{\mu}(\sigma^{\mu}\bar{\varepsilon})_{\alpha}\ a \\
&\quad + \frac{\sqrt{2}}{2}\psi_{\tilde{\nu}\alpha}[\bar{\psi}_{\tilde{\mu}}\bar{\sigma}^{\tilde{\nu}}\sigma^{\tilde{\mu}}\bar{\varepsilon}]\ a - i\psi_{\tilde{\mu}\alpha}[\rho\sigma^{\tilde{\mu}}\bar{\varepsilon}]\,, \\
\delta f &= \partial_{\tilde{\mu}}[a\bar{\psi}_{\tilde{\nu}}\bar{\sigma}^{\tilde{\mu}}\sigma^{\tilde{\nu}}\bar{\varepsilon} + i\sqrt{2}\bar{\varepsilon}\bar{\sigma}^{\tilde{\mu}}\rho]\,,
\end{aligned} \tag{5.21}$$

where the variation $\delta\rho$ has been simplified with the use of the Fierz identity (C.9). Note that these transformations were obtained in [Wess and Zumino (1978a); Wess and Bagger (1992)].

In order to define an invariant volume playing the rôle of E in the case of chiral actions, we now construct a special chiral density connected to the vielbein, denoted by $\mathcal{E}$. Its component can be expressed in terms of the components of the gravitation multiplet through their transformation laws. This derivation starts from the lowest order component which is assumed to be given by [Wess and Zumino (1978a); Wess and Bagger (1992)]

$$a = \det(e_{\tilde{\mu}}{}^{\mu}) \equiv e\ . \tag{5.22}$$

The variation of a is thus given, on the one hand, by (5.21), and on the other hand, by the definition of a itself, $\delta a = \delta e$. This last quantity can be obtained along the same lines as the computation of δE, from (5.7),

$$\begin{aligned}\delta e &= e(\delta e_{\tilde{\mu}}{}^{\mu})e_{\mu}{}^{\tilde{\mu}} \\ &= -ie[\varepsilon\sigma^{\tilde{\mu}}\bar{\psi}_{\tilde{\mu}} - \psi_{\tilde{\mu}}\sigma^{\tilde{\mu}}\bar{\varepsilon}]\,. \end{aligned} \tag{5.23}$$

The last equality follows from the transformation property of the vierbein given in (3.141). Identifying (5.23) with the variation δa in (5.21), the terms in $\bar{\varepsilon}$ cancel whilst those in ε allow to get an expression for ρ [Wess and Zumino (1978a); Wess and Bagger (1992)],

$$\rho_{\alpha} = \frac{\sqrt{2}}{2}ie(\sigma^{\tilde{\mu}}\bar{\psi}_{\tilde{\mu}})_{\alpha}\,. \tag{5.24}$$

By the same token evaluating the variation of ρ yields the expression of f in terms of the components of the gravitation supermultiplet. Inserting (5.22) and (5.24) into the variation of ρ (5.21), we obtain

$$\begin{aligned}\delta\rho_{\alpha} =&\ \sqrt{2}\varepsilon_{\alpha}[f - \frac{1}{3}M^{*}e] - i\sqrt{2}\mathcal{D}_{\tilde{\mu}}[e(\sigma^{\tilde{\mu}}\bar{\varepsilon})_{\alpha}] \\ &+ \sqrt{2}e\Big[\frac{1}{6}b_{\mu}(\sigma^{\mu}\bar{\varepsilon})_{\alpha} + 2\psi_{\tilde{\mu}\alpha}(\bar{\psi}_{\tilde{\nu}}\bar{\sigma}^{\tilde{\mu}\tilde{\nu}}\bar{\varepsilon})\,\Big]\,, \end{aligned} \tag{5.25}$$

after having introduced the matrices $\bar{\sigma}^{\tilde{\mu}\tilde{\nu}} = e_{\mu}{}^{\tilde{\mu}}e_{\nu}{}^{\tilde{\nu}}\bar{\sigma}^{\mu\nu}$. However, we can also compute the variation of ρ starting directly from its expression given in (5.24),

$$\begin{aligned}\delta\rho_{\alpha} &= i\frac{\sqrt{2}}{2}\sigma^{\mu}{}_{\alpha\dot{\alpha}}\delta(\bar{\psi}_{\tilde{\mu}}{}^{\dot{\alpha}}ee_{\mu}{}^{\tilde{\mu}}) \\ &= i\frac{\sqrt{2}}{2}\sigma^{\mu}{}_{\alpha\dot{\alpha}}\Big[(\delta\bar{\psi}_{\tilde{\mu}}{}^{\dot{\alpha}})ee_{\mu}{}^{\tilde{\mu}} + \bar{\psi}_{\tilde{\mu}}{}^{\dot{\alpha}}(\delta e)e_{\mu}{}^{\tilde{\mu}} + \bar{\psi}_{\tilde{\mu}}{}^{\dot{\alpha}}e(\delta e_{\mu}{}^{\tilde{\mu}})\Big]\,. \end{aligned}$$

The variations of the graviton and gravitino have been obtained in Sec. 3.5 whilst the one of e is given in (5.23). Therefore, separating the terms in ε from those in $\bar{\varepsilon}$, one gets

$$\begin{aligned}\delta\rho_{\alpha} =& \Big[\frac{\sqrt{2}}{2}e(\sigma^{\tilde{\mu}}\bar{\psi}_{\tilde{\mu}})_{\alpha}[\varepsilon\sigma^{\tilde{\nu}}\bar{\psi}_{\tilde{\nu}}] - \frac{\sqrt{2}}{2}e(\sigma^{\tilde{\nu}}\bar{\psi}_{\tilde{\mu}})_{\alpha}[\varepsilon\sigma^{\tilde{\mu}}\bar{\psi}_{\tilde{\nu}}] + \frac{\sqrt{2}}{6}eM^{*}(\sigma^{\tilde{\mu}}\bar{\sigma}_{\tilde{\mu}}\varepsilon)_{\alpha}\Big] \\ &+ \sqrt{2}e\Big[\frac{1}{2}(\sigma^{\tilde{\nu}}\bar{\psi}_{\tilde{\mu}})_{\alpha}[\psi_{\tilde{\nu}}\sigma^{\tilde{\mu}}\bar{\varepsilon}] - \frac{1}{2}(\sigma^{\tilde{\mu}}\bar{\psi}_{\tilde{\mu}})_{\alpha}[\psi_{\tilde{\nu}}\sigma^{\tilde{\nu}}\bar{\varepsilon}] - ie_{\mu}{}^{\tilde{\mu}}\mathcal{D}_{\tilde{\mu}}[(\sigma^{\mu}\bar{\varepsilon})_{\alpha}] \\ &+ \frac{1}{6}(\sigma^{\tilde{\mu}}\bar{\sigma}_{\tilde{\mu}}\sigma^{\tilde{\nu}}\bar{\varepsilon})_{\alpha}b_{\tilde{\nu}} - \frac{1}{2}(\sigma^{\tilde{\mu}}\bar{\varepsilon})_{\alpha}b_{\tilde{\mu}}\Big]\,, \end{aligned} \tag{5.26}$$

where we have introduced $b_{\tilde{\mu}} = e_{\tilde{\mu}}{}^{\mu} b_{\mu}$. The first two terms of the first line of the equation above can be rewritten, using the Fierz identity (C.9), as

$$\begin{aligned}
&\frac{\sqrt{2}}{2} e(\sigma^{\tilde{\mu}}\bar{\psi}_{\tilde{\mu}})_\alpha[\varepsilon\sigma^{\tilde{\nu}}\bar{\psi}_{\tilde{\nu}}] - \frac{\sqrt{2}}{2} e(\sigma^{\tilde{\nu}}\bar{\psi}_{\tilde{\mu}})_\alpha[\varepsilon\sigma^{\tilde{\mu}}\bar{\psi}_{\tilde{\nu}}] \\
&= \frac{\sqrt{2}}{2} e(\sigma^{\mu}\bar{\psi}_{\tilde{\mu}})_\alpha[\varepsilon\sigma^{\nu}\bar{\psi}_{\tilde{\nu}}][e_\nu{}^{\tilde{\nu}} e_\mu{}^{\tilde{\mu}} - e_\nu{}^{\tilde{\mu}} e_\mu{}^{\tilde{\nu}}] \\
&= \sqrt{2} e\varepsilon_\alpha[\bar{\psi}_{\tilde{\mu}}\bar{\sigma}^{\tilde{\mu}\tilde{\nu}}\bar{\psi}_{\tilde{\nu}}] ,
\end{aligned} \tag{5.27}$$

while similarly, the first two terms of the second line can be rewritten as

$$\begin{aligned}
&\frac{1}{2}(\sigma^{\tilde{\nu}}\bar{\psi}_{\tilde{\mu}})_\alpha[\psi_{\tilde{\nu}}\sigma^{\tilde{\mu}}\bar{\varepsilon}] - \frac{1}{2}(\sigma^{\tilde{\mu}}\bar{\psi}_{\tilde{\mu}})_\alpha[\psi_{\tilde{\nu}}\sigma^{\tilde{\nu}}\bar{\varepsilon}] \\
&= 2\psi_{\tilde{\mu}\alpha}[\bar{\psi}_{\tilde{\nu}}\bar{\sigma}^{\tilde{\mu}\tilde{\nu}}\bar{\varepsilon}] - \frac{1}{2}(\sigma^{\tilde{\mu}}\bar{\varepsilon})_\alpha[\psi_{\tilde{\nu}}\sigma^{\tilde{\nu}}\bar{\psi}_{\tilde{\mu}} - \psi_{\tilde{\mu}}\sigma^{\tilde{\nu}}\bar{\psi}_{\tilde{\nu}}] .
\end{aligned}$$

The last of these two terms can be rewritten in terms of derivatives of the graviton by introducing the torsion tensor,

$$\begin{aligned}
&-\frac{1}{2}(\sigma^{\tilde{\mu}}\bar{\varepsilon})_\alpha[\psi_{\tilde{\nu}}\sigma^{\tilde{\nu}}\bar{\psi}_{\tilde{\mu}} - \psi_{\tilde{\mu}}\sigma^{\tilde{\nu}}\bar{\psi}_{\tilde{\nu}}] \\
&= i(\sigma^{\tilde{\mu}}\bar{\varepsilon})_\alpha e_\rho{}^{\tilde{\nu}} T_{\tilde{\mu}\tilde{\nu}}{}^{\rho}\big| \\
&= -i(\sigma^{\mu}\bar{\varepsilon})_\alpha e_\mu{}^{\tilde{\mu}} e_\rho{}^{\tilde{\nu}}\Big[\mathcal{D}_{\tilde{\mu}} e_{\tilde{\nu}}{}^{\rho} - \mathcal{D}_{\tilde{\nu}} e_{\tilde{\mu}}{}^{\rho}\Big] \\
&= -i(\sigma^{\mu}\bar{\varepsilon})_\alpha\Big[e_\mu{}^{\tilde{\mu}}(\partial_{\tilde{\mu}} e_{\tilde{\nu}}{}^{\rho}) e_\rho{}^{\tilde{\nu}} - e_\mu{}^{\tilde{\mu}}(\partial_{\tilde{\nu}} e_{\tilde{\mu}}{}^{\rho}) e_\rho{}^{\tilde{\nu}} - \omega_{\tilde{\nu}\mu}{}^{\rho} e_\rho{}^{\tilde{\nu}}\Big] \\
&= -i(\sigma^{\mu}\bar{\varepsilon})_\alpha\Big[\frac{1}{e} e_\mu{}^{\tilde{\mu}}\mathcal{D}_{\tilde{\mu}} e + \mathcal{D}_{\tilde{\mu}} e_\mu{}^{\tilde{\mu}}\Big] .
\end{aligned} \tag{5.28}$$

This latter property will be fundamental proving that $\mathcal{E}$ is a chiral density. Moreover, it relates the torsion tensor to the (covariant) derivative of the vierbein as

$$\begin{aligned}
e e_\mu{}^{\tilde{\mu}} e_\rho{}^{\tilde{\nu}} T_{\tilde{\mu}\tilde{\nu}}{}^{\rho} &= e e_\mu{}^{\tilde{\mu}} e_\rho{}^{\tilde{\nu}}(-\mathcal{D}_{\tilde{\mu}} e_{\tilde{\nu}}{}^{\rho} + \mathcal{D}_{\tilde{\nu}} e_{\tilde{\mu}}{}^{\rho}) \\
&= -\mathcal{D}_{\tilde{\mu}}(e e_\mu{}^{\tilde{\mu}}) .
\end{aligned} \tag{5.29}$$

For the first equality in (5.28), we use the expression of the torsion tensor with curved indices computed from the one carrying flat indices, as shown in (3.118), whilst for the second equality, we write the torsion tensor following the definition given in Table 3.1. In the third equality, on the basis of (3.12), we expand the covariant derivative of the graviton in terms of the usual spacetime derivative and the superconnection, the latter being given by (3.104) following our gauge-fixing conditions and obeying the symmetry property (3.4). Finally, to validate the last equality, on the one hand, we use again the definition of the covariant derivative of the vielbein (see (3.12)) which leads to

$$-e_\mu{}^{\tilde\mu}(\partial_{\tilde\nu} e_{\tilde\mu}{}^\rho) e_\rho{}^{\tilde\nu} - \omega_{\tilde\nu\mu}{}^\rho e_\rho{}^{\tilde\nu} = -e_\mu{}^{\tilde\mu}\big[\partial_{\tilde\nu}(e_{\tilde\mu}{}^\rho e_\rho{}^{\tilde\nu}) - e_{\tilde\mu}{}^\rho(\partial_{\tilde\nu} e_\rho{}^{\tilde\nu})\big] - \omega_{\tilde\nu\mu}{}^\rho e_\rho{}^{\tilde\nu}$$
$$= (\partial_{\tilde\nu} e_\mu{}^{\tilde\nu}) - \omega_{\tilde\nu\mu}{}^\rho e_\rho{}^{\tilde\nu} = \mathcal{D}_{\tilde\nu} e_\mu{}^{\tilde\nu} ,$$

and, on the other and, we recall that, analogously to (5.9), we can compute $\mathcal{D}_{\tilde\mu} e$ as

$$\begin{aligned} \mathcal{D}_{\tilde\mu} e &= \partial_{\tilde\mu} e \\ &= e\,(\partial_{\tilde\mu} e_{\tilde\nu}{}^\nu) e_\nu{}^{\tilde\nu} . \end{aligned} \tag{5.30}$$

Therefore, the variation of ρ in (5.26) reads

$$\delta\rho_\alpha = \sqrt{2}\varepsilon_\alpha\Big[e\,[\bar\psi_{\tilde\mu}\bar\sigma^{\tilde\mu\tilde\nu}\bar\psi_{\tilde\nu}] + \frac{2}{3} e M^*\Big] - i\sqrt{2}\mathcal{D}_{\tilde\mu}[e(\sigma^{\tilde\mu}\bar\varepsilon)_\alpha]$$
$$+ \sqrt{2}e\Big[2\psi_{\tilde\mu\alpha}[\bar\psi_{\tilde\nu}\bar\sigma^{\tilde\mu\tilde\nu}\bar\varepsilon] + \frac{1}{6}(\sigma^{\tilde\mu}\bar\varepsilon)_\alpha b_{\tilde\mu}\Big] ,$$

due to (C.6). Comparing with (5.25), one derives the expression of f in terms of the gravitation multiplet [Wess and Zumino (1978a); Wess and Bagger (1992)],

$$f = eM^* + e\bar\psi_{\tilde\mu}\bar\sigma^{\tilde\mu\tilde\nu}\bar\psi_{\tilde\nu} . \tag{5.31}$$

Collecting all the previous results, one then establishes the complete expression of the chiral density $\mathcal{E}$,

$$\mathcal{E} = e\Big[1 + i\Theta\sigma^{\tilde\mu}\bar\psi_{\tilde\mu} - \Theta\cdot\Theta(M^* + \bar\psi_{\tilde\mu}\bar\sigma^{\tilde\mu\tilde\nu}\bar\psi_{\tilde\nu})\Big] . \tag{5.32}$$

However, to actually prove that $\mathcal{E}$ is a chiral density, we must still check that the variation of f in (5.21), which in this case read

$$\delta f = 4\partial_{\tilde\mu}[e\bar\psi_{\tilde\nu}\bar\sigma^{\tilde\mu\tilde\nu}\bar\varepsilon] , \tag{5.33}$$

satisfies the results obtained on the basis of (5.31) and the variation of the components of the gravitation supermultiplet calculated in Sec. 3.5. The computation of the latter starts from

$$\delta f = \delta\big(eM^* + e\bar\psi_{\tilde\mu}\bar\sigma^{\tilde\mu\tilde\nu}\bar\psi_{\tilde\nu}\big)$$
$$= (\delta e)M^* + e(\delta M^*) + (\delta e)\bar\psi_{\tilde\mu}\bar\sigma^{\tilde\mu\tilde\nu}\bar\psi_{\tilde\nu} + e(\delta\bar\psi_{\tilde\mu})\bar\sigma^{\tilde\mu\tilde\nu}\bar\psi_{\tilde\nu}$$
$$+ e\bar\psi_{\tilde\mu}\bar\sigma^{\tilde\mu\tilde\nu}(\delta\bar\psi_{\tilde\nu}) + e(\delta e_\mu{}^{\tilde\mu})\bar\psi_{\tilde\mu}\bar\sigma^{\mu\tilde\nu}\bar\psi_{\tilde\nu} + e(\delta e_\nu{}^{\tilde\nu})\bar\psi_{\tilde\mu}\bar\sigma^{\tilde\mu\nu}\bar\psi_{\tilde\nu} ,$$

where $\bar{\sigma}^{\mu\tilde{\nu}} = e_\nu{}^{\tilde{\nu}}\bar{\sigma}^{\mu\nu}$ and $\bar{\sigma}^{\tilde{\mu}\nu} = e_\mu{}^{\tilde{\mu}}\bar{\sigma}^{\mu\nu}$. After introducing the variation of the graviton, the gravitino and the auxiliary M field calculated in Sec. 3.5, this equation becomes,

$$\begin{aligned}
\delta f = ie\,&\Big[(\varepsilon\sigma^{\tilde{\rho}}\bar{\psi}_{\tilde{\rho}})(\bar{\psi}_{\tilde{\mu}}\bar{\sigma}^{\tilde{\nu}\tilde{\mu}}\bar{\psi}_{\tilde{\nu}}) + (\varepsilon\sigma^{\tilde{\mu}}\bar{\psi}_{\tilde{\rho}})(\bar{\psi}_{\tilde{\mu}}\bar{\sigma}^{\tilde{\rho}\tilde{\nu}}\bar{\psi}_{\tilde{\nu}}) + (\varepsilon\sigma^{\tilde{\nu}}\bar{\psi}_{\tilde{\rho}})(\bar{\psi}_{\tilde{\mu}}\bar{\sigma}^{\tilde{\mu}\tilde{\rho}}\bar{\psi}_{\tilde{\nu}})\Big]\\
&+ ie\Big[(\psi_{\tilde{\mu}}\sigma^{\tilde{\mu}}\bar{\varepsilon})(\bar{\psi}_{\tilde{\nu}}\bar{\sigma}^{\tilde{\nu}\tilde{\rho}}\bar{\psi}_{\tilde{\rho}}) + (\psi_{\tilde{\mu}}\sigma^{\tilde{\nu}}\bar{\varepsilon})(\bar{\psi}_{\tilde{\nu}}\bar{\sigma}^{\tilde{\rho}\tilde{\mu}}\bar{\psi}_{\tilde{\rho}}) + (\psi_{\tilde{\mu}}\sigma^{\tilde{\rho}}\bar{\varepsilon})(\bar{\psi}_{\tilde{\nu}}\bar{\sigma}^{\tilde{\mu}\tilde{\nu}}\bar{\psi}_{\tilde{\rho}})\Big]\\
&+ ieM^*\Big[\frac{1}{3}\varepsilon\sigma_{\tilde{\mu}}\bar{\sigma}^{\tilde{\mu}\tilde{\nu}}\bar{\psi}_{\tilde{\nu}} - \frac{1}{3}\bar{\psi}_{\tilde{\mu}}\bar{\sigma}^{\tilde{\mu}\tilde{\nu}}\bar{\sigma}_{\tilde{\nu}}\varepsilon - \varepsilon\sigma^{\tilde{\mu}}\bar{\psi}_{\tilde{\mu}} + \psi_{\tilde{\mu}}\sigma^{\tilde{\mu}}\bar{\varepsilon} + \overline{\varepsilon\sigma}^{\tilde{\mu}}\psi_{\tilde{\mu}}\Big]\\
&+ ieb^{\tilde{\mu}}\Big[\bar{\varepsilon}\cdot\bar{\psi}_{\tilde{\mu}} - \frac{1}{3}\overline{\varepsilon\sigma}_{\tilde{\mu}}\sigma_{\tilde{\rho}}\bar{\sigma}^{\tilde{\rho}\tilde{\nu}}\bar{\psi}_{\tilde{\nu}} + \overline{\varepsilon\sigma}_{\tilde{\mu}}{}^{\tilde{\nu}}\bar{\psi}_{\tilde{\nu}} - \frac{1}{3}\bar{\psi}_{\tilde{\rho}}\bar{\sigma}^{\tilde{\rho}\tilde{\nu}}\bar{\sigma}_{\tilde{\nu}}\sigma_{\tilde{\mu}}\bar{\varepsilon} + \bar{\psi}_{\tilde{\rho}}\bar{\sigma}^{\tilde{\rho}}{}_{\tilde{\mu}}\bar{\varepsilon}\Big]\\
&- 2e\Big[\overline{\varepsilon\sigma}^{\tilde{\mu}\tilde{\nu}}\bar{\psi}_{\tilde{\mu}\tilde{\nu}} + (\mathcal{D}_{\tilde{\mu}}\bar{\varepsilon})\bar{\sigma}^{\tilde{\mu}\tilde{\nu}}\bar{\psi}_{\tilde{\nu}} + \bar{\psi}_{\tilde{\mu}}\bar{\sigma}^{\tilde{\mu}\tilde{\nu}}(\mathcal{D}_{\tilde{\nu}}\bar{\varepsilon})\Big]\,. \qquad (5.34)
\end{aligned}$$

Analogously to the computation performed in (5.27) with the help of the Fierz identity (C.9), we can derive the property

$$\begin{aligned}
(\sigma^\mu\bar{\psi}_{\tilde{\rho}})_\alpha(\bar{\psi}_{\tilde{\mu}}\bar{\sigma}^{\nu\rho}\bar{\psi}_{\tilde{\nu}}) &= \frac{1}{4}\Big[(\sigma^\mu\bar{\psi}_{\tilde{\rho}})_\alpha(\bar{\psi}_{\tilde{\mu}}\bar{\sigma}^\nu)^\beta(\sigma^\rho\bar{\psi}_{\tilde{\nu}})_\beta - (\nu\leftrightarrow\rho)\Big]\\
&= -\frac{1}{4}\Big[(\sigma^\rho\bar{\psi}_{\tilde{\nu}})_\alpha(\bar{\psi}_{\tilde{\mu}}\bar{\sigma}^\nu\sigma^\mu\bar{\psi}_{\tilde{\rho}}) + (\sigma^\nu\bar{\psi}_{\tilde{\mu}})_\alpha(\bar{\psi}_{\tilde{\rho}}\bar{\sigma}^\mu\sigma^\rho\bar{\psi}_{\tilde{\nu}})\Big]\\
&\quad + \frac{1}{4}\Big[\nu\leftrightarrow\rho\Big]\,,
\end{aligned}$$

which implies that the first line of (5.34) identically vanishes. In the same fashion, we can show that

$$\begin{aligned}
\psi_{\tilde{\mu}\alpha}(\bar{\psi}_{\tilde{\nu}}\bar{\sigma}^{\nu\rho}\bar{\psi}_{\tilde{\rho}}) &= \frac{1}{4}\Big[\psi_{\tilde{\mu}\alpha}(\bar{\psi}_{\tilde{\nu}}\bar{\sigma}^\nu)^\beta(\sigma^\rho\bar{\psi}_{\tilde{\rho}})_\beta - (\nu\leftrightarrow\rho)\Big]\\
&= \frac{1}{4}\Big[(\sigma^\nu\bar{\psi}_{\tilde{\nu}})_\alpha(\psi_{\tilde{\mu}}\sigma^\rho\bar{\psi}_{\tilde{\rho}}) - (\sigma^\rho\bar{\psi}_{\tilde{\rho}})_\alpha(\bar{\psi}_{\tilde{\nu}}\sigma^\nu\bar{\psi}_{\tilde{\mu}})\Big] - \frac{1}{4}\Big[\nu\leftrightarrow\rho\Big]\,,
\end{aligned}$$

which allows to simplify the second line of (5.34) as

$$\begin{aligned}
ie(\psi_{\tilde{\mu}}&\sigma^{\tilde{\mu}}\bar{\varepsilon})(\bar{\psi}_{\tilde{\nu}}\bar{\sigma}^{\tilde{\nu}\tilde{\rho}}\bar{\psi}_{\tilde{\rho}}) + ie(\psi_{\tilde{\mu}}\sigma^{\tilde{\nu}}\bar{\varepsilon})(\bar{\psi}_{\tilde{\nu}}\bar{\sigma}^{\tilde{\rho}\tilde{\mu}}\bar{\psi}_{\tilde{\rho}}) + ie(\psi_{\tilde{\mu}}\sigma^{\tilde{\rho}}\bar{\varepsilon})(\bar{\psi}_{\tilde{\nu}}\bar{\sigma}^{\tilde{\mu}\tilde{\nu}}\bar{\psi}_{\tilde{\rho}})\\
&= 2ie(\overline{\varepsilon\sigma}^{\tilde{\nu}\tilde{\mu}}\bar{\psi}_{\tilde{\nu}})(\psi_{\tilde{\mu}}\sigma^{\tilde{\rho}}\bar{\psi}_{\tilde{\rho}}) + 2ie(\overline{\varepsilon\sigma}^{\tilde{\rho}\tilde{\nu}}\bar{\psi}_{\tilde{\nu}})(\psi_{\tilde{\mu}}\sigma^{\tilde{\mu}}\bar{\psi}_{\tilde{\rho}}) + 2ie(\overline{\varepsilon\sigma}^{\tilde{\mu}\tilde{\rho}}\bar{\psi}_{\tilde{\nu}})(\psi_{\tilde{\mu}}\sigma^{\tilde{\nu}}\bar{\psi}_{\tilde{\rho}})\\
&= -2ie(\overline{\varepsilon\sigma}^{\tilde{\mu}\tilde{\nu}}\bar{\psi}_{\tilde{\nu}})(\psi_{\tilde{\mu}}\sigma^\rho\bar{\psi}_{\tilde{\rho}} - \psi_{\tilde{\rho}}\sigma^\rho\bar{\psi}_{\tilde{\mu}})e_\rho{}^{\tilde{\rho}} + ie(\overline{\varepsilon\sigma}^{\tilde{\mu}\tilde{\rho}}\bar{\psi}_{\tilde{\nu}})(\psi_{\tilde{\mu}}\sigma^\rho\bar{\psi}_{\tilde{\rho}} - \psi_{\tilde{\rho}}\sigma^\rho\bar{\psi}_{\tilde{\mu}})e_\rho{}^{\tilde{\nu}}\\
&= 4(\overline{\varepsilon\sigma}^{\tilde{\mu}\tilde{\nu}}\bar{\psi}_{\tilde{\nu}})\;e\;e_\rho{}^{\tilde{\rho}}T_{\tilde{\mu}\tilde{\rho}}{}^\rho\Big| - 2(\overline{\varepsilon\sigma}^{\tilde{\mu}\tilde{\rho}}\bar{\psi}_{\tilde{\nu}})\;e\;e_\rho{}^{\tilde{\nu}}T_{\tilde{\mu}\tilde{\rho}}{}^\rho\Big|\\
&= -4(\overline{\varepsilon\sigma}^{\tilde{\mu}\tilde{\nu}}\bar{\psi}_{\tilde{\nu}})(\mathcal{D}_{\tilde{\mu}}e) - 4e(\overline{\varepsilon\sigma}^{\mu\tilde{\nu}}\bar{\psi}_{\tilde{\nu}})(\mathcal{D}_{\tilde{\mu}}e_\mu{}^{\tilde{\mu}}) - 4e(\overline{\varepsilon\sigma}^{\tilde{\mu}\nu}\bar{\psi}_{\tilde{\nu}})(\mathcal{D}_{\tilde{\mu}}e_\nu{}^{\tilde{\nu}})\,.
\end{aligned}$$

For the next-to-last equality, we have introduced the torsion tensor with curved indices as given by (3.118), whilst in the last equality, we refer to (5.29) for the first term and to a similar derivation as the one performed in (5.28) for the other terms, using in addition the antisymmetry properties of the $\bar{\sigma}^{\mu\nu}$ matrices. The two properties of the Pauli matrices (C.6) and (C.8) allow to show that the third and fourth lines of (5.34) both vanish. Therefore, replacing $\psi_{\tilde{\mu}\tilde{\nu}}$ by its definition in (3.117), one gets

$$\begin{aligned}
\delta f =& -4(\overline{\varepsilon\bar{\sigma}^{\tilde{\mu}\tilde{\nu}}\bar{\psi}_{\tilde{\nu}}})(\mathcal{D}_{\tilde{\mu}}e) - 4e(\overline{\varepsilon\bar{\sigma}^{\mu\tilde{\nu}}\bar{\psi}_{\tilde{\nu}}})(\mathcal{D}_{\tilde{\mu}}e_\mu{}^{\tilde{\mu}}) - 4e(\overline{\varepsilon\bar{\sigma}^{\tilde{\mu}\nu}\bar{\psi}_{\tilde{\nu}}})(\mathcal{D}_{\tilde{\mu}}e_\nu{}^{\tilde{\nu}}) \\
& - 4e\Big[\overline{\varepsilon\bar{\sigma}^{\tilde{\mu}\tilde{\nu}}(\mathcal{D}_{\tilde{\mu}}\bar{\psi}_{\tilde{\nu}})} + \overline{(\mathcal{D}_{\tilde{\mu}}\bar{\varepsilon})\bar{\sigma}^{\tilde{\mu}\tilde{\nu}}\bar{\psi}_{\tilde{\nu}}}\Big] \\
=& -4\mathcal{D}_{\tilde{\mu}}\big[e\overline{\varepsilon\bar{\sigma}^{\tilde{\mu}\tilde{\nu}}\bar{\psi}_{\tilde{\nu}}}\big] \\
=& \; 4\partial_{\tilde{\mu}}\big[e\bar{\psi}_{\tilde{\nu}}\bar{\sigma}^{\tilde{\mu}\tilde{\nu}}\bar{\varepsilon}\big]\,,
\end{aligned}$$

which agrees with (5.33) and which complete the proof that $\mathcal{E}$ is a chiral density. A geometrical interpretation of $\mathcal{E}$ is given in [Butter (2010)].

Using the invariant measure $\mathcal{E}$ we can rewrite the Lagrangian (5.11) in a purely chiral form,

$$\begin{aligned}
\mathcal{L} =& \frac{3}{8}\int \mathrm{d}^2\Theta\; \mathcal{E}(\overline{\mathcal{D}}\cdot\overline{\mathcal{D}} - 8\mathcal{R})e^{-\frac{1}{3}K(\Phi,\Phi^\dagger e^{-2gV})} \\
& + \int \mathrm{d}^2\Theta\; \mathcal{E}W(\Phi) + \frac{1}{16g^2}\int \mathrm{d}^2\Theta\; \mathcal{E}h_{ab}(\Phi)W^{a\alpha}W^b_\alpha + \text{h.c.}\,,
\end{aligned} \tag{5.35}$$

where the normalisation ensures a correct flat limit Lagrangian. We now address the explicit computation of the different terms appearing in (5.35). This requires several ingredients which have all been calculated in the previous sections and are summarised in Tables 5.1 and 5.2.

5.3 Super-Weyl rescaling and invariant actions

In this section, we derive the transformation properties of the action (5.35) under a super-Weyl rescaling, and show how the combination of a Howe–Tucker and a Kähler transformation leaves the action invariant.

5.3.1 *Transformation properties of the basic objects necessary to build the action*

In order to compute the variation of the action (5.35) under a Howe–Tucker transformation as defined in Table 3.5, we first calculate the transformation laws of the chiral density $\mathcal{E}$, those of a generic chiral superfield Φ as well as those of $(\overline{\mathcal{D}}\cdot\overline{\mathcal{D}} - 8\mathcal{R})V$ where V is a real superfield. For the sake of completeness, the variation of the Berezian E is also derived.

Table 5.1 Ingredients for an explicit construction of the chiral supergravity action.

The chiral form of the supergravity action is given by (5.35),

$$\mathcal{L} = \frac{3}{8}\int d^2\Theta\, \mathcal{E}(\overline{\mathcal{D}}\cdot\overline{\mathcal{D}} - 8\mathcal{R})e^{-\frac{1}{3}K(\Phi,\Phi^\dagger e^{-2gV})}$$
$$+ \int d^2\Theta\, \mathcal{E}W(\Phi) + \frac{1}{16g^2}\int d^2\Theta\, \mathcal{E}h(\Phi)_{ab}W^{a\alpha}W^b_\alpha + \text{h.c.}\,.$$

We recall that Φ being a generic chiral superfield, its expansion in terms of the Θ-variables reads

$$\Phi = \Phi\big| + \Theta\cdot(\mathcal{D}\Phi)\big| - \frac{1}{4}\Theta\cdot\Theta\,(\mathcal{D}\cdot\mathcal{D}\Phi)\big|\,,$$

where we use covariant derivatives with flat indices, or

$$\Phi = \phi + \sqrt{2}\Theta\cdot\chi - \Theta\cdot\Theta F\,,$$

when we introduce the component fields denoted by ϕ, χ and F. The computation of the first term of the action above involves the chiral superfield $\Xi = (\overline{\mathcal{D}}\cdot\overline{\mathcal{D}} - 8\mathcal{R})\Phi^\dagger$ associated to the antichiral superfield $\Phi^\dagger$. Its Θ-expansion can be directly derived from the results summarised in Table 4.1,

$$\begin{aligned}
\Xi = {} & 4F^\dagger + \frac{4}{3}M\phi^\dagger \\
& +\Theta\cdot\Big[-4i\sqrt{2}(\sigma^\mu\hat{D}_\mu\bar{\chi}) + \frac{2\sqrt{2}}{3}b_\mu(\sigma^\mu\bar{\chi}) + \phi^\dagger\Big(\frac{8}{3}(\sigma^{\mu\nu}\psi_{\mu\nu}) - \frac{4i}{3}(\sigma^\mu\bar{\psi}_\mu)M + \frac{4i}{3}\psi^\mu b_\mu\Big)\Big] \\
& -\Theta\cdot\Theta\Big[-4e_\mu{}^{\tilde{\mu}}\mathcal{D}_{\tilde{\mu}}\hat{D}^\mu\phi^\dagger + \frac{8i}{3}b^\mu\hat{D}_\mu\phi^\dagger + 2\sqrt{2}\bar{\psi}_\mu\cdot\hat{D}^\mu\bar{\chi} - \frac{2\sqrt{2}}{3}\bar{\chi}\bar{\sigma}^{\mu\nu}\bar{\psi}_{\mu\nu} - \frac{8}{3}M^*F^\dagger \\
& \quad + \sqrt{2}i\bar{\chi}\cdot\bar{\psi}_\mu b^\mu + \frac{2\sqrt{2}i}{3}\bar{\chi}\bar{\sigma}^{\nu\mu}\bar{\psi}_\mu b_\nu + \phi^\dagger\Big(-\frac{2}{3}e_\mu{}^{\tilde{\mu}}e_\nu{}^{\tilde{\nu}}R_{\tilde{\mu}\tilde{\nu}}{}^{\mu\nu}\big| - \frac{8}{9}MM^* + \frac{4}{9}b_\mu b^\mu \\
& \quad + \frac{4i}{3}e_\mu{}^{\tilde{\mu}}\mathcal{D}_{\tilde{\mu}}b^\mu + \frac{2}{3}\bar{\psi}_\mu\cdot\bar{\psi}^\mu M + \frac{2}{3}\psi_\nu\sigma^\nu\bar{\psi}_\mu b^\mu + \frac{4i}{3}\bar{\psi}^\mu\bar{\sigma}^\nu\psi_{\mu\nu} + \frac{1}{6}\varepsilon^{\mu\nu\rho\sigma}(\psi_\mu\sigma_\sigma\bar{\psi}_{\nu\rho} + \bar{\psi}_\mu\bar{\sigma}_\sigma\psi_{\nu\rho})\Big)\Big]
\end{aligned}$$

The gauge interactions included in the first term of the action above can be computed from the components of the superfield $\mathcal{X} = (\overline{\mathcal{D}}\cdot\overline{\mathcal{D}} - 8\mathcal{R})\Phi^\dagger e^{-2gV}$ introduced in Sec. 4.3. Adopting the Wess-Zumino gauge, defined in (4.40), the expressions of the component fields can be found in Table 4.4, and allow to write the Θ-expansion of $\mathcal{X}$ as

$$\begin{aligned}
\mathcal{X} = {} & \Xi + \Theta\cdot\Big[-4\sqrt{2}g(\sigma^\mu\bar{\chi})v_\mu + 8ig\phi^\dagger\lambda + 4ig\phi^\dagger v_\nu(\sigma^\mu\bar{\sigma}^\nu\psi_\mu)\Big] \\
& -\Theta\cdot\Theta\Big[8igv_\mu\hat{D}^\mu\phi^\dagger - 4\sqrt{2}ig\bar{\chi}\cdot\bar{\lambda} + 4ig\phi^\dagger\hat{D}_\mu v^\mu + \frac{8}{3}g\phi^\dagger v_\mu b^\mu - 4gD\phi^\dagger + 4g^2\phi^\dagger v_\mu v^\mu\Big]\,,
\end{aligned}$$

where the expansion of the superfield Ξ has been given above.

From (5.8) and the variations of the component of the supervierbein given in Table 3.5, one immediately obtains the variation of the Berezian,

$$\begin{aligned}
\delta_\Sigma E &= (-)^{|\tilde{M}|}E\delta_\Sigma E_{\tilde{M}}{}^M E_M{}^{\tilde{M}} \\
&= -(-)^{|M|}\delta_\Sigma E_M{}^{\tilde{M}} E_{\tilde{M}}{}^M \\
&= 2(\Sigma + \Sigma^\dagger)E\,.
\end{aligned}$$

Table 5.2 Ingredients for an explicit construction of the chiral supergravity action cont'd.

The kinetic and gauge interaction terms for the vector superfields, *i.e.*, the third term of the action above can be computed from the results of Table 4.3 in the case of non-abelian gauge groups. Starting from the expression of the components of the spinorial superfield $W_\alpha = -1/4(\overline{\mathcal{D}}\cdot\overline{\mathcal{D}} - 8\mathcal{R})e^{2gV}\mathcal{D}_\alpha e^{-2gV}$, its expansion in the Θ-variable reads

$$W_\alpha = -2g\Big[i\lambda_\alpha + \Big[i(\sigma^{\mu\nu}\Theta)_\alpha(\hat{F}_{\mu\nu}) + \Theta_\alpha D\Big] - \Theta\cdot\Theta\Big[(\sigma^\mu \hat{D}_\mu\bar{\lambda})_\alpha - \frac{i}{2}b_\mu(\sigma^\mu\bar{\lambda})_\alpha - \frac{i}{2}M^*\lambda_\alpha\Big]\Big],$$

with

$$\hat{F}_{\mu\nu} = F_{\mu\nu} - \frac{i}{2}\Big[\bar{\lambda}\bar{\sigma}_\nu\psi_\mu + \lambda\sigma_\nu\bar{\psi}_\mu - \bar{\lambda}\bar{\sigma}_\mu\psi_\nu - \lambda\sigma_\mu\bar{\psi}_\nu\Big].$$

We refer to Tables 4.1 and 4.3 for the explicit form of all the covariant derivatives of fields appearing in all the expressions above, and notice that the abelian case can be easily recovered from the non-abelian one (or from the results of Table 4.2). Finally, the superfield $\mathcal{R}$ can be read from Table 3.8,

$$\begin{aligned}\mathcal{R} = \frac{1}{6}\Big[&- M + \Theta\cdot\Big[-2(\sigma^{\mu\nu}\psi_{\mu\nu}) + i(\sigma^\mu\bar{\psi}_\mu)M - i\psi_\mu b^\mu\Big] - \Theta\cdot\Theta\Big[\frac{1}{2}e_\mu{}^{\tilde{\mu}}e_\nu{}^{\tilde{\nu}}R_{\tilde{\mu}\tilde{\nu}}{}^{\mu\nu}\Big| + \frac{2}{3}MM^* \\ &-\frac{1}{3}b_\mu b^\mu - ie_\mu{}^{\tilde{\mu}}\mathcal{D}_{\tilde{\mu}}b^\mu - \frac{1}{2}\bar{\psi}_\mu\cdot\bar{\psi}^\mu M - \frac{1}{2}\psi_\nu\sigma^\nu\bar{\psi}_\mu b^\mu \\ &-i\bar{\psi}^\mu\bar{\sigma}^\nu\psi_{\mu\nu} - \frac{1}{8}\varepsilon^{\mu\nu\rho\sigma}\Big[\psi_\mu\sigma_\sigma\bar{\psi}_{\nu\rho} + \bar{\psi}_\mu\bar{\sigma}_\sigma\psi_{\nu\rho}\Big]\Big]\Big],\end{aligned}$$

and the chiral density is given by Eq. (5.32),

$$\mathcal{E} = e\Big[1 + i\Theta\sigma^{\tilde{\mu}}\bar{\psi}_{\tilde{\mu}} - \Theta\cdot\Theta(M^* + \bar{\psi}_{\tilde{\mu}}\bar{\sigma}^{\tilde{\mu}\tilde{\nu}}\bar{\psi}_{\tilde{\nu}})\Big].$$

In order to derive the variation of the chiral density (5.32),

$$\mathcal{E} = e + ie\Theta\sigma^{\tilde{\mu}}\bar{\psi}_{\tilde{\mu}} - \Theta\cdot\Theta(eM^* + e\bar{\psi}_{\tilde{\mu}}\bar{\sigma}^{\tilde{\mu}\tilde{\nu}}\bar{\psi}_{\tilde{\nu}}), \tag{5.36}$$

we tackle the problem component by component and first compute the variations of the different objects included in (5.36), namely, e, $e_\mu{}^{\tilde{\mu}}$, $e_{\tilde{\mu}}{}^\mu$, $\bar{\psi}_{\tilde{\mu}}{}^{\dot{\alpha}}$, $\bar{\psi}_{\tilde{\mu}\dot{\alpha}}$, M^* and b_μ. From the results of Table 3.5 and Table 3.6, one gets the relations

$$\begin{aligned}\delta_\Sigma e_\mu{}^{\tilde{\mu}} &= \delta_\Sigma E_\mu{}^{\tilde{\mu}}\Big| = -(\Sigma + \Sigma^\dagger)\Big|\, e_\mu{}^{\tilde{\mu}}, \\ \delta_\Sigma e_{\tilde{\mu}}{}^\mu &= \delta_\Sigma E_{\tilde{\mu}}{}^\mu\Big| = (\Sigma + \Sigma^\dagger)\Big|e_{\tilde{\mu}}{}^\mu, \\ \delta_\Sigma\bar{\psi}_{\tilde{\mu}\dot{\alpha}} &= 2\delta_\Sigma E_{\tilde{\mu}\dot{\alpha}}\Big| = (2\Sigma - \Sigma^\dagger)\Big|\bar{\psi}_{\tilde{\mu}\dot{\alpha}} + ie_{\tilde{\mu}}{}^\mu\, \mathcal{D}^\alpha\Sigma\Big|\, \sigma_{\mu\alpha\dot{\alpha}}, \\ \delta_\Sigma\bar{\psi}_{\tilde{\mu}}{}^{\dot{\alpha}} &= (2\Sigma - \Sigma^\dagger)\Big|\bar{\psi}_{\tilde{\mu}}{}^{\dot{\alpha}} - ie_{\tilde{\mu}}{}^\mu\bar{\sigma}_\mu{}^{\dot{\alpha}\alpha}\mathcal{D}_\alpha\Sigma\Big|, \\ \delta_\Sigma\psi_{\tilde{\mu}}{}^\alpha &= 2\delta_\Sigma E_{\tilde{\mu}}{}^\alpha\Big| = (2\Sigma^\dagger - \Sigma)\Big|\psi_{\tilde{\mu}}{}^\alpha + ie_{\tilde{\mu}}{}^\mu\overline{\mathcal{D}}_{\dot{\alpha}}\Sigma^\dagger\Big|\bar{\sigma}_\mu{}^{\dot{\alpha}\alpha}, \\ \delta_\Sigma\psi_{\tilde{\mu}\alpha} &= (2\Sigma^\dagger - \Sigma)\Big|\psi_{\tilde{\mu}\alpha} - ie_{\tilde{\mu}}{}^\mu\, \sigma_{\mu\alpha\dot{\alpha}}\overline{\mathcal{D}}^{\dot{\alpha}}\Sigma^\dagger\Big|, \\ \delta_\Sigma M^* &= -6\delta_\Sigma\mathcal{R}^\dagger\Big| = 2(\Sigma - 2\Sigma^\dagger)\Big|M^* + \tfrac{3}{2}\mathcal{D}\cdot\mathcal{D}\Sigma\Big|, \\ \delta_\Sigma b_\mu &= -3\delta_\Sigma G_\mu\Big| = -(\Sigma + \Sigma^\dagger)\Big|b_\mu - 3i\mathcal{D}_\mu[\Sigma - \Sigma^\dagger]\Big|,\end{aligned} \tag{5.37}$$

which allow to compute the variations of the different components of $\mathcal{E}$,

$$\begin{aligned}
\delta_\Sigma e &= e(\delta_\Sigma e_{\tilde\mu}{}^{\mu}) e_\mu{}^{\tilde\mu} \\
&= 4(\Sigma + \Sigma^\dagger)\big|\, e\,, \\
\delta_\Sigma(ie\sigma^{\tilde\mu}\bar\psi_{\tilde\mu})_\alpha &= i\sigma^\mu{}_{\alpha\dot\alpha}\delta_\Sigma(ee_\mu{}^{\tilde\mu}\bar\psi_{\tilde\mu}{}^{\dot\alpha}) \\
&= ie(5\Sigma + 2\Sigma^\dagger)\big|(\sigma^{\tilde\mu}\bar\psi_{\tilde\mu})_\alpha + 4e\mathcal{D}_\alpha\Sigma\big|\,, \\
\delta_\Sigma(eM^*) &= 6e\Sigma\big|M^* + \frac{3}{2}e\mathcal{D}\cdot\mathcal{D}\Sigma\big|\,, \\
\delta_\Sigma(e\bar\psi_{\tilde\mu}\bar\sigma^{\tilde\mu\tilde\nu}\bar\psi_{\tilde\nu}) &= \bar\sigma^{\mu\nu\dot\beta}{}_{\dot\alpha}\delta_\Sigma(ee_\mu{}^{\tilde\mu}e_\nu{}^{\tilde\nu}\bar\psi_{\tilde\mu\dot\beta}\bar\psi_{\tilde\nu}{}^{\dot\alpha}) \\
&= 6e\Sigma\big|\bar\psi_{\tilde\mu}\bar\sigma^{\tilde\mu\tilde\nu}\bar\psi_{\tilde\nu} + 3ie\mathcal{D}\Sigma\big|\sigma^{\tilde\rho}\bar\psi_{\tilde\rho}\,.
\end{aligned}$$

In order to derive the relations above, we refer to (5.23) for the expression of the variation of e and to (C.6) for simplifying products of Pauli matrices. The full variation of $\mathcal{E}$, obtained by collecting the contributions derived above, can however be written under a more compact form by introducing the spinorial superfield [Wess and Bagger (1992)]

$$S^\alpha = \Theta^\alpha[2\Sigma^\dagger - \Sigma]\big| + \Theta\cdot\Theta\mathcal{D}^\alpha\Sigma\big|\,. \tag{5.38}$$

Indeed, a direct calculation gives

$$\delta_\Sigma\mathcal{E} = 6\Sigma\mathcal{E} + \partial_{\Theta^\alpha}(S^\alpha\mathcal{E})\,, \tag{5.39}$$

where we recall that the components of the superfield Σ read

$$\Sigma = \Sigma\big| + \Theta\cdot\mathcal{D}\Sigma\big| - \frac{1}{4}\Theta\cdot\Theta\mathcal{D}\cdot\mathcal{D}\Sigma\big|\,.$$

The super-Weyl transformations of chiral and vector superfields are defined in such a way as to preserve the respective constraints. Therefore, they can in general be written as

$$\begin{aligned}
\delta_\Sigma\Phi &= w\Sigma\Phi\,, \\
\delta_\Sigma V &= w'(\Sigma + \Sigma^\dagger)V\,,
\end{aligned} \tag{5.40}$$

where we have introduced the Weyl weights w and w' of the chiral and vector superfields. Switching to the hybrid Θ-variables, one has to account for the variations of the superderivatives (as shown in the first relations of Table 3.5). Recalling that the expansion of the chiral superfield Φ in terms of the Θ−variables is written as $\Phi = \phi + \sqrt{2}\Theta\cdot\chi - \Theta\cdot\Theta F$ we can deduce the variations of the component fields from (5.40) [Wess and Bagger (1992)],

$$\begin{aligned}\delta_\Sigma\phi &= \delta_\Sigma\Phi\big| \\ &= w\Sigma\Phi\big|\,, \\ \delta_\Sigma\chi_\alpha &= \delta_\Sigma\mathcal{D}_\alpha\Phi\big| \\ &= (\Sigma - 2\Sigma^\dagger)\big|\chi_\alpha + w\mathcal{D}_\alpha(\Sigma\Phi)\big|\,, \\ \delta_\Sigma F &= \frac{1}{4}\delta_\Sigma(\mathcal{D}\cdot\mathcal{D})\Phi\big| \\ &= 2(\Sigma - 2\Sigma^\dagger)\big|F + \sqrt{2}\mathcal{D}\Sigma\big|\cdot\chi + w\mathcal{D}\cdot\mathcal{D}(\frac{1}{4}\Sigma\Phi)\,,\end{aligned} \tag{5.41}$$

where the last relation has been derived with the help of (B.5). Collecting the results and introducing the spinorial superfield S^α, we get

$$\delta_\Sigma\Phi = w\Sigma\Phi - S^\alpha\partial_{\Theta^\alpha}\Phi\,. \tag{5.42}$$

Defining the chiral superfield $\Xi = (\overline{\mathcal{D}}\cdot\overline{\mathcal{D}} - 8\mathcal{R})V$ as the chiral projection of the vector superfield V, its transformation laws are given by [Wess and Bagger (1992)]

$$\begin{aligned}\delta_\Sigma\Xi &= \Big[(\delta_\Sigma\overline{\mathcal{D}})\cdot\overline{\mathcal{D}} + \overline{\mathcal{D}}\cdot(\delta_\Sigma\overline{\mathcal{D}}) - 8\delta_\Sigma\mathcal{R}\Big]V + w'(\overline{\mathcal{D}}\cdot\overline{\mathcal{D}} - 8\mathcal{R})\Big[(\Sigma + \Sigma^\dagger)V\Big] \\ &= (\overline{\mathcal{D}}\cdot\overline{\mathcal{D}} - 8\mathcal{R})\Big[2(\Sigma^\dagger - 2\Sigma)V + w'(\Sigma + \Sigma^\dagger)V\Big] - S^\alpha\partial_{\Theta^\alpha}\Xi\,,\end{aligned} \tag{5.43}$$

where, as for the chiral superfield Φ, the term in S^α is related to the usage of the superderivatives for defining the component fields and where we have employed (B.5) for the action of the Lorentz generators.

5.3.2 *Transformation of the action*

Using the results of the previous section, we can derive the behaviour of the action (5.35) under a super-Weyl rescaling. Imposing the Kähler potential to have a vanishing Weyl weight, the variation of the corresponding Lagrangian can be computed with the help of (5.39) and (5.43),

$$\begin{aligned}&\delta_\Sigma\Big[\mathcal{E}(\overline{\mathcal{D}}\cdot\overline{\mathcal{D}} - 8\mathcal{R})\exp\Big(-\frac{1}{3}K(\Phi, \Phi^\dagger e^{-2gV})\Big)\Big] \\ &= \mathcal{E}(\overline{\mathcal{D}}\cdot\overline{\mathcal{D}} - 8\mathcal{R})\Big[2(\Sigma + \Sigma^\dagger)\exp\Big(-\frac{1}{3}K(\Phi, \Phi^\dagger e^{-2gV})\Big)\Big] \\ &\quad + \partial_{\Theta^\alpha}\Big[S^\alpha\mathcal{E}(\overline{\mathcal{D}}\cdot\overline{\mathcal{D}} - 8\mathcal{R})\exp\Big(-\frac{1}{3}K(\Phi, \Phi^\dagger e^{-2gV})\Big)\Big]\,.\end{aligned} \tag{5.44}$$

In the same fashion, assuming the weight of the superpotential to be zero, (5.39) and (5.42) lead to

$$\delta_\Sigma(\mathcal{E}W) = 6\Sigma\mathcal{E}W + \partial_{\Theta^\alpha}[S^\alpha\mathcal{E}W] \,. \tag{5.45}$$

If the vector superfield V is also weightless under Weyl rescaling, the variation of the associated spinorial superfield (see Table 5.1) can be directly computed from the variation of the superderivatives,

$$\begin{aligned}\delta_\Sigma W_\alpha &= \Big[(\delta_\Sigma\overline{\mathcal{D}})\cdot\overline{\mathcal{D}} + \overline{\mathcal{D}}\cdot(\delta_\Sigma\overline{\mathcal{D}}) - 8(\delta_\Sigma\mathcal{R})\Big]e^{2gV}\mathcal{D}_a e^{-2gV} \\ &\quad + \Big[\overline{\mathcal{D}}\cdot\overline{\mathcal{D}} - 8\mathcal{R}\Big]e^{2gV}(\delta_\Sigma\mathcal{D}_a)e^{-2gV} - S^\beta\partial_{\Theta^\beta}W_\alpha \\ &= -3\Sigma W_\alpha - S^\beta\partial_{\Theta^\beta}W_\alpha \,. \end{aligned} \tag{5.46}$$

Using (5.39) and (5.42) and assuming that the gauge kinetic function h_{ab} is weightless too, we find

$$\delta_\Sigma[\mathcal{E}h_{ab}W^{\alpha a}W_\alpha{}^b] = \partial_{\Theta^\beta}\Big(\mathcal{E}S^\beta h_{ab}W^{\alpha a}W_\alpha{}^b\Big) \,. \tag{5.47}$$

Integrating over $d\Theta^2$, *i.e.*, dropping the derivatives with respect to Θ, and considering a finite transformation, (5.44), (5.45) and (5.47) imply that the Lagrangian (5.35) transforms as [Wess and Bagger (1992)]

$$\begin{aligned}\mathcal{L} \to \frac{3}{8}\int d^2\Theta\; \mathcal{E}(\overline{\mathcal{D}}\cdot\overline{\mathcal{D}} - 8\mathcal{R})e^{2(\Sigma+\Sigma^\dagger)}\exp\left[-\frac{1}{3}K(\Phi, \Phi^\dagger e^{-2gV})\right] \\ + \int d^2\Theta\; \mathcal{E}e^{6\Sigma}W(\Phi) + \frac{1}{16g^2}\int d^2\Theta\; \mathcal{E}h(\Phi)_{ab}W^{\alpha a}W^b_\alpha + \text{h.c.}\,.\end{aligned} \tag{5.48}$$

We observe that the last term of the action, describing the gauge sector, is invariant under a super-Weyl rescaling. In the same manner, the supergravity action (5.13) is also invariant under a Howe–Tucker transformation, and therefore leads to conformal supergravity. Identical invariance properties could also have been obtained with the action written in the non-chiral form, as in (5.11).

5.3.3 *Kähler invariance and Howe–Tucker transformations*

In global supersymmetry, the action is invariant under a Kähler transformation,

$$K(\Phi, \Phi^\dagger e^{-2gV}) \to K(\Phi, \Phi^\dagger e^{-2gV}) + F(\Phi) + F^\star(\Phi^\dagger) \,, \tag{5.49}$$

where $F(\Phi)$ ($F^\star(\Phi^\dagger)$) is a gauge-singlet holomorphic (antiholomorphic) function of the chiral (antichiral) superfield Φ ($\Phi^\dagger$). Therefore, it must be noted that $F(\Phi) = F(e^{-2gV}\Phi)$ and that $F^\star(\Phi^\dagger) = F^\star(\Phi^\dagger e^{-2gV})$.

The superpotential contribution to the final Lagrangian contains contribution from the Kähler potential (see Chapter 6, *e.g.* the covariant derivative $\mathcal{D}_iW$). In particular, the scalar potential is given by (6.69). So if we impose that under the Kähler transformation (5.49) we have

$$W \to e^{-F} W ,$$

for the superpotential, then we automatically have (see Chapter 6 for the definition of the covariant derivative $\mathcal{D}_i W = W_i + K_i W$)

$$\mathcal{D}_i W \to e^{-F} \mathcal{D}_i W ,$$

and the scalar potential is invariant (see (6.71)).

Inspecting the first two terms of (5.48), one observes that the variation of the Kähler part of the action under a Howe–Tucker transformation has exactly the form of a Kähler transformation. Therefore, combining the Howe–Tucker transformation of parameter Σ with a Kähler transformation of parameter F makes it possible to render the action invariant. Hence, omitting the dependence of the Kähler potential and the functions F and $F^\star$ for clarity [Wess and Bagger (1992)], we obtain

$$\begin{aligned} &\frac{3}{8}\int d^2\Theta\mathcal{E}(\overline{\mathcal{D}}\cdot\overline{\mathcal{D}} - 8\mathcal{R})e^{-\frac{1}{3}K} + \int d^2\Theta\mathcal{E}W \to \\ &\frac{3}{8}\int d^2\Theta\mathcal{E}(\overline{\mathcal{D}}\cdot\overline{\mathcal{D}} - 8\mathcal{R})e^{2(\Sigma+\Sigma^\dagger)}e^{-\frac{1}{3}K-\frac{1}{3}(F+F^\star)} + \int d^2\Theta\mathcal{E}e^{-F+6\Sigma}W . \end{aligned} \tag{5.50}$$

Therefore, fixing $F = 6\Sigma$ makes the action invariant under Kähler-super-Weyl transformations. This compensates its variations under Howe–Tucker transformations, and the full Lagrangian is consequently made invariant under the combined Kähler-super-Weyl transformations. In other words, a Kähler transformation can be seen as a Howe–Tucker transformation.

Therefore, super-Weyl rescaling and Kähler transformations are related. This observation has an interesting consequence for the fermionic fields, since Howe–Tucker transformations generate a (Weyl) rotation in the fermionic sector, and *vice versa*, Kähler transformations generate a $U(1)$ rotation of the fermionic fields[1]. Hence, Eqs. (5.37), (5.41) and (5.46) induce,

$$\begin{aligned} \chi^i &\to e^{(\Sigma-2\Sigma^\dagger)|}\chi^i , \\ \lambda_a &\to e^{-3\Sigma|}\lambda_a , \\ \bar\psi_{\tilde\mu} &\to e^{(2\Sigma-\Sigma^\dagger)|}\Big(\bar\psi_{\tilde\mu} - i\bar\sigma_{\tilde\mu}\mathcal{D}\Sigma\big|\Big) , \end{aligned} \tag{5.51}$$

where we recall that $\lambda_\alpha = -iW_\alpha\big|$ in the abelian case and $\lambda_\alpha = i/(2g)W_\alpha\big|$ in the non-abelian one. If we decompose now $\Sigma\big| = \sigma + i\eta$ with σ, η real we get (not

[1] Since the transformation parameters are complex, we consider more precisely the complexified of $U(1)$, *i.e.*, $U(1,\mathbb{C}) \cong \mathbb{C}^* \cong U(1) \oplus \mathbb{R}^*$.

taking into account the shift for the gravitino)

$$\chi^i \to e^{-\sigma} e^{3i\eta} \chi^i ,$$
$$\lambda_a \to e^{-3\sigma} e^{-3i\eta} \lambda_a ,$$
$$\bar{\psi}_{\tilde{\mu}} \to e^{-\sigma} e^{3i\eta} \bar{\psi}_{\tilde{\mu}} ,$$

showing explicitly that we have a dilatation together with a $U(1)$–rotation. The last relation indicates that Weyl rotations of the gravitino are always accompanied by an additional shift. Moreover, since the chiral superfield Φ is invariant, due to (5.51), we have $\Theta^\alpha \to e^{-(\Sigma-2\Sigma^\dagger)|}\Theta^\alpha$. Then, (5.46), together with the expression of the spinorial superfield W_α given in Table 5.1, result in

$$\begin{aligned} \hat{F}_{\mu\nu} &\to e^{-2(\Sigma+\Sigma^\dagger)|} \hat{F}_{\mu\nu} , \\ v_\mu^a &\to e^{-(\Sigma+\Sigma^\dagger)|} v_\mu^a , \\ D^a &\to e^{-2(\Sigma+\Sigma^\dagger)|} D^a . \end{aligned} \tag{5.52}$$

The second line above follows from the transformation property of the graviton which means that $F_{\tilde{\mu}\tilde{\nu}} \to F_{\tilde{\mu}\tilde{\nu}}$ and consequently $v_{\tilde{\mu}} \to v_{\tilde{\mu}}$ which in turn enables us to deduce the transformation of v_μ. We obtain only a pure dilatation for the gauge field and the auxiliary field. The fact that $v_{\tilde{\mu}}$ and $F_{\tilde{\mu}\tilde{\nu}}$ are invariant under a Weyl rescaling will be important in the sequel.

In order to have an invariant action under the transformations (5.51) and (5.49) (with $F = 6\Sigma$), the spacetime derivatives of the fermionic fields have to be replaced by derivatives that are covariant with respect to Weyl rotations. The substitution rules concerning the fermionic matter field can be directly computed by inspecting the corresponding Lagrangian which is proportional to $\chi^i\sigma^\mu\partial_\mu\bar{\chi}_{i^*} - \partial_\mu\chi^i\sigma^\mu\bar{\chi}_{i^*}$. Requiring the cancellation of the additional terms in this last expression through the Nœther procedure, one gets

$$\begin{aligned} \partial_\mu\chi^i &\to \Big[\partial_\mu - \frac{1}{4}(K_j\partial_\mu\phi^j - K^{j^*}\partial_\mu\phi^\dagger_{j^*})\Big]\chi^i , \\ \partial_\mu\bar{\chi}_{i^*} &\to \Big[\partial_\mu + \frac{1}{4}(K_j\partial_\mu\phi^j - K^{j^*}\partial_\mu\phi^\dagger_{j^*})\Big]\bar{\chi}_{i^*} . \end{aligned} \tag{5.53}$$

In a similar fashion, the gaugino Lagrangian is proportional to $\lambda\sigma^\mu\partial_\mu\bar{\lambda} - \partial_\mu\lambda\sigma^\mu\bar{\lambda}$ and becomes invariant after the substitution

$$\begin{aligned} \partial_\mu\lambda &\to \Big[\partial_\mu + \frac{1}{4}(K_i\partial_\mu\phi^i - K^{i^*}\partial_\mu\phi^\dagger_{i^*})\Big]\lambda , \\ \partial_\mu\bar{\lambda} &\to \Big[\partial_\mu - \frac{1}{4}(K_i\partial_\mu\phi^i - K^{i^*}\partial_\mu\phi^\dagger_{i^*})\Big]\bar{\lambda} , \end{aligned} \tag{5.54}$$

whilst the gravitino Lagrangian is proportional to $\varepsilon^{\mu\nu\rho\sigma}(\psi_\mu\sigma_\sigma\bar{\psi}_{\nu\rho} - \bar{\psi}_\mu\bar{\sigma}_\sigma\psi_{\nu\rho})$ and becomes invariant after

$$\begin{aligned}
\partial_\mu\psi_\nu &\to \Big[\partial_\mu + \frac{1}{4}(K_i\partial_\mu\phi^i - K^{i^*}\partial_\mu\phi^\dagger_{i^*})\Big]\psi_\nu\,, \\
\partial_\mu\bar{\psi}_\nu &\to \Big[\partial_\mu - \frac{1}{4}(K_i\partial_\mu\phi^i - K^{i^*}\partial_\mu\phi^\dagger_{i^*})\Big]\bar{\psi}_\nu\,.
\end{aligned} \tag{5.55}$$

In we perform a Kähler transformation $K(\Phi, \Phi^\dagger) \to K(\Phi, \Phi^\dagger) + F(\Phi) + F^\dagger(\Phi^\dagger)$, the metric, the Christofell symbols and the curvature stay invariant. However,

$$\frac{1}{4}\Big(K_i\partial_\mu\phi^i - K^{i^*}\partial_\mu\phi^\dagger_{i^*}\Big) \to \frac{1}{4}\Big(K_i\partial_\mu\phi^i - K^{i^*}\partial_\mu\phi^\dagger_{i^*}\Big) + \frac{i}{2}\partial_\mu\mathrm{Im}(F)\,,$$

where now $F = F(\phi)$ is understood as a function of the scalar fields ϕ^i. In order to have Kähler invariance, the Kähler transformation has to come along with a Weyl rotation [Witten and Bagger (1982)]

$$\begin{aligned}
\chi^i &\to e^{\frac{i}{2}\mathrm{Im}(F)}\chi^i\,, \\
\lambda &\to e^{-\frac{i}{2}\mathrm{Im}(F)}\lambda\,, \\
\psi_\mu &\to e^{-\frac{i}{2}\mathrm{Im}(F)}\psi_\mu\,.
\end{aligned}$$

In Chapter 6 we shall see that the derivatives (5.53),(5.54) and (5.55) will be generated naturally.

Before moving on, let us note that, when the superpotential in non-vanishing, a Kähler transformation of parameter $F = \ln(W)$ rescales the superpotential to one. In this case, the supergravity action can be rewritten in an equivalent way as

$$\begin{aligned}
\mathcal{L} = &\frac{3}{8}\int \mathrm{d}^2\Theta\, \mathcal{E}(\overline{\mathcal{D}}\cdot\overline{\mathcal{D}} - 8\mathcal{R})e^{-\frac{1}{3}\mathcal{G}(\Phi,\Phi^\dagger e^{-2gV})} \\
&+ \int \mathrm{d}^2\Theta\, \mathcal{E} + \frac{1}{16g^2}\int \mathrm{d}^2\Theta\, \mathcal{E}h(\Phi)_{ab}W^{a\alpha}W^b_\alpha + \text{h.c.}\,,
\end{aligned}$$

where we have introduced to so-called generalised Kähler potential $\mathcal{G} = K + \ln|W|^2$, thereby unifying the Kähler potential and the superpotential.

Chapter 6

Explicit computation of the supergravity action

In this chapter, we will expand the Lagrangian (5.35) for a generic supersymmetric theory. We will hence derive the most general expression of the supergravity Lagrangian in terms of the components of the gravitation multiplet and of those of the chiral and vector superfield of the theory. We specify the matter sector of the model by a set of chiral superfields denoted by Φ^i which are assumed to lie in a representation $\mathfrak{R}$ of a gauge group G. Subsequently, the antichiral superfields $\Phi^\dagger_{i^*}$ lie in the $\overline{\mathfrak{R}}$ representation of G (the complex conjugate representation). To the gauge group we associate a vector superfield V, lying in the adjoint representation of G, and we adopt the Wess-Zumino gauge (4.40). In order to facilitate the computation of the different terms appearing in the Lagrangian, we decompose the latter into different sectors, *i.e.*, the pure gravitation sector, the gravitational and gauge interactions of the matter content and those of the gauge superfields. This approach is complementary to historical approaches obtained by means of the tensor calculus [Cremmer et al. (1979, 1982, 1978b, 1983b)].

6.1 The pure supergravity action

As suggested by (5.12), the first term of the Lagrangian (5.35) contains the pure supergravity action, given by,

$$\mathcal{L}_{\text{pure sugra}} = -3 \int \mathrm{d}^2\Theta\, \mathcal{E}\mathcal{R} + \text{h.c.} = -3[\mathcal{E}\mathcal{R}]_{\Theta^2} + \text{h.c.}\,,$$

where the Planck mass is set to one and the notation $[\Phi]_{\Theta^2}$ stands for the $\Theta\cdot\Theta$-coefficient of the expansion of the superfield Φ in terms of the Θ-variables. From the results of Table 5.1 and the expression of the product of two chiral superfields computed in (2.4), one gets

$$-3[\mathcal{E}\mathcal{R}]_{\Theta^2} = \frac{1}{4}ee_\mu{}^{\tilde{\mu}}e_\nu{}^{\tilde{\nu}}R_{\tilde{\mu}\tilde{\nu}}{}^{\mu\nu}\Big| - \frac{1}{16}e\varepsilon^{\mu\nu\rho\sigma}\Big[\psi_\mu\sigma_\sigma\bar{\psi}_{\nu\rho} + 5\bar{\psi}_\mu\bar{\sigma}_\sigma\psi_{\nu\rho}\Big]$$
$$-\frac{1}{6}eMM^* - \frac{1}{6}eb_\mu b^\mu - \frac{i}{2}\mathcal{D}_{\tilde{\mu}}[eb^{\tilde{\mu}}]\,,$$

where we have used the property of the Pauli matrices (C.5) and the relation

$$e[\psi_{\tilde{\mu}}\sigma^\nu\bar{\psi}_{\tilde{\nu}} - \psi_{\tilde{\nu}}\sigma^\nu\bar{\psi}_{\tilde{\mu}}]e_\mu{}^{\tilde{\mu}}e_\nu{}^{\tilde{\nu}} = -2i\Big[e_\mu{}^{\tilde{\mu}}(\mathcal{D}_{\tilde{\mu}}e) + e(\mathcal{D}_{\tilde{\mu}}e_\mu{}^{\tilde{\mu}})\Big]\,, \tag{6.1}$$

given in (5.28) and (5.29). Consequently, after summing with the hermitian conjugate piece $[\mathcal{E}^\dagger\mathcal{R}^\dagger]_{\bar{\Theta}^2}$, one obtains,

$$\mathcal{L}_{\text{pure sugra}} = \frac{1}{2}ee_\mu{}^{\tilde{\mu}}e_\nu{}^{\tilde{\nu}}R_{\tilde{\mu}\tilde{\nu}}{}^{\mu\nu}\Big| + \frac{1}{4}e\varepsilon^{\mu\nu\rho\sigma}\Big[\psi_\mu\sigma_\sigma\bar{\psi}_{\nu\rho} - \bar{\psi}_\mu\bar{\sigma}_\sigma\psi_{\nu\rho}\Big] - \frac{1}{3}e[MM^* + b_\mu b^\mu]\,.$$

This Lagrangian contains the Einstein-Hilbert Lagrangian for the spin-two graviton and the Rarita-Schwinger Lagrangian describing the dynamics of the spin-3/2 gravitino coupled to gravity. The quadratic terms in the fields M and b^μ show that the latter do not propagate and will have to be eliminated from the Lagrangian through their equations of motion. The pure supergravity Lagrangian was obtained first in [Deser and Zumino (1976); Freedman et al. (1976)] without auxiliary fields and later in [Stelle and West (1978c); Ferrara and van Nieuwenhuizen (1978a)] the minimal set of auxiliary fields was identified.

6.2 Coupling matter and gauge sector to supergravity

We now turn to the calculation of the Lagrangian describing the dynamics of the matter content of the theory, after it is coupled to supergravity. The full Lagrangian can be split into two terms,

$$\begin{aligned}\mathcal{L}_{\text{matter}} &\equiv \mathcal{L}_{\text{K}} + \mathcal{L}_{\text{W}}\\ &= \int d^2\Theta\,\frac{3}{8}\mathcal{E}(\bar{\mathcal{D}}\cdot\bar{\mathcal{D}} - 8\mathcal{R})e^{-\frac{1}{3}K(\Phi,\Phi^\dagger e^{-2gV})}\\ &\quad + \int d^2\Theta\,\mathcal{E}W(\Phi) + \text{h.c.}\,,\end{aligned} \tag{6.2}$$

the first one being related to kinetic and gauge interactions for the chiral superfields and the second one to the interactions driven by the superpotential. In our computation we follow [Wess and Bagger (1992)] where the results were given without much detail. One can also see [Dragon et al. (1987)] for some indications. We mention again that the coupling of matter to supergravity was obtained first in [Ferrara et al. (1976b); Freedman and Schwarz (1977); Cremmer et al. (1979, 1982, 1978b, 1983b)] by means of the tensor calculus and later in [Bagger (1983); Bagger and Witten (1982); Witten and Bagger (1982)] using the superspace approach.

6.2.1 *Computation of the Lagrangian coming from the Kähler potential*

In order to compute the Lagrangian $\mathcal{L}_{\mathrm{K}}$, we introduce the superspace kinetic energy Ω[1],

$$\Omega(\Phi, \Phi^\dagger) = -3e^{-\frac{1}{3}K(\Phi,\Phi^\dagger)} = W_I(\Phi^\dagger)W^I(\Phi)\,, \tag{6.3}$$

where the functions $W_I(\Phi^\dagger)$ are anti-holomorphic and depend on the antichiral superfields $\Phi^\dagger_{i^*}$ and the functions $W^I(\Phi)$ are holomorphic and depend on the chiral superfields Φ^i. As in Sec. 2.4, $W^I(\Phi)$ and $W_I(\Phi^\dagger)$ can be written as

$$\begin{aligned}
W^I(\Phi) &= W^I + \sqrt{2}\Theta\cdot(W^I_i\chi^i) - \Theta\cdot\Theta\Big[W^I_iF^i + \frac{1}{2}W^I_{ij}\chi^i\cdot\chi^j\Big]\,,\\
W_I(\Phi^\dagger) &= W_I + \sqrt{2}\bar\Theta\cdot(W_I^{i^*}\bar\chi_{i^*}) - \bar\Theta\cdot\bar\Theta\Big[W_I^{i^*}F^\dagger_{i^*} + \frac{1}{2}W_I^{i^*j^*}\bar\chi_{i^*}\cdot\bar\chi_{j^*}\Big]\,,
\end{aligned} \tag{6.4}$$

where we denote the scalar, fermionic and auxiliary components of the superfields Φ^i by ϕ^i, χ^i and F^i and those of the superfields $\Phi^\dagger_{i^*}$ by $\phi^\dagger_{i^*}, \bar\chi_{i^*}$ and $F^\dagger_{i^*}$, respectively. In (6.4), we have introduced the notations $W^I \equiv W^I(\phi)$ and $W_I \equiv W_I(\phi^\dagger)$, where the arguments of the W functions are the scalar components of the chiral superfields instead of the superfields themselves. Finally, the derivatives of W_I and W^I (as function of the scalar fields ϕ and $\phi^\dagger$) read

$$\begin{aligned}
W^I_i &= \frac{\partial W^I(\phi)}{\partial\phi^i}\,,\\
W^I_{ij} &= \frac{\partial^2 W^I(\phi)}{\partial\phi^i\partial\phi^j}\,,\\
W_I^{i^*} &= \frac{\partial W_I(\phi^\dagger)}{\partial\phi^\dagger_{i^*}}\,,\\
W_I^{i^*j^*} &= \frac{\partial^2 W_I(\phi^\dagger)}{\partial\phi^\dagger_{i^*}\partial\phi^\dagger_{j^*}}\,.
\end{aligned}$$

[1]This expansion is not always possible. Since at the very end of the computation the result does not depend on this expansion, it can formally be extended to any cases.

In order to calculate the Lagrangian (6.2), we need to evaluate $(\overline{\mathcal{D}}\cdot\overline{\mathcal{D}} - 8\mathcal{R})\Omega(\Phi,\Phi^\dagger e^{-2gV})$. However, since W^I are holomorphic functions which depend on the chiral superfields, the projector operator only acts on the functions W_I depending on the antichiral superfields, and

$$\begin{aligned}(\overline{\mathcal{D}}\cdot\overline{\mathcal{D}} - 8\mathcal{R})\Omega(\Phi,\Phi^\dagger e^{-2gV}) &= \Big[(\overline{\mathcal{D}}\cdot\overline{\mathcal{D}} - 8\mathcal{R})W_I(\Phi^\dagger e^{-2gV})\Big]W^I(\Phi)\\ &\equiv \mathcal{X}_I W^I\,. \end{aligned}\tag{6.5}$$

Moreover, in the Wess-Zumino gauge, all terms which are at least trilinear in V vanish (as in Sec. 4.3). Therefore $\mathcal{X}_I$, reduces to

$$\begin{aligned}\mathcal{X}_I &= (\overline{\mathcal{D}}\cdot\overline{\mathcal{D}} - 8\mathcal{R})W_I(\Phi^\dagger e^{-2gV})\\ &= (\overline{\mathcal{D}}\cdot\overline{\mathcal{D}} - 8\mathcal{R})\Big[W_I(\Phi^\dagger)\\ &\qquad + \frac{\partial W_I(\Phi^\dagger)}{\partial\Phi^\dagger_{i^*}}\Big(-2g(\Phi^\dagger V)_{i^*} + 2g^2(\Phi^\dagger V^2)_{i^*}\Big)\\ &\qquad + 2g^2\frac{\partial^2 W_I(\Phi^\dagger)}{\partial\Phi^\dagger_{i^*}\partial\Phi^\dagger_{j^*}}(\Phi^\dagger V)_{i^*}(\Phi^\dagger V)_{j^*}\Big]\,,\end{aligned}\tag{6.6}$$

where the first and second-order derivatives of $W_I(\Phi^\dagger)$ are given by

$$\begin{aligned}\frac{\partial W_I(\Phi^\dagger)}{\partial\Phi^\dagger_{i^*}} &= W_I^{i^*} + \sqrt{2}\bar\Theta\cdot(W_I^{i^*j^*}\bar\chi_{j^*}) - \bar\Theta\cdot\bar\Theta\Big[W_I^{i^*j^*}F^\dagger_{j^*} + \frac{1}{2}W_I^{i^*j^*k^*}\bar\chi_{j^*}\cdot\bar\chi_{k^*}\Big]\,,\\ \frac{\partial^2 W_I(\Phi^\dagger)}{\partial\Phi^\dagger_{i^*}\partial\Phi^\dagger_{j^*}} &= W_I^{i^*j^*} + \sqrt{2}\bar\Theta\cdot(W_I^{i^*j^*k^*}\bar\chi_{k^*})\\ &\quad - \bar\Theta\cdot\bar\Theta\Big[W_I^{i^*j^*k^*}F^\dagger_{k^*} + \frac{1}{2}W_I^{i^*j^*k^*\ell^*}\bar\chi_{k^*}\cdot\bar\chi_{\ell^*}\Big]\,,\end{aligned}\tag{6.7}$$

with the following shortland notations for the third-order and fourth-order derivatives of the function W_I,

$$\begin{aligned}W_I^{i^*j^*k^*} &= \frac{\partial^3 W_I(\phi^\dagger)}{\partial\phi^\dagger_{i^*}\partial\phi^\dagger_{j^*}\partial\phi^\dagger_{k^*}}\,,\\ W_I^{i^*j^*k^*\ell^*} &= \frac{\partial^4 W_I(\phi^\dagger)}{\partial\phi^\dagger_{i^*}\partial\phi^\dagger_{j^*}\partial\phi^\dagger_{k^*}\partial\phi^\dagger_{\ell^*}}\,.\end{aligned}$$

The expansion of (6.6) in terms of the hybrid Grassmann variables can be calculated in several steps. We first start with the computation of the components of the superfield $\Xi_I = (\overline{\mathcal{D}}\cdot\overline{\mathcal{D}} - 8\mathcal{R})W_I(\Phi^\dagger)$. They are derived from those of the superfield

Ξ given in Table 5.1, after replacing, in the expansion of Ξ, the components of $\Phi^\dagger$ by those of the antichiral superfield W_I given in (6.4),

$$
\begin{aligned}
\Xi_I = & \left[4W_I^{i^*} F^\dagger_{i^*} + 2W_I^{i^* j^*} \bar{\chi}_{i^*} \cdot \bar{\chi}_{j^*} + \frac{4}{3} M W_I\right] \\
& + \Theta \cdot \Big\{ -4i\sqrt{2} W_I^{i^*} \sigma^\mu \hat{D}_\mu \bar{\chi}_{i^*} - 4i\sqrt{2} W_I^{i^* j^*} \sigma^\mu \bar{\chi}_{i^*} \partial_\mu \phi^\dagger_{j^*} + \frac{2\sqrt{2}}{3} W_I^{i^*} \sigma^\mu \bar{\chi}_{i^*} b_\mu \\
& \quad -2i W_I^{i^* j^*} \sigma^\mu \bar{\psi}_\mu \bar{\chi}_{i^*} \cdot \bar{\chi}_{j^*} + W_I \Big(\frac{8}{3} \sigma^{\mu\nu} \psi_{\mu\nu} - \frac{4}{3} i M \sigma^\mu \bar{\psi}_\mu + \frac{4}{3} i \psi^\mu b_\mu \Big) \Big\} \\
& - \Theta \cdot \Theta \Big\{ -4 e_\mu{}^{\tilde{\mu}} W_I^{i^*} \mathcal{D}_{\tilde{\mu}} \hat{D}^\mu \phi^\dagger_{i^*} + \frac{8i}{3} W_I^{i^*} b^\mu \hat{D}_\mu \phi^\dagger_{i^*} - 4 e_\mu{}^{\tilde{\mu}} W_I^{i^* j^*} \partial_{\tilde{\mu}} \phi^\dagger_{j^*} \hat{D}^\mu \phi^\dagger_{i^*} \\
& \quad + 2\sqrt{2} W_I^{i^*} (\bar{\psi}_\mu \cdot \hat{D}^\mu \bar{\chi}_{i^*}) + 2\sqrt{2} W_I^{i^* j^*} (\bar{\psi}^\mu \cdot \bar{\chi}_{i^*}) \partial_\mu \phi^\dagger_{j^*} + W_I^{i^* j^*} \bar{\psi}_\mu \cdot \bar{\psi}^\mu \bar{\chi}_{i^*} \cdot \bar{\chi}_{j^*} \\
& \quad - \frac{2\sqrt{2}}{3} W_I^{i^*} (\bar{\chi}_{i^*} \bar{\sigma}^{\mu\nu} \bar{\psi}_{\mu\nu}) - \frac{8}{3} M^* W_I^{i^*} F^\dagger_{i^*} - \frac{4}{3} M^* W_I^{i^* j^*} (\bar{\chi}_{i^*} \cdot \bar{\chi}_{j^*}) \\
& \quad + \sqrt{2} i W_I^{i^*} (\bar{\chi}_{i^*} \cdot \bar{\psi}_\mu) b^\mu - \frac{2\sqrt{2} i}{3} W_I^{i^*} (\bar{\chi}_{i^*} \bar{\sigma}^{\mu\nu} \bar{\psi}_\mu) b_\nu \\
& \quad + W_I \Big(- \frac{2}{3} e_\mu{}^{\tilde{\mu}} e_\nu{}^{\tilde{\nu}} R_{\tilde{\mu}\tilde{\nu}}{}^{\mu\nu} \Big| - \frac{8}{9} M M^* + \frac{4}{9} b_\mu b^\mu \\
& \qquad + \frac{4i}{3} e_\mu{}^{\tilde{\mu}} \mathcal{D}_{\tilde{\mu}} b^\mu + \frac{2}{3} \bar{\psi}_\mu \cdot \bar{\psi}^\mu M + \frac{2}{3} \psi_\nu \sigma^\nu \bar{\psi}_\mu b^\mu + \frac{4i}{3} \bar{\psi}^\mu \bar{\sigma}^\nu \psi_{\mu\nu} \\
& \qquad + \frac{1}{6} \varepsilon^{\mu\nu\rho\sigma} (\psi_\mu \sigma_\sigma \bar{\psi}_{\nu\rho} + \bar{\psi}_\mu \bar{\sigma}_\sigma \psi_{\nu\rho}) \Big) \Big\} .
\end{aligned}
$$

We now turn to the computation of the terms linear in g included in (6.6), which we denote by $\mathcal{X}_I^{(g)}$. Since

$$
\mathcal{X}_I^{(g)} = -2g\big(\bar{\mathcal{D}} \cdot \bar{\mathcal{D}} - 8\mathcal{R}\big) \Big[\frac{\partial W_I(\Phi^\dagger)}{\partial \Phi^\dagger_{i^*}} \Phi^\dagger_{j^*} \Big] V^a (T_a)^{j^*}{}_{i^*} ,
$$

the components of $\mathcal{X}_I^{(g)}$ can be deduced from those of the superfield $\mathcal{X}$ introduced in (4.71). Replacing the components of the superfield $\Phi^\dagger$ given in Table 5.1 by those of the superfields

$$
\begin{aligned}
\frac{\partial W_I(\Phi^\dagger)}{\partial \Phi^\dagger_{i^*}} \Phi^\dagger_{j^*} = & W_I^{i^*} \phi^\dagger_{j^*} + \sqrt{2} \bar{\Theta} \cdot \Big[W_I^{i^* k^*} \bar{\chi}_{k^*} \phi^\dagger_{j^*} + W_I^{i^*} \bar{\chi}_{j^*} \Big] \\
& - \bar{\Theta} \cdot \bar{\Theta} \Big[W_I^{i^*} F^\dagger_{j^*} + W_I^{i^* k^*} F^\dagger_{k^*} \phi^\dagger_{j^*} + \frac{1}{2} W_I^{i^* k^* \ell^*} \bar{\chi}_{k^*} \cdot \bar{\chi}_{\ell^*} \phi^\dagger_{j^*} + W_I^{i^* k^*} \bar{\chi}_{k^*} \cdot \bar{\chi}_{j^*} \Big] ,
\end{aligned}
$$

which can be deduced from the conjugate equations to (2.4) and (6.7), and, keeping only the terms linear in g, we get

$$\mathcal{X}_I^{(g)} = \Theta\cdot\Big[-4\sqrt{2}gW_I^{i^*k^*}(\sigma^\mu\bar{\chi}_{k^*})(\phi^\dagger T_a)_{i^*}v_\mu^a - 4\sqrt{2}gW_I^{i^*}(\sigma^\mu\bar{\chi}T_a)_{i^*}v_\mu^a$$
$$+ 8igW_I^{i^*}(\phi^\dagger T_a)_{i^*}\lambda^a + 4igW_I^{i^*}(\phi^\dagger T_a)_{i^*}v_\nu^a(\sigma^\mu\bar{\sigma}^\nu\psi_\mu)\Big]$$
$$-\Theta\cdot\Theta\Big[\ 8igW_I^{i^*k^*}\hat{D}^\mu\phi_{k^*}^\dagger(\phi^\dagger T_a)_{i^*}v_\mu^a + 8igW_I^{i^*}(\hat{D}^\mu\phi^\dagger T_a)_{i^*}v_\mu^a$$
$$-4\sqrt{2}igW_I^{i^*}(\bar{\chi}T_a)_{i^*}\cdot\bar{\lambda}^a - 4\sqrt{2}igW_I^{i^*k^*}(\bar{\chi}_{k^*}\cdot\bar{\lambda}^a)(\phi^\dagger T_a)_{i^*}$$
$$+ 4igW_I^{i^*}(\phi^\dagger T_a)_{i^*}\hat{D}_\mu v^{\mu a} + \frac{8}{3}gW_I^{i^*}(\phi^\dagger T_a)_{i^*}v_\mu^a b^\mu - 4gD^aW_I^{i^*}(\phi^\dagger T_a)_{i^*}\Big].$$

Finally, the two terms proportional to g^2 included in (6.6), which we label as $\mathcal{X}_I^{(g^2)}$,

$$\mathcal{X}_I^{(g^2)} = 2g^2\big(\bar{\mathcal{D}}\cdot\bar{\mathcal{D}} - 8\mathcal{R}\big)\Big[\frac{\partial W_I(\Phi^\dagger)}{\partial\Phi_{i^*}^\dagger}\Phi_{j^*}^\dagger(T_aT_b)^{j^*}{}_{i^*}V^aV^b$$
$$+\frac{\partial^2 W_I}{\partial\Phi_{i^*}^\dagger\partial\Phi_{j^*}^\dagger}\Phi_{\ell^*}^\dagger\Phi_{k^*}^\dagger(T_a)^{\ell^*}{}_{i^*}(T_b)^{k^*}{}_{j^*}V^aV^b\Big],$$

are derived in a similar fashion. Starting from the results for the g^2 components of the superfield $\mathcal{X}$ presented in Table 5.1, we replace the zeroth-order component of the superfield $\Phi^\dagger$ by

$$\frac{\partial W_I(\Phi^\dagger)}{\partial\Phi_{i^*}^\dagger}\Phi_{j^*}^\dagger\Big| = W_I^{i^*}\phi_{j^*}^\dagger\,,$$
$$\frac{\partial^2 W_I(\Phi^\dagger)}{\partial\Phi_{i^*}^\dagger\partial\Phi_{j^*}^\dagger}\Phi_{k^*}^\dagger\Phi_{\ell^*}^\dagger\Big| = W^{i^*j^*}\phi_{k^*}^\dagger\phi_{\ell^*}^\dagger\,,$$

for the first and second contributions to $\mathcal{X}_I^{(g^2)}$, respectively. This leads to

$$\mathcal{X}_I^{(g^2)} = -\Theta\cdot\Theta\Big[4g^2W_I^{i^*}(\phi^\dagger T_aT_b)_{i^*}v_\mu^a v^{\mu b} + 4g^2W_I^{i^*j^*}(\phi^\dagger T_a)_{i^*}(\phi^\dagger T_b)_{j^*}v_\mu^a v^{\mu b}\Big].$$

Collecting all the contributions, we obtain

$$
\begin{aligned}
\mathcal{X}_I &= \Big(\overline{\mathcal{D}}\cdot\overline{\mathcal{D}} - 8\mathcal{R}\Big)W_I(\Phi^\dagger e^{-2gV}) = \Xi_I + \mathcal{X}_I^{(g)} + \mathcal{X}_I^{(g^2)}\\
&= 4\Big[\Big[W_I^{i^*}F^\dagger_{i^*} + \frac{1}{2}W_I^{i^*j^*}\bar\chi_{i^*}\cdot\bar\chi_{j^*}\Big] + \frac{1}{3}MW_I\Big]\\
&\quad+\Theta\cdot\Big[-4i\sqrt{2}W_I^{i^*}(\sigma^\mu\hat D_\mu\bar\chi_{i^*}) - 4i\sqrt{2}W_I^{i^*j^*}(\sigma^\mu\bar\chi_{i^*})\partial_\mu\phi^\dagger_{j^*} - 2iW_I^{i^*j^*}\sigma^\mu\bar\psi_\mu\bar\chi_{i^*}\cdot\bar\chi_{j^*}\\
&\qquad+\frac{2\sqrt{2}}{3}W_I^{i^*}(\sigma^\mu\bar\chi_{i^*})b_\mu + W_I\Big[\frac{8}{3}(\sigma^{\mu\nu}\psi_{\mu\nu}) - \frac{4}{3}iM(\sigma^\mu\bar\psi_\mu) + \frac{4}{3}i\psi^\mu b_\mu\Big]\\
&\qquad-4\sqrt{2}gW_I^{i^*k^*}(\sigma^\mu\bar\chi_{k^*})(\phi^\dagger T_a)_{i^*}v^a_\mu - 4\sqrt{2}gW_I^{i^*}(\sigma^\mu\bar\chi T_a)_{i^*}v^a_\mu\\
&\qquad+8igW_I^{i^*}(\phi^\dagger T_a)_{i^*}\lambda^a + 4igW_I^{i^*}(\phi^\dagger T_a)_{i^*}v^a_\nu(\sigma^\mu\bar\sigma^\nu\psi_\mu)\Big]\\
&\quad-\Theta\cdot\Theta\Big[-4e_\mu{}^{\tilde\mu}W_I^{i^*}\mathcal{D}_{\tilde\mu}\hat D^\mu\phi^\dagger_{i^*} - 4e_\mu{}^{\tilde\mu}W_I^{i^*j^*}\partial_{\tilde\mu}\phi^\dagger_{j^*}\hat D^\mu\phi^\dagger_{i^*} + \frac{8}{3}iW_I^{i^*}\hat D_\mu\phi^\dagger_{i^*}b^\mu\\
&\qquad+2\sqrt{2}W_I^{i^*}(\bar\psi_\mu\cdot\hat D^\mu\bar\chi_{i^*}) + 2\sqrt{2}W_I^{i^*j^*}(\bar\psi^\mu\cdot\bar\chi_{i^*})\partial_\mu\phi^\dagger_{j^*} - \frac{2\sqrt{2}}{3}W_I^{i^*}(\bar\chi_{i^*}\bar\sigma^{\mu\nu}\psi_{\mu\nu})\\
&\qquad+W_I^{i^*j^*}\bar\psi_\mu\cdot\bar\psi^\mu\bar\chi_{i^*}\cdot\bar\chi_{j^*} - \frac{8}{3}M^*\Big[W_I^{i^*}F^\dagger_{i^*} + \frac{1}{2}W_I^{i^*j^*}(\bar\chi_{i^*}\cdot\bar\chi_{j^*})\Big]\\
&\qquad+\sqrt{2}iW_I^{i^*}(\bar\chi_{i^*}\cdot\bar\psi_\mu)b^\mu - \frac{2\sqrt{2}}{3}iW_I^{i^*}(\bar\chi_{i^*}\bar\sigma^{\mu\nu}\bar\psi_\mu)b_\nu\\
&\qquad+W_I\Big[-\frac{2}{3}e_\mu{}^{\tilde\mu}e_\nu{}^{\tilde\nu}R_{\tilde\mu\tilde\nu}{}^{\mu\nu}\Big| - \frac{8}{9}MM^* + \frac{4}{9}b_\mu b^\mu\\
&\qquad\quad+\frac{4i}{3}e_\mu{}^{\tilde\mu}\mathcal{D}_{\tilde\mu}b^\mu + \frac{2}{3}\bar\psi_\mu\cdot\bar\psi^\mu M + \frac{2}{3}\psi_\nu\sigma^\nu\bar\psi_\mu b^\mu + \frac{4i}{3}\bar\psi^\mu\bar\sigma^\nu\psi_{\mu\nu}\\
&\qquad\quad+\frac{1}{6}\varepsilon^{\mu\nu\rho\sigma}(\psi_\mu\sigma_\sigma\bar\psi_{\nu\rho} + \bar\psi_\mu\bar\sigma_\sigma\psi_{\nu\rho})\Big]\\
&\qquad+8igW_I^{i^*k^*}\hat D^\mu\phi^\dagger_{k^*}(\phi^\dagger T_a)_{i^*}v^a_\mu + 8igW_I^{i^*}\big(\hat D^\mu\phi^\dagger T_a\big)_{i^*}v^a_\mu\\
&\qquad-4\sqrt{2}igW_I^{i^*k^*}(\bar\chi_{k^*}\cdot\bar\lambda^a)(\phi^\dagger T_a)_{i^*} - 4\sqrt{2}igW_I^{i^*}(\bar\chi T_a)_{i^*}\cdot\bar\lambda^a\\
&\qquad+4igW_I^{i^*}(\phi^\dagger T_a)_{i^*}\hat D_\mu v^{\mu a} + \frac{8}{3}gW_I^{i^*}(\phi^\dagger T_a)_{i^*}v^a_\mu b^\mu - 4gD^aW_I^{i^*}(\phi^\dagger T_a)_{i^*}\\
&\qquad+4g^2W_I^{i^*}(\phi^\dagger T_aT_b)_{i^*}v^a_\mu v^{\mu b} + 4g^2W_I^{i^*j^*}(\phi^\dagger T_a)_{i^*}(\phi^\dagger T_b)_{j^*}v^a_\mu v^{\mu b}\Big]\,.
\end{aligned}
$$

Using this last result as well as the expression of $W^I(\Phi)$ in (6.4), we derive the action of the projector on the superspace kinetic energy, as defined in (6.5), to get

$$
\mathcal{X}_I W^I = (\overline{\mathcal{D}}\cdot\overline{\mathcal{D}} - 8\mathcal{R})\Omega(\Phi,\Phi^\dagger e^{-2gV})\,.
$$

Introducing notations similar to those of Sec. 2.4, such as $\Omega_i = W_I W^I_i$ or $\Omega^{i^*}{}_i = W_I^{i^*}W^I_i$, the product of the two chiral superfields $\mathcal{X}_I$ and W^I is obtained from (2.4) and reads

$$
\begin{aligned}
\mathcal{X}_I W^I = &\Big[4\Big(\Omega^{i^*} F^\dagger_{i^*} + \frac{1}{2}\Omega^{i^* j^*}\bar{\chi}_{i^*}\cdot\bar{\chi}_{j^*}\Big) + \frac{4}{3}\Omega M\Big]\\
&+\Theta\cdot\Big[-2i\Omega^{i^* j^*}\sigma^\mu\bar{\psi}_\mu\bar{\chi}_{i^*}\cdot\bar{\chi}_{j^*}\\
&\quad -4i\sqrt{2}\Omega^{i^*}\sigma^\mu\Big(\hat{D}_\mu\bar{\chi}_{i^*} - ig(\bar{\chi}T_a)_{i^*}v^a_\mu\Big) + \frac{2\sqrt{2}}{3}\Omega^{i^*}\sigma^\mu\bar{\chi}_{i^*}b_\mu\\
&\quad -4i\sqrt{2}\Omega^{i^* j^*}\sigma^\mu\bar{\chi}_{i^*}\Big(\partial_\mu\phi^\dagger_{j^*} - ig(\phi^\dagger T_a)_{j^*}v^a_\mu\Big) + \Omega\Big(\frac{8}{3}\sigma^{\mu\nu}\psi_{\mu\nu} - \frac{4}{3}Mi\sigma^\mu\bar{\psi}_\mu\\
&\quad +\frac{4}{3}i\psi^\mu b_\mu\Big) + 8ig\Omega^{i^*}(\phi^\dagger T_a)_{i^*}\lambda^a + 4ig\Omega^{i^*}(\phi^\dagger T_a)_{i^*}v^a_\nu\sigma^\mu\bar{\sigma}^\nu\psi_\mu\\
&\quad +4\sqrt{2}\Big(\Omega^{i^*}{}_i F^\dagger_{i^*}\chi^i + \frac{1}{2}\Omega^{i^*}{}_i{}^{j^*}(\bar{\chi}_{i^*}\cdot\bar{\chi}_{j^*})\chi^i\Big) + \frac{4\sqrt{2}}{3}\Omega_i M\chi^i\Big]\\
&-\Theta\cdot\Theta\Big[4\Big(\Omega^{i^*}{}_i F^\dagger_{i^*}F^i + \frac{1}{2}\Omega^{i^*}{}_i{}^{j^*}F^i\bar{\chi}_{i^*}\cdot\bar{\chi}_{j^*} + \frac{1}{2}\Omega_i{}^{i^*}{}_j F^\dagger_{i^*}\chi^i\cdot\chi^j\\
&\quad +\frac{1}{4}\Omega^{i^* j^*}{}_{ij}\bar{\chi}_{i^*}\cdot\bar{\chi}_{j^*}\chi^i\cdot\chi^j\Big) + \frac{4}{3}\Omega_i M F^i + \frac{2}{3}\Omega_{ij}M\chi^i\cdot\chi^j\\
&\quad -4i\Omega^{i^*}{}_i\chi^i\sigma^\mu\Big(\hat{D}_\mu\bar{\chi}_{i^*} - ig(\bar{\chi}T_a)_{i^*}v^a_\mu\Big) - \sqrt{2}i\Omega^{i^*}{}_i{}^{j^*}\chi^i\sigma^\mu\bar{\psi}_\mu\bar{\chi}_{i^*}\cdot\bar{\chi}_{j^*}\\
&\quad -4i\Omega^{i^*}{}_i{}^{j^*}\chi^i\sigma^\mu\bar{\chi}_{i^*}\Big(\partial_\mu\phi^\dagger_{j^*} - ig(\phi^\dagger T_a)_{j^*}v^a_\mu\Big)\\
&\quad +\Omega_i\Big(\frac{4\sqrt{2}}{3}\chi^i\sigma^{\mu\nu}\psi_{\mu\nu} - i\frac{2\sqrt{2}}{3}M\chi^i\sigma^\mu\bar{\psi}_\mu + \frac{2\sqrt{2}}{3}i\chi^i\cdot\psi_\mu b^\mu\Big)\\
&\quad +\frac{2}{3}\Omega^{i^*}{}_i\chi^i\sigma^\mu\bar{\chi}_{i^*}b_\mu + 4\sqrt{2}ig\Omega^{i^*}{}_i(\phi^\dagger T_a)_{i^*}\lambda^a\cdot\chi^i - 4e_\mu{}^{\tilde{\mu}}\Omega^{i^*}\mathcal{D}_{\tilde{\mu}}\hat{D}^\mu\phi^\dagger_{i^*}\\
&\quad +2\sqrt{2}ig\Omega^{i^*}{}_i(\phi^\dagger T_a)_{i^*}\chi^i\sigma^\mu\bar{\sigma}^\nu\psi_\mu v^a_\nu - 4e_\mu{}^{\tilde{\mu}}\Omega^{i^* j^*}\partial_{\tilde{\mu}}\phi^\dagger_{j^*}\hat{D}^\mu\phi^\dagger_{i^*}\\
&\quad +8ig\Omega^{i^*}(\hat{D}^\mu\phi^\dagger T_a)_{i^*}v^a_\mu + 8ig\Omega^{i^* j^*}\hat{D}^\mu\phi^\dagger_{j^*}(\phi^\dagger T_a)_{i^*}v^a_\mu + \Omega^{i^* j^*}\bar{\psi}_\mu\cdot\bar{\psi}^\mu\bar{\chi}_{i^*}\cdot\bar{\chi}_{j^*}\\
&\quad +\frac{8}{3}i\Omega^{i^*}\Big(\hat{D}_\mu\phi^\dagger_{i^*} - ig(\phi^\dagger T_a)_{i^*}v^a_\mu\Big)b^\mu + 2\sqrt{2}\Omega^{i^*}\bar{\psi}_\mu\cdot\hat{D}^\mu\bar{\chi}_{i^*} + 2\sqrt{2}\Omega^{i^* j^*}\bar{\psi}^\mu\cdot\bar{\chi}_{i^*}\partial_\mu\phi^\dagger_{j^*}\\
&\quad +4ig\Omega^{i^*}(\phi^\dagger T_a)_{i^*}\hat{D}_\mu v^{\mu a} - \frac{2\sqrt{2}}{3}\Omega^{i^*}\bar{\chi}_{i^*}\bar{\sigma}^{\mu\nu}\bar{\psi}_{\mu\nu} - \frac{8}{3}M^*\Big(\Omega^{i^*}F^\dagger_{i^*} + \frac{1}{2}\Omega^{i^* j^*}\bar{\chi}_{i^*}\cdot\bar{\chi}_{j^*}\Big)\\
&\quad +\sqrt{2}i\Omega^{i^*}\bar{\chi}_{i^*}\cdot\bar{\psi}_\mu b^\mu - \frac{2\sqrt{2}i}{3}\Omega^{i^*}\bar{\chi}_{i^*}\bar{\sigma}^{\mu\nu}\bar{\psi}_\mu b_\nu + \Omega\Big\{-\frac{2}{3}e_\mu{}^{\tilde{\mu}}e_\nu{}^{\tilde{\nu}}R_{\tilde{\mu}\tilde{\nu}}{}^{\mu\nu}\Big| - \frac{8}{9}MM^*\\
&\quad +\frac{4}{9}b_\mu b^\mu + \frac{4i}{3}e_\mu{}^{\tilde{\mu}}\mathcal{D}_{\tilde{\mu}}b^\mu + \frac{2}{3}\bar{\psi}_\mu\cdot\bar{\psi}^\mu M + \frac{2}{3}\psi_\nu\sigma^\nu\bar{\psi}_\mu b^\mu + \frac{4i}{3}\bar{\psi}^\mu\bar{\sigma}^\nu\psi_{\mu\nu}\\
&\quad +\frac{1}{6}\varepsilon^{\mu\nu\rho\sigma}(\psi_\mu\sigma_\sigma\bar{\psi}_{\nu\rho} + \bar{\psi}_\mu\bar{\sigma}_\sigma\psi_{\nu\rho})\Big\} - 4gD^a\Omega^{i^*}(\phi^\dagger T_a)_{i^*}\\
&\quad -4\sqrt{2}ig\Omega^{i^* j^*}\bar{\chi}_{j^*}\cdot\bar{\lambda}^a(\phi^\dagger T_a)_{i^*} - 4\sqrt{2}ig\Omega^{i^*}(\bar{\chi}T_a)_{i^*}\cdot\bar{\lambda}^a\\
&\quad +4g^2\Omega^{i^*}(\phi^\dagger T_a T_b)_{i^*}v^a_\mu v^{\mu b} + 4g^2\Omega^{i^* j^*}(\phi^\dagger T_a)_{i^*}(\phi^\dagger T_b)_{j^*}v^a_\mu v^{\mu b}\Big]. \qquad (6.8)
\end{aligned}
$$

This expression can be simplified however after introducing fully covariant derivatives, *i.e.*, derivatives covariant at the same time with respect to supergravity, gauge symmetry and the Kähler geometry,

$$\begin{aligned}
\widetilde{\mathcal{D}}_\mu \phi^i &= e_\mu{}^{\tilde{\mu}} \partial_{\tilde{\mu}} \phi^i + ig(T_a\phi)^i v_\mu^a \,, \\
\widetilde{\mathcal{D}}_\mu \phi^\dagger_{i^*} &= e_\mu{}^{\tilde{\mu}} \partial_{\tilde{\mu}} \phi^\dagger_{i^*} - ig(\phi^\dagger T_a)_{i^*} v_\mu^a \,, \\
\widetilde{\mathcal{D}}_\mu \chi^i &= e_\mu{}^{\tilde{\mu}} \mathcal{D}_{\tilde{\mu}} \chi^i + ig(T_a\chi)^i v_\mu^a + \Gamma_j{}^i{}_k \widetilde{\mathcal{D}}_\mu \phi^j \chi^k \,, \\
\widetilde{\mathcal{D}}_\mu \bar{\chi}_{i^*} &= e_\mu{}^{\tilde{\mu}} \mathcal{D}_{\tilde{\mu}} \bar{\chi}_{i^*} - ig(\bar{\chi} T_a)_{i^*} v_\mu^a + \Gamma^{j^*}{}_{i^*}{}^{k^*} \widetilde{\mathcal{D}}_\mu \phi^\dagger_{j^*} \bar{\chi}_{k^*} \,, \\
\mathcal{D}_i \Omega_j &= \Omega_{ij} - \Gamma_i{}^k{}_j \Omega_k \,, \\
\mathcal{D}^{i^*} \Omega^{j^*} &= \Omega^{i^* j^*} - \Gamma^{i^*}{}_{k^*}{}^{j^*} \Omega^{k^*} \,,
\end{aligned} \tag{6.9}$$

which are the only natural geometrical objects of the theory. The hatted derivatives $\hat{D}_\mu \phi^\dagger_{i^*}$ and $\hat{D}_\mu \bar{\chi}_{i^*}$ given in Table 4.1 can then be rewritten under the form

$$\begin{aligned}
\hat{D}_\mu \phi^\dagger_{i^*} &= \widetilde{\mathcal{D}}_\mu \phi^\dagger_{i^*} - \frac{\sqrt{2}}{2} \bar{\psi}_\mu \cdot \bar{\chi}_{i^*} + ig(\phi^\dagger T_a)_{i^*} v_\mu^a \,, \\
\hat{D}_\mu \bar{\chi}_{i^*} &= \widetilde{\mathcal{D}}_\mu \bar{\chi}_{i^*} + ig(\bar{\chi} T_a)_{i^*} v_\mu^a - \Gamma^{j^*}{}_{i^*}{}^{k^*} \widetilde{\mathcal{D}}_\mu \phi^\dagger_{j^*} \bar{\chi}_{k^*} + \frac{1}{\sqrt{2}} \bar{\psi}_\mu F^\dagger_{i^*} + \frac{i}{\sqrt{2}} \psi_\mu \sigma^\nu \widetilde{\mathcal{D}}_\nu \phi^\dagger_{i^*} \\
&\quad - \frac{g}{\sqrt{2}} \psi_\mu \sigma^\nu (\phi^\dagger T_a)_{i^*} v_\nu^a - \frac{i}{2} (\psi_\mu \sigma^\nu) \bar{\psi}_\nu \cdot \bar{\chi}_{i^*} \,,
\end{aligned}$$

which allows to simplify various terms in (6.8). Indeed, we can collect the terms labelled by $\widetilde{\Theta}$ included in the Θ-component of the superfield $X_I W^I$ under,

$$\begin{aligned}
\widetilde{\Theta}_{(1)} \equiv &- 4i\sqrt{2} \Omega^{i^*} \sigma^\mu \Big(\hat{D}_\mu \bar{\chi}_{i^*} - ig(\bar{\chi} T_a)_{i^*} v_\mu^a \Big) \\
= &- 4i\sqrt{2} \Omega^{i^*} \sigma^\mu \widetilde{\mathcal{D}}_\mu \bar{\chi}_{i^*} - 4i \Omega^{i^*} \sigma^\mu \bar{\psi}_\mu F^\dagger_{i^*} + 4i\sqrt{2} \Big[\Omega^{i^* j^*} - \mathcal{D}^{i^*} \Omega^{j^*} \Big] \sigma^\mu \bar{\chi}_{j^*} \widetilde{\mathcal{D}}_\mu \phi^\dagger_{i^*} \\
&- 4 \Omega^{i^*} (\sigma^\mu \bar{\sigma}^\nu \psi_\mu) \widetilde{\mathcal{D}}_\nu \phi^\dagger_{i^*} - 4ig \Omega^{i^*} (\sigma^\mu \bar{\sigma}^\nu \psi_\mu) (\phi^\dagger T_a)_{i^*} v_\nu^a \\
&+ 2\sqrt{2} \Omega^{i^*} (\sigma^\mu \bar{\sigma}^\nu \psi_\mu) \bar{\psi}_\nu \cdot \bar{\chi}_{i^*} \,, \\
\widetilde{\Theta}_{(2)} \equiv &- 4i\sqrt{2} \Omega^{i^* j^*} \sigma^\mu \bar{\chi}_{i^*} \bigg(\partial_\mu \phi^\dagger_{j^*} - ig(\phi^\dagger T_a)_{j^*} v_\mu^a \bigg) = -4i\sqrt{2} \Omega^{i^* j^*} \sigma^\mu \bar{\chi}_{i^*} \widetilde{\mathcal{D}}_\mu \phi^\dagger_{j^*} \,, \\
\widetilde{\Theta}_{(3)} \equiv &\ 4ig \Omega^{i^*} (\phi^\dagger T_a)_{i^*} v_\nu^a \sigma^\mu \bar{\sigma}^\nu \psi_\mu \,.
\end{aligned}$$

Their sum is given by

$$\begin{aligned}
\widetilde{\Theta} &= \widetilde{\Theta}_{(1)} + \widetilde{\Theta}_{(2)} + \widetilde{\Theta}_{(3)} \\
&= -4\sqrt{2} i \sigma^\mu \Big[\Omega^{i^*} \widetilde{\mathcal{D}}_\mu \bar{\chi}_{i^*} + \mathcal{D}^{i^*} \Omega^{j^*} \bar{\chi}_{i^*} \widetilde{\mathcal{D}}_\mu \phi^\dagger_{j^*} \Big] - 4i \Omega^{i^*} \sigma^\mu \bar{\psi}_\mu F^\dagger_{i^*} \\
&\quad - 4 \Omega^{i^*} (\sigma^\mu \bar{\sigma}^\nu \psi_\mu) \Big[\widetilde{\mathcal{D}}_\nu \phi^\dagger_{i^*} - \frac{1}{\sqrt{2}} \bar{\psi}_\nu \cdot \bar{\chi}_{i^*} \Big] \,.
\end{aligned}$$

Hence, even if the three contributions above are apparently not covariant, the missing terms are compensating among themselves, rendering the total result fully covariant.

In the same way we now address several terms (which we denote by $\widetilde{\Theta}^2$) included in the $\Theta\cdot\Theta$-component of the quantity $X_I W^I$ in (6.8), namely

$$\begin{aligned}
\widetilde{\Theta}^2_{(1)} &= -4e_\mu{}^{\tilde{\mu}}\Omega^{i^*}\mathcal{D}_{\tilde{\mu}}\hat{\mathcal{D}}^\mu\phi^\dagger_{i^*}\\
&= -4e_\mu{}^{\tilde{\mu}}\mathcal{D}_{\tilde{\mu}}\big(\Omega^{i^*}\widetilde{\mathcal{D}}^\mu\phi^\dagger_{i^*}\big)+4e_\mu{}^{\tilde{\mu}}\Omega^{i^*j^*}\partial_{\tilde{\mu}}\phi^\dagger_{j^*}\widetilde{\mathcal{D}}^\mu\phi^\dagger_{i^*}+4e_\mu{}^{\tilde{\mu}}\Omega^{i^*}{}_i\partial_{\tilde{\mu}}\phi^i\widetilde{\mathcal{D}}^\mu\phi^\dagger_{i^*}\\
&\quad-4ige_\mu{}^{\tilde{\mu}}\Omega^{i^*}\partial_{\tilde{\mu}}(\phi^\dagger T_a)_{i^*}v^{\mu a}-4ige_\mu{}^{\tilde{\mu}}\Omega^{i^*}(\phi^\dagger T_a)_{i^*}\mathcal{D}_{\tilde{\mu}}v^{\mu a}\\
&\quad+2\sqrt{2}e_\mu{}^{\tilde{\mu}}\Omega^{i^*}\mathcal{D}_{\tilde{\mu}}\bar{\psi}^\mu\cdot\bar{\chi}_{i^*}+2\sqrt{2}e_\mu{}^{\tilde{\mu}}\Omega^{i^*}\bar{\psi}^\mu\cdot\mathcal{D}_{\tilde{\mu}}\bar{\chi}_{i^*}\,,\\
\widetilde{\Theta}^2_{(2)} &= -4e_\mu{}^{\tilde{\mu}}\Omega^{i^*j^*}\partial_{\tilde{\mu}}\phi^\dagger_{j^*}\hat{\mathcal{D}}^\mu\phi^\dagger_{i^*}\\
&= -4e_\mu{}^{\tilde{\mu}}\Omega^{i^*j^*}\partial_{\tilde{\mu}}\phi^\dagger_{j^*}\widetilde{\mathcal{D}}^\mu\phi^\dagger_{i^*}+2\sqrt{2}e_\mu{}^{\tilde{\mu}}\Omega^{i^*j^*}\partial_{\tilde{\mu}}\phi^\dagger_{j^*}\bar{\psi}^\mu\cdot\bar{\chi}_{i^*}\\
&\quad-4ige_\mu{}^{\tilde{\mu}}\Omega^{i^*j^*}\partial_{\tilde{\mu}}\phi^\dagger_{j^*}(\phi^\dagger T_a)_{i^*}v^{\mu a}\,,\\
\widetilde{\Theta}^2_{(3)} &= 8ig\Omega^{i^*}(\hat{\mathcal{D}}^\mu\phi^\dagger T_a)_{i^*}v^a_\mu\\
&= 4ig\Omega^{i^*}(\widetilde{\mathcal{D}}^\mu\phi^\dagger T_a)_{i^*}v^a_\mu+4ige_\mu{}^{\tilde{\mu}}\Omega^{i^*}\partial_{\tilde{\mu}}(\phi^\dagger T_a)_{i^*}v^{\mu a}-4\sqrt{2}ig\Omega^{i^*}\bar{\psi}_\mu\cdot(\bar{\chi}T_a)_{i^*}v^{\mu a}\\
&\quad-4g^2\Omega^{i^*}(\phi^\dagger T_aT_b)_{i^*}v^a_\mu v^{\mu b}\,,\\
\widetilde{\Theta}^2_{(4)} &= 4g^2\Omega^{i^*}(\phi^\dagger T_aT_b)_{i^*}v^a_\mu v^{\mu b}\,,\\
\widetilde{\Theta}^2_{(5)} &= 8ig\Omega^{i^*j^*}\hat{\mathcal{D}}^\mu\phi^\dagger_{j^*}(\phi^\dagger T_a)_{i^*}v^a_\mu\\
&= 4ig\Omega^{i^*j^*}\widetilde{\mathcal{D}}^\mu\phi^\dagger_{j^*}(\phi^\dagger T_a)_{i^*}v^a_\mu+4ige_\mu{}^{\tilde{\mu}}\Omega^{i^*j^*}\partial_{\tilde{\mu}}\phi^\dagger_{j^*}(\phi^\dagger T_a)_{i^*}v^{\mu a}\\
&\quad-4\sqrt{2}ig\Omega^{i^*j^*}\bar{\psi}^\mu\cdot\bar{\chi}_{j^*}(\phi^\dagger T_a)_{i^*}v^a_\mu-4g^2\Omega^{i^*j^*}(\phi^\dagger T_a)_{i^*}(\phi^\dagger T_b)_{j^*}v^a_\mu v^{\mu b}\,,\\
\widetilde{\Theta}^2_{(6)} &= 4g^2\Omega^{i^*j^*}(\phi^\dagger T_a)_{i^*}(\phi^\dagger T_b)_{j^*}v^a_\mu v^{\mu b}\,,\\
\widetilde{\Theta}^2_{(7)} &= 4ig\Omega^{i^*}(\phi^\dagger T_a)_{i^*}\hat{D}_\mu v^{\mu a}\\
&= 4ig\Omega^{i^*}(\phi^\dagger T_a)_{i^*}e_\mu{}^{\tilde{\mu}}\mathcal{D}_{\tilde{\mu}}v^{\mu a}+2g\Omega^{i^*}(\phi^\dagger T_a)_{i^*}\big[\bar{\lambda}^a\bar{\sigma}^\mu\psi_\mu-\bar{\psi}_\mu\bar{\sigma}^\mu\lambda^a\big]\\
&\quad+2g\Omega^{i^*}(\phi^\dagger T_a)_{i^*}\psi^\mu\sigma^\nu\bar{\psi}_\mu v^a_\nu\,,\\
\widetilde{\Theta}^2_{(8)} &= 2\sqrt{2}\Omega^{i^*}\bar{\psi}_\mu\cdot\hat{\mathcal{D}}^\mu\bar{\chi}_{i^*}\\
&= 2\sqrt{2}\Big[\Omega^{i^*}\bar{\psi}_\mu\cdot\widetilde{\mathcal{D}}^\mu\bar{\chi}_{i^*}+ig\Omega^{i^*}\bar{\psi}_\mu\cdot(\bar{\chi}T_a)_{i^*}v^{\mu a}+[\mathcal{D}^{i^*}\Omega^{j^*}-\Omega^{i^*j^*}]\widetilde{\mathcal{D}}_\mu\phi^\dagger_{i^*}\bar{\psi}^\mu\cdot\bar{\chi}_{j^*}\\
&\quad+\frac{1}{\sqrt{2}}\Omega^{i^*}\bar{\psi}^\mu\cdot\bar{\psi}_\mu F^\dagger_{i^*}-\frac{i}{\sqrt{2}}\Omega^{i^*}\bar{\psi}_\mu\bar{\sigma}^\nu\psi^\mu\,\widetilde{\mathcal{D}}_\nu\phi^\dagger_{i^*}+\frac{g}{\sqrt{2}}\Omega^{i^*}\bar{\psi}_\mu\bar{\sigma}^\nu\psi^\mu(\phi^\dagger T_a)_{i^*}v^a_\nu\\
&\quad+\frac{i}{2}\Omega^{i^*}\bar{\psi}^\mu\bar{\sigma}^\nu\psi_\mu\,\bar{\psi}_\nu\cdot\bar{\chi}_{i^*}\Big]\,,\\
\widetilde{\Theta}^2_{(9)} &= 2\sqrt{2}\Omega^{i^*j^*}\bar{\psi}^\mu\cdot\bar{\chi}_{i^*}\partial_\mu\phi^\dagger_{j^*}=2\sqrt{2}\Omega^{i^*j^*}\bar{\psi}^\mu\cdot\bar{\chi}_{i^*}\Big[\widetilde{\mathcal{D}}_\mu\phi^\dagger_{j^*}+ig(\phi^\dagger T_a)_{j^*}v^a_\mu\Big]\,,
\end{aligned}$$

where we recall that for a scalar field, $\partial_{\tilde{\mu}}\phi = \mathcal{D}_{\tilde{\mu}}\phi$, and where we have used the definition of $\hat{D}_\mu v^{\mu a}$ given in Table 4.3 for the computation of $\widetilde{\Theta}^2_{(7)}$. Using in addition the gauge invariance property of the superspace kinetic energy, we find[2]

$$0 = \delta_a\left(\Omega^{i^*}\widetilde{\mathcal{D}}^\mu\phi^\dagger_{i^*}\right) = -\Omega^{i^*j^*}(\phi^\dagger T_a)_{j^*}\widetilde{\mathcal{D}}^\mu\phi^\dagger_{i^*} + \Omega^{i^*}{}_i(T_a\phi)^i\widetilde{\mathcal{D}}^\mu\phi^\dagger_{i^*} - \Omega^{i^*}(\widetilde{\mathcal{D}}^\mu\phi^\dagger T_a)_{i^*}\,,$$

we can sum over all the $\widetilde{\Theta}^2$ quantities and simplify the result as

$$\begin{aligned}
\widetilde{\Theta}^2_{(a)} &= \widetilde{\Theta}^2_{(1)} + \widetilde{\Theta}^2_{(2)} + \widetilde{\Theta}^2_{(3)} + \widetilde{\Theta}^2_{(4)} + \widetilde{\Theta}^2_{(5)} + \widetilde{\Theta}^2_{(6)} + \widetilde{\Theta}^2_{(7)} + \widetilde{\Theta}^2_{(8)} + \widetilde{\Theta}^2_{(9)} \\
&= -4e_\mu{}^{\tilde{\mu}}\mathcal{D}_{\tilde{\mu}}\left(\Omega^{i^*}\widetilde{\mathcal{D}}^\mu\phi^\dagger_{i^*}\right) + 4\Omega^{i^*}{}_i\widetilde{\mathcal{D}}_\mu\phi^\dagger_{i^*}\widetilde{\mathcal{D}}^\mu\phi^i + 2\Omega^{i^*}\bar{\psi}_\mu\cdot\bar{\psi}^\mu F^\dagger_{i^*} \\
&\quad + 2g\Omega^{i^*}(\phi^\dagger T_a)_{i^*}\left[\bar{\lambda}^a\bar{\sigma}^\mu\psi_\mu - \bar{\psi}_\mu\bar{\sigma}^\mu\lambda^a\right] + 2\sqrt{2}\Omega^{i^*}e_\mu{}^{\tilde{\mu}}\mathcal{D}_{\tilde{\mu}}\bar{\psi}^\mu\cdot\bar{\chi}_{i^*} \\
&\quad + 4\sqrt{2}\bar{\psi}^\mu\cdot\left[\Omega^{i^*}\widetilde{\mathcal{D}}_\mu\bar{\chi}_{i^*} + \mathcal{D}^i\Omega^{j^*}\widetilde{\mathcal{D}}_\mu\phi^\dagger_{i^*}\bar{\chi}_{j^*}\right] - 2i\Omega^{i^*}(\bar{\psi}_\mu\bar{\sigma}^\nu\psi^\mu)\widetilde{\mathcal{D}}_\nu\phi^\dagger_{i^*} \\
&\quad + \sqrt{2}i\Omega^{i^*}(\bar{\psi}_\mu\bar{\sigma}^\nu\psi^\mu)\,\bar{\psi}_\nu\cdot\bar{\chi}_{i^*}\,.
\end{aligned}$$

In a similar fashion, the three following terms,

$$\begin{aligned}
\widetilde{\Theta}^2_{(10)} &= -4i\Omega^{i^*}{}_i\chi^i\sigma^\mu\left(\hat{D}_\mu\bar{\chi}_{i^*} - ig(\bar{\chi}T_a)_{i^*}v^a_\mu\right) \\
&= -4i\Omega^{i^*}{}_i\chi^i\sigma^\mu\widetilde{\mathcal{D}}_\mu\bar{\chi}_{i^*} + 4i\Gamma^{i^*}{}_{k^*}{}^{j^*}\Omega^{k^*}{}_i\chi^i\sigma^\mu\bar{\chi}_{j^*}\widetilde{\mathcal{D}}_\mu\phi^\dagger_{i^*} \\
&\quad - 2\sqrt{2}i\Omega^{i^*}{}_i\chi^i\sigma^\mu\bar{\psi}_\mu F^\dagger_{i^*} - 2\sqrt{2}\Omega^{i^*}{}_i\chi^i\sigma^\mu\bar{\sigma}^\nu\psi_\mu\widetilde{\mathcal{D}}_\nu\phi^\dagger_{i^*} \\
&\quad - 2\sqrt{2}ig\Omega^{i^*}{}_i\chi^i\sigma^\mu\bar{\sigma}^\nu\psi_\mu(\phi^\dagger T_a)_{i^*}v^a_\nu + 2\Omega^{i^*}{}_i\chi^i\sigma^\mu\bar{\sigma}^\nu\psi_\mu\,\bar{\psi}_\nu\cdot\bar{\chi}_{i^*}\,, \\
\widetilde{\Theta}^2_{(11)} &= -4i\Omega^{i^*}{}_i{}^{j^*}\chi^i\sigma^\mu\bar{\chi}_{i^*}\left(\partial_\mu\phi^\dagger_{j^*} - ig(\phi^\dagger T_a)_{j^*}v^a_\mu\right) = -4i\Gamma^{i^*}{}_{k^*}{}^{j^*}\Omega^{k^*}{}_i\chi^i\sigma^\mu\bar{\chi}_{j^*}\,\widetilde{\mathcal{D}}_\mu\phi^\dagger_{i^*}\,, \\
\widetilde{\Theta}^2_{(12)} &= 2\sqrt{2}ig\Omega^{i^*}{}_i(\phi^\dagger T_a)_{i^*}\chi^i\sigma^\mu\bar{\sigma}^\nu\psi_\mu v^a_\nu\,,
\end{aligned}$$

can be grouped together, yielding

$$\begin{aligned}
\widetilde{\Theta}^2_{(b)} &= \widetilde{\Theta}^2_{(10)} + \widetilde{\Theta}^2_{(11)} + \widetilde{\Theta}^2_{(12)} \\
&= -4i\Omega^{i^*}{}_i\chi^i\sigma^\mu\widetilde{\mathcal{D}}_\mu\bar{\chi}_{i^*} - 2\sqrt{2}\Omega^{i^*}{}_i\left[\chi^i\sigma^\mu\bar{\sigma}^\nu\psi_\mu\widetilde{\mathcal{D}}_\nu\phi^\dagger_{i^*} + i\chi^i\sigma^\mu\bar{\psi}_\mu F^\dagger_{i^*}\right] \\
&\quad + 2\Omega^{i^*}{}_i\chi^i\sigma^\mu\bar{\sigma}^\nu\psi_\mu\,\bar{\psi}_\nu\cdot\bar{\chi}_{i^*}\,,
\end{aligned}$$

as well as

$$\widetilde{\Theta}^2_{(c)} = \frac{8}{3}i\Omega^{i^*}\left[\hat{D}_\mu\phi^\dagger_{i^*} - ig(\phi^\dagger T_a)_{i^*}v^a_\mu\right]b^\mu = \frac{8}{3}i\Omega^{i^*}\widetilde{\mathcal{D}}_\mu\phi^\dagger_{i^*}b^\mu - \frac{4\sqrt{2}}{3}i\Omega^{i^*}\bar{\psi}_\mu\cdot\bar{\chi}_{i^*}b^\mu\,.$$

Collecting all contributions, we can now rewrite the product $\mathcal{Y} = -1/4X_IW^I$ under a fully covariant form,

$$\mathcal{Y} = \mathcal{Y}^{(0)} + \Theta\cdot\mathcal{Y}^{(1)} + \Theta\cdot\Theta\mathcal{Y}^{(2)}\,, \tag{6.10}$$

[2] We recall that for the representation $\overline{\mathcal{R}}$, the generators of the Lie algebra are given by $-T^*_a$. Since the representation is unitary we have $T^*_a = T^t_a$.

where the components are given by

$$
\begin{aligned}
\mathcal{Y}^{(0)} =& -\frac{1}{2}\Omega^{i^* j^*}\bar{\chi}_{i^*}\cdot\bar{\chi}_{j^*} - \Omega^{i^*}F^{\dagger}_{i^*} - \frac{1}{3}\Omega M\,,\\
\mathcal{Y}^{(1)} =& \, i\sqrt{2}\sigma^{\mu}\Big[\Omega^{i^*}\widetilde{\mathcal{D}}_{\mu}\bar{\chi}_{i^*} + \mathcal{D}^{i^*}\Omega^{j^*}\bar{\chi}_{j^*}\widetilde{\mathcal{D}}_{\mu}\phi^{\dagger}_{i^*}\Big] - \frac{2}{3}\Omega\sigma^{\mu\nu}\psi_{\mu\nu}\\
&+ \Omega^{i^*}(\sigma^{\mu}\bar{\sigma}^{\nu}\psi_{\mu})\Big[\widetilde{\mathcal{D}}_{\nu}\phi^{\dagger}_{i^*} - \frac{1}{\sqrt{2}}\bar{\psi}_{\nu}\cdot\bar{\chi}_{i^*}\Big] + \frac{i}{2}\Omega^{i^* j^*}\sigma^{\mu}\bar{\psi}_{\mu}\bar{\chi}_{i^*}\cdot\bar{\chi}_{j^*}\\
&- 2ig\Omega^{i^*}(\phi^{\dagger}T_a)_{i^*}\lambda^a - \frac{\sqrt{2}}{2}\Omega^{i^*}{}_{i}{}^{j^*}\bar{\chi}_{i^*}\cdot\bar{\chi}_{j^*}\chi^{i} - \frac{1}{3\sqrt{2}}\Omega^{i^*}\sigma^{\mu}\bar{\chi}_{i^*}b_{\mu} - \sqrt{2}\Omega^{i^*}{}_{i}F^{\dagger}_{i^*}\chi^{i}\\
&- \frac{\sqrt{2}}{3}\Omega_{i}M\chi^{i} + i\Omega^{i^*}\sigma^{\mu}\bar{\psi}_{\mu}F^{\dagger}_{i^*} - \frac{i}{3}\Omega\Big[\psi^{\mu}b_{\mu} - M\sigma^{\mu}\bar{\psi}_{\mu}\Big]\,,\\
\mathcal{Y}^{(2)} =& -e_{\mu}{}^{\tilde{\mu}}\mathcal{D}_{\tilde{\mu}}(\Omega^{i^*}\widetilde{\mathcal{D}}^{\mu}\phi^{\dagger}_{i^*}) + \Omega^{i^*}{}_{i}\widetilde{\mathcal{D}}_{\mu}\phi^{\dagger}_{i^*}\widetilde{\mathcal{D}}^{\mu}\phi^{i} - i\Omega^{i^*}{}_{i}\chi^{i}\sigma^{\mu}\widetilde{\mathcal{D}}_{\mu}\bar{\chi}_{i^*}\\
&- \frac{\sqrt{2}}{2}\Omega^{i^*}{}_{i}\chi^{i}\sigma^{\mu}\bar{\sigma}^{\nu}\psi_{\mu}\widetilde{\mathcal{D}}_{\nu}\phi^{\dagger}_{i^*} + \frac{\sqrt{2}}{2}\Omega^{i^*}e_{\mu}{}^{\tilde{\mu}}\mathcal{D}_{\tilde{\mu}}\bar{\psi}^{\mu}\cdot\bar{\chi}_{i^*} - \frac{1}{2}i\Omega^{i^*}\bar{\psi}_{\mu}\bar{\sigma}^{\nu}\psi^{\mu}\widetilde{\mathcal{D}}_{\nu}\phi^{\dagger}_{i^*}\\
&+ \sqrt{2}\bar{\psi}^{\mu}\cdot\Big[\Omega^{i^*}\widetilde{\mathcal{D}}_{\mu}\bar{\chi}_{i^*} + \mathcal{D}^{i^*}\Omega^{j^*}\widetilde{\mathcal{D}}_{\mu}\phi^{\dagger}_{i^*}\bar{\chi}_{j^*}\Big] + \frac{1}{4}\Omega^{i^* j^*}\bar{\psi}_{\mu}\cdot\bar{\psi}^{\mu}\bar{\chi}_{i^*}\cdot\bar{\chi}_{j^*}\\
&- \frac{\sqrt{2}i}{4}\Omega^{i^*}{}_{i}{}^{j^*}\chi^{i}\sigma^{\mu}\bar{\psi}_{\mu}\bar{\chi}_{i^*}\cdot\bar{\chi}_{j^*} + \frac{1}{4}\Omega^{i^* j^*}{}_{ij}\bar{\chi}_{i^*}\cdot\bar{\chi}_{j^*}\chi^{i}\cdot\chi^{j} + \frac{\sqrt{2}}{3}\Omega_{i}\chi^{i}\sigma^{\mu\nu}\psi_{\mu\nu}\\
&+ \frac{1}{2}\Omega^{i^*}{}_{i}\chi^{i}\sigma^{\mu}\bar{\sigma}^{\nu}\psi_{\mu}\bar{\psi}_{\nu}\cdot\bar{\chi}_{i^*} + \frac{\sqrt{2}i}{4}\Omega^{i^*}\bar{\psi}_{\mu}\bar{\sigma}^{\nu}\psi^{\mu}\,\bar{\psi}_{\nu}\cdot\bar{\chi}_{i^*} + \sqrt{2}ig\Omega^{i^*}{}_{i}(\phi^{\dagger}T_a)_{i^*}\lambda^a\cdot\chi^{i}\\
&- \sqrt{2}ig\Omega^{i^*}(\bar{\chi}T_a)_{i^*}\cdot\bar{\lambda}^a - \sqrt{2}gi\,\Omega^{i^* j^*}\,\bar{\chi}_{j^*}\cdot\bar{\lambda}^a(\phi^{\dagger}T_a)_{i^*}\\
&+ \frac{1}{2}g\Omega^{i^*}(\phi^{\dagger}T_a)_{i^*}\Big[\bar{\lambda}^a\bar{\sigma}^{\mu}\psi_{\mu} - \bar{\psi}_{\mu}\bar{\sigma}^{\mu}\lambda^a\Big] - \frac{\sqrt{2}}{6}\Omega^{i^*}\bar{\chi}_{i^*}\bar{\sigma}^{\mu\nu}\bar{\psi}_{\mu\nu}\\
&+ \Omega\Big\{-\frac{1}{6}e_{\mu}{}^{\tilde{\mu}}e_{\nu}{}^{\tilde{\nu}}R_{\tilde{\mu}\tilde{\nu}}{}^{\mu\nu}\Big| + \frac{i}{3}\bar{\psi}^{\mu}\bar{\sigma}^{\nu}\psi_{\mu\nu} + \frac{1}{24}\varepsilon^{\mu\nu\rho\sigma}(\psi_{\mu}\sigma_{\sigma}\bar{\psi}_{\nu\rho} + \bar{\psi}_{\mu}\bar{\sigma}_{\sigma}\psi_{\nu\rho})\\
&- \frac{2}{9}MM^{*} + \frac{1}{9}b_{\mu}b^{\mu} + \frac{i}{3}e_{\mu}{}^{\tilde{\mu}}\mathcal{D}_{\tilde{\mu}}b^{\mu} + \frac{1}{6}\bar{\psi}_{\mu}\cdot\bar{\psi}^{\mu}M + \frac{1}{6}\psi_{\nu}\sigma^{\nu}\bar{\psi}_{\mu}b^{\mu}\Big\} + \Omega^{i^*}{}_{i}F^{\dagger}_{i^*}F^{i}\\
&- i\frac{\sqrt{2}}{2}\Omega^{i^*}{}_{i}\chi^{i}\sigma^{\mu}\bar{\psi}_{\mu}F^{\dagger}_{i^*} + \frac{2}{3}i\Omega^{i^*}\widetilde{\mathcal{D}}_{\mu}\phi^{\dagger}_{i^*}b^{\mu} - \frac{\sqrt{2}}{12}i\Omega^{i^*}\bar{\chi}_{i^*}\bar{\sigma}^{\mu}\sigma^{\nu}\bar{\psi}_{\mu}b_{\nu}\\
&+ \frac{1}{2}\Omega^{i^*}\bar{\psi}_{\mu}\cdot\bar{\psi}^{\mu}F^{\dagger}_{i^*} + \frac{1}{2}\Omega^{i^*}{}_{i}{}^{j^*}F^{i}\bar{\chi}_{i^*}\cdot\bar{\chi}_{j^*} + \frac{1}{2}\Omega_{i}{}^{i^*}{}_{j}F^{\dagger}_{i^*}\chi^{i}\cdot\chi^{j} - i\Omega_{i}\frac{\sqrt{2}}{6}M\chi^{i}\sigma^{\mu}\bar{\psi}_{\mu}\\
&+ \frac{1}{3}\Omega_{i}MF^{i} + \frac{1}{6}\Omega_{ij}M\chi^{i}\cdot\chi^{j} + \frac{1}{6}\Omega^{i^*}{}_{i}\chi^{i}\sigma^{\mu}\bar{\chi}_{i^*}b_{\mu} + \frac{\sqrt{2}}{6}i\Omega_{i}\chi^{i}\cdot\psi_{\mu}b^{\mu}\\
&- \frac{2}{3}M^{*}\Big[\Omega^{i^*}F^{\dagger}_{i^*} + \frac{1}{2}\Omega^{i^* j^*}\bar{\chi}_{i^*}\cdot\bar{\chi}_{j^*}\Big] - gD^{a}\Omega^{i^*}(\phi^{\dagger}T_a)_{i^*}\,.
\end{aligned}
$$

In this expression, we have ordered the terms such that the kinetic terms come first, followed by the interaction terms and then by all the terms involving auxiliary fields. Using (2.4), we now compute the product of the chiral superfield $\mathcal{Y}$ by $\mathcal{E}$,

$$\begin{aligned}\mathcal{Z} = \mathcal{E}\mathcal{Y} &= \mathcal{Z}^{(0)} + \Theta\cdot\mathcal{Z}^{(1)} + \Theta\cdot\Theta\mathcal{Z}^{(2)} \\ &= e\Big[\mathcal{Y}^{(0)} + \Theta\cdot(\mathcal{Y}^{(1)} + i\sigma^\mu\bar{\psi}_\mu\mathcal{Y}^{(0)}) \\ &\quad + \Theta\cdot\Theta\Big(\mathcal{Y}^{(2)} + \frac{i}{2}\bar{\psi}_\mu\bar{\sigma}^\mu\mathcal{Y}^{(1)} - \mathcal{Y}^{(0)}(M^* + \bar{\psi}_\mu\bar{\sigma}^{\mu\nu}\bar{\psi}_\nu)\Big)\Big]\,.\end{aligned}$$

The components of the superfield $\mathcal{Z}$ are then given by

$$\begin{aligned}
e^{-1}\mathcal{Z}^{(0)} &= -\frac{1}{2}\Omega^{i^*j^*}\bar{\chi}_{i^*}\cdot\bar{\chi}_{j^*} - \Omega^{i^*}F^\dagger_{i^*} - \frac{1}{3}\Omega M\,,\\
e^{-1}\mathcal{Z}^{(1)} &= i\sqrt{2}\sigma^\mu\Big[\Omega^{i^*}\widetilde{\mathcal{D}}_\mu\bar{\chi}_{i^*} + \mathcal{D}^{i^*}\Omega^{j^*}\bar{\chi}_{i^*}\widetilde{\mathcal{D}}_\mu\phi^\dagger_{j^*}\Big] - \frac{2}{3}\Omega\sigma^{\mu\nu}\psi_{\mu\nu}\\
&\quad + \Omega^{i^*}(\sigma^\mu\bar{\sigma}^\nu\psi_\mu)\Big[\widetilde{\mathcal{D}}_\nu\phi^\dagger_{i^*} - \frac{1}{\sqrt{2}}\bar{\psi}_\nu\cdot\bar{\chi}_{i^*}\Big] - 2ig\Omega^{i^*}(\phi^\dagger T_a)_{i^*}\lambda^a - \frac{i}{3}\Omega\psi^\mu b_\mu\\
&\quad - \frac{\sqrt{2}}{2}\Omega^{i^*}{}_i{}^{j^*}\bar{\chi}_{i^*}\cdot\bar{\chi}_{j^*}\chi^i - \frac{1}{3\sqrt{2}}\Omega^{i^*}\sigma^\mu\bar{\chi}_{i^*}b_\mu - \sqrt{2}\Omega^{i^*}{}_iF^\dagger_{i^*}\chi^i - \frac{\sqrt{2}}{3}\Omega_iM\chi^i\,,\\
e^{-1}\mathcal{Z}^{(2)} &= \Omega\Big[-\frac{1}{6}e_\mu{}^{\tilde{\mu}}e_\nu{}^{\tilde{\nu}}R_{\tilde{\mu}\tilde{\nu}}{}^{\mu\nu}\Big| + \frac{1}{24}\varepsilon^{\mu\nu\rho\sigma}(\psi_\mu\sigma_\sigma\bar{\psi}_{\nu\rho} + 5\bar{\psi}_\mu\bar{\sigma}_\sigma\psi_{\nu\rho})\Big]\\
&\quad - e_\mu{}^{\tilde{\mu}}\mathcal{D}_{\tilde{\mu}}(\Omega^{i^*}\widetilde{\mathcal{D}}^\mu\phi^\dagger_{i^*}) + \Omega^{i^*}{}_i\widetilde{\mathcal{D}}_\mu\phi^\dagger_{i^*}\widetilde{\mathcal{D}}^\mu\phi^i - i\Omega^{i^*}{}_i\chi^i\sigma^\mu\widetilde{\mathcal{D}}_\mu\bar{\chi}_{i^*}\\
&\quad + \frac{i}{2}\Omega^{i^*}\Big[\bar{\psi}_\mu\bar{\sigma}^\mu\psi^\rho - \bar{\psi}^\rho\bar{\sigma}^\mu\psi_\mu + i\varepsilon^{\mu\nu\rho\sigma}\bar{\psi}_\mu\bar{\sigma}_\sigma\psi_\nu\Big]\Big[\widetilde{\mathcal{D}}_\rho\phi^\dagger_{i^*} - \frac{1}{\sqrt{2}}\bar{\psi}_\rho\cdot\bar{\chi}_{i^*}\Big]\\
&\quad + \frac{\sqrt{2}}{2}\bar{\psi}_\mu\bar{\sigma}^\nu\sigma^\mu\Big[\Omega^{i^*}\widetilde{\mathcal{D}}_\nu\bar{\chi}_{i^*} + \mathcal{D}^{i^*}\Omega^{j^*}\widetilde{\mathcal{D}}_\nu\phi^\dagger_{i^*}\bar{\chi}_{j^*}\Big] + \frac{\sqrt{2}}{3}\Omega_i\chi^i\sigma^{\mu\nu}\psi_{\mu\nu}\\
&\quad - \frac{\sqrt{2}}{2}\Omega^{i^*}{}_i\chi^i\sigma^\mu\bar{\sigma}^\nu\psi_\mu\Big[\widetilde{\mathcal{D}}_\nu\phi^\dagger_{i^*} - \frac{1}{\sqrt{2}}\bar{\psi}_\nu\cdot\bar{\chi}_{i^*}\Big] + \frac{\sqrt{2}}{2}\Omega^{i^*}e_\mu{}^{\tilde{\mu}}\mathcal{D}_{\tilde{\mu}}\bar{\psi}^\mu\cdot\bar{\chi}_{i^*}\\
&\quad - \frac{\sqrt{2}}{6}\Omega^{i^*}\bar{\chi}_{i^*}\bar{\sigma}^{\mu\nu}\bar{\psi}_{\mu\nu} + \frac{1}{4}\Omega^{i^*j^*}{}_{ij}\bar{\chi}_{i^*}\cdot\bar{\chi}_{j^*}\chi^i\cdot\chi^j + \sqrt{2}ig\Omega^{i^*}{}_i(\phi^\dagger T_a)_{i^*}\lambda^a\cdot\chi^i\\
&\quad - \sqrt{2}ig\Omega^{i^*}(\bar{\chi}T_a)_{i^*}\cdot\bar{\lambda}^a - \sqrt{2}gi\,\Omega^{i^*j^*}\bar{\chi}_{j^*}\cdot\bar{\lambda}^a(\phi^\dagger T_a)_{i^*}\\
&\quad + \frac{1}{2}g\Omega^{i^*}(\phi^\dagger T_a)_{i^*}\Big[\bar{\lambda}^a\bar{\sigma}^\mu\psi_\mu + \bar{\psi}_\mu\bar{\sigma}^\mu\lambda^a\Big] + \Omega^{i^*}{}_iF^\dagger_{i^*}F^i + \frac{1}{9}\Omega MM^*\\
&\quad + \frac{1}{9}\Omega b_\mu b^\mu + \frac{1}{3}\Omega_iMF^i + \frac{1}{3}\Omega^{i^*}M^*F^\dagger_{i^*} + \frac{i}{3e}\Omega\mathcal{D}_{\tilde{\mu}}(ee_\mu{}^{\tilde{\mu}}b^\mu) + \frac{2}{3}i\Omega^{i^*}\widetilde{\mathcal{D}}_\mu\phi^\dagger_{i^*}b^\mu\\
&\quad + \frac{1}{6}\Omega^{i^*j^*}M^*\bar{\chi}_{i^*}\cdot\bar{\chi}_{j^*} + \frac{1}{6}\Omega_{ij}M\chi^i\cdot\chi^j + \frac{1}{6}\Omega^{i^*}{}_i\chi^i\sigma^\mu\bar{\chi}_{i^*}b_\mu\\
&\quad - \frac{\sqrt{2}}{6}i\Omega^{i^*}\bar{\chi}_{i^*}\cdot\bar{\psi}_\mu b^\mu + \frac{\sqrt{2}}{6}i\Omega_i\chi^i\cdot\psi_\mu b^\mu + \frac{1}{2}\Omega^{i^*}{}_i{}^{j^*}F^i\bar{\chi}_{i^*}\cdot\bar{\chi}_{j^*}\\
&\quad + \frac{1}{2}\Omega_i{}^{i^*}{}_jF^\dagger_{i^*}\chi^i\cdot\chi^j - gD^a\Omega^{i^*}(\phi^\dagger T_a)_{i^*}\,,
\end{aligned}$$

where, for the computation of the highest-order component, we have used the properties of the Pauli matrices (C.5) and (C.8) and those of the torsion tensor given in (3.118) and (5.29) leading to

$$\mathcal{D}_{\tilde{\mu}}(ee_\mu{}^{\tilde{\mu}}) = -\frac{ie}{2}(\psi_\nu\sigma^\nu\bar{\psi}_\mu - \psi_\mu\sigma^\nu\bar{\psi}_\nu)\,. \tag{6.11}$$

The $\mathcal{Z}^{(2)}$ component can be further simplified after performing two integrations by parts,

$$\begin{aligned}
\text{first}: \quad & -ee_\mu{}^{\tilde{\mu}}\mathcal{D}_{\tilde{\mu}}(\Omega^{i^*}\widetilde{\mathcal{D}}^\mu\phi^\dagger_{i^*}) = -\frac{ie}{2}[\bar{\psi}_\nu\bar{\sigma}^\nu\psi_\mu - \bar{\psi}_\mu\bar{\sigma}^\nu\psi_\nu][\Omega^{i^*}\widetilde{\mathcal{D}}^\mu\phi^\dagger_{i^*}] \\
& \qquad +\mathcal{D}_{\tilde{\mu}}\Big[-ee_\mu{}^{\tilde{\mu}}(\Omega^{i^*}\widetilde{\mathcal{D}}^\mu\phi^\dagger_{i^*})\Big]\,, \\
\text{second}: \quad & \frac{\sqrt{2}}{2}e\bar{\psi}_\mu\bar{\sigma}^\nu\sigma^\mu[\Omega^{i^*}\widetilde{\mathcal{D}}_\nu\bar{\chi}_{i^*} + \mathcal{D}^{i^*}\Omega^{j^*}\widetilde{\mathcal{D}}_\nu\phi^\dagger_{i^*}\bar{\chi}_{j^*}] \\
& = -\frac{\sqrt{2}}{2}e\Omega^{i^*}{}_i\,\bar{\chi}_{i^*}\bar{\sigma}^\mu\sigma^\nu\bar{\psi}_\mu\widetilde{\mathcal{D}}_\nu\phi^i + \frac{\sqrt{2}}{2}e\Omega^{i^*}\,\bar{\chi}_{i^*}\bar{\sigma}^{\mu\nu}\bar{\psi}_{\mu\nu} \\
& \quad -\frac{\sqrt{2}}{2}e\Omega^{i^*}e_\mu{}^{\tilde{\mu}}\mathcal{D}_{\tilde{\mu}}\bar{\psi}^\mu\cdot\bar{\chi}_{i^*} \\
& \quad +\frac{\sqrt{2}i}{4}e[\bar{\psi}_\rho\bar{\sigma}^\rho\psi_\nu - \bar{\psi}_\nu\bar{\sigma}^\rho\psi_\rho][\Omega^{i^*}\bar{\psi}_\mu\bar{\sigma}^\nu\sigma^\mu\bar{\chi}_{i^*}] \\
& \quad +\mathcal{D}_{\tilde{\nu}}\Big[\frac{\sqrt{2}}{2}ee_\nu{}^{\tilde{\nu}}\Omega^{i^*}\bar{\psi}_\mu\bar{\sigma}^\nu\sigma^\mu\bar{\chi}_{i^*}\Big] \\
& \quad +\frac{\sqrt{2}i}{2}e\psi_\nu\sigma^\rho\bar{\psi}_\mu\Big[\Omega^{i^*}\bar{\psi}_\rho\bar{\sigma}^{\nu\mu}\bar{\chi}_{i^*}\Big]\,.
\end{aligned} \tag{6.12}$$

In order to derive the last equality, we recall that $\bar{\psi}_{\mu\nu} = e_\mu{}^{\tilde{\mu}}e_\nu{}^{\tilde{\nu}}(\mathcal{D}_{\tilde{\mu}}\bar{\psi}_{\tilde{\nu}} - \mathcal{D}_{\tilde{\nu}}\bar{\psi}_{\tilde{\mu}})$ and that

$$\begin{aligned}
e_\nu{}^{\tilde{\nu}}\mathcal{D}_{\tilde{\nu}}e_\mu{}^{\tilde{\mu}}(\bar{\psi}_{\tilde{\mu}}\bar{\sigma}^{\nu\mu}) &= -\mathcal{D}_{\tilde{\nu}}e_{\tilde{\mu}}{}^\rho(\bar{\psi}_\rho\bar{\sigma}^{\tilde{\nu}\tilde{\mu}}) \\
&= -\frac{1}{2}T_{\tilde{\mu}\tilde{\nu}}{}^\rho\big|(\bar{\psi}_\rho\bar{\sigma}^{\tilde{\nu}\tilde{\mu}}) \\
&= -\frac{i}{4}\Big[\psi_\nu\sigma^\rho\bar{\psi}_\mu - \psi_\mu\sigma^\rho\bar{\psi}_\nu\Big](\bar{\psi}_\rho\bar{\sigma}^{\nu\mu})\,,
\end{aligned} \tag{6.13}$$

using the constraints (3.29) and the definition of the lowest order components of the vielbein given in Table 3.6. In addition, we must be careful with the curved or flat nature of the indices with respect to derivation, *e.g.*, $\mathcal{D}_{\tilde{\mu}}\psi^\mu = \mathcal{D}_{\tilde{\mu}}(e_{\tilde{\nu}}{}^\mu\psi^{\tilde{\nu}})$.

The integration of the superfield $\mathcal{Z}$ over the two-dimensional chiral superspace allows to obtain the pieces of the supergravity Lagrangian associated to the Kähler potential,

$$\mathcal{L}_K = \frac{1}{2}[\mathcal{Z}]_{\Theta^2} + \text{h.c.} = \mathcal{L}_{K,\text{kin}} + \mathcal{L}_{K,\text{int}} + \mathcal{L}_{K,\text{aux}} ,$$

where we are considering separately the kinetic terms ($\mathcal{L}_{K,\text{kin}}$), after being rendered covariant with respect to the gauge transformations, the supergravity transformations and the Kähler manifold, the interaction terms ($\mathcal{L}_{K,\text{int}}$) and the terms related to the auxiliary fields ($\mathcal{L}_{K,\text{aux}}$). The total derivatives appearing in (6.12) correspond to surface terms, *i.e.*,

$$\mathcal{D}_{\tilde{\mu}}[ee_{\mu}{}^{\tilde{\mu}}(\Omega^{i^*}\widetilde{\mathcal{D}}^{\mu}\phi^{\dagger}_{i^*})] = \partial_{\tilde{\mu}}[ee_{\mu}{}^{\tilde{\mu}}(\Omega^{i^*}\widetilde{\mathcal{D}}^{\mu}\phi^{\dagger}_{i^*})]$$

and

$$\mathcal{D}_{\tilde{\nu}}[ee_{\nu}{}^{\tilde{\nu}}\Omega^{i^*}\bar{\psi}_{\mu}\bar{\sigma}^{\nu}\sigma^{\mu}\bar{\chi}_{i^*}] = \partial_{\tilde{\nu}}[ee_{\nu}{}^{\tilde{\nu}}\Omega^{i^*}\bar{\psi}_{\mu}\bar{\sigma}^{\nu}\sigma^{\mu}\bar{\chi}_{i^*}] .$$

As mentioned in Sec. 3.1, the arguments of the derivatives are covariant with respect to the Lorentzian indices, but not with respect to the Einsteinian indices, the latter being irrelevant for this work. Omitting the surface terms from the kinetic Lagrangian and adding the hermitian conjugate pieces, we get

$$\begin{aligned}
\mathcal{L}_{K,\text{kin}} = &-\frac{1}{6}e\Omega e_{\mu}{}^{\tilde{\mu}}e_{\nu}{}^{\tilde{\nu}}R_{\tilde{\mu}\tilde{\nu}}{}^{\mu\nu}\Big| - \frac{1}{12}e\Omega\varepsilon^{\mu\nu\rho\sigma}\Big[\psi_{\mu}\sigma_{\sigma}\bar{\psi}_{\nu\rho} - \bar{\psi}_{\mu}\bar{\sigma}_{\sigma}\psi_{\nu\rho}\Big] \\
&+ \frac{\sqrt{2}}{3}e\Big[\Omega_{i}\chi^{i}\sigma^{\mu\nu}\psi_{\mu\nu} + \Omega^{i^*}\bar{\chi}_{i^*}\bar{\sigma}^{\mu\nu}\bar{\psi}_{\mu\nu}\Big] + e\Omega^{i^*}{}_{i}\widetilde{\mathcal{D}}_{\mu}\phi^{\dagger}_{i^*}\widetilde{\mathcal{D}}^{\mu}\phi^{i} \\
&- \frac{1}{4}e\varepsilon^{\mu\nu\rho\sigma}\bar{\psi}_{\mu}\bar{\sigma}_{\sigma}\psi_{\nu}\Big[\Omega^{i^*}\widetilde{\mathcal{D}}_{\rho}\phi^{\dagger}_{i^*} - \Omega_{i}\widetilde{\mathcal{D}}_{\rho}\phi^{i}\Big] \\
&- \frac{i}{2}e\Omega^{i^*}{}_{i}\Big[\chi^{i}\sigma^{\mu}\widetilde{\mathcal{D}}_{\mu}\bar{\chi}_{i^*} - \widetilde{\mathcal{D}}_{\mu}\chi^{i}\sigma^{\mu}\bar{\chi}_{i^*}\Big] \\
&- \frac{\sqrt{2}}{2}e\Omega^{i^*}{}_{i}\Big[\chi^{i}\sigma^{\mu}\bar{\sigma}^{\nu}\psi_{\mu}\widetilde{\mathcal{D}}_{\nu}\phi^{\dagger}_{i^*} + \bar{\chi}_{i^*}\bar{\sigma}^{\mu}\sigma^{\nu}\bar{\psi}_{\mu}\widetilde{\mathcal{D}}_{\nu}\phi^{i}\Big] ,
\end{aligned} \tag{6.14}$$

whilst all the remaining terms independent of the auxiliary fields are collected into the interaction Lagrangian, containing quartic interaction terms, together with gauge interactions terms not included in the covariant derivatives. After using (C.8) on the second relation of (6.12), the Lagrangian $\mathcal{L}_{K,\text{int}}$ reads

$$
\begin{aligned}
\mathcal{L}_{K,\text{int}} = &\ \frac{\sqrt{2}i}{4} e\psi_\nu\sigma^\rho\bar{\psi}_\mu\left[\Omega_i\psi_\rho\sigma^{\nu\mu}\chi^i + \Omega^{i^*}\bar{\psi}_\rho\bar{\sigma}^{\nu\mu}\bar{\chi}_{i^*}\right] \\
& - \frac{\sqrt{2}i}{4} e\left[\bar{\psi}_\mu\bar{\sigma}^\mu\psi_\rho - \bar{\psi}_\rho\bar{\sigma}^\mu\psi_\mu\right]\left[\Omega^{i^*}\bar{\psi}_\nu\bar{\sigma}^{\nu\rho}\bar{\chi}_{i^*} + \Omega_i\psi_\nu\sigma^{\nu\rho}\chi^i\right] \\
& + \frac{\sqrt{2}}{8} e\varepsilon^{\mu\nu\rho\sigma}\bar{\psi}_\mu\bar{\sigma}_\sigma\psi_\nu\left[\Omega^{i^*}\bar{\psi}_\rho\cdot\bar{\chi}_{i^*} - \Omega_i\psi_\rho\cdot\chi^i\right] \\
& + \frac{e}{4}\Omega^{i^*}{}_i\left[\chi^i\sigma^\mu\bar{\sigma}^\nu\psi_\mu\bar{\psi}_\nu\cdot\bar{\chi}_{i^*} + \bar{\chi}_{i^*}\bar{\sigma}^\mu\sigma^\nu\bar{\psi}_\mu\,\psi_\nu\cdot\chi^i\right] \\
& + \frac{e}{4}\Omega^{i^*j^*}{}_{ij}\bar{\chi}_{i^*}\cdot\bar{\chi}_{j^*}\chi^i\cdot\chi^j + \frac{\sqrt{2}}{2} ieg\Omega^{i^*}{}_i\left[(\phi^\dagger T_a)_{i^*}\lambda^a\cdot\chi^i - (T_a\phi)^i\bar{\lambda}^a\cdot\bar{\chi}_{i^*}\right] \\
& - \frac{\sqrt{2}}{2} ieg\left[\Omega^{i^*}(\bar{\chi}T_a)_{i^*}\cdot\bar{\lambda}^a - \Omega_i(T_a\chi)^i\cdot\lambda^a\right] \\
& - \frac{\sqrt{2}}{2} ieg\left[\Omega^{i^*j^*}\,\bar{\chi}_{j^*}\cdot\bar{\lambda}^a(\phi^\dagger T_a)_{i^*} - \Omega_{ij}\,\chi^j\cdot\lambda^a(T_a\phi)^i\right] \\
& + \frac{e}{4} g\left[\Omega_i(T_a\phi)^i + \Omega^{i^*}(\phi^\dagger T_a)_{i^*}\right]\left[\bar{\lambda}^a\bar{\sigma}^\mu\psi_\mu + \bar{\psi}_\mu\bar{\sigma}^\mu\lambda^a\right]. \qquad (6.15)
\end{aligned}
$$

All the remaining terms of the Lagrangian are either linear or bilinear in the auxiliary fields. From the definition of the superspace kinetic energy Ω (6.3), one may deduce the properties

$$
\begin{aligned}
\Omega_i &= (\ln\Omega)_i\Omega\,, \\
\Omega^{i^*} &= (\ln\Omega)^{i^*}\Omega\,, \\
\Omega^{i^*}{}_i &= (\ln\Omega)^{i^*}{}_i\Omega + (\ln\Omega)_i(\ln\Omega)^{i^*}\Omega\,,
\end{aligned}
$$

which allow to rewrite the auxiliary Lagrangian as

$$
\begin{aligned}
\mathcal{L}_{K,\text{aux}} = &\ \frac{1}{9} e\Omega\left|M + 3(\ln\Omega)^{i^*}F^\dagger_{i^*}\right|^2 + e\Omega(\ln\Omega)^{i^*}{}_i F^\dagger_{i^*}F^i + \frac{1}{9} e\Omega b_\mu b^\mu \\
& + \frac{1}{6} e\Omega_{ij}M\chi^i\cdot\chi^j + \frac{1}{6} e\Omega^{i^*j^*}M^*\bar{\chi}_{i^*}\cdot\bar{\chi}_{j^*} \\
& + \frac{1}{2} e\Omega_i{}^{i^*}{}_j F^\dagger_{i^*}\chi^i\cdot\chi^j + \frac{1}{2} e\Omega^{i^*}{}_i{}^{j^*}F^i\bar{\chi}_{i^*}\cdot\bar{\chi}_{j^*} \\
& - \frac{1}{2} egD^a\left[\Omega^{i^*}(\phi^\dagger T_a)_{i^*} + \Omega_i(T_a\phi)^i\right] \\
& + eb_\mu\Big[\frac{i}{3}\Omega^{i^*}\tilde{\mathcal{D}}^\mu\phi^\dagger_{i^*} - \frac{i}{3}\Omega_i\tilde{\mathcal{D}}^\mu\phi^i \\
& \qquad - \frac{\sqrt{2}i}{6}\Omega^{i^*}\bar{\chi}_{i^*}\cdot\bar{\psi}^\mu + \frac{\sqrt{2}i}{6}\Omega_i\chi^i\cdot\psi^\mu + \frac{1}{6}\Omega^{i^*}{}_i\chi^i\sigma^\mu\bar{\chi}_{i^*}\Big]. \qquad (6.16)
\end{aligned}
$$

6.2.2 Computation of the Lagrangian coming from the superpotential

The second Lagrangian $\mathcal{L}_W$, is related to the interactions coming from the superpotential:

$$\mathcal{L}_W = \int \mathrm{d}^2\Theta\ \mathcal{E}W(\Phi) + \text{h.c.}\,.$$

The superpotential being a holomorphic function of the matter superfields Φ^i, according to (6.4), it reads

$$W(\Phi) = W + \sqrt{2}\Theta\cdot(W_i\chi^i) - \Theta\cdot\Theta\Big[W_i F^i + \frac{1}{2}W_{ij}\chi^i\cdot\chi^j\Big]\,,$$

where the fields ϕ^i, χ^i and F^i denote the scalar, fermionic and auxiliary components of the chiral superfields Φ^i. Moreover, in our notations, the functions W, W_i and W_{ij} depend on the scalar components of the chiral superfields, and are defined as in (2.14) by

$$\begin{aligned} W_i &= \frac{\partial W(\phi)}{\partial \phi^i}\,, \\ W_{ij} &= \frac{\partial^2 W(\phi)}{\partial \phi^i \partial \phi^j}\,. \end{aligned} \tag{6.17}$$

The integrand of the integral above is the product of the superpotential with the chiral density:

$$\begin{aligned} \mathcal{E}W(\Phi) = {}& eW + \Theta\cdot\Big[\sqrt{2}eW_i\chi^i + ieW\sigma^\mu\bar{\psi}_\mu\Big] \\ & -\Theta\cdot\Theta\Big[eW\big(M^* + \bar{\psi}_\mu\bar{\sigma}^{\mu\nu}\bar{\psi}_\nu\big) + e\big(F^i W_i + \frac{1}{2}W_{ij}\chi^i\cdot\chi^j\big) - \frac{\sqrt{2}}{2}ieW_i\bar{\psi}_\mu\bar{\sigma}^\mu\chi^i\Big]\,. \end{aligned}$$

After integrating over the chiral superspace and adding the conjugate pieces, we get the Lagrangian

$$\begin{aligned} \mathcal{L}_W = {}& \frac{\sqrt{2}}{2}ie\big(W_i\bar{\psi}_\mu\bar{\sigma}^\mu\chi^i + W^{\star i^*}\psi_\mu\sigma^\mu\bar{\chi}_{i^*}\big) - eW\big(M^* + \bar{\psi}_\mu\bar{\sigma}^{\mu\nu}\bar{\psi}_\nu\big) \\ & - eW^\star\big(M + \psi_\mu\sigma^{\mu\nu}\psi_\nu\big) - e\big(F^i W_i + F^\dagger_{i^*}W^{\star i^*} + \frac{1}{2}W_{ij}\chi^i\cdot\chi^j + \frac{1}{2}W^{\star i^* j^*}\bar{\chi}_{i^*}\cdot\bar{\chi}_{j^*}\big)\,, \end{aligned} \tag{6.18}$$

where we have introduced the first and second order derivatives of the antiholomorphic function $W^\star$ defined by

$$\begin{aligned} W^{\star i^*} &= \frac{\partial W^\star(\phi^\dagger)}{\partial \phi^\dagger_{i^*}}\,, \\ W^{\star i^* j^*} &= \frac{\partial^2 W^\star(\phi^\dagger)}{\partial \phi^\dagger_{i^*}\partial \phi^\dagger_{j^*}}\,. \end{aligned}$$

6.2.3 Computation of the Lagrangian describing the gauge sector

The last term of the Lagrangian (5.35) is related to the gauge sector and contains kinetic and interaction terms for the components of the vector superfields,

$$\mathcal{L}_{\text{gauge}} = \frac{1}{16g^2}\int \mathrm{d}^2\Theta\mathcal{E}h_{ab}(\Phi)W^{\alpha a}W^b_\alpha + \text{h.c.}\,. \tag{6.19}$$

Using the results of Table 5.1 for the spinorial superfield W_α, we can calculate the product

$$\begin{aligned}\frac{1}{4g^2}W^{\alpha a}W^b_\alpha = &-\lambda^a\cdot\lambda^b + \Theta\sigma^{\mu\nu}\Big[\lambda^b\hat{F}^a_{\mu\nu} + \lambda^a\hat{F}^b_{\mu\nu}\Big] + i\Theta\cdot\Big[\lambda^a D^b + \lambda^b D^a\Big]\\ &+\Theta\cdot\Theta\Big[-\frac{1}{2}\hat{F}^a_{\mu\nu}\hat{F}^{\mu\nu b} - \frac{i}{2}\hat{F}^a_{\mu\nu}\,{}^*\hat{F}^{\mu\nu b} - \lambda^a\cdot\lambda^b M^* + D^a D^b\\ &-\frac{1}{2}\Big[\lambda^a\sigma^\mu\bar{\lambda}^b + \lambda^b\sigma^\mu\bar{\lambda}^a\Big]b_\mu - i\Big[\lambda^a\sigma^\mu\hat{D}_\mu\bar{\lambda}^b + \lambda^b\sigma^\mu\hat{D}_\mu\bar{\lambda}^a\Big]\Big]\,.\end{aligned}$$

This product must then be multiplied with the gauge kinetic function

$$h_{ab}(\Phi) = h_{ab} + \sqrt{2}\Theta\cdot\chi^i h_{abi} - \Theta\cdot\Theta\Big[h_{abi}F^i + \frac{1}{2}h_{abij}\chi^i\cdot\chi^j\Big]\,,$$

where the derivatives of h_{ab} are defined as in (2.25) by

$$\begin{aligned}h_{abi} &= \frac{\partial h_{ab}(\phi)}{\partial\phi^i}\,,\\ h_{abij} &= \frac{\partial^2 h_{ab}(\phi)}{\partial\phi^i\partial\phi^j}\,.\end{aligned}$$

Recalling that h_{ab} is a symmetric tensor, we obtain,

$$\begin{aligned}\frac{1}{4g^2}h_{ab}(\Phi)W^{\alpha a}W^b_\alpha = &-h_{ab}\lambda^a\cdot\lambda^b\\ &+\Theta\cdot\Big[-\sqrt{2}h_{abi}\lambda^a\cdot\lambda^b\chi^i + h_{ab}\Big(2\sigma^{\mu\nu}\lambda^a\hat{F}^b_{\mu\nu} + 2i\lambda^a D^b\Big)\Big]\\ &+\Theta\cdot\Theta\Big[\;h_{ab}\Big(-\frac{1}{2}\hat{F}^a_{\mu\nu}\hat{F}^{\mu\nu b} - \frac{i}{2}\hat{F}^a_{\mu\nu}\,{}^*\hat{F}^{\mu\nu b} - 2i\lambda^a\sigma^\mu\hat{D}_\mu\bar{\lambda}^b + D^a D^b\\ &-\lambda^a\sigma^\mu\bar{\lambda}^b b_\mu - \lambda^a\cdot\lambda^b M^*\Big) + \Big(h_{abi}F^i + \frac{1}{2}h_{abij}\chi^i\cdot\chi^j\Big)\lambda^a\cdot\lambda^b\\ &-\sqrt{2}h_{abi}\chi^i\sigma^{\mu\nu}\lambda^b\hat{F}^a_{\mu\nu} - \sqrt{2}ih_{abi}\chi^i\cdot\lambda^a D^b\Big]\,.\end{aligned}$$

We then deduce the expansion of

$$\frac{1}{4g^2}\mathcal{E}h_{ab}(\Phi)W^{\alpha a}W^b_\alpha = \mathcal{W}^{(0)} + \Theta\cdot\mathcal{W}^{(1)} + \Theta\cdot\Theta\mathcal{W}^{(2)}\,,$$

with

$$\mathcal{W}^{(0)} = -eh_{ab}\lambda^a\cdot\lambda^b\,,$$

$$\mathcal{W}^{(1)} = 2eh_{ab}\sigma^{\mu\nu}\lambda^a\hat{F}^b_{\mu\nu} - \sqrt{2}eh_{abi}\lambda^a\cdot\lambda^b\chi^i - ieh_{ab}\sigma^\mu\bar{\psi}_\mu\lambda^a\cdot\lambda^b + 2ieh_{ab}\lambda^a D^b\,,$$

$$\begin{aligned}\mathcal{W}^{(2)} =& -\frac{1}{2}eh_{ab}\Big[\hat{F}^a_{\mu\nu}\hat{F}^{\mu\nu b} + i\hat{F}^a_{\mu\nu}\,{}^*\hat{F}^{\mu\nu b}\Big] - \sqrt{2}eh_{abi}\chi^i\sigma^{\mu\nu}\lambda^b\hat{F}^a_{\mu\nu}\\ &+ ieh_{ab}\bar{\psi}_\mu\bar{\sigma}^\mu\sigma^{\nu\rho}\lambda^a\hat{F}^b_{\nu\rho} - 2ieh_{ab}\lambda^a\sigma^\mu\hat{D}_\mu\bar{\lambda}^b + eh_{ab}\lambda^a\cdot\lambda^b\bar{\psi}_\mu\bar{\sigma}^{\mu\nu}\bar{\psi}_\nu\\ &+\frac{1}{2}eh_{abij}\chi^i\cdot\chi^j\lambda^a\cdot\lambda^b - \frac{\sqrt{2}i}{2}eh_{abi}\bar{\psi}_\mu\bar{\sigma}^\mu\chi^i\lambda^a\cdot\lambda^b + eh_{abi}F^i\lambda^a\cdot\lambda^b + eh_{ab}D^aD^b\\ &- D^a\Big[eh_{ab}\bar{\psi}_\mu\bar{\sigma}^\mu\lambda^b + \sqrt{2}ieh_{abi}\chi^i\cdot\lambda^b\Big] - eh_{ab}\lambda^a\sigma^\mu\bar{\lambda}^b b_\mu\,.\end{aligned}$$

The expression of the highest-order component $\mathcal{W}^{(2)}$ can be further simplified after replacing the hatted derivatives by fully covariant derivatives, using (4.66) and (4.69), to get

$$\begin{aligned}\hat{F}_{\mu\nu} &= F_{\mu\nu} - \frac{i}{2}\Big[\bar{\lambda}\bar{\sigma}_\nu\psi_\mu - \bar{\psi}_\mu\bar{\sigma}_\nu\lambda - \bar{\lambda}\bar{\sigma}_\mu\psi_\nu + \bar{\psi}_\nu\sigma_\mu\bar{\lambda}\Big]\,,\\ \hat{D}_\mu\bar{\lambda} &= \widetilde{\mathcal{D}}_\mu\bar{\lambda} + \frac{1}{2}(\bar{\psi}_\mu\bar{\sigma}^{\nu\rho})\Big[F_{\nu\rho} - i(\lambda\bar{\sigma}_\rho\psi_\nu - \bar{\psi}_\nu\bar{\sigma}_\rho\lambda)\Big] - \frac{i}{2}D\bar{\psi}_\mu\,,\\ \widetilde{\mathcal{D}}_\mu\bar{\lambda} &= e_\mu{}^{\tilde{\mu}}\Big(\mathcal{D}_{\tilde{\mu}}\bar{\lambda} + ig[v_{\tilde{\mu}},\bar{\lambda}]\Big)\,.\end{aligned}$$

The first relation above allows to rewrite the kinetic terms of the vector field,

$$\begin{aligned}-\frac{1}{2}eh_{ab}\hat{F}^a_{\mu\nu}\hat{F}^{\mu\nu b} &= -\frac{1}{2}eh_{ab}F^a_{\mu\nu}F^{\mu\nu b}\\ &\quad+\frac{i}{2}eh_{ab}[\bar{\lambda}^a\bar{\sigma}_\nu\psi_\mu + \lambda^a\sigma_\nu\bar{\psi}_\mu][F^{\mu\nu b} + \hat{F}^{\mu\nu b}]\,,\\ -\frac{i}{2}eh_{ab}\hat{F}^a_{\mu\nu}\,{}^*\hat{F}^{\mu\nu b} &= -\frac{i}{2}eh_{ab}F^a_{\mu\nu}\,{}^*F^{\mu\nu b}\\ &\quad-\frac{1}{4}eh_{ab}\varepsilon_{\mu\nu\rho\sigma}[\bar{\lambda}^a\bar{\sigma}^\nu\psi^\mu + \lambda^a\sigma^\nu\bar{\psi}^\mu][F^{\rho\sigma b} + \hat{F}^{\rho\sigma b}]\,,\end{aligned}\tag{6.20}$$

whilst the second one enables to rewrite

$$\begin{aligned}&ieh_{ab}\Big[\bar{\psi}_\mu\bar{\sigma}^\mu\sigma^{\nu\rho}\lambda^a\hat{F}^b_{\nu\rho} - 2\lambda^a\sigma^\mu\hat{D}_\mu\bar{\lambda}^b\Big]\\ &\quad= -ieh_{ab}\Big[2\lambda^a\sigma^\mu\widetilde{\mathcal{D}}_\mu\bar{\lambda}^b + i\varepsilon^{\mu\nu\rho\sigma}\lambda^a\sigma_\nu\bar{\psi}_\mu\hat{F}^b_{\rho\sigma} - iD^a\lambda^b\sigma_\mu\bar{\psi}^\mu\Big]\\ &\quad= -ieh_{ab}\bigg[2\lambda^a\sigma^\mu\widetilde{\mathcal{D}}_\mu\bar{\lambda}^b - iD^a\lambda^b\sigma_\mu\bar{\psi}^\mu\\ &\qquad+\frac{i}{2}\varepsilon^{\mu\nu\rho\sigma}\lambda^a\sigma_\nu\bar{\psi}_\mu\Big[F^b_{\rho\sigma} + \hat{F}^b_{\rho\sigma} - i(\bar{\lambda}^b\bar{\sigma}_\sigma\psi_\rho + \lambda^b\sigma_\sigma\bar{\psi}_\rho)\Big]\bigg]\,,\end{aligned}\tag{6.21}$$

where we have simplified the products of Pauli matrices with the help of the relation (C.5). Adding all the terms linear in the field strength tensors $F_{\mu\nu}$ and $\hat{F}_{\mu\nu}$ from (6.20) and (6.21), one gets

$$
\begin{aligned}
&\frac{1}{2}eh_{ab}\Big[F^{\rho\sigma b}+\hat{F}^{\rho\sigma b}\Big]\Big[i(\lambda^a\sigma_\sigma\bar{\psi}_\rho+\bar{\lambda}^a\bar{\sigma}_\sigma\psi_\rho)+\frac{1}{2}\varepsilon_{\mu\nu\rho\sigma}(\lambda^a\sigma^\nu\bar{\psi}^\mu-\bar{\lambda}^a\bar{\sigma}^\nu\psi^\mu)\Big]\\
=&\frac{i}{2}eh_{ab}\Big[\lambda^a\sigma_\mu\bar{\sigma}_{\rho\sigma}\psi^\mu+\bar{\lambda}^a\bar{\sigma}_\mu\sigma_{\rho\sigma}\psi^\mu\Big]\Big[F^{\rho\sigma b}+\hat{F}^{\rho\sigma b}\Big]\,,
\end{aligned}
$$

using again the relation (C.5), whilst the quartic term of (6.21) can be rewritten with the help of the Fierz identity (C.9) and the duality property of the Pauli matrices (C.2) as

$$
\begin{aligned}
-\frac{i}{2}e\varepsilon^{\mu\nu\rho\sigma}h_{ab}\lambda^a\sigma_\nu\bar{\psi}_\mu\lambda^b\sigma_\sigma\bar{\psi}_\rho &= -\,eh_{ab}\lambda^a\cdot\lambda^b\bar{\psi}_\mu\bar{\sigma}^{\mu\nu}\bar{\psi}_\nu\,,\\
-\frac{i}{2}e\varepsilon^{\mu\nu\rho\sigma}h_{ab}\lambda^a\sigma_\nu\bar{\psi}_\mu\bar{\lambda}^b\bar{\sigma}_\sigma\psi_\rho &= \frac{i}{2}e\varepsilon^{\mu\nu\rho\sigma}h_{ab}\psi_\rho\sigma_\nu\bar{\psi}_\mu\bar{\lambda}^b\bar{\sigma}_\sigma\lambda^a+2eh_{ab}\bar{\psi}_\mu\sigma^{\mu\nu}\bar{\lambda}^b\;\lambda^a\cdot\psi_\nu\,.
\end{aligned}
$$

Collecting the results of the previous equations, we can rewrite $\mathcal{W}^{(2)}$ as

$$
\begin{aligned}
\mathcal{W}^{(2)}=&-\frac{1}{2}eh_{ab}\Big[F^a_{\mu\nu}F^{\mu\nu b}+iF^a_{\mu\nu}\,{}^*F^{\mu\nu b}\Big]-2ieh_{ab}\lambda^a\sigma^\mu\tilde{\bar{\mathcal{D}}}_\mu\bar{\lambda}^b\\
&+\frac{i}{2}eh_{ab}\Big[\bar{\lambda}^a\bar{\sigma}^\mu\sigma^{\nu\rho}\psi_\mu+\lambda^a\sigma^\mu\bar{\sigma}^{\nu\rho}\bar{\psi}_\mu\Big]\Big[F^b_{\nu\rho}+\hat{F}^b_{\nu\rho}\Big]\\
&-\sqrt{2}eh_{abi}\chi^i\sigma^{\mu\nu}\lambda^bF^a_{\mu\nu}+\sqrt{2}ieh_{abi}\chi^i\sigma^{\mu\nu}\lambda^b\Big[\lambda^a\sigma_\nu\bar{\psi}_\mu+\bar{\lambda}^a\bar{\sigma}_\nu\psi_\mu\Big]\\
&+\frac{1}{2}eh_{abij}\chi^i\cdot\chi^j\lambda^a\cdot\lambda^b-\frac{\sqrt{2}i}{2}eh_{abi}\bar{\psi}_\mu\bar{\sigma}^\mu\chi^i\lambda^a\cdot\lambda^b+\frac{i}{2}eh_{ab}\varepsilon^{\mu\nu\rho\sigma}\psi_\rho\sigma_\nu\bar{\psi}_\mu\bar{\lambda}^b\bar{\sigma}_\sigma\lambda^a\\
&+2eh_{ab}\bar{\psi}_\mu\bar{\sigma}^{\mu\nu}\bar{\lambda}^b\lambda^a\cdot\psi_\nu+eh_{ab}D^aD^b\\
&-\sqrt{2}ieh_{abi}\chi^i\cdot\lambda^aD^b-eh_{ab}\lambda^a\sigma^\mu\bar{\lambda}^bb_\mu+eh_{abi}F^i\lambda^a\cdot\lambda^b\,.
\end{aligned}
$$

We can now perform the integration over the chiral superspace and, after adding the conjugate pieces, we obtain the Lagrangian associated to the gauge sector:

$$
\mathcal{L}_{\text{gauge}}=\frac{1}{4}[\mathcal{W}]_{\Theta^2}+\text{h.c.}=\mathcal{L}_{\text{gauge, kin}}+\mathcal{L}_{\text{gauge, int}}+\mathcal{L}_{\text{gauge, aux}}\,,
$$

where we have split the Lagrangian into (gauge invariant and supergravity invariant) kinetic terms ($\mathcal{L}_{\text{gauge, kin}}$), quartic interaction terms ($\mathcal{L}_{\text{gauge, int}}$) and the remaining terms, depending on the auxiliary fields ($\mathcal{L}_{\text{gauge, aux}}$). The different pieces of the Lagrangian are given by

$$\begin{aligned}
\mathcal{L}_{\text{gauge, kin}} &= -\frac{1}{4}eh^R_{ab}F^a_{\mu\nu}F^{\mu\nu b} + \frac{1}{4}eh^I_{ab}F^a_{\mu\nu}{}^*F^{\mu\nu b} - \frac{i}{2}eh^R_{ab}(\lambda^a\sigma^\mu\widetilde{\mathcal{D}}_\mu\bar{\lambda}^b - \widetilde{\mathcal{D}}_\mu\lambda^a\sigma^\mu\bar{\lambda}^b) \\
&\quad + \frac{1}{2}eh^I_{ab}\widetilde{\mathcal{D}}_\mu(\lambda^a\sigma^\mu\bar{\lambda}^b) + \frac{i}{4}eh^R_{ab}\left[\lambda^a\sigma_\mu\bar{\sigma}_{\nu\rho}\bar{\psi}^\mu + \bar{\lambda}^a\bar{\sigma}_\mu\sigma_{\nu\rho}\psi^\mu\right]\left[F^{\nu\rho b} + \hat{F}^{\nu\rho b}\right] \\
&\quad - \frac{\sqrt{2}}{4}e\left[h_{abi}\chi^i\sigma^{\mu\nu}\lambda^b + h^{\star\,i^*}_{ab}\bar{\chi}_{i^*}\bar{\sigma}^{\mu\nu}\bar{\lambda}^b\right]F^a_{\mu\nu}\,, \\
\mathcal{L}_{\text{gauge, int}} &= \frac{1}{8}e\left[h_{abij}\chi^i\cdot\chi^j\lambda^a\cdot\lambda^b + h^{\star\,i^*j^*}_{ab}\bar{\chi}_{i^*}\cdot\bar{\chi}_{j^*}\bar{\lambda}^a\cdot\bar{\lambda}^b\right] \\
&\quad - \frac{\sqrt{2}i}{8}e\left[h_{abi}\bar{\psi}_\mu\bar{\sigma}^\mu\chi^i\lambda^a\cdot\lambda^b + h^{\star\,i^*}_{ab}\psi_\mu\sigma^\mu\bar{\chi}_{i^*}\bar{\lambda}^a\cdot\bar{\lambda}^b\right] \\
&\quad + \frac{\sqrt{2}i}{4}e\left[h_{abi}\chi^i\sigma^{\mu\nu}\lambda^a + h^{\star\,i^*}_{ab}\bar{\chi}_{i^*}\bar{\sigma}^{\mu\nu}\bar{\lambda}^a\right]\left[\lambda^b\sigma_\nu\bar{\psi}_\mu + \bar{\lambda}^b\bar{\sigma}_\nu\psi_\mu\right] \\
&\quad + \frac{1}{4}eh^I_{ab}\varepsilon^{\mu\nu\rho\sigma}\bar{\psi}_\mu\bar{\sigma}_\nu\lambda^a\psi_\rho\sigma_\sigma\bar{\lambda}^b\,, \\
\mathcal{L}_{\text{gauge, aux}} &= \frac{1}{2}eh^R_{ab}D^aD^b - \frac{\sqrt{2}i}{4}eD^a\left[h_{abi}\chi^i\cdot\lambda^b - h^{\star\,i^*}_{ab}\bar{\chi}_{i^*}\cdot\bar{\lambda}^b\right] - \frac{1}{2}eh^R_{ab}\lambda^a\sigma^\mu\bar{\lambda}^b b_\mu \\
&\quad + \frac{1}{4}e\left[h_{abi}F^i\lambda^a\cdot\lambda^b + h^{\star\,i^*}_{ab}F^\dagger_{i^*}\bar{\lambda}^a\cdot\bar{\lambda}^b\right],
\end{aligned} \tag{6.22}$$

where $h^R_{ab} = \text{Re}(h_{ab})$ and $h^I_{ab} = \text{Im}(h_{ab})$ and the derivatives of the conjugate gauge kinetic function $h^\star_{ab}$ are defined as

$$\begin{aligned}
h^{\star\,i^*}_{ab} &= \frac{\partial h^\star_{ab}(\phi^\dagger)}{\partial\phi^\dagger_{i^*}}\,, \\
h^{\star\,i^*j^*}_{ab} &= \frac{\partial^2 h^\star_{ab}(\phi^\dagger)}{\partial\phi^\dagger_{i^*}\partial\phi^\dagger_{j^*}}\,.
\end{aligned}$$

For the kinetic term of the gauginos, we have used the relation,

$$\begin{aligned}
-\frac{i}{2}h_{ab}\lambda^a\sigma^\mu\widetilde{\mathcal{D}}_\mu\bar{\lambda}^b + \frac{i}{2}h^\star_{ab}\widetilde{\mathcal{D}}_\mu\lambda^a\sigma^\mu\bar{\lambda}^b &= -\frac{i}{2}eh^R_{ab}(\lambda^a\sigma^\mu\widetilde{\mathcal{D}}_\mu\bar{\lambda}^b - \widetilde{\mathcal{D}}_\mu\lambda^a\sigma^\mu\bar{\lambda}^b) \\
&\quad + \frac{1}{2}h^I_{ab}\widetilde{\mathcal{D}}_\mu(\lambda^a\sigma^\mu\bar{\lambda}^b)\,,
\end{aligned}$$

whilst the interaction terms are simplified by using the property

$$\begin{aligned}
&\frac{i}{4}e\varepsilon^{\mu\nu\rho\sigma}h^R_{ab}\psi_\rho\sigma_\sigma\bar{\psi}_\mu\bar{\lambda}^b\bar{\sigma}_\nu\lambda^a + \frac{1}{2}eh_{ab}\bar{\psi}_\mu\bar{\sigma}^{\mu\nu}\bar{\lambda}^b\,\lambda^a\cdot\psi_\nu + \frac{1}{2}eh^\star_{ab}\psi_\mu\sigma^{\mu\nu}\lambda^b\bar{\lambda}^a\cdot\bar{\psi}_\nu \\
&= -\frac{1}{4}eh^I_{ab}\varepsilon^{\mu\nu\rho\sigma}\bar{\psi}_\mu\bar{\sigma}_\nu\lambda^a\psi_\rho\sigma_\sigma\bar{\lambda}^b\,.
\end{aligned}$$

6.2.4 *Elimination of the auxiliary fields*

The auxiliary fields M, b_μ, F^i and D^a having non-propagating equations of motion, they can be eliminated from the Lagrangian in favour of the physical degrees of freedom of the theory. Collecting the auxiliary pieces of the Lagrangians $\mathcal{L}_{K,\text{aux}}$ (6.16), $\mathcal{L}_W$(6.18) and $\mathcal{L}_{\text{gauge,aux}}$ (6.22), one gets,

$$
\begin{aligned}
\mathcal{L}_{\text{aux}} = & \frac{1}{9}e\Omega NN^* + eN\Big[\frac{1}{6}\Omega_{ij}\chi^i\cdot\chi^j - W^\star\Big] + eN^*\Big[\frac{1}{6}\Omega^{i^*j^*}\bar{\chi}_{i^*}\cdot\bar{\chi}_{j^*} - W\Big] \\
& + \frac{1}{2}eh^R_{ab}D^aD^b + e\Omega(\ln\Omega)^{i^*}{}_iF^\dagger_{i^*}F^i + \frac{1}{9}e\Omega b_\mu b^\mu \\
& + eF^i\Big[\frac{1}{2}(\Omega^{i^*}{}_i{}^{j^*} - \Omega^{i^*j^*}(\ln\Omega)_i)\bar{\chi}_{i^*}\cdot\bar{\chi}_{j^*} + 3(\ln\Omega)_iW - W_i + \frac{1}{4}h_{abi}\lambda^a\cdot\lambda^b\Big] \\
& + eF^\dagger_{i^*}\Big[\frac{1}{2}(\Omega_i{}^{i^*}{}_j - \Omega_{ij}(\ln\Omega)^{i^*})\chi^i\cdot\chi^j + 3(\ln\Omega)^{i^*}W^\star - W^{\star i^*} + \frac{1}{4}h^{\star}_{ab}{}^{i^*}\bar{\lambda}^a\cdot\bar{\lambda}^b\Big] \\
& + eb_\mu\Big[\frac{i}{3}\Omega^{i^*}\tilde{\mathcal{D}}^\mu\phi^\dagger_{i^*} - \frac{i}{3}\Omega_i\tilde{\mathcal{D}}^\mu\phi^i - \frac{\sqrt{2}i}{6}\Omega^{i^*}\bar{\chi}_{i^*}\cdot\bar{\psi}^\mu + \frac{\sqrt{2}i}{6}\Omega_i\chi^i\cdot\psi^\mu \\
& \qquad + \frac{1}{6}\Omega^{i^*}{}_i\chi^i\sigma^\mu\bar{\chi}_{i^*} - \frac{1}{2}h^R_{ab}\lambda^a\sigma^\mu\bar{\lambda}^b\Big] \\
& + eD^a\Big[-\frac{g}{2}(\Omega^{i^*}(\phi^\dagger T_a)_{i^*} + \Omega_i(T_a\phi)^i) - \frac{\sqrt{2}}{4}i(h_{abi}\chi^i\cdot\lambda^b - h^{\star}_{ab}{}^{i^*}\bar{\chi}_{i^*}\cdot\bar{\lambda}^b)\Big],
\end{aligned}
\tag{6.23}
$$

where we have replaced the auxiliary field M of the gravitation supermultiplet by a new scalar field

$$
N = M + 3(\ln\Omega)^{i^*}F^\dagger_{i^*}\,, \tag{6.24}
$$

allowing to decouple the mixing term between the fields F^i and M. The equations of motion for the auxiliary fields are easily deduced from the Lagrangian above,

$$
\frac{\partial\mathcal{L}_{\text{aux}}}{\partial N} = \frac{\partial\mathcal{L}_{\text{aux}}}{\partial N^*} = \frac{\partial\mathcal{L}_{\text{aux}}}{\partial F^i} = \frac{\partial\mathcal{L}_{\text{aux}}}{\partial F^\dagger_{i^*}} = \frac{\partial\mathcal{L}_{\text{aux}}}{\partial b_\mu} = \frac{\partial\mathcal{L}_{\text{aux}}}{\partial D^a} = 0\,,
$$

which leads to

$$
\begin{aligned}
N = & \ \Omega^{-1}\Big(9W - \frac{3}{2}\Omega^{i^*j^*}\bar{\chi}_{i^*}\cdot\bar{\chi}_{j^*}\Big), \\
F^i = & \ \Omega^{-1}((\ln\Omega)^{-1})_{i^*}{}^i\Big[\frac{1}{2}(\Omega_{jk}(\ln\Omega)^{i^*} - \Omega_j{}^{i^*}{}_k)\chi^j\cdot\chi^k \\
& -3(\ln\Omega)^{i^*}W^\star + W^{\star i^*} - \frac{1}{4}h^{\star}_{ab}{}^{i^*}\bar{\lambda}^a\cdot\bar{\lambda}^b\Big], \\
b_\mu = & \ \Omega^{-1}\Big[-\frac{3}{2}i(\Omega^{i^*}\tilde{\mathcal{D}}_\mu\phi^\dagger_{i^*} - \Omega_i\tilde{\mathcal{D}}_\mu\phi^i) + \frac{3}{4}\sqrt{2}i(\Omega^{i^*}\bar{\chi}_{i^*}\cdot\bar{\psi}_\mu - \Omega_i\chi^i\cdot\psi_\mu) \\
& -\frac{3}{4}\Omega^{i^*}{}_i\chi^i\sigma_\mu\bar{\chi}_{i^*} + \frac{9}{4}h^R_{ab}\lambda^a\sigma_\mu\bar{\lambda}^b\Big], \\
D^a = & \ \frac{1}{2}(h^{-1})^{Rab}\Big[g(\Omega^{i^*}(\phi^\dagger T_b)_{i^*} + \Omega_i(T_b\phi)^i) + \frac{\sqrt{2}}{2}i(h_{bci}\chi^i\cdot\lambda^c - h^{\star}_{bc}{}^{i^*}\bar{\chi}_{i^*}\cdot\bar{\lambda}^c)\Big].
\end{aligned}
\tag{6.25}
$$

Inserting these solutions into the Lagrangian (6.23), one obtains additional kinetic and interaction terms for the physical fields. Considering the terms depending on N, F, b and D separately, one respectively derives

$$\begin{aligned}
\mathcal{L}_{\text{aux},N} &= -e\Omega^{-1}\Big\{\frac{1}{4}\Omega_{ij}\Omega^{i^*j^*}\chi^i\cdot\chi^j\bar{\chi}_{i^*}\cdot\bar{\chi}_{j^*} - \frac{3}{2}\Omega_{ij}W\chi^i\cdot\chi^j \\
&\qquad -\frac{3}{2}\Omega^{i^*j^*}W^\star\bar{\chi}_{i^*}\cdot\bar{\chi}_{j^*} + 9WW^\star\Big\}, \\
\mathcal{L}_{\text{aux},F} &= -e\Omega^{-1}((\ln\Omega)^{-1})^i{}_{i^*}\Big\{ + 9(\ln\Omega)_i(\ln\Omega)^{i^*}WW^\star + W_iW^{\star i^*} \\
&\quad +\frac{1}{4}\big[\Omega^{j^*}{}_i{}^{k^*} - \Omega^{j^*k^*}(\ln\Omega)_i\big]\big[\Omega_j{}^{i^*}{}_k - \Omega_{jk}(\ln\Omega)^{i^*}\big]\chi^j\cdot\chi^k\bar{\chi}_{j^*}\cdot\bar{\chi}_{k^*} \\
&\quad +\frac{1}{16}h_{abi}h^{\star}_{cd}{}^{i^*}\lambda^a\cdot\lambda^b\bar{\lambda}^c\cdot\bar{\lambda}^d - \frac{1}{4}h_{abi}W^{\star i^*}\lambda^a\cdot\lambda^b - \frac{1}{4}h^{\star}_{ab}{}^{i^*}W_i\bar{\lambda}^a\cdot\bar{\lambda}^b \\
&\quad +\frac{1}{2}\big[\Omega_j{}^{i^*}{}_k - \Omega_{jk}(\ln\Omega)^{i^*}\big]\big[3(\ln\Omega)_iW - W_i + \frac{1}{4}h_{abi}\lambda^a\cdot\lambda^b\big]\chi^j\cdot\chi^k \\
&\quad +\frac{1}{2}\big[\Omega^{j^*}{}_i{}^{k^*} - \Omega^{j^*k^*}(\ln\Omega)_i\big]\big[3(\ln\Omega)^{i^*}W^\star - W^{\star i^*} + \frac{1}{4}h^{\star}_{ab}{}^{i^*}\bar{\lambda}^a\cdot\bar{\lambda}^b\big]\bar{\chi}_{j^*}\cdot\bar{\chi}_{k^*} \\
&\quad -3(\ln\Omega)_iW\big[W^{\star i^*} - \frac{1}{4}h^{\star}_{ab}{}^{i^*}\bar{\lambda}^a\cdot\bar{\lambda}^b\big] - 3(\ln\Omega)^{i^*}W^\star\big[W_i - \frac{1}{4}h_{abi}\lambda^a\cdot\lambda^b\big]\Big\}, \\
\mathcal{L}_{\text{aux},b} &= e\Omega^{-1}\Big\{\frac{1}{4}\big[\Omega^{i^*}\widetilde{\mathcal{D}}_\mu\phi^\dagger_{i^*} - \Omega_i\widetilde{\mathcal{D}}_\mu\phi^i\big]\big[\Omega^{j^*}\widetilde{\mathcal{D}}^\mu\phi^\dagger_{j^*} - \Omega_j\widetilde{\mathcal{D}}^\mu\phi^j\big] \\
&\quad -\frac{9}{8}h^R_{ab}h^R_{cd}\lambda^a\cdot\lambda^c\bar{\lambda}^b\cdot\bar{\lambda}^d - \frac{1}{8}\Omega^{i^*}{}_i\Omega^{j^*}{}_j\chi^i\cdot\chi^j\bar{\chi}_{i^*}\cdot\bar{\chi}_{j^*} \qquad (6.26)\\
&\quad \frac{\sqrt{2}i}{8}\big[\Omega^{i^*}\bar{\chi}_{i^*}\cdot\bar{\psi}_\mu - \Omega_i\chi^i\cdot\psi_\mu\big]\big[\Omega^{j^*}{}_j\chi^j\sigma^\mu\bar{\chi}_{j^*} - 3h^R_{ab}\lambda^a\sigma^\mu\bar{\lambda}^b\big] \\
&\quad +\frac{1}{8}\big[\Omega^{i^*}\bar{\chi}_{i^*}\cdot\bar{\psi}_\mu - \Omega_i\chi^i\cdot\psi_\mu\big]\big[\Omega^{j^*}\bar{\chi}_{j^*}\cdot\bar{\psi}^\mu - \Omega_j\chi^j\cdot\psi^\mu\big] \\
&\quad -\frac{1}{4}\big[\Omega^{i^*}\widetilde{\mathcal{D}}_\mu\phi^\dagger_{i^*} - \Omega_i\widetilde{\mathcal{D}}_\mu\phi^i\big]\big[\sqrt{2}(\Omega^{j^*}\bar{\chi}_{j^*}\cdot\bar{\psi}^\mu - \Omega_j\chi^j\cdot\psi^\mu) + i\Omega^{j^*}{}_j\chi^j\sigma^\mu\bar{\chi}_{j^*} \\
&\quad -3ih^R_{ab}\lambda^a\sigma^\mu\bar{\lambda}^b\big] + \frac{3}{4}\Omega^{i^*}{}_ih^R_{ab}\chi^i\cdot\lambda^a\bar{\chi}_{i^*}\cdot\bar{\lambda}^b\Big\}, \\
\mathcal{L}_{\text{aux},D} &= -eh^{Rab}\Big\{\frac{g^2}{8}\big[\Omega^{i^*}(\phi^\dagger T_a)_{i^*} + \Omega_i(T_a\phi)^i\big]\big[\Omega^{j^*}(\phi^\dagger T_b)_{j^*} + \Omega_j(T_b\phi)^j\big] \\
&\quad -\frac{1}{16}\big[h_{aci}\chi^i\cdot\lambda^c - h^{\star}_{ac}{}^{i^*}\bar{\chi}_{i^*}\cdot\bar{\lambda}^c\big]\big[h_{bdj}\chi^j\cdot\lambda^d - h^{\star}_{bd}{}^{j^*}\bar{\chi}_{j^*}\cdot\bar{\lambda}^d\big] \\
&\quad +\frac{\sqrt{2}i}{8}g\big[\Omega^{i^*}(\phi^\dagger T_a)_{i^*} + \Omega_i(T_a\phi)^i\big]\big[h_{bdj}\chi^j\cdot\lambda^d - h^{\star}_{bd}{}^{j^*}\bar{\chi}_{j^*}\cdot\bar{\lambda}^d\big]\Big\},
\end{aligned}$$

where h^{Rab} is the inverse matrix of h^R_{ab}. Hence, the auxiliary fields contribute to the total Lagrangian with derivative terms rendering the fermionic part of the Lagrangian covariant with respect to the Weyl rotations (see Sec. 5.3.3) and with interaction terms. The scalar potential V is deduced from these last terms by collecting all scalar interactions, and can be rewritten in a more standard form by replacing the derivatives of the superspace kinetic energy by the Kähler potential K and its derivatives, namely

$$\begin{aligned}
\Omega &= -3e^{-\frac{1}{3}K} , \\
\Omega_i &= -\frac{1}{3}K_i\Omega , \\
\Omega^{i^*} &= -\frac{1}{3}K^{i^*}\Omega , \\
(\ln\Omega)_i &= -\frac{1}{3}K_i , \\
(\ln\Omega)^{i^*} &= -\frac{1}{3}K^{i^*} , \\
(\ln\Omega^{-1})^i{}_{i^*} &= -3K^i{}_{i^*} ,
\end{aligned} \tag{6.27}$$

where $K^i{}_{i^*} \equiv (K^{-1})^i{}_{i^*}$ and $(\ln\Omega^{-1})^i{}_{i^*}$ is the inverse of the matrix $(\ln\Omega)^{i^*}{}_i$. With these substitution, the scalar potential reads

$$\begin{aligned}
V = ee^{\frac{1}{3}K}\Big[&K^i{}_{i^*}\mathcal{D}_iW\mathcal{D}^{i^*}W^\star - 3\big|W\big|^2\Big] \\
&+\frac{g^2}{8}ee^{-\frac{2}{3}K}h^{Rab}\Big[K^{i^*}(\phi^\dagger T_a)_{i^*} + K_i(T_a\phi)^i\Big]\Big[K^{j^*}(\phi^\dagger T_b)_{j^*} + K_j(T_b\phi)^j\Big] ,
\end{aligned} \tag{6.28}$$

where we have defined $\mathcal{D}_iW = W_i + K_iW$. In contrast to the global supersymmetric case, the potential is not positive. Consequently, a non-vanishing vacuum expectation value of the potential is not a signature of supergravity breaking, as will be shown in the next chapter. Collecting the bilinear terms involving two fermions included in the Lagrangians (6.18) and (6.26), one derives

$$\begin{aligned}
\mathcal{L}_{\chi\chi} &= -\frac{1}{2}e\Big[W_{ij} + K_{ij}W + \frac{1}{3}K_iW_j + \frac{1}{3}K_jW_i \\
&\qquad + \frac{1}{3}K_iK_jW - \Gamma_i{}^k{}_j\mathcal{D}_kW\Big]\chi^i\cdot\chi^j + \text{h.c.} , \qquad\qquad (6.29) \\
\mathcal{L}_{\lambda\lambda} &= \frac{1}{4}ee^{\frac{1}{3}K}K^i{}_{i^*}h_{abi}\mathcal{D}^{i^*}W^\star\lambda^a\cdot\lambda^b + \text{h.c.} ,
\end{aligned}$$

where we have used the properties

$$
\begin{aligned}
\Omega^{i^*j^*} &= \Big[-\frac{1}{3}K^{i^*j^*} + \frac{1}{9}K^{i^*}K^{j^*}\Big]\Omega\,,\\
\Omega_{ij} &= \Big[-\frac{1}{3}K_{ij} + \frac{1}{9}K_iK_j\Big]\Omega\,,\\
\Omega^{i^*}{}_i{}^{j^*} &= \Big[-\frac{1}{3}K^{i^*}{}_i{}^{j^*} + \frac{1}{9}K^{i^*j^*}K_i + \frac{1}{9}K^{i^*}{}_iK^{j^*}\\
&\quad + \frac{1}{9}K^{i^*}K^{j^*}{}_i - \frac{1}{27}K^{i^*}K^{j^*}K_i\Big]\Omega\,,\\
\Omega_i{}^{i^*}{}_j &= \Big[-\frac{1}{3}K_i{}^{i^*}{}_j + \frac{1}{9}K_{ij}K^{i^*} + \frac{1}{9}K^{i^*}{}_iK_j\\
&\quad + \frac{1}{9}K_iK^{i^*}{}_j - \frac{1}{27}K_iK_jK^{i^*}\Big]\Omega\,,
\end{aligned}
\tag{6.30}
$$

and recall that $K^{i^*}{}_i{}^{j^*} = K^{\ell^*}{}_i\Gamma^{i^*}{}_{\ell^*}{}^{j^*}$. In a similar fashion, the four-fermion self-interaction terms of the fermionic field χ can be rewritten as,

$$
\begin{aligned}
\mathcal{L}_{\chi^4} = -\frac{1}{12}e\Omega\Big[&R_i{}^{i^*}{}_j{}^{j^*} - \frac{1}{2}K^{i^*}{}_iK^{j^*}{}_j\\
&-\frac{1}{9}K^{i^*}{}_iK^{j^*}K_j + \frac{1}{54}K_iK_jK^{i^*}K^{j^*}\Big]\chi^i\cdot\chi^i\bar{\chi}_{i^*}\cdot\bar{\chi}_{j^*}\,,
\end{aligned}
\tag{6.31}
$$

where

$$
\begin{aligned}
\Omega^{i^*}{}_i &= \Big[-\frac{1}{3}K^{i^*}{}_i + \frac{1}{9}K^{i^*}K_i\Big]\Omega\,,\\
\Omega^{i^*j^*}{}_{ij} &= \Big[-\frac{1}{3}K^{i^*j^*}{}_{ij} + \frac{1}{9}K_i{}^{i^*}{}_jK^{j^*} + \frac{1}{9}K_i{}^{j^*}{}_jK^{i^*} + \frac{1}{9}K_iK^{i^*}{}_j{}^{j^*}\\
&\quad + \frac{1}{9}K^{i^*}{}_i{}^{j^*}K_j + \frac{1}{9}K_{ij}K^{i^*j^*} + \frac{1}{9}K^{i^*}{}_iK^{j^*}{}_j + \frac{1}{9}K^{j^*}{}_iK^{i^*}{}_j\\
&\quad - \frac{1}{27}K^{i^*}{}_iK_jK^{j^*} - \frac{1}{27}K_{ij}K^{i^*}K^{j^*} - \frac{1}{27}K_iK^{i^*}{}_jK^{j^*} - \frac{1}{27}K^{j^*}{}_iK_jK^{i^*}\\
&\quad -\frac{1}{27}K_iK^{j^*}{}_jK^{i^*} - \frac{1}{27}K_iK_jK^{i^*j^*} + \frac{1}{81}K_iK_jK^{i^*}K^{j^*}\Big]\Omega\,,
\end{aligned}
\tag{6.32}
$$

with $R_i{}^{i^*}{}_J{}^{j^*}$ the curvature tensor of the Kähler manifold defined by $R_i{}^{i^*}{}_J{}^{j^*} = K^{i^*J^*}{}_{ij} - K^{\ell}{}_{\ell^*}\Gamma^{i^*}{}_{\ell}{}^{j^*}\Gamma_i{}^{\ell^*}{}_j$. From the Lagrangians (6.14), (6.22) and (6.26) the kinetic terms of the full supergravity Lagrangian can be written as

$$
\begin{aligned}
\mathcal{L}_{\text{kin}} =& -\frac{1}{6}e\Omega e_\mu{}^{\tilde\mu} e_\nu{}^{\tilde\nu} R_{\tilde\mu\tilde\nu}{}^{\mu\nu}\Big| - \frac{1}{12}e\Omega\varepsilon^{\mu\nu\rho\sigma}\Big[\psi_\mu\sigma_\sigma\bar\psi_{\nu\rho} - \bar\psi_\mu\bar\sigma_\sigma\psi_{\nu\rho}\Big]\\
&+\frac{\sqrt{2}}{3}e\Big[\Omega_i\chi^i\sigma^{\mu\nu}\psi_{\mu\nu} + \Omega^{i^*}\bar\chi_{i^*}\bar\sigma^{\mu\nu}\bar\psi_{\mu\nu}\Big] + e\Omega^{i^*}{}_i\widetilde{\mathcal{D}}_\mu\phi^\dagger_{i^*}\widetilde{\mathcal{D}}^\mu\phi^i\\
&+\frac{1}{4}e\Omega^{-1}(\Omega^{i^*}\widetilde{\mathcal{D}}_\mu\phi^\dagger_{i^*} - \Omega_i\widetilde{\mathcal{D}}_\mu\phi^i)(\Omega^{j^*}\widetilde{\mathcal{D}}^\mu\phi^\dagger_{j^*} - \Omega_j\widetilde{\mathcal{D}}^\mu\phi^j)\\
&-\frac{1}{4}e\varepsilon^{\mu\nu\rho\sigma}\bar\psi_\mu\bar\sigma_\sigma\psi_\nu\Big[\Omega^{i^*}\widetilde{\mathcal{D}}_\rho\phi^\dagger_{i^*} - \Omega_i\widetilde{\mathcal{D}}_\rho\phi^i\Big]\\
&-\frac{\sqrt{2}}{4}e\Omega^{-1}(\Omega^{i^*}\widetilde{\mathcal{D}}_\mu\phi^\dagger_{i^*} - \Omega_i\widetilde{\mathcal{D}}_\mu\phi^i)(\Omega^{j^*}\bar\chi_{j^*}\cdot\bar\psi^\mu - \Omega_j\chi^j\cdot\psi^\mu)\\
&-\frac{i}{2}e\Omega^{i^*}{}_i\Big[\chi^i\sigma^\mu\widetilde{\mathcal{D}}_\mu\bar\chi_{i^*} - \widetilde{\mathcal{D}}_\mu\chi^i\sigma^\mu\bar\chi_{i^*}\Big]\\
&-\frac{i}{4}e\Omega^{-1}\Omega^{j^*}{}_j(\Omega^{i^*}\widetilde{\mathcal{D}}_\mu\phi^\dagger_{i^*} - \Omega_i\widetilde{\mathcal{D}}_\mu\phi^i)\chi^j\sigma^\mu\bar\chi_{j^*}\\
&-\frac{\sqrt{2}}{2}e\Omega^{i^*}{}_i\Big[\chi^i\sigma^\mu\bar\sigma^\nu\psi_\mu\widetilde{\mathcal{D}}_\nu\phi^\dagger_{i^*} + \bar\chi_{i^*}\bar\sigma^\mu\sigma^\nu\bar\psi_\mu\widetilde{\mathcal{D}}_\nu\phi^i\Big]\\
&-\frac{1}{4}eh^R_{ab}F^a_{\mu\nu}F^{\mu\nu b} + \frac{1}{4}eh^I_{ab}F^a_{\mu\nu}{}^*F^{\mu\nu b} - \frac{i}{2}eh^R_{ab}(\lambda^a\sigma^\mu\widetilde{\mathcal{D}}_\mu\bar\lambda^b - \widetilde{\mathcal{D}}_\mu\lambda^a\sigma^\mu\bar\lambda^b)\\
&+\frac{3i}{4}e\Omega^{-1}h^R_{ab}(\Omega^{i^*}\widetilde{\mathcal{D}}_\mu\phi^\dagger_{i^*} - \Omega^i\widetilde{\mathcal{D}}_\mu\phi^i)\lambda^a\sigma^\mu\bar\lambda^b + \frac{1}{2}eh^I_{ab}\widetilde{\mathcal{D}}_\mu(\lambda^a\sigma^\mu\bar\lambda^b)\\
&+\frac{i}{4}eh^R_{ab}\Big[\lambda^a\sigma_\mu\bar\sigma_{\nu\rho}\bar\psi^\mu + \bar\lambda^a\bar\sigma_\mu\sigma_{\nu\rho}\psi^\mu\Big]\Big[F^{\nu\rho b} + \hat F^{\nu\rho b}\Big]\\
&-\frac{\sqrt{2}}{4}e\Big[h_{abi}\chi^i\sigma^{\mu\nu}\lambda^b + h^{\star}_{ab}{}^{i^*}\bar\chi_{i^*}\bar\sigma^{\mu\nu}\bar\lambda^b\Big]F^a_{\mu\nu}\,,
\end{aligned}
\tag{6.33}
$$

whilst the interaction terms are given by (see (6.15), (6.18) (6.22), (6.28), (6.29) and (6.31)),

$$
\begin{aligned}
\mathcal{L}_{\text{int}} =& \frac{\sqrt{2}i}{4}e\psi_\nu\sigma^\rho\bar\psi_\mu\Big[\Omega_i\psi_\rho\sigma^{\nu\mu}\chi^i + \Omega^{i^*}\bar\psi_\rho\bar\sigma^{\nu\mu}\bar\chi_{i^*}\Big]\\
&-\frac{\sqrt{2}i}{4}e\Big[\bar\psi_\mu\bar\sigma^\mu\psi_\rho - \bar\psi_\rho\bar\sigma^\mu\psi_\mu\Big]\Big[\Omega^{i^*}\bar\psi_\nu\bar\sigma^{\nu\rho}\bar\chi_{i^*} + \Omega_i\psi_\nu\sigma^{\nu\rho}\chi^i\Big]\\
&+\frac{\sqrt{2}}{8}e\varepsilon^{\mu\nu\rho\sigma}\bar\psi_\mu\bar\sigma_\sigma\psi_\nu\Big[\Omega^{i^*}\bar\psi_\rho\cdot\bar\chi_{i^*} - \Omega_i\psi_\rho\cdot\chi^i\Big]\\
&+\frac{e}{4}\Omega^{i^*}{}_i\Big[\chi^i\sigma^\mu\bar\sigma^\nu\psi_\mu\bar\psi_\nu\cdot\bar\chi_{i^*} + \bar\chi_{i^*}\bar\sigma^\mu\sigma^\nu\bar\psi_\mu\,\psi_\nu\cdot\chi^i\Big]\\
&+\frac{e}{8}\Omega^{-1}\Big[\Omega^{i^*}\bar\chi_{i^*}\cdot\bar\psi_\mu - \Omega_i\chi^i\cdot\psi_\mu\Big]\\
&\qquad\times\Big[\Omega^{j^*}\bar\chi_{j^*}\cdot\bar\psi^\mu - \Omega_j\chi^j\cdot\psi^\mu + \sqrt{2}i\Omega^{j^*}{}_j\chi^j\sigma^\mu\bar\chi_{j^*} - 3\sqrt{2}ih^R_{ab}\lambda^a\sigma^\mu\bar\lambda^b\Big]
\end{aligned}
$$

$$
\begin{aligned}
&-\frac{e}{12}\Omega\Big[R_i{}^{i^*}{}_j{}^{j^*}-\frac{1}{2}K^{i^*}{}_iK^{j^*}{}_j-\frac{1}{9}K^{i^*}{}_iK^{j^*}K_j+\frac{1}{54}K_iK_jK^{i^*}K^{j^*}\Big]\chi^i\cdot\chi^j\bar\chi_{i^*}\cdot\bar\chi_{j^*}\\
&+\frac{\sqrt{2}}{2}ieg\Omega^{i^*}{}_i\Big[(\phi^\dagger T_a)_{i^*}\lambda^a\cdot\chi^i-(T_a\phi)^i\bar\lambda^a\cdot\bar\chi_{i^*}\Big]\\
&-\frac{\sqrt{2}}{2}ieg\Big[\Omega^{i^*j^*}\bar\chi_{j^*}\cdot\bar\lambda^a(\phi^\dagger T_a)_{i^*}-\Omega_{ij}\chi^j\cdot\lambda^a(T_a\phi)^i\Big]\\
&-\frac{\sqrt{2}}{2}ieg\Big[\Omega^{i^*}(\bar\chi T_a)_{i^*}\cdot\bar\lambda^a-\Omega_i(T_a\chi)^i\cdot\lambda^a\Big]\\
&+\frac{e}{4}g\Big[\Omega_i(T_a\phi)^i+\Omega^{i^*}(\phi^\dagger T_a)_{i^*}\Big]\Big[\bar\lambda^a\bar\sigma^\mu\psi_\mu+\bar\psi_\mu\bar\sigma^\mu\lambda^a\Big]\\
&+\frac{\sqrt{2}}{2}ie\Big(W_i\bar\psi_\mu\bar\sigma^\mu\chi^i+W^{\star i^*}\psi_\mu\sigma^\mu\bar\chi_{i^*}\Big)-eW\bar\psi_\mu\bar\sigma^{\mu\nu}\bar\psi_\nu-eW^\star\psi_\mu\sigma^{\mu\nu}\psi_\nu\\
&-\frac{1}{2}e\Big[W_{ij}+K_{ij}W+\frac{1}{3}K_iW_j+\frac{1}{3}K_jW_i+\frac{1}{3}K_iK_jW-\Gamma_i{}^k{}_j\mathcal{D}_kW\Big]\chi^i\cdot\chi^j\\
&-\frac{1}{2}e\Big[W^{\star i^*j^*}+K^{i^*j^*}W^\star+\frac{1}{3}K^{i^*}W^{\star j^*}+\frac{1}{3}K^{j^*}W^{\star i^*}\\
&\qquad+\frac{1}{3}K^{i^*}K^{j^*}W^\star-\Gamma^{i^*}{}_{k^*}{}^{j^*}\mathcal{D}^{k^*}W^\star\Big]\bar\chi_{i^*}\cdot\bar\chi_{j^*}\\
&+\frac{1}{4}ee^{\frac{1}{3}K}K^i{}_{i^*}\Big[h_{abi}\mathcal{D}^{i^*}W^\star\lambda^a\cdot\lambda^b+h^\star_{ab}{}^{i^*}\mathcal{D}_iW\bar\lambda^a\cdot\bar\lambda^b\Big]\\
&+\frac{3}{4}e\Omega^{-1}\Omega^{i^*}{}_ih^R_{ab}\chi^i\cdot\lambda^a\bar\chi_{i^*}\cdot\bar\lambda^b\\
&+\frac{1}{8}e\Big[h_{abij}-\Big(\Gamma_i{}^k{}_jh_{abk}-\frac{1}{3}K_ih_{abj}-\frac{1}{3}K_jh_{abi}\Big)\Big]\chi^i\cdot\chi^j\lambda^a\cdot\lambda^b\\
&+\frac{1}{8}e\Big[h^\star_{ab}{}^{i^*j^*}-\Big(\Gamma^{i^*}{}_{k^*}{}^{j^*}h^\star_{ab}{}^{k^*}-\frac{1}{3}K^{i^*}h^\star_{ab}{}^{j^*}-\frac{1}{3}K^{j^*}h^\star_{ab}{}^{i^*}\Big)\Big]\bar\chi_{i^*}\cdot\bar\chi_{j^*}\bar\lambda^a\cdot\bar\lambda^b\\
&+\frac{1}{16}eh^{Rab}\Big[h_{aci}\chi^i\cdot\lambda^c-h^\star_{ac}{}^{i^*}\bar\chi_{i^*}\cdot\bar\lambda^c\Big]\Big[h_{bdj}\chi^j\cdot\lambda^d-h^\star_{bd}{}^{j^*}\bar\chi_{j^*}\cdot\bar\lambda^d\Big]\\
&-\frac{9}{8}e\Omega^{-1}h^R_{ab}h^R_{cd}\lambda^a\cdot\lambda^c\bar\lambda^b\cdot\bar\lambda^d\\
&-\frac{1}{16}ee^{K/3}K^i{}_{i^*}h_{abi}h^\star_{cd}{}^{i^*}\lambda^a\cdot\lambda^b\bar\lambda^c\cdot\bar\lambda^d\\
&-\frac{\sqrt{2}i}{8}e\Big[h_{abi}\bar\psi_\mu\bar\sigma^\mu\chi^i\lambda^a\cdot\lambda^b+h^\star_{ab}{}^{i^*}\psi_\mu\sigma^\mu\bar\chi_{i^*}\bar\lambda^a\cdot\bar\lambda^b\Big]\\
&+\frac{\sqrt{2}i}{4}e\Big[h_{abi}\chi^i\sigma^{\mu\nu}\lambda^a+h^\star_{ab}{}^{i^*}\bar\chi_{i^*}\bar\sigma^{\mu\nu}\lambda^a\Big]\Big[\lambda^b\sigma_\nu\bar\psi_\mu+\bar\lambda^b\bar\sigma_\nu\psi_\mu\Big]\\
&+\frac{1}{4}eh^I_{ab}\varepsilon^{\mu\nu\rho\sigma}\psi_\rho\sigma_\sigma\bar\lambda^b\,\bar\psi_\mu\bar\sigma_\nu\lambda^a\\
&-\frac{\sqrt{2}i}{8}eh^{Rab}\Big[\Omega^{i^*}(\phi^\dagger T_a)_{i^*}+\Omega_i(T_a\phi)^i\Big]\Big[h_{bdj}\chi^j\cdot\lambda^d-h^\star_{bd}{}^{j^*}\bar\chi_{j^*}\cdot\bar\lambda^d\Big]
\end{aligned}
$$

$$- \frac{g^2}{8} e e^{-\frac{2}{3}K} h^{Rab} \Big[K^{i^*} (\phi^\dagger T_a)_{i^*} + K_i (T_a \phi)^i \Big] \Big[K^{j^*} (\phi^\dagger T_b)_{j^*} + K_j (T_b \phi)^j \Big]$$
$$- e e^{\frac{1}{3}K} \Big[K^i{}_{i^*} \mathcal{D}_i W \mathcal{D}^{i^*} W^\star - 3 |W|^2 \Big] , \qquad (6.34)$$

respectively.

This Lagrangian is however not properly normalised. For instance, it has a Brans-Dicke type normalisation for the Einstein action of gravity. To remedy this issue, one needs to perform a dilatation of all fermionic fields together with an additional shift of the gravitino. After this field redefinition, it turns out that all the kinetic terms are correctly normalised. The fact that the Lagrangian is not correctly normalised is not specific to the superspace approach. Indeed in the historical papers [Cremmer et al. (1979, 1982, 1978b, 1983b)] they were faced with the same problem.

6.2.5 *Brans-Dicke and Einstein action: general features*

As stated above, the normalisation of the kinetic terms in (6.33) is in an unconventional Brans-Dicke form. Historically, Brans and Dicke have considered such a type of gravitational action and tried to make the Newton constant spacetime-dependent [Brans and Dicke (1961)]. Later, it has been realised that a field-dependent Weyl rescaling allows to recast the action from the Brans-Dicke form into the more standard Einstein-Hilbert form. Under a Weyl rescaling, the graviton transforms as

$$e_\mu{}^{\tilde{\mu}} \to \exp(-\lambda)\, e_\mu{}^{\tilde{\mu}} ,$$
$$e \to \exp(4\lambda)\, e ,$$

where λ is the transformation parameter. In the more general context of the super-Weyl transformations, we can then set $\lambda \equiv (\Sigma + \Sigma^\dagger)|$ as suggested by (5.37). Hence, the gravitational part of the action (6.33) becomes

$$- \frac{1}{6} e \Omega e_\mu{}^{\tilde{\mu}} e_\nu{}^{\tilde{\nu}} R_{\tilde{\mu}\tilde{\nu}}{}^{\mu\nu} \Big| \to -\frac{1}{6} e^{2\lambda} e \Omega e_\mu{}^{\tilde{\mu}} e_\nu{}^{\tilde{\nu}} R_{\tilde{\mu}\tilde{\nu}}{}^{\mu\nu} \Big| + \cdots , \qquad (6.35)$$

where the dots stand for additional terms related to the variation of the spin-connection. The complete calculation will be undertaken in due course. It is then clear that choosing[3]

$$\exp(2\lambda) = -\frac{3}{\Omega} ,$$

[3]From now on $\exp(\lambda)$ is denoted e^λ when there will be no ambiguity.

restores the correct normalisation factor for the Hilbert-Einstein Lagrangian. Similarly, with the choice above for the Weyl dilatation, the kinetic fermion terms become (the fermions transformation under a Weyl rescaling is given in Table 6.1)

$$\begin{aligned}
& -\frac{1}{12}e\Omega\varepsilon^{\mu\nu\rho\sigma}\Big[\psi_\mu\sigma_\sigma\bar{\psi}_{\nu\rho} - \bar{\psi}_\mu\bar{\sigma}_\sigma\psi_{\nu\rho}\Big] \\
& +\frac{\sqrt{2}}{3}e\Big[\Omega_i\chi^i\sigma^{\mu\nu}\psi_{\mu\nu} + \Omega^{i^*}\bar{\chi}_{i^*}\bar{\sigma}^{\mu\nu}\bar{\psi}_{\mu\nu}\Big] \\
& -\frac{i}{2}e\Omega^{i^*}{}_i\Big[\chi^i\sigma^\mu\widetilde{\mathcal{D}}_\mu\bar{\chi}_{i^*} - \widetilde{\mathcal{D}}_\mu\chi^i\sigma^\mu\bar{\chi}_{i^*}\Big] \\
\to\ & \frac{1}{4}e\varepsilon^{\mu\nu\rho\sigma}\Big[\psi_\mu\sigma_\sigma\bar{\psi}_{\nu\rho} - \bar{\psi}_\mu\bar{\sigma}_\sigma\psi_{\nu\rho}\Big] \\
& +\frac{\sqrt{2}}{3}e\Big[K_i\chi^i\sigma^{\mu\nu}\psi_{\mu\nu} + K^{i^*}\bar{\chi}_{i^*}\bar{\sigma}^{\mu\nu}\bar{\psi}_{\mu\nu}\Big] \\
& -\frac{i}{2}e\Big(K^{i^*}{}_i - \frac{1}{3}K^{i^*}K_i\Big)\Big[\chi^i\sigma^\mu\widetilde{\mathcal{D}}_\mu\bar{\chi}_{i^*} - \widetilde{\mathcal{D}}_\mu\chi^i\sigma^\mu\bar{\chi}_{i^*}\Big] + \cdots\ ,
\end{aligned}$$

where the dots stand for additional terms that will be undertaken in due course. In order to diagonalise the kinetic energy, *i.e.*, to eliminate the fermion-gravitino mixing term above, and to have a correctly normalised action for fermions, the Weyl dilatation must be followed by the gravitino shift:

$$\psi_{\tilde{\mu}} \to \psi_{\tilde{\mu}} + \frac{\sqrt{2}i}{2}\Omega^{-1}\Omega^{i^*}\sigma_{\tilde{\mu}}\bar{\chi}_{i^*}\ ,$$

as we will see explicitly later on. From Table 3.5 and App. H, and in order to recast these two transformations in a single Howe–Tucker transformation, we can choose the (superfield) transformation parameters to be

$$\begin{aligned}
\Sigma &= \frac{1}{2}\lambda + \sqrt{2}\Theta\cdot\lambda_i\chi^i - \Theta\cdot\Theta\mathcal{F}\ , \\
\Sigma^\dagger &= \frac{1}{2}\lambda + \sqrt{2}\bar{\Theta}\cdot\bar{\lambda}^{i^*}\bar{\chi}_{i^*} - \bar{\Theta}\cdot\bar{\Theta}\mathcal{F}^\dagger\ ,
\end{aligned} \tag{6.36}$$

namely

$$\begin{aligned}
\Sigma\Big| &= \frac{1}{2}\lambda\ , \\
\mathcal{D}\Sigma\Big| &= \lambda_i\chi^i = -\frac{1}{2}\Omega^{-1}\Omega_i\chi^i\ .
\end{aligned}$$

The components $\mathcal{F}$ and $\mathcal{F}^\dagger$ of Σ and $\Sigma^\dagger$ contribute to the transformation of M, and thus they are irrelevant at this stage since the auxiliary fields have been eliminated through their equations of motion. Therefore, we keep $\mathcal{F}$ and $\mathcal{F}^\dagger$ unspecified in

(6.36) and leave their explicit form for Sec. 6.7. For further use note

$$
\begin{aligned}
\lambda_i &= \frac{\partial\lambda}{\partial\phi^i} = -\frac{1}{2}\Omega^{-1}\Omega_i\,,\\
\lambda^{i^*} &= \frac{\partial\lambda}{\partial\phi^\dagger_{i^*}} = -\frac{1}{2}\Omega^{-1}\Omega^{i^*}\,,\\
\lambda_{ij} &= \frac{\partial\lambda}{\partial\phi^i\partial\phi^j} = -\frac{1}{2}\Omega^{-1}\Omega_{ij} + \frac{1}{2}\Omega^{-2}\Omega_i\Omega_j\,,\\
\lambda^{i^*j^*} &= \frac{\partial\lambda}{\partial\phi^\dagger_{i^*}\partial\phi^\dagger_{j^*}} = -\frac{1}{2}\Omega^{-1}\Omega^{i^*j^*} + \frac{1}{2}\Omega^{-2}\Omega^{i^*}\Omega^{j^*}\,.
\end{aligned}
\tag{6.37}
$$

On the basis of the results of (5.37), one immediately derives the variations of the graviton[4],

$$
\begin{aligned}
e_{\tilde\mu}{}^{\mu} &\to \exp(\lambda)\, e_{\tilde\mu}{}^{\mu}\,,\\
e_{\mu}{}^{\tilde\mu} &\to \exp(-\lambda)\, e_{\mu}{}^{\tilde\mu}\,,\\
e &\to \exp(4\lambda)\, e\,,
\end{aligned}
\tag{6.38}
$$

as well as those of the metric introduced in Sec. 3.3.2

$$
\begin{aligned}
g_{\tilde\mu\tilde\nu} &= e_{\tilde\mu}{}^{\mu} e_{\tilde\nu}{}^{\nu}\eta_{\mu\nu} \to e^{2\lambda} g_{\tilde\mu\tilde\nu}\,,\\
g^{\tilde\mu\tilde\nu} &= e_{\mu}{}^{\tilde\mu} e_{\nu}{}^{\tilde\nu}\eta_{\mu\nu} \to e^{-2\lambda} g^{\tilde\mu\tilde\nu}\,.
\end{aligned}
$$

The transformation laws of the gravitino (see (5.37) and (5.51)) contain two pieces, a pure Weyl-rescaling together with a shift,

$$
\begin{aligned}
\psi_{\tilde\mu} &\to e^{\frac{1}{2}\lambda}\Big[\psi_{\tilde\mu} + \frac{\sqrt{2}i}{2}\Omega^{-1}\Omega^{i^*}\sigma_{\tilde\mu}\bar\chi_{i^*}\Big]\,,\\
\bar\psi_{\tilde\mu} &\to e^{\frac{1}{2}\lambda}\Big[\bar\psi_{\tilde\mu} + \frac{\sqrt{2}i}{2}\Omega^{-1}\Omega_i\bar\sigma_{\tilde\mu}\chi^i\Big]\,,
\end{aligned}
\tag{6.39}
$$

where we have replaced the derivatives of λ by their values as given in (6.37). As for the graviton, the dilatation allows to recover a standard normalisation for the kinetic terms of the gravitino, whilst the shift of the field plays a key rôle in order to restore appropriate fermionic kinetic terms, as will be shown below. Eq. (5.37) also allows us to compute the transformations of the auxiliary components of the gravitation multiplet,

$$
\begin{aligned}
b_\mu &\to e^{-\lambda}\Big[b_\mu - \frac{3\sqrt{2}i}{4}\Omega^{-1}(\Omega_i\chi^i\cdot\psi_\mu - \Omega^{i^*}\bar\chi_{i^*}\cdot\bar\psi_\mu)\Big]\,,\\
M &\to e^{-\lambda}\Big[M + 6\mathcal{F}^\dagger\Big]\,,\\
M^\star &\to e^{-\lambda}\Big[M^\star + 6\mathcal{F}\Big]\,.
\end{aligned}
\tag{6.40}
$$

[4]In principle, one should use different notations for the transformed objects. However, for the sake of clarity, we use the same notations before and after the rescaling.

Even if the auxiliary fields have already been eliminated from the Lagrangian[5], these expressions (and in particular the transformation of b_μ) will be useful below to avoid a lengthy computation of the variations of the Lagrangian describing the dynamics of the gauge sector.

The transformation laws of the physical components of (anti)chiral superfields Φ^i ($\Phi^\dagger_{i^*}$) are obtained from (5.41) and (5.51)

$$\begin{aligned}
\phi^i &\to \phi^i\,, \\
\phi^\dagger_{i^*} &\to \phi^\dagger_{i^*}\,, \\
\chi^i &\to e^{-\frac{1}{2}\lambda}\chi^i\,, \\
\bar{\chi}_{i^*} &\to e^{-\frac{1}{2}\lambda}\bar{\chi}_{i^*}\,,
\end{aligned} \tag{6.41}$$

while the same equations allow to compute the variations of the auxiliary fields F and $F^\dagger$,

$$\begin{aligned}
F^i &\to e^{-\lambda}\Big[F^i - \Omega^{-1}\Omega_j\chi^i\cdot\chi^j\Big]\,, \\
F^\dagger_{i^*} &\to e^{-\lambda}\Big[F_{i^*} - \Omega^{-1}\Omega^{j^*}\bar{\chi}_{i^*}\cdot\bar{\chi}_{j^*}\Big]\,.
\end{aligned} \tag{6.42}$$

Turning to vector superfields, Howe–Tucker transformations of the physical components and of the auxiliary D-field are given by (5.51) and (5.52),

$$\begin{aligned}
v^a_{\tilde{\mu}} &\to v^a_{\tilde{\mu}}\,, \\
\lambda_a &\to e^{-\frac{3}{2}\lambda}\lambda_a\,, \\
\bar{\lambda}_a &\to e^{-\frac{3}{2}\lambda}\bar{\lambda}_a\,, \\
D^a &\to e^{-2\lambda}D^a\,.
\end{aligned} \tag{6.43}$$

The Lagrangian (6.33) contains several covariant derivatives, in particular, of the fermionic fields. In order to derive their variations under a Howe–Tucker transformation, one must calculate the transformation laws of the components of the connection. However, the calculation of the transformation laws of the different parts of the action can be simplified on the basis of the results of App. H (Eq.[H.12]). Indeed, in the following we perform the rescaling in two steps, applying first a pure dilation of the fields, and second a pure gravitino shift. Therefore, from the result of Table 3.5, we derive first the variation of the connection under

[5]It is also possible to proceed with the super-Weyl rescaling of the Lagrangian before eliminating the auxiliary fields. On the one hand, this leads to a modification of the equations of motion for the auxiliary fields, in particular due to the shifts for the F-, b- and M-fields. On the other hand, these shifts also leads to additional interaction terms in the Lagrangian, not depending on the auxiliary fields. Both effects compensate each other so that one can equivalently eliminate the auxiliary fields before or after the rescaling. In particular, this implies that the shifted Lagrangian is independent of the quantity $\mathcal{F}$ introduced in (6.36). See also the transformation of the auxiliary field N give in Sec. 6.7.

a pure dilatation,

$$
\begin{aligned}
\omega_{\tilde{\mu}\mu\nu} &\to \omega_{\tilde{\mu}\mu\nu} - \frac{1}{2}\Omega^{-1}\Big[e_{\tilde{\mu}\nu}e_{\mu}{}^{\tilde{\rho}} - e_{\tilde{\mu}\mu}e_{\nu}{}^{\tilde{\rho}}\Big]\partial_{\tilde{\rho}}\Omega\,, \\
\omega_{\tilde{\mu}\alpha\beta} &\to \omega_{\tilde{\mu}\alpha\beta} + \frac{1}{2}\Omega^{-1}e_{\tilde{\mu}\nu}(\sigma^{\nu\rho})_{\beta}{}^{\gamma}\varepsilon_{\alpha\gamma}\partial_{\rho}\Omega\,, \\
\omega_{\tilde{\mu}\dot{\alpha}\dot{\beta}} &\to \omega_{\tilde{\mu}\dot{\alpha}\dot{\beta}} + \frac{1}{2}\Omega^{-1}e_{\tilde{\mu}\nu}\varepsilon_{\dot{\alpha}\dot{\gamma}}(\bar{\sigma}^{\nu\rho})^{\dot{\gamma}}{}_{\dot{\beta}}\partial_{\rho}\Omega\,,
\end{aligned}
\tag{6.44}
$$

and then the variation induced by a gravitino shift,

$$
\begin{aligned}
\omega_{\tilde{\mu}\mu\nu} \to \omega_{\tilde{\mu}\mu\nu} + \frac{\sqrt{2}}{4}\Omega^{-1}\Big[&2\Omega_i\psi_{\tilde{\mu}}\sigma_{\mu\nu}\chi^i + 2\Omega^{i^*}\bar{\psi}_{\tilde{\mu}}\bar{\sigma}_{\mu\nu}\bar{\chi}_{i^*} \\
&+\Omega_i\Big(e_{\tilde{\mu}\nu}\psi_{\mu} - e_{\tilde{\mu}\mu}\psi_{\nu}\Big)\cdot\chi^i + \Omega^{i^*}\Big(e_{\tilde{\mu}\nu}\bar{\psi}_{\mu} - e_{\tilde{\mu}\mu}\bar{\psi}_{\nu}\Big)\cdot\bar{\chi}_{i^*}\Big]\,, \\
\omega_{\tilde{\mu}\alpha\beta} \to \omega_{\tilde{\mu}\alpha\beta} &- \frac{\sqrt{2}}{4}\Omega^{-1}\Omega_i\Big[\psi_{\tilde{\mu}\alpha}\chi^i_{\beta} + \psi_{\tilde{\mu}\beta}\chi^i_{\alpha}\Big] \\
&- \frac{\sqrt{2}}{4}\Omega^{-1}e_{\tilde{\mu}\nu}(\sigma^{\nu\rho})_{\beta}{}^{\gamma}\varepsilon_{\alpha\gamma}\Big[\Omega^{i^*}\bar{\psi}_{\rho}\cdot\bar{\chi}_{i^*} + \Omega_i\psi_{\rho}\cdot\chi^i\Big]\,, \\
\omega_{\tilde{\mu}\dot{\alpha}\dot{\beta}} \to \omega_{\tilde{\mu}\dot{\alpha}\dot{\beta}} &- \frac{\sqrt{2}}{4}\Omega^{-1}\Omega^{i^*}\Big[\bar{\psi}_{\tilde{\mu}\dot{\alpha}}\bar{\chi}_{i^*\dot{\beta}} + \bar{\psi}_{\tilde{\mu}\dot{\beta}}\bar{\chi}_{i^*\dot{\alpha}}\Big] \\
&- \frac{\sqrt{2}}{4}\Omega^{-1}e_{\tilde{\mu}\nu}\varepsilon_{\dot{\alpha}\dot{\gamma}}(\bar{\sigma}^{\nu\rho})^{\dot{\gamma}}{}_{\dot{\beta}}\Big[\Omega^{i^*}\bar{\psi}_{\rho}\cdot\bar{\chi}_{i^*} + \Omega_i\psi_{\rho}\cdot\chi^i\Big]\,.
\end{aligned}
\tag{6.45}
$$

In order to obtain the expressions above, we refer to (4.5) and (4.7) for the complete expressions of the derivatives $\mathcal{D}_{\mu}\Sigma$ and $\mathcal{D}_{\mu}\Sigma^{\dagger}$. We can rewrite the covariant derivatives of the fermions (6.9) and the covariant derivative of the gravitino by explicitly indicating the pieces depending on the connection, following (3.5),

$$
\begin{aligned}
\widetilde{\mathcal{D}}_{\tilde{\mu}}\chi^{i\alpha} &= \mathcal{D}_{\tilde{\mu}}\chi^{i\alpha} + \chi^{i\beta}\omega_{\tilde{\mu}\beta}{}^{\alpha}\,, \\
\widetilde{\mathcal{D}}_{\tilde{\mu}}\bar{\chi}^{\dot{\alpha}}_{i^*} &= \mathcal{D}_{\tilde{\mu}}\bar{\chi}^{\dot{\alpha}}_{i^*} - \omega_{\tilde{\mu}}{}^{\dot{\alpha}}{}_{\dot{\beta}}\bar{\chi}^{\dot{\beta}}_{i^*}\,, \\
\widetilde{\mathcal{D}}_{\tilde{\mu}}\psi_{\tilde{\nu}\alpha} &= \partial_{\tilde{\mu}}\psi_{\tilde{\nu}\alpha} - \omega_{\tilde{\mu}\alpha}{}^{\beta}\psi_{\tilde{\nu}\beta}\,, \\
\widetilde{\mathcal{D}}_{\tilde{\mu}}\bar{\psi}_{\tilde{\nu}}{}^{\dot{\alpha}} &= \partial_{\tilde{\mu}}\bar{\psi}_{\tilde{\nu}}{}^{\dot{\alpha}} - \omega_{\tilde{\mu}}{}^{\dot{\alpha}}{}_{\dot{\beta}}\bar{\psi}_{\tilde{\nu}}{}^{\dot{\beta}}\,,
\end{aligned}
\tag{6.46}
$$

where in the notations above, the derivation operator $\mathcal{D}_{\tilde{\mu}}$ is still covariant with respect to the gauge group. Since the gauge bosons are invariant under a super-Weyl rescaling (see (5.52)), we obtain the transformation laws of the derivatives

(6.46) under a pure dilatation,

$$
\begin{aligned}
\widetilde{\mathcal{D}}_{\tilde{\mu}}\chi^{i\alpha} &\to e^{-\frac{1}{2}\lambda}\Big[\widetilde{\mathcal{D}}_{\tilde{\mu}}\chi^{i\alpha} + \frac{1}{4}\Omega^{-1}\partial_{\tilde{\mu}}\Omega\chi^{i\alpha} + \frac{1}{2}e_{\tilde{\mu}\nu}\,\Omega^{-1}\partial_\rho\Omega\,(\chi^i\sigma^{\nu\rho})^\alpha\Big]\,,\\
\widetilde{\mathcal{D}}_{\tilde{\mu}}\bar{\chi}^{\dot{\alpha}}_{i^*} &\to e^{-\frac{1}{2}\lambda}\Big[\widetilde{\mathcal{D}}_{\tilde{\mu}}\bar{\chi}^{\dot{\alpha}}_{i^*} + \frac{1}{4}\Omega^{-1}\partial_{\tilde{\mu}}\Omega\bar{\chi}^{\dot{\alpha}}_{i^*} - \frac{1}{2}e_{\tilde{\mu}\nu}\,\Omega^{-1}\partial_\rho\Omega\,(\bar{\sigma}^{\nu\rho}\bar{\chi}_{i^*})^{\dot{\alpha}}\Big]\,,\\
\widetilde{\mathcal{D}}_{\tilde{\mu}}\psi_{\tilde{\nu}\alpha} &\to e^{\frac{1}{2}\lambda}\Big[\widetilde{\mathcal{D}}_{\tilde{\mu}}\psi_{\tilde{\nu}\alpha} - \frac{1}{4}\Omega^{-1}\partial_{\tilde{\mu}}\Omega\psi_{\tilde{\nu}\alpha} - \frac{1}{2}\Omega^{-1}\partial_\rho\Omega\,e_{\tilde{\mu}\nu}(\sigma^{\nu\rho}\psi_{\tilde{\nu}})_\alpha\Big]\,,\\
\widetilde{\mathcal{D}}_{\tilde{\mu}}\bar{\psi}_{\tilde{\nu}}{}^{\dot{\alpha}} &\to e^{\frac{1}{2}\lambda}\Big[\widetilde{\mathcal{D}}_{\tilde{\mu}}\bar{\psi}_{\tilde{\nu}}{}^{\dot{\alpha}} - \frac{1}{4}\Omega^{-1}\partial_{\tilde{\mu}}\Omega\bar{\psi}_{\tilde{\nu}}{}^{\dot{\alpha}} - \frac{1}{2}\Omega^{-1}\partial_\rho\Omega\,e_{\tilde{\mu}\nu}(\bar{\sigma}^{\nu\rho}\bar{\psi}_{\tilde{\nu}})^{\dot{\alpha}}\Big]\,,
\end{aligned}
\tag{6.47}
$$

and under a gravitino shift

$$
\begin{aligned}
\widetilde{\mathcal{D}}_{\tilde{\mu}}\chi^{i\alpha} \to\; & \widetilde{\mathcal{D}}_{\tilde{\mu}}\chi^{i\alpha} - \frac{\sqrt{2}}{4}\Omega^{-1}\Omega_j\Big[\chi^i\cdot\psi_{\tilde{\mu}}\,\chi^{j\alpha} - \chi^i\cdot\chi^j\,\psi_{\tilde{\mu}}{}^\alpha\Big]\\
& -\frac{\sqrt{2}}{4}e_{\tilde{\mu}\nu}\Omega^{-1}(\chi^i\sigma^{\nu\rho})^\alpha\Big[\Omega^{j^*}\bar{\psi}_\rho\cdot\bar{\chi}_{j^*} + \Omega_j\psi_\rho\cdot\chi^j\Big]\,,\\
\widetilde{\mathcal{D}}_{\tilde{\mu}}\bar{\chi}^{\dot{\alpha}}_{i^*} \to\; & \widetilde{\mathcal{D}}_{\tilde{\mu}}\bar{\chi}^{\dot{\alpha}}_{i^*} - \frac{\sqrt{2}}{4}\Omega^{-1}\Omega^{j^*}\Big[\bar{\chi}_{i^*}\cdot\bar{\psi}_{\tilde{\mu}}\,\bar{\chi}^{\dot{\alpha}}_{j^*} - \bar{\chi}_{i^*}\cdot\bar{\chi}_{j^*}\,\bar{\psi}_{\tilde{\mu}}{}^{\dot{\alpha}}\Big]\\
& +\frac{\sqrt{2}}{4}e_{\tilde{\mu}\nu}\Omega^{-1}(\bar{\sigma}^{\nu\rho}\bar{\chi}_{i^*})^{\dot{\alpha}}\Big[\Omega^{j^*}\bar{\psi}_\rho\cdot\bar{\chi}_{j^*} + \Omega_j\psi_\rho\cdot\chi^j\Big]\,,\\
\widetilde{\mathcal{D}}_{\tilde{\mu}}\psi_{\tilde{\nu}\alpha} \to\; & \widetilde{\mathcal{D}}_{\tilde{\mu}}\psi_{\tilde{\nu}\alpha} + \frac{\sqrt{2}i}{2}\widetilde{\mathcal{D}}_{\tilde{\mu}}\Big[\Omega^{-1}\Omega^{i^*}(\sigma_{\tilde{\nu}}\bar{\chi}_{i^*})_\alpha\Big]\\
& +\frac{\sqrt{2}}{4}\Omega^{-1}\Omega_i\Big(\psi_{\tilde{\mu}\alpha}\,\chi^i\cdot\psi_{\tilde{\nu}} - \chi^i_\alpha\,\psi_{\tilde{\mu}}\cdot\psi_{\tilde{\nu}}\Big)\\
& +\Big[\frac{\sqrt{2}}{4}\Omega^{-1}e_{\tilde{\mu}\nu}(\sigma^{\nu\rho}\psi_{\tilde{\nu}})_\alpha + \frac{i}{4}\Omega^{-2}\Omega^{i^*}e_{\tilde{\mu}\nu}(\sigma^{\nu\rho}\sigma_{\tilde{\nu}}\bar{\chi}_{i^*})_\alpha\Big]\times\\
& \qquad\times\Big[\Omega^{j^*}\bar{\psi}_\rho\cdot\bar{\chi}_{j^*} + \Omega_j\psi_\rho\cdot\chi^j\Big]\\
& +\frac{i}{4}\Omega^{-2}\Omega_i\Omega^{i^*}\Big(\psi_{\tilde{\mu}\alpha}\chi^i\sigma_{\tilde{\nu}}\bar{\chi}_{i^*} - \chi^i_\alpha\psi_{\tilde{\mu}}\sigma_{\tilde{\nu}}\bar{\chi}_{i^*}\Big)\,,\\
\widetilde{\mathcal{D}}_{\tilde{\mu}}\bar{\psi}_{\tilde{\nu}}{}^{\dot{\alpha}} \to\; & \widetilde{\mathcal{D}}_{\tilde{\mu}}\bar{\psi}_{\tilde{\nu}}{}^{\dot{\alpha}} + \frac{\sqrt{2}i}{2}\widetilde{\mathcal{D}}_{\tilde{\mu}}\Big[\Omega^{-1}\Omega_i(\bar{\sigma}_{\tilde{\nu}}\chi^i)^{\dot{\alpha}}\Big]\\
& +\frac{\sqrt{2}}{4}\Omega^{-1}\Omega^{i^*}\Big(\bar{\psi}_{\tilde{\mu}}{}^{\dot{\alpha}}\,\bar{\chi}_{i^*}\cdot\bar{\psi}_{\tilde{\nu}} - \bar{\chi}^{\dot{\alpha}}_{i^*}\,\bar{\psi}_{\tilde{\mu}}\cdot\bar{\psi}_{\tilde{\nu}}\Big)\\
& +\Big[\frac{\sqrt{2}}{4}\Omega^{-1}e_{\tilde{\mu}\nu}(\bar{\sigma}^{\nu\rho}\bar{\psi}_{\tilde{\nu}})^{\dot{\alpha}} + \frac{i}{4}\Omega^{-2}\Omega_i e_{\tilde{\mu}\nu}(\bar{\sigma}^{\nu\rho}\bar{\sigma}_{\tilde{\nu}}\chi^i)^{\dot{\alpha}}\Big]\times\\
& \qquad\Big[\Omega^{j^*}\bar{\psi}_\rho\cdot\bar{\chi}_{j^*} + \Omega_j\psi_\rho\cdot\chi^j\Big]\\
& +\frac{i}{4}\Omega^{-2}\Omega_i\Omega^{i^*}\Big(\bar{\psi}_{\tilde{\mu}}{}^{\dot{\alpha}}\bar{\chi}_{i^*}\bar{\sigma}_{\tilde{\nu}}\chi^i - \bar{\chi}^{\dot{\alpha}}_{i^*}\bar{\psi}_{\tilde{\mu}}\bar{\sigma}_{\tilde{\nu}}\chi^i\Big)\,.
\end{aligned}
\tag{6.48}
$$

Table 6.1 Field redefinition.

To get a correctly normalised Lagrangian, a Weyl rescaling followed by a gravitino shift must be done. The field dependant rescaling is specified by

$$\exp(2\lambda) = -\frac{3}{\Omega}\,,$$

and is given by

$$\begin{aligned}
&e_{\tilde{\mu}}{}^{\mu} \to \exp(\lambda)\, e_{\tilde{\mu}}{}^{\mu}\,,\ \ e_{\mu}{}^{\tilde{\mu}} \to \exp(-\lambda)\, e_{\mu}{}^{\tilde{\mu}}\,,\ \ e \to \exp(4\lambda)\, e\,,\\
&\psi_{\tilde{\mu}} \to e^{\frac{1}{2}\lambda}\psi_{\tilde{\mu}}\,,\ \ \bar{\psi}_{\tilde{\mu}} \to e^{\frac{1}{2}\lambda}\bar{\psi}_{\tilde{\mu}}\,,\\
&\phi^i \to \phi^i\,,\ \ \phi^\dagger_{i^*} \to \phi^\dagger_{i^*}\,,\\
&\chi^i \to e^{-\frac{1}{2}\lambda}\chi^i\,,\ \ \bar{\chi}_{i^*} \to e^{-\frac{1}{2}\lambda}\bar{\chi}_{i^*}\,,\\
&v_{\tilde{\mu}}^a \to v_{\tilde{\mu}}^a\,,\\
&\lambda^a \to e^{-\frac{3}{2}\lambda}\lambda^a\,,\ \ \bar{\lambda}^a \to e^{-\frac{3}{2}\lambda}\bar{\lambda}^a\,.
\end{aligned}$$

From the transformation of the connection given in (6.44) (page 182), the covariant derivatives transform as

$$\begin{aligned}
\widetilde{\mathcal{D}}_{\tilde{\mu}}\chi^{i\alpha} &\to e^{-\frac{1}{2}\lambda}\Big[\widetilde{\mathcal{D}}_{\tilde{\mu}}\chi^{i\alpha} + \tfrac{1}{4}\Omega^{-1}\partial_{\tilde{\mu}}\Omega\chi^{i\alpha} + \tfrac{1}{2}e_{\tilde{\mu}\nu}\,\Omega^{-1}\partial_\rho\Omega\,(\chi^i\sigma^{\nu\rho})^\alpha\Big]\,,\\
\widetilde{\mathcal{D}}_{\tilde{\mu}}\bar{\chi}^{\dot{\alpha}}_{i^*} &\to e^{-\frac{1}{2}\lambda}\Big[\widetilde{\mathcal{D}}_{\tilde{\mu}}\bar{\chi}^{\dot{\alpha}}_{i^*} + \tfrac{1}{4}\Omega^{-1}\partial_{\tilde{\mu}}\Omega\bar{\chi}^{\dot{\alpha}}_{i^*} - \tfrac{1}{2}e_{\tilde{\mu}\nu}\,\Omega^{-1}\partial_\rho\Omega\,(\bar{\sigma}^{\nu\rho}\bar{\chi}_{i^*})^{\dot{\alpha}}\Big]\,,\\
\widetilde{\mathcal{D}}_{\tilde{\mu}}\psi_{\tilde{\nu}\alpha} &\to e^{\frac{1}{2}\lambda}\Big[\widetilde{\mathcal{D}}_{\tilde{\mu}}\psi_{\tilde{\nu}\alpha} - \tfrac{1}{4}\Omega^{-1}\partial_{\tilde{\mu}}\Omega\psi_{\tilde{\nu}\alpha} - \tfrac{1}{2}\Omega^{-1}\partial_\rho\Omega\, e_{\tilde{\mu}\nu}(\sigma^{\nu\rho}\psi_{\tilde{\nu}})_\alpha\Big]\,,\\
\widetilde{\mathcal{D}}_{\tilde{\mu}}\bar{\psi}_{\tilde{\nu}}{}^{\dot{\alpha}} &\to e^{\frac{1}{2}\lambda}\Big[\widetilde{\mathcal{D}}_{\tilde{\mu}}\bar{\psi}_{\tilde{\nu}}{}^{\dot{\alpha}} - \tfrac{1}{4}\Omega^{-1}\partial_{\tilde{\mu}}\Omega\bar{\psi}_{\tilde{\nu}}{}^{\dot{\alpha}} - \tfrac{1}{2}\Omega^{-1}\partial_\rho\Omega\, e_{\tilde{\mu}\nu}(\bar{\sigma}^{\nu\rho}\bar{\psi}_{\tilde{\nu}})^{\dot{\alpha}}\Big]\,.
\end{aligned}$$

The gravitino shift is also field dependant:

$$\psi_{\tilde{\mu}} \to \psi_{\tilde{\mu}} + \frac{\sqrt{2}i}{2}\Omega^{-1}\Omega^{i^*}\sigma_{\tilde{\mu}}\bar{\chi}_{i^*}\,,$$

and the transformation of the covariant derivatives under the gravitino shift are given in Eq.[6.48] (page 183) using the transformation of the connection given in (6.45) (page 182).

from those of the connection. The dilation parametrised by $\Sigma\big|$ and the gravitino shift associated to $\mathcal{D}\Sigma\big|$ leading to the corresponding field redefinitions were already considered in [Ferrara et al. (1976b); Freedman and Schwarz (1977); Cremmer et al. (1979, 1982, 1978b, 1983b)] and later in [Wess and Bagger (1992)] in order to cast the action in a conventionally normalised form.

We summarise in Table 6.1 the field redefinition of the various fields.

6.2.6 *Field redefinition: the gauge sector*

The conformal invariance of the gauge sector ensures its invariance with respect to Howe–Tucker transformations (5.48). However, even if the complete contribution is left unchanged by a super-Weyl rescaling,

$$\delta_\Sigma\Big(\mathcal{L}_{\text{gauge,kin}} + \mathcal{L}_{\text{gauge,int}} + \mathcal{L}_{\text{gauge,aux}}\Big) = 0\,, \tag{6.49}$$

the kinetic terms $\mathcal{L}_{\text{gauge,kin}}$, interaction terms $\mathcal{L}_{\text{gauge,int}}$ and the terms depending on the auxiliary fields $\mathcal{L}_{\text{gauge,aux}}$ are separately not invariant. We refer to (6.22) for the expressions of the three parts of the gauge sector. However, one has to be careful with the terms arising from the insertion of the solutions of the equations of motion for the auxiliary fields as they are mixing chiral and vector superfields. We adopt the choice to rescale them explicitly in Sec. 6.2.8, so that we now have to compute the variation of $\mathcal{L}_{\text{gauge,kin}}$ and $\mathcal{L}_{\text{gauge,int}}$. From (6.49), one has

$$\delta_\Sigma\Big(\mathcal{L}_{\text{gauge,kin}} + \mathcal{L}_{\text{gauge,int}}\Big) = -\delta_\Sigma\mathcal{L}_{\text{gauge,aux}}\,, \tag{6.50}$$

where

$$\begin{aligned}\mathcal{L}_{\text{gauge, aux}} = &\ \frac{1}{2}eh^R_{ab}D^aD^b - \frac{\sqrt{2}i}{4}eD^a\Big[h_{abi}\chi^i\cdot\lambda^b - h^\star_{ab}{}^{i^*}\overline{\chi}_{i^*}\cdot\overline{\lambda}^b\Big]\\ &-\frac{1}{2}eh^R_{ab}\lambda^a\sigma^\mu\overline{\lambda}^b b_\mu + \frac{1}{4}e\Big[h_{abi}F^i\lambda^a\cdot\lambda^b + h^\star_{ab}{}^{i^*}F^\dagger_{i^*}\overline{\lambda}^a\cdot\overline{\lambda}^b\Big]\,.\end{aligned} \tag{6.51}$$

Since (6.51) is invariant under a pure dilatation of the fields, according to (6.40), (6.41), (6.42) and (6.43), it is sufficient to shift the auxiliary fields. Therefore, the Lagrangian related to the gauge sector becomes,

$$\begin{aligned}\mathcal{L}_{\text{gauge,kin}} + \mathcal{L}_{\text{gauge,int}} \to &\ \mathcal{L}_{\text{gauge,kin}} + \mathcal{L}_{\text{gauge,int}}\\ &+\frac{\sqrt{2}i}{8}eh^R_{ab}\lambda^a\sigma^\mu\overline{\lambda}^b\Big[K_i\chi^i\cdot\psi_\mu - K^{i^*}\overline{\chi}_{i^*}\cdot\overline{\psi}_\mu\Big]\\ &-\frac{1}{12}e\Big[h_{abi}K_j\chi^i\cdot\chi^j\lambda^a\cdot\lambda^b + h^\star_{ab}{}^{i^*}K^{i^*}\overline{\chi}_{i^*}\cdot\overline{\chi}_{j^*}\overline{\lambda}^a\cdot\overline{\lambda}^b\Big]\,,\end{aligned} \tag{6.52}$$

after shifting the b- and F-fields and after replacing the superspace kinetic energy and its derivatives by the Kähler potential and its derivatives, using (6.27).

Concerning the kinetic terms related to the gaugino field, a remark is in order. Inspecting the fourth and third terms starting from the end of (6.33), we realise that they can both be grouped together after introducing a new derivative $\check{\mathcal{D}}_\mu$. This derivation has been made covariant with respect to the $U(1)$-connection of the Kähler manifold, as illustrated in (5.54), in addition to the (already considered) covariance with respect to gauge transformations, supergravity transformations and the Kähler manifold as included in the covariant derivative $\widetilde{\mathcal{D}}_\mu$,

$$\begin{aligned}&-\frac{i}{2}eh^R_{ab}\Big[\lambda^a\sigma^\mu\widetilde{\mathcal{D}}_\mu\overline{\lambda}^b - \widetilde{\mathcal{D}}_\mu\lambda^a\sigma^\mu\lambda^b - \frac{3}{2}\Omega^{-1}(\Omega^{i^*}\widetilde{\mathcal{D}}_\mu\phi^\dagger_{i^*} - \Omega_i\widetilde{\mathcal{D}}_\mu\phi^i)\lambda^a\sigma^\mu\overline{\lambda}^b\Big]\\ &= -\frac{i}{2}eh^R_{ab}\Big(\lambda^a\sigma^\mu\check{\mathcal{D}}_\mu\overline{\lambda}^b - \check{\mathcal{D}}_\mu\lambda^a\sigma^\mu\overline{\lambda}^b\Big)\,.\end{aligned}$$

The full covariant derivative of the gaugino field is then defined as in (5.54),

$$\check{\mathcal{D}}_{\tilde{\mu}}\lambda^a = \widetilde{\mathcal{D}}_{\tilde{\mu}}\lambda^a + \frac{1}{4}(K_i\widetilde{\mathcal{D}}_\mu\phi^i - K_{i^*}\widetilde{\mathcal{D}}_\mu\phi^\dagger_{i^*})\lambda^a$$
$$\check{\mathcal{D}}_{\tilde{\mu}}\bar{\lambda}^a = \widetilde{\mathcal{D}}_{\tilde{\mu}}\bar{\lambda}^a - \frac{1}{4}(K_i\widetilde{\mathcal{D}}_\mu\phi^i - K_{i^*}\widetilde{\mathcal{D}}_\mu\phi^\dagger_{i^*})\bar{\lambda}^a .$$

Observe also that

$$\check{\mathcal{D}}_\mu(\lambda^a\sigma^\mu\bar{\lambda}^b) = \widetilde{\mathcal{D}}_\mu(\lambda^a\sigma^\mu\bar{\lambda}^b) .$$

Collecting the different contributions, the pure kinetic terms are given by

$$\begin{aligned}\mathcal{L}^{(\text{gauge})}_{\text{kin}} &= -\frac{1}{4}eh^R_{ab}F^a_{\mu\nu}F^{\mu\nu b} + \frac{1}{4}eh^I_{ab}F^a_{\mu\nu}{}^*F^{\mu\nu b}\\ &\quad - \frac{i}{2}eh^R_{ab}\Big(\lambda^a\sigma^\mu\check{\mathcal{D}}_\mu\bar{\lambda}^b - \check{\mathcal{D}}_\mu\lambda^a\sigma^\mu\bar{\lambda}^b\Big) + \frac{1}{2}eh^I_{ab}\check{\mathcal{D}}_\mu(\lambda^a\sigma^\mu\bar{\lambda}^b)\\ &\quad + \frac{i}{4}eh^R_{ab}\Big[\lambda^a\sigma_\mu\bar{\sigma}_{\nu\rho}\bar{\psi}^\mu + \bar{\lambda}^a\bar{\sigma}_\mu\sigma_{\nu\rho}\psi^\mu\Big]\Big[F^{\nu\rho b} + \hat{F}^{\nu\rho b}\Big]\\ &\quad - \frac{\sqrt{2}}{4}e\Big[h_{abi}\chi^i\sigma^{\mu\nu}\lambda^b + h^\star_{ab}{}^{i^*}\bar{\chi}_{i^*}\bar{\sigma}^{\mu\nu}\bar{\lambda}^b\Big]F^a_{\mu\nu}\end{aligned} \tag{6.53}$$

while the interaction terms read

$$\begin{aligned}\mathcal{L}^{(\text{gauge})}_{\text{int}} &= \frac{1}{8}e\Big[h_{abij}\chi^i\cdot\chi^j\lambda^a\cdot\lambda^b + h^\star_{ab}{}^{i^*j^*}\bar{\chi}_{i^*}\cdot\bar{\chi}_{j^*}\bar{\lambda}^a\cdot\bar{\lambda}^b\Big]\\ &\quad - \frac{\sqrt{2}i}{8}e\Big[h_{abi}\bar{\psi}_\mu\bar{\sigma}^\mu\chi^i\lambda^a\cdot\lambda^b + h^\star_{ab}{}^{i^*}\psi_\mu\sigma^\mu\bar{\chi}_{i^*}\bar{\lambda}^a\cdot\bar{\lambda}^b\Big]\\ &\quad + \frac{\sqrt{2}i}{4}e\Big[h_{abi}\chi^i\sigma^{\mu\nu}\lambda^a + h^\star_{ab}{}^{i^*}\bar{\chi}_{i^*}\bar{\sigma}^{\mu\nu}\bar{\lambda}^a\Big]\Big[\lambda^b\sigma_\nu\bar{\psi}_\mu + \bar{\lambda}^b\bar{\sigma}_\nu\psi_\mu\Big]\\ &\quad + \frac{1}{4}eh^I_{ab}\varepsilon^{\mu\nu\rho\sigma}\bar{\psi}_\mu\bar{\sigma}_\nu\lambda^a\psi_\rho\sigma_\sigma\bar{\lambda}^b\\ &\quad + \frac{\sqrt{2}i}{8}eh^R_{ab}\lambda^a\sigma^\mu\bar{\lambda}^b\Big[K_i\chi^i\cdot\psi_\mu - K^{i^*}\bar{\chi}_{i^*}\cdot\bar{\psi}_\mu\Big]\\ &\quad - \frac{1}{12}e\Big[h_{abi}K_j\chi^i\cdot\chi^j\lambda^a\cdot\lambda^b + h^\star_{ab}{}^{i^*}K^{i^*}\bar{\chi}_{i^*}\cdot\bar{\chi}_{j^*}\bar{\lambda}^a\cdot\bar{\lambda}^b\Big] .\end{aligned} \tag{6.54}$$

Observe that the Lagrangians above do not include the additional terms arising from the auxiliary fields. The latter will be treated in Sec. 6.2.8. Let us note however that we have included the new interacting terms appearing in (6.52).

6.2.7 *Field redefinition: kinetic terms for the matter sector*

We now address the field redefinition terms-by-terms focusing first on the kinetic terms. As stated previously, the calculation of the transformation laws of the different parts will be performed in two steps, applying first a pure dilation of the fields, followed by a pure gravitino shift.

Rescaling of the graviton kinetic terms.

As mentioned above, Weyl rescaling allows for getting a correct normalisation of the gravitational action, as sketched in (6.35). However, additional terms appear due to the variation of the connection since the curvature tensor $R_{\tilde{\mu}\tilde{\nu}}{}^{\mu\nu}$ depends on it. Under a pure dilation, the connection undergoes the transformation (6.44) and consequently, the gravitational action becomes

$$\mathcal{L}_{\text{grav}} = -\frac{1}{6}e\Omega e_\mu{}^{\tilde{\mu}} e_\nu{}^{\tilde{\nu}} R_{\tilde{\mu}\tilde{\nu}}{}^{\mu\nu}\Big| \to \frac{1}{2}ee_\mu{}^{\tilde{\mu}} e_\nu{}^{\tilde{\nu}} R_{\tilde{\mu}\tilde{\nu}}{}^{\mu\nu}\Big| + \delta_e^{(1)} , \tag{6.55}$$

with

$$\begin{aligned}
\delta_e^{(1)} = \frac{1}{2}ee_\mu{}^{\tilde{\mu}} e_\nu{}^{\tilde{\nu}} \bigg\{ & -\frac{1}{2}\partial_{\tilde{\mu}}\Big[\Omega^{-1}\partial_{\tilde{\rho}}\Omega(e_{\tilde{\nu}}{}^{\nu} e^{\mu\tilde{\rho}} - e_{\tilde{\nu}}{}^{\mu} e^{\nu\tilde{\rho}})\Big] \\
& + \frac{1}{2}\partial_{\tilde{\nu}}\Big[\Omega^{-1}\partial_{\tilde{\rho}}\Omega(e_{\tilde{\mu}}{}^{\nu} e^{\mu\tilde{\rho}} - e_{\tilde{\mu}}{}^{\mu} e^{\nu\tilde{\rho}})\Big] \\
& - \frac{1}{2}\Omega^{-1}\partial_{\tilde{\rho}}\Omega\Big[\omega_{\tilde{\nu}}{}^{\mu\rho}(e_{\tilde{\mu}}{}^{\nu} e_\rho{}^{\tilde{\rho}} - e_{\tilde{\mu}\rho} e^{\nu\tilde{\rho}}) + \omega_{\tilde{\mu}\rho}{}^{\nu}(e_{\tilde{\nu}}{}^{\rho} e^{\mu\tilde{\rho}} - e_{\tilde{\nu}}{}^{\mu} e^{\rho\tilde{\rho}}) - (\tilde{\mu} \leftrightarrow \tilde{\nu})\Big] \\
& + \frac{1}{4}\Omega^{-2}\partial_{\tilde{\rho}}\Omega\partial_{\tilde{\sigma}}\Omega\Big[(e_{\tilde{\nu}}{}^{\rho} e^{\mu\tilde{\rho}} - e_{\tilde{\nu}}{}^{\mu} e^{\rho\tilde{\rho}})(e_{\tilde{\mu}}{}^{\nu} e_\rho{}^{\tilde{\sigma}} - e_{\tilde{\mu}\rho} e^{\nu\tilde{\sigma}}) \\
& \qquad - (e_{\tilde{\mu}}{}^{\rho} e^{\mu\tilde{\rho}} - e_{\tilde{\mu}}{}^{\mu} e^{\rho\tilde{\rho}})(e_{\tilde{\nu}}{}^{\nu} e_\rho{}^{\tilde{\sigma}} - e_{\tilde{\nu}\rho} e^{\nu\tilde{\sigma}})\Big]\bigg\} \\
= & -\frac{3}{2}\partial_{\tilde{\mu}}(e\Omega^{-1}\partial^{\tilde{\mu}}\Omega) + \Omega^{-1}\partial^\mu\Omega\partial_{\tilde{\mu}}(ee_\mu{}^{\tilde{\mu}}) + \Omega^{-1}\partial_\rho\Omega e_\mu{}^{\tilde{\mu}}\omega_{\tilde{\mu}}{}^{\mu\rho} + \frac{3}{4}e\Omega^{-2}\partial_\mu\Omega\partial^\mu\Omega \\
= & -\frac{3}{2}\partial_{\tilde{\mu}}(e\Omega^{-1}\partial^{\tilde{\mu}}\Omega) - \frac{i}{2}e\Omega^{-1}\partial_\rho\Omega\Big[\psi_\mu\sigma^\mu\bar{\psi}^\rho - \psi^\rho\sigma^\mu\bar{\psi}_\mu\Big] + \frac{3}{4}e\Omega^{-2}\partial_\mu\Omega\partial^\mu\Omega .
\end{aligned} \tag{6.56}$$

The first two terms of the right-hand side of the second equality have been deduced from the first two lines of the right-hand side of the first equality using the definition of the metric in curved space in terms of the vierbein (see Sec. 3.3.2) together with the spacetime derivative of e given in (5.30) and after an integration by parts. The third line of the right-hand side of the first equality can be simplified to the third term of the right-hand side of the second equality whilst the terms in Ω^{-2} appearing in the first equality can be reduced to the last term of the right-hand side of the second equality, using again the expression of the metric in curved space in terms of the vierbein. The last equality is derived from the second one after collecting contributions to the covariant derivatives and using (6.11). Let us recall the relation $e_{\tilde{\mu}\mu} = e_{\mu\tilde{\mu}}$ and that care must be taken when rising and lowering curved indices, since the spacetime derivative of the (curved) metric does not vanish.

We now perform a pure gravitino shift of (6.55) and (6.56). The graviton kinetic term becomes

$$\frac{1}{2}ee_\mu{}^{\tilde{\mu}} e_\nu{}^{\tilde{\nu}} R_{\tilde{\mu}\tilde{\nu}}{}^{\mu\nu}\Big| \to \frac{1}{2}ee_\mu{}^{\tilde{\mu}} e_\nu{}^{\tilde{\nu}} R_{\tilde{\mu}\tilde{\nu}}{}^{\mu\nu}\Big| + \delta_e^{(2)} ,$$

where the variation is computed from the shift of the connection (6.45),

$$
\begin{aligned}
\delta_e^{(2)} = \frac{1}{2} e\Big[e_\mu{}^{\tilde\mu} e_\nu{}^{\tilde\nu} - e_\nu{}^{\tilde\mu} e_\mu{}^{\tilde\nu}\Big]\Bigg\{ & \frac{\sqrt{2}}{2}\partial_{\tilde\mu}\Big[\Omega^{-1}(\Omega_i\psi_{\tilde\nu}\sigma^{\mu\nu}\chi^i + \Omega^{i^*}\bar\psi_{\tilde\nu}\bar\sigma^{\mu\nu}\bar\chi_{i^*})\Big] \\
& + \frac{\sqrt{2}}{4}\partial_{\tilde\mu}\Big[\Omega^{-1}(e_{\tilde\nu}{}^\nu\psi^\mu - e_{\tilde\nu}{}^\mu\psi^\nu)\cdot\Omega_i\chi^i\Big] + \frac{\sqrt{2}}{4}\partial_{\tilde\mu}\Big[\Omega^{-1}(e_{\tilde\nu}{}^\nu\bar\psi^\mu - e_{\tilde\nu}{}^\mu\bar\psi^\nu)\cdot\Omega^{i^*}\bar\chi_{i^*}\Big] \\
& + \frac{\sqrt{2}}{2}\Omega^{-1}\Big[\Omega_i\psi_{\tilde\mu}\sigma_\rho{}^\nu\chi^i + \Omega^{i^*}\bar\psi_{\tilde\mu}\bar\sigma_\rho{}^\nu\bar\chi_{i^*} + \frac{1}{2}(e_{\tilde\mu}{}^\nu\psi_\rho - e_{\tilde\mu\rho}\psi^\nu)\cdot\Omega_i\chi^i \\
& \qquad\qquad + \frac{1}{2}(e_{\tilde\mu}{}^\nu\bar\psi_\rho - e_{\tilde\mu\rho}\bar\psi^\nu)\cdot\Omega^{i^*}\bar\chi_{i^*}\Big]\omega_{\tilde\nu}{}^{\mu\rho} \\
& + \frac{\sqrt{2}}{2}\Omega^{-1}\Big[\Omega_i\psi_{\tilde\nu}\sigma^{\mu\rho}\chi^i + \Omega^{i^*}\bar\psi_{\tilde\nu}\bar\sigma^{\mu\rho}\bar\chi_{i^*} + \frac{1}{2}(e_{\tilde\nu}{}^\rho\psi^\mu - e_{\tilde\nu}{}^\mu\psi^\rho)\cdot\Omega_i\chi^i \\
& \qquad\qquad + \frac{1}{2}(e_{\tilde\nu}{}^\rho\bar\psi^\mu - e_{\tilde\nu}{}^\mu\bar\psi^\rho)\cdot\Omega^{i^*}\bar\chi_{i^*}\Big]\omega_{\tilde\mu\rho}{}^\nu \\
& + \frac{1}{2}\Omega^{-2}\Big[\Omega_i\psi_{\tilde\nu}\sigma^{\mu\rho}\chi^i + \Omega^{i^*}\bar\psi_{\tilde\nu}\bar\sigma^{\mu\rho}\bar\chi_{i^*}\Big]\Big[\Omega_j\psi_{\tilde\mu}\sigma_\rho{}^\nu\chi^j + \Omega^{j^*}\bar\psi_{\tilde\mu}\bar\sigma_\rho{}^\nu\bar\chi_{j^*}\Big] \\
& + \frac{1}{4}\Omega^{-2}\Big[\Omega_i\psi_{\tilde\nu}\sigma^{\mu\rho}\chi^i + \Omega^{i^*}\bar\psi_{\tilde\nu}\bar\sigma^{\mu\rho}\bar\chi_{i^*}\Big] \\
& \qquad \times\Big[(e_{\tilde\mu}{}^\nu\psi_\rho - e_{\tilde\mu\rho}\psi^\nu)\cdot\Omega_j\chi^j + (e_{\tilde\mu}{}^\nu\bar\psi_\rho - e_{\tilde\mu\rho}\bar\psi^\nu)\cdot\Omega^{j^*}\bar\chi_{j^*}\Big] \\
& + \frac{1}{4}\Omega^{-2}\Big[\Omega_i\psi_{\tilde\mu}\sigma_\rho{}^\nu\chi^i + \Omega^{i^*}\bar\psi_{\tilde\mu}\bar\sigma_\rho{}^\nu\bar\chi_{i^*}\Big] \\
& \qquad \times\Big[(e_{\tilde\nu}{}^\rho\psi^\mu - e_{\tilde\nu}{}^\mu\psi^\rho)\cdot\Omega_j\chi^j + (e_{\tilde\nu}{}^\rho\bar\psi^\mu - e_{\tilde\nu}{}^\mu\bar\psi^\rho)\cdot\Omega^{j^*}\bar\chi_{j^*}\Big] \\
& + \frac{1}{8}\Omega^{-2}\Big[(e_{\tilde\mu}{}^\nu\psi_\rho - e_{\tilde\mu\rho}\psi^\nu)\cdot\Omega_i\chi^i + (e_{\tilde\mu}{}^\nu\bar\psi_\rho - e_{\tilde\mu\rho}\bar\psi^\nu)\cdot\Omega^{i^*}\bar\chi_{i^*}\Big] \\
& \qquad \times\Big[(e_{\tilde\nu}{}^\rho\psi^\mu - e_{\tilde\nu}{}^\mu\psi^\rho)\cdot\Omega_j\chi^j + (e_{\tilde\nu}{}^\rho\bar\psi^\mu - e_{\tilde\nu}{}^\mu\bar\psi^\rho)\cdot\Omega^{j^*}\bar\chi_{j^*}\Big]\Bigg\} \\
= & -\frac{\sqrt{2}}{2}\Omega^{-1}\partial_{\tilde\mu}(ee_\mu{}^{\tilde\mu})\Big[\Omega_i\psi_\nu\sigma^{\mu\nu}\chi^i + \Omega^{i^*}\bar\psi_\nu\bar\sigma^{\mu\nu}\bar\chi_{i^*}\Big] \\
& + \frac{\sqrt{2}}{2}\Omega^{-1}e\partial_{\tilde\mu}e_{\tilde\nu}{}^\rho\Big[\Omega_i\psi_\rho\sigma^{\tilde\mu\tilde\nu}\chi^i + \Omega^{i^*}\bar\psi_\rho\bar\sigma^{\tilde\mu\tilde\nu}\bar\chi_{i^*}\Big] \\
& - \frac{\sqrt{2}}{2}\Omega^{-1}\partial_{\tilde\mu}(ee_\mu{}^{\tilde\mu})\Big[\Omega_i\psi^\mu\cdot\chi^i + \Omega^{i^*}\bar\psi^\mu\cdot\bar\chi_{i^*}\Big] \\
& + \frac{\sqrt{2}}{2}e\Omega^{-1}\Big[\Omega_i\psi_\nu\sigma^{\mu\nu}\chi^i + \Omega^{i^*}\bar\psi_\nu\bar\sigma^{\mu\nu}\bar\chi_{i^*}\Big]\omega_{\tilde\mu\mu}{}^\rho e_\rho{}^{\tilde\mu} \\
& + \frac{\sqrt{2}}{2}e\Omega^{-1}\Big[\Omega_i\psi_{\tilde\nu}\sigma^{\mu\nu}\chi^i + \Omega^{i^*}\bar\psi_{\tilde\nu}\bar\sigma^{\mu\nu}\bar\chi_{i^*}\Big]e_\mu{}^{\tilde\mu}\omega_{\tilde\mu\nu}{}^\rho e_\rho{}^{\tilde\nu} \\
& + \frac{\sqrt{2}}{2}e\Omega^{-1}\Big[\Omega_i\psi^\mu\cdot\chi^i + \Omega^{i^*}\bar\psi^\mu\cdot\bar\chi_{i^*}\Big]\omega_{\tilde\mu\mu}{}^\rho e_\rho{}^{\tilde\mu}
\end{aligned}
$$

$$
\begin{aligned}
&-\frac{1}{2}e\Omega^{-2}\Omega_i\Omega^{i^*}\Big[\psi_\mu\sigma^{\mu\nu}\chi^i\bar{\psi}_\nu\cdot\bar{\chi}_{i^*}+\bar{\psi}_\mu\bar{\sigma}^{\mu\nu}\bar{\chi}_{i^*}\psi_\nu\cdot\chi^i\Big]\\
&+\frac{3}{8}e\Omega^{-2}\Big[\Omega_i\psi^\mu\cdot\chi^i+\Omega^{i^*}\bar{\psi}^\mu\cdot\bar{\chi}_{i^*}\Big]\Big[\Omega_j\psi_\mu\cdot\chi^j+\Omega^{j^*}\bar{\psi}_\mu\cdot\bar{\chi}_{j^*}\Big]\\
&+\frac{\sqrt{2}}{2}\partial_{\tilde{\mu}}\Big[e\Omega^{-1}(\Omega_i\psi_\nu\sigma^{\tilde{\mu}\nu}\chi^i+\Omega^{i^*}\bar{\psi}_\nu\bar{\sigma}^{\tilde{\mu}\nu}\bar{\chi}_i)\Big]\\
&+\frac{3\sqrt{2}}{4}\partial_{\tilde{\mu}}\Big[e\Omega^{-1}(\Omega_i\psi^{\tilde{\mu}}\cdot\chi^i+\Omega i^*\bar{\psi}^{\tilde{\mu}}\cdot\bar{\chi}_{i^*})\Big]\,.
\end{aligned}
\tag{6.57}
$$

In the equation above, the relation between the right-hand side of the first and second equality is as follows. The first line of the first equality leads to the two first lines of the second equality, up to a total derivative given as the next-to-last term of the expression above. For the computation of the second term, we refer to a procedure similar to the one followed in (6.13). The second line of the first equality gives rise to the third line of the second equality, up to a total derivative shown as the last term of the above expression. This computation requires the expression of the spacetime derivative of e given in (5.30)). Lines three to six of the first equality correspond to the three terms depending on the connection appearing on the right-hand side of the second equality. These terms are immediately derived from the properties of the vierbein. Finally, the terms in Ω^{-2} can be simplified after using (C.6) and (C.8) and the Fierz identity (C.9). Collecting terms to form covariant derivatives, the variation $\delta_e^{(2)}$ can be rewritten in a more compact form as

$$
\begin{aligned}
\delta_e^{(2)} = {}& \frac{\sqrt{2}i}{4}e\Omega^{-1}\Big[\psi_\rho\sigma^\rho\bar{\psi}_\mu-\psi_\mu\sigma^\rho\bar{\psi}_\rho\Big]\Big[\Omega_i\psi_\nu\sigma^{\mu\nu}\chi^i+\Omega^{i^*}\bar{\psi}_\nu\bar{\sigma}^{\mu\nu}\bar{\chi}_{i^*}\Big]\\
&+\frac{\sqrt{2}i}{4}e\Omega^{-1}\psi_\mu\sigma^\rho\bar{\psi}_\nu\Big[\Omega_i\psi_\rho\sigma^{\mu\nu}\chi^i+\Omega^{i^*}\bar{\psi}_\rho\bar{\sigma}^{\mu\nu}\bar{\chi}_{i^*}\Big]\\
&+\frac{\sqrt{2}i}{4}e\Omega^{-1}\Big[\psi_\rho\sigma^\rho\bar{\psi}_\mu-\psi_\mu\sigma^\rho\bar{\psi}_\rho\Big]\Big[\Omega_i\psi^\mu\cdot\chi^i+\Omega^{i^*}\bar{\psi}^\mu\cdot\bar{\chi}_{i^*}\Big]\\
&-\frac{1}{2}e\Omega^{-2}\Omega_i\Omega^{i^*}\Big[\psi_\mu\sigma^{\mu\nu}\chi^i\bar{\psi}_\nu\cdot\bar{\chi}_{i^*}+\bar{\psi}_\mu\bar{\sigma}^{\mu\nu}\bar{\chi}_{i^*}\psi_\nu\cdot\chi^i\Big]\\
&+\frac{3}{8}e\Omega^{-2}\Big[\Omega_i\psi^\mu\cdot\chi^i+\Omega^{i^*}\bar{\psi}^\mu\cdot\bar{\chi}_{i^*}\Big]\Big[\Omega_j\psi_\mu\cdot\chi^j+\Omega^{j^*}\bar{\psi}_\mu\cdot\bar{\chi}_{j^*}\Big]\,,
\end{aligned}
\tag{6.58}
$$

where the derivatives of the vierbein have been simplified according to (6.11) and (6.13) and where we have omitted the total derivatives.

Finally, the gravitino shift must also be applied to the second term of (6.55), *i.e.*, to the variation $\delta_e^{(1)}$,

$$
\delta_e^{(1)} \to \delta_e^{(1)} + \delta_e^{(2,1)}\,. \tag{6.59}
$$

and the expression of $\delta_e^{(2,1)}$ can be directly derived from the transformation laws of the gravitino field

$$\begin{aligned}\delta_e^{(2,1)} &= -\frac{3\sqrt{2}}{4}e\Omega^{-2}\partial_\rho\Omega\left[\Omega_i\psi^\rho\cdot\chi^i + \Omega^{i^*}\bar{\psi}^\rho\cdot\bar{\chi}_{i^*}\right] \\ &\quad - \frac{\sqrt{2}}{2}e\Omega^{-2}\partial_\mu\Omega\left[\Omega_i\psi_\nu\sigma^{\mu\nu}\chi^i + \Omega^{i^*}\bar{\psi}_\nu\bar{\sigma}^{\mu\nu}\bar{\chi}_{i^*}\right],\end{aligned}$$

where only the second term of (6.56) contributes.

Collecting all the contributions while omitting all the total derivatives, the transformed gravitational Lagrangian reads

$$\begin{aligned}\mathcal{L}_{\text{grav}} \to\ & \frac{1}{2}ee_\mu{}^{\tilde{\mu}}e_\nu{}^{\tilde{\nu}}R_{\tilde{\mu}\tilde{\nu}}{}^{\mu\nu}\Big| + \frac{3}{4}e\Omega^{-2}\partial_\mu\Omega\partial^\mu\Omega - \frac{i}{2}e\Omega^{-1}\partial_\rho\Omega\left[\psi_\mu\sigma^\mu\bar{\psi}^\rho - \psi^\rho\sigma^\mu\bar{\psi}_\mu\right] \\ & -\frac{3\sqrt{2}}{4}e\Omega^{-2}\partial_\mu\Omega\left[\Omega_i\psi^\mu\cdot\chi^i + \Omega^{i^*}\bar{\psi}^\mu\cdot\bar{\chi}_{i^*}\right] \\ & -\frac{\sqrt{2}}{2}e\Omega^{-2}\partial_\mu\Omega\left[\Omega_i\psi_\nu\sigma^{\mu\nu}\chi^i + \Omega^{i^*}\bar{\psi}_\nu\bar{\sigma}^{\mu\nu}\bar{\chi}_{i^*}\right] \\ & +\frac{3}{8}e\Omega^{-2}\left[\Omega_i\psi^\mu\cdot\chi^i + \Omega^{i^*}\bar{\psi}^\mu\cdot\bar{\chi}_{i^*}\right]\left[\Omega_j\psi_\mu\cdot\chi^j + \Omega^{j^*}\bar{\psi}_\mu\cdot\bar{\chi}_{j^*}\right] \\ & +\frac{\sqrt{2}i}{4}e\Omega^{-1}\psi_\mu\sigma^\rho\bar{\psi}_\nu\left[\Omega_i\psi_\rho\sigma^{\mu\nu}\chi^i + \Omega^{i^*}\bar{\psi}_\rho\bar{\sigma}^{\mu\nu}\bar{\chi}_{i^*}\right] \\ & -\frac{1}{2}e\Omega^{-2}\Omega_i\Omega^{i^*}\left[\psi_\mu\sigma^{\mu\nu}\chi^i\bar{\psi}_\nu\cdot\bar{\chi}_{i^*} + \bar{\psi}_\mu\bar{\sigma}^{\mu\nu}\bar{\chi}_{i^*}\psi_\nu\cdot\chi^i\right] \\ & +\frac{\sqrt{2}i}{4}e\Omega^{-1}\left[\psi_\rho\sigma^\rho\bar{\psi}_\mu - \psi_\mu\sigma^\rho\bar{\psi}_\rho\right]\times \\ & \qquad \times\left[\Omega_i\psi^\mu\cdot\chi^i + \Omega^{i^*}\bar{\psi}^\mu\cdot\bar{\chi}_{i^*} + \Omega_i\psi_\nu\sigma^{\mu\nu}\chi^i + \Omega^{i^*}\bar{\psi}_\nu\bar{\sigma}^{\mu\nu}\bar{\chi}_{i^*}\right].\end{aligned}$$

Rescaling of the gravitino kinetic terms.

Under a pure dilation, the transformation laws of the first derivatives of the gravitino read as (6.47) while those of gravitino are given in (6.39). Together with the Howe–Tucker transformations of the vielbein (6.38), one derives the variation of the gravitino Lagrangian,

$$\begin{aligned}\mathcal{L}_{3/2} &= -\frac{1}{12}e\Omega\varepsilon^{\mu\nu\rho\sigma}\left[\psi_\mu\sigma_\sigma\bar{\psi}_{\nu\rho} - \bar{\psi}_\mu\bar{\sigma}_\sigma\psi_{\nu\rho}\right] \\ &\to \frac{1}{4}e\varepsilon^{\mu\nu\rho\sigma}\left[\psi_\mu\sigma_\sigma\bar{\psi}_{\nu\rho} - \bar{\psi}_\mu\bar{\sigma}_\sigma\psi_{\nu\rho}\right] + \delta_{3/2}^{(1)}\end{aligned} \qquad (6.60)$$

where the kinetic terms describing the dynamics of the gravitino is now standardly normalised. The variation $\delta^{(1)}_{3/2}$ is computed as

$$\delta^{(1)}_{3/2} = \frac{1}{2}e\varepsilon^{\mu\nu\rho\sigma}e_\mu{}^{\tilde\mu}e_\nu{}^{\tilde\nu}e_\rho{}^{\tilde\rho}\Big[-\frac{1}{4}\Omega^{-1}\partial_{\tilde\nu}\Omega\psi_{\tilde\mu}\sigma_\sigma\bar\psi_{\tilde\rho} - \frac{1}{2}\Omega^{-1}\partial_\kappa\Omega\, e_{\tilde\nu\eta}\,\psi_{\tilde\mu}\sigma_\sigma\bar\sigma^{\eta\kappa}\bar\psi_{\tilde\rho}\Big] + \text{h.c.}$$
$$= -\frac{i}{2}e\Omega^{-1}\partial_\mu\Omega\big(\psi^\mu\sigma^\nu\bar\psi_\nu + \bar\psi^\mu\bar\sigma^\nu\psi_\nu\big)\,, \qquad (6.61)$$

where we have used (C.5) and (C.12). The full variation under Howe–Tucker transformation is obtained after applying a gravitino shift to the expression above, *i.e.*, form (6.48) and (6.39). The Rarita-Schwinger Lagrangian, hence becomes

$$\frac{1}{4}e\varepsilon^{\mu\nu\rho\sigma}\Big[\psi_\mu\sigma_\sigma\bar\psi_{\nu\rho} - \bar\psi_\mu\bar\sigma_\sigma\psi_{\nu\rho}\Big] \to \frac{1}{4}e\varepsilon^{\mu\nu\rho\sigma}\Big[\psi_\mu\sigma_\sigma\bar\psi_{\nu\rho} - \bar\psi_\mu\bar\sigma_\sigma\psi_{\nu\rho}\Big] + \delta^{(2)}_{3/2}\,.$$

The computation of the variation $\delta^{(2)}_{3/2}$ proceeds as

$$\begin{aligned}
\delta^{(2)}_{3/2} = \frac{1}{2}e\varepsilon^{\mu\nu\rho\sigma}\Big[&\frac{\sqrt{2}i}{2}\psi_\mu\sigma_\sigma e_\rho{}^{\tilde\rho}\widetilde{\mathcal{D}}_\nu\big(\Omega^{-1}\Omega_i\bar\sigma_{\tilde\rho}\chi^i\big)\\
&+\Omega^{-1}\Omega^{i^*}\bar\chi_{i^*}\bar\sigma_\mu\sigma_\sigma\Big(\frac{1}{2}e_\rho{}^{\tilde\rho}\widetilde{\mathcal{D}}_\nu[\Omega^{-1}\Omega_i\bar\sigma_{\tilde\rho}\chi^i] - \frac{\sqrt{2}i}{4}\psi_{\nu\rho}\Big)\Big]\\
&+\frac{\sqrt{2}}{8}e\Omega^{-1}\varepsilon^{\mu\nu\rho\sigma}\Big[\Omega^{i^*}\big(\psi_\mu\sigma_\sigma\bar\psi_\nu\,\bar\chi_{i^*}\cdot\bar\psi_\rho - \psi_\mu\sigma_\sigma\bar\chi_{i^*}\,\bar\psi_\nu\cdot\bar\psi_\rho\big)\\
&+\psi_\mu\sigma_\sigma\bar\sigma_{\nu\kappa}\bar\psi_\rho\big(\Omega^{i^*}\bar\psi^\kappa\cdot\bar\chi_{i^*} + \Omega_i\psi^\kappa\cdot\chi^i\big)\Big]\\
&+\frac{i}{8}e\Omega^{-2}\varepsilon^{\mu\nu\rho\sigma}\Big[\Omega^{i^*}\Omega_i\psi_\mu\sigma_\sigma\big(\bar\psi_\nu\bar\chi_{i^*}\bar\sigma_\rho\chi^i - \bar\chi_{i^*}\bar\psi_\nu\bar\sigma_\rho\chi^i\big)\\
&-\Omega^{i^*}\Omega^{j^*}\bar\chi_{i^*}\bar\sigma_\mu\sigma_\sigma\big(\bar\psi_\nu\bar\chi_{j^*}\cdot\bar\psi_\rho - \bar\chi_{j^*}\bar\psi_\nu\cdot\bar\psi_\rho\big)\Big]\\
&+\frac{i}{8}e\Omega^{-2}\varepsilon^{\mu\nu\rho\sigma}\Big[\Omega_i\psi_\mu\sigma_\sigma\bar\sigma_{\nu\kappa}\bar\sigma_\rho\chi^i - \Omega^{i^*}\bar\chi_{i^*}\bar\sigma_\mu\sigma_\sigma\bar\sigma_{\nu\kappa}\bar\psi_\rho\Big]\\
&\qquad\times\Big[\Omega^{j^*}\bar\psi^\kappa\cdot\bar\chi_{j^*} + \Omega_j\psi^\kappa\cdot\chi^j\Big]\\
&+\frac{\sqrt{2}}{16}e\Omega^{-3}\Omega^{i^*}\Omega_i\varepsilon^{\mu\nu\rho\sigma}\Big[\Omega^{j^*}\bar\chi_{j^*}\bar\sigma_\mu\sigma_\sigma\big(\bar\psi_\nu\bar\chi_{i^*}\bar\sigma_\rho\chi^i - \bar\chi_{i^*}\bar\psi_\nu\bar\sigma_\rho\chi^i\big)\\
&+\bar\chi_{i^*}\bar\sigma_\mu\sigma_\sigma\bar\sigma_{\nu\kappa}\bar\sigma_\rho\chi^i\big(\Omega^{j^*}\bar\psi^\kappa\cdot\bar\chi_{j^*} + \Omega_j\psi^\kappa\cdot\chi^j\big)\Big] + \text{h.c.}\\
= -\sqrt{2}\partial_{\tilde\nu}&\Big[e\Omega^{-1}e_\nu{}^{\tilde\nu}(\Omega_i\psi_\mu\sigma^{\mu\nu}\chi^i + \Omega^{i^*}\bar\psi_\mu\bar\sigma^{\mu\nu}\bar\chi_{i^*})\Big]\\
&-\frac{\sqrt{2}i}{2}e\Omega^{-1}\Big[\psi_\rho\sigma^\rho\bar\psi_\nu - \psi_\nu\sigma^\rho\bar\psi_\rho\Big]\Big[\Omega_i\psi_\mu\sigma^{\mu\nu}\chi^i + \Omega^{i^*}\bar\psi_\mu\bar\sigma^{\mu\nu}\bar\chi_{i^*}\Big]\\
&-\frac{\sqrt{2}i}{2}e\Omega^{-1}\psi_\nu\sigma^\rho\bar\psi_\mu\Big[\Omega_i\psi_\rho\sigma^{\mu\nu}\chi^i + \Omega^{i^*}\bar\psi_\rho\bar\sigma^{\mu\nu}\bar\chi_{i^*}\Big]
\end{aligned}$$

$$
\begin{aligned}
&-\frac{\sqrt{2}}{4}e\Omega^{-1}\varepsilon^{\mu\nu\rho\sigma}\bar{\psi}_\mu\bar{\sigma}_\sigma\psi_\nu\Big[\Omega_i\chi^i\cdot\psi_\rho-\Omega^{i^*}\bar{\chi}_{i^*}\cdot\bar{\psi}_\rho\Big]\\
&+\sqrt{2}e\Omega^{-1}\Big[\Omega^{i^*}\bar{\chi}_{i^*}\bar{\sigma}^{\mu\nu}\bar{\psi}_{\mu\nu}+\Omega_i\chi^i\sigma^{\mu\nu}\psi_{\mu\nu}\Big]\\
&+\frac{3i}{2}e\Omega^{-2}\Big[\Omega_i\chi^i\sigma^\mu\widetilde{\mathcal{D}}_\mu(\Omega^{i^*}\bar{\chi}_{i^*})-\Omega^{i^*}\widetilde{\mathcal{D}}_\mu(\Omega_i\chi^i)\sigma^\mu\bar{\chi}_{i^*}\Big]\\
&+e\Omega^{-2}\Omega_i\Omega^{i^*}\Big[\chi^i\sigma^{\mu\nu}\psi_\nu\bar{\chi}_{i^*}\cdot\bar{\psi}_\mu+\bar{\chi}_{i^*}\bar{\sigma}^{\mu\nu}\bar{\psi}_\nu\chi^i\cdot\psi_\mu\Big]\\
&-\frac{\sqrt{2}}{8}e\Omega^{-1}\varepsilon^{\mu\nu\rho\sigma}\bar{\psi}_\mu\bar{\sigma}_\sigma\psi_\nu\Big[\Omega_i\chi^i\cdot\psi_\rho-\Omega^{i^*}\bar{\chi}_{i^*}\cdot\bar{\psi}_\rho\Big]\\
&+\frac{\sqrt{2}i}{4}e\Omega^{-1}\Big[\psi_\mu\sigma^\nu\bar{\psi}_\nu-\psi_\nu\sigma^\nu\bar{\psi}_\mu\Big]\Big[\Omega_i\chi^i\cdot\psi^\mu+\Omega^{i^*}\bar{\chi}_{i^*}\cdot\bar{\psi}^\mu\Big]\\
&-\frac{3}{4}e\Omega^{-2}\Big[\Omega_i\chi^i\cdot\psi^\mu+\Omega^{i^*}\bar{\chi}_{i^*}\cdot\bar{\psi}^\mu\Big]\Big[\Omega_j\chi^j\cdot\psi_\mu+\Omega^{j^*}\bar{\chi}_{j^*}\cdot\bar{\psi}_\mu\Big]\\
&+e\Omega^{-2}\Omega_i\Omega^{i^*}\Big[\chi^i\sigma^{\mu\nu}\psi_\nu\bar{\chi}_{i^*}\cdot\bar{\psi}_\mu+\bar{\chi}_{i^*}\bar{\sigma}^{\mu\nu}\bar{\psi}_\nu\chi^i\cdot\psi_\mu\Big]\\
&-\frac{3\sqrt{2}i}{16}e\Omega^{-3}\Omega_i\Omega^{i^*}\Big[\Omega_j\chi^i\cdot\chi^j\bar{\chi}_{i^*}\bar{\sigma}^\nu\psi_\nu+\Omega^{j^*}\bar{\chi}_{i^*}\cdot\bar{\chi}_{j^*}\chi^i\sigma^\nu\bar{\psi}_\nu\Big]\,.
\end{aligned}
$$

In the two first lines of the right-hand side of the first equality above, we have collected all the derivative terms. Together with their (not explicitly written) hermitian conjugate counterparts and after integrating by parts, these terms give rise to the first seven contributions of the right-hand side of the second equality. Simplifications have been performed with the help of the relations among the Pauli matrices and the antisymmetric tensors of rank four (C.2), (C.5), (C.6) and (C.12) and of the two properties (6.11) and (6.13). The terms inversely proportional to Ω, *i.e.*, the second line of the right-hand side of the first equality (with the corresponding hermitian conjugate terms), give rise to the eighth and ninth terms of the second equality after accounting for (C.5) and (C.12). Using again the properties of App. C, the terms in Ω^{-2} appearing on the right-hand side of the first equality simplify to those in Ω^{-2} of the second equality, while those in Ω^{-3} lead to the last term of the second equality. Collecting terms of the same nature, one finally gets,

$$
\begin{aligned}
\delta^{(2)}_{3/2}=&-\sqrt{2}\partial_{\tilde{\nu}}\Big[e\Omega^{-1}e_\nu{}^{\tilde{\nu}}(\Omega_i\psi_\mu\sigma^{\mu\nu}\chi^i+\Omega^{i^*}\bar{\psi}_\mu\bar{\sigma}^{\mu\nu}\bar{\chi}_{i^*})\Big]\\
&-\frac{\sqrt{2}i}{2}e\Omega^{-1}\psi_\nu\sigma^\rho\bar{\psi}_\mu\Big[\Omega_i\psi_\rho\sigma^{\mu\nu}\chi^i+\Omega^{i^*}\bar{\psi}_\rho\bar{\sigma}^{\mu\nu}\bar{\chi}_{i^*}\Big]\\
&+\sqrt{2}e\Omega^{-1}\Big[\Omega^{i^*}\bar{\chi}_{i^*}\bar{\sigma}^{\mu\nu}\bar{\psi}_{\mu\nu}+\Omega_i\chi^i\sigma^{\mu\nu}\psi_{\mu\nu}\Big]\\
&-\frac{\sqrt{2}i}{4}e\Omega^{-1}\Big[\psi_\rho\sigma^\rho\bar{\psi}_\nu-\psi_\nu\sigma^\rho\bar{\psi}_\rho\Big]\Big[\Omega_i\psi_\mu\sigma^\mu\bar{\sigma}^\nu\chi^i+\Omega^{i^*}\bar{\psi}_\mu\bar{\sigma}^\mu\sigma^\nu\bar{\chi}_{i^*}\Big]
\end{aligned}
$$

$$
\begin{aligned}
&-\frac{3\sqrt{2}}{8}e\Omega^{-1}\varepsilon^{\mu\nu\rho\sigma}\bar{\psi}_\mu\bar{\sigma}_\sigma\psi_\nu\big[\Omega_i\chi^i\cdot\psi_\rho-\Omega^{i^*}\bar{\chi}_{i^*}\cdot\bar{\psi}_\rho\big]\\
&+\frac{3i}{2}e\Omega^{-2}\big[\Omega_i\chi^i\sigma^\mu\widetilde{\mathcal{D}}_\mu(\Omega^{i^*}\bar{\chi}_{i^*})-\Omega^{i^*}\widetilde{\mathcal{D}}_\mu(\Omega_i\chi^i)\sigma^\mu\bar{\chi}_{i^*}\big]\\
&-\frac{3}{4}e\Omega^{-2}\big[\Omega_i\chi^i\cdot\psi^\mu+\Omega^{i^*}\bar{\chi}_{i^*}\cdot\bar{\psi}^\mu\big]\big[\Omega_j\chi^j\cdot\psi_\mu+\Omega^{j^*}\bar{\chi}_{j^*}\cdot\bar{\psi}_\mu\big]\\
&+2e\Omega^{-2}\Omega_i\Omega^{i^*}\big[\chi^i\sigma^{\mu\nu}\psi_\nu\bar{\chi}_{i^*}\cdot\bar{\psi}_\mu+\bar{\chi}_{i^*}\bar{\sigma}^{\mu\nu}\bar{\psi}_\nu\chi^i\cdot\psi_\mu\big]\\
&-\frac{3\sqrt{2}i}{16}e\Omega^{-3}\Omega_i\Omega^{i^*}\big[\Omega_j\chi^i\cdot\chi^j\bar{\chi}_{i^*}\bar{\sigma}^\nu\psi_\nu+\Omega^{j^*}\bar{\chi}_{i^*}\cdot\bar{\chi}_{j^*}\chi^i\sigma^\nu\bar{\psi}_\nu\big]\,.
\end{aligned}
$$

In contrast, the computation of the variation of the quantity $\delta^{(1)}_{3/2}$ under a gravitino shift

$$
\delta^{(1)}_{3/2}\to\delta^{(1)}_{3/2}+\delta^{(2,1)}_{3/2}\,, \tag{6.62}
$$

is easier to compute and the expression of $\delta^{(2,1)}_{3/2}$ can be immediately simplified to

$$
\begin{aligned}
\delta^{(2,1)}_{3/2}=&\ \frac{3\sqrt{2}}{4}e\Omega^{-2}\partial_\mu\Omega\big[\Omega_i\chi^i\cdot\psi^\mu+\Omega^{i^*}\bar{\chi}_{i^*}\cdot\bar{\psi}^\mu\big]\\
&-\frac{\sqrt{2}}{2}e\Omega^{-2}\partial_\mu\Omega\big[\Omega_i\chi^i\sigma^{\mu\nu}\psi_\nu+\Omega^{i^*}\bar{\chi}_{i^*}\bar{\sigma}^{\mu\nu}\bar{\psi}_\nu\big]\,,
\end{aligned}
$$

from (C.6) and (C.8).

Collecting all the contributions while omitting the total derivative, the transformed gravitino Lagrangian reads,

$$
\begin{aligned}
\mathcal{L}_{3/2}\to&\ \frac{1}{4}e\varepsilon^{\mu\nu\rho\sigma}\big[\psi_\mu\sigma_\sigma\bar{\psi}_{\nu\rho}-\bar{\psi}_\mu\bar{\sigma}_\sigma\psi_{\nu\rho}\big]\\
&-\frac{\sqrt{2}i}{2}e\Omega^{-1}\psi_\nu\sigma^\rho\bar{\psi}_\mu\big[\Omega_i\psi_\rho\sigma^{\mu\nu}\chi^i+\Omega^{i^*}\bar{\psi}_\rho\bar{\sigma}^{\mu\nu}\bar{\chi}_{i^*}\big]\\
&+\sqrt{2}e\Omega^{-1}\big[\Omega^{i^*}\bar{\chi}_{i^*}\bar{\sigma}^{\mu\nu}\bar{\psi}_{\mu\nu}+\Omega_i\chi^i\sigma^{\mu\nu}\psi_{\mu\nu}\big]\\
&-\frac{\sqrt{2}i}{4}e\Omega^{-1}\big[\psi_\rho\sigma^\rho\bar{\psi}_\nu-\psi_\nu\sigma^\rho\bar{\psi}_\rho\big]\big[\Omega_i\psi_\mu\sigma^\mu\bar{\sigma}^\nu\chi^i+\Omega^{i^*}\bar{\psi}_\mu\bar{\sigma}^\mu\sigma^\nu\bar{\chi}_{i^*}\big]\\
&-\frac{3\sqrt{2}}{8}e\Omega^{-1}\varepsilon^{\mu\nu\rho\sigma}\bar{\psi}_\mu\bar{\sigma}_\sigma\psi_\nu\big[\Omega_i\chi^i\cdot\psi_\rho-\Omega^{i^*}\bar{\chi}_{i^*}\cdot\bar{\psi}_\rho\big]\\
&+\frac{3i}{2}e\Omega^{-2}\big[\Omega_i\chi^i\sigma^\mu\widetilde{\mathcal{D}}_\mu(\Omega^{i^*}\bar{\chi}_{i^*})-\Omega^{i^*}\widetilde{\mathcal{D}}_\mu(\Omega_i\chi^i)\sigma^\mu\bar{\chi}_{i^*}\big]\\
&-\frac{3}{4}e\Omega^{-2}\big[\Omega_i\chi^i\cdot\psi^\mu+\Omega^{i^*}\bar{\chi}_{i^*}\cdot\bar{\psi}^\mu\big]\big[\Omega_j\chi^j\cdot\psi_\mu+\Omega^{j^*}\bar{\chi}_{j^*}\cdot\bar{\psi}_\mu\big]\\
&+2e\Omega^{-2}\Omega_i\Omega^{i^*}\big[\chi^i\sigma^{\mu\nu}\psi_\nu\bar{\chi}_{i^*}\cdot\bar{\psi}_\mu+\bar{\chi}_{i^*}\bar{\sigma}^{\mu\nu}\bar{\psi}_\nu\chi^i\cdot\psi_\mu\big]
\end{aligned}
$$

$$
\begin{aligned}
&-\frac{3\sqrt{2}i}{16}e\Omega^{-3}\Omega_i\Omega^{i^*}\left[\Omega_j\chi^i\cdot\chi^j\bar{\chi}_{i^*}\bar{\sigma}^\nu\psi_\nu+\Omega^{j^*}\bar{\chi}_{i^*}\cdot\bar{\chi}_{j^*}\chi^i\sigma^\nu\bar{\psi}_\nu\right]\\
&+\frac{3\sqrt{2}}{4}e\Omega^{-2}\partial_\mu\Omega\left[\Omega_i\chi^i\cdot\psi^\mu+\Omega^{i^*}\bar{\chi}_{i^*}\cdot\bar{\psi}^\mu\right]\\
&-\frac{\sqrt{2}}{2}e\Omega^{-2}\partial_\mu\Omega\left[\Omega_i\chi^i\sigma^{\mu\nu}\psi_\nu+\Omega^{i^*}\bar{\chi}_{i^*}\bar{\sigma}^{\mu\nu}\bar{\psi}_\nu\right]\\
&-\frac{i}{2}e\Omega^{-1}\partial_\mu\Omega\left(\psi^\mu\sigma^\nu\bar{\psi}_\nu+\bar{\psi}^\mu\bar{\sigma}^\nu\psi_\nu\right).
\end{aligned}
$$

Rescaling of the fermionic kinetic terms.

As before, the variation of the fermionic kinetic term is computed in two steps on the basis of the results of App. H. Under a pure dilatation, this term transforms as,

$$
\begin{aligned}
\mathcal{L}_\chi &= -\frac{i}{2}e\Omega^{i^*}{}_i\left[\chi^i\sigma^\mu\widetilde{\mathcal{D}}_\mu\bar{\chi}_{i^*}-\widetilde{\mathcal{D}}_\mu\chi^i\sigma^\mu\bar{\chi}_{i^*}\right]\\
&\to \frac{3i}{2}e\Omega^{-1}\Omega^{i^*}{}_i\left[\chi^i\sigma^\mu\widetilde{\mathcal{D}}_\mu\bar{\chi}_{i^*}-\widetilde{\mathcal{D}}_\mu\chi^i\sigma^\mu\bar{\chi}_{i^*}\right].
\end{aligned}
$$

It can be noted that the terms proportional to $\partial_\mu\Omega$ resulting from the dilatation of the derivatives $\widetilde{\mathcal{D}}_\mu\chi^i$ are cancelled by those induced by the variation of $\widetilde{\mathcal{D}}_\mu\bar{\chi}_{i^*}$ (see (6.47)). After using (C.6) and the Fierz identity (C.9), the gravitino shift of the derivative factors in (6.48) leads to the transformation law

$$
\begin{aligned}
\mathcal{L}_\chi \to\ & \frac{3i}{2}e\Omega^{-1}\Omega^{i^*}{}_i\left[\chi^i\sigma^\mu\widetilde{\mathcal{D}}_\mu\bar{\chi}_{i^*}-\widetilde{\mathcal{D}}_\mu\chi^i\sigma^\mu\bar{\chi}_{i^*}\right]\\
&+\frac{3\sqrt{2}}{4}ie\Omega^{-2}\Omega^{i^*}{}_i\left[\Omega^{j^*}\chi^i\sigma^\mu\bar{\psi}_\mu\bar{\chi}_{i^*}\cdot\bar{\chi}_{j^*}-\Omega_j\psi_\mu\sigma^\mu\bar{\chi}_{i^*}\chi^i\cdot\chi^j\right]\\
&+\frac{3\sqrt{2}}{8}ie\Omega^{-2}\Omega^{i^*}{}_i\chi^i\sigma^\mu\bar{\chi}_{i^*}\left[\Omega^{j^*}\bar{\chi}_{j^*}\cdot\bar{\psi}_\mu-\Omega_j\chi^j\cdot\psi^\mu\right].
\end{aligned}
$$

Rescaling of the gravitino-fermion kinetic mixing terms.

Under a pure dilatation, the transformation law of the gravitino-fermion kinetic mixing terms can be calculated from (6.47),

$$
\begin{aligned}
\mathcal{L}_{\chi\psi} &= \frac{\sqrt{2}}{3}e\left[\Omega_i\chi^i\sigma^{\mu\nu}\psi_{\mu\nu}+\Omega^{i^*}\bar{\chi}_{i^*}\bar{\sigma}^{\mu\nu}\bar{\psi}_{\mu\nu}\right]\\
&\to -\sqrt{2}e\Omega^{-1}\left[\Omega_i\chi^i\sigma^{\mu\nu}\psi_{\mu\nu}+\Omega^{i^*}\bar{\chi}_{i^*}\bar{\sigma}^{\mu\nu}\bar{\psi}_{\mu\nu}\right]+\delta^{(1)}_{\chi\psi}\,,
\end{aligned}
\tag{6.63}
$$

with

$$\begin{aligned}
\delta^{(1)}_{\chi\psi} &= \sqrt{2}e\Omega^{-1}\Omega_i\chi^i\sigma^{\mu\nu}\Big[\frac{1}{2}\Omega^{-1}\partial_\mu\Omega\psi_\nu + \Omega^{-1}\partial^\rho\Omega\,\sigma_{\mu\rho}\psi_\nu\Big] + \text{h.c.}\\
&= -\frac{3\sqrt{2}}{4}e\Omega^{-2}\partial_\mu\Omega\Big[\Omega_i\psi_\nu\sigma^\mu\bar{\sigma}^\nu\chi^i + \Omega^{i^*}\bar{\psi}_\nu\bar{\sigma}^\mu\sigma^\nu\bar{\chi}_{i^*}\Big]\,,
\end{aligned}$$

where, for the last equality, we have used (C.2) and (C.6). Performing the gravitino shift, the pure mixing term of (6.63) becomes

$$\begin{aligned}
&-\sqrt{2}e\Omega^{-1}\Big[\Omega_i\chi^i\sigma^{\mu\nu}\psi_{\mu\nu} + \Omega^{i^*}\bar{\chi}_{i^*}\bar{\sigma}^{\mu\nu}\bar{\psi}_{\mu\nu}\Big]\\
\to &-\sqrt{2}e\Omega^{-1}\Big[\Omega_i\chi^i\sigma^{\mu\nu}\psi_{\mu\nu} + \Omega^{i^*}\bar{\chi}_{i^*}\bar{\sigma}^{\mu\nu}\bar{\psi}_{\mu\nu}\Big] + \delta^{(2)}_{\chi\psi}\,,
\end{aligned}$$

where the variation can be calculated according to (6.48) as

$$\begin{aligned}
\delta^{(2)}_{\chi\psi} &= -2ie\Omega^{-1}\Omega_i\chi^i\sigma^{\mu\nu}e_\nu{}^{\tilde{\nu}}\mathcal{D}_\mu\big(\Omega^{-1}\Omega^{i^*}\bar{\sigma}_{\tilde{\nu}}\bar{\chi}_{i^*}\big)\\
&\quad -e\Omega^{-2}\Omega_i\chi^i\sigma^{\mu\nu}\sigma_{\mu\rho}\psi_\nu\Big[\Omega_j\psi^\rho\cdot\chi^j+\Omega^{j^*}\bar{\psi}^\rho\cdot\bar{\chi}_{j^*}\Big]\\
&\quad -e\Omega^{-2}\Omega_i\Omega_j\chi^i\sigma^{\mu\nu}\psi_\mu\chi^j\cdot\psi_\nu\\
&\quad -\frac{\sqrt{2}i}{2}e\Omega^{-3}\Omega_i\Omega^{i^*}\chi^i\sigma^{\mu\nu}\sigma_{\mu\rho}\sigma_\nu\bar{\chi}_{i^*}\Big[\Omega_j\psi^\rho\cdot\chi^j+\Omega^{j^*}\bar{\psi}^\rho\cdot\bar{\chi}_{j^*}\Big]\\
&\quad -\frac{\sqrt{2}i}{2}e\Omega^{-3}\Omega_i\Omega_j\Omega^{i^*}\big(\chi^i\sigma^{\mu\nu}\psi_\mu\chi^j\sigma_\nu\bar{\chi}_{i^*}+\chi^i\sigma^{\mu\nu}\chi^j\psi_\mu\sigma_\nu\bar{\chi}_{i^*}\big) + \text{h.c.}\\
&= -3ie\Omega^{-2}\Big[\Omega_i\chi^i\sigma^\mu\tilde{\mathcal{D}}_\mu(\Omega^{i^*}\bar{\chi}_{i^*}) - \Omega^{i^*}\tilde{\mathcal{D}}_\mu(\Omega_i\chi^i)\sigma^\mu\bar{\chi}_{i^*}\Big]\\
&\quad +2e\Omega^{-2}\Omega_i\Omega^{i^*}\Big[\chi^i\sigma^{\mu\nu}\psi_\nu\bar{\chi}_{i^*}\cdot\bar{\psi}_\mu+\bar{\chi}_{i^*}\bar{\sigma}^{\mu\nu}\bar{\psi}_\nu\chi^i\cdot\psi_\mu\Big]\\
&\quad +\frac{3}{4}e\Omega^{-2}\Big[\Omega_i\chi^i\cdot\psi^\mu+\Omega^{i^*}\bar{\chi}_{i^*}\cdot\bar{\psi}^\mu\Big]\Big[\Omega_j\chi^j\cdot\psi_\mu+\Omega^{j^*}\bar{\chi}_{j^*}\cdot\bar{\psi}_\mu\Big]\\
&\quad -3e\Omega^{-2}\Omega_i\Omega^{i^*}\Big[\chi^i\sigma^{\mu\nu}\psi_\nu\bar{\chi}_{i^*}\cdot\bar{\psi}_\mu+\bar{\chi}_{i^*}\bar{\sigma}^{\mu\nu}\bar{\psi}_\nu\chi^i\cdot\psi_\mu\Big]\\
&\quad +\frac{3\sqrt{2}i}{8}e\Omega^{-3}\Omega_i\Omega^{i^*}\Big[\Omega_j\chi^i\cdot\chi^j\bar{\chi}_{i^*}\bar{\sigma}^\mu\psi_\mu + \Omega^{j^*}\bar{\chi}_{i^*}\cdot\bar{\chi}_{j^*}\chi^i\sigma^\mu\bar{\psi}_\mu\Big]\,.
\end{aligned}$$

For the simplification of the derivative term, *i.e.*, the first term of the right-hand side of the first equality, we have used (C.6) and (C.8) and followed a procedure similar to the procedure introduced in (6.13). This leads to the contributions written in the two first lines of the second equality. The non-derivative terms in Ω^{-2} can immediately be simplified by applying (C.6). Concerning the gathering of the terms in Ω^{-3}, we have used the Fierz identity (C.9) and the relation (C.6). Moreover, it can be easily shown that $\delta^{(1)}_{\chi\psi}$ is invariant under a gravitino shift with the help of (C.6). Therefore, the full variation of $\mathcal{L}_{\chi\psi}$ under the Howe–Tucker reads

$$\begin{aligned}\mathcal{L}_{\chi\psi} \to & -\sqrt{2}e\Omega^{-1}\Big[\Omega_i\chi^i\sigma^{\mu\nu}\psi_{\mu\nu} + \Omega^{i^*}\bar{\chi}_{i^*}\bar{\sigma}^{\mu\nu}\bar{\psi}_{\mu\nu}\Big] \\ & -3ie\Omega^{-2}\Big[\Omega_i\chi^i\sigma^\mu\widetilde{\mathcal{D}}_\mu(\Omega^{i^*}\bar{\chi}_{i^*}) - \Omega^{i^*}\widetilde{\mathcal{D}}_\mu(\Omega_i\chi^i)\sigma^\mu\bar{\chi}_{i^*}\Big] \\ & +\frac{3}{4}e\Omega^{-2}\Big[\Omega_i\chi^i\cdot\psi^\mu + \Omega^{i^*}\bar{\chi}_{i^*}\cdot\bar{\psi}^\mu\Big]\Big[\Omega_j\chi^j\cdot\psi_\mu + \Omega^{j^*}\bar{\chi}_{j^*}\cdot\bar{\psi}_\mu\Big] \\ & -3e\Omega^{-2}\Omega_i\Omega^{i^*}\Big[\chi^i\sigma^{\mu\nu}\psi_\nu\bar{\chi}_{i^*}\cdot\bar{\psi}_\mu + \bar{\chi}_{i^*}\bar{\sigma}^{\mu\nu}\bar{\psi}_\nu\chi^i\cdot\psi_\mu\Big] \\ & -\frac{3\sqrt{2}}{4}e\Omega^{-2}\partial_\mu\Omega\Big[\Omega_i\psi_\nu\sigma^\mu\bar{\sigma}^\nu\chi^i + \Omega^{i^*}\bar{\psi}_\nu\bar{\sigma}^\mu\sigma^\nu\bar{\chi}_{i^*}\Big] \\ & +\frac{3\sqrt{2}i}{8}e\Omega^{-3}\Omega_i\Omega^{i^*}\Big[\Omega_j\chi^i\cdot\chi^j\bar{\chi}_{i^*}\bar{\sigma}^\mu\psi_\mu + \Omega^{j^*}\bar{\chi}_{i^*}\cdot\bar{\chi}_{j^*}\chi^i\sigma^\mu\bar{\psi}_\mu\Big]\,.\end{aligned}$$

Rescaling of the scalar kinetic terms.

The variation under Howe–Tucker transformations of the scalar kinetic term in (6.33) corresponds to the variation under a dilation alone. After recalling that

$$\partial_\mu\Omega\partial^\mu\Omega = \Omega^{i^*}\Omega^{j^*}\widetilde{\mathcal{D}}^\mu\phi^\dagger_{i^*}\widetilde{\mathcal{D}}_\mu\phi^\dagger_{j^*} + \Omega_i\Omega_j\widetilde{\mathcal{D}}_\mu\phi^i\widetilde{\mathcal{D}}^\mu\phi^j + 2\Omega_i\Omega^{i^*}\widetilde{\mathcal{D}}_\mu\phi^i\widetilde{\mathcal{D}}^\mu\phi^\dagger_{i^*}\,, \tag{6.64}$$

one gets

$$\begin{aligned}\mathcal{L}_\phi &= e\Omega^{i^*}{}_i\widetilde{\mathcal{D}}_\mu\phi^\dagger_{i^*}\widetilde{\mathcal{D}}^\mu\phi^i + \frac{1}{4}e\Omega^{-1}\Big[\Omega^{i^*}\widetilde{\mathcal{D}}_\mu\phi^\dagger_{i^*} - \Omega_i\widetilde{\mathcal{D}}_\mu\phi^i\Big]\Big[\Omega^{j^*}\widetilde{\mathcal{D}}^\mu\phi^\dagger_{j^*} - \Omega_j\widetilde{\mathcal{D}}^\mu\phi^j\Big] \\ &\to eK^{i^*}{}_i\widetilde{\mathcal{D}}_\mu\phi^i\widetilde{\mathcal{D}}^\mu\phi^\dagger_{i^*} - \frac{3}{4}e\Omega^{-2}\partial_\mu\Omega\partial^\mu\Omega\,.\end{aligned} \tag{6.65}$$

To obtain the above expression, we have expressed the superspace kinetic energy and its derivatives in terms of the Kähler potential and its derivatives, as given in (6.27) and (6.32).

Rescaling of the terms depending on the derivatives of the scalar fields.

Finally, we now turn to the terms in (6.33) that are linear in the derivatives of the scalar fields,

$$\begin{aligned}\mathcal{L}_{\mathcal{D}\phi} = & -\frac{1}{4}e\varepsilon^{\mu\nu\rho\sigma}\bar{\psi}_\mu\bar{\sigma}_\sigma\psi_\nu\Big[\Omega^{i^*}\widetilde{\mathcal{D}}_\rho\phi^\dagger_{i^*} - \Omega_i\widetilde{\mathcal{D}}_\rho\phi^i\Big] \\ & -\frac{\sqrt{2}}{4}e\Omega^{-1}(\Omega^{i^*}\widetilde{\mathcal{D}}_\mu\phi^\dagger_{i^*} - \Omega_i\widetilde{\mathcal{D}}_\mu\phi^i)(\Omega^{j^*}\bar{\chi}_{j^*}\cdot\bar{\psi}^\mu - \Omega_j\chi^j\cdot\psi^\mu) \\ & -\frac{i}{4}e\Omega^{-1}\Omega^{j^*}{}_j(\Omega^{i^*}\widetilde{\mathcal{D}}_\mu\phi^\dagger_{i^*} - \Omega_i\widetilde{\mathcal{D}}_\mu\phi^i)\chi^j\sigma^\mu\bar{\chi}_{j^*} \\ & -\frac{\sqrt{2}}{2}e\Omega^{i^*}{}_i\Big[\chi^i\sigma^\mu\bar{\sigma}^\nu\psi_\mu\widetilde{\mathcal{D}}_\nu\phi^\dagger_{i^*} + \bar{\chi}_{i^*}\bar{\sigma}^\mu\sigma^\nu\bar{\psi}_\mu\widetilde{\mathcal{D}}_\nu\phi^i\Big]\,.\end{aligned}$$

Their transformation law under a pure dilatation can be readily derived from (6.39) and (6.41),

$$\begin{aligned}
\mathcal{L}_{\mathcal{D}\phi} \rightarrow\ & \frac{3}{4}e\Omega^{-1}\varepsilon^{\mu\nu\rho\sigma}\bar{\psi}_\mu\bar{\sigma}_\sigma\psi_\nu\Big[\Omega^{i^*}\widetilde{\mathcal{D}}_\rho\phi^\dagger_{i^*}-\Omega_i\widetilde{\mathcal{D}}_\rho\phi^i\Big] \\
&+\frac{3\sqrt{2}}{4}e\Omega^{-2}(\Omega^{i^*}\widetilde{\mathcal{D}}_\mu\phi^\dagger_{i^*}-\Omega_i\widetilde{\mathcal{D}}_\mu\phi^i)(\Omega^{j^*}\bar{\chi}_{j^*}\cdot\bar{\psi}^\mu-\Omega_j\chi^j\cdot\psi^\mu) \\
&+\frac{3i}{4}e\Omega^{-2}\Omega^{j^*}{}_j(\Omega^{i^*}\widetilde{\mathcal{D}}_\mu\phi^\dagger_{i^*}-\Omega_i\widetilde{\mathcal{D}}_\mu\phi^i)\chi^j\sigma^\mu\bar{\chi}_{j^*} \\
&+\frac{3\sqrt{2}}{2}e\Omega^{-1}\Omega^{i^*}{}_i\Big[\chi^i\sigma^\mu\bar{\sigma}^\nu\psi_\mu\widetilde{\mathcal{D}}_\nu\phi^\dagger_{i^*}+\bar{\chi}_{i^*}\bar{\sigma}^\mu\sigma^\nu\bar{\psi}_\mu\widetilde{\mathcal{D}}_\nu\phi^i\Big]\,.
\end{aligned}$$

With the help of (C.2), (C.6) and (C.12), the variation under a gravitino shift derived from (6.39) is computed as

$$\begin{aligned}
\mathcal{L}_{\mathcal{D}\phi} \rightarrow\ & \frac{3}{4}e\Omega^{-1}\varepsilon^{\mu\nu\rho\sigma}\bar{\psi}_\mu\bar{\sigma}_\sigma\psi_\nu\Big[\Omega^{j^*}\widetilde{\mathcal{D}}_\rho\phi^\dagger_{j^*}-\Omega_j\widetilde{\mathcal{D}}_\rho\phi^j\Big] \\
&+3ie\Omega^{-2}\chi^j\sigma^\mu\bar{\chi}_{j^*}\Big[\Omega^{j^*}{}_i\Omega_j\widetilde{\mathcal{D}}_\mu\phi^i-\Omega^{i^*}{}_j\Omega^{j^*}\widetilde{\mathcal{D}}_\mu\phi^\dagger_{i^*}\Big] \\
&+\frac{3}{4}e\Big[\Omega^{j^*}\widetilde{\mathcal{D}}_\rho\phi^\dagger_{j^*}-\Omega_j\widetilde{\mathcal{D}}_\rho\phi^j\Big]\Big[i(\Omega^{-3}\Omega_i\Omega^{i^*}+\Omega^{-2}\Omega^{i^*}{}_i)\chi^i\sigma^\rho\bar{\chi}_{i^*} \\
&\qquad-\sqrt{2}\Omega^{-2}\Big(\Omega_i\chi^i\sigma^\mu\bar{\sigma}^\rho\psi_\mu-\Omega^{i^*}\bar{\chi}_{i^*}\bar{\sigma}^\mu\sigma^\rho\bar{\psi}_\mu\Big)\Big] \\
&+\frac{3\sqrt{2}}{2}e\Omega^{-1}\Omega^{i^*}{}_i\Big[\chi^i\sigma^\mu\bar{\sigma}^\nu\psi_\mu\widetilde{\mathcal{D}}_\nu\phi^\dagger_{i^*}+\bar{\chi}_{i^*}\bar{\sigma}^\mu\sigma^\nu\bar{\psi}_\mu\widetilde{\mathcal{D}}_\nu\phi^i\Big]\,.
\end{aligned}$$

Rescaling of all the kinetic terms.

We now collect all the results derived in this section. After inspecting the equations above, one observes that a super-Weyl rescaling of the kinetic part of the Lagrangian (6.33) leads to standardly normalised kinetic terms, together with new interaction terms. The pure kinetic part of the rescaled Lagrangian, $\mathcal{L}^{(\rm kin)}_{\rm kin}$, reads,

$$\begin{aligned}
\mathcal{L}^{(\rm kin)}_{\rm kin} =\ & \frac{1}{2}ee_\mu{}^{\tilde\mu}e_\nu{}^{\tilde\nu}R_{\tilde\mu\tilde\nu}{}^{\mu\nu}\Big| + \frac{1}{4}e\varepsilon^{\mu\nu\rho\sigma}\Big[\psi_\mu\sigma_\sigma\check{\bar{\psi}}_{\nu\rho}-\bar{\psi}_\mu\bar{\sigma}_\sigma\check{\psi}_{\nu\rho}\Big] \\
&-\frac{i}{2}eK^{i^*}{}_i\Big[\chi^i\sigma^\mu\check{\mathcal{D}}_\mu\bar{\chi}_{i^*}-\check{\mathcal{D}}_\mu\chi^i\sigma^\mu\bar{\chi}_{i^*}\Big]+eK^{i^*}{}_i\widetilde{\mathcal{D}}_\mu\phi^i\widetilde{\mathcal{D}}^\mu\phi^\dagger_{i^*} \\
&-\frac{\sqrt{2}}{2}eK^{i^*}{}_i\Big[\chi^i\sigma^\mu\bar{\sigma}^\nu\psi_\mu\widetilde{\mathcal{D}}_\nu\phi^\dagger_{i^*}+\bar{\chi}_{i^*}\bar{\sigma}^\mu\sigma^\nu\bar{\psi}_\mu\widetilde{\mathcal{D}}_\nu\phi^i\Big]\,,
\end{aligned} \tag{6.66}$$

where according to the results obtained in Sec. 5.3.3, we have introduced the fully covariant derivatives

$$\begin{aligned}
\check{\psi}_{\mu\nu} &= e_\mu{}^{\tilde{\mu}} e_\nu{}^{\tilde{\nu}} \big(\check{\mathcal{D}}_{\tilde{\mu}}\psi_{\tilde{\nu}} - \check{\mathcal{D}}_{\tilde{\nu}}\psi_{\tilde{\mu}}\big)\,, \\
\check{\mathcal{D}}_{\tilde{\mu}}\psi_{\tilde{\nu}} &= \mathcal{D}_{\tilde{\mu}}\psi_{\tilde{\nu}} + \frac{1}{4}\big(K_i \widetilde{\mathcal{D}}_{\tilde{\mu}}\phi^i - K^{i^*}\widetilde{\mathcal{D}}_{\tilde{\mu}}\phi^\dagger_{i^*}\big)\psi_{\tilde{\nu}}\,, \\
\check{\bar{\psi}}_{\mu\nu} &= e_\mu{}^{\tilde{\mu}} e_\nu{}^{\tilde{\nu}} \big(\check{\mathcal{D}}_{\tilde{\mu}}\bar{\psi}_{\tilde{\nu}} - \check{\mathcal{D}}_{\tilde{\nu}}\bar{\psi}_{\tilde{\mu}}\big)\,, \\
\check{\mathcal{D}}_{\tilde{\mu}}\bar{\psi}_{\tilde{\nu}} &= \mathcal{D}_{\tilde{\mu}}\bar{\psi}_{\tilde{\nu}} - \frac{1}{4}\big(K_i \widetilde{\mathcal{D}}_{\tilde{\mu}}\phi^i - K^{i^*}\widetilde{\mathcal{D}}_{\tilde{\mu}}\phi^\dagger_{i^*}\big)\psi_{\tilde{\nu}}\,, \\
\check{\mathcal{D}}_\mu \chi^i &= \widetilde{\mathcal{D}}_\mu \chi^i - \frac{1}{4}(K_j \mathcal{D}_\mu \phi^j - K^{j^*}\mathcal{D}_\mu \phi^\dagger_{j^*})\,\chi^i\,, \\
\check{\mathcal{D}}_\mu \bar{\chi}_{i^*} &= \widetilde{\mathcal{D}}_\mu \bar{\chi}_{i^*} + \frac{1}{4}(K_j \mathcal{D}_\mu \phi^j - K^{j^*}\mathcal{D}_\mu \phi^\dagger_{j^*})\,\bar{\chi}_{i^*}\,.
\end{aligned} \tag{6.67}$$

A remark is in order with respect to the simplification of the fermionic terms. Originally, the derivatives were covariant with respect to the kinetic energy Ω. However, in (6.66), they have been rendered covariant with respect to the Kähler manifold,

$$\Omega^{i^*}{}_i \widetilde{\mathcal{D}}^{(\Omega)}_\mu \chi^i = \Omega^{i^*}{}_i \widetilde{\mathcal{D}}^{(K)}_\mu \chi^i + \Omega_j{}^{i^*}{}_k \widetilde{\mathcal{D}}_\mu \phi^j \chi^k - \Omega^{i^*}{}_i \Gamma^{(K)}{}_j{}^i{}_k \widetilde{\mathcal{D}}_\mu \phi^j \chi^k\,. \tag{6.68}$$

In the expression above, we have explicitly indicated the dependence in Ω or K for the derivatives and the connection and we recall that $\Omega^{i^*}{}_i{}^{j^*} = \Omega^{\ell^*}{}_i \Gamma^{(\Omega) i^*}{}_{\ell^*}{}^{j^*}$, where the third order derivative of Ω is given in (6.30).

Adding the gauge contribution, we obtain

$$\begin{aligned}
\mathcal{L}_{\text{kin}} = {}& \frac{1}{2} e e_\mu{}^{\tilde{\mu}} e_\nu{}^{\tilde{\nu}} R_{\tilde{\mu}\tilde{\nu}}{}^{\mu\nu}\big| + \frac{1}{4} e \varepsilon^{\mu\nu\rho\sigma}\Big[\psi_\mu \sigma_\sigma \check{\bar{\psi}}_{\nu\rho} - \bar{\psi}_\mu \bar{\sigma}_\sigma \check{\psi}_{\nu\rho}\Big] \\
& - \frac{1}{4} e h^R_{ab} F^a_{\mu\nu} F^{\mu\nu b} + \frac{1}{4} e h^I_{ab} F^a_{\mu\nu}{}^* F^{\mu\nu b} \\
& - \frac{i}{2} e h^R_{ab} (\lambda^a \sigma^\mu \check{\mathcal{D}}_\mu \bar{\lambda}^b - \check{\mathcal{D}}_\mu \lambda^a \sigma^\mu \bar{\lambda}^b) + \frac{1}{2} e h^I_{ab} \check{\mathcal{D}}_\mu (\lambda^a \sigma^\mu \bar{\lambda}^b) \\
& - \frac{i}{2} e K^{i^*}{}_i \Big[\chi^i \sigma^\mu \check{\mathcal{D}}_\mu \bar{\chi}_{i^*} - \check{\mathcal{D}}_\mu \chi^i \sigma^\mu \bar{\chi}_{i^*}\Big] + e K^{i^*}{}_i \widetilde{\mathcal{D}}_\mu \phi^i \widetilde{\mathcal{D}}^\mu \phi^\dagger_{i^*} \\
& - \frac{\sqrt{2}}{2} e K^{i^*}{}_i \Big[\chi^i \sigma^\mu \bar{\sigma}^\nu \psi_\mu \widetilde{\mathcal{D}}_\nu \phi^\dagger_{i^*} + \bar{\chi}_{i^*} \bar{\sigma}^\mu \sigma^\nu \bar{\psi}_\mu \widetilde{\mathcal{D}}_\nu \phi^i\Big] \\
& + \frac{i}{4} e h^R_{ab} \Big[\lambda^a \sigma_\mu \bar{\sigma}_{\nu\rho} \bar{\psi}^\mu + \bar{\lambda}^a \bar{\sigma}_\mu \sigma_{\nu\rho} \psi^\mu\Big]\Big[F^{\nu\rho b} + \hat{F}^{\nu\rho b}\Big] \\
& - \frac{\sqrt{2}}{4} e \Big[h_{abi} \chi^i \sigma^{\mu\nu} \lambda^b + h^\star_{ab}{}^{i^*} \bar{\chi}_{i^*} \bar{\sigma}^{\mu\nu} \bar{\lambda}^b\Big] F^a_{\mu\nu}\,.
\end{aligned}$$

Note that we have used the relationship $\check{\mathcal{D}}_\mu(\lambda^a\sigma^\mu\bar{\lambda}^b) = \widetilde{\mathcal{D}}(\lambda^a\sigma^\mu\bar{\lambda}^b)$ with

$$\check{\mathcal{D}}_{\tilde{\mu}}\lambda^{\alpha a} = e_\mu{}^{\tilde{\mu}}\Big(\partial_\mu\lambda^{\alpha a} + \lambda^{\beta a}\omega_{\tilde{\mu}\beta}{}^\alpha - g f_{bc}{}^a v_{\tilde{\mu}}^b\lambda^{\alpha c} + \frac{1}{4}(K_i\widetilde{\mathcal{D}}_{\tilde{\mu}}\phi^i - K_{i^*}\widetilde{\mathcal{D}}_{\tilde{\mu}}\phi^\dagger_{i^*})\lambda^{\alpha a}\Big),$$

$$\check{\mathcal{D}}_{\tilde{\mu}}\bar{\lambda}^a_{\dot{\alpha}} = e_\mu{}^{\tilde{\mu}}\Big(\partial_\mu\bar{\lambda}^a_{\dot{\alpha}} + \bar{\lambda}^a_{\dot{\beta}}\omega_{\tilde{\mu}}{}^{\dot{\beta}}{}_{\dot{\alpha}} - g f_{bc}{}^a v_{\tilde{\mu}}^b\bar{\lambda}^c_{\dot{\alpha}} - \frac{1}{4}(K_i\widetilde{\mathcal{D}}_{\tilde{\mu}}\phi^i - K_{i^*}\widetilde{\mathcal{D}}_{\tilde{\mu}}\phi^\dagger_{i^*})\bar{\lambda}^a_{\dot{\alpha}}\Big).$$

Finally, the interaction part of the rescaled Lagrangian, $\mathcal{L}^{(\text{kin})}_{\text{int}}$, is given by

$$\begin{aligned}
\mathcal{L}^{(\text{kin})}_{\text{int}} &= \frac{\sqrt{2}}{8}e\varepsilon^{\mu\nu\rho\sigma}\bar{\psi}_\mu\bar{\sigma}_\sigma\psi_\nu\Big[K_i\chi^i\cdot\psi_\rho - K^{i^*}\bar{\chi}_{i^*}\cdot\bar{\psi}_\rho\Big] \\
&\quad -\frac{\sqrt{2}i}{4}e\Big[\psi_\rho\sigma^\rho\bar{\psi}_\mu - \psi_\mu\sigma^\rho\bar{\psi}_\rho\Big]\Big[K_i\psi_\nu\sigma^{\mu\nu}\chi^i + K^{i^*}\bar{\psi}_\nu\bar{\sigma}^{\mu\nu}\bar{\chi}_{i^*}\Big] \\
&\quad -\frac{\sqrt{2}i}{4}e\psi_\mu\sigma^\rho\bar{\psi}_\nu\Big[K_i\psi_\rho\sigma^{\mu\nu}\chi^i + K^{i^*}\bar{\psi}_\rho\bar{\sigma}^{\mu\nu}\bar{\chi}_{i^*}\Big] \\
&\quad +\frac{1}{24}e\Big[K_iK_j\psi^\mu\cdot\chi^i\psi_\mu\cdot\chi^j + K^{i^*}K^{j^*}\bar{\psi}^\mu\cdot\bar{\chi}_{i^*}\bar{\psi}_\mu\cdot\bar{\chi}_{j^*}\Big] \\
&\quad +\frac{1}{12}eK_iK^{i^*}\Big[\psi^\mu\cdot\chi^i\bar{\psi}_\mu\cdot\bar{\chi}_{i^*} - 2\psi_\mu\sigma^{\mu\nu}\chi^i\bar{\psi}_\nu\cdot\bar{\chi}_{i^*} - 2\psi_\nu\cdot\chi^i\bar{\psi}_\mu\bar{\sigma}^{\mu\nu}\bar{\chi}_{i^*}\Big] \\
&\quad +\frac{\sqrt{2}i}{24}eK^{i^*}{}_i\Big[K_j(2\bar{\chi}_{i^*}\bar{\sigma}^\mu\psi_\mu\chi^i\cdot\chi^j - \chi^i\sigma^\mu\bar{\chi}_{i^*}\psi_\mu\cdot\chi^j) \\
&\qquad +K^{j^*}(2\chi^i\sigma^\mu\bar{\psi}_\mu\bar{\chi}_{i^*}\cdot\bar{\chi}_{j^*} + \chi^i\sigma^\mu\bar{\chi}_{i^*}\bar{\psi}_\mu\cdot\bar{\chi}_{j^*})\Big] \\
&\quad -\frac{\sqrt{2}i}{36}eK^{i^*}K_i\Big[K_j\bar{\chi}_{i^*}\bar{\sigma}^\mu\psi_\mu\chi^i\cdot\chi^j + K^{j^*}\chi^i\sigma^\mu\bar{\psi}_\mu\bar{\chi}_{i^*}\cdot\bar{\chi}_{j^*}\Big].
\end{aligned}$$

6.2.8 *Field redefinition: interacting terms*

We address now the super-Weyl rescaling of the interaction terms in (6.34). As the interactions arising from the gauge sector, *i.e.*, $\mathcal{L}_{\text{gauge,int}}$ in (6.22), have already been treated in Sec. 6.2.6, they will be omitted here. For the sake of clarity, below we will address the various terms one by one, following their sequence in (6.34). As usual, their variation under Howe–Tucker transformations is computed in two steps. The dilatation of the fields is performed first, followed by the shift of the gravitino.

Under a dilatation, the calculation of the variation of the first term is immediate and reads,

$$\begin{aligned}
&\frac{\sqrt{2}i}{4}e\psi_\nu\sigma^\rho\bar{\psi}_\mu\Big[\Omega_i\psi_\rho\sigma^{\nu\mu}\chi^i + \Omega^{i^*}\bar{\psi}_\rho\bar{\sigma}^{\nu\mu}\bar{\chi}_{i^*}\Big] \\
\to\ \mathcal{L}_{\text{int},1} &= \frac{\sqrt{2}i}{4}e\psi_\nu\sigma^\rho\bar{\psi}_\mu\Big[K_i\psi_\rho\sigma^{\nu\mu}\chi^i + K^{i^*}\bar{\psi}_\rho\bar{\sigma}^{\nu\mu}\bar{\chi}_{i^*}\Big],
\end{aligned}$$

where we have used (6.38), (6.39) and (6.41) for the rescaling of the vierbein, the gravitino and the fermionic fields, respectively, and replaced the derivatives of the kinetic energy by derivatives of the Kähler potential. The variation of the Lagrangian term $\mathcal{L}_{\text{int},1}$ under a gravitino shift reads

$$\begin{aligned}
\mathcal{L}_{\text{int},1} \rightarrow\ & \frac{\sqrt{2}i}{4}e\Big[\psi_\nu\sigma_\rho\bar{\psi}_\mu + \frac{\sqrt{2}i}{6}K^{j^*}\bar{\chi}_{j^*}\bar{\sigma}_\nu\sigma_\rho\bar{\psi}_\mu - \frac{\sqrt{2}i}{6}K_j\psi_\nu\sigma_\rho\bar{\sigma}_\mu\chi^j \\
&\quad + \frac{1}{18}K_jK^{j^*}\bar{\chi}_{j^*}\bar{\sigma}_\nu\sigma_\rho\bar{\sigma}_\mu\chi^j\Big]\Big[K_i\psi^\rho\sigma^{\nu\mu}\chi^i + K^{i^*}\bar{\psi}^\rho\bar{\sigma}^{\nu\mu}\bar{\chi}_{i^*} \\
&\quad + \frac{\sqrt{2}i}{6}K_iK^{i^*}\bar{\chi}_{i^*}\bar{\sigma}^\rho\sigma^{\nu\mu}\chi^i + \frac{\sqrt{2}i}{6}K_iK^{i^*}\chi^i\sigma^\rho\bar{\sigma}^{\nu\mu}\bar{\chi}_{i^*}\Big] \\
=\ & \frac{\sqrt{2}i}{4}e\Big[\psi_\nu\sigma_\rho\bar{\psi}_\mu + \frac{\sqrt{2}i}{6}K^{j^*}\bar{\chi}_{j^*}\bar{\sigma}_\nu\sigma_\rho\bar{\psi}_\mu - \frac{\sqrt{2}i}{6}K_j\psi_\nu\sigma_\rho\bar{\sigma}_\mu\chi^j \\
&\quad + \frac{i}{18}\varepsilon_{\mu\nu\rho\eta}K_jK^{j^*}\bar{\chi}_{j^*}\bar{\sigma}^\eta\chi^j\Big]\Big[K_i\psi^\rho\sigma^{\nu\mu}\chi^i + K^{i^*}\bar{\psi}^\rho\bar{\sigma}^{\nu\mu}\bar{\chi}_{i^*} \\
&\quad + \frac{\sqrt{2}}{6}\varepsilon^{\mu\nu\rho\sigma}K_iK^{i^*}\bar{\chi}_{i^*}\bar{\sigma}_\sigma\chi^i\Big] \\
=\ & \mathcal{L}_{\text{int},1} - \frac{\sqrt{2}i}{16}eK_iK^{i^*}\Big[K_j\chi^i\cdot\chi^j\psi_\rho\sigma^\rho\bar{\chi}_{i^*} + K^{j^*}\bar{\chi}_{i^*}\cdot\bar{\chi}_{j^*}\bar{\psi}_\rho\bar{\sigma}^\rho\chi^i\Big] \\
&\quad - \frac{1}{12}eK_iK^{i^*}\Big[\bar{\chi}_{i^*}\bar{\sigma}^\nu\psi_\nu\chi^i\sigma^\mu\bar{\psi}_\mu + \bar{\chi}_{i^*}\bar{\sigma}^\mu\psi_\nu\chi^i\sigma^\nu\bar{\psi}_\mu + \chi^i\cdot\psi_\mu\bar{\psi}^\mu\cdot\bar{\chi}_{i^*} \\
&\qquad - \chi^i\cdot\psi_\mu\bar{\psi}_\nu\bar{\sigma}^{\nu\mu}\bar{\chi}_{i^*} - \bar{\chi}_{i^*}\cdot\bar{\psi}_\mu\psi_\nu\sigma^{\nu\mu}\chi^i\Big] \\
&\quad + \frac{1}{8}eK_iK_j\Big[\chi^j\cdot\psi_\mu\psi^\mu\cdot\chi^i + 2\chi^j\cdot\psi_\mu\psi_\nu\sigma^{\mu\nu}\chi^i\Big] \\
&\quad + \frac{1}{8}eK^{i^*}K^{j^*}\Big[\bar{\chi}_{j^*}\cdot\bar{\psi}_\mu\bar{\psi}^\mu\cdot\bar{\chi}_{i^*} + 2\bar{\chi}_{j^*}\cdot\bar{\psi}_\mu\bar{\psi}_\nu\bar{\sigma}^{\mu\nu}\bar{\chi}_{i^*}\Big] \\
&\quad + \frac{1}{18}eK_iK_jK^{i^*}K^{j^*}\chi^i\cdot\chi^j\bar{\chi}_{i^*}\cdot\bar{\chi}_{j^*}\,,
\end{aligned}$$

where, for the first equality, we have used the properties (C.5) of the Pauli matrices and the antisymmetry property of the second brackets under the exchange of the Lorentz indices $\mu \leftrightarrow \nu$. The second equality is derived after grouping terms with the same structure in the derivatives of the Kähler potential and applying (C.5), (C.6), the Fierz identity (C.9) and the duality properties (C.2).

After the dilation of all the fields, the second term of (6.34) transforms as

$$\begin{aligned}
& -\frac{\sqrt{2}i}{4}e\bar{\psi}_\mu\bar{\sigma}^\mu\psi_\rho\Big[\Omega_i\psi_\nu\sigma^{\nu\rho}\chi^i + \Omega^{i^*}\bar{\psi}_\nu\bar{\sigma}^{\nu\rho}\bar{\chi}_{i^*}\Big] + \text{h.c.} \\
\rightarrow\ & \mathcal{L}_{\text{int},2} = -\frac{\sqrt{2}i}{4}e\bar{\psi}_\mu\bar{\sigma}^\mu\psi_\rho\Big[K_i\psi_\nu\sigma^{\nu\rho}\chi^i + K^{i^*}\bar{\psi}_\nu\bar{\sigma}^{\nu\rho}\bar{\chi}_{i^*}\Big] + \text{h.c.}\,,
\end{aligned}$$

where we have once again used (6.38), (6.39) and (6.41) for the rescaling of the different fields. Applying a gravitino shift to $\mathcal{L}_{\text{int},2}$, one gets

$$\begin{aligned}
\mathcal{L}_{\text{int},2} \to & -\frac{\sqrt{2}i}{4}e\Big[\big(\bar{\psi}_\mu\bar{\sigma}^\mu\psi_\rho - \bar{\psi}_\rho\bar{\sigma}^\mu\psi_\mu\big) - \frac{\sqrt{2}i}{6}K^{j^*}\big(\bar{\psi}_\mu\bar{\sigma}^\mu\sigma_\rho\bar{\chi}_{j^*} - \bar{\psi}_\rho\bar{\sigma}^\mu\sigma_\mu\bar{\chi}_{j^*}\big) \\
& + \frac{\sqrt{2}i}{6}K_j\big(\chi^j\sigma_\mu\bar{\sigma}^\mu\psi_\rho - \chi^j\sigma_\rho\bar{\sigma}^\mu\psi_\mu\big)\Big]\Big[K_i\psi_\nu\sigma^{\nu\rho}\chi^i + K^{i^*}\bar{\psi}_\nu\bar{\sigma}^{\nu\rho}\bar{\chi}_{i^*}\Big] \\
= & \ \mathcal{L}_{\text{int},2} + \frac{1}{12}eK_iK^{i^*}\big[\bar{\chi}_{i^*}\bar{\sigma}^\nu\psi_\nu\chi^i\sigma^\mu\bar{\psi}_\mu + \bar{\chi}_{i^*}\bar{\sigma}^\mu\psi_\nu\chi^i\sigma^\nu\bar{\psi}_\mu + \chi^i\cdot\psi_\mu\bar{\psi}^\mu\cdot\bar{\chi}_{i^*} \\
& + 3\chi^i\cdot\psi_\mu\bar{\psi}_\nu\bar{\sigma}^{\nu\mu}\bar{\chi}_{i^*} + 3\bar{\chi}_{i^*}\cdot\bar{\psi}_\mu\psi_\nu\sigma^{\nu\mu}\chi^i\big] \\
& - \frac{1}{24}eK_iK_j\big[3\chi^j\cdot\psi_\mu\psi^\mu\cdot\chi^i - 2\chi^j\cdot\psi_\mu\psi_\nu\sigma^{\nu\mu}\chi^i\big] \\
& - \frac{1}{24}eK^{i^*}K^{j^*}\big[3\bar{\chi}_{j^*}\cdot\bar{\psi}_\mu\bar{\psi}^\mu\cdot\bar{\chi}_{i^*} - 2\bar{\chi}_{j^*}\cdot\bar{\psi}_\mu\bar{\psi}_\nu\bar{\sigma}^{\nu\mu}\bar{\chi}_{i^*}\big],
\end{aligned}$$

where we have used (C.5) for the simplifications.

Turning to the third term of (6.34), its change under a pure dilatation of the fields reads, using the same relations as those mentioned for the first two terms,

$$\begin{aligned}
& -\frac{\sqrt{2}}{8}e\varepsilon^{\mu\nu\rho\sigma}\bar{\psi}_\mu\bar{\sigma}_\sigma\psi_\nu\big[\Omega_i\psi_\rho\cdot\chi^i - \Omega^{i^*}\bar{\psi}_\rho\cdot\bar{\chi}_{i^*}\big] \\
& \to \ \mathcal{L}_{\text{int},3} = -\frac{\sqrt{2}}{8}e\varepsilon^{\mu\nu\rho\sigma}\bar{\psi}_\mu\bar{\sigma}_\sigma\psi_\nu\big[K_i\psi_\rho\cdot\chi^i - K^{i^*}\bar{\psi}_\rho\cdot\bar{\chi}_{i^*}\big],
\end{aligned}$$

which, under a gravitino shift, becomes

$$\begin{aligned}
\mathcal{L}_{\text{int},3} \to & -\frac{\sqrt{2}}{8}e\varepsilon^{\mu\nu\rho\sigma}\Big[\bar{\psi}_\mu\bar{\sigma}_\sigma\psi_\nu + \frac{\sqrt{2}i}{6}K_j\chi^j\sigma_\mu\bar{\sigma}_\sigma\psi_\nu \\
& - \frac{\sqrt{2}i}{6}K^{j^*}\bar{\psi}_\mu\bar{\sigma}_\sigma\sigma_\nu\bar{\chi}_{j^*} + \frac{1}{18}K^{j^*}K_j\chi^j\sigma_\mu\bar{\sigma}_\sigma\sigma_\nu\bar{\chi}_{j^*}\Big] \\
& \times\Big[K_i\psi_\rho\cdot\chi^i + \frac{\sqrt{2}i}{6}K^{i^*}K_i\bar{\chi}_{i^*}\bar{\sigma}_\rho\chi^i - K^{i^*}\bar{\psi}_\rho\cdot\bar{\chi}_{i^*} - \frac{\sqrt{2}i}{6}K^{i^*}K_i\chi^i\sigma_\rho\bar{\chi}_{i^*}\Big] \\
= & \ \mathcal{L}_{\text{int},3} - \frac{i}{12}e\varepsilon^{\mu\nu\rho\sigma}K^{i^*}K_i\bar{\psi}_\mu\bar{\sigma}_\sigma\psi_\nu\bar{\chi}_{i^*}\bar{\sigma}_\rho\chi^i \\
& - e\Big[K_i\psi_\rho\cdot\chi^i - K^{i^*}\bar{\psi}_\rho\cdot\bar{\chi}_{i^*} + \frac{\sqrt{2}i}{3}K^{i^*}K_i\bar{\chi}_{i^*}\bar{\sigma}_\rho\chi^i\Big] \\
& \times\Big[\frac{1}{6}K_j\chi^j\sigma^{\nu\rho}\psi_\nu + \frac{1}{6}K^{j^*}\bar{\psi}_\mu\bar{\sigma}^{\mu\rho}\bar{\chi}_{j^*} + \frac{\sqrt{2}i}{24}K^{j^*}K_j\chi^j\sigma^\rho\bar{\chi}_{j^*}\Big]
\end{aligned}$$

$$= \mathcal{L}_{\text{int},3} + \frac{1}{6}eK_iK_j\psi_\rho\cdot\chi^i\psi_\nu\sigma^{\nu\rho}\chi^j + \frac{1}{6}eK^{i^*}K^{j^*}\bar{\psi}_\rho\cdot\bar{\chi}_{i^*}\bar{\psi}_\nu\bar{\sigma}^{\nu\rho}\bar{\chi}_{j^*}$$

$$+\frac{\sqrt{2}i}{16}eK_iK^{i^*}K_j\chi^i\cdot\chi^j\psi_\rho\sigma^\rho\bar{\chi}_{i^*} + \frac{\sqrt{2}i}{16}eK_iK^{i^*}K^{j^*}\bar{\chi}_{i^*}\cdot\bar{\chi}_{j^*}\bar{\psi}_\rho\bar{\sigma}^\rho\chi^i$$

$$-\frac{1}{18}eK_iK_jK^{i^*}K^{j^*}\bar{\chi}_{i^*}\cdot\bar{\chi}_{j^*}\chi^i\cdot\chi^j\,.$$

The second equality is deduced after using the identities (C.5), the duality properties (C.2) and (C.12). The last equality arises from the Fierz identities (C.9).

After the dilatation of the fourth term of (6.34),

$$\frac{e}{4}\Omega^{i^*}{}_i\big[\chi^i\sigma^\mu\bar{\sigma}^\nu\psi_\mu\bar{\psi}_\nu\cdot\bar{\chi}_{i^*} + \bar{\chi}_{i^*}\bar{\sigma}^\mu\sigma^\nu\bar{\psi}_\mu\psi_\nu\cdot\chi^i\big]$$

$$\to \mathcal{L}_{\text{int},4} = \frac{e}{4}\Big[K^{i^*}{}_i - \frac{1}{3}K^{i^*}K_i\Big]\big[\chi^i\sigma^\mu\bar{\sigma}^\nu\psi_\mu\bar{\psi}_\nu\cdot\bar{\chi}_{i^*} + \bar{\chi}_{i^*}\bar{\sigma}^\mu\sigma^\nu\bar{\psi}_\mu\psi_\nu\cdot\chi^i\big]\,,$$

the gravitino shift leads to the following transformation law

$$\mathcal{L}_{\text{int},4} \to \mathcal{L}_{\text{int},4} + \frac{e}{4}\Big[K^{i^*}{}_i - \frac{1}{3}K^{i^*}K_i\Big]$$

$$\times\Big[\frac{1}{18}K_jK^{j^*}(\chi^i\sigma^\mu\bar{\sigma}^\nu\sigma_\mu\bar{\chi}_{j^*}\chi^j\sigma_\nu\bar{\chi}_{i^*} + \bar{\chi}_{i^*}\bar{\sigma}^\mu\sigma^\nu\bar{\sigma}_\mu\chi^j\bar{\chi}_{j^*}\bar{\sigma}_\nu\cdot\chi^i)$$

$$+\frac{\sqrt{2}i}{6}K^{j^*}(\bar{\chi}_{i^*}\bar{\sigma}^\mu\sigma^\nu\bar{\psi}_\mu\bar{\chi}_{j^*}\bar{\sigma}_\nu\chi^i - \chi^i\sigma^\mu\bar{\sigma}^\nu\sigma_\mu\bar{\chi}_{j^*}\bar{\psi}_\nu\cdot\bar{\chi}_{i^*})$$

$$+\frac{\sqrt{2}i}{6}K_j(\chi^i\sigma^\mu\bar{\sigma}^\nu\psi_\mu\chi^j\sigma_\nu\bar{\chi}_{i^*} - \bar{\chi}_{i^*}\bar{\sigma}^\mu\sigma^\nu\bar{\sigma}_\mu\chi^j\psi_\nu\cdot\chi^i)\Big]$$

$$= \mathcal{L}_{\text{int},4} + e\Big[K^{i^*}{}_i - \frac{1}{3}K^{i^*}K_i\Big]$$

$$\times\Big[\frac{\sqrt{2}i}{12}(K^{j^*}\bar{\chi}_{i^*}\cdot\bar{\chi}_{j^*}\bar{\psi}_\nu\bar{\sigma}^\nu\chi^i + K_j\chi^i\cdot\chi^j\psi_\nu\sigma^\nu\bar{\chi}_{i^*}) - \frac{1}{9}K_jK^{j^*}\chi^i\cdot\chi^j\bar{\chi}_{j^*}\cdot\bar{\chi}_{i^*}\Big]\,,$$

after using again the relations of App. C.

We now proceed to the fifth term of the interaction Lagrangian,

$$\mathcal{L}_{\text{int},5} = \frac{e}{8}\Omega^{-1}\big[\Omega^{i^*}\bar{\chi}_{i^*}\cdot\bar{\psi}_\mu - \Omega_i\chi^i\cdot\psi_\mu\big]$$

$$\times\big[\Omega^{j^*}\bar{\chi}_{j^*}\cdot\bar{\psi}^\mu - \Omega_j\chi^j\cdot\psi^\mu + \sqrt{2}i\Omega^{j^*}{}_j\chi^j\sigma^\mu\bar{\chi}_{j^*} - 3\sqrt{2}ih^R_{ab}\lambda^a\sigma^\mu\bar{\lambda}^b\big]\,,$$

where we recall that the last contribution arises from the auxiliary fields, as can be seen in the Lagrangian $\mathcal{L}_{\text{aux},b}$ (see (6.26)). Under a pure dilatation, the Lagrangian $\mathcal{L}_{\text{int},5}$ becomes

$$\mathcal{L}_{\text{int},5} \to -\frac{e}{24}\big[K^{i^*}\bar{\chi}_{i^*}\cdot\bar{\psi}_\mu - K_i\chi^i\cdot\psi_\mu\big]$$

$$\times\Big[K^{j^*}\bar{\chi}_{j^*}\cdot\bar{\psi}^\mu - K_j\chi^j\cdot\psi^\mu + \sqrt{2}i(K^{j^*}{}_j - \frac{1}{3}K^{j^*}K_j)\chi^j\sigma^\mu\bar{\chi}_{j^*} - 3\sqrt{2}ih^R_{ab}\lambda^a\sigma^\mu\bar{\lambda}^b\Big]\,.$$

This last expression acquires additional contributions from the shift of the gravitino field since

$$K^{i^*}\bar{\chi}_{i^*}\cdot\bar{\psi}_\mu - K_i\chi^i\cdot\psi_\mu \to K^{i^*}\bar{\chi}_{i^*}\cdot\bar{\psi}_\mu - K_i\chi^i\cdot\psi_\mu + \frac{\sqrt{2}i}{3}K_iK^{i^*}\chi^i\sigma_\mu\bar{\chi}_{i^*}\ .$$

Therefore, the transformation law of $\mathcal{L}_{\text{int},5}$ under a general Howe–Tucker transformation reads

$$\begin{aligned}\mathcal{L}_{\text{int},5} \to\ & \frac{e}{48}\Big[K^{i^*}K^{j^*}\bar{\chi}_{i^*}\cdot\bar{\chi}_{j^*}\bar{\psi}^\mu\cdot\bar{\psi}_\mu + K_iK_j\chi^i\cdot\chi^j\psi^\mu\cdot\psi_\mu + 4K_iK^{i^*}\bar{\chi}_{i^*}\cdot\bar{\psi}_\mu\chi^i\cdot\psi^\mu\Big]\\ & -\frac{\sqrt{2}i}{24}eK^{i^*}{}_i\Big[K^{j^*}\bar{\chi}_{j^*}\cdot\bar{\psi}_\nu\chi^i\sigma^\nu\bar{\chi}_{i^*} - K_j\chi^j\cdot\psi_\nu\chi^i\sigma^\nu\bar{\chi}_{i^*}\\ & +\frac{2\sqrt{2}i}{3}K_jK^{j^*}\chi^i\cdot\chi^j\bar{\chi}_{j^*}\cdot\bar{\chi}_{i^*}\Big] + \frac{\sqrt{2}i}{8}eh^R_{ab}\lambda^a\sigma^\mu\bar{\lambda}^b\Big[K^{i^*}\bar{\chi}_{i^*}\cdot\bar{\psi}_\mu - K_i\chi^i\cdot\psi_\mu\Big]\\ & -\frac{1}{6}eK_iK^{i^*}h^R_{ab}\bar{\chi}_{i^*}\cdot\bar{\lambda}^b\chi^i\cdot\lambda^a - \frac{\sqrt{2}i}{72}eK_iK^{i^*}\Big[K^{j^*}\bar{\chi}_{j^*}\cdot\bar{\psi}_\mu - K_j\chi^j\cdot\psi_\mu\Big]\chi^i\sigma^\mu\bar{\chi}_{i^*}\ ,\end{aligned}$$

after using (C.6) and (C.9) for the simplifications.

The sixth, seventh, eighth and ninth terms of (6.34) being independent of the gravitino, they are only affected by a dilatation. Collecting them, the associated transformation law is given by

$$\begin{aligned}\mathcal{L}_{\text{int},6-9} =\ & -\frac{e}{12}\Omega\Big[R_i{}^{i^*}{}_j{}^{j^*} - \frac{1}{2}K^{i^*}{}_iK^{j^*}{}_j - \frac{1}{9}K^{i^*}{}_iK^{j^*}K_j + \frac{1}{54}K_iK_jK^{i^*}K^{j^*}\Big]\chi^i\cdot\chi^j\bar{\chi}_{i^*}\cdot\bar{\chi}_{j^*}\\ & +\frac{\sqrt{2}}{2}ieg\Omega^{i^*}{}_i\Big[(\phi^\dagger T_a)_{i^*}\lambda^a\cdot\chi^i - (T_a\phi)^i\bar{\lambda}^a\cdot\bar{\chi}_{i^*}\Big]\\ & -\frac{\sqrt{2}}{2}ieg\Big[\Omega^{i^*j^*}\bar{\chi}_{j^*}\cdot\bar{\lambda}^a(\phi^\dagger T_a)_{i^*} - \Omega_{ij}\chi^j\cdot\lambda^a(T_a\phi)^i\Big]\\ & -\frac{\sqrt{2}}{2}ieg\Big[\Omega^{i^*}(\bar{\chi}T_a)_{i^*}\cdot\bar{\lambda}^a - \Omega_i(T_a\chi)^i\cdot\lambda^a\Big]\\ \to\ & \frac{e}{4}\Big[R_i{}^{i^*}{}_j{}^{j^*} - \frac{1}{2}K^{i^*}{}_iK^{j^*}{}_j - \frac{1}{9}K^{i^*}{}_iK^{j^*}K_j + \frac{1}{54}K_iK_jK^{i^*}K^{j^*}\Big]\chi^i\cdot\chi^j\bar{\chi}_{i^*}\cdot\bar{\chi}_{j^*}\\ & +\frac{\sqrt{2}i}{2}eg\Big[2K^{i^*}{}_i - \frac{1}{3}K_iK^{i^*}\Big]\Big[(\phi^\dagger T_a)_{i^*}\lambda^a\cdot\chi^i - (T_a\phi)^i\bar{\lambda}^a\cdot\bar{\chi}_{i^*}\Big]\\ & +\frac{\sqrt{2}i}{6}eg\Big[K^{i^*}K^{j^*}\bar{\chi}_{j^*}\cdot\bar{\lambda}^a(\phi^\dagger T_a)_{i^*} - K_iK_j\chi^j\cdot\lambda^a(T_a\phi)^i\Big]\ .\end{aligned}$$

The simplifications performed here are based on the gauge invariance identities

$$\begin{aligned}0 &= \delta_a(K_i\chi^i) = K_{ij}(T_a\phi)^j\chi^i - K^{i^*}{}_i(\phi^\dagger T_a)_{i^*}\chi^i + K_i(T_a\chi)^i\ ,\\ 0 &= \delta_a(K^{i^*}\bar{\chi}_{i^*}) = -K^{i^*j^*}(\phi^\dagger T_a)_{j^*}\bar{\chi}_{i^*} + K^{i^*}{}_i(T_a\phi)^i\bar{\chi}_{i^*} - K^{i^*}(\bar{\chi}T_a)_{i^*}\ .\end{aligned}$$

In contrast, the tenth term of (6.34), arising from the Kähler potential, transforms both under a pure dilatation and a gravitino shift. The dilatation contributions reads

$$\begin{aligned}&\frac{e}{4}g\big[\Omega_i(T_a\phi)^i+\Omega^{i^*}(\phi^\dagger T_a)_{i^*}\big]\big[\bar\lambda^a\bar\sigma^\mu\psi_\mu+\bar\psi_\mu\bar\sigma^\mu\lambda^a\big]\\ \to\ &\mathcal{L}_{\text{int},10}=\frac{e}{4}g\big[K_i(T_a\phi)^i+K^{i^*}(\phi^\dagger T_a)_{i^*}\big]\big[\bar\lambda^a\bar\sigma^\mu\psi_\mu+\bar\psi_\mu\bar\sigma^\mu\lambda^a\big]\,,\end{aligned}$$

whilst after a gravitino shift, using (C.6), one immediately gets

$$\begin{aligned}\mathcal{L}_{\text{int},10}\to\mathcal{L}_{\text{int},10}+\frac{\sqrt{2}i}{6}eg\Big[&K_iK_j\chi^j\cdot\lambda^a(T_a\phi)^i-K^{i^*}K^{j^*}\bar\chi_{j^*}\cdot\bar\lambda^a(\phi^\dagger T_a)_{i^*}\\ &-K_iK^{i^*}(T_a\phi)^i\bar\lambda^a\cdot\bar\chi_{i^*}+K_iK^{i^*}(\phi^\dagger T_a)_{i^*}\lambda^a\cdot\chi^i\Big]\,.\end{aligned}$$

The next terms of (6.34) concern the superpotential. The first three of these terms,

$$\mathcal{L}_{\text{int},11-13}=\frac{\sqrt{2}}{2}ie\big(W_i\bar\psi_\mu\bar\sigma^\mu\chi^i+W^{\star i^*}\psi_\mu\sigma^\mu\bar\chi_{i^*}\big)-eW\bar\psi_\mu\bar\sigma^{\mu\nu}\bar\psi_\nu-eW^\star\psi_\mu\sigma^{\mu\nu}\psi_\nu\,,$$

are modified both under a dilatation and a gravitino shift. Applying the dilatation part of the Howe–Tucker transformations leads to

$$\mathcal{L}_{\text{int},11-13}\to ee^{K/2}\Big[\frac{\sqrt{2}}{2}i\big(W_i\bar\psi_\mu\bar\sigma^\mu\chi^i+W^{\star i^*}\psi_\mu\sigma^\mu\bar\chi_{i^*}\big)-W\bar\psi_\mu\bar\sigma^{\mu\nu}\bar\psi_\nu-W^\star\psi_\mu\sigma^{\mu\nu}\psi_\nu\Big]\,,$$

which, after including in addition the gravitino shift, becomes

$$\begin{aligned}\mathcal{L}_{\text{int},11-13}\to ee^{K/2}\Big\{&\frac{\sqrt{2}i}{2}\big[W_i\bar\psi_\mu\bar\sigma^\mu\chi^i+W^{\star i^*}\psi_\mu\sigma^\mu\bar\chi_{i^*}-K_iW\chi^i\sigma^\mu\bar\psi_\mu-K^{i^*}W^\star\bar\chi_{i^*}\bar\sigma^\mu\psi_\mu\big]\\ &-W\bar\psi_\mu\bar\sigma^{\mu\nu}\bar\psi_\nu-W^\star\psi_\mu\sigma^{\mu\nu}\psi_\nu\\ &-\frac{1}{3}\big[K_iW_j+K_jW_i+K_iK_jW\big]\chi^i\cdot\chi^j\\ &-\frac{1}{3}\big[K^{i^*}W^{\star j^*}+K^{j^*}W^{\star i^*}+K^{i^*}K^{j^*}W^\star\big]\bar\chi_{i^*}\cdot\bar\chi_{j^*}\Big\}\,.\end{aligned}$$

The next terms are related to the superpotential, and we have

$$\begin{aligned}\mathcal{L}_{\text{int},14-16}=&-\frac{1}{2}e\Big[W_{ij}+K_{ij}W+\frac{1}{3}K_iW_j+\frac{1}{3}K_jW_i+\frac{1}{3}K_iK_jW-\Gamma_i{}^k{}_j\mathcal{D}_kW\Big]\chi^i\cdot\chi^j\\ &-\frac{1}{2}e\Big[W^{\star i^*j^*}+K^{i^*j^*}W^\star+\frac{1}{3}K^{i^*}W^{\star j^*}+\frac{1}{3}K^{j^*}W^{\star i^*}\end{aligned}$$

$$
\begin{aligned}
&+\frac{1}{3}K^{i^*}K^{j^*}W^\star - \Gamma^{i^*}{}_{k^*}{}^{j^*}\mathcal{D}^{k^*}W^\star\Big]\bar{\chi}_{i^*}\cdot\bar{\chi}_{j^*} \\
&+\frac{1}{4}ee^{\frac{1}{3}K}K^i{}_{i^*}\Big[h_{abi}\mathcal{D}^{i^*}W^\star\lambda^a\cdot\lambda^b + h^{\star}_{ab}{}^{i^*}\mathcal{D}_iW\bar{\lambda}^a\cdot\bar{\lambda}^b\Big].
\end{aligned}
$$

They do not depend on the gravitino and are therefore only affected by the dilatation part of the Hower-Tucker transformations. This gives,

$$
\begin{aligned}
\mathcal{L}_{\text{int},14-16} \;\to\; e^{K/2}\Bigg\{ & -\frac{1}{2}\Big[W_{ij} + K_{ij}W + \frac{1}{3}K_iW_j + \frac{1}{3}K_jW_i \\
&\qquad + \frac{1}{3}K_iK_jW - \Gamma_i{}^k{}_j\mathcal{D}_kW\Big]\chi^i\cdot\chi^j \\
& -\frac{1}{2}\Big[W^{\star i^*j^*} + K^{i^*j^*}W^\star + \frac{1}{3}K^{i^*}W^{\star j^*} + \frac{1}{3}K^{j^*}W^{\star i^*} \\
&\qquad + \frac{1}{3}K^{i^*}K^{j^*}W^\star - \Gamma^{i^*}{}_{k^*}{}^{j^*}\mathcal{D}^{k^*}W^\star\Big]\bar{\chi}_{i^*}\cdot\bar{\chi}_{j^*} \\
& + \frac{1}{4}K^i{}_{i^*}\Big[h_{abi}\mathcal{D}^{i^*}W^\star\lambda^a\cdot\lambda^b + h^{\star}_{ab}{}^{i^*}\mathcal{D}_iW\bar{\lambda}^a\cdot\bar{\lambda}^b\Big]\Bigg\}.
\end{aligned}
$$

Among the remaining terms in the interaction Lagrangian, only the 21rd, the 22th and the 26th terms are not invariant under Howe–Tucker transformations[6], their variation being given by

$$
\begin{aligned}
\mathcal{L}_{\text{int},21-22,26} \;=\; & -\frac{9}{8}e\Omega^{-1}h^R_{ab}h^R_{cd}\lambda^a\cdot\lambda^c\bar{\lambda}^b\cdot\bar{\lambda}^d - \frac{1}{16}ee^{K/3}K^i{}_{i^*}h_{abi}h^{\star}_{cd}{}^{i^*}\lambda^a\cdot\lambda^b\bar{\lambda}^c\cdot\bar{\lambda}^d \\
& -\frac{\sqrt{2}i}{8}eh^{Rab}\Big[\Omega^{i^*}(\phi^\dagger T_a)_{i^*} + \Omega_i(T_a\phi)^i\Big]\Big[h_{bdj}\chi^j\cdot\lambda^d - h^{\star}_{bd}{}^{j^*}\bar{\chi}_{j^*}\cdot\bar{\lambda}^d\Big] \\
\to\; & +\frac{3}{8}eh^R_{ab}h^R_{cd}\lambda^a\cdot\lambda^c\bar{\lambda}^b\cdot\bar{\lambda}^d - \frac{1}{16}eK^i{}_{i^*}h_{abi}h^{\star}_{cd}{}^{i^*}\lambda^a\cdot\lambda^b\bar{\lambda}^c\cdot\bar{\lambda}^d\,. \\
& -\frac{\sqrt{2}i}{8}eh^{Rab}\Big[K^{i^*}(\phi^\dagger T_a)_{i^*} + K_i(T_a\phi)^i\Big]\Big[h_{bdj}\chi^j\cdot\lambda^d - h^{\star}_{bd}{}^{j^*}\bar{\chi}_{j^*}\cdot\bar{\lambda}^d\Big].
\end{aligned}
$$

Indeed, we recall that all the other terms are trivially invariant under a dilatation. Concerning the shift part of the Howe–Tucker transformations, the only terms depending on the gravitino are driven by the gauge interaction part of Lagrangian (6.22). Therefore, conformal invariance of the gauge sector ensures that their variation is exactly compensated by the one of the gauge kinetic terms, as mentioned above, and that it variation can be safely ignored in the computations[7].

[6]More precisely, the variation of the terms 23rd, 24th and 25th have already been considered (see Eq.[6.50]).

[7]The eD^aD_a terms being both invariant under a dilatation and a gravitino shift, their treatment can be safely decoupled from the one of the kinetic and interaction terms, which is the guideline implicitly followed in this section.

Finally, we still have to compute the variation of the scalar potential under a dilatation. The result is given by

$$\begin{aligned}
V &= \frac{g^2}{8} e e^{-\frac{2}{3}K} h^{Rab}\Big[K^{i^*}(\phi^\dagger T_a)_{i^*} + K_i(T_a\phi)^i\Big]\Big[K^{j^*}(\phi^\dagger T_b)_{j^*} + K_j(T_b\phi)^j\Big] \\
&\quad + e e^{\frac{1}{3}K}\Big[K^i{}_{i^*}\mathcal{D}_i W \mathcal{D}^{i^*} W^\star - 3\big|W\big|^2\Big] \\
&\to \frac{g^2}{8} e h^{Rab}\Big[K^{i^*}(\phi^\dagger T_a)_{i^*} + K_i(T_a\phi)^i\Big]\Big[K^{j^*}(\phi^\dagger T_b)_{j^*} + K_j(T_b\phi)^j\Big] \\
&\quad + e e^{K}\Big[K^i{}_{i^*}\mathcal{D}_i W \mathcal{D}^{i^*} W^\star - 3\big|W\big|^2\Big]. \qquad (6.69)
\end{aligned}$$

Interactions.

We are now ready to collect all the interaction terms resulting from the rescaling procedure. Then, we have to consider the non-derivative terms of the rescaled kinetic Lagrangian (6.67) together with all the terms computed from the interaction Lagrangian. Hence, one gets

$$\begin{aligned}
\mathcal{L}_{\text{int}} &= \sqrt{2}ie g K^{i^*}{}_i\Big[(\phi^\dagger T_a)_{i^*}\lambda^a\cdot\chi^i - (T_a\phi)^i\bar\lambda^a\cdot\bar\chi_{i^*}\Big] \\
&\quad -\frac{\sqrt{2}i}{8} e h^{Rab}\Big[K^{i^*}(\phi^\dagger T_a)_{i^*} + K_i(T_a\phi)^i\Big]\Big[h_{bdj}\chi^j\cdot\lambda^d - h^\star_{bd}{}^{j^*}\bar\chi_{j^*}\cdot\bar\lambda^d\Big] \\
&\quad e e^{K/2}\Big\{ - W\bar\psi_\mu\bar\sigma^{\mu\nu}\bar\psi_\nu - W^\star\psi_\mu\sigma^{\mu\nu}\psi_\nu \\
&\qquad -\frac{1}{2}\mathcal{D}_i\mathcal{D}_j W\chi^i\cdot\chi^j - \frac{1}{2}\overline{\mathcal{D}}^{i^*}\overline{\mathcal{D}}^{j^*}W^*\bar\chi_{i^*}\cdot\bar\chi_{j^*} \\
&\qquad +\frac{1}{4}K^i{}_{i^*}\Big[h_{abi}\mathcal{D}^{i^*}W^\star\lambda^a\cdot\lambda^b + h^\star_{ab}{}^{i^*}\mathcal{D}_i W\bar\lambda^a\cdot\bar\lambda^b\Big] \\
&\qquad +\frac{\sqrt{2}i}{2}\Big[\mathcal{D}_i W\bar\psi_\mu\bar\sigma^\mu\chi^i + \overline{\mathcal{D}}^{i^*}W^*\psi_\mu\sigma^\mu\bar\chi_{i^*}\Big]\Big\} \\
&\quad +\frac{e}{4}g\Big[K_i(T_a\phi)^i + K^{i^*}(\phi^\dagger T_a)_{i^*}\Big]\Big[\bar\lambda^a\bar\sigma^\mu\psi_\mu + \bar\psi_\mu\bar\sigma^\mu\lambda^a\Big] \\
&\quad -\frac{g^2}{8} e h^{Rab}\Big[K^{i^*}(\phi^\dagger T_a)_{i^*} + K_i(T_a\phi)^i\Big]\Big[K^{j^*}(\phi^\dagger T_b)_{j^*} + K_j(T_b\phi)^j\Big] \\
&\quad - e e^{K}\Big[K^i{}_{i^*}\mathcal{D}_i W\mathcal{D}^{i^*}W^\star - 3\big|W\big|^2\Big] \\
&\quad +\frac{e}{4}\Big[R_i{}^{i^*}{}_j{}^{j^*} - \frac{1}{2}K^{i^*}{}_i K^{j^*}{}_j - \frac{1}{3}K^{i^*}{}_i K^{j^*}K_j + \frac{1}{6}K_i K_j K^{i^*}K^{j^*}\Big]\chi^i\cdot\chi^j\bar\chi_{i^*}\cdot\bar\chi_{j^*} \\
&\quad -\frac{\sqrt{2}i}{72} e K_i K^{i^*}\Big[K^{j^*}\bar\chi_{j^*}\cdot\bar\psi_\mu - K_j\chi^j\cdot\psi_\mu\Big]\chi^i\sigma^\mu\bar\chi_{i^*} \\
&\quad +\frac{e}{4}K^{i^*}{}_i\chi^i\sigma_\mu\bar\chi_{i^*}\Big[\psi_\nu\sigma^\mu\bar\psi^\nu - i\varepsilon^{\mu\nu\rho\sigma}\psi_\nu\sigma_\rho\bar\psi_\sigma\Big]
\end{aligned}$$

$$
\begin{aligned}
&+\frac{e}{8}\Big[\mathcal{D}_i\mathcal{D}_j h_{ab}\chi^i\cdot\chi^j\lambda^a\cdot\lambda^b+\overline{\mathcal{D}}^{i^*}\overline{\mathcal{D}}^{j^*}h^*_{ab}\bar\chi_{i^*}\cdot\bar\chi_{j^*}\bar\lambda^a\cdot\bar\lambda^b\Big]\\
&-e\frac{1}{4}(K^{i^*}{}_i+\frac{1}{3}K^{i^*}K_i)h^R_{ab}\chi^i\cdot\lambda^a\bar\chi_{i^*}\cdot\bar\lambda^b\\
&+\frac{1}{16}eh^{Rab}\Big[h_{aci}\chi^i\cdot\lambda^c-h^{\star}_{ac}{}^{i^*}\bar\chi_{i^*}\cdot\bar\lambda^c\Big]\Big[h_{bdj}\chi^j\cdot\lambda^d-h^{\star}_{bd}{}^{j^*}\bar\chi_{j^*}\cdot\bar\lambda^d\Big]\\
&+\frac{3}{8}eh^R_{ab}h^R_{cd}\lambda^a\cdot\lambda^c\bar\lambda^b\cdot\bar\lambda^d-\frac{1}{16}eK^i{}_{i^*}h_{abi}h^{\star}_{cd}{}^{i^*}\lambda^a\cdot\lambda^b\bar\lambda^c\cdot\bar\lambda^d\\
&-\frac{\sqrt{2}i}{8}e\Big[h_{abi}\bar\psi_\mu\bar\sigma^\mu\chi^i\lambda^a\cdot\lambda^b+h^{\star}_{ab}{}^{i^*}\psi_\mu\sigma^\mu\bar\chi_{i^*}\bar\lambda^a\cdot\bar\lambda^b\Big]\\
&+\frac{\sqrt{2}i}{4}e\Big[h_{abi}\chi^i\sigma^{\mu\nu}\lambda^a+h^{\star}_{ab}{}^{i^*}\bar\chi_{i^*}\bar\sigma^{\mu\nu}\lambda^a\Big]\Big[\lambda^b\sigma_\nu\bar\psi_\mu+\bar\lambda^b\bar\sigma_\nu\psi_\mu\Big]\\
&-\frac{1}{4}eh^I_{ab}\varepsilon^{\mu\nu\rho\sigma}\psi_\rho\sigma_\sigma\bar\lambda^b\,\bar\psi_\mu\bar\sigma_\nu\lambda^a\,.
\end{aligned}\tag{6.70}
$$

Let us note that in order to simplify the terms in $K^{i^*}{}_i$, we apply several times the relations of App. C. We have also introduced the notations

$$
\begin{aligned}
\mathcal{D}_iW&=W_i+K_iW\,,\\
\mathcal{D}_i\mathcal{D}_iW&=W_{ij}+K_{ij}W+K_i\mathcal{D}_jW+K_j\mathcal{D}_iW-K_iK_jW-\Gamma_i{}^k{}_j\mathcal{D}_kW\,,\\
\mathcal{D}_ih_{ab}&=h_{abi}\,,\\
\mathcal{D}_i\mathcal{D}_jh_{ab}&=h_{abij}-\Gamma_i{}^k{}_jh_{abk}\,.
\end{aligned}
$$

These derivatives are fully covariant. This is obvious with respect to the Kähler manifold, thanks to the Christofell symbol. But it is also true with respect to Kähler transformation. Recalling that under a Kähler transformation $K\to K+F+F^\dagger$ the superpotential transforms like $W\to e^{-F}W$, as simple computation shows that likewise we have

$$
\begin{aligned}
W&\to e^{-K}W\,,\\
\mathcal{D}_iW&\to e^{-K}\mathcal{D}_iW\,,\\
\mathcal{D}_i\mathcal{D}_jW&\to e^{-K}\mathcal{D}_i\mathcal{D}_jW\,.
\end{aligned}\tag{6.71}
$$

Note that the final Lagrangian has four additional terms in the four-fermion interactions, with respect to Wess and Bagger (1992) (two more $\chi\chi\bar\chi\bar\chi$ terms, one more $\chi\lambda\bar\chi\bar\lambda$ term and one more $\chi\chi\bar\chi\bar\psi_\mu$+h.c. term).

6.3 Lagrangian of matter coupled to gauge interactions and supergravity

In this section, we recap the results in Tables 6.2 and 6.3 and analyse the content of the Lagrangian.

Table 6.2 Lagrangian of matter coupled to gauge interactions and supergravity: the kinetic part.

The Kinetic part of the Lagrangian of matter coupled with gauge interactions and supergravity reads,

$$\begin{aligned}
\mathcal{L}_{\rm kin} = & \frac{1}{2}ee_\mu{}^{\tilde\mu}e_\nu{}^{\tilde\nu}R_{\tilde\mu\tilde\nu}{}^{\mu\nu}\Big| + \frac{1}{4}e\varepsilon^{\mu\nu\rho\sigma}\Big[\psi_\mu\sigma_\sigma\bar{\check\psi}_{\nu\rho} - \bar\psi_\mu\bar\sigma_\sigma\check\psi_{\nu\rho}\Big] \\
& -\frac{1}{4}eh^R_{ab}F^a_{\mu\nu}F^{\mu\nu b} + \frac{1}{4}eh^I_{ab}F^a_{\mu\nu}{}^*F^{\mu\nu b} \\
& -\frac{i}{2}eh^R_{ab}(\lambda^a\sigma^\mu\check{\mathcal{D}}_\mu\bar\lambda^b - \check{\mathcal{D}}_\mu\lambda^a\sigma^\mu\bar\lambda^b) + \frac{1}{2}eh^I_{ab}\check{\mathcal{D}}_\mu(\lambda^a\sigma^\mu\bar\lambda^b) \\
& -\frac{i}{2}eK^{i^*}{}_i\Big[\chi^i\sigma^\mu\check{\mathcal{D}}_\mu\bar\chi_{i^*} - \check{\mathcal{D}}_\mu\chi^i\sigma^\mu\bar\chi_{i^*}\Big] + eK^{i^*}{}_i\widetilde{\mathcal{D}}_\mu\phi^i\widetilde{\mathcal{D}}^\mu\phi^\dagger_{i^*} \\
& -\frac{\sqrt2}{2}eK^{i^*}{}_i\Big[\chi^i\sigma^\mu\bar\sigma^\nu\psi_\mu\widetilde{\mathcal{D}}_\nu\phi^\dagger_{i^*} + \bar\chi_{i^*}\bar\sigma^\mu\sigma^\nu\bar\psi_\mu\widetilde{\mathcal{D}}_\nu\phi^i\Big] \\
& +\frac{i}{4}eh^R_{ab}\Big[\lambda^a\sigma_\mu\bar\sigma_{\nu\rho}\bar\psi^\mu + \bar\lambda^a\bar\sigma_\mu\sigma_{\nu\rho}\psi^\mu\Big]\Big[F^{\nu\rho b} + \hat F^{\nu\rho b}\Big] \\
& -\frac{\sqrt2}{4}e\Big[h_{abi}\chi^i\sigma^{\mu\nu}\lambda^b + h^\star_{ab}{}^{i^*}\bar\chi_{i^*}\bar\sigma^{\mu\nu}\bar\lambda^b\Big]F^a_{\mu\nu}\,.
\end{aligned}$$

Where the covariant derivatives are defined by

$$\begin{aligned}
\widetilde{\mathcal{D}}_\mu\phi^i &= e_\mu{}^{\tilde\mu}\Big[\partial_{\tilde\mu}\phi^i + igv^a_{\tilde\mu}(\phi T_a)^i\Big]\,, \\
\check{\mathcal{D}}_\mu\chi^{i\alpha} &= e_\mu{}^{\tilde\mu}[\partial_{\tilde\mu}\chi^{i\alpha} + \chi^{i\beta}\omega_{\tilde\mu\beta}{}^\alpha + igv^a_{\tilde\mu}(T_a\chi^\alpha)^i + \Gamma_j{}^i{}_k\chi^{\alpha j}\widetilde{\mathcal{D}}_{\tilde\mu}\phi^j - \frac{1}{4}(K_j\widetilde{\mathcal{D}}_{\tilde\mu}\phi^j - K^{j^*}\widetilde{\mathcal{D}}_{\tilde\mu}\phi^\dagger_{i^*})\chi^{\alpha i}\Big]\,, \\
\check{\mathcal{D}}_{\tilde\mu}\lambda^{\alpha a} &= e_\mu{}^{\tilde\mu}\Big[\partial_\mu\lambda^{\alpha a} + \lambda^{\beta a}\omega_{\tilde\mu\beta}{}^\alpha - gf_{bc}{}^a v^b_{\tilde\mu}\lambda^{\alpha c} + \frac{1}{4}(K_i\widetilde{\mathcal{D}}_{\tilde\mu}\phi^i - K_{i^*}\widetilde{\mathcal{D}}_{\tilde\mu}\phi^\dagger_{i^*})\lambda^{\alpha a}\Big]\,, \\
\check{\mathcal{D}}_\mu\psi_{\tilde\nu}{}^\alpha &= e_{\tilde\mu}{}^\mu\Big[\partial_\mu\psi_{\tilde\nu}{}^\alpha + \psi_{\tilde\nu}{}^\beta\omega_{\tilde\mu\beta}{}^\alpha + \frac{1}{4}(K_j\widetilde{\mathcal{D}}_{\tilde\mu}\phi^j - K^{j^*}\widetilde{\mathcal{D}}_{\tilde\mu}\phi^\dagger_{i^*})\psi_{\tilde\nu}{}^\alpha\Big]\,,
\end{aligned}$$

the Christofell symbol and the spin connection by

$$\begin{aligned}
\Gamma_i{}^j{}_k &= K^k{}_{k^*}K_i{}^{k^*}{}_j\,, \\
\omega_{\tilde\mu\tilde\nu\tilde\rho} &= -\frac{1}{2}e_{\mu\tilde\rho}(\partial_{\tilde\mu}e_{\tilde\nu}{}^\mu - \partial_{\tilde\nu}e_{\tilde\mu}{}^\mu) + \frac{1}{2}e_{\mu\tilde\mu}(\partial_{\tilde\nu}e_{\tilde\rho}{}^\mu - \partial_{\tilde\rho}e_{\tilde\nu}{}^\mu) - \frac{1}{2}e_{\mu\tilde\nu}\big(\partial_{\tilde\rho}e_{\tilde\mu}{}^\mu - \partial_{\tilde\mu}e_{\tilde\rho}{}^\mu\big) \\
&\quad +\frac{i}{4}e_{\mu\tilde\rho}(\psi_{\tilde\mu}\sigma^\mu\bar\psi_{\tilde\nu} - \psi_{\tilde\nu}\sigma^\mu\bar\psi_{\tilde\mu}) - \frac{i}{4}e_{\mu\tilde\mu}(\psi_{\tilde\nu}\sigma^\mu\bar\psi_{\tilde\rho} - \psi_{\tilde\rho}\sigma^\mu\bar\psi_{\tilde\nu}) + \frac{i}{4}e_{\mu\tilde\nu}(\psi_{\tilde\rho}\sigma^\mu\bar\psi_{\tilde\mu} - \psi_{\tilde\mu}\sigma^\mu\bar\psi_{\tilde\rho})\,,
\end{aligned}$$

and

$$\begin{aligned}
\check\psi_{\nu\rho} &= e_\mu{}^{\tilde\mu}e_\nu{}^{\tilde\nu}(\check{\mathcal{D}}_{\tilde\mu}\psi_{\tilde\nu} - \check{\mathcal{D}}_{\tilde\nu}\psi_{\tilde\mu})\,, \\
\hat F_{\mu\nu} &= e_\mu{}^{\tilde\mu}e_\nu{}^{\tilde\nu}\big(F_{\tilde\mu\tilde\nu} - \frac{i}{2}[\bar\lambda\bar\sigma_{\tilde\nu}\psi_{\tilde\mu} - \bar\psi_{\tilde\mu}\bar\sigma_{\tilde\nu}\lambda - \bar\lambda\bar\sigma_{\tilde\mu}\psi_{\tilde\nu} + \bar\psi_{\tilde\nu}\sigma_{\tilde\mu}\bar\lambda]\big)\,.
\end{aligned}$$

In Table 6.2 we have regrouped the kinetic terms. The four first lines are the standard kinetic terms for spin $2, 3/2, 1, 1/2$ and 0 particles. More precisely, the first line corresponds to the Einstein-Hilbert Lagrangian for gravity and to the Rarita-Swchinger Lagrangian for the spin $3/2$ gravitino. The second and third lines are the standard gauge part of the Lagrangian (for the gauge boson and

Table 6.3 Lagrangian of matter coupled to gauge interactions and supergravity: the interacting part.

Interacting part of the Lagrangian of matter coupled to gauge interactions and supergravity

$$
\begin{aligned}
\mathcal{L}_{\text{int}} = & \frac{e}{4}g\Big[K_i(T_a\phi)^i+K^{i^*}(\phi^\dagger T_a)_{i^*}\Big]\Big[\bar{\lambda}^a\bar{\sigma}^\mu\psi_\mu+\bar{\psi}_\mu\bar{\sigma}^\mu\lambda^a\Big] \\
& +ee^{\frac{K}{2}}\Big\{\frac{\sqrt{2}}{2}i\Big[\mathcal{D}_iW\bar{\psi}_\mu\bar{\sigma}^\mu\chi^i+\overline{\mathcal{D}}^{i^*}W^*\psi_\mu\sigma^\mu\bar{\chi}_{i^*}\Big] \\
& \qquad -W\bar{\psi}_\mu\bar{\sigma}^{\mu\nu}\bar{\psi}_\nu-W^\star\psi_\mu\sigma^{\mu\nu}\psi_\nu-\frac{1}{2}\mathcal{D}_i\mathcal{D}_jW\chi^i\cdot\chi^j-\frac{1}{2}\overline{\mathcal{D}}^{i^*}\overline{\mathcal{D}}^{j^*}W^*\bar{\chi}_{i^*}\cdot\bar{\chi}_{j^*} \\
& \qquad +\frac{1}{4}K^i{}_{i^*}\Big[h_{abi}\mathcal{D}^{i^*}W^\star\lambda^a\cdot\lambda^b+h^{\star}_{ab}{}^{i^*}\mathcal{D}_iW\bar{\lambda}^a\cdot\bar{\lambda}^b\Big]\Big\} \\
& +\sqrt{2}iegK^{i^*}{}_i\Big[(\phi^\dagger T_a)_{i^*}\lambda^a\cdot\chi^i-(T_a\phi)^i\bar{\lambda}^a\cdot\bar{\chi}_{i^*}\Big] \\
& -\frac{\sqrt{2}i}{8}eh^{Rab}\Big[K^{i^*}(\phi^\dagger T_a)_{i^*}+K_i(T_a\phi)^i\Big]\Big[h_{bdj}\chi^j\cdot\lambda^d-h^{\star}_{bd}{}^{j^*}\bar{\chi}_{j^*}\cdot\bar{\lambda}^d\Big] \\
& -\frac{g^2}{8}eh^{Rab}\Big[K^{i^*}(\phi^\dagger T_a)_{i^*}+K_i(T_a\phi)^i\Big]\Big[K^{j^*}(\phi^\dagger T_b)_{j^*}+K_j(T_b\phi)^j\Big] \\
& -ee^K\Big[K^i{}_{i^*}\mathcal{D}_iW\mathcal{D}^{i^*}W^\star-3|W|^2\Big] \\
& +\frac{e}{4}\Big[R_i{}^{i^*}{}_j{}^{j^*}-\frac{1}{2}K^{i^*}{}_iK^{j^*}{}_j-\frac{1}{3}K^{i^*}{}_iK^{j^*}K_j+\frac{1}{6}K_iK_jK^{i^*}K^{j^*}\Big]\chi^i\cdot\chi^j\bar{\chi}_{i^*}\cdot\bar{\chi}_{j^*} \\
& -\frac{\sqrt{2}i}{72}eK_iK^{i^*}\Big[K^{j^*}\bar{\chi}_{j^*}\cdot\bar{\psi}_\mu-K_j\chi^j\cdot\psi_\mu\Big]\chi^i\sigma^\mu\bar{\chi}_{i^*} \\
& +\frac{e}{4}K^{i^*}{}_i\chi^i\sigma_\mu\bar{\chi}_{i^*}\Big[\psi_\nu\sigma^\mu\bar{\psi}^\nu-i\varepsilon^{\mu\nu\rho\sigma}\psi_\nu\sigma_\rho\bar{\psi}_\sigma\Big]-e\frac{1}{4}(K^{i^*}{}_i+\frac{1}{3}K^{i^*}K_i)h^R_{ab}\chi^i\cdot\lambda^a\bar{\chi}_{i^*}\cdot\bar{\lambda}^b \\
& +\frac{1}{16}eh^{Rab}\Big[h_{aci}\chi^i\cdot\lambda^c-h^{\star}_{ac}{}^{i^*}\bar{\chi}_{i^*}\cdot\bar{\lambda}^c\Big]\Big[h_{bdj}\chi^j\cdot\lambda^d-h^{\star}_{bd}{}^{j^*}\bar{\chi}_{j^*}\cdot\bar{\lambda}^d\Big] \\
& +\frac{e}{8}\Big[\mathcal{D}_i\mathcal{D}_jh_{ab}\chi^i\cdot\chi^j\lambda^a\cdot\lambda^b+\overline{\mathcal{D}}^{i^*}\overline{\mathcal{D}}^{j^*}h^*_{ab}\bar{\chi}_{i^*}\cdot\bar{\chi}_{j^*}\bar{\lambda}^a\cdot\bar{\lambda}^b\Big]+\frac{3}{8}eh^R_{ab}h^R_{cd}\lambda^a\cdot\lambda^c\bar{\lambda}^b\cdot\bar{\lambda}^d \\
& -\frac{1}{16}eK^i{}_{i^*}h_{abi}h^{\star}_{cd}{}^{i^*}\lambda^a\cdot\lambda^b\bar{\lambda}^c\cdot\bar{\lambda}^d-\frac{\sqrt{2}i}{8}e\Big[h_{abi}\bar{\psi}_\mu\bar{\sigma}^\mu\chi^i\lambda^a\cdot\lambda^b+h^{\star}_{ab}{}^{i^*}\psi_\mu\sigma^\mu\bar{\chi}_{i^*}\bar{\lambda}^a\cdot\bar{\lambda}^b\Big] \\
& +\frac{\sqrt{2}i}{4}e\Big[h_{abi}\chi^i\sigma^{\mu\nu}\lambda^a+h^{\star}_{ab}{}^{i^*}\bar{\chi}_{i^*}\bar{\sigma}^{\mu\nu}\lambda^a\Big]\Big[\lambda^b\sigma_\nu\bar{\psi}_\mu+\bar{\lambda}^b\bar{\sigma}_\nu\psi_\mu\Big] \\
& -\frac{1}{4}eh^I_{ab}\varepsilon^{\mu\nu\rho\sigma}\psi_\rho\sigma_\sigma\bar{\lambda}^b\bar{\psi}_\mu\bar{\sigma}_\nu\lambda^a\,,
\end{aligned}
$$

the gauginos), and, finally, the fourth line is the standard matter part of the Lagrangian, describing spin $1/2$ matter and its scalar partner. Some remarks are in order however. Indeed, since the spin-connection is given by (3.119), we are in presence of torsion. If we develop the connection in the form

$$\omega(e,\psi)_{\tilde{\mu}\mu\nu}=\omega(e)_{\tilde{\mu}\mu\nu}+K_{\tilde{\mu}\mu\nu}(\psi)\,,$$

where the tensor $K_{\tilde{\mu}\mu\nu}(\psi)$, which is called the contortion, depends only on the gravitino, we see that the first term is torsion-free and only the second induces a

Table 6.4 Lagrangian of matter coupled to gauge interactions and supergravity: the interacting part cont'd.

where the various covariant derivatives are

$$\begin{aligned}
\mathcal{D}_i W &= W_i + K_i W \,, \\
\mathcal{D}_i \mathcal{D}_i W &= W_{ij} + K_{ij} W + K_i \mathcal{D}_j W + K_j \mathcal{D}_i W - K_i K_j W - \Gamma_i{}^k{}_j \mathcal{D}_k W \,, \\
\mathcal{D}_i h_{ab} &= h_{abi} \,, \\
\mathcal{D}_i \mathcal{D}_j h_{ab} &= h_{abij} - \Gamma_i{}^k{}_j h_{abk} \,,
\end{aligned}$$

the curvature tensor of the Kähler manifold is given by

$$R_i{}^{i^*}{}_j{}^{j^*} = K^{i^* j^*}{}_{ij} - K^{m^*}{}_m \Gamma^{i^*}{}_{m^*}{}^{j^*} \Gamma_i{}^m{}_j \,,$$

and the Christofell symbols by

$$\Gamma_i{}^k{}_j = K^k{}_{k^*} K_i{}^{k^*}{}_j \,.$$

torsion. In particular if we write the curvature tensor in the form

$$R(e, \psi) = R(e) + \text{torsion terms} \,,$$

the torsion terms contribute to four-gravitino interactions. In the same manner the covariant derivatives of the gravitino and fermions can be decomposed in the form

$$\check{\mathcal{D}}_{\tilde{\mu}}(e, \psi) = \check{\mathcal{D}}_{\tilde{\mu}}(e) + \text{torsion terms} \,,$$

and here the torsion terms generate two-gravitino-two-gaugino, two-gravitino-two-fermion or four-gravitino interactions.

In the same manner, if we decompose the covariant derivatives of the gravitino and fermions in the form

$$\check{\mathcal{D}}_{\tilde{\mu}} = \widetilde{\mathcal{D}}_{\tilde{\mu}} + \text{Kähler terms} \,,$$

we observe that we generate interactions of the fermions (gravitino, matter and gauginos) with the scalars. Finally the fifth and sixth lines of Table 6.2 together with the first and second lines of Table 6.3 correspond to the contraction of the gravitino with the conserved current (for a general expression of the conserved current in supersymmetric theories one can see *e.g.*, [Fuks and Rausch de Traubenberg (2011)]).

Now we address the terms of Table 6.3. The third to the fifth lines correspond to the mass terms in the fermionic sector. They will be analysed in Chapter 7. The first and second lines correspond to mixing terms of the gravitino with the goldstino. They will play a key rôle in the super-Higgs effect, *i.e.*, the Higgs effect, but in the context of supergravity. In Sec. 7.2 we will show how they cooperate together to eliminate the goldstino from the spectrum giving a mass to the gravitino.

The seventh and the eighth lines are minus the potential. This means in contrast to supersymmetry, in supergravity the potential is no-longer positive. This is of importance, notably is cosmology. Indeed, supergravity being a theory of gravitation, there a non-vanishing value for the potential at the minimum corresponds to a non-vanishing cosmological constant. In particular, at the minimum, we have (see Sec. 7.1),

$$\langle V \rangle = \langle K^{i^*}{}_i F^i F_{i^*} \rangle + \frac{1}{2}\langle h_{Rab} D^a D^b \rangle - \frac{1}{3}\langle N N^* \rangle .$$

Here the two first terms are the usual $F-$ and $D-$terms. The last one will be central in breaking supergravity while making the cosmological constant vanishing[8,9] (see Chapter 7 Sec. 7.3). This form of the potential is essential in gravity induced supersymmetry breaking, and in particular it generates the so-called soft-supersymmetric breaking terms (see Chapter 7). Finally, the remaining terms of the Lagrangian are the four-fermion interactions.

At this point, we would like to comment our Lagrangian, and in particular the covariant derivatives of the fermions. In our construction we were assuming from the very beginning that our superfields were in a unitary representation $\mathcal{R}$ of the gauge group G and thus that K and W are gauge singlets. Indeed, we have imposed the invariance of the Lagrangian under the gauge transformation

$$\begin{aligned} \delta_a \phi^i &= (T_a \phi)^i , \\ \delta_a \phi^\dagger_{i^*} &= -(\phi^\dagger T_a)_{i^*} , \end{aligned}$$

where T_a (resp. $T^*_{\underline{a}} = T^t_a$) are the generators of the Lie algebra of G in the representation $\mathcal{R}$ (resp. $\overline{\mathcal{R}}$). In the language of App. F the Killing vectors associated to the gauge symmetry are given by (see Eq.[F.11])

$$\begin{aligned} k^i_a(\phi) &= -i(T_a \phi)^i , \\ k^\dagger_{i^* a}(\phi^\dagger) &= i(\phi^\dagger T_a)_{i^*} . \end{aligned}$$

Consequently, in our construction, we were positing from the very beginning that the symmetries of the Kähler manifold are known and simply correspond to the gauge group G. Note however that there is another approach to determine the symmetry of the Kähler manifold without any reference to an *a priori* Lie group. In this approach, the symmetries are encoded in the Killing equation (F.7) and correspond to the transformation that preserve the Kähler metric. Then, it turns out that the symmetries are not necessary linear in the fields and could be non-linear. This means, in particular that the (gauge part of the) covariant derivatives

[8]If we assume that there are no fermion condensations, *i.e.*, $\langle \lambda \cdot \lambda \rangle \neq 0$ *etc*, at the minimum the potential is expressed in terms of the vacuum expectation value of the $F-$, $D-$ and $N-$terms.

[9]This is very different from supersymmetry where the signature of supersymmetry breaking is just the positivity of the potential.

has to be modified in the form of Eq.[F.14]. This approach has been followed in [Wess and Bagger (1992); Bagger (1983)]. Yet even more general Lagrangians can be obtained. In our construction of the final Lagrangian we have assumed (as in [Wess and Bagger (1992)]) that the Kähler potential is gauge invariant, *i.e.*, that under a gauge transformation we have

$$\delta_a K = K_i (T_a \phi)^i - K^{i^*} (\phi^\dagger T_a)_{i^*} = 0 \, .$$

However this assumption is too strong and we could have simply assumed that the Kähler potential is gauge invariant up to a Kähler transformation

$$\delta_a K = r_a(\phi) + r_a^\dagger(\phi^\dagger) \, ,$$

with $r_a(\phi)$ holomorphic functions. Within this assumption additional terms are present in the covariant derivatives for all the fermions. Note that in the reference of Wess and Bagger (1992) these terms are not present (as in our construction). On can consult *e.g.* [Kallosh et al. (2000); Freedman and Van Proeyen (2012)] where these additional terms are considered and have a natural superconformal origin (see Sec. 6.7).

6.4 The generalised Kähler potential

Another interesting presentation of supergravity is obtained performing the Kähler transformation of Sec. 5.3.3 with $F = \ln W$ [10]

$$\begin{aligned} K &\to \mathcal{G} = K + \ln |W|^2 \, , \\ K_i &\to \mathcal{G}_i \, , \\ W &\to 1 \, , \\ \mathcal{D}_i \mathcal{D}_j W &\to \mathcal{D}_i \mathcal{D}_j \mathcal{G} = \mathcal{G}_{ij} + \mathcal{G}_i \mathcal{G}_j - \Gamma_i{}^k{}_j \mathcal{G}_k \\ &\qquad = \mathcal{D}_i \mathcal{G}_j + \mathcal{G}_i \mathcal{G}_j \, . \end{aligned} \tag{6.72}$$

The metric $K^{i^*}{}_i = \mathcal{G}^{i^*}{}_i$, the Christofell symbols $\Gamma_i{}^k{}_j$ and the curvature remain unchanged. This transformation simply means that in the context of supergravity, the Kähler potential and the superpotential unify in the so-called generalised Kähler potential $\mathcal{G}$. Thus, only two basic functions are needed: the generalised Kähler potential and the gauge kinetic function.

We have seen in Sec. 5.3.3 that a specific combination of Kähler transformation and Howe–Tucker transformation leaves the action invariant. So, equivalently

[10] It should be noted that this transformation is possible if and only if the superpotential is not equal to zero.

the Kähler transformation above can be seen as a field redefinition, or a Howe–Tucker transformation with superfield

$$\Xi = -\frac{1}{6}\ln W\,.$$

Now, from the substitutions (6.72), the Lagrangian can easily be made to the form [Cremmer et al. (1983b)],

$$\begin{aligned}
\mathcal{L} = &\frac{1}{2}ee_\mu{}^{\tilde\mu}e_\nu{}^{\tilde\nu}R_{\tilde\mu\tilde\nu}{}^{\mu\nu}\Big| + \frac{1}{4}e\varepsilon^{\mu\nu\rho\sigma}\left[\psi_\mu\sigma_\sigma\check{\bar\psi}_{\nu\rho} - \bar\psi_\mu\bar\sigma_\sigma\check\psi_{\nu\rho}\right]\\
&-\frac{1}{4}eh^R_{ab}F^a_{\mu\nu}F^{\mu\nu b} + \frac{1}{4}eh^I_{ab}F^a_{\mu\nu}{}^*F^{\mu\nu b}\\
&-\frac{i}{2}eh^R_{ab}\left(\lambda^a\sigma^\mu\check{\mathcal{D}}_\mu\bar\lambda^b - \check{\mathcal{D}}_\mu\lambda^a\sigma^\mu\bar\lambda^b\right) + \frac{1}{2}eh^I_{ab}\check{\mathcal{D}}_\mu(\lambda^a\sigma^\mu\bar\lambda^b)\\
&-\frac{i}{2}e\mathcal{G}^{i^*}{}_i\left[\chi^i\sigma^\mu\check{\mathcal{D}}_\mu\bar\chi_{i^*} - \check{\mathcal{D}}_\mu\chi^i\sigma^\mu\bar\chi_{i^*}\right] + e\mathcal{G}^{i^*}{}_i\widetilde{\mathcal{D}}_\mu\phi^i\widetilde{\mathcal{D}}^\mu\phi^\dagger_{i^*}\\
&-\frac{\sqrt2}{2}e\mathcal{G}^{i^*}{}_i\left[\chi^i\sigma^\mu\bar\sigma^\nu\psi_\mu\widetilde{\mathcal{D}}_\nu\phi^\dagger_{i^*} + \bar\chi_{i^*}\bar\sigma^\mu\sigma^\nu\bar\psi_\mu\widetilde{\mathcal{D}}_\nu\phi^i\right]\\
&+\frac{i}{4}eh^R_{ab}\left[\lambda^a\sigma_\mu\bar\sigma_{\nu\rho}\bar\psi^\mu + \bar\lambda^a\bar\sigma_\mu\sigma_{\nu\rho}\psi^\mu\right]\left[F^{\nu\rho b} + \hat F^{\nu\rho b}\right]\\
&-\frac{\sqrt2}{4}e\left[h_{abi}\chi^i\sigma^{\mu\nu}\lambda^b + h^\star_{ab}{}^{i^*}\bar\chi_{i^*}\bar\sigma^{\mu\nu}\bar\lambda^b\right]F^a_{\mu\nu}\\
&+\frac{e}{4}g\left[\mathcal{G}_i(T_a\phi)^i + \mathcal{G}^{i^*}(\phi^\dagger T_a)_{i^*}\right]\left[\bar\lambda^a\bar\sigma^\mu\psi_\mu + \bar\psi_\mu\bar\sigma^\mu\lambda^a\right]\\
&+ee^{\frac{\mathcal{G}}{2}}\Bigg\{\frac{\sqrt2}{2}i\left[\mathcal{G}_i\bar\psi_\mu\bar\sigma^\mu\chi^i + \mathcal{G}^{i^*}\psi_\mu\sigma^\mu\bar\chi_{i^*}\right] - \bar\psi_\mu\bar\sigma^{\mu\nu}\bar\psi_\nu - \psi_\mu\sigma^{\mu\nu}\psi_\nu\\
&\qquad-\frac{1}{2}(\mathcal{D}_i\mathcal{G}_j + \mathcal{G}_i\mathcal{G}_j)\chi^i\cdot\chi^j - \frac{1}{2}(\overline{\mathcal{D}}^{i^*}\mathcal{G}^{j^*} + \mathcal{G}^{i^*}\mathcal{G}^{j^*})\bar\chi_{i^*}\cdot\bar\chi_{j^*}\\
&\qquad+\frac{1}{4}\mathcal{G}^i{}_{i^*}\left[h_{abi}\mathcal{G}^{i^*}\lambda^a\cdot\lambda^b + h^\star_{ab}{}^{i^*}\mathcal{G}_i\bar\lambda^a\cdot\bar\lambda^b\right]\Bigg\}\\
&+\sqrt2 ieg\mathcal{G}^{i^*}{}_i\left[(\phi^\dagger T_a)_{i^*}\lambda^a\cdot\chi^i - (T_a\phi)^i\bar\lambda^a\cdot\bar\chi_{i^*}\right]\\
&-\frac{\sqrt2 i}{8}eh^{Rab}\left[\mathcal{G}^{i^*}(\phi^\dagger T_a)_{i^*} + \mathcal{G}_i(T_a\phi)^i\right]\left[h_{bdj}\chi^j\cdot\lambda^d - h^\star_{bd}{}^{j^*}\bar\chi_{j^*}\cdot\bar\lambda^d\right]\\
&-\frac{g^2}{8}eh^{Rab}\left[\mathcal{G}^{i^*}(\phi^\dagger T_a)_{i^*} + \mathcal{G}_i(T_a\phi)^i\right]\left[\mathcal{G}^{j^*}(\phi^\dagger T_b)_{j^*} + \mathcal{G}_j(T_b\phi)^j\right]\\
&-ee^{\mathcal{G}}\left[\mathcal{G}^i{}_{i^*}\mathcal{D}_i\mathcal{G}\mathcal{D}^{i^*}\mathcal{G} - 3\right]\\
&+\frac{e}{4}\left[R_i{}^{i^*}{}_j{}^{j^*} - \frac{1}{2}\mathcal{G}^{i^*}{}_i\mathcal{G}^{j^*}{}_j - \frac{1}{3}\mathcal{G}^{i^*}{}_i\mathcal{G}^{j^*}\mathcal{G}_j + \frac{1}{6}\mathcal{G}_i\mathcal{G}_j\mathcal{G}^{i^*}\mathcal{G}^{j^*}\right]\chi^i\cdot\chi^j\bar\chi_{i^*}\cdot\bar\chi_{j^*}\\
&-\frac{\sqrt2 i}{72}e\mathcal{G}_i\mathcal{G}^{i^*}\left[\mathcal{G}^{j^*}\bar\chi_{j^*}\cdot\bar\psi_\mu - \mathcal{G}_j\chi^j\cdot\psi_\mu\right]\chi^i\sigma^\mu\bar\chi_{i^*}
\end{aligned}$$

$$
\begin{aligned}
&+\frac{e}{4}\mathcal{G}^{i^*}{}_i\chi^i\sigma_\mu\bar{\chi}_{i^*}\left[\psi_\nu\sigma^\mu\bar{\psi}^\nu - i\varepsilon^{\mu\nu\rho\sigma}\psi_\nu\sigma_\rho\bar{\psi}_\sigma\right]\\
&-\frac{1}{4}e(\mathcal{G}^{i^*}{}_i + \frac{1}{3}\mathcal{G}^{i^*}\mathcal{G}_i)h^R_{ab}\chi^i\cdot\lambda^a\bar{\chi}_{i^*}\cdot\bar{\lambda}^b\\
&+\frac{1}{16}eh^{Rab}\left[h_{aci}\chi^i\cdot\lambda^c - h^{\star}_{ac}{}^{i^*}\bar{\chi}_{i^*}\cdot\bar{\lambda}^c\right]\left[h_{bdj}\chi^j\cdot\lambda^d - h^{\star}_{bd}{}^{j^*}\bar{\chi}_{j^*}\cdot\bar{\lambda}^d\right]\\
&+\frac{e}{8}\left[\mathcal{D}_i\mathcal{D}_j h_{ab}\chi^i\cdot\chi^j\lambda^a\cdot\lambda^b + \overline{\mathcal{D}}^{i^*}\overline{\mathcal{D}}^{j^*}h^*_{ab}\bar{\chi}_{i^*}\cdot\bar{\chi}_{j^*}\bar{\lambda}^a\cdot\bar{\lambda}^b\right]\\
&+\frac{3}{8}eh^R_{ab}h^R_{cd}\lambda^a\cdot\lambda^c\bar{\lambda}^b\cdot\bar{\lambda}^d - \frac{1}{16}e\mathcal{G}^i{}_{i^*}h_{abi}h^{\star}_{cd}{}^{i^*}\lambda^a\cdot\lambda^b\bar{\lambda}^c\cdot\bar{\lambda}^d\\
&-\frac{\sqrt{2}i}{8}e\left[h_{abi}\bar{\psi}_\mu\bar{\sigma}^\mu\chi^i\lambda^a\cdot\lambda^b + h^{\star}_{ab}{}^{i^*}\psi_\mu\sigma^\mu\bar{\chi}_{i^*}\bar{\lambda}^a\cdot\bar{\lambda}^b\right]\\
&+\frac{\sqrt{2}i}{4}e\left[h_{abi}\chi^i\sigma^{\mu\nu}\lambda^a + h^{\star}_{ab}{}^{i^*}\bar{\chi}_{i^*}\bar{\sigma}^{\mu\nu}\lambda^a\right]\left[\lambda^b\sigma_\nu\bar{\psi}_\mu + \bar{\lambda}^b\bar{\sigma}_\nu\psi_\mu\right]\\
&-\frac{1}{4}eh^I_{ab}\varepsilon^{\mu\nu\rho\sigma}\psi_\rho\sigma_\sigma\bar{\lambda}^b\,\bar{\psi}_\mu\bar{\sigma}_\nu\lambda^a\,,
\end{aligned}
$$

where the various covariant derivatives are given in Table 6.2 with the substitution $K \to \mathcal{G}$.

Two remarks are in order now. First, the D–part of the potential now reads,

$$
V_D = \frac{g^2}{8}eh^{Rab}\left[\mathcal{G}^{i^*}(\phi^\dagger T_a)_{i^*} + \mathcal{G}_i(T_a\phi)^i\right]\left[\mathcal{G}^{j^*}(\phi^\dagger T_b)_{j^*} + \mathcal{G}_j(T_b\phi)^j\right],
$$

but

$$
\mathcal{G}_i(T_a\phi)^i = \left(K_i + \frac{W_i}{W}\right)\left(T_a\phi\right)^i = K_i\left(T_a\phi\right)^i + \frac{1}{W}\delta_a W = K_i\left(T_a\phi\right)^i
$$

because $\delta_a W = W_i(T_a\phi)^i = 0$ since W is gauge invariant. Second, in Chapter 7 we will analyse supergravity breaking and show that when supergravity is broken the gravitino gets massive. Expressed in terms of K and W, the mass-term comes from

$$
e\,\exp\left[\frac{1}{2}K\right]\left(W\bar{\psi}_\mu\sigma^{\mu\nu}\bar{\psi}_\nu + \text{h.c.}\right), \tag{6.73}
$$

and leads to

$$
m_{\frac{3}{2}} = \langle W\rangle\exp\left[\frac{1}{2}\langle K\rangle\right],
$$

whereas expressed in terms of $\mathcal{G}$ it comes from

$$
e\,\exp\left[\frac{1}{2}\mathcal{G}\right]\left(\bar{\psi}_\mu\sigma^{\mu\nu}\bar{\psi}_\nu + \text{h.c.}\right), \tag{6.74}
$$

and reduces to

$$m_{\frac{3}{2}} = \exp\left[\frac{1}{2}\langle\mathcal{G}\rangle\right].$$

In the notation above, $\langle \cdots \rangle$ means that the fields are replaced by their vacuum expectation values. In contrast to Chapter 7, here we have taken $m_p = 1$ for the Planck mass. Now, since W is holomorphic, in general $\big\langle W\big\rangle = \big\langle |W|\big\rangle e^{i\alpha}$ and the gravitino mass is complex when computed with K and W. Unlike, the computation with $\mathcal{G}$, a real function, leads to a real mass for the gravitino. It should be emphasise that the mass parameters which appear either in (6.73) or in (6.74) are not the physical masses. The physical mass-square is given by $m^\dagger_{\frac{3}{2}} m_{\frac{3}{2}}$ which is obviously positive and the same in both cases. The fact that $m_{\frac{3}{2}}$ is complex or real comes form the field redefinition. To simplify the discussion we denote by ψ^K_μ the gravitino in (6.73) and by $\psi^{\mathcal{G}}_\mu$ the gravitino in (6.74). Nonetheless, we have seen that to obtain the action with $\mathcal{G}$ we have to perform a field redefinition (or equivalently a $U(1)$-rotation). In particular, without the gravitino shift we have the dilatation (see (5.51))

$$\bar{\psi}^K_\mu \to e^{\left(2\Xi-\Xi^\dagger\right)|}\bar{\psi}^{\mathcal{G}}_\mu ,$$

but

$$\begin{aligned} 2\Xi - \Xi^\dagger &= -\frac{1}{6}\Big(2\ln|W| + 2i\alpha - (\ln W - i\alpha)\Big) \\ &= -\frac{1}{6}\ln|W| - \frac{1}{2}i\alpha , \end{aligned}$$

which resolves the apparent contradiction. To be complete, we need to mention also that when expending the action expressed in term of $\mathcal{G}$ (following exactly the same lines as in Secs. 6.2.1, 6.2.2 and 6.2.3), the final result is not properly normalised, and field redefinitions analogous to Sec. 6.2.5 must be performed. However, the parameter of the dilation being real, no additional phase factor appears in $\psi^{\mathcal{G}}$.

Now, if we attempt to recover the flat limit of our Lagrangian (expressed either in terms of K and W or in terms of $\mathcal{G}$), we have first to reintroduce the mass terms, *i.e.*, the Planck mass m_P by dimensional analysis and then take the limit where $m_p \to \infty$. In this limit, the gravity and Rarita-Schwinger parts decouple, the covariant derivatives of the various fields become the covariant derivatives of Chapter 2. Furthermore all interactions with the gravitino are Planck suppressed. Finally some of the four-fermions interactions are also Planck suppressed. At the very end, we are left with the Lagrangian of Chapter 2.

6.5 Supergravity limit of renormalisable supersymmetric theories

There is another limit which is interesting to study, namely when the Kähler manifold is flat, the gauge kinetic function is vanishing, and the superpotential involves only renormalisable interactions. In this limit, it is not necessary to have specific notations for anti-chiral superfields. This means that from now on, every time we are considering this limit, an anti-chiral superfield will be denoted $\Phi^\dagger_i$. Thus the three basic functions read

$$\begin{aligned} W(\Phi) &= \frac{1}{6}\lambda_{ijk}\Phi^i\Phi^j + \frac{1}{2}m_{ij}\Phi^i\Phi^j \ , \\ K(\Phi,\Phi^\dagger) &= \Phi^i\Phi^\dagger_i \ , \\ h_{ab}(\Phi) &= \delta_{ab} \ . \end{aligned}$$

In this limit the Lagrangian given in Tables 6.2 and 6.3 simplifies drastically and reduces to

$$\begin{aligned} \mathcal{L} = & \frac{1}{2}e e_\mu{}^{\tilde\mu} e_\nu{}^{\tilde\nu} R_{\tilde\mu\tilde\nu}{}^{\mu\nu}\Big| + \frac{1}{4}e\varepsilon^{\mu\nu\rho\sigma}\Big[\psi_\mu\sigma_\sigma\check{\bar\psi}_{\nu\rho} - \bar\psi_\mu\bar\sigma_\sigma\check\psi_{\nu\rho}\Big] - \frac{1}{4}e\delta_{ab}F^a_{\mu\nu}F^{\mu\nu b} \\ & -\frac{i}{2}e\delta_{ab}(\lambda^a\sigma^\mu\check{\mathcal{D}}_\mu\bar\lambda^b - \check{\mathcal{D}}_\mu\lambda^a\sigma^\mu\bar\lambda^b) - \frac{i}{2}e\Big[\chi^i\sigma^\mu\check{\mathcal{D}}_\mu\bar\chi_i - \check{\mathcal{D}}_\mu\chi^i\sigma^\mu\bar\chi_i\Big] + e\widetilde{\mathcal{D}}_\mu\phi^i\widetilde{\mathcal{D}}^\mu\phi^\dagger_i \\ & -\frac{\sqrt2}{2}e\Big[\chi^i\sigma^\mu\bar\sigma^\nu\psi_\mu\widetilde{\mathcal{D}}_\nu\phi^\dagger_i + \bar\chi_i\bar\sigma^\mu\sigma^\nu\bar\psi_\mu\widetilde{\mathcal{D}}_\nu\phi^i\Big] \\ & +\frac{i}{4}e\delta_{ab}\Big[\lambda^a\sigma_\mu\bar\sigma_{\nu\rho}\bar\psi^\mu + \bar\lambda^a\bar\sigma_\mu\sigma_{\nu\rho}\psi^\mu\Big]\Big[F^{\nu\rho b} + \hat F^{\nu\rho b}\Big] \\ & +\frac{e}{2}g(\phi^\dagger T_a\phi)^i\Big[\bar\lambda^a\bar\sigma^\mu\psi_\mu + \bar\psi_\mu\bar\sigma^\mu\lambda^a\Big] \\ & +ee^{\frac12\phi^i\phi^\dagger_i}\Big\{-W\bar\psi_\mu\bar\sigma^{\mu\nu}\bar\psi_\nu - W^\star\psi_\mu\sigma^{\mu\nu}\psi_\nu + \frac{\sqrt2}{2}i\Big[\mathcal{D}_iW\bar\psi_\mu\bar\sigma^\mu\chi^i + \overline{\mathcal{D}}^i W^*\psi_\mu\sigma^\mu\bar\chi_i\Big] \\ & \qquad -\frac{1}{2}\mathcal{D}_i\mathcal{D}_jW\chi^i\cdot\chi^j - \frac{1}{2}\overline{\mathcal{D}}^i\overline{\mathcal{D}}^jW^*\bar\chi_i\cdot\bar\chi_j\Big\} \\ & +\sqrt2 ieg\Big[(\phi^\dagger T_a)_i\lambda^a\cdot\chi^i - (T_a\phi)^i\bar\lambda^a\cdot\bar\chi_i\Big] \\ & -\frac{g^2}{2}e\delta^{ab}(\phi^\dagger T_a\phi)(\phi^\dagger T_b\phi) - ee^{\phi^i\phi^\dagger_i}\Big[\mathcal{D}_iW\mathcal{D}^iW^\star - 3|W|^2\Big] \\ & -\frac{e}{8}\Big[\delta^i{}_j\delta^k{}_\ell + \frac{2}{3}\delta^i{}_j\phi^k\phi^\dagger_\ell - \frac{1}{3}\phi^i\phi^k\phi^\dagger_j\phi^\dagger_\ell\Big]\chi^j\cdot\chi^\ell\bar\chi_i\cdot\bar\chi_k \\ & -\frac{\sqrt2 i}{72}e\phi^j\phi^\dagger_i\Big[\phi^k\bar\chi_k\cdot\bar\psi_\mu - \phi^\dagger_k\chi^k\cdot\psi_\mu\Big]\chi^i\sigma^\mu\bar\chi_j \\ & +\frac{e}{4}\chi^i\sigma_\mu\bar\chi_i\Big[\psi_\nu\sigma^\mu\bar\psi^\nu - i\varepsilon^{\mu\nu\rho\sigma}\psi_\nu\sigma_\rho\bar\psi_\sigma\Big] - \frac{1}{4}e(\delta^j{}_i + \frac{1}{3}\phi^j\phi^\dagger_i)\delta_{ab}\chi^i\cdot\lambda^a\bar\chi_j\cdot\bar\lambda^b \\ & +\frac{3}{8}e\delta_{ab}\delta_{cd}\lambda^a\cdot\lambda^c\bar\lambda^b\cdot\bar\lambda^d \end{aligned}$$

6.6 The Fayet-Iliopoulos term

It is known that in (global) supersymmetry the Lagrangian (2.26) can be extended slightly. Indeed, for an abelian gauge theory one can add the so-called Fayet-Iliopoulos term [Fayet and Iliopoulos (1974)]

$$\mathcal{L}_{\text{F.I.}} = \xi \int \mathrm{d}^2\theta \mathrm{d}^2\bar{\theta} V ,$$

with V the corresponding vector superfield. However, this term does not straightforwardly generalise to supergravity since *e.g.*

$$\xi \int \mathrm{d}^2\Theta \mathcal{E}(\overline{\mathcal{D}} \cdot \overline{\mathcal{D}} - 8\mathcal{R})V ,$$

is no longer gauge invariant. The Fayet-Iliopoulos term was first introduced in supergravity in [Freedman (1977); Das et al. (1977); de Wit and van Nieuwenhuizen (1978); Barbieri et al. (1982a); Stelle and West (1978a); Ferrara et al. (1983)], more recently in relation to $R-$invariance in [Chamseddine and Dreiner (1995); Castano et al. (1996)] and was completely clarified in [Binetruy et al. (2004)] and related to the superconformal approach of supergravity (see Sec. 6.7).

Following App. F and in particular Equations (F.10), (F.11), (F.12) and (F.13) the symmetries of the Kähler manifold are related to the Killing equations and the corresponding moment map. In particular, this shows that for an abelian gauge group the moment map can be shifted from $\mathcal{P}$ to $\mathcal{P} - \frac{1}{2}\xi$ and consequently the $D-$part of the potential for an abelian gauge group is shifted correspondingly as shown in [Wess and Bagger (1992)]. In our approach, since the symmetries are known from the very beginning this corresponds to the shift

$$\begin{aligned} \mathcal{P} &= -\frac{1}{2}\Big[K_i(Q\phi)^i + K^{i^*}(\phi^\dagger Q)_{i^*}\Big] \\ \to \mathcal{P} - \frac{1}{2}\xi &= -\frac{1}{2}\Big[K_i(Q\phi)^i + K^{i^*}(\phi^\dagger Q)_{i^*} + \xi\Big] , \end{aligned}$$

and the $D-$part of potential for an abelian gauge group takes the form

$$V_D = \frac{g^2}{8}\Big[K_i(Q\phi)^i + K^{i^*}(\phi^\dagger Q)_{i^*} + \xi\Big]\Big[K_j(Q\phi)^j + K^{j^*}(\phi^\dagger Q)_{j^*} + \xi\Big] . \tag{6.75}$$

We have denoted the $U(1)-$generator by Q. Observe that this term can be easily generated in the superspace language by adding

$$-\frac{1}{8}g\xi \int \mathrm{d}^2\Theta \mathcal{E}\Theta \cdot \Theta \mathcal{D}^a W_a + \text{h.c.} , \tag{6.76}$$

to the Lagrangian. Observe finally that the Fayet-Iliopoulos term has a more deeper origin in the superconformal approach and following the results of [Binetruy et al. (2004)] and the discussion concerning the covariant derivatives of the previous section, the Fayet-Iliopoulos term will not only modify the potential of the theory, but also contribute to a new term in the covariant derivatives of all the fermions.

6.7 Compensating fields and superconformal invariance

In Sec. 6.2.5 we have seen that appropriate field redefinitions enable us to cast the supergravity action in a form with conventionally normalised kinetic terms. More precisely, in order to get the full supergravity action, we first compute the component fields of the action (5.35) and then perform a super-Weyl rescaling.

However, we can proceed otherwise to compute the full supergravity action. One possible alternative is to use the so-called compensating fields. The key observation in this method, is that the action (5.35) is not invariant under a superconformal transformation. Thus, the idea is to introduce a compensating field in order to have a superconformal invariant action. Such a mechanism was invented by Stuckelberg in order to construct a massive theory with an enhanced $U(1)$ invariance [Stueckelberg (1938a,b,c)]. This mechanism can be applied in the context of supergravity see *e.g.* [Gates et al. (1983); Buchbinder and Kuzenko (1998)]. For that purpose, we introduce a chiral superfield Φ^0 of conformal weight -2 together with the usual chiral superfields Φ^i of conformal weight zero describing matter. Within the new chiral superfield, the superconformal invariant action takes the form

$$\begin{aligned}\mathcal{L} = \frac{3}{8}&\int \mathrm{d}^2\Theta\ \mathcal{E}(\overline{\mathcal{D}}\cdot\overline{\mathcal{D}} - 8\mathcal{R})\left\{[\Phi^0\tilde{\Phi}^\dagger_{0^*}\exp\left[-\frac{1}{3}K(\Phi,\Phi^\dagger e^{-2gV})\right]\right\}\\ &+\int \mathrm{d}^2\Theta\ \mathcal{E}(\Phi^0)^3 W(\Phi)\\ &+\frac{1}{16g^2}\int \mathrm{d}^2\Theta\ \mathcal{E}h(\Phi)_{ab}W^{a\alpha}W^b_\alpha + \text{h.c.}\,.\end{aligned} \tag{6.77}$$

Indeed, using the results of Sec. 5.3.2 together with the transformation

$$\begin{aligned}\Phi^0 &\to e^{-2\Sigma}\Phi^0\,,\\ \Phi^\dagger_{0^*} &\to e^{-2\Sigma^\dagger}\Phi^\dagger_{0^*}\,,\end{aligned}$$

we observe that (6.77) is manifestly superconformally invariant. The computation of the component terms of the action proceeds exactly along the same lines as in

Sec. 6.2 with the substitution

$$\begin{aligned}
\Phi^i &\to \Phi^I = (\Phi^0, \Phi^i)\,, \\
\Phi^\dagger_{i^*} &\to \Phi^\dagger_{I^*} = (\Phi^\dagger_{0^*}, \Phi^\dagger_{i^*})\,, \\
\Omega &\to \tilde{\Omega} = \Phi^0 \Phi^\dagger_{0^*} \Omega\,, \\
W &\to \tilde{W} = (\Phi^0)^3 W\,,
\end{aligned}$$

and

$$\begin{aligned}
\tilde{\Omega}_I &= (\Phi^\dagger_{0^*}\Omega, \Phi^0\Phi^\dagger_{0^*}\Omega_i)\,, \\
\tilde{W}_I &= (3(\Phi^0)^2 W, (\Phi^0)^3 W_i)\,,
\end{aligned}$$

etc. In this substitution process, some care has to be taken with respect to the elimination of the auxiliary fields. Indeed, the part of the Lagrangian involving F^0 and $F^\dagger_{0^*}$ vanishes identically since $\ln(\tilde{\Omega})^{0^*}{}_I = \ln(\tilde{\Omega})^{I^*}{}_0 = 0$ and the coefficients of F^0 (resp. $F^\dagger_{0^*}$) in (6.23) are equal to zero. This means that F^0 and $F^\dagger_{0^*}$ do not contribute to the final Lagrangian. Once the component action is obtained, we break this artificial superconformal invariance by choosing the compensating field such that the physical fields are normalised with standard kinetic terms , *i.e.*, choosing Φ^0 such that

$$\Phi^0 \bar{\Phi}^\dagger_{0^*} \Omega\big| = -3\,. \tag{6.78}$$

Promoting Φ^0 at the level of superfield and using (5.50), we set $\Phi^0 = \exp(2\Sigma)$

$$\Sigma = \frac{1}{2}\lambda + \sqrt{2}\Theta\cdot\omega\; - \Theta\cdot\Theta\mathcal{F}\,,$$

with

$$\begin{aligned}
e^\lambda &= \sqrt{\frac{-3}{\Omega}}\,, \\
\omega &= \lambda_i \chi^i = -\frac{1}{2}\Omega^{-1}\Omega_i\chi^i\,.
\end{aligned}$$

To identify $\mathcal{F}$ we use on the one hand the transformation of M and $F^\dagger_{i^*}$ given in (6.40) and (6.42) and the transformation rule of N which can be deduced from (6.26):

$$N \to e^{-\lambda} N\,,$$

and the other hand the relationship between N and M (6.24). We thus obtain

$$\Sigma = \frac{1}{2}\lambda - \frac{\sqrt{2}}{2}\Theta\cdot\left(\Omega^{-1}\Omega_i\chi^i\right) - \frac{1}{2}\Theta\cdot\Theta\Omega^{-2}\Omega_i\Omega_j\chi^i\cdot\chi^j\,,$$

and

$$\begin{aligned}\Phi^0 &= \phi^0 + \sqrt{2}\Theta \cdot \chi^0 - \Theta \cdot \Theta F^0 \\ &= \exp(2\Sigma) \\ &= \exp\left[\lambda - \sqrt{2}\Theta \cdot (\Omega^{-1}\Omega_i \chi^i) - \Theta \cdot \Theta \Omega^{-2}\Omega_i \Omega_j \chi^i \cdot \chi^j\right] \\ &= \sqrt{\frac{-3}{\Omega}}\Big(1 - \sqrt{2}\Omega^{-1}\Theta \cdot (\Omega_i \chi^i) - \frac{3}{2}\Theta \cdot \Theta \Omega^{-2}\Omega_i \Omega_j \chi^i \cdot \chi^j\Big)\,.\end{aligned} \tag{6.79}$$

Now, we just have to substitute (6.79) into (6.77) using the results of Sec. 6.2 to obtain the Lagrangian. Note that, as in Sec. 6.2.5 we have to take care of the covariant derivatives with respect to the Kähler manifold. In particular we have to substitute $\mathcal{D}(\tilde{\Omega})$ by $\mathcal{D}(K)$. Thus, choosing (6.79) breaks explicitly the enhanced conformal invariance and leads to a conventionally normalised action. For instance, the pure gravitational part of the action reduces to

$$\mathcal{L}_{\text{gravitation}} = -\frac{1}{6}e_\mu{}^{\tilde{\mu}} e_\nu{}^{\tilde{\nu}} \tilde{\Omega} R_{\tilde{\mu}\tilde{\nu}}{}^{\mu\nu}\Big| = \frac{1}{2} e e_\mu{}^{\tilde{\mu}} e_\nu{}^{\tilde{\nu}} R_{\tilde{\mu}\tilde{\nu}}{}^{\mu\nu}\Big|\,.$$

In the same way, the gravitino kinetic part of the action reads

$$\begin{aligned}\mathcal{L}_{\text{kin. gravitino}} &= -\frac{1}{12} e\tilde{\Omega}\varepsilon^{\mu\nu\rho\sigma}\Big(\psi_\mu \sigma_\sigma \bar{\psi}_{\nu\rho} - \bar{\psi}_\mu \bar{\sigma}_\sigma \psi_{\nu\rho}\Big) \\ &\quad + \frac{\sqrt{2}}{3} e\Big(\tilde{\Omega}_I \chi^I \sigma^{\mu\nu}\psi_{\mu\nu} + \tilde{\Omega}^{I^*} \bar{\chi}_{I^*} \bar{\sigma}^{\mu\nu}\bar{\psi}_{\mu\nu}\Big) \\ &= \frac{1}{4}\varepsilon^{\mu\nu\rho\sigma}\Big(\psi_\mu \sigma_\sigma \bar{\psi}_{\nu\rho} - \bar{\psi}_\mu \bar{\sigma}_\sigma \psi_{\nu\rho}\Big)\,,\end{aligned}$$

because

$$\begin{aligned}\tilde{\Omega}_I \chi^I &= \bar{\phi}_{0^*}\Omega\chi^0 + \phi^0 \bar{\phi}_{0^*}\Omega\chi^i \\ &= 0\,,\end{aligned}$$

using (6.79). Furthermore since Φ^0 is a gauge singlet

$$(T_a\phi)^I = (T_a\phi)^i\,.$$

Similarly, the computation of the on other terms give the same results as the ones obtained in Sec. 6.2.5 so we will not reproduce the calculation. Note however, that unlike the field redefinition where we first perform the expansion of the components of the Lagrangian and then proceed to a field redefinition, here, we proceed in the other way around: we first chose an appropriate compensating field and then proceed to an expansion of the corresponding Lagrangian.

To end this section we would like to mention two closely related methods to obtain actions in supergravity. The first method [Binetruy et al. (2001)] is geometrical. Enlarging the usual superspace to a $U(1)$−superspace and considering a super-Weyl rescaling which does not preserve the constraints (3.29), it has been shown that the action of matter coupled to supergravity (written in the non-chiral form – see Sec. 5.1 –) takes the pure geometrical form

$$\begin{aligned}\mathcal{L}_{\text{matter}} &= -3\int \mathrm{d}^4\theta E\exp(-1/3K) \\ &\to -3\int \mathrm{d}^4\theta E' .\end{aligned}$$

The second method is based on superconformal methods. In Poincaré supergravity, the structure group is the Poincaré supergroup. The conformal methods [Ferrara et al. (1977b); Kaku et al. (1978); Kaku and Townsend (1978); Townsend and van Nieuwenhuizen (1979); Ferrara et al. (1978)] are based on an enlargement of the structure group to the superconformal group (see App. H). Within this approach the full superconformal group is gauged, and following analogous technique as the ones developed in this book, invariant actions are constructed. Then, at the very end the extra-symmetries are gauge fixed and Poincaré supergravity is recovered. Note that in [Kugo and Uehara (1983b)] the authors use a (compensating) conformal superfield. For a detailed analysis in terms of tensor calculus one can see [Freedman and Van Proeyen (2012)] or in terms of superfields one can see [Butter (2010)].

To give more details of the conformal methods would go much beyond the scope of this book. We would like however, to mention some points which are enlightening concerning the superconformal methods. Firstly, the Howe–Tucker transformations of Sec. 3.2.1 can be seen as a Weyl rescaling accompanied by an extra modification (*e.g.* the gravitino is dilated and shifted). This has a natural origin in conformal supergravity (see *e.g.* [Butter (2010)] and App. H). Secondly, conformal supergravity provides some insights on the Kähler manifold, which is in fact a projective Kähler manifold . More precisely if we assume to have chiral superfields $\Phi^I, I = 0, \cdots, n$ of conformal weight Δ_I (with at least one being different from zero, say Δ_0), the superfields Φ^I transform non-trivially under the superconformal transformation

$$\begin{aligned}\Phi^I &\to e^{\Delta_I \Sigma}\Phi^I , \\ \Phi^\dagger_{I^*} &\to e^{\Delta_I \Sigma^\dagger}\Phi^\dagger_{I^*} ,\end{aligned}$$

with Σ a chiral superfield. By a field redefinition we can set

$$\begin{aligned}\Phi^0 &\to \tilde\Phi^0 = (\Phi^0)^{\frac{2}{\Delta_0}} , \\ \Phi^i &\to \Xi^i = \frac{\Phi^i}{(\tilde\Phi^0)^{\frac{\Delta_i}{2}}} .\end{aligned}$$

Now the superfields have conformal weight $\tilde{\Delta}_0 = 2, \tilde{\Delta}_i = 0$. Since the Kähler potential and superpotential have conformal weight 2 and 6 respectively they transform under the dilatation like

$$\begin{aligned} \Omega &\to e^{2(\Sigma+\Sigma^\dagger)}\Omega \,, \\ K &\to e^{6\Sigma} W \,, \end{aligned}$$

and by dimensional arguments we automatically have

$$\begin{aligned} -\frac{1}{8}\tilde{\Omega}(\Phi^I, \Phi^\dagger_{I^*}) &= -\frac{1}{8}\tilde{\Phi}^0\tilde{\Phi}^\dagger_{0^*}\Omega(\Xi^i, \Xi_{i^*}) = \frac{3}{8}\tilde{\Phi}^0\tilde{\Phi}^\dagger_{0^*}e^{-\frac{1}{3}K(\Xi^i,\Xi_{i^*})} \,, \\ \tilde{W}(\Phi^I) &= \left(\tilde{\Phi}^0\right)^3 W(\Xi^i) \,. \end{aligned}$$

The second equality in the first equation is simply a consequence of the fact that the kinetic energy (that is -3Ω) has to be positive. Since the basic fields are now $\Xi^i = \Phi^i/(\tilde{\Phi}^0)^{\frac{\Delta_i}{2}}$, the Kähler manifold is a projective Kähler manifold. This observation was used *e.g.* in [Freedman and Van Proeyen (2012)] to simplify *a priori* some computations. Another parametrisation of the projective Kähler manifold can be obtained by a Kähler transformation. For instance, if we assume $\Delta_1 \neq 0$ setting $\tilde{\Phi}'^1 = (\Phi^1)^{2/\Delta_1}$ we define the Kähler transformation

$$K \to K + F + \bar{F} \,, W \to e^{-F} W \,,$$

with

$$F = 3\ln\Big(\frac{\tilde{\Phi}'^1}{\tilde{\Phi}^0}\Big) \,.$$

This transformation can equivalently be seen as a conformal transformation with parameter $\Sigma = \frac{1}{2}\ln\Big(\frac{\tilde{\Phi}'^1}{\tilde{\Phi}^0}\Big)$ so that

$$\begin{aligned} \tilde{\Phi}^0 &\to \tilde{\Phi}'^1 \,, \\ \Phi^i &\to \Phi'^i = \Big(\frac{\tilde{\Phi}'^1}{\tilde{\Phi}^0}\Big)^{\frac{\Delta_i}{2}}\Phi^i \,, \\ \Xi^i &\to \Xi'^i = \frac{\Phi'^i}{(\tilde{\Phi}'^1)^{\frac{2}{\Delta_i}}} \,, \\ K(\Xi^i, \Xi^\dagger_i) &\to K(\Xi'^i, \Xi'^\dagger_i) \,, \\ W(\Xi^i) &\to W(\Xi'^i) \,, \end{aligned}$$

where now $\tilde{\Phi}'^1$ is of conformal weight 2, the fields Φ'^i are of conformal weight Δ_i and Ξ'^i are of conformal weight 0. Thus, we finally obtain

$$\begin{aligned} \tilde{\Phi}^0\tilde{\Phi}^\dagger_{0^*}e^{-\frac{1}{3}K(\Xi^i,\Xi_{i^*})} &\to \tilde{\Phi}^1\tilde{\Phi}^\dagger_{1^*}e^{-\frac{1}{3}K(\Xi'^i,\Xi'_{i^*})} \,, \\ \left(\tilde{\Phi}^0\right)^3 W(\Xi^i) &\to \left(\tilde{\Phi}^1\right)^3 W(\Xi'^i) \,, \end{aligned}$$

with the projective superfields now given by Ξ'^i. This corresponds to another chart (or system of coordinates) for the projective Kähler manifold. The conformal methods to compute the supergravity action were introduced first by Kugo and Uehara (1985), but similarly to the method introduced in this book they are very involved. This method was also considered in [Cheung et al. (2011)] in order to obtain part of the supergravity Lagrangian (but not all) without to many technical difficulties.

Finally one can see [Butter (2010)] for comparisons between the different approaches to construct actions in supergravity.

6.8 Supergravity in Jordan frame

When expanding (6.2) the final action was obtained in a Brans-Dicke form (see Eqs. [6.33] and [6.34]). This is a specific Jordan frame with the frame function given by

$$Z(\phi, \phi^\dagger) = \Omega(\phi, \phi^\dagger) ,$$

because as mentioned earlier, the Einstein-Hilbert action is multiplied by $-1/3Z > 0$. This function Z is called a frame function, $Z = -3$ corresponding to the usual Einstein frame, since the action is canonically normalised. Generically, a Jordan frame is a frame where the Einstein-Hilbert part of the action is not canonically normalised.

It turns out that it might be interesting to construct the $N = 1, d = 4$ supergravity action in an arbitrary Jordan frame, especially for cosmological purpose. To construct Jordan frame supergravity we introduce, as usual, the Kähler potential, the superpotential and the gauge kinetic function, together with an arbitrary real frame function $Z(\Phi, \Phi^\dagger)$. The full action in a Jordan frame were obtained in [Ferrara et al. (2010, 2011)] by superconformal techniques.

Supergravity action in an arbitrary Jordan frame can be equivalently obtained along the lines of Secs. 6.2.6, 6.2.7 and 6.2.8 by an appropriate field redefinition. However, in this section we reproduce the various steps, but only for the scalar part starting from supergravity in an Einstein frame (here, for simplicity, we do not consider the coupling of scalar fields to gauge fields since it doesn't change anything in the discussion):

$$\mathcal{L}_{\text{scal}} = \frac{1}{2} e e_\mu{}^{\tilde\mu} e_\nu{}^{\tilde\nu} R_{\tilde\mu\tilde\nu}{}^{\mu\nu} + e K^{i^*}{}_i g^{\tilde\mu\tilde\nu} \partial_{\tilde\mu}\phi^i \partial_{\tilde\nu}\phi^\dagger_{i^*} - e V_E ,$$

where V_E (see Eq. [6.69]) is the scalar potential in Einstein frame.

Introducing now an arbitrary real frame function, $Z(\phi, \phi^\dagger)$ such that $-1/3Z > 0$, the metric in a Jordan frame $g_J^{\tilde\mu\tilde\nu}$ is related to the metric in the Einstein frame

$g_E^{\tilde{\mu}\tilde{\nu}}$ through

$$g_J^{\tilde{\mu}\tilde{\nu}} = \Big(-\frac{1}{3}Z(\phi,\phi^\dagger)\Big)g_E^{\tilde{\mu}\tilde{\nu}}\ ,$$

this corresponds to the field redefinition

$$g^{\tilde{\mu}\tilde{\nu}} \to \Big(-\frac{1}{3}Z\Big)^{-1}g^{\tilde{\mu}\tilde{\nu}}\ .$$

The results of Secs. 6.2.6, 6.2.7 and 6.2.8 can be directly used with the substitution

$$\Omega = \frac{9}{Z}\ . \tag{6.80}$$

As said above, here we only go ahead with the scalar part.

For the Einstein-Hilbert part of the action we obtain (using (6.80))

$$\begin{aligned}
-3\Omega^{-1}e\Big(\frac{1}{6}\Omega e_\mu{}^{\tilde{\mu}}e_\nu{}^{\tilde{\nu}}R_{\tilde{\mu}\tilde{\nu}}{}^{\mu\nu}\Big) &\to -3\Omega^{-1}e\Big(\frac{1}{2}e_\mu{}^{\tilde{\mu}}e_\nu{}^{\tilde{\nu}}R_{\tilde{\mu}\tilde{\nu}}{}^{\tilde{\mu}\tilde{\nu}} + \frac{3}{4}\Omega^{-2}\partial_{\tilde{\mu}}\Omega\partial^{\tilde{\mu}}\Omega + \cdots\Big)\\
&= e\Big(-\frac{1}{6}Ze_\mu{}^{\tilde{\mu}}e_\nu{}^{\tilde{\nu}}R_{\tilde{\mu}\tilde{\nu}}{}^{\mu\nu} - \frac{1}{4}\frac{1}{Z}\partial_{\tilde{\mu}}Z\partial^{\tilde{\mu}}Z + \cdots\Big)\\
&= e\Big(-\frac{1}{6}Ze_\mu{}^{\tilde{\mu}}e_\nu{}^{\tilde{\nu}}R_{\tilde{\mu}\tilde{\nu}}{}^{\mu\nu} + \frac{1}{Z}Z_iZ^{i^*}\partial_{\tilde{\mu}}\phi^i\partial^{\tilde{\mu}}\phi^\dagger_{i^*}\\
&\quad -\frac{1}{4}\frac{1}{Z}\Big[Z_i\partial_{\tilde{\mu}}\phi^i - Z^{i^*}\partial_{\tilde{\mu}}\phi^\dagger_{i^*}\Big]\Big[Z_j\partial^{\tilde{\mu}}\phi^j - Z^{j^*}\partial^{\tilde{\mu}}\phi^\dagger_{j^*}\Big] + \cdots\Big)
\end{aligned}$$

where the dots correspond to the gravitino part and a total derivative, both being irrelevant for our analysis. For the kinetic part of the scalar fields we obtain

$$eK^{i^*}{}_i g^{\tilde{\mu}\tilde{\nu}}\partial_{\tilde{\mu}}\phi^i\partial_{\tilde{\nu}}\phi^\dagger_{i^*} \to -e\frac{1}{3}ZK^{i^*}{}_i\partial_{\tilde{\mu}}\phi^i\partial^{\tilde{\mu}}\phi^\dagger_{i^*}\ .$$

And finally the scalar potential rescales as

$$eV_E \to e\frac{Z^2}{9}V_E\ .$$

Collecting all terms, and omitting the doted part, we obtain finally

$$\begin{aligned}
\mathcal{L}_{\text{scal}} = e\bigg[&-\frac{1}{6}Ze_\mu{}^{\tilde{\mu}}e_\nu{}^{\tilde{\nu}}R_{\tilde{\mu}\tilde{\nu}}{}^{\mu\nu} + \Big(-\frac{1}{3}ZK^{i^*}{}_i + \frac{Z_iZ^{i^*}}{Z}\Big)\partial_{\tilde{\mu}}\phi^i\partial^{\tilde{\mu}}\phi^\dagger_{i^*}\\
&-\frac{1}{4}\frac{1}{Z}\Big[Z_i\partial_{\tilde{\mu}}\phi^i - Z^{i^*}\partial_{\tilde{\mu}}\phi^\dagger_{i^*}\Big]\Big[Z_j\partial^{\tilde{\mu}}\phi^j - Z^{j^*}\partial^{\tilde{\mu}}\phi^\dagger_{j^*}\Big] - \frac{Z^2}{9}V_E\bigg]\ .
\end{aligned} \tag{6.81}$$

If we assume now that the relationship between the frame function and the Kähler potential is

$$K = -3\ln\Big[-\frac{1}{3}Z\Big]\ , \tag{6.82}$$

then

$$\mathcal{L}_{\rm scal} = e\Bigg[-\frac{1}{6}Ze_\mu{}^{\tilde\mu}e_\nu{}^{\tilde\nu}R_{\tilde\mu\tilde\nu}{}^{\mu\nu} + Z^{i^*}{}_i\partial_{\tilde\mu}\phi^i\partial^{\tilde\mu}\phi^\dagger_{i^*}$$
$$-\frac{1}{4}\frac{1}{Z}\Big[Z_i\partial_{\tilde\mu}\phi^i - Z^{i^*}\partial_{\tilde\mu}\phi^\dagger_{i^*}\Big]\Big[Z_j\partial^{\tilde\mu}\phi^j - Z^{j^*}\partial^{\tilde\mu}\phi^\dagger_{j^*}\Big] - \frac{Z^2}{9}V_E\Bigg].$$

In particular, we recover the Lagrangian given in (6.33) and (6.34) before the field redefinition. Correspondingly the Lagrangian of Sec. 6.2.4 is given in a specific Jordan frame where (6.82) holds.

Chapter 7

Supergravity breaking

This chapter is devoted to some applications of supergravity. Supergravity cannot be an exact symmetry and has to be broken. Indeed, since $P_\mu P^\mu$ is a Casimir operator in a given multiplet, matter, *i.e.* chiral superfields, or gauge, *i.e.* vector superfields, the fermions and bosons are degenerated in mass. In particular, this means that if supergravity is a symmetry in particle physics we should have observed experimentally such degenerate pairs, *i.e.*, a scalar and a fermion with the same mass and quantum numbers, *e.g.*, a neutral massless Majorana fermion (associated to the photon) or a light charged scalar (associated to the electron). This is obviously not the case. Consequently, supergravity has to be spontaneously broken. This feature, is not only a property of supergravity but also of supersymmetry. The first mechanisms of supersymmetry breaking involved chiral superfields [O'Raifeartaigh (1975)] or a vector superfield associated to a $U(1)-$gauge symmetry [Fayet and Iliopoulos (1974); Fayet (1975)]. However, neither are realistic since after supersymmetry breaking the spectrum contains supersymmetric particles lighter than the standard particles. Then it was realised that it is more natural to break supersymmetry but in the context of supergravity. There are several ways to break supergravity, but all share a common feature, that is, supergravity is broken in a hidden sector. The interaction (gravitational, gauge, *etc*) of the hidden sector induces supergravity breaking within the observable sector. In the first types of mechanisms, called gravity mediated supersymmetry breaking [Chamseddine et al. (1982); Barbieri et al. (1982b); Ibañez (1982); Ohta (1983); Ellis et al. (1983); Alvarez-Gaumé et al. (1983); Polonyi (1977)], the supersymmetry breaking is communicated to the observable sector *via* gravitational interactions. In this mechanism the Kähler potential and the scalar potential play an essential rôle. In the second types of mechanisms, called gauge mediated supersymmetry breaking, supersymmetry is broken by a singlet field (called the spurion) which couples (*via* the superpotential) to a field called the messenger which is in a non-trivial representation of the gauge group. Quantum loop corrections involving the messenger break supersymmetry in the observable sector [Nilles (1984); Dimopoulos and Raby (1981); Dine et al. (1981); Derendinger and Savoy (1982); Fayet (1978); Dine and Fischler (1982); Nappi and Ovrut (1982); Alvarez-Gaumé et al. (1982); Dine and Nelson (1993); Dine et al. (1995, 1996); Giudice and Rattazzi (1999); Arkani-Hamed et al. (1998)]. In the third type of mechanisms, called anomaly mediated supersymmetry breaking, the introduction of compensating fields having a conformal (super-Weyl) anomaly induces supersymmetry breaking by purely quantum effects [Randall and Sundrum (1999); Arkani-Hamed et al. (1998); Bagger et al. (2000)]. Several other mechanisms to break supergravity have been proposed, as in string theory for instance *etc.*

In broken supersymmetry a massless fermion appears (the goldstone fermion or goldstino). In the O'Raifeartaigh or the Fayet-Illiopoulos mechanisms this massless particle remains in the spectrum. Furthermore, the signature of broken supersymmetry is a strictly positive potential, since some auxiliary fields (an F−term in the O'Raifeartaigh mechanism or a D−term in Fayet-Illiopoulos models) develop a non-zero vacuum expectation value. We recall that in supersymmetry the potential is positive. In supergravity things change drastically. Firstly, the potential is no-longer positive, thus positivity of the potential is not a signature of a broken supergravity. Secondly, the goldstino does not remain in the spectrum but is eaten by the gravitino, which thus acquire its $\pm 1/2$ helicity states in such a way it becomes massive. In fact by selecting a well-chosen gauge, one can eliminate the goldstino through an appropriate supergravity transformation. This is the analogue of the Brout-Englet-Higgs mechanism in supergravity and is called super-Brout-Englert-Higgs mechanism [Cremmer et al. (1979, 1978b, 1983b)] (see also [Wess and Zumino (1974a); Iliopoulos and Zumino (1974)]). This mechanism will be explicitly studied in the following.

In the first part of this chapter we study gravity induced supersymmetry breaking in details. It is shown that when supergravity is broken in the limit where gravity is negligible a supersymmetric theory with additional terms which break supersymmetry explicitly (but in a soft way) is obtained. The general structure of these terms will be computed. In supergravity the order parameter is the gravitino mass $m_{3/2}$ (at least when the cosmological constant vanishes). Thus, for a vanishing cosmological constant, a non-vanishing $m_{3/2}$ is a signature of broken supergravity whereas in the general case at least one auxiliary field F^i (or D^a) must develop a non-vanishing vacuum expectation value. For general properties of broken supergravity one can see [Witten (1981a,b); Fayet and Iliopoulos (1974); Intriligator and Seiberg (2007)]. In this chapter, since we would like to study the effects of gravitation on the system we reintroduce the mass parameters by dimensional analysis. Indeed, a chiral superfield has the dimension of a mass, and $[K] = M^2, [W] = M^3$ and $[h_{ab}] = M^0$.

The problem with all mechanisms which break supersymmetry is the fact that the cosmological constant is not vanishing unless some fine tuning in the parameters are involved. Here, however a class of supergravity theories will be introduced for which the cosmological constant vanishes naturally due to geometrical properties of the Kähler manifold. This class of supergravity models, called no-scale supergravity is the subject of a section at the end of this Chapter. Finally we will turn to the computation of mass matrices in broken supergravity.

7.1 Goldstino and super-Brout-Englert-Higgs effect in supergravity

Before studying the super-Brout-Englert-Higgs effect in supergravity, we would like to recall some results concerning supersymmetry breaking and in particular the rôle of the goldstino in supersymmetry. Starting either from the flat limit of the supergravity Lagrangian of Sec.6.2.8 (see also Tables 6.2 and 6.3) or from the flat Lagrangian of Chapter 2, and after having eliminated the auxiliary fields F^i and D^a, we obtain

$$\begin{aligned}
F^i &= K^i{}_{i^*} W^{\star i^*} - \frac{1}{2}\Gamma_j{}^i{}_k \chi^j \cdot \chi^k - \frac{1}{4} K^i{}_{i^*} h^{\star}_{ab}{}^{i^*} \bar{\lambda}^a \cdot \bar{\lambda}^b \,, \\
D^a &= \frac{g}{2} h^{Rab}[K^{j^*}(\phi^\dagger T_b)_{j^*} + K_j(T_b\phi)^j] + i\frac{\sqrt{2}}{4} h^{Rab} h_{bci}\chi^i \cdot \lambda^c \\
&\quad - i\frac{\sqrt{2}}{4} h^{Rab} h^{\star}_{bc}{}^{i^*} \bar{\chi}_{i^*} \cdot \bar{\lambda}^c \,.
\end{aligned} \tag{7.1}$$

The quadratic fermion-terms then reads

$$\begin{aligned}
e^{-1}\mathcal{L}_{\text{ferm.}} &= -\frac{1}{2} D_i D_j W \chi^i \cdot \chi^j + \frac{1}{4} K^i{}_{i^*} W^{\star i^*} h_{abi} \lambda^a \cdot \lambda^b + ig\sqrt{2} K^{i^*}{}_i (\phi^\dagger T_a)_{i^*} \chi^i \cdot \lambda^a \\
&\quad - i\frac{\sqrt{2}}{8} h^{Rab}[K^{i^*}(\phi^\dagger T_a)_{i^*} + K_i(T_a\phi)^i] h_{bcj}\chi^j \cdot \lambda^c + \text{h.c.}\,,
\end{aligned}$$

and the potential is expressed by

$$\begin{aligned}
V = K^i{}_{i^*} W^{\star i^*} W_i &+ \frac{g^2}{8} h^{Rab}[K^{i^*}(\phi^\dagger T_a)_{i^*} + K_i(T_a\phi)^i] \\
&\times [K^{j^*}(\phi^\dagger T_b)_{j^*} + K_j(T_b\phi)^j].
\end{aligned} \tag{7.2}$$

If we now assume that supersymmetry is broken and correspondingly some of the auxiliary fields F^i, D^a and some of the scalar fields ϕ^i take a vacuum expectation value, we obtain the fermionic mass matrix,

$$\mathcal{L}_{\text{mass}} = -\frac{1}{2}\left(\chi^i \; i\sqrt{2}\lambda^a\right)$$
$$\times\begin{pmatrix} \langle D_i D_j W\rangle & \frac{1}{4}\langle D^c h_{cbi}\rangle - g\langle K^{i^*}{}_i(\phi^\dagger T_b)_{i^*}\rangle \\ \frac{1}{4}\langle D^d h_{daj}\rangle - g\langle K^{i^*}{}_j(\phi^\dagger T_a)_{i^*}\rangle & \frac{1}{4}\langle h_{abk}F^k\rangle \end{pmatrix}\begin{pmatrix}\chi^j \\ i\sqrt{2}\lambda^b\end{pmatrix}.$$

To proceed further let us consider (i) the minimisation of the potential

$$\partial_k V = 0\,,$$

and (ii) the gauge invariance of the superpotential

$$\delta_a W = 0\,.$$

The equation of minimisation of the potential can be simplified

-first by

$$\begin{aligned} \partial_k K^i{}_{j^*} &= -K^i{}_{i^*} K_k{}^{i^*}{}_j K^j{}_{j^*} = -K^j{}_{j^*}\Gamma_k{}^i{}_j\,, \\ \partial_k h^{Rab} &= -h^{Rac} h^R_{cdk} h^{Rbd} = -\frac{1}{2} h^{Rac} h^{Rbd} h_{cdk}\,, \end{aligned} \tag{7.3}$$

which can be deduced on the one hand from $\partial_k(K^{i^*}{}_j K^j{}_{j^*}) = 0$ and on the other hand from $\partial_k(h^{Rab}h^R_{bc}) = 0$ and $h^R_{ab} = 1/2(h_{ab} + h^\star_{ab})$ with $h^\star_{ab}$ the anti-holomorphic counter part of h_{ab};

-second by the gauge invariance of the Kähler potential

$$0 = \delta_a K = K_i(T_a\phi)^i - K^{i^*}(\phi^\dagger T_a)_{i^*}\,.$$

Next we use the gauge invariance of the superpotential $W^\star$ together with the equations (7.1). Finally the minimisation of the potential and the gauge invariance of the superpotential can be unified into

$$\begin{pmatrix} \langle D_i D_j W\rangle & \frac{1}{4}\langle D^c h_{cbi}\rangle - g\langle K^{i^*}{}_i(\phi^\dagger T_b)_{i^*}\rangle \\ \frac{1}{4}\langle D^d h_{daj}\rangle - g\langle K^{i^*}{}_j(\phi^\dagger T_a)_{i^*}\rangle & \frac{1}{4}\langle h_{abk}F^k\rangle \end{pmatrix}\begin{pmatrix}\langle F^j\rangle \\ -\langle D^b\rangle\end{pmatrix} = 0\,.$$

This implies that the mass matrix has a zero mode, and in particular that

$$\begin{aligned} \psi_G &= \frac{1}{\sqrt{2}}\left(\langle K^{i^*}{}_i F^\dagger_{i^*}\rangle \; -\langle h_{Rab}D^b\rangle\right)\begin{pmatrix}\chi^i \\ i\sqrt{2}\lambda^a\end{pmatrix} \\ &= \frac{1}{\sqrt{2}}\langle K^{i^*}{}_i F^\dagger_{i^*}\rangle\chi^i - i\langle h_{Rab}D^b\rangle\lambda^a\,, \end{aligned} \tag{7.4}$$

is a massless particle, the goldstone fermion written in the basis $(\chi^i, i\sqrt{2}\lambda^a)$. This massless particle is just the manifestation of broken supersymmetric theories. The emergence of a massless goldstone fermion where first shown in [Iliopoulos and Zumino (1974); Salam and Strathdee (1974a)].

In supergravity things change as we will see now. First of all the two-fermion interactions are given by (see Table 6.3)

$$\begin{aligned}
e^{-1}\mathcal{L}_{\text{ferm.}} &= e^{\frac{1}{2}\frac{K}{m_P^2}}\left\{-\frac{1}{2}e^{\frac{1}{2}\frac{K}{m_P^2}}\mathcal{D}_i\mathcal{D}_jW\chi^i\cdot\chi^j + \frac{1}{4}K^i{}_{i^*}h_{abi}\mathcal{D}^{i^*}W^\star\lambda^a\cdot\lambda^b\right\} \\
&+\sqrt{2}igK^{i^*}{}_i(\phi^\dagger T_a)_{i^*}\lambda^a\cdot\chi^i \\
&-i\frac{\sqrt{2}}{8}h^{Rab}[K^{i^*}(\phi^\dagger T_a)_{i^*} + K_i(T_a\phi)^i]h_{bcj}\chi^j\cdot\lambda^c + \text{h.c.}\,,
\end{aligned}$$

and the potential (see Table 6.3) by

$$\begin{aligned}
V &= e^{\frac{K}{m_P^2}}\left(K^i{}_{i^*}\mathcal{D}_iW\mathcal{D}^{i^*}W^\star - \frac{3}{m_m^2}|W|^2\right) \\
&+\frac{g^2}{8}h^{Rab}[K^{i^*}(\phi^\dagger T_a)_{i^*} + K_i(T_a\phi)^i][K^{j^*}(\phi^\dagger T_b)_{j^*} + K_j(T_b\phi)^j]\,.
\end{aligned}$$

From supergravity transformations of the fermions χ^i, λ^a given in (4.78) and (4.79), performing a super-Weyl transformation (6.41), (6.42) and (6.43) (be careful: we express the new fields with respect to the old fields) with the substitution $\varepsilon \to e^{1/2\lambda}\varepsilon$ we get

$$\begin{aligned}
\delta\chi^i &= \sqrt{2}\varepsilon F^i + \cdots\,, \\
\delta\lambda^a &= i\varepsilon D^a + \cdots\,,
\end{aligned}$$

where the dots represent derivative in the fields and terms involving fermions. Now the auxiliary fields, using (6.25) are given by (after the Weyl rescaling)

$$\begin{aligned}
F^i &= e^{\frac{1}{2}\frac{K}{m_P^2}}K^i{}_{i^*}\mathcal{D}^{i^*}W^\star + \cdots\,, \\
D^a &= h^{Rab}(K^{i^*}(\phi^\dagger T_b)_{i^*} + K_i(\phi T_b)^i) + \cdots\,, \\
N &= -3e^{\frac{1}{2}\frac{K}{m_P^2}}W + \cdots\,,
\end{aligned} \tag{7.5}$$

where the dots involves fermions, *i.e.*, terms that vanish when vacuum expectation values are taken[1]. These relations are of interest for several reasons. If we now assume that the auxiliary fields develop a vacuum expectation value we have on the one hand for the potential

[1] We do not allow fermion condensation in our analysis.

$$
\begin{aligned}
\langle V\rangle &= \langle K^{i^*}{}_i F^i F_{i^*}\rangle + \frac{1}{2}\langle h_{Rab} D^a D^b\rangle - \frac{1}{3}\langle NN^\star\rangle \\
&= \langle e^{\frac{K}{m_p^2}} K^i{}_{i^*} \mathcal{D}_i W \mathcal{D}^{i^*} W^\star\rangle \\
&\quad + \frac{1}{8} g^2 \langle (K^{i^*}(\phi^\dagger T_a)_{i^*} + K_i (T_a\phi)^i)(K^{j^*}(\phi^\dagger T_b)_{j^*} + K_j (T_a\phi)^j)\rangle \\
&\quad - 3\langle e^{\frac{K}{m_p^2}} |W|^2\rangle \\
&= \langle V_+\rangle + \langle V_-\rangle ,
\end{aligned} \tag{7.6}
$$

where $< V_+ >$ is strictly positive and $< V_- > 0$ strictly negative. On the other hand, using (7.4) and (7.6), the variation of the goldstino under a supergravity transformation is given by

$$
\delta\psi_G = \langle V_+\rangle \varepsilon + \cdots ,
$$

where the dots involves derivatives of fields and terms involving fermions. It is crucial to observe that the goldstino is shifted by a supergravity transformation and more importantly that $< V_+ >$ never vanishes. This has the very important consequence that one can now chose a gauge which eliminates the goldstino. More precisely, considering the second and third lines of Table 6.3, one can perform a shift of the goldstino such that it disappears from the Lagrangian (see Eqs. [7.10-7.12] below). It is important to emphasise that the remaining part of the transformation, *i.e.*, the dots, are necessary in order to interpret the transformation as a supergravity transformation. In particular, this means that the gravitino becomes massive with mass

$$
m_{\frac{3}{2}} = \frac{1}{m_p^2}\langle e^{\frac{1}{2}\frac{K}{m_p^2}} W\rangle , \tag{7.7}
$$

see Sec. 6.2.8 and the helicities $\pm 1/2$ are given by the goldstino which is eaten by the gravitino. This is the analogue of the Brout-Englert-Higgs mechanism in supergravity [Deser and Zumino (1977); Cremmer et al. (1979)].

7.2 Gravity induced supersymmetry breaking

As mentioned previously, supergravity is broken in a hidden sector. The problem is (and this is the case for any theory containing a hidden sector) that there are no indications upon the hidden sector (content in fields, interactions, *etc.*). We only have to ensure that when the hidden sector communicates in one way or another with the observable sector it does not lead to some predictions which are not in accordance with experimental results. Despite this point we will see that gravity mediated supersymmetry breaking shares some generic features[2].

7.2.1 *The very first model*

In this section we consider the historical model which induces a supergravity breaking [Barbieri et al. (1982b); Polonyi (1977)].

We denote Φ^i the chiral superfields of the observable sector and $Z = (z, \chi_z, F_z)$ the chiral superfield of the hidden sector. We further assume that the superpotential takes the form

$$\begin{aligned} W(\Phi^i, Z) &= W_m(\Phi^i) + W_h(Z)\ , \\ W_h(Z) &= \mu m_p(Z + \beta)\ , \end{aligned} \tag{7.8}$$

with m_p the Planck mass and μ, β two dimensionfull parameters. We also assume that the Kähler manifold is flat

$$K(\Phi^i, \Phi^\dagger_i, Z, Z^\dagger) = \Phi^i \Phi^\dagger_i + ZZ^\dagger\ ,$$

and that the superpotential $W_m(\phi)$ is such that the fields ϕ^i do not develop a vacuum expectation value. Since the Kähler potential is minimal the anti-chiral multiplet will be written using the notations of Sec. 6.5. Moreover, in the limit $m_p \to \infty$ the two sectors (hidden and observable) decouple. It is indeed the coupling to supergravity which leads to interactions between the two sectors. Remembering that we have introduced $\mathcal{D}_i W = W_i + \frac{1}{m_p^2} K_i W$, we get

$$\begin{aligned} \mathcal{D}_z W &= \mu m_p + \frac{1}{m_p^2} z^\dagger \left[\mu m_p (z + \beta) + W_m\right]\ , \\ \mathcal{D}_i W &= \frac{\partial W_m}{\partial \phi^i} + \frac{1}{m_p^2} \phi^\dagger_i \left[\mu m_p (z + \beta) + W_m\right]\ . \end{aligned}$$

The relevant part of the potential (6.69) becomes (here we do not consider the gauge part since it has no effect on the analysis)

$$\begin{aligned} V(z, z^\dagger, \phi^i, \phi^\dagger_i) = e^{\frac{zz^\dagger + \phi^i \phi^\dagger_i}{m_p^2}} \bigg\{ &\Big|\mu m_p + \frac{1}{m_p^2} z^\dagger (W_h + W_m)\Big|^2 \\ &+ \sum_i \Big|\frac{\partial W_m}{\partial \phi^i} + \frac{1}{m_p^2} \phi^\dagger_i (W_h + W_m)\Big|^2 - \frac{3}{m_p^2} \Big|W_h + W_m\Big|^2 \bigg\}\ . \end{aligned}$$

[2]These general features are also shared by the other mechanisms of supergravity breaking.

Since supergravity is broken in the hidden sector, we can assume $\phi^i = 0$ at the minimum of V and, without loss of generality, that z is real. Then,

$$V(z) = e^{\frac{z^2}{m_p^2}}\left[\left(\mu m_p + \frac{\mu}{m_p}z(z+\beta)\right)^2 - 3\mu^2(z+\beta)^2\right],$$

$$\frac{\mathrm{d}V}{\mathrm{d}z} = 2\frac{\mu^2}{m_p^2}e^{\frac{z^2}{m_p^2}}\left[\frac{1}{m_p^2}z^3 + \frac{\beta}{m_p^2}z^2 - 2\beta\right]\left[z^2 + z\beta + m_p^2\right].$$

We are looking for a minimum $< z >= v$ such that the cosmological constant vanishes, that is $V(v) = 0$. This leads to

$$v = (\sqrt{3} \pm 1)m_p\,,$$
$$\beta = (\mp 2 - \sqrt{3})m_p\,.$$

More precisely, when $\beta = -(2 + \sqrt{3})m_p$ the minimisation equation has five real roots but the one corresponding to $v = (\sqrt{3} + 1)m_p$ is not the absolute minimum, whereas when $\beta = (2 - \sqrt{3})m_p$ the minimisation equation has only one real root $v = (\sqrt{3} - 1)m_p$. Thus, among the two solutions we select

$$v = (\sqrt{3} - 1)m_p\,,$$
$$\beta = (2 - \sqrt{3})m_p\,.$$

This solution is not natural since some fine tuning is needed in order to have a vanishing cosmological constant. In particular, a precise value of β is needed.

Now we are ready to study the consequences of the fact that the field z develops a vacuum expectation value. First of all from (see Table 6.3)

$$\frac{1}{m_p^2}e^{\frac{1}{m_p^2}\frac{K}{2}}\left(W\psi_\mu\sigma^{\mu\nu}\psi_\nu + W^\star\bar{\psi}_\mu\bar{\sigma}^{\mu\nu}\bar{\psi}_\nu\right)\Big|_{\substack{z\,\to\, v,\\ \phi^i\,\to\, 0}} = \mu e^{2-\sqrt{3}}\left(\psi_\mu\sigma^{\mu\nu}\psi_\nu + \bar{\psi}_\mu\bar{\sigma}^{\mu\nu}\bar{\psi}_\nu\right)\,,$$

the gravitino becomes massive with mass

$$m_{\frac{3}{2}} = \mu e^{2-\sqrt{3}}\,. \tag{7.9}$$

In particular the scale at which supergravity is broken is controlled by μ which can be much bellow the Planck scale.

Secondly we study the hidden sector and take the limit $m_p \to \infty$. Performing the shift $z \to z+v$, Taylor expending the exponential, and taking the limit $m_p \to \infty$,

we obtain

$$
\begin{aligned}
V(z+v) &= e^{\frac{|z+v|^2}{m_p^2}}\left[\left|\mu m_p + \frac{1}{m_p^2} z^\dagger W_h(v+z)\right|^2 - \frac{3}{m_p^2}\left|W_h(v+z)\right|^2\right] \\
&\sim e^{\frac{v}{m_p^2}}\Bigg\{\left[\left|\mu m_p + \frac{1}{m_p^2} z^\dagger W_h(z+v)\right|^2 - \frac{3}{m_p^2}\left|W_h(z+v)\right|^2\right] \\
&\quad + \left(\frac{|z|^2}{v^2} + \frac{z+z^\dagger}{v}\right)\left[\left|\mu m_p + \frac{1}{m_p^2} z^\dagger W_h(z)\right|^2 - \frac{3}{m_p^2}\left|W_h(z)\right|^2\right]\Bigg\} + o\left(\frac{1}{m_p^2}\right) \\
&\sim e^{\frac{v^2}{m_p^2}}\left[\left|\mu m_p + \frac{1}{m_p^2} z^\dagger W_h(v+z)\right|^2 - \frac{3}{m_p^2}\left|W_h(v+z)\right|^2\right],
\end{aligned}
$$

where we have used the fact that the cosmological constant vanishes and that we are at a minimum of the potential *i.e.* $V(v) = 0, V_z(v) = 0$ in the third and second lines respectively. Introducing the gravitino mass, and neglecting the vanishing terms (that is terms like $z^2 z^{\dagger 2}, (z^2 z^\dagger + z^{\dagger 2} z)$) we finally obtain

$$
\begin{aligned}
V(z) &= 2m_{\frac{3}{2}}^2 z z^\dagger + (\sqrt{3}-1)m_{\frac{3}{2}}^2(z^2 + z^{\dagger 2}) \\
&= \sqrt{3} m_{\frac{3}{2}}^2 X^2 + (2-\sqrt{3}) m_{\frac{3}{2}}^2 Y^2 .
\end{aligned}
$$

where $z = 1/\sqrt{2}(X+iY)$. We thus have two massive scalars with masses square $m_X^2 = \sqrt{3} m_{\frac{3}{2}}^2$ and $m_Y^2 = (2-\sqrt{3})m_{\frac{3}{2}}^2$. For the fermionic sector we recall that the bilinear terms are (see Table 6.3)

$$
\frac{1}{2} e^{\frac{1}{m_p^2}\frac{K}{2}}\left[\mathcal{D}_i \mathcal{D}_j W \chi^i \cdot \chi^j + \mathcal{D}^i \mathcal{D}^j W^\star \bar{\chi}_i \cdot \bar{\chi}_j\right]
$$

with

$$
\begin{aligned}
\mathcal{D}_z \mathcal{D}_z W &= W_{zz} + 1/m_p^2 K_{zz} W + 1/m_p^2 K_z \mathcal{D}_z W + 1/m_p^2 K_z \mathcal{D}_z W \\
&\quad -1/m_p^4 K_z K_z W - \Gamma_z{}^z{}_z \mathcal{D}_z W \\
&= \frac{2}{m_p^2} K_z \mathcal{D}_z W - \frac{1}{m_p^4} K_z K_z W \\
&= 2\mu .
\end{aligned}
$$

Thus, the mass of the fermion χ_z is $m_{\chi_z} = m_{3/2}$ and χ_z is the goldstino. This is legitimated by the fact that supergravity is broken in the direction of the chiral superfield Z. To be more precise, from Sec. 6.2.8 the coupling of the goldstino with the gravitino reads (see Table 6.3)

$$\begin{aligned}
\mathcal{L}_{\text{gold-gravit.}} &= -e^{\frac{1}{m_p^2}\frac{K}{2}}\Big(\frac{1}{m_p^2}W\psi_\mu\sigma^{\mu\nu}\psi_\nu + \frac{1}{m_p^2}W^\star\bar{\psi}_\mu\bar{\sigma}^{\mu\nu}\bar{\psi}_\nu \\
&\quad + \frac{1}{2}\mathcal{D}_z\mathcal{D}_z W\chi_z\cdot\chi_z + \frac{1}{2}\overline{\mathcal{D}}_{z^\dagger}\overline{\mathcal{D}}_{z^\dagger}W^\star\bar{\chi}_z\cdot\bar{\chi}_z \\
&\quad - \frac{i}{m_p}\frac{\sqrt{2}}{2}\mathcal{D}_z W\bar{\psi}_\mu\bar{\sigma}^\mu\chi_z - \frac{i}{m_p}\frac{\sqrt{2}}{2}\overline{\mathcal{D}}_{z^\dagger}W^\star\psi_\mu\sigma^\mu\bar{\chi}_z\Big) \\
&= -m_{\frac{3}{2}}\Big(\psi_\mu\sigma^{\mu\nu}\psi_\nu + \bar{\psi}_\mu\bar{\sigma}^{\mu\nu}\bar{\psi}_\nu \frac{\sqrt{6}}{2}i\bar{\psi}_\mu\bar{\sigma}^\mu\chi_z \\
&\quad - \frac{\sqrt{6}}{2}i\psi_\mu\sigma^\mu\bar{\chi}_z + \chi_z\cdot\chi_z + \bar{\chi}_z\cdot\bar{\chi}_z\Big)\,.
\end{aligned} \tag{7.10}$$

Now if we perform the gravitino shift

$$\psi_\mu \to \psi_\mu - \frac{\sqrt{6}}{6}i\sigma_\mu\bar{\chi}_z + \cdots\,, \tag{7.11}$$

the mass term (7.10) becomes

$$\mathcal{L}_{\text{gold-gravit.}} = -m_{\frac{3}{2}}\Big(\psi^\mu\sigma^{\mu\nu}\psi_\nu + \bar{\psi}^\mu\bar{\sigma}^{\mu\nu}\bar{\psi}_\nu\Big)\,, \tag{7.12}$$

and the goldstino has been explicitly eliminated by an appropriate supergravity transformation. Note however, that the dots in (7.11) represent some terms (derivatives) [Cremmer et al. (1979)] or [Freedman and Van Proeyen (2012)] that are needed in order to interpret the shift as a supergravity transformation. This is an illustration of the general principle stated in the previous section.

Now we turn to the study of the matter sector: the substitution $z \to v$ gives

$$\begin{aligned}
V(\phi^i,\phi_i^\dagger) &= e^{\frac{\phi^i\phi_i^\dagger}{m_p^2}}\Big\{\Big|\sqrt{3}m_p m_{\frac{3}{2}} + \frac{1}{m_p}(\sqrt{3}-1)W\Big|^2 \\
&\quad + \sum_i\Big|\frac{\partial W}{\partial\phi^i} + m_{\frac{3}{2}}\phi_i^\dagger + \frac{1}{m_p^2}\phi_i^\dagger W\Big|^2 - 3\Big|m_{\frac{3}{2}}m_p + \frac{1}{m_p}W\Big|^2\Big\}\,,
\end{aligned}$$

where we have introduce $W = e^{\frac{1}{2}\frac{v^2}{m_p^2}}W_m$ and the gravitino mass $m_{3/2}$. Taking the limit where $m_p \to \infty$ and keeping the non-vanishing terms, the potential reduces to

$$V(\phi) = W_i W^{\star i} + m_{\frac{3}{2}}^2\phi^i\phi_i^\dagger + m_{\frac{3}{2}}\Big(-\sqrt{3}W + \phi^i W_i + \text{h.c.}\Big)\,.$$

In particular if we consider a gauge invariant renormalisable superpotential

$$W(\Phi) = \alpha_i\Phi^i + \frac{1}{2}m_{ij}\Phi^i\Phi^j + \frac{1}{6}\lambda_{ijk}\Phi^i\Phi^j\Phi^k\,,$$

we obtain

$$V = W_i W^{\star i} + V_{\text{soft}} ,$$

with

$$V_{\text{soft}} = m^2_{\frac{3}{2}} \phi^i \phi^\dagger_i + m_{\frac{3}{2}} \Big[\frac{1}{6}(3 - \sqrt{3}) \lambda_{ijk} \phi^i \phi^j \phi^k$$
$$+ \frac{1}{2}(2 - \sqrt{3}) m_{ij} \phi^i \phi^j + (1 - \sqrt{3}) \alpha_i \phi^i + \text{h.c.} \Big] . \qquad (7.13)$$

The potential V_{soft} contains the so-called soft-supersymmetric breaking terms. These terms are soft in the sense that they preserve the nice renormalisability properties of supersymmetric theories, *i.e.* they generate only logarithmic divergent terms coming from loop diagrams. Indeed, a general study of the terms which explicitly break supersymmetry without reintroducing quadratic divergences was performed in [Girardello and Grisaru (1982)]. It is interesting to observe that these terms appear naturally in the context of gravity mediated supersymmetry breaking. Four types of terms are automatically generated:

(1) a universal scalar mass term : $m_0^2 = m_{3/2}^2$;

(2) a universal trilinear coupling $A_0 \lambda_{ijk}, A_0 = (3 - \sqrt{3}) m_{3/2}$;

(3) a universal bilinear term $B_0 m_{ij}, B_0 = (2 - \sqrt{3}) m_{3/2}$;

(4) a universal linear term $C_0 \alpha_i, C_0 = (1 - \sqrt{3}) m_{m/2}$.

Several observations are in order. First of all, the soft-supersymmetric breaking terms are related to the gravitino mass. This is a simple consequence of the superpotential we have chosen in the hidden sector. For different superpotentials we could obtain universal A_0, B_0 and C_0 which are fully independent. Secondly, we observe that all the terms are universal. This is due to the special form of our Kähler potential. Non-minimal Kähler potential would have given non-universal soft supersymmetric breaking terms. This will be the subject of the next section.

Finally, we investigate the gauge sector of the theory. Recall that the bilinear terms for the gauginos are given by (see Table 6.3)

$$\frac{1}{4} e^{\frac{1}{2}\frac{K}{m_P^2}} \Big(\mathcal{D}^i W^\star h_{abi} \lambda^a \cdot \lambda^b + \mathcal{D}_i W h^{\star i}{}_{ab} \bar{\lambda}^a \cdot \bar{\lambda}^b \Big) .$$

If we assume that the gauge kinetic function does not couple the hidden and the matter sectors, and if we remember that z is a gauge singlet, we have

$$h_{ab}(\phi, z) = \delta_{ab} h^h(z) + h^m_{ab}(\phi) ,$$

and only the first term contributes to the gaugino mass to give

$$M_{\frac{1}{2}} = \frac{1}{2} e^{\frac{1}{2}\frac{v^2}{m_p^2}} \left\{ \sqrt{3} m_p \mu \Big(\frac{\mathrm{d}h^h(z)}{\mathrm{d}z} \Big)_{|z=v} \right\} = \frac{\sqrt{3}}{2} \alpha m_p m_{\frac{3}{2}} \, .$$

Since $[\alpha] = M^{-1}$ it is natural to assume that $\alpha \sim m_p^{-1}$, and the gaugino mass is of the order of the gravitino mass. As for the matter part, the gravity induced supersymmetry breaking generates an universal gaugino mass. Note however that for semisimple Lie groups we can in principle take a different $h^h(z)$ for each simple factors.

7.2.2 *General form of soft-supersymmetric breaking terms*

In this section we address the most general possible form of the soft supersymmetric breaking terms. We assume, as in the previous section, to have a hidden sector and an observable sector. Let us denote by $X^I, I = 1, \cdots, m+n$ the chiral superfields. We set $X^I = (H^i, \Phi^a), i = 1, \cdots, m, a = 1, \cdots, n$ where the H^i represent superfields in the hidden sector and the Φ^a superfields in the observable sector. By a superfield in the observable sector we mean a superfield describing a supersymmetric extension (minimal or not, involving GUT or not) of the standard model (see *e.g.* Sec. 8.1.2). We assume further that supersymmetry is broken in the hidden sector by a mechanism similar to the mechanism of Sec. 7.2.1. So, at the minimum of the potential some of the fields in the hidden sector develop a vacuum expectation value of the order of the Planck mass. We also suppose that some of the fields in the observable sector can develop much smaller vacuum expectation value(s) as the electroweak scale or possibly as an intermediate scales corresponding to some extensions of the Standard Model. The scalar potential (6.69) (we do not consider the gauge part of the scalar potential in this analysis)

$$V = e^{\frac{1}{m_p^2} K(h, h^\dagger, \phi, \phi^\dagger)} \Big(\mathcal{D}_I W \mathcal{D}^{I^*} W^\star K^I{}_{I^*} - \frac{3}{m_p^2} |W|^2 \Big) \, ,$$

needs to be expended after supergravity breaking. Since some of the hs develop a vacuum expectation value of the order of the Planck mass m_p, the expansion of the scalar potential might generate dangerous terms, *i.e.*, terms where some fields in the visible sector have a coupling that diverges when the Planck mass goes to infinity (for instance a term like $m_p^2 \phi^1 \phi^1$ is a dangerous term). We thus restrict the form of W and K in such a way that low energy supergravity is consistent with particle physics, *i.e.*, when there are no dangerous terms. This problem has been solved in Soni and Weldon (1983) and the possible forms of K and W are given by

$$\begin{aligned} K(H,H^\dagger,\Phi,\Phi^\dagger) &= m_p^2 K_2(H,H^\dagger) + m_p K_1(H,H^\dagger) + K_0(H,H^\dagger,\Phi,\Phi^\dagger) \\ W(H,\Phi) &= m_p^2 W_2(H) + m_p W_1(H) + W_0(H,\Phi)\,, \end{aligned} \tag{7.14}$$

where W_2, W_1 are arbitrary holomorphic functions of H; K_2, K_1 are arbitrary real functions of $(H, H^\dagger)$; W_0 is an arbitrary holomorphic function of (H, Φ) and K_0 is an arbitrary real function of $(H, H^\dagger, \Phi, \Phi^\dagger)$.

Recently this analysis was reconsidered [Moultaka et al. (2019a)] and new solutions were obtained (see Sec. 7.2.3). We do not consider these new solutions in this section, and identify the most general soft supersymmetric breaking terms under the hypothesis of (7.14).

Following [Giudice and Masiero (1988); Brignole et al. (1998)] we define

$$\begin{aligned} K(H,H^\dagger,\Phi,\Phi^\dagger) &= \hat{K}(H,H^\dagger) + \Phi^\dagger_{a^*}\Lambda^{a^*}{}_a(H,H^\dagger)\Phi^a + \Big(Z(H,H^\dagger)g_2(\Phi) + \text{h.c.}\Big) \\ &\equiv \hat{K}(H,H^\dagger) + \widetilde{K}(H,H^\dagger,\Phi,\Phi^\dagger) \\ W(H,\Phi) &= \hat{W}(H) + W_0(H)g_3(\Phi) \\ &\equiv \hat{W}(H) + \widetilde{W}(H,\Phi)\,, \end{aligned} \tag{7.15}$$

where $\hat{K}$ and $\hat{W}$ depend only on the hidden sector whereas $\widetilde{K}$ and $\widetilde{W}$ depend on the visible and hidden sectors. We also assume the orders of magnitude in accordance with Soni and Weldon (1983)

$$\hat{K} \sim m_p^2\,, \quad \widetilde{K} \sim M^2\,, \quad \hat{W} \sim m_p^2 M\,, \quad \widetilde{W} \sim M^3\,, \tag{7.16}$$

with $M \ll m_p$ an intermediate scale. We only assume one intermediate scale for simplicity, say M, since multiple ones will not alter the final result.

Before performing the computation of V we would like to comment on the generality of (7.15) within the hypothesis of Soni and Weldon (1983). The form of $\widetilde{W} = W_0(H)g_3(\Phi)$ and of $\widetilde{K} = \Phi^\dagger_{a^*}\Lambda^{a^*}{}_a(H,H^\dagger)\Phi^a + \Big(Z(H,H^\dagger)g_2(\Phi) + \text{h.c.}\Big)$ is not the most general since *a priori* we could have considered these two functions as arbitrary functions of the hidden and the observable sectors. This limitation is in fact not a problem since the formula obtained within our hypothesis can easily be extended to consider the more general case. For instance if $Z(H,H^\dagger)g_2(\Phi)$ is substituted by $\Gamma(H,H^\dagger,\Phi)$, the term $\partial^{i^*}Zg_2$ in (7.20) becomes $\partial^{i^*}\Gamma$. Next, we could have considered, as in [Giudice and Masiero (1988)]

$$\begin{aligned} Z(H,H^\dagger)g_2(\Phi) &\to Z_k(H,H^\dagger)g_2^k(\Phi)\,, \\ W_0(H)g_3(\Phi) &\to W_{0k}(H)g_3^k(\Phi)\,, \end{aligned}$$

i.e. several functions g_2^k and g_3^k. This in fact does not constitute an obstacle either since, as we will see, our formula can be easily extended to this case. In the second part of this section we explicitly extend our computation to the more general renormalisable theory [Brignole et al. (1998)] with the substitution

$$\begin{aligned} Z(H,H^\dagger)g_2(\Phi) &\to \frac{1}{2}Z_{ab}(H,H^\dagger)\Phi^a\Phi^b \,, \\ W_0(H)g_3(\Phi) &\to \frac{1}{2}m_{ab}(H)\Phi^a\Phi^b + \frac{1}{6}\lambda_{abc}(H)\Phi^a\Phi^b\Phi^c \,. \end{aligned} \tag{7.17}$$

When some of the fields in the hidden sector acquire a vacuum expectation value, some of the F–fields associated with the hidden sector (7.5) develop a vacuum expectation value

$$F^I = \left\langle e^{\frac{1}{2}\frac{\hat{K}}{m_P^2}} \hat{K}^I{}_{I^*} \mathcal{D}^{I^*} W^* \right\rangle , \tag{7.18}$$

with $\hat{K}^I{}_{I^*}$ the inverse of $\hat{K}^{I^*}{}_I$. Since the vacuum expectation values of the fields in the observable sector are negligible with respect to the vacuum expectation values of the fields in the hidden sector, only the latter contribute. In other words we do not consider the contributions of the former in the following computation.

The first step in the determination of the scalar potential is a pertubative expansion of the Kähler metric. We write the Kähler metric in block diagonal form

$$\begin{aligned} K^{I^*}{}_I &= \frac{\partial^2 K}{\partial X^I \partial X^\dagger_{I^*}} \\ &= \begin{pmatrix} K^{a^*}{}_a & K^{a^*}{}_i \\ K^{i*}{}_a & K^{i^*}{}_i \end{pmatrix} \\ &= \begin{pmatrix} A & B \\ C & D \end{pmatrix} \\ &= \begin{pmatrix} A_0 & B_1 \\ C_1 & D_0 + D_2 \end{pmatrix} , \end{aligned}$$

with

$$\begin{aligned} D_0{}^{i^*}{}_i &= \hat{K}^{i^*}{}_i \,, \\ D_2{}^{i^*}{}_i &= \phi^\dagger_{a^*}\partial_i\partial^{i^*}\Lambda^{a^*}{}_a\phi^a + \left(\partial_i\partial^{i^*}Zg_2 + \text{h.c.}\right) , \\ A_0{}^{a^*}{}_a &= \Lambda^{a^*}{}_a \,, \\ B^{a^*}{}_i &= \partial_i\Lambda^{a^*}{}_a\phi^a + \partial_i Z^\star \partial^{a^*} g_2^\star \,, \\ C^{i^*}{}_a &= \phi^\dagger_{a^*}\partial^{i^*}\Lambda^{a^*}{}_a + \partial^{i^*} Z\partial_a g_2 \,. \end{aligned}$$

• The superscript indicates the order of magnitude in the Planck mass. For instance $D_0 \sim m_p^0, D_2 \sim m_p^{-2}$. Using (5.5), the inverse of the Kähler metric is given by

$$K^I{}_{I^*} = \begin{pmatrix} (A - BD^{-1}C)^{-1} & -(A - BD^{-1}C)^{-1}BD^{-1} \\ -D^{-1}C(A - BD^{-1}C)^{-1} & D^{-1}C(A - BD^{-1}C)^{-1}BD^{-1} + D^{-1} \end{pmatrix} .$$

Inverting the metric pertubatively , using (7.16), we obtain

$$K^I{}_{I^*} = \begin{pmatrix} K^a{}_{a^*} & K^a{}_{i^*} \\ K^i{}_{a^*} & K^i{}_{i^*} \end{pmatrix}$$

with

$$\begin{aligned}
K^i{}_{i^*} &= \hat{K}^i{}_{i^*} + \hat{K}^i{}_{j^*}\Big([\phi^\dagger_{a^*}\partial^{j^*}\Lambda^{a^*}{}_a + \partial^{j^*}Z\partial_a g_2](\Lambda^{-1})^a{}_{b^*}[\partial_j\Lambda^{b^*}{}_b\phi^b + \partial_j Z^\star\partial^{b^*}g_2^\star] \\
&\qquad -\phi^\dagger_{a^*}\partial_j\partial^{j^*}\Lambda^{a^*}{}_a\phi^a - [\partial_j\partial^{j^*}Zg_2 + \text{h.c.}]\Big)\hat{K}^j{}_{i^*} \\
&\equiv \hat{K}^i{}_{i^*} + \widetilde{K}^i{}_{i^*} \\
&\sim m_p^0 + m_p^{-2} , \\
K^a{}_{a^*} &= \ (\Lambda^{-1})^a{}_{b^*}[\partial_i\Lambda^{b^*}{}_b\phi^b + \partial_i Z^\star\partial^{b^*}g_2^\star]\hat{K}^i{}_{i^*}[\phi^\dagger_{c^*}\partial^{i^*}\Lambda^{c^*}{}_c + \partial^{i^*}Z\partial_c g_2](\Lambda^{-1})^c{}_{a^*} , \\
&\quad +(\Lambda^{-1})^a{}_{a^*} \\
&\sim m_p^{-2} + m_p^0 , \\
K^a{}_{i^*} &= -(\Lambda^{-1})^a{}_{a^*}[\partial_i\Lambda^{a^*}{}_b\phi^b + \partial_i Z^\star\partial^{a^*}g_2^\star]\hat{K}^i{}_{i^*} \\
&\sim m_p^{-1} , \\
K^i{}_{a^*} &= -K^i{}_{i^*}[\phi^\dagger_{b^*}\partial^{i^*}\Lambda^{b^*}{}_a + \partial^{i^*}Z\partial_a g_2](\Lambda^{-1})^a{}_{a^*} \\
&\sim m_p^{-1} .
\end{aligned}$$

We have also indicated the order of magnitude of the various terms. Introducing the two functions (of order m_p^{-1})

$$\begin{aligned}
\rho_i &= \partial_i\Big(\frac{1}{m_p^2}\hat{K} + \ln\frac{\hat{W}}{m_p^3}\Big) , \\
\tilde{\rho}_i &= \partial_i\Big(\frac{1}{m_p^2}\hat{K} + \ln\frac{\widetilde{W}}{m_p^3}\Big) ,
\end{aligned}$$

which leads to the relationships

$$\begin{aligned}
F^i &= m_p^2 m^\dagger_{\frac{3}{2}} K^i{}_{i^*}\rho^{*i^*} , \\
F^\star_{i^*} &= m_p^2 m_{\frac{3}{2}} K^i{}_{i^*}\rho_i ,
\end{aligned}$$

the covariant derivatives take the form

$$\begin{aligned}
\mathcal{D}_i W &= \hat{W}\rho_i + \widetilde{W}\tilde{\rho}_i + \frac{1}{m_p^2}\hat{W}\widetilde{K}_i + \frac{1}{m_p^2}\widetilde{W}\widetilde{K}_i \\
&\sim m_p \ + m_p^{-1} + \ m_p^{-1} \ + \ m_p^{-3}, \\
\mathcal{D}_a W &= \partial_a \tilde{W} + \frac{1}{m_p^2}\hat{W}\widetilde{K}_a + \frac{1}{m_p^2}\widetilde{W}\widetilde{K}_a \\
&\sim m_p^0 \ + \ m_p^0 \ + \ m_p^{-2}.
\end{aligned}$$

We have indicated again the orders of magnitude of the various terms in view of the expansion we are about to develop.

Introducing the gravitino mass

$$m_{\frac{3}{2}} = \frac{1}{m_p^2}\langle \hat{W}\rangle e^{\frac{1}{2}\frac{\langle \hat{K}\rangle}{m_p^2}}$$

we obtain

$$\begin{aligned}
e^{\frac{1}{m_p^2}K}\mathcal{D}_i W K^i{}_{i^*}\mathcal{D}^{i^*}W^\star &= e^{\frac{1}{m_p^2}K}\Bigg(|\hat{W}|^2\rho_i\rho^{\star i^*}\hat{K}^i{}_{i^*} + [\hat{W}\widetilde{W}^\star\rho_i\tilde{\rho}^{\star i^*} + \text{h.c.}]\hat{K}^i{}_{i^*} \\
&\quad + \frac{1}{m_p^2}|\hat{W}|^2[\rho_i\widetilde{K}^{i^*} + \text{h.c.}]\hat{K}^i{}_{i^*} + |\hat{W}|^2\rho_i\rho^{\star i^*}\widetilde{K}^i{}_{i^*}\Bigg), \\
e^{\frac{1}{m_p^2}K}\mathcal{D}_a W K^a{}_{a^*}\mathcal{D}^{a^*}W^\star &= e^{\frac{1}{m_p^2}K}\Bigg(\partial_a\widetilde{W}\partial^{a^*}\widetilde{W}^\star(\Lambda^{-1})^a{}_{a^*} \\
&\quad + \frac{1}{m_p^2}[\partial_a\widetilde{W}\hat{W}^\star\widetilde{K}^{a^*} + \text{h.c.}](\Lambda^{-1})^a{}_{a^*} \\
&\quad + \frac{1}{m_p^4}|\hat{W}|^2\widetilde{K}_a\widetilde{K}^{a^*}(\Lambda^{-1})^a{}_{a^*}\Bigg), \\
e^{\frac{1}{m_p^2}K}\mathcal{D}_a W K^a{}_{i^*}\mathcal{D}^{i^*}W^\star &= e^{\frac{1}{m_p^2}K}\Bigg(\partial_a\widetilde{W}\hat{W}^\star\rho^{\star i^*}K^a{}_{i^*} + \frac{1}{m_p^2}|\hat{W}|^2\widetilde{K}_a\rho^{\star i^*}K^a{}_{i^*}\Bigg).
\end{aligned}$$

We have stopped the expansion at the appropriate order so that the Planck suppressed terms in the potential are not included (see (7.16)), *i.e.*, we have only kept the terms up to the order m_p^0. In the expansion above, we have to be careful since the gravitino mass is in $1/m_p^2$. Now we substitute the fields h^i by their vacuum expectation values. Developing the expression above is a little involved since there are many terms. However, regrouping the various terms, being careful not to forget the $|W|^2$-term, following the basic functions

$$\begin{aligned}
S^{a^*}{}_a &= \langle \Lambda^{a^*}{}_a \rangle + \frac{1}{m^2_{\frac{3}{2}}} \langle F^\dagger_{i^*} \rangle [\langle \partial^{i^*} \Lambda^{a^*}{}_b \rangle \langle (\Lambda^{-1})^b{}_{b^*} \rangle \langle \partial_i \Lambda^{b^*}{}_a \rangle - \langle \partial_i \partial^{i^*} \Lambda^{a^*}{}_a \rangle] \langle F^i \rangle \,, \\
R^{a^*}{}_a &= \langle \Lambda^{a^*}{}_a \rangle - \frac{1}{m^\dagger_{\frac{3}{2}}} \langle F^i \rangle \langle \partial_i \Lambda^{a^*}{}_a \rangle \,, \\
A &= \frac{1}{m^\dagger_{\frac{3}{2}}} \langle F^i \rangle \langle \tilde{\rho}_i \rangle \,, \\
\mu &= m_{\frac{3}{2}} \langle Z \rangle - \langle F^\dagger_{i^*} \rangle \langle \partial^{i^*} Z \rangle \,, \\
B &= \frac{1}{m^\dagger_{\frac{3}{2}}} \frac{\langle (2m^2_{\frac{3}{2}} + m_{\frac{3}{2}} F^i \partial_i - m^\dagger_{\frac{3}{2}} F^\dagger_{i^*} \partial^{i^*} - F^i F^\dagger_{i^*} \partial_i \partial^{i^*}) Z \rangle}{\langle (m_{\frac{3}{2}} - F^\dagger_{i^*} \partial^{i^*}) Z \rangle}
\end{aligned} \tag{7.19}$$

enable us to regroup them in a compact form. We also introduce the superpotential of the observable sector

$$W_m = \hat{g}_3 + \mu g_2 \tag{7.20}$$

where

$$\hat{g}_3 = \langle W_0 \rangle e^{\frac{1}{2} \frac{\langle \hat{K} \rangle}{m_p^2}} g_3 \,.$$

Finally we define

$$\Lambda = \frac{1}{m_p^2} \langle F^\dagger_{i^*} \rangle \langle \hat{K}^{i^*}{}_i \rangle \langle F^i \rangle - 3 m^2_{\frac{3}{2}} \,, \tag{7.21}$$

the cosmological constant. Regrouping all terms and keeping only the non Planck-suppressed terms we obtain [Giudice and Masiero (1988)]

$$\begin{aligned}
V &= m_p^2 \Lambda + \partial_a W_m \langle (\Lambda^{-1})^a{}_{a^*} \rangle \partial^{a^*} W^\star_m + \phi^\dagger_{a^*} \Big(m^2_{\frac{3}{2}} S^{a^*}{}_a + \Lambda \langle \Lambda^{a^*}{}_a \rangle \Big) \phi^a \\
&\quad + m^\dagger_{\frac{3}{2}} \Big\{ \langle (\Lambda^{-1})^a{}_{a^*} \rangle R^{a^*}{}_b \phi^b \partial_a W_m + \big(A - 3 \big) \hat{g}_3 + \Big[(B-2)\mu + \frac{1}{m^\dagger_{\frac{3}{2}}} \Lambda \langle Z \rangle \Big] g_2 \Big\} + \text{h.c.} \,.
\end{aligned} \tag{7.22}$$

Note that since in the expansion of $(\mathcal{D}_i W (K^{-1})^i{}_{i^*} \mathcal{D}^{i^*} W^\star$ and in the expansion of $-3/m_p^2 |W|^2$ there are terms of order m_p^2 (see the cosmological constant) there is an additional contribution coming from the expansion of the exponential factor (the term $1/m_p^2 \widetilde{K}$ being of order m_p^{-2}). Note that in general it is assumed that $\Lambda = 0$ in order to have a vanishing cosmological constant. In this case these additional terms vanish. The form of the potential above can be decomposed into two parts:

(1) $V_{\rm susy} = \partial_a W_m \langle (\Lambda^{-1})^a{}_{a^*}\rangle \partial^{a^*} W^\star_m$ corresponding to the potential of a flat supersymmetric model with superpotential $W_m = \hat{g}_3 + \mu g_2$ and Kähler potential $K = \Phi^\dagger \langle \Lambda \rangle \Phi$.

(2) Softly broken terms:

$$V_{\rm soft} = \phi^\dagger_{a^*}\Big(m^2_{\frac{3}{2}} S^{a^*}{}_a + \Lambda \langle \Lambda^{a^*}{}_a\rangle\Big)\phi^a$$
$$+ m^\dagger_{\frac{3}{2}}\Big\{\langle (\Lambda^{-1})^a{}_{a^*}\rangle R^{a^*}{}_b \phi^b \partial_a W_m + \Big(A-3\Big)\hat{g}_3 + (B-2)\mu + \frac{1}{m^\dagger_{\frac{3}{2}}} \Lambda Z) g_2\Big\} + \text{h.c.}$$

i.e., terms which explicitly break supersymmetry.

A direct inspection of the potential shows that there are no dangerous coupling terms.

We turn now to the more general renormalisable theory. With the substitution (7.17) the functions (7.19) now read

$$\begin{aligned}
\mu &\to m_{\frac{3}{2}}\langle Z_{ab}\rangle - \langle F^\dagger_{i^*}\rangle \langle \partial^{i^*} Z_{ab}\rangle\,,\\
A &\to \frac{1}{m^\dagger_{\frac{3}{2}}} F^i\Big(\frac{1}{m_p^2}\langle \partial_i \hat{K}\rangle + \frac{\frac{1}{6}\langle \partial_i \lambda_{abc}\rangle \phi^a\phi^b\phi^c + \frac{1}{2}\langle \partial_i m_{ab}\rangle \phi^a \phi^b}{\frac{1}{6}\langle \lambda_{def}\rangle \phi^d\phi^e\phi^f + \frac{1}{2}\langle m_{de}\rangle \phi^d\phi^e}\Big)
\end{aligned} \tag{7.23}$$

$$(B-2)\mu \to \frac{1}{m^\dagger_{\frac{3}{2}}}\Big(m^\dagger_{\frac{3}{2}}\langle F^\dagger_{i^*}\rangle\langle \partial^{i^*} Z_{ab}\rangle + m_{\frac{3}{2}}\langle F^i\rangle \langle \partial_i Z_{ab}\rangle - \langle F^i F^\dagger_{i^*}\rangle \langle \partial_i \partial^{i^*} Z_{ab}\rangle\Big)\,,$$

with S and R as in (7.19). Within this substitution the scalar potential is easily obtained from (7.22) [Brignole et al. (1998)] in the form

$$V = m_p^2 \Lambda + \partial_a W_m (\Lambda^{-1})^a{}_{a^*} \partial_{a^*} W^\star_m + V_{\rm soft}\,,$$

where the cosmological constant is given in (7.21). The superpotential for unbroken supersymmetry W_m is given by

$$W_m = \frac{1}{6}\hat{\lambda}_{abc}\phi^a\phi^b\phi^c + \frac{1}{2}(\hat{m}_{ab} + m_{\frac{3}{2}}\langle Z_{ab}\rangle - F^\dagger_{i^*}\langle \partial^{i^*} Z_{ab}\rangle)\phi^a\phi^b\,,$$

with

$$\begin{aligned}
\hat{\lambda}_{abc} &= e^{\frac{1}{2m_p^2}\langle \hat{K}\rangle} \langle \lambda_{abc}\rangle\,,\\
\hat{m}_{ab} &= e^{\frac{1}{2m_p^2}\langle \hat{K}\rangle} \langle m_{ab}\rangle\,,
\end{aligned}$$

and the soft supersymmetric breaking terms by

$$V_{\rm soft} = \phi^\dagger_{a^*}(m^2_{\frac{3}{2}} S^{a^*}{}_a + \Lambda\langle \Lambda^a{}_{a^*}\rangle)\phi^a + \frac{1}{6}A_{abc}\phi^a\phi^b\phi^c + \frac{1}{2}B_{ab}\phi^a\phi^b + \text{h.c.}\,,$$

with

$$A_{abc} = e^{\frac{1}{2}\frac{\langle\hat{K}\rangle}{m_p^2}} F^i \Bigg[\frac{1}{m_p^2} \langle \partial_i \hat{K} \rangle \langle \lambda_{abc} \rangle + \langle \partial_i \lambda_{abc} \rangle$$
$$- \Big(\langle (\Lambda^{-1})^d{}_{a^*} \partial_i \Lambda^{a^*}{}_a \lambda_{dbc} \rangle + (a \leftrightarrow b) + (a \leftrightarrow c) \Big) \Bigg] , \tag{7.24}$$

and

$$B_{ab} = e^{\frac{1}{2}\frac{\langle\hat{K}\rangle}{m_p^2}} F^i \Bigg[\frac{1}{m_p^2} \langle \partial_i \hat{K} \rangle \langle m_{ab} \rangle + \langle \partial_i m_{ab} \rangle - \Big(\langle (\Lambda^{-1})^c{}_{a^*} \partial_i \Lambda^{a^*}{}_b m_{ac} \rangle + (a \leftrightarrow b) \Big) \Bigg]$$
$$- m^\dagger_{\frac{3}{2}} e^{\frac{1}{2}\frac{\langle\hat{K}\rangle}{m_p^2}} \langle m_{ab} \rangle + (2 m^2_{\frac{3}{2}} + \Lambda) \langle Z_{ab} \rangle - m^\dagger_{\frac{3}{2}} F^\star_{i^*} \langle \partial^{i^*} Z_{ab} \rangle$$
$$+ m_{\frac{3}{2}} F^i \Bigg[\langle \partial_i Z_{ab} \rangle - \Big(\langle (\Lambda^{-1})^c{}_{a^*} \partial_i \Lambda^{a^*}{}_b Z_{ac} \rangle + (a \leftrightarrow b) \Big) \Bigg]$$
$$- F^i F^\star_{i^*} \Bigg[\langle \partial_i \partial^{i^*} Z_{ab} \rangle - \Big(\langle (\Lambda^{-1})^c{}_{a^*} \partial_i \Lambda^{a^*}{}_b \partial^{i^*} Z_{ac} \rangle + (a \leftrightarrow b) \Big) \Bigg] . \tag{7.25}$$

The general form of renormalisable softly broken supersymmetry is summarised in Tables 7.1 and 7.2.

The last soft supersymmetric breaking terms are the gaugino masses which come from the term (we denote A the gauge indices not to be confused with the observable sector indices)

$$\frac{1}{4} e^{\frac{1}{2}\frac{K}{m_p^2}} (K^{-1})^I{}_{I^*} \Big(\mathcal{D}^{I^*} W^\star \frac{\partial f_{AB}}{\partial X^I} \lambda^A \cdot \lambda^B + \text{h.c.} \Big) ,$$

where $f_{AB}(H, \Phi)$ is the gauge kinetic function. The gaugino mass turns out to be

$$m_{AB} = \frac{1}{2} F^i \langle \partial_i f_{AB} \rangle |_{\phi^a = 0} .$$

If the gauge group decomposes into semisimple factors $\mathfrak{g} = \oplus_\alpha \mathfrak{g}_\alpha$ in order not to break gauge symmetry we further assume $f_{AB} \to f_\alpha$, *i.e.*, one kinetic function for each semisimple part and the corresponding gaugino mass of each semisimple factor is given by

$$m_\alpha = \frac{1}{2} F^i \langle \partial_i f_\alpha \rangle |_{\phi^i = 0} . \tag{7.26}$$

To conclude this section we would like to make several remarks. Indeed, it should be emphasised that any mechanism of supergravity breaking leads to soft-supersymmetric breaking terms[3]: scalar and gauginos masses, trilinear, bilinear

[3] In fact recently new mechanisms leading to hard breaking terms were discovered (see Sec. 7.2.3)

Table 7.1 The general form of supersymmetric soft breaking terms.

For a renormalisable theory the general form of the Kähler potential and superpotential are given by

$$K(H,H^\dagger,\Phi,\Phi^\dagger) = \hat{K}(H,H^\dagger) + \Phi^\dagger_{a^*}\Lambda^{a^*}{}_a(H,H^\dagger)\Phi^a + \Big(\frac{1}{2}Z_{ab}(H,H^\dagger)\Phi^a\Phi^b + \text{h.c.}\Big),$$
$$W(H,\Phi) = \hat{W}(H) + \frac{1}{2}m_{ab}(H)\Phi^a\Phi^b + \frac{1}{6}\lambda_{abc}(H)\Phi^a\Phi^b\Phi^c\,,$$

where Φ are the fields of the observable sector and H the fields of the hidden sector. After supersymmetry breaking in the hidden sector, the scalar potential takes the form

$$V = m_p^2\Lambda + V_{\text{susy}} + V_{\text{soft}}\,.$$

The cosmological constant reads:

$$\Lambda = \frac{1}{m_p^2}\langle F^\dagger_{i^*}\rangle\langle \hat{K}^{i^*}{}_i\rangle\langle F^i\rangle - 3m^2_{\frac{3}{2}}\,,$$

the superpotential for unbroken supersymmetry reduces to

$$V_{\text{susy}} = \frac{1}{6}\hat{\lambda}_{abc}\phi^a\phi^b\phi^c + \frac{1}{2}(\hat{m}_{ab} + m_{\frac{3}{2}}\langle Z_{ab}\rangle - F^\dagger_{i^*}\langle\partial^{i^*}Z_{ab}\rangle)\phi^a\phi^b\,,$$

with

$$\hat{\lambda}_{abc} = e^{\frac{1}{2m_p^2}\langle\hat{K}\rangle}\langle\lambda_{abc}\rangle\,,$$
$$\hat{m}_{ab} = e^{\frac{1}{2m_p^2}\langle\hat{K}\rangle}\langle m_{ab}\rangle\,.$$

and linear terms for the scalar potential. Moreover, these terms are in direct correspondence with the various functions of the hidden sector appearing in the superpotential and in the Kähler potential. Finally, it is important to mention that all the soft-supersymmetric breaking terms are obtained at high energy, namely at the energy where supergravity is broken. At low energy their values can be computed using the renormalisation group equations (see App. G).

7.2.3 *Soft supersymmetric breaking terms versus hard supersymmetric breaking terms*

We have analysed gravity induced supersymmetry breaking in detail and obtained the general form of soft supersymmetric breaking terms. Considering other mechanisms which break supersymmetry in the same manner, leads automatically to:

$$\lim_{m_p\to\infty}\Big(\mathcal{L}_{\text{sugra}} - \mathcal{L}_{2,3/2}\Big) = \mathcal{L}_{\text{susy}} + \mathcal{L}_{\text{soft}}\,, \tag{7.27}$$

in the limit when the Planck mass goes to infinity. That is in this limit we obtain a theory of pure supergravity (involving the graviton and the gravitino $\mathcal{L}_{2,3/2}$ which

Table 7.2 The general form of supersymmetric soft breaking terms cont'd.

The soft supersymmetric breaking terms are:

$$V_{\text{soft}} = \phi^\dagger_{a^*}(m^2_{\frac{3}{2}} S^{a^*}{}_a + \Lambda\langle\Lambda^a{}_{a^*}\rangle)\phi^a + \frac{1}{6}A_{abc}\phi^a\phi^b\phi^c + \frac{1}{2}B_{ab}\phi^a\phi^b + \text{h.c.}\,,$$

with

$$\begin{aligned}
A_{abc} &= e^{\frac{1}{2}\frac{\langle \hat{K}\rangle}{m_p^2}}\langle F^i\rangle\Bigg[\frac{1}{m_p^2}\langle\partial_i\hat{K}\rangle\langle\lambda_{abc}\rangle + \langle\partial_i\lambda_{abc}\rangle \\
&\qquad -\Big(\langle(\Lambda^{-1})^d{}_{a^*}\partial_i\Lambda^{a^*}{}_a\lambda_{dbc}\rangle + (a\leftrightarrow b) + (a\leftrightarrow c)\Big)\Bigg]\,,\\
B_{ab} &= e^{\frac{1}{2}\frac{\langle \hat{K}\rangle}{m_p^2}}\langle F^i\rangle\Bigg[\frac{1}{m_p^2}\langle\partial_i\hat{K}\rangle\langle m_{ab}\rangle + \langle\partial_i m_{ab}\rangle - \Big(\langle(\Lambda^{-1})^c{}_{a^*}\partial_i\Lambda^{a^*}{}_b m_{ac}\rangle + (a\leftrightarrow b)\Big)\Bigg]\\
&\quad -m^\dagger_{\frac{3}{2}}e^{\frac{1}{2}\frac{\langle \hat{K}\rangle}{m_p^2}}\langle m_{ab}\rangle + (2m^2_{\frac{3}{2}} + \Lambda)\langle Z_{ab}\rangle - m^\dagger_{\frac{3}{2}}\langle F^\star_{i^*}\rangle\langle\partial^{i^*}Z_{ab}\rangle\\
&\quad +m_{\frac{3}{2}}\langle F^i\rangle\Big[\langle\partial_i Z_{ab}\rangle - \Big(\langle(\Lambda^{-1})^c{}_{a^*}\partial_i\Lambda^{a^*}{}_b Z_{ac}\rangle + (a\leftrightarrow b)\Big)\Big]\\
&\quad -\langle F^i F^\star_{i^*}\rangle\Big[\langle\partial_i\partial^{i^*}Z_{ab}\rangle - \Big(\langle(\Lambda^{-1})^c{}_{a^*}\partial_i\Lambda^{a^*}{}_b\partial^{i^*}Z_{ac}\rangle + (a\leftrightarrow b)\Big)\Big]\,,\\
S^{a^*}{}_a &= \langle\Lambda^{a^*}{}_a\rangle + \frac{1}{m^2_{\frac{3}{2}}}\langle F^\dagger_{i^*}\rangle[\langle\partial^{i^*}\Lambda^{a^*}{}_b\rangle\langle(\Lambda^{-1})^b{}_{b^*}\rangle\langle\partial_i\Lambda^{b^*}{}_a\rangle - \langle\partial_i\partial^{i^*}\Lambda^{a^*}{}_a\rangle]\langle F^i\rangle\,.
\end{aligned}$$

decouples to a supersymmetric theory $\mathcal{L}_{\text{susy}}$ including additional soft supersymmetric breaking terms $\mathcal{L}_{\text{soft}}$, *i.e.*, terms which break supersymmetry explicitly, but *only in a soft way.*

This low energy limit softly broken supersymmetry is obtained, as said previously, by considering a hidden sector, where supergravity is broken, and an observable sector (describing a supersymmetric extension of the Standard Model, see *e.g.*, Chapter 8). The interaction of the hidden sector with the observable sector communicates supersymmetry breaking to the observable sector, and a Lagrangian of the form (7.27) is automatically generated as a consequence. An example of such a mechanism was studied in previous sections.

However, considering (7.27) it is not obvious that the limit is consistent with particle physics. It could happen that the coupling of the hidden sector to the observable sector leads to dangerous terms. For instance, all terms involving coupling in the observable sector must have a finite limit when $m_p \to \infty$. This strong requirement has been analysed in Soni and Weldon (1983). In their analysis they first made a clear distinction between the hidden sector with potential vacuum expectation values of the order of the Planck mass and the observable sector with potential vacuum expectation values much below the Planck scale (for instance of the GUT, or of the electroweak scale). Assuming to have other scales than the Planck scale, Soni and Weldon performed an expansion of the Kähler potential and the superpotential

$$\begin{aligned} K(H,\Phi,H^\dagger,\Phi^\dagger) &= \sum_{n=1}^{N} m_p^n K_n(H,\Phi,H^\dagger,\Phi^\dagger)\,, \\ W(H,\Phi) &= \sum_{m=1}^{M} m_p^m W_m(H,\Phi)\,. \end{aligned} \tag{7.28}$$

Recall that H denotes the fields in the hidden sector and Φ the fields in the observable sector. Imposing the physical requirement stated above, they obtained the general form for the expansion of the superpotential and the Kähler potential

$$\begin{aligned} K(H,\Phi,H^\dagger,\Phi^\dagger) &= m_p^2 K_2(H,H^\dagger) + m_p K_1(H,H^\dagger) + K_0(H,\Phi,H^\dagger,\Phi^\dagger)\,, \\ W(H,\Phi) &= m_p^2 W_2(H) + m_p W_1(H) + W_0(H,\Phi)\,, \end{aligned} \tag{7.29}$$

namely that the observable sector fields do appear only in the m_p to the power zero components. This form automatically forbids dangerous coupling terms when taking the limit $m_p \to \infty$. Moreover, after breaking supergravity the form of the superpotential and Kähler potential in (7.29) leads only to a softly broken supersymmetry. At this point this was seen as an advantage in the construction of physically acceptable phenomenological models, since after supersymmetry breaking the masses of the supersymmetric partners of the particles of the Standard Model (see Chapter 8) are of the order of the TeV scale, and so does not spoil the solution to the hierarchy problem. Then all pioneer works were based upon (7.28) [Chamseddine et al. (1982); Barbieri et al. (1982b); Ibañez (1982); Ohta (1983); Ellis et al. (1983); Alvarez-Gaumé et al. (1983); Polonyi (1977); Cremmer et al. (1983a); Nilles et al. (1983); Nilles (1983); Soni and Weldon (1983); Giudice and Masiero (1988); Nilles (1984); Hall et al. (1983)].

However, the absence of direct experimental signatures of a low energy broken supersymmetry may ultimately question, if not its existence, at least the way this symmetry is realised in Nature. In particular, since no deviations from the Standard Model predictions have been obtained in colliders, the analysis of Soni Weldon has been revisited with a critical eye and was taken up again in Moultaka et al. (2019a). In this analysis the physical requirement of having no dangerous terms translates into a tower of differential equations. Solving these equations by taking into consideration the holomorphy condition of the superpotential, makes it possible to classify all possible solutions. Of course the Soni and Weldon solutions were re-obtained, but a bunch of new solutions were discovered. In particular, the full classification of the solutions compatible with low energy physics was then obtained, at least in the case of a flat Kähler manifold, and a couple of new solutions (but not the full classification) were identified for curved Kähler manifold (some of the new solutions share interesting properties with no-scale supergravity, –See Sec. 7.3) [Moultaka et al. (2019a)]. Some of the solutions lead to a new class of models where a third type of hybrid superfield was identified, which is neither in the hidden sector, nor in the observable sector. This class of models bears several notable differences compared to the usual class:

(1) there are new types of terms breaking supersymmetry explicitly called usually "hard-breaking terms" since they lead to quadratic divergences if included in loop corrections . These new terms couple the new-fields to the observable sector fields, but in fact these terms are parametrically small;
(2) the gravitino mass, which is the typical scale of supersymmetry breaking, can be much smaller compared to the usual gravity mediated scenario;
(3) the usual hidden sector fields have large masses, the new (hybrid) type of fields have weak-scale masses and can couple to the observable fields.

Whereas this new class of theories spoils the solution to the hierarchy problem because of the presence of "hard-breaking terms", they are worth to be examined since they have potentially different phenomenologies and experimental signatures. In particular these types of models could explain some features which are not currently explained (or at least only with difficulty) within softly broken supersymmetry.

7.3 No-scale supergravity

As we have seen in the previous section, gravity mediated supersymmetry breaking suffers from a fine-tuning problem. Indeed, we need to adjust some parameters in order to have a vanishing cosmological constant (see Eq. [7.21]). There are however some supergravity models, termed no-scale supergravity models, where the vanishing of the cosmological constant is just a geometric constraint upon the Kähler potential. In fact, the $F-$part of the scalar potential vanishes identically. This means that the potential presents some flat directions, and consequently the gravitino mass is not specified classically but only at the quantum level. These models were introduced in [Cremmer et al. (1983c); Ellis et al. (1984b); Lahanas and Nanopoulos (1987)] with some applications in the construction of physical models. In section 5.3.3 we have seen that a Howe–Tucker transformation allows the substitutions of the Kähler potential and of the superpotential

$$\begin{aligned} K &\to \mathcal{G} = K + m_p^2 \ln \frac{|W|^2}{m_p^6}\,, \\ W &\to m_p^3\,. \end{aligned} \tag{7.30}$$

This transformation implies also that

$$\begin{aligned} \mathcal{D}_i W &\to m_p \mathcal{G}_i\,, \\ \mathcal{D}_i \mathcal{D}_j W &\to m_p \mathcal{G}_{ij} - \frac{1}{m_p}\mathcal{G}_i \mathcal{G}_j - m_p \mathcal{G}_k \mathcal{G}^k{}_{k^*} \mathcal{G}_i{}^{k^*}{}_j\,. \end{aligned} \tag{7.31}$$

The generalised Kähler potential is now a combination of the Kähler potential and the superpotential. It is a simple computation to show that the $F-$part of the potential takes the form

$$V = m_p^2 e^{\frac{1}{m_p^2}\mathcal{G}} \Big(\mathcal{G}_i \mathcal{G}^i{}_{i^*} \mathcal{G}^{i^*} - 3m_p^2 \Big)\,,$$

where we have used the notations $(\mathcal{G}^{-1})^i{}_{i^*} \equiv \mathcal{G}^i{}_{i^*}$. To ensure that the cosmological constant vanishes, we impose that for any field configuration the potential vanishes, *i.e.*,

$$V = 0 \quad \text{for all field configuration}$$

whose solution is

$$\mathcal{G}_i \mathcal{G}^i{}_{i^*} \mathcal{G}^{i^*} = 3m_p^2\,. \tag{7.32}$$

We have thus substituted the (not natural) fine tuning problem into a (natural) geometrical condition. Since in such supergravity theories there are no mass scale but the Planck mass, these types of models are called no-scale models. Using the transformation (7.30), the gravitino mass reads (see also Sec. 6.4)

$$m_{\frac{3}{2}} = m_p \exp\Big(\frac{1}{2}\frac{1}{m_p^2}\mathcal{G}\Big)\,.$$

Similarly, using (7.4) and (7.5) the goldstino takes the form (we assume that only F-terms develop vacuum expectation values)

$$\psi_G = m_p \frac{1}{\sqrt{2}} \exp\Big(\frac{1}{2}\frac{1}{m_p^2}\mathcal{G}\Big)\, \mathcal{G}_i \chi^i\,. \tag{7.33}$$

A signature of broken supersymmetry then is a non-vanishing $\mathcal{G}_i$ at the minimum of the potential.

We consider now an example of a Kähler manifold having the property (7.32). Just like in the previous section, it is constituted by a hidden sector and an observable sector. We denote $Z = \Phi^0$ the chiral superfield of the hidden sector and

$\Phi^i, i = 1, \cdots, n$ the chiral superfields of the observable sector. We introduce the notation $I = (0, i)$. For the generalised Kähler potential we chose [Ellis et al. (1984b); Lahanas and Nanopoulos (1987)],

$$\mathcal{G} = -3m_p^2 \ln\left[\frac{\phi^0 + \phi_{0^*}^\dagger}{m_p} - \frac{h(\phi^i, \phi_{i^*}^\dagger)}{m_p^2}\right],$$

where h is an unspecified real function related to the interactions in the observable sector. In order to facilitate the computation, from now on we set again $m_p = 1$. Remembering that $[\Phi] = M, [h] = M^2$, the Planck mass can be reintroduced easily by dimensional arguments. A simple computation shows that

$$\begin{aligned} \mathcal{G}_0 &= -3e^{\frac{1}{3}\mathcal{G}}\,, \\ \mathcal{G}_i &= 3h_i e^{\frac{1}{3}\mathcal{G}} = -h_i \mathcal{G}_0\,, \end{aligned} \tag{7.34}$$

and

$$\begin{aligned} \mathcal{G}^{0^*}{}_0 &= 3e^{\frac{2}{3}\mathcal{G}}\,, \\ \mathcal{G}^{0^*}{}_i &= -3h_i e^{\frac{2}{3}\mathcal{G}}\,, \\ \mathcal{G}^{i^*}{}_0 &= -3h^{i^*} e^{\frac{2}{3}\mathcal{G}}\,, \\ \mathcal{G}^{i^*}{}_i &= 3h_i h^{i^*} e^{\frac{2}{3}\mathcal{G}} + 3h^{i^*}{}_i e^{\frac{1}{3}\mathcal{G}}\,. \end{aligned}$$

Thus, we have

$$\mathcal{G}^{I^*}{}_I = 3e^{\frac{2}{3}\mathcal{G}} \begin{pmatrix} 1 & 0 \\ -h^{i^*} & -h^{i^*}{}_j \end{pmatrix} \begin{pmatrix} 1 & 0 \\ 0 & e^{-\frac{1}{3}\mathcal{G}} h^j{}_{j^*} \end{pmatrix} \begin{pmatrix} 1 & -h_i \\ 0 & -h^{j^*}{}_i \end{pmatrix}.$$

Recalling that for

$$P = \begin{pmatrix} 1 & v^t \\ 0 & M \end{pmatrix},$$

we have

$$P^{-1} = \begin{pmatrix} 1 & -v^t M^{-1} \\ 0 & M^{-1} \end{pmatrix},$$

we obtain,

$$\mathcal{G}^I{}_{I^*} = \frac{1}{3} e^{-\frac{2}{3}\mathcal{G}} \begin{pmatrix} 1 & -h_k h^k{}_{j^*} \\ 0 & -h^i{}_{j^*} \end{pmatrix} \begin{pmatrix} 1 & 0 \\ 0 & e^{\frac{1}{3}\mathcal{G}} h^{j^*}{}_j \end{pmatrix} \begin{pmatrix} 1 & 0 \\ -h^{k^*} h^j{}_{k^*} & -h^j{}_{i^*} \end{pmatrix}, \tag{7.35}$$

and

$$
\begin{aligned}
\mathcal{G}^0{}_{0^*} &= \frac{1}{3}e^{-\frac{2}{3}\frac{1}{m_P^2}\mathcal{G}} + \frac{1}{3}h_j h^{j^*} h^j{}_{j^*} e^{-\frac{1}{3}\frac{1}{m_P^2}\mathcal{G}} \\
&= 3\mathcal{G}_0^{-2} - \mathcal{G}_j \mathcal{G}^{j^*} h^j{}_{j^*} \mathcal{G}_0^{-3} , \\
\mathcal{G}^0{}_{i^*} &= \frac{1}{3}h_j h^j{}_{i^*} e^{-\frac{1}{3}\frac{1}{m_P^2}\mathcal{G}} = h^j{}_{i^*}\mathcal{G}_j \mathcal{G}_0^{-2} , \\
\mathcal{G}^i{}_{0^*} &= \frac{1}{3}h^{j^*} h^i{}_{j^*} e^{-\frac{1}{3}\frac{1}{m_P^2}\mathcal{G}} = h^i{}_{j^*}\mathcal{G}^{j^*}\mathcal{G}_0^{-2} , \\
\mathcal{G}^i{}_{i^*} &= -\frac{1}{3}h^i{}_{i^*} e^{-\frac{1}{3}\frac{1}{m_P^2}\mathcal{G}} = -h^i{}_{i^*}\mathcal{G}_0^{-1} ,
\end{aligned}
\tag{7.36}
$$

where $(h^{-1})^i{}_{i^*} \equiv h^i{}_{i^*}$ since there is no ambiguity. Thus, we get,

$$
\mathcal{G}_I G^I{}_{I^*} = -e^{-\frac{1}{3}\mathcal{G}}\delta^{0^*}{}_{I^*} ,
$$

and then from (7.34) the desired condition (7.32). Consequently, the generalised Kähler potential (7.30) satisfies the requirement of no-scale supergravity. Now it is of interest to study the mass matrix of the fermions. It is given by (see Table 6.3 with (7.31))

$$
\begin{aligned}
\mathcal{L}_{\text{mass}} &= e e^{\frac{1}{2}\mathcal{G}}\Big([\mathcal{G}_{IJ} + \mathcal{G}_I\mathcal{G}_J - \mathcal{G}_K\mathcal{G}^K{}_{K^*}\mathcal{G}_I{}^{K^*}{}_J]\chi^I\cdot\chi^J + \text{h.c.}\Big) \\
&= e e^{\frac{1}{2}\mathcal{G}}\Big(m_{IJ}\chi^I\cdot\chi^J + \text{h.c.}\Big) .
\end{aligned}
\tag{7.37}
$$

From

$$
\begin{aligned}
\mathcal{G}_{00} &= \frac{1}{3}\mathcal{G}_0\mathcal{G}_0 , \\
\mathcal{G}_{i0} &= \frac{1}{3}\mathcal{G}_0\mathcal{G}_i , \\
\mathcal{G}_{ij} &= \frac{1}{3}\mathcal{G}_i\mathcal{G}_j - h_{ij}\mathcal{G}_0 ,
\end{aligned}
$$

and

$$
\begin{aligned}
\mathcal{G}_0{}^{0^*}{}_0 &= \frac{2}{9}\mathcal{G}_0\mathcal{G}_0\mathcal{G}^{0^*} , \\
\mathcal{G}_0{}^{i^*}{}_0 &= \frac{2}{9}\mathcal{G}_0\mathcal{G}_0\mathcal{G}^{i^*} , \\
\mathcal{G}_0{}^{0^*}{}_i &= \frac{2}{9}\mathcal{G}_0\mathcal{G}_i\mathcal{G}^{0^*} , \\
\mathcal{G}_0{}^{j^*}{}_i &= \frac{2}{9}\mathcal{G}_0\mathcal{G}_i\mathcal{G}^{j^*} - \frac{1}{3}h^{j^*}{}_i\mathcal{G}_0\mathcal{G}_0 , \\
\mathcal{G}_i{}^{0^*}{}_j &= \frac{2}{9}\mathcal{G}_i\mathcal{G}_j\mathcal{G}^{0^*} - \frac{1}{3}h_{ij}\mathcal{G}_0\mathcal{G}^{0^*} , \\
\mathcal{G}_i{}^{k^*}{}_j &= \frac{2}{9}\mathcal{G}_i\mathcal{G}_j\mathcal{G}^{k^*} - \frac{1}{3}[h_{ij}\mathcal{G}^{k^*} + h^{k^*}{}_i\mathcal{G}_j + h^{k^*}{}_j\mathcal{G}_i]\mathcal{G}_0 - h_i{}^{k^*}{}_j\mathcal{G}_0 ,
\end{aligned}
\tag{7.38}
$$

together with (7.36) rewritten in a more appropriate form

$$\begin{aligned}
\mathcal{G}^0{}_{0^*} &= 3(\mathcal{G}^0)^{-1}(\mathcal{G}_{0^*})^{-1} - \mathcal{G}_j \mathcal{G}^{j^*} h^j{}_{j^*}(\mathcal{G}_0)^{-3} \,, \\
\mathcal{G}^0{}_{i^*} &= h^j{}_{i^*} G_j (\mathcal{G}_0)^{-2} \,, \\
\mathcal{G}^i{}_{0^*} &= h^i{}_{j^*} \mathcal{G}^{j^*}(\mathcal{G}_0)^{-2} \,, \\
\mathcal{G}^i{}_{i^*} &= -h^i{}_{i^*}(\mathcal{G}_0)^{-1} \,,
\end{aligned} \tag{7.39}$$

and (7.34) (noting that $\mathcal{G}_0 = \mathcal{G}^{0^*}$) we obtain after a bit of calculus

$$m_{IJ} = \frac{2}{3}\mathcal{G}_I \mathcal{G}_J \,, \tag{7.40}$$

and thus

$$\begin{aligned}
\mathcal{L}_{\text{mass}} &= \frac{1}{3} e^{\frac{1}{2}\mathcal{G}}\big(\mathcal{G}_I \mathcal{G}_J \chi^I \cdot \chi^J + \text{h.c.}\big) \\
&= \frac{2}{3} e^{-\frac{1}{2}\mathcal{G}}\big(\psi_G \cdot \psi_G + \bar{\psi}_G \cdot \bar{\psi}_G\big) \,,
\end{aligned}$$

where ψ_G is the goldstino given in (7.33). This means that the fermions-fermions interaction can be eliminated by an appropriate supergravity transformation (see Sec. 7.1). In addition this implies that there are no mass splittings between the fermions and the bosons when supersymmetry is broken. Consequently, these types of model are not appropriate in particles physics.

To obtain more realistic models, we now define the generalised Kähler potential

$$G = \mathcal{G} + F + F^\dagger = -3\ln\left[\Phi^0 + \Phi_{0^*} - h(\Phi, \Phi^\dagger)\right] + F(\Phi^i) + F^\dagger(\Phi_i^\dagger) \,.$$

Note that this new definition of G is *not a Kähler transformation*. The holomorphic functions F just reproduces interactions. However, when performing the Kähler transformation $G \to G - F - F^\star = \mathcal{G}$, this function can now be understood as the superpotential

$$W(\Phi) = e^{F(\Phi)} \,.$$

We consider the case of a diagonal function h and use the notations of Sec. 6.5 for the anti-chiral superfields. We also assume that the function F depends only on the observable sector. We note that $G^I{}_J = \mathcal{G}^I{}_J$, $(G^{-1})^I{}_J = (\mathcal{G}^{-1})^I{}_J$, $G_I = \mathcal{G}_I + F_I$ with $F_0 = 0$ and that $G_{IJ} = \mathcal{G}_{IJ} + F_{IJ}$ with $F_{0I} = 0$. The potential is then given by

$$\begin{aligned}
e^G\big(G_I(\mathcal{G}^{-1})^I{}_J G^J - 3\big) &= e^G\big(F_I(\mathcal{G}^{-1})^I{}_J F^{\dagger J} + F_I(\mathcal{G}^{-1})^I{}_J \mathcal{G}^J + \mathcal{G}_I(\mathcal{G}^{-1})^I{}_J F^{\dagger J}\big) \\
&= \frac{1}{3} e^G e^{-\frac{1}{3}\mathcal{G}} F_i F^{\dagger i} \,.
\end{aligned}$$

Here we have used the vanishing condition (7.32). Now, if we add the D−terms coming from the gauge sector we get

$$V = \frac{1}{3}e^{G}e^{-\frac{1}{3}\mathcal{G}}F_iF^{\dagger i} + \frac{1}{2}h^{Rab}D_aD_b \,.$$

The potential is therefore positive. If supersymmetry is broken in a hidden sector we have $< F_i >=< D_a >= 0$ and the cosmological constant vanishes. To obtain the mass matrix of the fermionic sector, we use (7.40) to get

$$\begin{aligned} m_{IJ} &= \frac{2}{3}\mathcal{G}_IF_J + F_{IJ} + F_I\mathcal{G}_J + F_J\mathcal{G}_I + F_IF_J - F_k(\mathcal{G}^{-1})^k{}_{K^*}\mathcal{G}_I{}^{K^*}{}_J \\ &= \frac{2}{3}(\mathcal{G}_I + F_I)(\mathcal{G}_J + F_J) + \frac{1}{3}F_IF_J + F_{IJ} \,. \end{aligned}$$

Here we have used (7.38) and (7.39). Subtracting the goldstino

$$\eta = (\mathcal{G}_I + F_I)\chi^I \,,$$

along the lines of the next section the fermion mass-term can be obtained [Ellis et al. (1984b)]:

$$m_{IJ} = e^{\frac{1}{2}G}F_{IJ} \,.$$

Up to now we did not study the kinetic part of the scalars and of the fermions. To proceed, now we examine

$$h(\phi,\phi^\dagger) = \frac{1}{3}\phi^i\phi^\dagger_i \,.$$

For the scalar fields we have

$$\begin{aligned} \frac{1}{e}\mathcal{L}_{\text{scal.}} &= \mathcal{G}^J{}_I\partial_\mu\phi^\dagger_J\partial^\mu\phi^I \\ &= 3e^{\frac{2}{3}\mathcal{G}}\Big[3\partial_\mu\phi^\dagger_0\partial^\mu\phi^0 - \phi^i\partial_\mu\phi^\dagger_i\partial^\mu\phi^0 - \phi^\dagger_i\partial_\mu\phi^i\partial^\mu\phi^\dagger_0 - \phi^i\phi^\dagger_j\partial_\mu\phi^\dagger_i\partial^\mu\phi^j\Big] \\ &\quad + e^{\frac{1}{3}\mathcal{G}}\partial_\mu\phi^\dagger_i\partial^\mu\phi^i \\ &= e^{\frac{1}{3}\mathcal{G}}\partial_\mu\phi^i\partial^\mu\phi^\dagger_i + \frac{1}{12}\partial_\mu\mathcal{G}\partial^\mu\mathcal{G} \\ &\quad + \frac{3}{4}e^{\frac{2}{3}\mathcal{G}}\Big[\partial_\mu(\phi^0 - \phi^\dagger_0) - \frac{1}{3}(\phi^\dagger_i\partial_\mu\phi^i - \phi^i\partial_\mu\phi^\dagger_i)\Big]\times \\ &\qquad \times\Big[\partial_\mu(\phi^\dagger_0 - \phi^0) - (\phi^j\partial^\mu\phi^\dagger_j - \phi^\dagger_j\partial^\mu\phi^j)\Big] \,. \end{aligned}$$

Assuming that ϕ^0 is in the hidden sector and develops a vacuum expectation value whereas the field in the observable sector, say ϕ^i does not after supersymmetry breaking, and keeping only non-Planck suppressed term, if we take

$$Y_0^R = \sqrt{\frac{1}{6}}\mathcal{G}\,,$$
$$Y_0^I = -i\sqrt{\frac{3}{2}}e^{\frac{1}{3}\langle\mathcal{G}\rangle}\big[\phi^0 - \phi_0^\dagger\big]\,,$$
$$\varphi^i = e^{\frac{1}{6}\langle\mathcal{G}\rangle}\phi^i\,,$$

the kinetic terms are correctly normalised:

$$\frac{1}{e}\mathcal{L}_{\text{scal.}} = \partial_\mu\varphi^i\partial^\mu\varphi_i^\dagger + \frac{1}{2}\partial_\mu Y^R\partial^\mu Y^R + \frac{1}{2}\partial_\mu Y^I\partial^\mu Y^I\,.$$

Likewise for the fermionic sector we have

$$\begin{aligned}
\frac{1}{e}\mathcal{L}_{\text{ferm.}} &= -i\mathcal{G}^I{}_J\chi^J\sigma^\mu\mathcal{D}_\mu\bar{\chi}_J \\
&= -ie^{\frac{2}{3}\mathcal{G}}\Big\{3\chi^0\sigma^\mu\mathcal{D}_\mu\bar{\chi}_0 - \phi^i\chi^0\sigma^\mu\mathcal{D}_\mu\bar{\chi}_i - \phi_i^\dagger\chi^i\sigma^\mu\mathcal{D}_\mu\bar{\chi}_0 + \phi^i\phi_j^\dagger\chi^j\sigma^\mu\mathcal{D}_\mu\bar{\chi}_i\Big\} \\
&\quad -ie^{\frac{1}{3}\mathcal{G}}\chi^i\sigma^\mu\mathcal{D}_\mu\bar{\chi}_i \\
&= -i\frac{1}{3}\big(\mathcal{G}_I\chi^I\big)\mathcal{G}^J\sigma^\mu\mathcal{D}_\mu\bar{\chi}_J - ie^{\frac{1}{3}\mathcal{G}}\chi^i\sigma^\mu\mathcal{D}_\mu\bar{\chi}_i\,.
\end{aligned}$$

After supersymmetry breaking, the first term reduces to (up to the Planck-suppressed terms)

$$\mathcal{G}_I\chi^I \to \langle\mathcal{G}_0\rangle\chi^0\,,$$

i.e., to the Golstino, which can be eliminated, as shown in the next section. Meanwhile the second term, after the appropriate renormalisation

$$\lambda^i = e^{\left\langle\frac{1}{6}\mathcal{G}\right\rangle}\chi^i\,,$$

describes the matter part which is correctly normalised.

These models where studied in the context of SU(5) grand-Unified theories in [Ellis et al. (1984b); Lahanas and Nanopoulos (1987)] or in the context of the Standard Model [Ellis et al. (1984c)]. One can see these references for a deeper analysis.

7.4 Mass matrices

Throughout this section we also implicitly assume that all masses will be given in the Planck units with $m_p = 1$. Now, we consider a generic supergravity theory where supersymmetry is broken by a mechanism or another. We simply assume that some of the auxiliary fields F^i and some of the scalar fields ϕ^i develop a vacuum expectation value (we do not suppose that the D^a's take vev.'s). We further assume a vanishing cosmological constant. Remembering that the potential is given by

$$\begin{aligned} V &= e^{\mathcal{G}}(\mathcal{G}^i\mathcal{G}_i - 3) + \frac{g^2}{8}h^{Rab}\Big(\mathcal{G}^{i^*}(\phi^\dagger T_a)_{i^*} + \mathcal{G}_i(T_a\phi)^i\Big)\Big(\mathcal{G}^{j^*}(\phi^\dagger T_b)_{j^*} + \mathcal{G}_j(T_b\phi)^j\Big) \\ &\equiv V_F + V_D \,, \end{aligned}$$

with $\mathcal{G}^i = \mathcal{G}^i{}_{i^*}\mathcal{G}^{i^*}$, the vanishing of the D–term and of the cosmological constant lead to

$$\begin{aligned} \langle \mathcal{G}^i\mathcal{G}_i\rangle &= 3\,, \\ \langle D^a\rangle &= \frac{g}{2}\langle h^{Rab}(\mathcal{G}^{i^*}(\phi^\dagger T_b)_{i^*} + \mathcal{G}_i(T_b\phi)^i)\rangle = 0\,. \end{aligned} \tag{7.41}$$

Recalling the relations (7.3) that will be used throughout the computation of the derivatives of V_F and V_D, we have that

$$\begin{aligned} \partial_k\mathcal{G}^i{}_{i^*} &= -\mathcal{G}^j{}_{i^*}\Gamma_j{}^i{}_k\,, \\ \partial_k h^{Rab} &= -\frac{1}{2}h^{Rac}h^{Rbd}h_{cdk}\,, \end{aligned} \tag{7.42}$$

and that $\mathcal{G}$ is a gauge singlet (see Sec. 7.1) the minimisation of the potential gives

$$\langle \partial_k V\rangle = \langle e^{\mathcal{G}}(\mathcal{G}^i\mathcal{D}_k\mathcal{G}_i + \mathcal{G}_k)\rangle = 0\,. \tag{7.43}$$

We have explicitly used (7.41). For further use we also mention that

$$\partial_k V_D = g\mathcal{G}^{i^*}{}_k(\phi^\dagger T_a)_{i^*}D^a - \frac{1}{4}h_{abk}D^aD^b\,. \tag{7.44}$$

Now, we compute the scalar mass matrix by taking the second derivative of the scalar potential. We first compute the second derivative of V_F and then of V_D. Using the first equation of (7.41), (7.42) and adding the vanishing term $\langle \Gamma_\ell{}^m{}_k\partial_m V\rangle = 0$ we get

$$\langle \partial_\ell\partial_k V_F\rangle = \langle(\partial_\ell\partial_k - \Gamma_\ell{}^m{}_k\partial_m)V_F\rangle = \langle e^{\mathcal{G}}(\mathcal{G}^i\mathcal{D}_\ell\mathcal{D}_k\mathcal{G}_i + \mathcal{D}_k\mathcal{G}_\ell + \mathcal{D}_\ell\mathcal{G}_k)\rangle\,,$$

and

$$\begin{aligned} \langle \partial^{\ell^*}\partial_k V\rangle &= \langle e^{\mathcal{G}}((\partial^{\ell^*}\mathcal{G}^i)\mathcal{D}_k\mathcal{G}_i + \mathcal{G}^i\partial^{\ell^*}\mathcal{D}_k\mathcal{G}_i + \mathcal{G}^{\ell^*}{}_k)\rangle \\ &= \langle e^{\mathcal{G}}(\mathcal{D}^{\ell^*}\mathcal{G}^i\mathcal{D}_k\mathcal{G}_i - \mathcal{G}^i\mathcal{G}_{i^*}R_k{}^{\ell^*}{}_i{}^{i^*} + \mathcal{G}^{\ell^*}{}_k)\rangle\,. \end{aligned}$$

To simplify the equation above, we have used the expression of the curvature tensor (2.20) and the property $\partial^{\ell^*}\mathcal{G}^i = \mathcal{D}^{\ell^*}\mathcal{G}^i$. The second derivatives of V_D are easier to compute. Using (7.44), the second relation of (7.42) and $\langle D^a\rangle = 0$ we obtain

$$\begin{aligned} \langle \partial_\ell\partial_k V_D\rangle &= g^2\langle h^{Rab}\mathcal{G}^{i^*}{}_k\mathcal{G}^{j^*}{}_\ell(\phi^\dagger T_a)_{i^*}(\phi^\dagger T_b)_{j^*}\rangle\,, \\ \langle \partial^{\ell^*}\partial_k V_D\rangle &= g^2\langle h^{Rab}\mathcal{G}^{i^*}{}_k\mathcal{G}^{\ell^*}{}_i(\phi^\dagger T_a)_{i^*}(T_b\phi)^i\rangle\,. \end{aligned}$$

Adding both contributions and writing the scalar mass matrix in the form

$$\mathcal{M}_0^2 = \begin{pmatrix} (\mathcal{M}_0^2)_i{}^{j^*} & \mathcal{M}_0^2)_{ij} \\ (\mathcal{M}_0^2)^{i^*j^*} & (\mathcal{M}_0^2)^{i^*}{}_j \end{pmatrix},$$

we get

$$(\mathcal{M}_0^2)_{ij} = \langle e^{\mathcal{G}}(\mathcal{G}^k \mathcal{D}_i \mathcal{D}_j \mathcal{G}_k + \mathcal{D}_i \mathcal{G}_j + \mathcal{D}_j \mathcal{G}_i) + g^2 h^{Rab} \mathcal{G}^{k^*}{}_i \mathcal{G}^{\ell^*}{}_j (\phi^\dagger T_a)_{k^*} (\phi^\dagger T_b)_{\ell^*} \rangle ,$$
$$(\mathcal{M}_0^2)_i{}^{i^*} = \langle e^{\mathcal{G}}(\mathcal{D}^{i^*} \mathcal{G}^k \mathcal{D}_i \mathcal{G}_k - \mathcal{G}^k \mathcal{G}_{k^*} R_i{}^{i^*}{}_k{}^{k^*} + \mathcal{G}^{i^*}{}_i) + g^2 h^{Rab} \mathcal{G}^{k^*}{}_i \mathcal{G}^{i^*}{}_k (\phi^\dagger T_a)_{k^*} (T_b \phi)^k \rangle .$$

The trace of $\mathcal{M}_0^2$ is easily obtained as

$$\begin{aligned} \mathrm{Tr}(\mathcal{M}_0^2) &= 2\langle \mathcal{G}^i{}_{i^*} \rangle (\mathcal{M}_0^2)^i{}_{i^*} \\ &= 2m_{\frac{3}{2}}^2 \Big(n + \langle \mathcal{G}^i{}_{i^*} \mathcal{D}^{i^*} \mathcal{G}^j \mathcal{D}_i \mathcal{G}_j \rangle - \langle \mathcal{G}^i \mathcal{G}_{i^*} R_i{}^{i^*} \rangle\Big) \\ &\quad + 2g^2 \langle h^{Rab} \mathcal{G}^{i^*}{}_i (\phi^\dagger T_a)_{i^*} (T_b \phi)^i \rangle , \end{aligned} \tag{7.45}$$

with n the number of chiral multiplets (or scalar fields) and $R_i{}^{i^*} = \mathcal{G}^j{}_{j^*} R_j{}^{j^*}{}_i{}^{i^*}$.

The mass matrix for the gauge bosons is readily obtained from the covariant derivatives as

$$(\mathcal{M}_1^2)_{ab} = 2g^2 \langle \mathcal{G}^{i^*}{}_i (T_a \phi)^i (\phi^\dagger T_b)_{i^*} \rangle ,$$

and its trace reduces to

$$\mathrm{Tr}(\mathcal{M}_1^2) = \langle h^{Rab} \rangle (\mathcal{M}_1^2)_{ab} = 2\langle g^2 h^{Rab} \mathcal{G}^{i^*}{}_i (\phi^\dagger T_a)_{i^*} (T_b \phi)^i \rangle . \tag{7.46}$$

Now we turn to the fermionic mass matrix. To obtain the fermionic mass matrix we have, as we have seen previously, to eliminate the goldstino. The goldstino is given by

$$\Psi_G = \frac{1}{\sqrt{2}} \langle e^{\frac{1}{2}\mathcal{G}} \mathcal{G}_i \rangle \chi^i - i \langle h_{Rab} D^b \rangle \lambda^a = \frac{1}{\sqrt{2}} \langle e^{\frac{1}{2}\mathcal{G}} \mathcal{G}_i \rangle \chi^i .$$

The coupling of the goldstino to the gravitino reads

$$\begin{aligned} -\frac{\sqrt{2}}{2} i \langle e^{\frac{1}{2}\mathcal{G}} \mathcal{G}_i \rangle \chi^i \sigma^\mu \bar{\psi}_\mu + \text{h.c.} &= -i \langle e^{\frac{1}{2}\mathcal{G}} \rangle \Big(\langle e^{-\frac{1}{2}\mathcal{G}} \rangle \Psi_G \sigma^\mu \bar{\psi}_\mu + \text{h.c.} \Big) \\ &= -\frac{\sqrt{2}}{2} i \langle e^{\frac{1}{2}\mathcal{G}} \rangle \Big(\eta \sigma^\mu \bar{\psi}_\mu + \text{h.c.} \Big) , \end{aligned}$$

where we have introduced

$$\eta = \sqrt{2} e^{-\frac{1}{2}\mathcal{G}} \Psi_G .$$

Since the goldstino transformation is given by

$$\begin{aligned}\delta\eta &= \langle \mathcal{G}_i \rangle \delta\chi^i + \cdots \\ &= \sqrt{2}\varepsilon \mathcal{G}_i F^i + \cdots \\ &= \sqrt{2} e^{\frac{1}{2}\mathcal{G}} \mathcal{G}^i \mathcal{G}_i \varepsilon + \cdots \\ &= 3\sqrt{2} m_{\frac{3}{2}} \varepsilon + \cdots ,\end{aligned}$$

choosing $\varepsilon = -\frac{\sqrt{2}}{6} m_{\frac{3}{2}}^{-1} \eta$ explicitly eliminates the goldstino [Cremmer et al. (1979, 1978b)]. In particular this means that the goldstino vanishes from the Lagrangian. This is the analogue of the "unitary gauge" of gauge theories.

It is however possible to diagonalise the mass matrix directly by performing the field redefinition [Wess and Bagger (1992); Freedman and Van Proeyen (2012)]

$$\psi_\mu \to \psi_\mu - \frac{\sqrt{2}}{6} i\bar{\eta}\bar{\sigma}_\mu - \frac{\sqrt{2}}{3} m_{\frac{3}{2}}^{-1} \partial_\mu \eta \,, \tag{7.47}$$

i.e., by not being in a "unitary gauge". Since this computation is more involved but also instructive, we will now give some details. Recall the relations $\sigma_\mu \bar{\sigma}^{\mu\nu} = 3/2\sigma^\nu, \sigma_\mu \bar{\sigma}^{\mu\nu} \bar{\sigma}_\nu = 6$, using $m_{\frac{3}{2}} = \langle e^{\mathcal{G}} \rangle$ we obtain

$$\begin{aligned}&-m_{\frac{3}{2}} \Big[\bar{\psi}_\mu \sigma^{\mu\nu} \bar{\psi}_\nu + \text{h.c.} \Big] \\ &\to -m_{\frac{3}{2}} \Big[\bar{\psi}_\mu \sigma^{\mu\nu} \bar{\psi}_\nu + \frac{\sqrt{2}}{2} i\psi_\mu \sigma^\mu \bar{\eta} + \frac{2\sqrt{2}}{3} m_{\frac{3}{2}}^{-1} \psi_\mu \sigma^{\mu\nu} \partial_\nu \eta \\ &\quad + \frac{1}{3} \bar{\eta} \cdot \bar{\eta} - \frac{i}{3} m_{\frac{3}{2}}^{-1} \bar{\eta} \bar{\sigma}^\mu \partial_\mu \eta + \frac{2}{9} m_{\frac{3}{2}}^{-2} \partial_\mu \eta \sigma^{\mu\nu} \partial_\nu \eta + \text{h.c.} \Big] ,\end{aligned}$$

$$\begin{aligned}&-m_{\frac{3}{2}} \Big[i\frac{\sqrt{2}}{2} \bar{\eta} \bar{\sigma}^\mu \psi_\mu + \text{h.c.} \Big] \\ &\to -m_{\frac{3}{2}} \Big[i\frac{\sqrt{2}}{3} \bar{\eta} \bar{\sigma}^\mu \psi_\mu - \frac{2}{3} \bar{\eta} \cdot \bar{\eta} + \frac{i}{3} m_{\frac{3}{2}}^{-1} \bar{\eta} \bar{\sigma}^\mu \partial_\mu \eta + \text{h.c.} \Big]\end{aligned}$$

and (omitting the covariant derivative for the gravitino for simplicity)

$$\begin{aligned}&\frac{1}{2} \epsilon^{\mu\nu\rho\sigma} \Big[\psi_\mu \sigma_\sigma \partial_\nu \bar{\psi}_\rho + \text{h.c.} \Big] \\ &\to \frac{1}{2} \epsilon^{\mu\nu\rho\sigma} \Big[\psi_\mu \sigma_\sigma \partial_\nu \bar{\psi}_\rho + \text{h.c.} \Big] \\ &\quad \Big\{ -\frac{\sqrt{3}}{2} \Big[\psi_\mu \sigma^{\mu\nu} \partial_\nu \eta - \bar{\psi}_\mu \bar{\sigma}^{\mu\nu} \partial_\nu \bar{\eta} \Big] + \frac{i}{6} \bar{\eta} \bar{\sigma}^\mu \partial_\mu \eta \\ &\quad -\frac{2}{9} m_{\frac{3}{2}}^{-1} \partial_\mu \eta \sigma^{\mu\nu} \partial_\nu \eta + \text{h.c.} \Big\} .\end{aligned}$$

Collecting all contributions we finally obtain

$$
\begin{aligned}
&\frac{1}{2}\epsilon^{\mu\nu\rho\sigma}\Big[\psi_\mu\sigma_\sigma\partial_\nu\bar{\psi}_\rho - \bar{\psi}_\mu\sigma_\sigma\partial_\nu\psi_\rho\Big] \\
&-m_{\frac{3}{2}}\Big[\bar{\psi}_\mu\bar{\sigma}^{\mu\nu}\bar{\psi}_\nu + \psi_\mu\sigma^{\mu\nu}\psi_\nu + i\frac{\sqrt{2}}{2}\bar{\eta}\bar{\sigma}^\mu\psi_\mu + i\frac{\sqrt{2}}{2}\eta\bar{\sigma}^\mu\psi_\mu\Big] \\
\to\ &\frac{1}{2}\epsilon^{\mu\nu\rho\sigma}\Big[\psi_\mu\sigma_\sigma\partial_\nu\bar{\psi}_\rho - \bar{\psi}_\mu\sigma_\sigma\partial_\nu\psi_\rho\Big] - m_{\frac{3}{2}}\Big[\bar{\psi}_\mu\bar{\sigma}^{\mu\nu}\bar{\psi}_\nu + \psi_\mu\sigma^{\mu\nu}\psi_\nu\Big] \\
&+\frac{1}{6}\Big[\bar{\eta}\bar{\sigma}^\mu\partial_\mu\eta - \eta\sigma^\mu\partial_\mu\bar{\eta}\Big] - \frac{1}{3}m_{\frac{3}{2}}\Big[\eta\cdot\eta + \bar{\eta}\cdot\bar{\eta}\Big]
\end{aligned} \tag{7.48}
$$

for the gravitino/goldstino term and thus the mass matrix is diagonal, as expected.

Coming back to the mass term, we have

$$
\begin{aligned}
e^{-1}\mathcal{L}^{\text{mass}}_{\frac{1}{2}} = -\langle e^{\frac{1}{2}\mathcal{G}}\rangle\bigg[&\psi_\mu\sigma^{\mu\nu}\psi_\nu + \bar{\psi}_\mu\bar{\sigma}^{\mu\nu}\bar{\psi}_\nu + \frac{1}{2}\Big(\langle\mathcal{D}_i\mathcal{G}_j\rangle + \langle\mathcal{G}_i\mathcal{G}_j\rangle\Big)\chi^i\cdot\chi^j \\
&+\frac{1}{2}\Big(\langle\mathcal{D}^{i^*}\mathcal{G}^{j^*}\rangle + \langle\mathcal{G}^{i^*}\mathcal{G}^{j^*}\rangle\Big)\bar{\chi}_{i^*}\cdot\bar{\chi}_{j^*} + i\frac{\sqrt{2}}{2}\eta\sigma^\mu\bar{\psi}_\mu + i\frac{\sqrt{2}}{2}\bar{\eta}\bar{\sigma}^\mu\psi_\mu \\
&+\frac{1}{4}\langle\mathcal{G}^i{}_{i^*}h_{abi}\mathcal{G}^{i^*}\rangle\lambda^a\cdot\lambda^b + \frac{1}{4}\langle\mathcal{G}^i{}_{i^*}h^{\star}_{ab}{}^{i^*}\mathcal{G}_i\rangle\bar{\lambda}^a\cdot\bar{\lambda}^b\bigg] \\
&+i\sqrt{2}g\langle\mathcal{G}^{i^*}{}_i(\phi^\dagger T_a)_{i^*}\rangle\chi^i\cdot\lambda^a - i\sqrt{2}g\langle\mathcal{G}^{i^*}{}_i(T_a\phi)^i\rangle\bar{\chi}_{i^*}\cdot\bar{\lambda}^a
\end{aligned}
$$

Subtracting the Goldstino [Wess and Bagger (1992); Freedman and Van Proeyen (2012)] induces new contributions to the fermions mass matrix. Here, since just F−terms take a non-vanishing value only

$$
\frac{1}{2}\Big(\langle\mathcal{D}_i\mathcal{G}_j\rangle + \langle\mathcal{G}_i\mathcal{G}_j\rangle\Big)\chi^i\cdot\chi^j\,,
$$

have additional mass term, coming from (7.4):

$$
\frac{1}{2}\Big(\langle\mathcal{D}_i\mathcal{G}_j\rangle + \langle\mathcal{G}_i\mathcal{G}_j\rangle\Big)\chi^i\cdot\chi^j - \frac{1}{3}\mathcal{G}_i\mathcal{G}_j\chi^i\cdot\chi^j = \frac{1}{2}\Big(\langle\mathcal{D}_i\mathcal{G}_j\rangle + \frac{1}{3}\langle\mathcal{G}_i\mathcal{G}_j\rangle\Big)\chi^i\cdot\chi^j\,,
$$

and

$$
\begin{aligned}
e^{-1}\mathcal{L}^{\text{mass}}_{\frac{1}{2}} \to &-m_{\frac{3}{2}}\big(\psi_\mu\sigma^{\mu\nu}\psi_\nu+\bar{\psi}_\mu\bar{\sigma}^{\mu\nu}\bar{\psi}_\nu\big)-\frac{1}{2}(\mathcal{M}_{\frac{1}{2}})_{ij}\chi^i\cdot\chi^i-\frac{1}{2}(\mathcal{M}^\dagger_{\frac{1}{2}})^{i^*j^*}\bar{\chi}_{i^*}\cdot\bar{\chi}_{j^*} \\
&-\frac{1}{2}(\mathcal{M}_{\frac{1}{2}})_{ab}\lambda^a\cdot\lambda^b-\frac{1}{2}(\mathcal{M}^\dagger_{\frac{1}{2}})_{ab}\bar{\lambda}^a\cdot\bar{\lambda}^b-(\mathcal{M}_{\frac{1}{2}})_{ai}\lambda^a\cdot\chi^i-(\mathcal{M}^\dagger_{\frac{1}{2}})_a{}^{i^*}\bar{\lambda}^a\cdot\bar{\chi}_{i^*}\,.
\end{aligned}
$$

Thus, we get,

$$
\begin{aligned}
m_{\frac{3}{2}} &= \langle e^{\frac{1}{2}\mathcal{G}}\rangle\,,\\
(\mathcal{M}_{\frac{1}{2}})_{ij} &= \langle e^{\frac{1}{2}\mathcal{G}}\rangle(\langle\mathcal{D}_i\mathcal{G}_j\rangle + \frac{1}{3}\langle\mathcal{G}_i\mathcal{G}_j\rangle)\,,\\
(\mathcal{M}_{\frac{1}{2}})_{ab} &= \frac{1}{2}\langle e^{\frac{1}{2}\mathcal{G}}\rangle\langle\mathcal{G}^i{}_{i^*}h_{abi}\mathcal{G}^{i^*}\rangle\,,\\
(\mathcal{M}_{\frac{1}{2}})_{ia} &= i\sqrt{2}g\langle\mathcal{G}^{i^*}{}_i(\phi^\dagger T_a)_{i^*}\rangle\,,\\
(\mathcal{M}_{\frac{1}{2}})^{i^*}{}_a &= -i\sqrt{2}g\langle\mathcal{G}^{i^*}{}_i(T_a\phi)^i\rangle\,.
\end{aligned}
$$

Due to the elimination of the Goldstino the rank of the fermions mass matrix is reduced by one. Since the square mass matrix of the fermions is then given by

$$
(\mathcal{M}^2_{\frac{1}{2}})_{(i,a),(j,b)} = \begin{pmatrix} (\mathcal{M}^\dagger_{\frac{1}{2}})^{i^*k^*} & (\mathcal{M}^\dagger_{\frac{1}{2}})^{i^*}{}_c \\ (\mathcal{M}^\dagger_{\frac{1}{2}})_a{}^{k^*} & (\mathcal{M}^\dagger_{\frac{1}{2}})_{ac}\end{pmatrix}\begin{pmatrix}\mathcal{G}^\ell{}_{k^*} & 0 \\ 0 & h^{Rcd}\end{pmatrix}\begin{pmatrix}(\mathcal{M}_{\frac{1}{2}})_{\ell j} & (\mathcal{M}_{\frac{1}{2}})_{\ell b}\\ (\mathcal{M}_{\frac{1}{2}})_{dj} & (\mathcal{M}_{\frac{1}{2}})_{db}\end{pmatrix}\,,
$$

and using (7.41), (7.43) and $\mathcal{D}_i\mathcal{G}^{j^*}{}_j = 0$, we obtain

$$
\begin{aligned}
(\mathcal{M}^2_{\frac{1}{2}})^{i^*}{}_i &= (\mathcal{M}^\dagger_{\frac{1}{2}})^{i^*j^*}\langle\mathcal{G}^j{}_{j^*}\rangle(\mathcal{M}_{\frac{1}{2}})_{ji} + (\mathcal{M}^\dagger_{\frac{1}{2}})^{i^*}{}_a\langle h^{Rab}\rangle(\mathcal{M}^\dagger_{\frac{1}{2}})_{bj}\\
&= \langle e^{\mathcal{G}}\rangle\langle\mathcal{G}^j{}_{j^*}\rangle\Big[\langle\mathcal{D}_i\mathcal{G}_j\rangle + \frac{1}{3}\langle\mathcal{G}_i\mathcal{G}_j\rangle\Big]\Big[\langle\mathcal{D}^{i^*}\mathcal{G}^{j^*}\rangle + \frac{1}{3}\langle\mathcal{G}^{i^*}\mathcal{G}^{j^*}\rangle\Big]\\
&\quad +2g^2\langle\mathcal{G}^{i^*}{}_j\mathcal{G}^{j^*}{}_i h^{Rab}(\phi^\dagger T_a)_{j^*}(T_b\phi)^j\rangle\\
&= m^2_{\frac{3}{2}}\Big(\langle\mathcal{G}^j{}_{j^*}\mathcal{D}_i\mathcal{G}_j\mathcal{D}^{i^*}\mathcal{G}^j\rangle - \frac{1}{3}\langle\mathcal{G}_i\mathcal{G}^{i^*}\rangle\Big) + 2g^2\langle\mathcal{G}^{i^*}{}_j\mathcal{G}^{j^*}{}_i h^{Rab}(\phi^\dagger T_a)_{j^*}(T_b\phi)^j\rangle\,,\\
(\mathcal{M}^2_{\frac{1}{2}})_{ab} &= (\mathcal{M}^\dagger_{\frac{1}{2}})_a{}^{i^*}\langle\mathcal{G}^i{}_{i^*}\rangle(\mathcal{M}_{\frac{1}{2}})_{ib} + (\mathcal{M}^\dagger_{\frac{1}{2}})_{ac}\langle h^{Rcd}\rangle(\mathcal{M}_{\frac{1}{2}})_{db}\\
&= 2g^2\langle\mathcal{G}^i{}_{i^*}(\phi^\dagger T_a)_{i^*}(T_b\phi)^i\rangle + \frac{1}{4}m^2_{\frac{3}{2}}\Big(\langle\mathcal{G}^i{}_{i^*}\mathcal{G}^j{}_{j^*}h^{Rcd}h_{aci}h^{\star}_{bd}{}^{j^*}\mathcal{G}^{i^*}\mathcal{G}_j\rangle\Big)\,.
\end{aligned}
$$

Where we have simplified the first calculation using the minimisation of the potential and the vanishing of the cosmological constant.

Taking the trace gives

$$
\begin{aligned}
\mathrm{Tr}(\mathcal{M}^2_{\frac{1}{2}}) &= \langle\mathcal{G}^i{}_{i^*}\rangle(\mathcal{M}^2_{\frac{1}{2}})^{i^*}{}_i + \langle h^{Rab}\rangle(\mathcal{M}^2_{\frac{1}{2}})_{ab}\\
&= m^2_{\frac{3}{2}}\Big(\langle\mathcal{G}^i{}_{i^*}\mathcal{D}_i\mathcal{G}_j\mathcal{D}^{i^*}\mathcal{G}^j\rangle - 1 + \frac{1}{4}\langle\mathcal{G}^i{}_{i^*}\mathcal{G}^j{}_{j^*}h^{Rab}h^{Rcd}h_{aci}h^{\star}_{bd}{}^{j^*}\mathcal{G}^{i^*}\mathcal{G}_j\rangle\Big)\\
&\quad +4g^2\langle h^{Rab}\mathcal{G}^i{}_{i^*}(\phi^\dagger T_a)_{i^*}(T_b\phi)^i\rangle\,.
\end{aligned}\tag{7.49}
$$

Finally adding (7.45), (7.46) and (7.49) we obtain

$$\begin{aligned}\mathrm{Str}(\mathcal{M}^2) &= \sum_j (-)^{2j}(2j+1)\mathrm{Tr}(\mathcal{M})_j^2 = \mathrm{Tr}(\mathcal{M})_0^2 - 2\mathrm{Tr}(\mathcal{M})_{\frac{1}{2}}^2 + 3\mathrm{Tr}(\mathcal{M})_1^2 - 4m_{\frac{3}{2}}^2 \\ &= m_{\frac{3}{2}}^2\left[2(n-1) - 2\langle \mathcal{G}^i \mathcal{G}_{i^*} R_i{}^{i^*}\rangle - \frac{1}{2}\langle \mathcal{G}^i{}_{i^*} \mathcal{G}^j{}_{j^*} h^{Rab} h^{Rcd} h_{aci} h^{\star}_{bd}{}^{j^*} \mathcal{G}^{i^*} \mathcal{G}_j\rangle\right].\end{aligned}$$

This explicitly shows that the gravitino mass is an order parameter for broken supergravity.

Chapter 8

Supergravity in particles physics and cosmology

In this chapter we investigate two applications of supergravity in Physics.

One application which has been intensively studied, for many decades, by any means of formal, theoretical, phenomenological or experimental aspects, is the application to particle physics. There are several ways to embed the Standard Model (or one of its Grand Unification extensions) in supergravity. Further, as we have seen in Chapter 7, supergravity or supersymmetry has to be broken for phenomenological reasons. Also there are many ways to break supergravity. One extension which figures most prominently in the literature is the Minimal Supersymmetric Standard Model (MSSM). This model, or at least the model were supersymmetry is broken in a universal way (see Sec. 7.2.1), is almost excluded experimentally. Nevertheless in this chapter the MSSM will be introduced, and in particular it will be established that considering specific hidden sectors it is possible to naturally reproduce the different models studied in the literature (such as the cMSSM, or any possible choice for the soft supersymmetric breaking terms considered so far in any phenomenological analysis). Since this study is generic it can be easily extended to different specific models.

Supergravity has at least two applications in cosmology. First, in certain models the lightest supersymmetric particle is stable. If it is, colourless and electrically neutral, this particle can then be seen as a natural candidate for dark matter in the Universe. This possibility will not be studied in this book. Second, in supergravity there are many scalar fields and many flat directions in the potential. So inherently supergravity has many ways to trigger an inflationary phase in the Universe with a scalar field. This property was rapidly seen as an asset in order to build supergravity models having an inflationary phase. However, the second application of supergravity we will be considering in this chapter, is more recent. Introducing a class of Models, called superconformal Supergravity Models, it will be briefly showed how the Higgs inflation introduced in the Standard Model can be extended to Supergravity. The specific model that will consider is a slight extension of the MSSM, where a gauge singlet chiral superfield is introduced.

8.1 Supergravity in particles physics

To apply supergravity to particles physics the matter/interaction content of the theory and the fundamental functions must be specified. The interaction content is constituted by a compact Lie group G (it can be the Standard Model gauge group or any Grand-Unified Theory gauge group) and its corresponding compact Lie algebra $\mathfrak{g}$. Thus, for the gauge sector a vector superfied is associated to any generator of the Lie algebra $\mathfrak{g}$. Stated differently, the vector superfield lies in the adjoint representation of $\mathfrak{g}$ which is always real since the Lie algebra $\mathfrak{g}$ is a real Lie algebra. On the other hand the matter content (fermionic and scalar fields) is described by chiral (anti-chirals) superfields in a given unitary representation $\mathcal{R}$ ($\bar{\mathcal{R}}$) of G. Next, the fundamental functions of the theory, namely the Kähler potential, the superpotential and the gauge kinetic function are introduced. However, this is not the end of the story. Indeed, as we have seen in Chapter 7, supergravity has to be broken. Two options can be considered. Either one introduces, in addition to the matter content, a hidden sector such that supergravity is broken in this sector as done in Sec.7.2. The interaction of the hidden sector with the observable sector then generates the soft-supersymmetric breaking terms. Or more pragmatically, one introduces by hand the so-called soft-breaking terms, having in mind that we know that they could be generated by an *ad hoc* hidden sector. The first option is more appropriate if one considers supergravity as an effective theory[1] of a more fundamental theory, like string theory for instance, where *in principle* everything could be predicted. The second approach is more efficient for a phenomenological purpose, in particular, within this last approach, phenomenological analyses just reduce and constraint the parameter space of the theory leading to predictions that could be tested at Colliders.

8.1.1 *A quick reminder of the Standard Model*

The Standard Model of particle physics is a gauge theory describing almost all interactions at the infinitesimal level. There are many books devoted to the Standard Model giving lot of details. For a review of Particle Physics see [Tanabashi et al. (2018)]. It is not our purpose to give a detailed description of this model here, but just to introduce briefly its field content, in order to fix the notations. The interested reader can see [Campoamor-Stursberg and Rausch de Traubenberg (2019)]

[1] Supergravity is unavoidably an effective theory since like gravity, it is non-renormalisable.

for instance which uses the notations of this book. In particular the Lagrangian will not be introduced. The Standard Model Gauge group is

$$G_{\text{S.M}} = SU(3)_c \times SU(2)_L \times U(1)_Y \ ,$$

where $SU(3)_c$ describes the strong interaction between quarks and $SU(2)_L \times U(1)_Y$ the electroweak sector. The gauge bosons of strong interaction or Quantum Chromodynamics are the eight gluons and the gauge bosons of the electroweak sector are the three W-bosons and the B-boson.

The matter sector of the Standard model is constituted of quarks and leptons. There are three families of quarks and leptons. Right handed fermions do not feel the weak interaction and consequently are singlets under $SU(2)_L$ whereas left-handed fermions are $SU(2)_L$-doublets. The precise content is thus

(1) for the left-handed quarks: $q^f = \begin{pmatrix} u^f_L \\ d^f_L \end{pmatrix} = (\underline{\mathbf{3}}, \underline{\mathbf{2}}, \frac{1}{6})$;

(2) for the left-handed anti-quarks $u^{fc}_L = (\underline{\bar{\mathbf{3}}}, \underline{\mathbf{1}}, -\frac{2}{3})$ and $d^{fc}_L = (\underline{\bar{\mathbf{3}}}, \underline{\mathbf{1}}, \frac{1}{3})$;

(3) for the left-handed leptons: $\ell^f = \begin{pmatrix} \nu^f_L \\ e^f_L \end{pmatrix} = (\underline{\mathbf{1}}, \underline{\mathbf{2}}, -\frac{1}{2})$;

(4) for the left-handed anti-leptons: $e^{fc}_L = (\underline{\bar{\mathbf{1}}}, \underline{\mathbf{1}}, 1)$ and $n^{fc}_L = (\underline{\bar{\mathbf{1}}}, \underline{\mathbf{1}}, 0)$;

where $f = 1, 2, 3$ is a family index and ψ^c_L is the charge conjugate of the right-handed fermion ψ_R which is left-handed and describes the anti-fermion. For

- $f = 1$: $u^1 = u, d^1 = d$ are respectively the up, down quarks and $e^1 = e, n^1 = \nu_e$ the electron and its associated neutrino;

- $f = 2$: $u^2 = c, d^2 = s$ are respectively the charm, strange quarks and $e^2 = \mu, n^2 = \nu_\mu$ the muon and its associated neutrino;

- $f = 3$: $u^3 = t, d^3 = b$ are respectively the top, bottom quarks and $e^3 = \tau, n^3 = \nu_\tau$ the tau and its associated neutrino.

Finally a scalar doublet dubbed the Higgs boson is necessary for the consistency of the model:

$$H = (\underline{\mathbf{1}}, \underline{\mathbf{2}}, -\frac{1}{2}) \ .$$

At 100 GeV the Higgs boson by developing a non-vanishing vacuum expectation value breaks spontaneously the electroweak sector $SU(2)_L \times U(1)_Y$ down to the electromagnetic interactions $U(1)_{e.m.}$ conferring a mass to all the particles except the photon and the eight gluons. Below 100 GeV the gauge group is therefore reduced to $SU(3)_c \times U(1)_{e.m.}$.

8.1.2 *The Standard Model in supergravity*

There are various possible supersymmetric extensions of the Standard Model of particles physics. The so-called Minimal Supersymmetry Standard Model (MSSM) is the simplest supersymmetric version of the Standard Model [Nilles (1984); Fayet (1976, 1977)]. The gauge group is

$$G_{\text{S.M.}} = SU(3)_c \times SU(2)_L \times U(1)_Y \,.$$

Thus, for each simple factor of $G_{\text{S.M.}}$ one introduces a vector superfield in the adjoint representation

$$\begin{aligned} V_3 &= (\underset{\sim}{\mathbf{8}}, \underset{\sim}{\mathbf{1}}, 0)\,, \text{ for } SU(3)_c\,, \\ V_2 &= (\underset{\sim}{\mathbf{1}}, \underset{\sim}{\mathbf{3}}, 0)\,, \text{ for } SU(2)_L\,, \\ V_1 &= (\underset{\sim}{\mathbf{1}}, \underset{\sim}{\mathbf{1}}, 0)\,, \text{ for } U(1)_Y\,. \end{aligned}$$

The matter sector of the MSSM contains three generations of chiral superfields ($f = 1, 2, 3$)

$$\begin{aligned} Q^f_L &= (\underset{\sim}{\mathbf{3}}, \underset{\sim}{\mathbf{2}}, \ \tfrac{1}{6})\,, \\ U^f_L &= (\bar{\underset{\sim}{\mathbf{3}}}, \underset{\sim}{\mathbf{1}}, -\tfrac{2}{3})\,, \\ D^f_L &= (\bar{\underset{\sim}{\mathbf{3}}}, \underset{\sim}{\mathbf{1}}, \tfrac{1}{3})\,, \\ L^f_L &= (\underset{\sim}{\mathbf{1}}, \underset{\sim}{\mathbf{2}}, -\tfrac{1}{2})\,, \\ E^f_L &= (\underset{\sim}{\mathbf{1}}, \underset{\sim}{\mathbf{2}}, \ 1)\,, \\ N^f_L &= (\underset{\sim}{\mathbf{1}}, \underset{\sim}{\mathbf{1}}, 0)\,, \end{aligned} \tag{8.1}$$

describing the usual quarks and lepton with their associated supersymmetric partners the squarks and the sleptons. Finally, the Higgs sector of the theory contains two Higgs doublets and is embedded in the chiral superfields

$$\begin{aligned} H_D &= (\underset{\sim}{\mathbf{1}}, \underset{\sim}{\mathbf{2}}, -\tfrac{1}{2})\,, \\ H_U &= (\underset{\sim}{\mathbf{1}}, \underset{\sim}{\mathbf{2}}, \tfrac{1}{2})\,. \end{aligned}$$

Two Higgs doublets (instead of one doublet for the Standard Model) are needed for at least two reasons. Firstly, to cancel the chiral anomaly coming from the fermionic component of say, H_U one has to introduce another Higgs

Table 8.1 The particle content of the MSSM.

supe-rmultiplet	name	particle	supersymmetric partner	repre-sentation
Q^f_L U^f_L D^f_L	quaks/ squarks	$q^f_L = \begin{pmatrix} u^f_L \\ d^f_L \end{pmatrix}$ u^{fc}_L d^{fc}_L	$\tilde{q}^f_L = \begin{pmatrix} \tilde{u}^f_L \\ \tilde{d}^f_L \end{pmatrix}$ $\tilde{u}^{f\dagger}_R$ $\tilde{d}^{f\dagger}_R$	$(\mathbf{3}, \mathbf{2}, \frac{1}{6})$ $(\bar{\mathbf{3}}, \mathbf{1}, -\frac{2}{3})$ $(\bar{\mathbf{3}}, \mathbf{1}, \frac{1}{3})$
L^f_R E^f_L N^f_L	leptons/ slpetons	$\ell^f_L = \begin{pmatrix} \nu^f_L \\ e^f_L \end{pmatrix}$ e^{fc}_L ν^{fc}_L	$\tilde{\ell}^f_L = \begin{pmatrix} \tilde{\nu}^f_L \\ \tilde{e}^f_L \end{pmatrix}$ $\tilde{e}^{f\dagger}_R$ $\tilde{\nu}^{f\dagger}_R$	$(\mathbf{1}, \mathbf{2}, -\frac{1}{2})$ $(\mathbf{1}, \mathbf{1}, 1)$ $(\mathbf{1}, \mathbf{1}, 0)$
H_D H_U	Higgs/ Higgsino	$h_D = \begin{pmatrix} h_{D0} \\ h_{D-} \end{pmatrix}$ $h_U = \begin{pmatrix} h_{U+} \\ h_{U0} \end{pmatrix}$	$\tilde{h}_D = \begin{pmatrix} \tilde{h}_{D0} \\ \tilde{h}_{D-} \end{pmatrix}$ $\tilde{h}_U = \begin{pmatrix} \tilde{h}_{U+} \\ \tilde{h}_{U0} \end{pmatrix}$	$(\mathbf{1}, \mathbf{2}, -\frac{1}{2})$ $(\mathbf{1}, \mathbf{2}, \frac{1}{2})$
V_1	boson-B bino	B	$\tilde{B}$	$(\mathbf{1}, \mathbf{1}, 0)$
V_2	bosons-W winos	W	$\tilde{W}$	$(\mathbf{1}, \mathbf{3}, 0)$
V_3	gluons gluinos	g	$\tilde{g}$	$(\mathbf{8}, \mathbf{1}, 0)$

The superpotential invariant under $R-$parity takes the form

$$W(\Phi) = -y_{eff'} L^f_L \cdot H_D E^{f'}_L - y_{dff'} Q^f_L \cdot H_D D^{f'}_L + y_{nff'} L^f_L \cdot H_U N^{f'}_L + y_{uff'} Q^f_L \cdot H_U U^{f'}_L + \frac{1}{2} m_{ff'} N^f_L N^{f'}_L + \mu H_D \cdot H_U \,,$$

where the contribution of the hidden sector is not taken into account.

field namely H_D with opposite hypercharge. Secondly, since the superpotential is a holomorphic function, two Higgs doublets are needed to give a mass term to both up-type and down-type fermions. In Table 8.1 we summarise the particle content of the MSSM. Note also that the electric charge Q is given by the Gell-Mann-Nishijima relation $Q = Y + t_3$ where Y is the hypercharge and t_3 the Cartan generator of $\mathfrak{su}(2)$. The particles of the Standard Model are denoted without a tilde whereas their corresponding supersymmetric partners with a tilde. For instance the chiral superfied Q contains the quark q and the squark $\tilde{q}$ and for SU(2) B represents the $B-$boson and $\tilde{B}$ the bino, its fermionic superpartner.

In addition to the field content, in supergravity a model is specified by a choice of the three fundamental functions: the Kähler potential K, the superpotential W and the gauge kinetic function h. In this section we denote Z^i the fields in the hidden sector and $\Phi^A = (Q^f_L, U^f_L, D^f_L, L^f_L, E^f_L, N^f_L, H_U, H_D), f = 1,2,3$ the twenty chiral superfields of the MSSM and we take the notations of Sec. 6.5 for the anti-chiral superfields. We also denote generically $\Phi^{fa} = (Q^f_L, U^f_L, D^f_L, L^f_L, E^f_L, N^f_L), a = Q, U, D, L, E, N$, the superfields of the f−th generation of quarks and leptons.

Using the results of Sec. 7.2.2 the general form for the Kähler potential and superpotential is given by Eq. [7.14] where now the various functions have to be invariant under the Standard Model gauge group $G_{\text{S.M.}}$. To define the superpotential we assume, in order to forbid a fast proton decay, that the R−parity defined by $R = (-)^{2S+3B+L}$ (with S the spin, B the baryon number and L the lepton number) is conserved[2]. With the notations of (7.14) the most general invariant (under the gauge group and R−parity) superpotential reads[3]

$$\begin{aligned} W(Z,\Phi) = \hat{W}(Z,Z^\dagger) &- y_{eff'}(Z)L^f_L \cdot H_D E^{f'}_L - y_{dff'}(Z)Q^f_L \cdot H_D D^{f'}_L \\ &+ y_{nff'}(Z)L^f_L \cdot H_U N^{f'}_L + y_{uff'}(Z)Q^f_L \cdot H_U U^{f'}_L \\ &+ \frac{1}{2} m_{ff'}(Z) N^f_L N^{f'}_L + \mu(Z) H_D \cdot H_U \, , \end{aligned} \tag{8.2}$$

where the Yukawa matrices $y_{ff'}(Z)$, describing the Yukawa interactions, as well as the μ− term $\mu(Z)$ depend on the hidden sector. We have denoted by a dot the $SU(2)$−invariant product *e.g.* $H_U \cdot H_D = \epsilon_{ij} H^i_U H^j_D$. This notation will be used throughout this section.

Now, using (7.23), (7.24) and (7.25), after having specified the Kähler potential, we obtain the scalar potential. *A priori* if we do not make any restriction on the different functions a phenomenological analysis would be rather complicated not to say intractable. Thus, to simplify the form of the different functions, an argument coming from experimental data is now invoked. Indeed, the fermions of each family have the same content, and in particular the same quantum numbers (see (8.1)) but their masses increase with f. For instance we have $m_{u^3}(= m_t) > m_{u^2}(= m_s) > m_{u^1}(= m_u)$. This observation suggests to take the Kähler potential in the form[4]

---

[2]The R− parity is a discrete symmetry associated to R− symmetry (see App. H).

[3]Since H_D has the same quantum number as L^f_L, a term like $L^f \cdot L^f E$ is invariant under the gauge group $G_{\text{S.M.}}$. However all particles of the Standard Model have an even R−parity whereas all supersymmetric partners have an odd R−parity. Thus, $L^f \cdot L^f E$ is not R−parity invariant.

[4]More general Kähler potentials and superpotentials, and perhaps more interesting ones, are considered at the end of this section.

$$K(Z,\Phi,Z^\dagger,\Phi^\dagger) = \hat{K}(Z,Z^\dagger) + \Lambda_h(Z,Z^\dagger)\Big[H_U^\dagger H_U + H_D^\dagger H_D\Big] + \sum_{f=1,2,3} \Lambda_f(Z,Z^\dagger) \sum_a \Phi^\dagger_{fa}\Phi^{fa} + \Big[\Gamma(Z,Z^\dagger) H_U \cdot H_D + \text{h.c.}\Big] \,, \tag{8.3}$$

and to make the superpotential reads

$$W(Z,\Phi) = \hat{W}(Z) + \sum_{f=1,2,3} \Big[H_f(Z) G_3^f(\Phi) + H'_f(Z) G_2^f(\Phi)\Big] \,, \tag{8.4}$$

with

$$G_3^f(\Phi) = -y_{ef} L_L^f \cdot H_D E_L^f - y_{df} Q_L^f \cdot H_D D_L^f + y_{nf} L_L^f \cdot H_U N_L^f + y_{uf} Q_L^f \cdot H_U U_L^f \,,$$
$$G_2^f(\Phi) = \frac{1}{2} m_f (N_L^f)^2 \,,$$

where the parameters appearing in G_3^f and in G_2^f are constant. Note that if G_2^f is not present W and K contain only dimensionless parameters. We have introduced two basic functions H_f, H'_f, for the each family in the superpotential, a basic function Λ_f for each family and a basic function Λ_h for the Higgs part in the Kähler potential. The form of the matter part of the superpotential suggests at least two remarks. (I) the Yukawa functions are diagonal and consequently flavour physics is neglected. (II) The $\mu-$term is set to zero, and by the non-renormalisation theorem [Grisaru et al. (1979); Seiberg (1993); Weinberg (2000)] no terms in $H_U \cdot H_D$ is generated by quantum corrections to W to all orders in perturbation theory. However, supergravity breaking generates a $\mu-$term in the superpotential W_m, but as pointed out not in W! (see (8.6) below and Sec. 7.2.2), this is the Giudice and Masiero (1988) mechanism.

Finally we introduce a gauge kinetic function for each simple factor of $G_{\text{S.M}}$ denoted h_1, h_2 and h_3.

8.1.3 *Supergravity breaking*

Along the lines of Sec. 7.2.2 we assume to have gravity mediated supersymmetry breaking. Thus, supergravity is broken in the hidden sector where some of the Z-fields develop a vacuum expectation value and the interaction of the hidden sector with the observable sector communicates supersymmetry breaking to the observable sector through the generation of softly breaking supersymmetric terms.

Using the explicit expression of F^i given in (7.18), the cosmological constant (7.21) takes the form

$$\Lambda = \frac{1}{m_p^2} e^{\frac{1}{m_p^2}\langle \hat{K}\rangle} \Big\{ \Big\langle \mathcal{D}^{i^*} \hat{W}^\star \mathcal{D}_i \hat{W} \hat{K}^i{}_{i^*} \Big\rangle - \frac{3}{m_p^2} \Big\langle |\hat{W}|^2 \Big\rangle \Big\} .$$

Now, if we assume that the superpotential $\hat{W}$ and the Kähler potential $\hat{K}$ of the hidden sector are subject to a no-scale condition (see Sec. 7.3)

$$\mathcal{D}^{i^*} \hat{W}^\star \mathcal{D}_i \hat{W} \hat{K}^i{}_{i^*} - \frac{3}{m_p^2} |\hat{W}|^2 = 0 , \tag{8.5}$$

the cosmological constant vanishes automatically. For instance if we suppose that there is only one field in the hidden sector we can take

$$\begin{aligned} \hat{K}(Z, Z^\dagger) &= -3 m_p^2 \ln \frac{Z + Z^\dagger}{m_p} , \\ \hat{W}(Z) &= m_p^3 . \end{aligned}$$

When the cosmological constant vanishes the order parameter for broken supergravity is the mass of the gravitino

$$m_{\frac{3}{2}} = \frac{1}{m_p^2} e^{\frac{1}{2} \frac{\langle \hat{K}\rangle}{m_p^2}} \langle \hat{W} \rangle .$$

For latter use we also introduce dimensionless ρ−functions defined by

$$\rho_{i^*} = \partial_i \Big(\frac{1}{m_p} \hat{K} + m_p \ln \frac{\hat{W}}{m_p^3} \Big) \hat{K}^i{}_{i^*} .$$

As we have seen in section 7.2.2, after supergravity is broken in the hidden sector, the potential of the observable sector (not considering the D−terms) reduces to

$$V = V_{\text{SUSY}} + V_{\text{Soft}} ,$$

where V_{SUSY} is the scalar potential of unbroken supersymmetry and V_{Soft} the potential which breaks (softly) supersymmetry explicitly. The different terms of the scalar potential can be directly deduces form (7.20),(7.22),(7.23),(7.24) and (7.25) or Tables 7.1 and 7.2. Before using the results of Sec. 7.2.2 some care must be exercised. Indeed, after supersymmetry breaking the kinetic part of the scalar component φ and of the fermion component χ of the chiral superfield Φ are not correctly normalised. Actually, the factor $\Lambda \Phi^\dagger \Phi$ in the Kähler potential leads to

$$\begin{aligned} L_{\text{kin scal.}} &= \langle \Lambda \rangle \partial_\mu \varphi \partial^\mu \varphi^\dagger , \\ L_{\text{kin ferm.}} &= -\frac{i}{2} \langle \Lambda \rangle \big(\chi \sigma^\mu \partial_\mu \bar{\chi} - \partial_\mu \chi \sigma^\mu \bar{\chi} \big) . \end{aligned}$$

Thus $\langle\Lambda\rangle$ must be positive, and in order to have correctly normalised fields, a field redefinition must be performed

$$\varphi \to \frac{1}{\sqrt{\langle\Lambda\rangle}}\varphi\,,$$
$$\chi \to \frac{1}{\sqrt{\langle\Lambda\rangle}}\chi\,.$$

After this field redefinition, and using the results of Sec. 7.2.2 we obtain (see Tables 7.1 and 7.2):

$$W_m = \sum_{f=1,2,3} W_f + m_{\frac{3}{2}} b_\mu H_U \cdot H_D\,, \tag{8.6}$$

where

$$W_f = -\hat{y}_{ef} L_L^f \cdot H_D E_L^f - \hat{y}_{df} Q_L^f \cdot H_D D_L^f + \hat{y}_{nf} L_L^f \cdot H_U N_L^f + \hat{y}_{uf} Q_L^f \cdot H_U U_L^f$$
$$+\frac{1}{2}\hat{m}_f (N_L^f)^2\,.$$

The various Yukawa constants are given by

$$\left.\begin{aligned}
\hat{y}_{ef} &= e^{\frac{1}{2}\frac{\langle\hat{K}\rangle}{m_p^2}} \frac{\langle H_f\rangle}{\langle\Lambda_f\sqrt{\Lambda_h}\rangle} y_{ef}\\
\hat{y}_{nf} &= e^{\frac{1}{2}\frac{\langle\hat{K}\rangle}{m_p^2}} \frac{\langle H_f\rangle}{\langle\Lambda_f\sqrt{\Lambda_h}\rangle} y_{nf}\\
\hat{y}_{df} &= e^{\frac{1}{2}\frac{\langle\hat{K}\rangle}{m_p^2}} \frac{\langle H_f\rangle}{\langle\Lambda_f\sqrt{\Lambda_h}\rangle} y_{df}\\
\hat{y}_{uf} &= e^{\frac{1}{2}\frac{\langle\hat{K}\rangle}{m_p^2}} \frac{\langle H_f\rangle}{\langle\Lambda_f\sqrt{\Lambda_h}\rangle} y_{uf}
\end{aligned}\right\} = \hat{y}_{If}\,,\quad I = e, n, d, u \tag{8.7}$$

the neutrino mass terms are

$$\hat{m}_f = e^{\frac{1}{2}\frac{\langle\hat{K}\rangle}{m_p^2}} \frac{\langle H'_f\rangle}{\langle\Lambda_f\rangle} m_f\,,$$

and the μ–term reads $\mu = m_{\frac{3}{2}} b_\mu$ with

$$b_\mu = \frac{1}{\langle\Lambda_h\rangle}\Big[\langle\Gamma\rangle - m_p\langle\rho_{i^*}\partial^{i^*}\Gamma\rangle\Big]\,. \tag{8.8}$$

Similarly we have

$$V_{\rm Soft} = |m_{\frac{3}{2}}|^2\Big[\sum_{f=1,2,3} S_f \sum_a \varphi^\dagger_{fa}\varphi^{fa} + S_h\big(h^\dagger_U h_U + h^\dagger_D h_D\big)\Big] + \Big[|m_{\frac{3}{2}}|^2\, b\, h_U\cdot h_D + \text{h.c.}\Big]$$
$$+ m^\dagger_{\frac{3}{2}}\Big[c_1\tilde{B}\cdot\tilde{B} + 32c_2\tilde{W}\cdot\tilde{W} + c_3\tilde{g}\cdot\tilde{g} + \text{h.c.}\Big] + \text{h.c.}$$
$$+ m^\dagger_{\frac{3}{2}}\sum_{f=1,2,3}\Big[-A_{ef}\tilde{\ell}^f_L\cdot h_D\tilde{e}^f_L - A_{df}\tilde{q}^f_L\cdot h_D\tilde{d}^f_L + A_{nf}\tilde{\ell}^f_L\cdot h_U\tilde{n}^f_L$$
$$+A_{uf}\tilde{q}^f_L\cdot h_U\tilde{u}^f_L + B_f(\tilde{n}^f)^2\Big] + \text{h.c.}\,. \tag{8.9}$$

The different soft-supersymmetric terms depend on the gravitino mass and the various functions (see Tables 7.1 and 7.2):

(1) Gaugino masses (see Eq. [7.26]):

$$c_\ell = \frac{1}{2}m_p e^{\frac{1}{2}\frac{\langle\hat{K}\rangle}{m_p^2}}\langle\rho^i\partial_i h_\ell\rangle\,. \tag{8.10}$$

(2) Scalar masses:

$$\begin{aligned} S_h &= 1 + m_p^2\Big\langle\rho_{i^*}\Big(\frac{\partial^{i^*}\Lambda_h\partial_i\Lambda_h}{\Lambda_h^2} - \frac{\partial^{i^*}\partial_i\Lambda_h}{\Lambda_h}\Big)\rho^i\Big\rangle\,,\\ S_f &= 1 + m_p^2\Big\langle\rho_{i^*}\Big(\frac{\partial^{i^*}\Lambda_f\partial_i\Lambda_f}{\Lambda_f^2} - \frac{\partial^{i^*}\partial_i\Lambda_f}{\Lambda_f}\Big)\rho^i\Big\rangle\,. \end{aligned} \tag{8.11}$$

(3) b–term:

$$b = \frac{1}{\langle\Lambda_h\rangle}\Big[m_p\langle\rho_{i^*}\partial^{i^*}\Gamma\rangle + m_p\langle\rho^i\partial_i\Gamma\rangle - m_p^2\langle\rho_{i^*}\rho^i\partial^{i^*}\partial_i\Gamma\rangle\Big]\,. \tag{8.12}$$

(4) Trilinear terms:

$$\begin{aligned} A_{If} &= \Big[\frac{1}{m_p}\langle\partial_i\hat{K}\rangle\hat{y}_{If} + m_p\langle\partial_i\hat{y}_{If}\rangle - 2m_p\frac{\langle\partial_i\Lambda_f\rangle}{\langle\Lambda_f\rangle}\hat{y}_{If} - m_p\frac{\langle\partial_i\Lambda_h\rangle}{\langle\Lambda_h\rangle}\hat{y}_{If}\Big]\langle\rho^i\rangle\\ &\equiv a_f y_{If}\,,\quad I = e,n,d,u\,, \end{aligned} \tag{8.13}$$

where $\langle\partial_i\hat{y}_{If}\rangle$ is a shorthand notation for $e^{\frac{1}{2}\frac{\langle\hat{K}\rangle}{m_p^2}}\big\langle\frac{\partial_i H_f}{\Lambda_f\sqrt{\Lambda_H}}\big\rangle y_{fI}$.

(5) Bilinear term:

$$\begin{aligned} B_f &= \Big[\frac{1}{m_p}\langle\partial_i\hat{K}\rangle\hat{m}_f + m_p\langle\partial_i\hat{m}_f\rangle - 2m_p\frac{\langle\partial_i\Lambda_f\rangle}{\langle\Lambda_f\rangle}\hat{m}_f\Big]\langle\rho^i\rangle - \hat{m}_f\,,\\ &\equiv b_f m_f\,, \end{aligned} \tag{8.14}$$

where $\langle\partial_i\hat{m}_f\rangle$ is a shorthand notation for $e^{\frac{1}{2}\frac{\langle\hat{K}\rangle}{m_p^2}}\big\langle\frac{\partial_i H'_f}{\Lambda_f\sqrt{\Lambda_f}}\big\rangle m_f$.

8.1.4 *Phenomenological analysis*

The final Lagrangian obtained when we neglect gravity, *i.e.*, when $m_p \to \infty$, is the Lagrangian (2.26) obtained in Chapter 2, with field content given in Table 8.1. The superpotential W_m is given in (8.6), the Kähler potential is canonical and the gauge kinetic functions are trivial. This Lagrangian describes the MSSM. Since supergravity has been broken in a hidden sector we also have the softly broken supersymmetric potential (8.9). We now make some general comments concerning the scalar potential V_{Soft} and the superpotential W_m.

(a) When supergravity is broken in the hidden sector some of the scalar fields Z develop a vacuum expectation value of the order of the Planck mass. Since the functions $b_\mu, c_\ell, S_h, S_f, a_f, b, H_h, H_{f'}$ and Λ_f, Λ_h are dimensionless these functions are automatically of the order of one. Moreover, the derivative ∂_i of each of these functions is of the order of m_p^{-1}.

(b) The mechanism which generates all dimensionfull parameters (of dimensions mass and mass^2) implies that all these parameters are controlled by the mass of the gravitino (see (8.8) and (8.9)). For the μ–parameter of W_m this is precisely the Giudici-Masiero mechanism:

$$\mu = m_{\frac{3}{2}} b_\mu \ .$$

For the functions $B_f = b_f \hat{m}_f$, the corresponding terms in the Lagrangian are multiplied by a factor $m^\dagger_{\frac{3}{2}}$.

The precise expression of the different terms involves the basic functions of the hidden sector which, as we have pointed out, are of the order of one.

(1) **Scalar soft supersymmetric breaking terms V_{Soft}:**

(c) The functions $S_f, S_h, f = 1, 2, 3$ for the scalar masses (8.11), $c_\ell, \ell = 1, 2, 3$ for the gaugino masses (8.10), $A_{fI}, f = 1, 2, 3$ for the trilinear terms (8.13), and b for the bilinear term (8.12) are thus all of the order of one. This means that almost all supersymmetric breaking terms are of the order of $m_{\frac{3}{2}}$. In contrast, due to the m_f–term, the function $b_f, f = 1, 2, 3$ for the bilinear terms (8.14), are of order $m_{\frac{3}{2}} m_p$.

(2) **Superpotential W_m:**

(a) The Yukawa constants (8.7) have an interesting expression in terms of the functions H_f and Λ_f. A specific choice of these functions, controlled by the hidden sector, could play a rôle in the hierarchy of the masses of the families of quarks and leptons: the third family is heavier than the second and the second family is heavier than the first.

(b) The neutrino mass terms m_f appearing in G_2^f ($f = 1, 2, 3$) are the only mass parameters in the Kähler potential (8.3) and in the superpotential (8.4). Before supergravity is broken we have only one mass scale: the Planck mass. So it seems natural to expect these parameters to be of the order of the Planck mass. This observation could play a rôle in the neutrino puzzle (why only left handed neutrinos are observed).

At the energy where supergravity is broken, the model we have constructed so far contains fifteen mass parameters:

(1) Four parameters S_f, S_h for scalar mass^2 terms:

$m^2_{af} = |m_{\frac{3}{2}}|^2 S_f$, $a = Q, U, D, L, E, N$, $f = 1, 2, 3$ and $m^2_U = m^2_D = |m_{\frac{3}{2}}|^2 S_h$;

(2) Three parameters c_ℓ for gaugino masses: $m_\ell = m^\dagger_{\frac{3}{2}} c_\ell, \ell = 1, 2, 3$;

(3) Three parameters a_f for trilinear couplings: $m^\dagger_{\frac{3}{2}} A_{fI} = m^\dagger_{\frac{3}{2}} a_f y_{fI}, I = e, n, d, u$;

(4) Three parameters b_f for bilinear couplings: $m^\dagger_{\frac{3}{2}} B_f = m^\dagger_{\frac{3}{2}} b_f m_f, f = 1, 2, 3$;

(5) One parameter for $\mu = m_{\frac{3}{2}} b_\mu$;

(6) One parameter for b;

and twelve dimensionless parameters:

(7) The Yukawa couplings $y_{fI}, f = 1, 2, 3, I = e, n, d, u$.

All these parameters are given at high energy, say at 10^{16}GeV where supergravity is broken. Using the renormalisation group equations given in App. G, we are able to compute these parameters at low energy, say at 8Tev. Knowing all parameters at low energy some phenomenological predictions can be compared to experimental data. However, this method fails *a priori* since we have no information upon the hidden sector or upon the basic functions needed to get all the physical parameters. Thus, to compare our model with experimental data we must proceed with a phenomenological analysis.

We will not perform a deep phenomenological analysis here, because it would go well beyond the scope of this book, but only sketch the main lines or ideas. The interested reader can see [Nilles (1984); Martin (1997)] for a phenomenological analysis or [Fuks and Rausch de Traubenberg (2011)] for an explicit computation of the mass matrices. To simplify the discussion we also will not consider the neutrino contribution (y_{fn}, m_f). A phenomenological analysis is a back-and-forth between low energy 100GeV and high energy 10^{16}GeV. Some parameters are directly given at low energy while some are given at high energy. At low energy, say 100GeV the electroweak symmetry is broken and the neutral part of the Higgs fields develops vacuum expectations value v_U and v_D. Since the mass of the Z-boson is known the value of $v_U^2 + v_D^2$ is fixed to $(174\text{Gev})^2$. However the ratio v_U/v_D, in general denoted $\tan\beta$, remains as a free parameter. Once $\tan\beta$ is fixed the Yukawa constant y_{fI} can be deduced from the known masses of the quarks and leptons. Furthermore, at low energy, the value of the gauge coupling constants of the $U(1)_Y, SU(2)_L$ and $SU(3)_c$ gauge groups are also known and consequently are fixed. The remaining parameters are given at high energy. However, since no information upon the hidden sector, and, correspondingly, upon the soft supersymmetric breaking terms, is available, we parametrise our ignorance by scanning the set of parameters at high energy. In our model we have twelve mass parameters, but using the definition of the vacuum[5] at 100GeV we can deduce $\mu = m_{\frac{3}{2}} b_\mu$ (up to a sign) and $B = |m_{\frac{3}{2}}|^2 b$ from the minimisation of the scalar potential. Thus, in the model we have considered, a phenomenological analysis can be performed by scanning an ten-dimensional parameters space, corresponding to all (ten) dimensionfull parameters except $\mu = m_{\frac{3}{2}} b_\mu$ and $B = |m_{\frac{3}{2}}|^2 b$ given at high energy, a choice for the sign of μ together with $\tan\beta$ given at low energy. This scanning only reproduces a class of models described by the Kähler potential (8.3) and superpotential (8.4). Of course such an analysis necessitates computer tool like, for instance, SuSpect [Djouadi et al. (2007)].

One popular model, simpler than the one we have been considering, is the so-called cMSSM [Martin (1997); Nilles (1984); Kane et al. (1994)] and is a very specific model where we have universality of the soft supersymmetric breaking terms. To recover the cMSSM from our model we must consider:

(1) A flat Kähler potential $\Lambda_1 = \Lambda_2 = \Lambda_3 = \Lambda_h = 1$ implying that all scalar masses are equal.

(2) A superpotential with (i) $H_1 = H_2 = H_3 = H$ implying that all trilinear terms are equal and (ii) $H'_1 = H'_2 = H'_3 = 0$ meaning that there are no mass term for the neutrinos.

[5]The vacuum is defined by minimising the scalar potential with respect to the two neutral components of the Higgs fields h_U and h_D. See [Martin (1997); Fuks and Rausch de Traubenberg (2011)] for an explicit computation of the minimisation equations.

(3) Gauge kinetic functions such that $h_1 = h_2 = h_3 = h$ leading to an equal mass for all gauginos.

A deep phenomenological analysis of this model versus experimental data reduces drastically the allowed region of the parameter space of the cMSSM, not to say almost excludes this simplest model [Bechtle et al. (2015); Han et al. (2017)].

Of course the cMSSM is not the only softly broken extension of the MSSM which is possible. It is interesting to observe that all models (with gravity induced supersymmetry breaking) could be obtained by an adequate hidden sector and their corresponding fundamental functions. The more general form of superpotential (when R–parity is conserved) is given by (8.2) thus all models can be deduced from this general form.

We now conclude considering a slight generalisation of (8.3) and (8.4) although not as general as (8.2):

$$\begin{aligned}
K(Z,\Phi,Z^\dagger,\Phi^\dagger) &= \hat{K}(Z,Z^\dagger) + \Lambda_U(Z,Z^\dagger)H_U^\dagger H_U + \Lambda_D(Z,Z^\dagger)H_D^\dagger H_D \\
&\quad + \sum_{f=1,2,3}\sum_a \Lambda_{af}(Z,Z^\dagger)\Phi^\dagger_{fa}\Phi^{fa} \\
&\quad + \Big[\Gamma(Z,Z^\dagger)H_U\cdot H_D + \text{h.c.}\Big]\,, \\
W(Z,\Phi) &= \hat{W}(Z) + \sum_{f=1,2,3}\Big(- H_{ef}(Z)\, y_{ef} L_L^f\cdot H_D E_L^f - H_{df}(Z)\, y_{df} Q_L^f \cdot H_D D_L^f \\
&\quad + H_{nf}(Z)\, y_{nf} L_L^f \cdot H_U N_L^f + H_{uf}(Z)\, y_{uf} Q_L^f\cdot H_U U_L^f \\
&\quad + \frac{1}{2}H'_f(Z)\, m_f (N_L^f)^2\Big)\,,
\end{aligned} \tag{8.15}$$

with $\hat{K}$ and $\hat{W}$ subject to a no-scale condition (8.5). Here, more $\Lambda-$ and $H-$functions need to be introduced. Consequently, the number of mass parameters in V_{Soft} is larger. However, the corresponding softly broken part of the scalar potential can be easily deduced from (8.9) and (8.11-8.13). For instance we get

$$\begin{aligned}
A_{ef} &= \Bigg[\frac{1}{m_p}\langle\partial_i\hat{K}\rangle\hat{y}_{ef} + m_p\partial_i\hat{y}_{ef} - m_p\frac{\langle\partial_i\Lambda_{Lf}\rangle}{\langle\Lambda_{Lf}\rangle}\hat{y}_{ef} \\
&\quad - m_p\frac{\langle\partial_i\Lambda_{Ef}\rangle}{\langle\Lambda_{Ef}\rangle}\hat{y}_{ef} - m_p\frac{\langle\partial_i\Lambda_D\rangle}{\langle\Lambda_D\rangle}\hat{y}_{ef}\Bigg]\langle\rho^i\rangle \\
&= a_{ef}y_{ef}\,,
\end{aligned}$$

with

$$\hat{y}_{ef} = e^{\frac{1}{2}\frac{\langle \hat{\mathcal{K}} \rangle}{m_p^2}} \frac{\langle H_{ef} \rangle}{\sqrt{\Lambda_D \Lambda_{ef} \Lambda_{\ell f}}} y_{ef} ,$$

$$\partial_i \hat{y}_{ef} = e^{\frac{1}{2}\frac{\langle \hat{\mathcal{K}} \rangle}{m_p^2}} \frac{\langle \partial_i H_{ef} \rangle}{\sqrt{\Lambda_D \Lambda_{ef} \Lambda_{\ell f}}} y_{ef} .$$

This ansatz makes it possible to reproduce all gravity mediated supergravity models considered in [Djouadi et al. (1998)].

The fact that supersymmetry has not been discovered in colliders does not mean that supersymmetry or supergravity are experimentally excluded. In fact the MSSM is the minimal supersymmetric extension of the Standard Model, and, in particular, there exist many possible generalisations of the MSSM. Secondly, the mechanism that breaks supergravity is essential for the phenomenological predictions of the model. In this book we have only concentrated on one mechanism to break supergravity, but there exist several other mechanisms.

8.2 Supergravity in cosmology

In this section we address the second application of supergravity in physics, namely in cosmology. Some basic general points concerning inflation are briefly recalled and the rôle of scalar fields is put forward. Recent results concerning Higgs inflation are also analysed. In the second part superconformal supergravity, in the Jordan frame, is introduced. A specific superconformal supergravity model with the MSSN content (see Table 8.1) and including an extra gauge singlet is introduced in relation to an inflationary phase in the Universe.

8.2.1 *Some basic elements on inflation*

For more details related to inflation and cosmology, one can see *e.g.* [Baumann (2011)]. Following isotropy and homogeneity principles, the metric describing the spacetime of the Universe is the Friedmann-Lemaître-Robertson-Walker metric

$$\mathrm{d}s^2 = \mathrm{d}t^2 - a(t)^2 \Big(\frac{\mathrm{d}r^2}{1 - \mathcal{K} r^2} + r^2 (\mathrm{d}\theta^2 + \sin\theta^2 \mathrm{d}\varphi^2) \Big) ,$$

where $a(t)$ is the scale factor and $\mathcal{K} = -1, 0, 1$ for space with negative, null, or positive constant curvature.

Consider a scalar field Φ minimally coupled to gravity:

$$S = \int \mathrm{d}^4 x \, \sqrt{-g} \Big(\frac{1}{2} R + \frac{1}{2} g^{\tilde{\mu}\tilde{\nu}} \partial_{\tilde{\mu}} \Phi \partial_{\tilde{\nu}} \Phi - V(\Phi) \Big) ,$$

with R the scalar curvature and $V(\Phi)$ the scalar potential. The energy-momentum tensor of the scalar field is

$$T_{\tilde{\mu}\tilde{\nu}} = \frac{2}{\sqrt{-g}} \frac{\delta S}{\delta g^{\tilde{\mu}\tilde{\nu}}} .$$

From

$$\det g = g = \sum_{\tilde{\nu}} g_{\tilde{\mu}\tilde{\nu}} \Delta^{\tilde{\nu}\tilde{\mu}} = \sum_{\tilde{\nu}} g_{\tilde{\mu}\tilde{\nu}} g g^{\tilde{\mu}\tilde{\nu}} ,$$

with $\Delta^{\tilde{\mu}\tilde{\nu}}$ the cofactor of the matrix $g_{\tilde{\mu}\tilde{\nu}}$ we obtain

$$T_{\tilde{\mu}\tilde{\nu}} = \partial_{\tilde{\mu}}\Phi\partial_{\tilde{\nu}}\Phi - g_{\tilde{\mu}\tilde{\nu}}\Big(\frac{1}{2}\partial^{\tilde{\rho}}\Phi\partial_{\tilde{\rho}}\Phi - V(\Phi)\Big) .$$

If we assume that the scalar field is constant, *i.e.*, if we assume that it does not depends on x^i but only on $x^0 = t$, the energy-momentum tensor simplifies and takes the form of a perfect fluid with pression P and density ρ

$$\begin{aligned} T_{00} &= \rho = \frac{1}{2}\dot{\Phi}^2 + V(\Phi) , \\ T_{\tilde{i}\tilde{j}} &= -g_{\tilde{i}\tilde{j}}\, P = -g_{\tilde{i}\tilde{j}} \Big(\frac{1}{2}\dot{\Phi}^2 - V(\Phi)\Big) . \end{aligned}$$

The Einstein equations reduce to

$$\begin{aligned} H^2 &= \frac{1}{3}\rho - \frac{\mathcal{K}}{a^2} , \\ \frac{\ddot{a}}{a} &= -\frac{1}{6}(\rho + 3P) , \end{aligned} \tag{8.16}$$

and the Euler-Lagrange equations of the scalar field are given by

$$\ddot{\Phi} + 3H\dot{\Phi} + \partial_\Phi V = 0 , \tag{8.17}$$

with

$$H = \frac{\dot{a}}{a} ,$$

the Hubble parameter. The Hubble parameter characterises the expansion of the Universe. Thus if $H < 0$ the Universe is in contraction whereas if $H > 0$ the Universe is in expansion.

Now, if we define

$$\varepsilon = -\frac{\dot{H}}{H^2} ,$$

from

$$\frac{\ddot{a}}{a} = H^2(1 - \varepsilon) , \tag{8.18}$$

if

$$\varepsilon = -\frac{\dot{H}}{H^2} < 1\,, \tag{8.19}$$

we have a Universe in acceleration or in inflation. In order to have a slow-roll inflation (see below), more conditions must be imposed.

On the one hand we assume that the scalar density dominates the curvature term. Then the first equation of (8.16) becomes

$$H^2 = \frac{1}{3}\rho = \frac{1}{3}\Big(\frac{1}{2}\dot{\Phi}^2 + V(\Phi)\Big)\,.$$

Further, using the second equation of (8.16), we obtain from (8.18)

$$\frac{1}{3}(1-\varepsilon)\rho = -\frac{1}{6}(\rho + 3P)\,,$$

and

$$\begin{aligned}\varepsilon &= \frac{3}{2}\Big(\frac{P}{\rho} + 1\Big)\\ &= \frac{3}{2}\Bigg(\frac{\frac{1}{2}\dot{\Phi}^2 - V}{\frac{1}{2}\dot{\Phi}^2 + V} + 1\Bigg)\\ &= \frac{1}{2}\frac{\dot{\Phi}^2}{H^2}\,.\end{aligned}$$

The limit where $\rho \to -P$ is then equivalent to the limit where $\varepsilon \to 0$ and this is equivalent to

$$\dot{\Phi}^2 \ll V\,.$$

On the other hand we suppose

$$\big|\ddot{\Phi}\big| \ll 3\big|H\dot{\Phi}\big|\,,\ \ \big|\partial_\Phi V\big|\,.$$

This last assumption can be cast into the form

$$\eta = -\frac{\ddot{\Phi}}{H\dot{\Phi}} \ll 1\,. \tag{8.20}$$

The condition (8.19) and (8.20) are known as the slow-roll conditions. Under these conditions, using (8.17), the first equation of (8.16) becomes

$$\begin{aligned}H^2 &= \frac{1}{3}V(\Phi) \sim \text{const.} = H_0^2\,,\\ \dot{\Phi} &\sim -\frac{\partial_\Phi V}{3H}\,.\end{aligned}$$

The first of these equations leads to

$$\dot{a} = H_0 a$$

whose solution is

$$a(t) = H_0 e^{tH} ,$$

and the Universe is in a slow-roll inflationary phase, *i.e.*, its size increases exponentially.

It can be shown that the two slow-roll conditions can be written in an approximate form as [Baumann (2011)]

$$\varepsilon_V = \frac{1}{2}\left(\frac{\partial_\Phi V}{V}\right)^2 < 1 ,$$

$$\eta_V = \frac{\partial^2_\Phi V}{V} \ll 1 .$$

8.2.2 *Higgs inflation*

In the previous section, we have seen that under certain conditions, a scalar field can induce an inflationary phase in the Universe. Such a field is consequently called an inflaton. It is then natural to ask if the *only* scalar field of the Standard Model of Particle Physics, namely the Higgs boson, can play the rôle of the inflaton. This question finds an answer in [Bezrukov and Shaposhnikov (2008); Bezrukov et al. (2009)]. To that purpose, they introduced the coupling of the Higgs field with gravity in a Jordan frame

$$S = m_p^2 \int \mathrm{d}x^4 \sqrt{-g}\left[\frac{1+\xi h^2}{2} R + \frac{1}{2} g^{\tilde{\mu}\tilde{\nu}} \partial_{\tilde{\mu}} h \partial_{\tilde{\nu}} h - \frac{\lambda}{4} m_p^2 \left(h^2 - \frac{v^2}{m_p^2}\right)^2\right] , \quad (8.21)$$

where h is the Higgs boson (in dimensionless units), ξ is a constant and v/m_p is the vacuum expectation value of h. Note that we have reintroduced the Planck mass for this analysis.

The gravitational part of the action is multiplied by the frame function

$$\Phi = -3(1 + \xi h^2) .$$

We have seen in Sec. 6.8 that a conformal transformation

$$g^{\tilde{\mu}\tilde{\nu}} \to (1 + \xi h^2) g^{\tilde{\mu}\tilde{\nu}} ,$$

enables one to obtain the action in an Einstein frame

$$S = m_p^2 \int \mathrm{d}x^4 \sqrt{-g}\left[\frac{1}{2} R + \frac{1}{2}\frac{1 + \xi(1+6\xi)h^2}{(1+\xi h^2)^2} \partial_{\tilde{\mu}} h \partial^{\tilde{\mu}} h - \frac{\lambda}{4} m_p^2 \left(\frac{h^2 - \frac{v^2}{m_p^2}}{1+\xi h^2}\right)^2\right] .$$

At this stage, the field h is not canonically normalised. To obtain a correctly normalised field, we introduce Ψ, the solution of the equation

$$\frac{\mathrm{d}\Psi}{\mathrm{d}h} = \frac{\sqrt{1+\xi(1+6\xi)h^2}}{1+\xi h^2} m_p \,,$$

and obtain

$$\mathcal{S} = \int \mathrm{d}x^4 \sqrt{-g} \left[\frac{m_p^2}{2} R + \frac{1}{2} \partial_{\tilde{\mu}} \Psi \partial^{\tilde{\mu}} \Psi - \frac{\lambda}{4} m_p^4 \left(\frac{h^2(\Psi) - \frac{v^2}{m_p^2}}{1+\xi h(\Psi)^2} \right)^2 \right],$$

where h is now a function of Ψ

$$\frac{\Psi}{m_p} = \sqrt{6 + \frac{1}{\xi}} \operatorname{arcsinh}\left[h \sqrt{\xi(1+6\xi)} \right] - \sqrt{6} \operatorname{arctanh}\left[\frac{\sqrt{6}\xi h}{\sqrt{1+\xi(1+6\xi)h^2}} \right]. \quad (8.22)$$

At least two limiting cases are worthy of discussion.

Low energy limit.

In this limit, we are at low energy, *e.g.*, at the electroweak scale. When $h \sim 0$ we obtain

$$\Psi \sim m_p h \,,$$

and

$$\frac{\lambda}{4} m_p^2 \left(\frac{h^2(\Psi) - \frac{v^2}{m_p^2}}{1+\xi h(\Psi)^2} \right)^2 \sim \frac{\lambda}{4} (\Psi^2 - v^2)^2 \,,$$

and we reproduce the Lagrangian of the Higgs boson canonically normalised with the standard Higgs potential. It should be observed that in this limit the Einstein frame and the Jordan frame approximately coincide.

High energy limit.

Here we assume $\xi h^2 \gg 1$ and $\xi \gg 1$. In this situation, expanding (8.22) we obtain

$$\Psi = m_p \sqrt{\frac{3}{2}} \ln(\xi h^2) \,,$$

and the potential reads

$$V = \frac{\lambda}{4} \frac{m_p^4}{\xi} \left(1 - e^{-\sqrt{\frac{2}{3}} \frac{\Psi}{m_p}} \right)^2 .$$

Next the slow-roll parameters become

$$\varepsilon_V = \frac{m_P^2}{2}\Big(\frac{V'}{V}\Big)^2 = \frac{4}{3}\frac{1}{\xi^2 h^4} < 1\,,$$

$$\eta_V = m_P^2 \frac{V''}{V} = -\frac{4}{3}\frac{1}{h^2 \xi}\,,$$

and the conditions for slow-roll inflation are satisfied.

8.2.3 *Superconformal supergravity*

Before studying inflation in supergravity, we would like to introduce a class of supergravity models called superconformal supergravity [Ferrara et al. (2011)]. Let us denote Z^i the chiral multiplets and $\Phi(Z, \bar{Z})$, $W(Z)$ the kinetic function (or the frame function) and the superpotential. Further, let us introduce a compensating chiral multiplet X^0, denote $Z^I = (X^0, Z^i)$ and define the chiral superfields

$$U^i = \frac{\sqrt{3} Z^i}{X^0}\,.$$

A superconformal supergravity model does satisfy the three following assumptions:

(1) The superpotential is homogeneous of degree three. We may thus define

$$\tilde{W}(Z^I) = \frac{1}{3\sqrt{3}}(X^0)^3 W(U^i) = W(Z^i)\,.$$

(2) If we define

$$\Omega(Z^I, Z_I^\dagger) = \frac{1}{3} X^0 X_0^\dagger \Phi(U^i, U_{i^*}^\dagger)\,,$$

then Ω is homogeneous of degree one in its holomorphic and anti-holomorphic parts (the U–fields have zero degree of homogeneity).

(3) The compensating field X^0 has a conformal weight -2 and the fields U^i have a conformal weight 0. Consequently the action is conformal invariant (see Sec. 6.7). This artificial conformal invariance can be broken by fixing the gauge

$$X^0 = X_0^\dagger = \sqrt{3}\,,$$
$$U^i = Z^i\,,$$
$$U_{i^*}^\dagger = Z_{i^*}^\dagger\,.$$

If we define the Kähler potential as usual by

$$K(Z, Z^\dagger) = -3\ln\Big(-\frac{\Phi}{3}\Big)\,,$$

we obtain

$$\begin{aligned}
K_i &= \frac{\partial K}{\partial Z^i} = -3\frac{\Phi_i}{\Phi}\,,\\
K^{i^*} &= \frac{\partial K}{\partial Z^\dagger_{i^*}} = -3\frac{\Phi^{i^*}}{\Phi}\,,\\
K^{i^*}{}_j &= \frac{\partial^2 K}{\partial Z^j \partial Z^\dagger_{i^*}} = -3\frac{\Phi^{i^*}{}_j}{\Phi} + 3\frac{\Phi^{i^*}\Phi_j}{\Phi^2}\,.
\end{aligned} \tag{8.23}$$

In the Jordan frame the scalar potential is given by

$$\begin{aligned}
V_J &= \frac{\Phi^2}{9} e^K \Big(\mathcal{D}_i W (K^{-1})^i{}_{j^*} \mathcal{D}^{j^*} W^\star - 3|W|^2\Big)\\
&= -\frac{3}{\Phi}\Big(\mathcal{D}_i W (K^{-1})^i{}_{j^*} \mathcal{D}^{j^*} W^\star - 3|W|^2\Big)\,.
\end{aligned} \tag{8.24}$$

However, as we will now show, due to conformal invariance the scalar potential takes on another expression in terms of the kinetic function Ω, and its associated metric:

$$\Omega^{I^*}{}_J = \frac{\partial^2 \Omega}{\partial X^J \partial X^\dagger_{I^*}} = \begin{cases}
\Omega^0{}_0 = \frac{1}{3}\Phi\\
\Omega^{i^*}{}_0 = \frac{1}{3} X^\dagger_0 \Phi^{i^*}\\
\Omega^0{}_i = \frac{1}{3} X^0 \Phi_i\\
\Omega^{i^*}{}_j = \frac{1}{3} X^0 X^\dagger_0 \Phi^{i^*}{}_j
\end{cases}\,,$$

which has the form

$$\begin{aligned}
\Omega^{I^*}{}_J &= \begin{pmatrix} \Omega^{i^*}{}_i & \Omega^{i^*}{}_0\\ \Omega^0{}_i & \Omega^0{}_0 \end{pmatrix}\\
&= \begin{pmatrix} A & B\\ C & D\end{pmatrix}\,.
\end{aligned}$$

Using (5.5), the inverse of the metric reduces to

$$(\Omega^{-1})^I{}_{I^*} = \begin{pmatrix} (A - BD^{-1}C)^{-1} & -(A - BD^{-1}C)^{-1}BD^{-1}\\ -D^{-1}C(A-BD^{-1}C)^{-1} & D^{-1}C(A-BD^{-1}C)^{-1}BD^{-1} + D^{-1}\end{pmatrix}\,.$$

This can also be directly checked by multiplying the two matrices. Since

$$(A - BD^{-1}C)^{i^*}{}_i = -\frac{1}{9}\Phi X^0 X^\dagger_0 K^{i^*}{}_i\,,$$

where $K^{i^*}{}_i$ is given in (8.23) we easily obtain

$$(\Omega^{-1})^I{}_{J^*} = \begin{cases}
(\Omega^{-1})^0{}_0 = -\frac{9}{\Phi^3}\Phi_i (K^{-1})^i{}_{j^*}\Phi^{j^*} + \frac{3}{\Phi} = -\frac{1}{\Phi}K_i (K^{-1})^i{}_{j^*} K^{j^*} + \frac{3}{\Phi}\\
(\Omega^{-1})^i{}_0 = \frac{9}{\Phi^2}\frac{1}{X^0}(K^{-1})^i{}_{j^*}\Phi^{j^*} = -\frac{3}{\Phi}\frac{1}{X^0}(K^{-1})^i{}_{j^*}K^{j^*}\\
(\Omega^{-1})^0{}_{j^*} = \frac{9}{\Phi^2}\frac{1}{X^\dagger_0}\Phi_i (K^{-1})^i{}_{j^*} = -\frac{3}{\Phi}\frac{1}{X^\dagger_0}K_i (K^{-1})^i{}_{j^*}\\
(\Omega^{-1})^i{}_{j^*} = -\frac{9}{\Phi}\frac{1}{X^0 X^\dagger_0}(K^{-1})^i{}_{j^*}
\end{cases}\,. \tag{8.25}$$

Therefore, an alternative expression of the scalar potential in the Jordan frame can be derived:

$$V_J = \partial_I \tilde{W} (\Omega^{-1})^I{}_{J^*} \partial^{J^*} \tilde{W}^\star \big|_{X^0 = X_0^\dagger = \sqrt{3}} \,. \tag{8.26}$$

From now on we were considering generic models with an unspecified frame function. We now introduce

$$\Phi(Z, Z^\dagger) = -3 + Z^i Z_i^\dagger - h(Z) - h^\star(Z^\dagger) \,,$$

where h is an arbitrary holomorphic function. Note that due to the canonical form of the frame function we use the notations of Sec. 6.5 for anti-chiral superfields. In this case a compact expression for the scalar potential can be obtained either from (8.24) or from (8.26) since the various metrics can be easily inverted.

In the first case, *i.e.*, computing the scalar potential directly, for the Kähler metric and its inverse we have

$$\begin{aligned}
K^i{}_j &= -3\frac{\delta^i{}_j}{\Phi} + 3\frac{(Z^i - h^{\star i})(Z_j^\dagger - h_j)}{\Phi^2} \,, \\
(K^{-1})^i{}_j &= \frac{1}{3}\frac{\Phi^3}{\Phi - (Z^k - h^{\star k})(Z_k^\dagger - h_k)} \\
&\quad \times \Big[-\frac{\delta^i{}_j}{\Phi} + \delta^i{}_j \frac{(Z^k - h^{\star k})(Z_k^\dagger - h_k)}{\Phi^2} - \frac{(Z^i - h^{\star i})(Z_j^\dagger - h_j)}{\Phi^2} \Big] \,.
\end{aligned}$$

In the second case, *i.e.*, using superconformal methods, we have

$$\Omega = Z^0 Z_0^\dagger \Big[-1 + \frac{1}{3} U^i U_i^\dagger - \frac{1}{3} h(U) - \frac{1}{3} h^\star(U^\dagger) \Big] \,,$$

and the metric and its inverse are given by

$$\begin{aligned}
\Omega^0{}_0 &= -1 + \frac{1}{3} U^i U_i^\dagger - \frac{1}{3} h - \frac{1}{3} h^\star \,, \\
\Omega^i{}_0 &= \frac{1}{3} X_0^\dagger (U^i - h^{\star i}) \,, \\
\Omega^0{}_i &= \frac{1}{3} X^0 (U_i^\dagger - h_i) \,, \\
\Omega^i{}_j &= \frac{1}{3} \delta^i{}_j X^0 X_0^\dagger \,,
\end{aligned}$$

and

$$(\Omega^{-1})^0{}_0 = \frac{1}{D} ,$$
$$(\Omega^{-1})^0{}_i = -\frac{1}{D}\frac{1}{X_0^\dagger}(U_i^\dagger - h_i) ,$$
$$(\Omega^{-1})^i{}_0 = -\frac{1}{D}\frac{1}{X^0}(U^i - h^{\star i}) ,$$
$$(\Omega^{-1})^i{}_j = 3\frac{\delta^i{}_j}{X^0 X_0^\dagger} + \frac{1}{D}\frac{1}{X^0 X_0^\dagger}(U^i - h^{\star i})(U_i^\dagger - h_i) ,$$

where

$$D = -1 + \frac{1}{3}U^k U_k^\dagger - \frac{1}{3}h - \frac{1}{3}h^\star - \frac{1}{3}(U^k - h^{\star k})(U_k^\dagger - h_k) .$$

Note that

$$X^0 X_0^\dagger D\Big|_{X^0 = X_0^\dagger = \sqrt{3}} = \Phi - (Z^k - h^{\star k})(Z_k^\dagger - h_k) .$$

A direct computation then gives (using $Z^i \partial_i W = 3W$ because W is cubic)

$$\begin{aligned} V_J &= \frac{\Phi^2}{9} e^K \Big(\mathcal{D}_i W (K^{-1})^i{}_j \mathcal{D}^j W^\star - 3|W|^2 \Big) \\ &= \partial_I \tilde{W} (\Omega^{-1})^I{}_J \partial^J \tilde{W}^\star \Big|_{X^0 = X_0^\dagger = \sqrt{3}} \\ &= \partial_i W \partial^i W^\star + \frac{\Big(h^{\star i}\partial_i W\Big)\Big(h_j \partial^j W^\star\Big)}{\Phi - (Z^k - h^{\star k})(Z_k^\dagger - h_k)} . \end{aligned}$$

We now assume that the function h is quadratic, *i.e.*,

$$\begin{aligned} W &= \frac{1}{6}\lambda_{ijk} Z^i Z^j Z^k , \\ \Phi &= -3 + Z^i Z_i^\dagger - \frac{1}{2}\chi_{ij} Z^i Z^j - \frac{1}{2}\chi^{ij} Z_i^\dagger Z_j^\dagger . \end{aligned}$$

The scalar potential thus reduces to

$$\begin{aligned} V_J &= \partial_i W \partial^i W^\star + \frac{\Big(\chi^{j\ell}\partial_j W Z_\ell^\dagger\Big)\Big(\chi_{ik}\partial^i W^\star z^k\Big)}{\Phi - (Z^i - \chi^{ik} Z_k^\dagger)(Z_i^\dagger - \chi_{i\ell} Z^\ell)} \\ &= \partial_i W \partial^i W^\star + \frac{1}{4}\frac{(\chi^{j\ell}\lambda_{jmn} Z^m Z^n Z_\ell^\dagger)(\chi_{ik}\lambda^{irs} Z_r^\dagger Z_s^\dagger Z^k)}{\Phi - (Z^k - \chi^{ik} Z_i^\dagger)(Z_k^\dagger - \chi_{jk} Z^j)} . \end{aligned}$$

It has been pointed out in [Ferrara et al. (2011)] that if the frame function is canonical, *i.e.*, if $h = 0$ then

$$\Phi = -3 + Z^i Z_i^\dagger \,,$$

and

$$V_J = \partial_i W \partial^i W^\star \,.$$

So, the scalar potential takes exactly the same form as in supersymmetry. Such models are called canonical superconformal supergravity.

One model that will be studied in the next section, is the Next to Minimal Supersymmetric-Standard Model (NMSSM). The field content of the NMSSM is the field content of the MSSM (see Table 8.1) with an additional gauge singlet denoted S. This model was studied by many authors [Fayet (1975); Ellwanger et al. (2010, 1997)] (and references therein). In this model, the superpotential is simply given by the superpotential of the MSSM (see Table 8.1) where the μ-term (*i.e.*, the $H_U \cdot H_D$-term) is replaced by a cubic term:

$$\begin{aligned} W = &-y_{eff'} L_L^f \cdot H_D E_L^{f'} - y_{dff'} Q_L^f \cdot H_D D_L^{f'} + y_{nff'} L_L^f \cdot H_U N_L^{f'} + y_{uff'} Q_L^f \cdot H_U U_L^{f'} \\ &+ \frac{1}{2} m_{ff'} N_L^f N_L^{f'} + \lambda S\, H_D \cdot H_U + \frac{1}{3} S^3 \,. \end{aligned}$$

In particular, the Higgs part of the superpotential takes the form

$$W_{\text{Higgs}} = \lambda S\, H_U \cdot H_D + \frac{1}{3} \rho S^3 \,.$$

The Jordan frame function is now taken to be[6]

$$\Phi = -3 + |S|^2 + H_U^\dagger H_U + H_D^\dagger H_D - \frac{3}{2}\chi_1\big(H_U \cdot H_D + H_U^\dagger \cdot H_D^\dagger\big) - \frac{3}{2}\chi_2\big(S^2 + (S^\dagger)^2\big) \,.$$

This model is a specific superconformal supergravity model with Kähler potential (reintroducing the Planck mass)

$$\begin{aligned} K &= -3m_p^2 \ln(-\Phi/3m_p^2) \\ &= -3m_p^2 \ln\Bigg[1 - \frac{1}{3m_p^2}\big(|S|^2 + H_U^\dagger H_U + H_D^\dagger H_D\big) \\ &\qquad + \frac{1}{2m_p^2}\chi_1\big(H_U \cdot H_D + H_U^\dagger \cdot H_D^\dagger\big) + \frac{1}{2m_p^2}\chi_2\big(S^2 + (S^\dagger)^2\big)\Bigg] \,. \end{aligned}$$

[6]Recall that H_U and H_D are doublets (see Table 8.1.)

At low energy, *i.e.*, when the fields are much below the Planck mass, we have

$$K \sim |S|^2 + H_U^\dagger H_U + H_D^\dagger H_D$$
$$-\frac{3}{2}\chi_1\left(H_U \cdot H_D + H_U^\dagger \cdot H_D^\dagger\right) - \frac{3}{2}\chi_2\left(S^2 + (S^\dagger)^2\right), \qquad (8.27)$$

and the fields are canonically normalised.

To make a connection with Sec. 8.1.2, we assume that there exists a hidden sector and that supergravity is broken in this sector. We consider $\tilde{W} = W + W_{\text{hidden}}$, $\tilde{K} = K + K_{\text{hidden}}$ with $K = -3\ln(-\Phi/3)$ where W, and K are the superpotential and the Kähler potential of the NMSSM whereas W_{hidden}, and K_{hidden} are the superpotential and the Kähler potential of the hidden sector. The scalar potential can be obtained using (8.24) where the summation must now include the fields in the hidden sector. Furthermore, if we assume that supergravity is broken in the hidden sector, and if we develop the fields of the hidden sector around their vacuum expectation value, the full potential is multiplied by $e^{\langle K_{\text{hidden}}\rangle/m_p^2}$. It is not our goal to reproduce here the full computation along the lines of Sec. 8.1.3, but we can easily analyse the consequences of the χ_1, χ_2 terms in (8.27). Firstly, when supergravity is broken the effective superpotential is redefined by

$$\tilde{W}_{\text{eff}} = e^{\frac{1}{2}\frac{1}{m_p^2}\langle K_{\text{hidden}}\rangle}\tilde{W},$$

and secondly, the gravitino becomes massive

$$m_{\frac{3}{2}} = \frac{1}{m_p^2}\langle W_{\text{hidden}}\rangle e^{\frac{1}{2}\frac{1}{m_p^2}\langle K_{\text{hidden}}\rangle}.$$

If we perform a Kähler transformation with

$$F = \frac{3}{2}\chi_1\, H_U \cdot H_D + \frac{3}{2}\chi_2\, S^2,$$

and assuming to be at low energy, we get

$$\tilde{K} \to \tilde{K} + F + F^\star = K_{\text{hidden}} + |S|^2 + H_U^\dagger H_U + H_D^\dagger H_D$$
$$\sim |S|^2 + H_U^\dagger H_U + H_D^\dagger H_D, \qquad (8.28)$$
$$\tilde{W} \to e^{-F/m_p^2}\tilde{W} \sim \tilde{W}_{\text{eff}} - \frac{\langle W_{\text{hidden,eff}}\rangle}{m_p^2}F$$
$$\sim W_{\text{eff}} + \frac{3}{2}m_{\frac{3}{2}}\chi_1\, H_U \cdot H_D + \frac{3}{2}m_{\frac{3}{2}}\chi_2 S^2,$$

we are lead to a canonical Kähler potential at low energy, and the quadratic terms are transfered to the superpotential and are controled by the gravitino mass. This is an analogue of the Giudice-Masiero mechanism studied previously. This mechanism is welcomed since when S takes a vacuum expectation value, it induces naturally an effective quadratic coupling between the two Higgs boson:

$$\Big(\frac{3}{2}m_{\frac{3}{2}}\chi_1 + \lambda\langle S\rangle\Big)H_U \cdot H_D = \mu_{\text{eff}} H_u \cdot H_D \ ,$$

and generates the last term in (8.2). In contrast to the original MSSM where this term is expected to be of the order of the Planck mass, here it is controlled by the gravitino mass and the vacuum expectation value of S and consequently $\mu_{\text{eff}} \ll m_p$. This is a possible solution of the $\mu-$problem of the MSSM [Ellwanger et al. (2010, 1997)].

We end this section with a general observation concerning superconformal models. One of the basic assumptions in order to have superconformal invariance pertains to the superpotential. In particular W must be homogeneous of degree three in the superfields:

$$W(Z) = \frac{1}{6}\lambda_{ijk}Z^iZ^jZ^k \ .$$

Consequently, W is invariant under a discrete $\mathbb{Z}_3$ symmetry:

$$Z^i \to e^{\frac{2i\pi}{3}} Z^i \ .$$

If this invariance is not broken such models suffer from domain wall problems. Indeed, if we have a minimum of the potential where the scalar part of the superfields takes a vacuum expectation value $\langle Z^i\rangle = v^i$, due to the $\mathbb{Z}_3-$symmetry automatically there exist three different solutions $\langle Z^i\rangle = e^{k\frac{2i\pi}{3}}v^i, k = 0, 1, 2$. Since the symmetry is discrete these three solutions cannot be continuously connected and the Universe has three different disconnected vacua. This is the domain wall problem and it affects the NMSSM. There are several ways to cure it. However, since the frame function contains an additional holomorphic function $h(Z)$, when this function is quadratic, this term explicitly breaks the discrete $\mathbb{Z}_3$ symmetry and solves the domain wall problem.

To conclude there are at least five reasons to consider superconformal supergravity, and in particular to consider the NMSSM in this context:

(1) In the Jordan frame the kinetic terms are canonically normalised and at low energy the Jordan frame coincides with the Einstein frame;

(2) The superpotential is cubic and the coupling constant (Yukawa terms) are dimensionless;

(3) After choosing the gauge $X^0 = X_0^\dagger = \sqrt{3}$ the conformal invariance is broken;

(4) At low energy, when supergravity is broken, the additional quadratic terms in the frame functions can be transfered, *via* a Kähler transformation analogous to (8.28), to quadratic terms in the effective superpotential. However the mass parameters so generated are much below the Planck mass and controlled by the gravitino mass;

(5) At high energy the quadratic terms in the frame function may play a rôle in the inflation in the early Universe.

The last point above is addressed briefly in the next section.

8.2.4 *Higgs inflation in supergravity*

A priori supergravity has all the ingredients to induce inflation naturally, since there exists many flat directions in the scalar potential. Several inflationary models induced by supergravity have been studied. More to the point, in [Einhorn and Jones (2010); Ferrara et al. (2010, 2011)] inflationary models were proposed in supergravity where the rôle of the inflaton is played by a Higgs boson. In the first paper, it was shown that within the MSSM we cannot have inflation whereas in the three papers it was argued that the presence of the singlet field of the NMSSM considerably improves the situation. Cosmology in Jordan and Einstein frames has been analysed in some detail in the last two papers.

We now compute the scalar potential. So far we have only considered the F-part of the potential, but now we also have to introduce the D-terms. Recall that the Higgs fields are doublets:

$$H_U = \begin{pmatrix} H_U^+ \\ H_U^0 \equiv H_u \end{pmatrix},$$

$$H_D = \begin{pmatrix} H_D^0 \equiv H_d \\ H_D^- \end{pmatrix},$$

where the superscript indicates the electromagnetic charge of the corresponding field. Here, we will only consider the neutral part of the Higgs boson, say H_u and H_d. For the F−terms, form (8.26) and taking into account that

$$\chi_{ud} = \chi_{du} = \frac{3}{2}\chi_1 \ , \quad \chi_{ss} = 3\chi_2 \ ,$$

the frame function becomes

$$\Phi = -3 + |S|^2 + H_U^\dagger H_U + H_D^\dagger H_D - \frac{3}{2}\chi_1(H_U \cdot H_D + H_U^\dagger \cdot H_D^\dagger) - \frac{3}{2}\chi_2(S^2 + \bar{S}^2) \, .$$

Considering only the neutral part of the Higgs bosons H_u, H_d, the potential in the Jordan frame takes the form:

$$V_J^F = |\lambda|^2(|H_u|^2 + |H_d|^2)|S|^2 + |\lambda H_u H_d + \rho S^2|^2$$
$$\frac{9\left|\frac{1}{2}\lambda\chi_1(|H_u|^2 + |H_d|^2)S + \chi_2\bar{S}(\lambda H_u H_d + \rho S^2)\right|^2}{-3 + \frac{3}{2}\chi_1(H_u H_d + \bar{H}_u\bar{H}_d) + \frac{3}{2}\chi_2(S^2 + \bar{S}^2) - \frac{9}{4}\chi_1^2(|H_u|^2 + |H_d|^2) - 9\chi_2^2|S|^2}\,.$$

To obtain the potential in the the Einstein frame we only have to multiply V_J by $9/\Phi^2$, thus

$$V_E^D = \frac{9}{\left[-3+|H_u|^2+|H_d|^2+|S|^2-\frac{3}{2}\chi_1(H_u H_d+\bar{H}_u\bar{H}_d)-\frac{3}{2}\chi_2\left(S^2+\bar{S}^2\right)\right]^2}$$
$$\times\left\{\lambda^2(|H_u|^2+|H_d|^2)|S|^2+|\lambda H_u H_d+\rho S^2|^2\right.$$
$$\left.+\frac{9\left|\frac{1}{2}\lambda\chi_1(|H_u|^2+|H_d|^2)S+\chi_2\bar{S}(\lambda H_u H_d+\rho S^2)\right|^2}{-3+\frac{3}{2}\chi_1(H_u H_d+\bar{H}_u\bar{H}_d)+\frac{3}{2}\chi_2(S^2+\bar{S}^2)-\frac{9}{4}\chi_1^2(|H_u|^2+|H_d|^2)-9\chi_2^2|S|^2}\right\}.$$

Since S is a singlet, only H_U and H_D contribute to the D–terms. From their quantum numbers given in Table 8.1 we obtain

$$V_J^D = \frac{\Phi^2}{9}\left[\frac{1}{8}g_2^2\left(K_U\sigma_i H_U + K_D\sigma_i H_D\right)\left(K_U\sigma_j H_U + K_D\sigma_j H_D\right)\delta^{ij}\right.$$
$$\left.+\frac{1}{8}g_1^2(K_U H_U - K_D H_D)^2\right],$$

where g_1 and g_2 are the coupling constants of the gauge groups $U(1)_Y$ and $SU(2)_L$ respectively. Since

$$K_U = -\frac{3H_U^\dagger}{\Phi},$$
$$K_D = -\frac{3H_D^\dagger}{\Phi},$$

considering only the neutral part of the Higgs fields we obtain

$$V_J^D = \frac{1}{8}(g_1^2 + g_2^2)\left(|H_u|^2 - |H_d|^2\right)^2,$$

in the Jordan frame and

$$V_E^D = \frac{9}{8}\frac{(g_1^2 + g_2^2)\left(|H_u|^2 - |H_d|^2\right)^2}{\left[-3 + |H_u|^2 + |H_d|^2 + |S|^2 - \frac{3}{2}\chi_1(H_u H_d + \bar{H}_u\bar{H}_d) - \frac{3}{2}\chi_2\left(S^2 + \bar{S}^2\right)\right]^2},$$

in the Einstein frame. The potential in the Jordan frame is then given by (see Sec. 6.8)

$$V_J = V_J^E + V_J^D \,.$$

According to Sec. 8.2.2, the bosonic part of the action in the Jordan frame takes the form (not taking into account the covariant derivatives for simplicity)

$$\begin{aligned} S &= \int \sqrt{-g}\Big[-\Phi/6R + \partial_{\tilde\mu} S\, \partial^{\tilde\mu} \bar S + \partial_{\tilde\mu} H_u \partial^{\tilde\mu} \bar H_u + \partial_{\tilde\mu} H_d \partial^{\tilde\mu} \bar H_d - \frac{1}{4}\frac{1}{\Phi}\mathcal{A}_{\tilde\mu}\mathcal{A}^{\tilde\mu} - V_J\Big] \\ &= \int \sqrt{-g}\Big[\Big\{\frac{1}{2} - \frac{1}{6}|S|^2 - \frac{1}{6}|H_u|^2 - \frac{1}{6}|H_d|^2 \\ &\qquad + \frac{1}{4}\chi_1(H_u H_d + \text{h.c.}) + \frac{1}{4}\chi_2(S^2 + \text{h.c.})\Big\}R \\ &\qquad - \frac{1}{4}\frac{1}{\Phi}\mathcal{A}_{\tilde\mu}\mathcal{A}^{\tilde\mu} + \partial_{\tilde\mu} S\, \partial^{\tilde\mu} \bar S + \partial_{\tilde\mu} H_u \partial^{\tilde\mu} \bar H_u + \partial_{\tilde\mu} H_d \partial^{\tilde\mu} \bar H_d - V_J\Big]\,, \end{aligned}$$

with

$$-\frac{1}{4}\frac{1}{\Phi}\mathcal{A}_\mu\mathcal{A}^\mu = -\frac{1}{4}\frac{1}{\Phi}\Big[\Phi_i \partial_{\tilde\mu} Z^i - \Phi^i \partial_{\tilde\mu} Z_i^\dagger\Big]\Big[\Phi_j \partial^{\tilde\mu} Z^j - \Phi^j \partial^{\tilde\mu} Z_j^\dagger\Big]\,,$$

where $Z^i = H_u, H_d, S, Z_i^\dagger = \bar S, \bar H_u, \bar H_d$. Note that if the fields Z^i are real $\mathcal{A}_\mu = 0$.

Before moving on, in order to make a comparison with Sec. 8.2.2 we will assume that $S = 0$ and $H_u = H_d = H/\sqrt{2}$, so that the bosonic part of the action reduces to

$$\begin{aligned} S = \int \sqrt{-g}\Big[&\Big\{\frac{1}{2} + \frac{1}{4}\Big(-\frac{1}{3} + \chi_1\Big)\big(H + \bar H\big)^2 + \frac{1}{4}\Big(\frac{1}{3} + \chi_1\Big)\big(H - \bar H\big)^2\Big\}R \\ &- \frac{1}{4}\frac{1}{\Phi}\mathcal{A}_{\tilde\mu}\mathcal{A}^{\tilde\mu} + \partial_{\tilde\mu} H \partial^{\tilde\mu} \bar H - \frac{\lambda}{4}(H\bar H)^2\Big]\,, \end{aligned}$$

which reproduces (8.21) in the limit where $H = \bar H$ or $H = -\bar H$, *i.e.*, when H is real or when H is purely imaginary. Furthermore, if we assume H to be real, then the action in the Jordan frame turns out to be canonically normalised since $\mathcal{A}_{\tilde\mu} = 0$. This motivates the following analysis.

As we have seen in Sec. 6.8, in the Einstein frame we have

$$S = \int \sqrt{-g}\Big[\frac{1}{2}R + K^i{}_j \partial_{\tilde\mu} Z^j \partial^{\tilde\mu} Z_i^\dagger - V_E\Big]\,.$$

As stated previously (see Eq.[8.27]) at low energy, the Kähler metric reduces to the canonical Kähler metric $K^i{}_j = \delta^i{}_j$.

The scalar potential, both in the Jordan and Einstein frames has a complicated shape. To perform out the analysis we have to select a specific configuration for the various fields. We now discuss the presence of the $\mathcal{A}_{\tilde\mu}\mathcal{A}^{\tilde\mu}$ term in the Jordan frame. In Sec. 8.2.2 we had to deal with a similar term, but in the Einstein frame. However, in this case since there was only one scalar, say the Higgs boson, this term has been absorbed by a field redefinition. In contrast, here we have more than one field in the definition of the $\mathcal{A}_{\tilde{\mu}}$ term, thus we cannot reabsorb this term by a field redefinition. Another possible solution to eliminate the $\mathcal{A}_{\tilde{\mu}}\mathcal{A}^{\tilde{\mu}}$ term, is to posit that there is a valley where all the fields are real such that $\mathcal{A}_\mu = 0$. This choice has been done in [Ferrara et al. (2010, 2011)]. The corresponding real fields were denoted h_u, h_d and s. When $\chi_2 = 0$ this regime is a minimum if the parameters satisfy some constraints [Ferrara et al. (2010)]. Further since the D–term has the flat direction:

$$V_D\Big|_{H_u=H_d} = 0\,,$$

by selecting the valley $h_u = h_d = h$ it is possible to show that the inflationary regime can be attained when $s = 0$. However, this regime turns out to be instable in the s-direction [Ferrara et al. (2010)], unless we add the term $-\zeta(\bar{S}S)^2$ to the frame function to stabilise the instability [Lee (2010); Ferrara et al. (2011)]. So an inflationary regime can be implemented in the NMSSM.

Other possibilities within the framework of the NMSSM are analysed in [Moultaka et al. (2019b)].

Appendix A

Conventions

In this Appendix, we collect the conventions adopted throughout this book. Lorentz indices in flat space are taken to be untilded $M = (\mu, \alpha, \dot{\alpha})$, whilst Einstein indices in curved space are taken to be tilded $\tilde{M} = (\tilde{\mu}, \tilde{\alpha}, \widetilde{\dot{\alpha}})$. We split the Greek alphabet so that vectors carry indices denoted by letters from the middle of the alphabet onward (μ, ν, *etc.*) and spinors carry indices denoted by letters from the beginning of the alphabet (α, β, *etc.*) and ($\dot{\alpha}, \dot{\beta}$, *etc.*).

The metric is given by

$$\eta_{\mu\nu} = \mathrm{diag}(1, -1, -1, -1)\,,$$

and the fully antisymmetric tensor of rank four is defined by

$$\varepsilon_{0123} = 1\,.$$

In the $\mathfrak{sl}(2, \mathbb{C})$ notations for dotted and undotted indices related to two-dimensional spinors, we use the following conventions to lower and raise the indices

$$\begin{aligned}
\psi_\alpha &= \varepsilon_{\alpha\beta}\psi^\beta\,,\\
\psi^\alpha &= \varepsilon^{\alpha\beta}\psi_\beta\,,\\
\bar{\psi}_{\dot{\alpha}} &= \varepsilon_{\dot{\alpha}\dot{\beta}}\bar{\psi}^{\dot{\beta}}\,,\\
\bar{\psi}^{\dot{\alpha}} &= \varepsilon^{\dot{\alpha}\dot{\beta}}\bar{\psi}_{\dot{\beta}}\,,
\end{aligned}$$

where ψ_α is a left-handed Weyl spinor, $(\psi_\alpha)^* = \bar{\psi}_{\dot{\alpha}}$ is a right-handed Weyl spinor, and where

$$\begin{aligned}
\varepsilon_{12} &= \varepsilon_{\dot{1}\dot{2}} = 1\,,\\
\varepsilon^{12} &= \varepsilon^{\dot{1}\dot{2}} = -1\,.
\end{aligned}$$

A Dirac spinor is given by

$$\Psi_D = \begin{pmatrix} \psi_\alpha \\ \bar{\lambda}^{\dot{\alpha}} \end{pmatrix}$$

and a Majorana spinor reduces to

$$\Psi_M = \begin{pmatrix} \psi_\alpha \\ \bar{\psi}^{\dot{\alpha}} \end{pmatrix} .$$

The Dirac matrices are taken to be

$$\Gamma^\mu = \begin{pmatrix} 0 & \sigma^\mu \\ \bar{\sigma}^\mu & 0 \end{pmatrix} ,$$

with

$$\begin{aligned} \sigma^\mu{}_{\alpha\dot{\alpha}} &= (1, \sigma^i) , \\ \bar{\sigma}^{\mu\dot{\alpha}\alpha} &= (1, -\sigma^i) , \end{aligned} \tag{A.1}$$

where σ^i $(i = 1, 2, 3)$ are the Pauli matrices,

$$\sigma^1 = \begin{pmatrix} 0 & 1 \\ 1 & 0 \end{pmatrix} ,$$

$$\sigma^2 = \begin{pmatrix} 0 & -i \\ i & 0 \end{pmatrix} ,$$

$$\sigma^3 = \begin{pmatrix} 1 & 0 \\ 0 & -1 \end{pmatrix} .$$

The generators of the Lorentz algebra in the spinor representation, are given by

$$\Gamma^{\mu\nu} = \frac{1}{4}[\Gamma^\mu, \Gamma^\nu] = \begin{pmatrix} \sigma^{\mu\nu} & 0 \\ 0 & \bar{\sigma}^{\mu\nu} \end{pmatrix} ,$$

with

$$(\sigma^{\mu\nu})_\alpha{}^\beta = \frac{1}{4}\left(\sigma^\mu{}_{\alpha\dot{\alpha}}\bar{\sigma}^{\nu\dot{\alpha}\beta} - \sigma^\nu{}_{\alpha\dot{\alpha}}\bar{\sigma}^{\mu\dot{\alpha}\beta}\right) ,$$

$$(\bar{\sigma}^{\mu\nu})^{\dot{\alpha}}{}_{\dot{\beta}} = \frac{1}{4}\left(\bar{\sigma}^{\mu\dot{\alpha}\alpha}\sigma^\nu{}_{\alpha\dot{\beta}} - \bar{\sigma}^{\nu\dot{\alpha}\alpha}\sigma^\mu{}_{\alpha\dot{\beta}}\right) .$$

Finally, scalar products of spinors are defined by:

$$\psi \cdot \lambda = \psi^\alpha \lambda_\alpha ,$$

$$\bar{\psi} \cdot \bar{\lambda} = \bar{\psi}_{\dot{\alpha}} \bar{\lambda}^{\dot{\alpha}} ,$$

where ψ and λ are two left-handed Weyl spinors.

A point in the superspace is defined by adjoining to the spacetime coordinates x^μ the Grassmann variables $(\theta^\alpha, \bar{\theta}_{\dot{\alpha}})$ corresponding to a Majorana spinor. The derivatives with respect to θ and $\bar{\theta}$ are defined by

$$\left\{\frac{\partial}{\partial\theta^\alpha}, \theta^\beta\right\} = \delta_\alpha{}^\beta ,$$

$$\left\{\frac{\partial}{\partial\bar{\theta}_{\dot{\alpha}}}, \bar{\theta}_{\dot{\beta}}\right\} = \delta^{\dot{\alpha}}{}_{\dot{\beta}} ,$$

or equivalently,

$$\left\{\frac{\partial}{\partial\theta_\alpha}, \theta_\beta\right\} = -\delta^\alpha{}_\beta ,$$

$$\left\{\frac{\partial}{\partial\overline{\theta}^{\dot\alpha}}, \overline{\theta}^{\dot\beta}\right\} = -\delta_{\dot\alpha}{}^{\dot\beta} .$$

This allows to define the variables conjugate to θ^α and $\overline{\theta}_{\dot\alpha}$, which satisfy

$$\{\partial_\alpha, \theta^\beta\} = \delta_\alpha{}^\beta ,$$

$$\{\overline{\partial}^{\dot\alpha}, \overline{\theta}_{\dot\beta}\} = \delta^{\dot\alpha}{}_{\dot\beta} ,$$

and are thus given by

$$\partial_\alpha \equiv \frac{\partial}{\partial\theta^\alpha} ,$$

$$\overline{\partial}^{\dot\alpha} \equiv \frac{\partial}{\partial\overline{\theta}_{\dot\alpha}} .$$

We also recall integration rules upon Grassmann variables

$$\int \mathrm{d}\theta = 0 ,$$

$$\int \mathrm{d}\theta\theta = 1 .$$

So for

$$f(\theta) = f_0 + \theta f_1 ,$$

we have

$$\int \mathrm{d}\theta f(\theta) = f_1 .$$

Extending these rules to our superspace leads to

$$\mathrm{d}^2\theta = -\frac{1}{4}\mathrm{d}\theta^\alpha \mathrm{d}\theta_\alpha ,$$

$$\mathrm{d}^2\overline{\theta} = -\frac{1}{4}\mathrm{d}\overline{\theta}_{\dot\alpha}\mathrm{d}\overline{\theta}^{\dot\alpha} ,$$

$$\mathrm{d}^4\theta = \mathrm{d}^2\theta \mathrm{d}^2\overline{\theta} .$$

For those readers who want to consult the book of Wess and Bagger (1992), we recall the minor differences in conventions we have adopted throughout this book:

$$
\begin{aligned}
\eta_{\mu\nu}^{\mathrm{WB}} &= -\eta_{\mu\nu} \,, \\
\epsilon_{\alpha\beta}^{\mathrm{WB}} &= -\epsilon_{\alpha\beta} \,, \\
\epsilon_{\dot\alpha\dot\beta}^{\mathrm{WB}} &= -\epsilon_{\dot\alpha\dot\beta} \,, \\
\text{Torsion tensor}^{\mathrm{WB}} &= -\text{Torsion tensor} \,.
\end{aligned}
$$

This leads to some trivial sign differences. For instance, some of our kinetic terms are of opposite sign with the corresponding Wess and Bagger kinetic terms, or for the auxiliary field of a chiral superfield

$$
F^{\mathrm{WB}} = -F \,,
$$

etc.

Appendix B

Some properties of the Lorentz group

In supergravity, the structure group is the Lorentz group, which means, that at any point of the curved superspace, there exists a tangent superspace on which the Lorentz transformations acts. In calculations undertaken in this tangent superspace, vector and spinor representations are central quantities. Therefore, we dedicate this Appendix to some of their basic properties.

Let us consider a Lorentz transformation with parameter $\omega_{\mu\nu} = -\omega_{\nu\mu}$. We define

$$\begin{aligned} \omega_\alpha{}^\beta &= \frac{1}{4}\omega_{\mu\nu}(\sigma^\mu\bar{\sigma}^\nu)_\alpha{}^\beta\,, \\ \omega^{\dot\alpha}{}_{\dot\beta} &= \frac{1}{4}\omega_{\mu\nu}(\bar{\sigma}^\mu\sigma^\nu)^{\dot\alpha}{}_{\dot\beta}\,. \end{aligned} \tag{B.1}$$

From the antisymmetry property of $\omega_{\mu\nu}$, one obtains $\omega_{\alpha\beta} = \omega_{\beta\alpha}$ and $\omega_{\dot\alpha\dot\beta} = \omega_{\dot\beta\dot\alpha}$, which allow to rewrite the action of a Lorentz transformation on the vector field V^μ and on the left-handed and right-handed spinors ψ^α and $\bar\psi_{\dot\alpha}$ as,

$$\begin{aligned} \delta V^\mu &= V^\nu\omega_\nu{}^\mu = \frac{1}{2}V^\nu\omega_{\sigma\rho}\Big(\eta^{\mu\rho}\delta^\sigma{}_\nu - \eta^{\mu\sigma}\delta^\rho{}_\nu\Big) = \frac{1}{2}V^\nu\omega_{\sigma\rho}(J^{\rho\sigma})_\nu{}^\mu\,, \\ \delta\psi^\alpha &= \psi^\beta\omega_\beta{}^\alpha = \frac{1}{2}\psi^\beta\omega_{\delta\gamma}\Big(\varepsilon^{\alpha\gamma}\delta^\delta{}_\beta + \varepsilon^{\alpha\delta}\delta^\gamma{}_\beta\Big) = \frac{1}{2}\psi^\beta\omega_{\delta\gamma}(J^{\gamma\delta})_\beta{}^\alpha\,, \\ \delta\bar\psi_{\dot\alpha} &= \bar\psi_{\dot\beta}\omega^{\dot\beta}{}_{\dot\alpha} = \frac{1}{2}\bar\psi_{\dot\beta}\omega^{\dot\delta\dot\gamma}\Big(\varepsilon_{\dot\alpha\dot\gamma}\delta_{\dot\delta}{}^{\dot\beta} + \varepsilon_{\dot\alpha\dot\delta}\delta_{\dot\gamma}{}^{\dot\beta}\Big) = \frac{1}{2}\bar\psi_{\dot\beta}\omega^{\dot\delta\dot\gamma}(J_{\dot\gamma\dot\delta})^{\dot\beta}{}_{\dot\alpha}\,, \end{aligned} \tag{B.2}$$

where one observes that $\omega^\dagger_{\alpha\beta} = -\omega_{\dot\alpha\dot\beta}$. Let us note that one should expect a minus sign for the variations of the spinors. Indeed, the basic objects of the Poincaré algebra are Dirac spinors $\Psi_D = (\psi_\alpha, \bar\psi^{\dot\alpha})^t$, lying in the representation $S_L \oplus S_R$ of the Lorentz algebra (where S denotes the spinor representation), and vectors V^μ lying in the vector representation. This means in principle that the transformation of $x^\mu, \theta_\alpha, \bar{\theta}^{\dot{\alpha}}$ are obtained by a left action[1]. In contrast, the fermionic part of a superspace coordinate $z^M = (x^\mu, \theta^\alpha, \bar{\theta}_{\dot{\alpha}})$ lies in the dual representation $S_L^* \oplus S_R^*$ of the Lorentz algebra so a minus sign is expected. However since the action of the Lorentz algebra on vectors is a left action the antisymmetric property of $\omega_{\mu\nu}$ leads to:

$$\delta X^\mu = \omega^\mu{}_\nu X^\nu = -V^\nu \omega_\nu{}^\mu \,.$$

This means in fact that all transformations in (B.2) are the inverse transformations.

Introducing the metric in the flat superspace $\eta^{MN} = (\eta^{\mu\nu}, \varepsilon^{\alpha\beta}, \varepsilon_{\dot{\alpha}\dot{\beta}})$ acting on $X^M = (V^\mu, \psi^\alpha, \bar{\psi}_{\dot{\alpha}})$, we can define the Lorentz generators as

$$(J^{MN})_P{}^Q = \eta^{QM}\delta^N{}_P - (-1)^{|M||N|}\eta^{QN}\delta^M{}_P \,, \tag{B.3}$$

where $|M|$ denotes the grading of the index M, *i.e.*, $|\mu| = 0$, and $|\alpha| = |\dot{\alpha}| = 1$. Hence, all the relations of (B.2) unify to,

$$\delta X^M = \frac{1}{2} X^N \omega_{QP} (J^{PQ})_N{}^M = X^N \omega_N{}^M \,,$$

and a direct calculation gives the well-known commutation relation between objects of the same nature,

$$[J^{MN}, J^{PQ}] = -\eta^{NP} J^{MQ} + \eta^{PM} J^{NQ} - \eta^{NQ} J^{PM} + \eta^{QM} J^{PN} \,, \tag{B.4}$$

since the gradings fulfil $|M| = |P| = |N| = |Q|$. Furthermore, the action (to the right) of the generators J^{MN} on a vectors V^Q is given by[2],

$$[V^Q, J^{MN}] = V^P (J^{MN})_P{}^Q = \eta^{QM} V^N - (-)^{|M||N|} \eta^{QN} V^M \,. \tag{B.5}$$

Since the scalar product $V^Q V_Q$ is a Lorentz scalar, we can derive the action of the generators on V_Q,

$$[V_Q, J^{MN}] = -(J^{MN})_Q{}^P V_P = -(-)^{|M||Q|} V^M \delta^N{}_Q + V^N \delta^M{}_Q \,. \tag{B.6}$$

In supergravity, the superconnection $\Omega_{\tilde{M}MN}$ and the curvature tensor R_{PQMN} play the rôle of the transformation parameters ω_{MN}. As a consequence, one would recover a Lorentz transformation under the condition that the indices M and N have the same gradings, *i.e.*, either they are both vector indices, or both left- or right-handed spin indices. Hence, as an example, the only non-vanishing components of $\Omega_{\tilde{M}MN}$ are $\Omega_{\tilde{M}\mu\nu}$, $\Omega_{\tilde{M}\alpha\beta}$ and $\Omega_{\tilde{M}}{}^{\dot{\alpha}\dot{\beta}}$.

[1] For a Lie algebra if we have a representation where the matrices M_a act by a left action on a vector space E like $\delta_a V = M_a V$ then the right action on the dual vector space is given by $\delta_a f = -f M_a$. In other words if M_a is a representation of $\mathfrak{g}$ then $-M_a^t$ (where M_a^t is the transpose of the matrices M_a) is also a representation.

[2] As in (B.2) there is no minus sign in the RHS.

From the isomorphism between bispinors and vectors, we can represent a bispinor as a vector and *vice versa*,

$$\begin{aligned} v_{\alpha\dot{\alpha}} &= \sigma^{\mu}{}_{\alpha\dot{\alpha}} v_\mu \,, \\ v_\mu &= \frac{1}{2}\bar{\sigma}_\mu{}^{\dot{\alpha}\alpha} v_{\alpha\dot{\alpha}} \,, \end{aligned} \tag{B.7}$$

where v_μ is a vector and $v_{\alpha\dot{\alpha}}$ a bispinor by using the relation

$$\bar{\sigma}^{\mu\dot{\beta}\beta}\sigma_{\mu\alpha\dot{\alpha}} = 2\delta^{\beta}{}_{\alpha}\delta^{\dot{\beta}}{}_{\dot{\alpha}} \,.$$

This allows to relate the parameters entering into (B.1),

$$\begin{aligned} \omega_{\alpha\dot{\alpha}\beta\dot{\beta}} &= \sigma^{\mu}{}_{\alpha\dot{\alpha}}\sigma^{\nu}{}_{\beta\dot{\beta}}\,\omega_{\mu\nu} \\ &= 2\Big[\varepsilon_{\alpha\beta}\omega_{\dot{\alpha}\dot{\beta}} - \varepsilon_{\dot{\alpha}\dot{\beta}}\omega_{\alpha\beta}\Big] \,. \end{aligned} \tag{B.8}$$

This identity can be easily proven, since $\omega_{\mu\nu}$ is antisymmetric under the exchange of the Lorentz indices $\mu \leftrightarrow \nu$. Therefore, $\omega_{\alpha\dot{\alpha}\beta\dot{\beta}}$ is antisymmetric under the exchange $(\alpha,\dot{\alpha}) \leftrightarrow (\beta,\dot{\beta})$. Decomposing this tensor into irreducible representations of $SL(2,\mathbb{C})$, we have

$$\omega_{\alpha\dot{\alpha}\beta\dot{\beta}} = \varepsilon_{\alpha\beta}x_{\dot{\alpha}\dot{\beta}} + \varepsilon_{\dot{\alpha}\dot{\beta}}x'_{\alpha\beta} \,,$$

where the tensors x and x' are symmetric. Using (B.1), one directly derives the form of the x and x' tensors, leading to (B.8). This last identity has a crucial consequence, *e.g.*, when we identify the curvature tensor corresponding to certain transformation parameters. Indeed, assuming the tensors $R_{\alpha\beta\gamma\delta}$ and $R_{\alpha\beta\dot{\gamma}\dot{\delta}}$ are non-vanishing, $R_{\alpha\beta\mu\nu}$ is non-vanishing too, since

$$R_{\alpha\beta\mu\nu} = \frac{1}{2}\bar{\sigma}_\mu{}^{\dot{\gamma}\gamma}\bar{\sigma}_\nu{}^{\dot{\delta}\delta}\Big(\varepsilon_{\gamma\delta}R_{\alpha\beta\dot{\gamma}\dot{\delta}} - \varepsilon_{\dot{\gamma}\dot{\delta}}R_{\alpha\beta\gamma\delta}\Big) \,. \tag{B.9}$$

This relation between spinor and vector representations also holds at the level of the generators of the Lorentz algebra. Hence, one can express the generators in the spinor representation in terms of those in the vector representation, and *vice versa*,

$$\begin{aligned} J_\alpha{}^\beta &= (\sigma^{\nu\mu})_\alpha{}^\beta J_{\mu\nu} \,, \\ J^{\dot{\alpha}}{}_{\dot{\beta}} &= (\bar{\sigma}^{\nu\mu})^{\dot{\alpha}}{}_{\dot{\beta}} J_{\mu\nu} \,, \\ J_{\mu\nu} &= \frac{1}{4}\bar{\sigma}_\nu{}^{\dot{\alpha}\alpha}\bar{\sigma}_\mu{}^{\dot{\beta}\beta}\Big(\varepsilon_{\alpha\beta}J_{\dot{\alpha}\dot{\beta}} - \varepsilon_{\dot{\alpha}\dot{\beta}}J_{\alpha\beta}\Big) \,. \end{aligned} \tag{B.10}$$

Let us note that it is straightforward to show that these last three relations are fully compatible with the Lorentz algebra (B.4), and that they emphasise the fact that the operators $J_\alpha{}^\beta$ and $J^{\dot{\alpha}}{}_{\dot{\beta}}$ are the self-dual and antiself-dual components of $J^{\mu\nu}$. Consequently, we have

$$\frac{1}{2}\omega_{\nu\rho}J^{\rho\nu} = \frac{1}{2}\omega_{\alpha\beta}J^{\beta\alpha} + \frac{1}{2}\omega_{\dot{\alpha}\dot{\beta}}J^{\dot{\beta}\dot{\alpha}} \,, \tag{B.11}$$

showing that the operators $J^{\mu\nu}$, $J^{\alpha\beta}$, $J^{\dot{\alpha}\dot{\beta}}$ are not linearly independent.

Appendix C

Useful identities

In this Appendix, we collect a set of fundamental identities useful for the derivation of all the results presented in this book. The Pauli matrices satisfy a series of fundamental identities:

$$\bar{\sigma}^{\mu\dot{\beta}\beta} = \sigma^{\mu}{}_{\alpha\dot{\alpha}}\varepsilon^{\beta\alpha}\varepsilon^{\dot{\beta}\dot{\alpha}} , \tag{C.1}$$

$$\begin{aligned} \frac{1}{2}\varepsilon_{\mu\nu\rho\sigma}\sigma^{\rho\sigma} &= -i\sigma_{\mu\nu} \\ \frac{1}{2}\varepsilon_{\mu\nu\rho\sigma}\bar{\sigma}^{\rho\sigma} &= i\bar{\sigma}_{\mu\nu} , \end{aligned} \tag{C.2}$$

$$\begin{aligned} \bar{\sigma}^{\mu}\sigma^{\nu}\bar{\sigma}^{\rho} + \bar{\sigma}^{\rho}\sigma^{\nu}\bar{\sigma}^{\mu} &= 2\left(\eta^{\mu\nu}\bar{\sigma}^{\rho} + \eta^{\nu\rho}\bar{\sigma}^{\mu} - \eta^{\mu\rho}\bar{\sigma}^{\nu}\right) , \\ \sigma^{\mu}\bar{\sigma}^{\nu}\sigma^{\rho} + \sigma^{\rho}\bar{\sigma}^{\nu}\sigma^{\mu} &= 2\left(\eta^{\mu\nu}\sigma^{\rho} + \eta^{\nu\rho}\sigma^{\mu} - \eta^{\mu\rho}\sigma^{\nu}\right) , \end{aligned} \tag{C.3}$$

$$\begin{aligned} \bar{\sigma}^{\mu}\sigma^{\nu}\bar{\sigma}^{\rho} - \bar{\sigma}^{\rho}\sigma^{\nu}\bar{\sigma}^{\mu} &= 2i\varepsilon^{\mu\nu\rho\sigma}\bar{\sigma}_{\sigma} , \\ \sigma^{\mu}\bar{\sigma}^{\nu}\sigma^{\rho} - \sigma^{\rho}\bar{\sigma}^{\nu}\sigma^{\mu} &= -2i\varepsilon^{\mu\nu\rho\sigma}\sigma_{\sigma} , \end{aligned} \tag{C.4}$$

$$\begin{aligned} \bar{\sigma}^{\mu}\sigma^{\nu}\bar{\sigma}^{\rho} &= i\varepsilon^{\mu\nu\rho\sigma}\bar{\sigma}_{\sigma} + \eta^{\mu\nu}\bar{\sigma}^{\rho} + \eta^{\nu\rho}\bar{\sigma}^{\mu} - \eta^{\mu\rho}\bar{\sigma}^{\nu} , \\ \sigma^{\mu}\bar{\sigma}^{\nu}\sigma^{\rho} &= -i\varepsilon^{\mu\nu\rho\sigma}\sigma_{\sigma} + \eta^{\mu\nu}\sigma^{\rho} + \eta^{\nu\rho}\sigma^{\mu} - \eta^{\mu\rho}\sigma^{\nu} , \end{aligned} \tag{C.5}$$

$$\begin{aligned} \sigma^{\mu}{}_{\alpha\dot{\alpha}}\bar{\sigma}_{\mu}{}^{\dot{\beta}\beta} &= 2\delta_{\alpha}{}^{\beta}\delta_{\dot{\alpha}}{}^{\dot{\beta}} , \\ \bar{\sigma}^{\mu\dot{\alpha}\alpha}\bar{\sigma}_{\mu}{}^{\dot{\beta}\beta} &= 2\varepsilon^{\dot{\alpha}\dot{\beta}}\varepsilon^{\alpha\beta} , \\ \sigma^{\mu}{}_{\alpha\dot{\alpha}}\sigma_{\mu\beta\dot{\beta}} &= 2\varepsilon_{\alpha\beta}\varepsilon_{\dot{\alpha}\dot{\beta}} , \end{aligned} \tag{C.6}$$

$$\sigma^{\mu}_{\alpha\dot{\alpha}}\bar{\sigma}^{\nu\dot{\alpha}\alpha} = \mathrm{tr}(\sigma^{\mu}\bar{\sigma}^{\nu}) = 2\eta^{\mu\nu} , \tag{C.7}$$

$$(\bar{\sigma}^\mu \sigma^\nu)^{\dot{\alpha}}{}_{\dot{\beta}} = \eta^{\mu\nu} \delta^{\dot{\alpha}}{}_{\dot{\beta}} + 2(\bar{\sigma}^{\mu\nu})^{\dot{\alpha}}{}_{\dot{\beta}} ,$$
$$(\sigma^\mu \bar{\sigma}^\nu)_\alpha{}^\beta = \eta^{\mu\nu} \delta_\alpha{}^\beta + 2(\sigma^{\mu\nu})_\alpha{}^\beta . \tag{C.8}$$

Given three left-handed spinors χ_1, χ_2 and χ_3 we have the Fierz identity

$$\chi_{1\alpha}(\chi_2 \cdot \chi_3) = -\chi_{2\alpha}(\chi_3 \cdot \chi_1) - \chi_{3\alpha}(\chi_1 \cdot \chi_2) , \tag{C.9}$$

which can be put into a more symmetrical form

$$\chi_{1\alpha}(\chi_2 \cdot \chi_3) + \chi_{2\alpha}(\chi_3 \cdot \chi_1) + \chi_{3\alpha}(\chi_1 \cdot \chi_2) = 0 . \tag{C.10}$$

Moreover, any left-handed spinor χ satisfies the relation

$$\chi^\alpha \chi^\beta = -\frac{1}{2} \varepsilon^{\alpha\beta} \chi \cdot \chi . \tag{C.11}$$

Finally, for the Levi-Civita tensor, we have

$$\frac{1}{2} \varepsilon_{\mu\nu\rho\sigma} \varepsilon^{\mu\nu\eta\kappa} = \delta_\sigma{}^\eta \delta_\rho{}^\kappa - \delta_\sigma{}^\kappa \delta_\rho{}^\eta . \tag{C.12}$$

Appendix D

Supermultiplets: conventions and main results

In this Appendix we collect some results concerning the basic multiplets considered throughout this book.

The graviton multiplet is associated to the vielbein and the various torsion and curvature tensors

$$
\begin{aligned}
e_{\tilde{\mu}}{}^{\mu} &= E_{\tilde{\mu}}{}^{\mu}\Big| \,, \\
\psi_{\tilde{\mu}}{}^{\alpha} &= 2E_{\tilde{\mu}}{}^{\alpha}\Big| \,, \\
\bar{\psi}_{\tilde{\mu}\dot{\alpha}} &= 2E_{\tilde{\mu}\dot{\alpha}}\Big| \,, \\
M &= -6\mathcal{R}\Big| \,, \\
M^{\star} &= -6\mathcal{R}^{\dagger}\Big| \,, \\
b_{\mu} &= -3G_{\mu}\Big| \,,
\end{aligned}
$$

we recall the notation $X| = X|_{\Theta=0}$. The last field of the gravitation multiplet, the connection, is not an elementary field but is composite and comes from the constraint $T_{\mu\nu}{}^{\rho} = 0$ which leads to

$$
\begin{aligned}
\omega_{\tilde{\mu}\tilde{\nu}\tilde{\rho}} = &-\frac{1}{2}e_{\mu\tilde{\rho}}(\partial_{\tilde{\mu}}e_{\tilde{\nu}}{}^{\mu} - \partial_{\tilde{\nu}}e_{\tilde{\mu}}{}^{\mu}) + \frac{1}{2}e_{\mu\tilde{\mu}}(\partial_{\tilde{\nu}}e_{\tilde{\rho}}{}^{\mu} - \partial_{\tilde{\rho}}e_{\tilde{\nu}}{}^{\mu}) - \frac{1}{2}e_{\mu\tilde{\nu}}\Big(\partial_{\tilde{\rho}}e_{\tilde{\mu}}{}^{\mu} - \partial_{\tilde{\mu}}e_{\tilde{\rho}}{}^{\mu}\Big) \\
&+ \frac{i}{4}e_{\mu\tilde{\rho}}(\psi_{\tilde{\mu}}\sigma^{\mu}\bar{\psi}_{\tilde{\nu}} - \psi_{\tilde{\nu}}\sigma^{\mu}\bar{\psi}_{\tilde{\mu}}) - \frac{i}{4}e_{\mu\tilde{\mu}}(\psi_{\tilde{\nu}}\sigma^{\mu}\bar{\psi}_{\tilde{\rho}} - \psi_{\tilde{\rho}}\sigma^{\mu}\bar{\psi}_{\tilde{\nu}}) \\
&+ \frac{i}{4}e_{\mu\tilde{\nu}}(\psi_{\tilde{\rho}}\sigma^{\mu}\bar{\psi}_{\tilde{\mu}} - \psi_{\tilde{\mu}}\sigma^{\mu}\bar{\psi}_{\tilde{\rho}}) \,.
\end{aligned}
$$

The transformation of the gravitation multiplet under supergravity transformations is given by

$$\delta e_\mu{}^{\tilde\mu} = ie_\nu{}^{\tilde\mu} e_\mu{}^{\tilde\rho}\Big[\varepsilon(x)\sigma^\nu\bar\psi_{\tilde\rho} - \psi_{\tilde\rho}\sigma^\nu\bar\varepsilon(x)\Big]\,,$$

$$\delta\psi_{\tilde\mu}{}^\alpha = -2\mathcal{D}_{\tilde\mu}\varepsilon^\alpha(x) + \frac{i}{3}e_{\tilde\mu}{}^\nu(\bar\varepsilon(x)\bar\sigma_\nu)^\alpha M + \frac{i}{3}e_{\tilde\mu}{}^\nu\Big[(\varepsilon(x)\sigma_\rho\bar\sigma_\nu)^\alpha b^\rho - 3\varepsilon^\alpha(x)\, b_\nu\Big]\,,$$

$$\delta\bar\psi_{\tilde\mu\dot\alpha} = -2\mathcal{D}_{\tilde\mu}\bar\varepsilon_{\dot\alpha}(x) + \frac{i}{3}e_{\tilde\mu}{}^\nu(\varepsilon(x)\sigma_\nu)_{\dot\alpha} M^\star - \frac{i}{3}e_{\tilde\mu}{}^\nu\Big[(\bar\varepsilon(x)\bar\sigma_\rho\sigma_\nu)_{\dot\alpha} b^\rho - 3\bar\varepsilon_{\dot\alpha}(x)\, b_\nu\Big]\,,$$

$$\delta M = -2(\varepsilon\sigma^{\mu\nu}\psi_{\mu\nu}) + i(\varepsilon\sigma^\mu\bar\psi_\mu)M - i(\varepsilon\cdot\psi_\mu)b^\mu\,,$$

$$\delta M^\star = -2(\bar\varepsilon\bar\sigma^{\mu\nu}\bar\psi_{\mu\nu}) + i(\bar\varepsilon\bar\sigma^\mu\psi_\mu)M^\star + i(\bar\varepsilon\cdot\bar\psi_\mu)b^\mu\,,$$

$$\delta b^\mu = \varepsilon\sigma^\nu\bar\psi_\nu{}^\mu - \bar\varepsilon\bar\sigma^\nu\psi_\nu{}^\mu - \frac{i}{4}\varepsilon^{\mu\nu\rho\sigma}\Big[\varepsilon\sigma_\sigma\bar\psi_{\nu\rho} + \bar\varepsilon\bar\sigma_\sigma\psi_{\nu\rho}\Big] + \frac{i}{2}\Big[\varepsilon\cdot\psi^\mu M^\star - \bar\varepsilon\cdot\bar\psi^\mu M\Big]$$

$$+\frac{i}{2}\Big[\varepsilon\sigma^\nu\bar\psi_\nu + \bar\varepsilon\bar\sigma^\nu\psi_\nu\Big]b^\mu + \frac{1}{4}\varepsilon^{\mu\nu\rho\sigma}\Big[\varepsilon\sigma_\sigma\bar\psi_\nu - \bar\varepsilon\bar\sigma_\sigma\psi_\nu\Big]b_\rho\,.$$

The transformation under a Howe–Tucker transformation defined by

$$\Sigma = (\sigma + i\alpha) + \Theta\cdot\omega - \Theta\cdot\Theta\mathcal{F}\,, \tag{D.1}$$

is given by

$$\begin{aligned}
\delta_\Sigma e_{\tilde\mu}{}^\mu &= 2\sigma e_{\tilde\mu}{}^\mu\,,\\
\delta_\Sigma\psi_{\tilde\mu} &= (\sigma - 3i\alpha)\psi_{\tilde\mu} + ie_{\tilde\mu}{}^\mu\bar\omega\bar\sigma_\mu\,,\\
\delta_\Sigma\bar\psi_{\tilde\mu} &= (\sigma + 3i\alpha)\bar\psi_{\tilde\mu} + ie_{\tilde\mu}{}^\mu\omega\sigma_\mu\,,\\
\delta_\Sigma M &= -2(\sigma + 3i\alpha)M + 6\mathcal{F}^\dagger\,,\\
\delta_\Sigma M^\star &= -2(\sigma - 3i\alpha)M^\star + 6\mathcal{F}\,,\\
\delta_\Sigma b_\mu &= -2\sigma b_\mu + 6e_\mu{}^{\tilde\mu}\partial_{\tilde\mu}\alpha + \frac{3}{2}i(\psi_\mu\cdot\omega - \bar\psi_\mu\cdot\bar\omega)\,,
\end{aligned} \tag{D.2}$$

and the transformation under a special conformal transformation (see App. H) by

$$\begin{aligned}
\delta_K e_{\tilde\mu}{}^\mu &= 0\,,\\
\delta_K\psi_{\tilde\mu} &= 0\,,\\
\delta_K\bar\psi_{\tilde\mu} &= 0\,,\\
\delta_K M &= 0\,,\\
\delta_K M^\star &= 0\,,\\
\delta_K b_\mu &= 0\,.
\end{aligned}$$

Associated to the gravitino are several types of covariant derivatives.

(1) Covariant derivatives with respect to general reparametrisations,

$$\begin{aligned}
\mathcal{D}_{\tilde{\mu}}\psi_{\tilde{\nu}}{}^{\alpha} &= \partial_{\tilde{\mu}}\psi_{\tilde{\nu}}{}^{\alpha} + \psi_{\tilde{\nu}}{}^{\beta}\omega_{\tilde{\mu}\beta}{}^{\alpha} , \\
\mathcal{D}_{\tilde{\mu}}\bar{\psi}_{\tilde{\nu}\dot{\alpha}} &= \partial_{\tilde{\mu}}\bar{\psi}_{\tilde{\nu}\dot{\alpha}} + \bar{\psi}_{\tilde{\nu}\dot{\beta}}\omega_{\tilde{\mu}}{}^{\dot{\beta}}{}_{\dot{\alpha}} , \\
\psi_{\mu\nu} &= e_{\mu}{}^{\tilde{\mu}}e_{\nu}{}^{\tilde{\nu}}(\mathcal{D}_{\tilde{\mu}}\psi_{\tilde{\nu}} - \mathcal{D}_{\tilde{\nu}}\psi_{\tilde{\mu}}) \\
&= e_{\mu}{}^{\tilde{\mu}}e_{\nu}{}^{\tilde{\nu}}\psi_{\tilde{\mu}\tilde{\nu}} , \\
\bar{\psi}_{\mu\nu} &= e_{\mu}{}^{\tilde{\mu}}e_{\nu}{}^{\tilde{\nu}}(\mathcal{D}_{\tilde{\mu}}\bar{\psi}_{\tilde{\nu}} - \mathcal{D}_{\tilde{\nu}}\bar{\psi}_{\tilde{\mu}}) \\
&= e_{\mu}{}^{\tilde{\mu}}e_{\nu}{}^{\tilde{\nu}}\bar{\psi}_{\tilde{\mu}\tilde{\nu}} .
\end{aligned}$$

(2) Covariant derivatives with respect to general reparametrisations and the $U(1)$–connection of the Kähler manifold

$$\begin{aligned}
\check{\mathcal{D}}_{\tilde{\mu}}\psi_{\tilde{\nu}} &= \mathcal{D}_{\tilde{\mu}}\psi_{\tilde{\nu}} + \frac{1}{4}\Big(K_i\widetilde{\mathcal{D}}_{\tilde{\mu}}\phi^i - K^{i^*}\widetilde{\mathcal{D}}_{\tilde{\mu}}\phi^{\dagger}_{i^*}\Big)\psi_{\tilde{\nu}} , \\
\check{\mathcal{D}}_{\tilde{\mu}}\bar{\psi}_{\tilde{\nu}} &= \mathcal{D}_{\tilde{\mu}}\bar{\psi}_{\tilde{\nu}} - \frac{1}{4}\Big(K_i\widetilde{\mathcal{D}}_{\tilde{\mu}}\phi^i - K^{i^*}\widetilde{\mathcal{D}}_{\tilde{\mu}}\phi^{\dagger}_{i^*}\Big)\psi_{\tilde{\nu}} , \\
\check{\psi}_{\mu\nu} &= e_{\mu}{}^{\tilde{\mu}}e_{\nu}{}^{\tilde{\nu}}(\check{\mathcal{D}}_{\tilde{\mu}}\psi_{\tilde{\nu}} - \check{\mathcal{D}}_{\tilde{\nu}}\psi_{\tilde{\mu}}) \\
&= e_{\mu}{}^{\tilde{\mu}}e_{\nu}{}^{\tilde{\nu}}\check{\psi}_{\tilde{\mu}\tilde{\nu}} , \\
\check{\bar{\psi}}_{\mu\nu} &= e_{\mu}{}^{\tilde{\mu}}e_{\nu}{}^{\tilde{\nu}}(\check{\mathcal{D}}_{\tilde{\mu}}\bar{\psi}_{\tilde{\nu}} - \check{\mathcal{D}}_{\tilde{\nu}}\bar{\psi}_{\tilde{\mu}}) \\
&= e_{\mu}{}^{\tilde{\mu}}e_{\nu}{}^{\tilde{\nu}}\check{\bar{\psi}}_{\tilde{\mu}\tilde{\nu}} ,
\end{aligned}$$

(see below for the definition of $\tilde{\mathcal{D}}\phi$).

Chiral (resp. anti-chiral) superfields are defined by

$$\begin{aligned}
\overline{\mathcal{D}}_{\dot{\alpha}}\Phi^i &= 0 , \\
\mathcal{D}_{\alpha}\Phi^{\dagger}_{i^*} &= 0 .
\end{aligned}$$

Their expansion in the Θ–variables are given by,

$$\begin{aligned}
\Phi^i &= \phi^i + \sqrt{2}\Theta\cdot\chi^i - \Theta\cdot\Theta F^i , \\
\Phi^{\dagger}_{i^*} &= \phi^{\dagger}_{i^*} + \sqrt{2}\bar{\Theta}\cdot\bar{\chi}_{i^*} - \bar{\Theta}\cdot\bar{\Theta}F^{\dagger}_{i^*} .
\end{aligned}$$

The chiral superfiefd associated to $\Phi^\dagger$ may be expressed as

$$
\begin{aligned}
\Xi &= (\overline{\mathcal{D}}\cdot\overline{\mathcal{D}} - 8\mathcal{R})\Phi^\dagger \\
&= 4F^\dagger + \frac{4}{3}M\phi^\dagger \\
&\quad +\Theta\cdot\bigg[-4i\sqrt{2}(\sigma^\mu \hat{D}_\mu \bar{\chi}) + \frac{2\sqrt{2}}{3}b_\mu(\sigma^\mu\bar{\chi}) \\
&\qquad +\phi^\dagger\Big(\frac{8}{3}(\sigma^{\mu\nu}\psi_{\mu\nu}) - \frac{4i}{3}(\sigma^\mu\bar{\psi}_\mu)M + \frac{4i}{3}\psi^\mu b_\mu\Big)\bigg] \\
&\quad -\Theta\cdot\Theta\bigg[-4e_\mu{}^{\tilde{\mu}}\mathcal{D}_{\tilde{\mu}}\hat{D}^\mu\phi^\dagger + \frac{8i}{3}b^\mu\hat{D}_\mu\phi^\dagger + 2\sqrt{2}\bar{\psi}_\mu\cdot\hat{D}^\mu\bar{\chi} \\
&\qquad -\frac{2\sqrt{2}}{3}\bar{\chi}\bar{\sigma}^{\mu\nu}\bar{\psi}_{\mu\nu} - \frac{8}{3}M^\star F^\dagger + \sqrt{2}i\bar{\chi}\cdot\bar{\psi}_\mu b^\mu + \frac{2\sqrt{2}i}{3}\bar{\chi}\bar{\sigma}^{\nu\mu}\bar{\psi}_\mu b_\nu \\
&\qquad +\phi^\dagger\Big(-\frac{2}{3}e_\mu{}^{\tilde{\mu}}e_\nu{}^{\tilde{\nu}}R_{\tilde{\mu}\tilde{\nu}}{}^{\mu\nu}\Big| - \frac{8}{9}MM^\star + \frac{4}{9}b_\mu b^\mu + \frac{4i}{3}e_\mu{}^{\tilde{\mu}}\mathcal{D}_{\tilde{\mu}}b^\mu + \frac{2}{3}\bar{\psi}_\mu\cdot\bar{\psi}^\mu M \\
&\qquad +\frac{2}{3}\psi_\nu\sigma^\nu\bar{\psi}_\mu b^\mu + \frac{4i}{3}\bar{\psi}^\mu\bar{\sigma}^\nu\psi_{\mu\nu} + \frac{1}{6}\varepsilon^{\mu\nu\rho\sigma}(\psi_\mu\sigma_\sigma\bar{\psi}_{\nu\rho} + \bar{\psi}_\mu\bar{\sigma}_\sigma\psi_{\nu\rho})\Big)\bigg]\,.
\end{aligned}
$$

The transformation of the chiral multiplet under supergravity transformations is given by,

$$
\begin{aligned}
\delta\phi^i &= -\sqrt{2}\varepsilon\cdot\chi^i\,, & \delta\phi^\dagger_{i^*} &= -\sqrt{2}\bar{\varepsilon}\cdot\bar{\chi}_{i^*}\,,\\
\delta\chi^i_\alpha &= \sqrt{2}\left[\varepsilon_\alpha F^i - i(\sigma^\mu\bar{\varepsilon})_\alpha\hat{\hat{D}}_\mu\phi^i\right], & \delta\bar{\chi}_{\dot{\alpha}i^*} &= \sqrt{2}\left[\bar{\varepsilon}_{\dot{\alpha}}F^\dagger_{i^*} + i(\varepsilon\sigma^\mu)_{\dot{\alpha}}\hat{\hat{D}}_\mu\phi^\dagger_{i^*}\right],\\
\delta F^i &= \frac{\sqrt{2}}{3}\varepsilon\cdot\chi^i M^\star - i\sqrt{2}\hat{\hat{D}}_\mu\chi^i\sigma^\mu\bar{\varepsilon} & \delta F^\dagger_{i^*} &= \frac{\sqrt{2}}{3}\bar{\varepsilon}\cdot\bar{\chi}_{i^*}M + i\sqrt{2}\varepsilon\sigma^\mu\hat{\hat{D}}_\mu\bar{\chi}_{i^*}\\
&\quad -\frac{\sqrt{2}}{6}b_\mu\chi^i\sigma^\mu\bar{\varepsilon} + 2ig\bar{\varepsilon}\cdot\bar{\lambda}^a(T_a\phi)^i\,, & &\quad -\frac{\sqrt{2}}{6}b_\mu\varepsilon\sigma^\mu\bar{\chi}_{i^*} - 2ig\varepsilon\cdot\lambda^a(\phi^\dagger T_a)_{i^*}\,.
\end{aligned}
$$

Its transformation under a Howe–Tucker transformation is given by

$$
\begin{aligned}
\delta_\Sigma\phi^i &= 0\,, & \delta_\Sigma\phi^\dagger_{i^*} &= 0\,,\\
\delta_\Sigma\chi^i &= (-\sigma + 3i\alpha)\chi^i\,, & \delta_\Sigma\chi^\dagger_{i^*} &= (-\sigma - 3i\alpha)\chi^\dagger_{i^*}\,,\\
\delta_\Sigma F^i &= 2(-\sigma + 3i\alpha)F^i + \sqrt{2}\omega\cdot\chi^i\,, & \delta_\Sigma F^\dagger_{i^*} &= 2(-\sigma - 3i\alpha)F^\dagger_{i^*} + \sqrt{2}\omega\cdot\chi^\dagger_{i^*}\,,
\end{aligned}
\tag{D.3}
$$

and under the special conformal transformations (see App. H) by

$$
\begin{aligned}
&\delta_K\phi^i = 0\,,\ \delta_K\phi^\dagger_{i^*} = 0\,,\\
&\delta_K\chi^i = 0\,,\ \delta_\Sigma\chi^\dagger_{i^*} = 0\,,\\
&\delta_K F^i = 0\,,\ \delta_K F^\dagger_{i^*} = 0\,.
\end{aligned}
$$

Several types of covariant derivatives can be defined.

(1) Covariant derivatives with respect to supergravity transformations

$$\hat{D}_\mu \phi^i = e_\mu{}^{\tilde\mu}\Big(\partial_{\tilde\mu}\phi^i - \frac{\sqrt{2}}{2}\psi_{\tilde\mu}\cdot\chi^i\Big)\,,$$

$$\hat{D}_\mu \chi^{\alpha i} = e_\mu{}^{\tilde\mu}\Big[\mathcal{D}_{\tilde\mu}\chi^{\alpha i} + \frac{1}{\sqrt{2}}\psi_{\tilde\mu}{}^\alpha F^i + \frac{i}{\sqrt{2}}(\bar\psi_{\tilde\mu}\bar\sigma^\nu)^\alpha \hat{D}_\nu\phi^i\Big]\,,$$

$$\hat{D}_\mu \phi^\dagger_{i^*} = e_\mu{}^{\tilde\mu}\Big(\partial_{\tilde\mu}\phi^\dagger_{i^*} - \frac{\sqrt{2}}{2}\bar\psi_{\tilde\mu}\cdot\bar\chi_{i^*}\Big)\,,$$

$$\hat{D}_\mu \bar\chi_{\dot\alpha i^*} = e_\mu{}^{\tilde\mu}\Big[\mathcal{D}_{\tilde\mu}\bar\chi_{\dot\alpha i^*} + \frac{1}{\sqrt{2}}\bar\psi_{\tilde\mu\dot\alpha}F^\dagger_{i^*} + \frac{i}{\sqrt{2}}(\psi_{\tilde\mu}\sigma^\nu)_{\dot\alpha}\hat{D}_\nu\phi^\dagger_{i^*}\Big]\,.$$

(2) Covariant derivatives with respect to supergravity transformations and gauge invariance

$$\hat{\hat{D}}_\mu\phi^i = \hat{D}_\mu\phi^i + ig(T_a\phi)^i v_\mu^a\,,$$

$$\hat{\hat{D}}_\mu\phi^\dagger_{i^*} = \hat{D}_\mu\phi^\dagger_{i^*} - ig(\phi^\dagger T_a)_{i^*} v_\mu^a\,,$$

$$\hat{\hat{D}}_\mu\chi^i = \hat{D}_\mu\chi^i + ig(T_a\chi)^i v_\mu^a\,,$$

$$\hat{\hat{D}}_\mu\bar\chi^\dagger_{i^*} = \hat{D}_\mu\bar\chi^\dagger_{i^*} - ig(\bar\chi^\dagger T_a)_{i^*} v_\mu^a\,.$$

(3) Covariant derivatives with respect to the Kähler manifold

$$\nabla_\mu\chi^i = \partial_\mu\chi^i + \Gamma_j{}^i{}_k\partial_\mu\phi^j\chi^k\,,$$

$$\nabla_\mu\bar\chi_{i^*} = \partial_\mu\bar\chi_{i^*} + \Gamma^{j^*}{}_{i^*}{}^{k^*}\partial_\mu\phi^\dagger_{j^*}\bar\chi_{k^*}\,,$$

with

$$\Gamma_i{}^k{}_j = (K^{-1})^k{}_{k^*}\frac{\partial^3 K}{\partial\phi^i\partial\phi^j\partial\phi^\dagger_{k^*}}\,,$$

$$\Gamma_{i^*}{}^{k^*}{}_{j^*} = (K^{-1})^k{}_{k^*}\frac{\partial^3 K}{\partial\phi^\dagger_{i^*}\partial\phi^\dagger_{j^*}\partial\phi^k}\,.$$

(4) Covariant derivatives with respect to general reparametrisations

$$\begin{aligned}
\mathcal{D}_\mu\phi^i &= e_\mu{}^{\tilde\mu}\partial_{\tilde\mu}\phi^i\,,\\
\mathcal{D}_\mu\phi^{i^*} &= e_\mu{}^{\tilde\mu}\partial_{\tilde\mu}\phi^\dagger_{i^*}\,,\\
\mathcal{D}_\mu\chi^{i\alpha} &= e_\mu{}^{\tilde\mu}\Big(\partial_{\tilde\mu}\chi^{i\alpha} + \chi^{i\beta}\omega_{\tilde\mu\beta}{}^\alpha\Big)\,,\\
\mathcal{D}_\mu\bar\chi_{i^*\dot\alpha} &= e_\mu{}^{\tilde\mu}\Big(\partial_{\tilde\mu}\bar\chi_{i^*\dot\alpha} + \bar\chi_{i^*\dot\beta}\omega_{\tilde\mu}{}^{\dot\beta}{}_{\dot\alpha}\Big)\,.
\end{aligned}\qquad\text{(D.4)}$$

(5) Covariant derivatives with respect to general, gauge symmetries and to the Kähler manifold

$$\begin{aligned}
\widetilde{\mathcal{D}}_\mu\phi^i &= e_\mu{}^{\tilde\mu}\partial_{\tilde\mu}\phi^i + ig(T_a\phi)^i v_\mu^a\,,\\
\widetilde{\mathcal{D}}_\mu\phi^\dagger_{i^*} &= e_\mu{}^{\tilde\mu}\partial_{\tilde\mu}\phi^\dagger_{i^*} - ig(\phi^\dagger T_a)_{i^*} v_\mu^a\,,\\
\widetilde{\mathcal{D}}_\mu\chi^i &= e_\mu{}^{\tilde\mu}\mathcal{D}_{\tilde\mu}\chi^i + ig(T_a\chi)^i v_\mu^a + \Gamma_j{}^i{}_k\widetilde{\mathcal{D}}_\mu\phi^j\chi^k\,,\\
\widetilde{\mathcal{D}}_\mu\bar\chi_{i^*} &= e_\mu{}^{\tilde\mu}\mathcal{D}_{\tilde\mu}\bar\chi_{i^*} - ig(\bar\chi T_a)_{i^*} v_\mu^a + \Gamma^{j^*}{}_{i^*}{}^{k^*}\widetilde{\mathcal{D}}_\mu\phi^\dagger_{j^*}\bar\chi_{k^*}\,.
\end{aligned}$$

(6) Covariant derivatives with respect to general, gauge symmetries, Kähler manifold and the $U(1)$–connection of the Kähler manifold

$$\begin{aligned}
\check{\mathcal{D}}_{\tilde\mu}\chi^i &= \widetilde{\mathcal{D}}_{\tilde\mu}\chi^i - \frac{1}{4}(K_j\tilde{\mathcal{D}}_{\tilde\mu}\phi^j - K^{j^*}\tilde{\mathcal{D}}_{\tilde\mu}\phi^\dagger_{j^*})\,\chi^i\,,\\
\check{\mathcal{D}}_{\tilde\mu}\bar\chi_{i^*} &= \widetilde{\mathcal{D}}_{\tilde\mu}\bar\chi_{i^*} + \frac{1}{4}(K_j\tilde{\mathcal{D}}_{\tilde\mu}\phi^j - K^{j^*}\tilde{\mathcal{D}}_{\tilde\mu}\phi^\dagger_{j^*})\,\bar\chi_{i^*}\,.
\end{aligned}$$

A vector superfield V^aT_a is defined by $V^\dagger = V$. In the Wess-Zumino gauge we have

$$\begin{aligned}
V\Big| &= 0\,,\\
\mathcal{D}_a V\Big| &= 0\,, \qquad \overline{\mathcal{D}}_{\dot\alpha}V\Big| = 0\,,\\
\mathcal{D}\cdot\mathcal{D}V\Big| &= 0\,,\ \ \overline{\mathcal{D}}\cdot\overline{\mathcal{D}}V\Big| = 0\,,\\
v_\mu &= \frac{1}{2}\bar\sigma_\mu{}^{\dot\alpha\alpha}\mathcal{D}_\alpha\overline{\mathcal{D}}_{\dot\alpha}V\Big|\\
&= -\frac{1}{2}\bar\sigma_\mu{}^{\dot\alpha\alpha}\overline{\mathcal{D}}_{\dot\alpha}\mathcal{D}_\alpha V\Big|\,.
\end{aligned}$$

The chiral (anti-chiral) spinor superfields are defined by

$$\begin{aligned}
W_\alpha &= -\frac{1}{4}\Big[\overline{\mathcal{D}}\cdot\overline{\mathcal{D}} - 8\mathcal{R}\Big]e^{2gV}\mathcal{D}_\alpha e^{-2gV}\\
&= -2g\Big[i\lambda_\alpha + \Big[i(\sigma^{\mu\nu}\Theta)_\alpha(\hat F_{\mu\nu}) + \Theta_\alpha D\Big]\\
&\qquad -\Theta\cdot\Theta\Big[(\sigma^\mu\hat D_\mu\bar\lambda)_\alpha - \frac{i}{2}b_\mu(\sigma^\mu\bar\lambda)_\alpha - \frac{i}{2}M^\star\lambda_\alpha\Big]\Big]\,,\\
\overline{W}_{\dot\alpha} &= \frac{1}{4}\Big[\mathcal{D}\cdot\mathcal{D} - 8\mathcal{R}^\dagger\Big]e^{-2gV}\overline{\mathcal{D}}_{\dot\alpha}e^{2gV}\\
&= -2g\Big[-i\bar\lambda_{\dot\alpha} + \Big[i(\bar\Theta\bar\sigma^{\mu\nu})_{\dot\alpha}(\hat F_{\mu\nu}) + \bar\Theta_{\dot\alpha}D\Big]\\
&\qquad -\bar\Theta\cdot\bar\Theta\Big[(\hat D_\mu\lambda\sigma^\mu)_{\dot\alpha} + \frac{i}{2}b_\mu(\lambda\sigma^\mu)_{\dot\alpha} + \frac{i}{2}M^\star\bar\lambda_{\dot\alpha}\Big]\Big]\,.
\end{aligned}$$

The transformation law of the vector multiplet under supergravity transformations is given by

$$\begin{aligned}
\delta v_\mu &= i(\varepsilon\sigma_\mu\bar{\lambda} - \lambda\sigma_\mu\bar{\varepsilon}) - \frac{i}{2}\left(\psi_\mu\sigma^\nu\bar{\varepsilon} - \varepsilon\sigma^\nu\bar{\psi}_\mu\right)v_\nu\,,\\
\delta\lambda_\alpha &= (\sigma^{\mu\nu}\varepsilon)_\alpha\hat{F}_{\mu\nu} + i\varepsilon_\alpha D\,,\\
\delta\bar{\lambda}_{\dot{\alpha}} &= -(\bar{\varepsilon}\bar{\sigma}^{\mu\nu})_{\dot{\alpha}}\hat{F}_{\mu\nu} - i\bar{\varepsilon}_{\dot{\alpha}} D\,,\\
\delta D &= -\varepsilon\sigma^\mu\hat{D}_\mu\bar{\lambda} - \hat{D}_\mu\lambda\sigma^\mu\bar{\varepsilon} + \frac{i}{2}b_\mu\left(\varepsilon\sigma^\mu\bar{\lambda} - \lambda\sigma^\mu\bar{\varepsilon}\right) + \frac{i}{2}\left(M^\star\varepsilon\cdot\lambda - M\bar{\varepsilon}\cdot\bar{\lambda}\right).
\end{aligned}$$

The transformation under a Howe–Tucker transformation is given by

$$\begin{aligned}
\delta_\Sigma v^a_{\tilde{\mu}} &= 0\,,\\
\delta_\Sigma\lambda^a &= -3(\sigma + i\alpha)\lambda^a\,,\\
\delta_\Sigma\bar{\lambda}^a &= -3(\sigma - i\alpha)\bar{\lambda}^a\,,\\
\delta_\Sigma D^a &= -4\sigma D^a\,,
\end{aligned}\tag{D.5}$$

and the transformation under a special conformal transformation (see App. H) by

$$\begin{aligned}
\delta_K v^a_{\tilde{\mu}} &= 0\,,\\
\delta_K\lambda^a &= 0\,,\\
\delta_\Sigma\bar{\lambda}^a &= 0\,,\\
\delta_\Sigma D^a &= 0\,.
\end{aligned}$$

Several types of covariant derivatives can be defined.

(1) Covariant derivatives with respect to supergravity transformations and gauge invariance

$$\begin{aligned}
\hat{D}_\mu v_\nu &= e_\mu{}^{\tilde{\mu}}(\partial_{\tilde{\mu}}v_\nu - \omega_{\tilde{\mu}\nu}{}^\rho v_\rho) - \frac{i}{2}\left[\bar{\lambda}\bar{\sigma}_\nu\psi_\mu - \bar{\psi}_\mu\bar{\sigma}_\nu\lambda\right] - \frac{i}{2}(\psi_\nu\sigma^\rho\bar{\psi}_\mu)v_\rho\,,\\
\hat{F}_{\mu\nu} &= e_\mu{}^{\tilde{\mu}}e_\nu{}^{\tilde{\nu}}\left\{\hat{D}_{\tilde{\mu}}v_{\tilde{\nu}} - \hat{D}_{\tilde{\nu}}v_{\tilde{\nu}} + ig[v_{\tilde{\mu}}, v_{\tilde{\nu}}]\right\}\\
&= e_\mu{}^{\tilde{\mu}}e_\nu{}^{\tilde{\nu}}\left\{F_{\tilde{\mu}\tilde{\nu}} - \frac{i}{2}\left[\bar{\lambda}\bar{\sigma}_{\tilde{\nu}}\psi_{\tilde{\mu}} + \lambda\sigma_{\tilde{\nu}}\bar{\psi}_{\tilde{\mu}} - \bar{\lambda}\bar{\sigma}_{\tilde{\mu}}\psi_{\tilde{\nu}} - \lambda\sigma_{\tilde{\mu}}\bar{\psi}_{\tilde{\nu}}\right]\right\},\\
\hat{D}_\mu\lambda_\alpha &= e_\mu{}^{\tilde{\mu}}\left(\widetilde{\mathcal{D}}_{\tilde{\mu}}\lambda_\alpha + ig[v_{\tilde{\mu}}, \lambda_\alpha]\right) - \frac{1}{2}(\sigma^{\nu\rho}\psi_\mu)_\alpha\hat{F}_{\nu\rho} + \frac{i}{2}D\psi_{\mu\alpha}\,,\\
\hat{D}_\mu\bar{\lambda}_{\dot{\alpha}} &= e_\mu{}^{\tilde{\mu}}\left(\widetilde{\mathcal{D}}_{\tilde{\mu}}\bar{\lambda}_{\dot{\alpha}} + ig[v_{\tilde{\mu}}, \bar{\lambda}_{\dot{\alpha}}]\right) + \frac{1}{2}(\bar{\psi}_\mu\bar{\sigma}^{\nu\rho})_{\dot{\alpha}}\hat{F}_{\nu\rho} - \frac{i}{2}D\bar{\psi}_{\mu\dot{\alpha}}\,.
\end{aligned}$$

(2) Covariant derivatives with respect to general reparametrisations

$$F_{\mu\nu} = e_\mu{}^{\tilde{\mu}} e_\nu{}^{\tilde{\nu}} \Big(\partial_{\tilde{\mu}} v_{\tilde{\nu}} - \partial_{\tilde{\nu}} v_{\tilde{\mu}} + ig[v_{\tilde{\mu}}, v_{\tilde{\nu}}] \Big)$$

$$= e_\mu{}^{\tilde{\mu}} e_\nu{}^{\tilde{\nu}} F_{\tilde{\mu}\tilde{\nu}} ,$$

$$\mathcal{D}_\mu \lambda^\alpha = e_\mu{}^{\tilde{\mu}} \, \mathcal{D}_{\tilde{\mu}} \lambda^\alpha = e_\mu{}^{\tilde{\mu}} \Big(\partial_{\tilde{\mu}} \lambda^\alpha + \lambda^\beta \omega_{\tilde{\mu}\beta}{}^\alpha \Big) ,$$

$$\mathcal{D}_\mu \bar{\lambda}_{\dot{\alpha}} = e_\mu{}^{\tilde{\mu}} \Big(\partial_{\tilde{\mu}} \bar{\lambda}_{\dot{\alpha}} + \bar{\lambda}_{\dot{\beta}} \omega_{\tilde{\mu}}{}^{\dot{\beta}}{}_{\dot{\alpha}} \Big) .$$

(3) Covariant derivatives with respect to general reparametrisations and gauge invariance

$$\widetilde{\mathcal{D}}_\mu \lambda = \mathcal{D}_\mu \lambda + ig[v_{\tilde{\mu}}, \lambda] ,$$

$$\widetilde{\mathcal{D}}_\mu \bar{\lambda} = \mathcal{D}_\mu \bar{\lambda} + ig[v_{\tilde{\mu}}, \bar{\lambda}] .$$

(4) Covariant derivatives with respect to general reparametrisations, gauge invariance and the $U(1)$–connection of the Kähler manifold

$$\check{\mathcal{D}}_{\tilde{\mu}} \lambda^a = \widetilde{\mathcal{D}}_{\tilde{\mu}} \lambda^a + \frac{1}{4} (K_i \widetilde{\mathcal{D}}_{\tilde{\mu}} \phi^i - K_{i^*} \widetilde{\mathcal{D}}_{\tilde{\mu}} \phi^\dagger_{i^*}) \lambda^a ,$$

$$\check{\mathcal{D}}_{\tilde{\mu}} \bar{\lambda}^a = \widetilde{\mathcal{D}}_{\tilde{\mu}} \bar{\lambda}^a - \frac{1}{4} (K_i \widetilde{\mathcal{D}}_{\tilde{\mu}} \phi^i - K_{i^*} \widetilde{\mathcal{D}}_{\tilde{\mu}} \phi^\dagger_{i^*}) \bar{\lambda}^a .$$

Appendix E

How to construct an action in supergravity

In this Appendix we summarise the main steps to construct an invariant action in supergravity. There are several ways to proceed, but we give the essential stages within the superspace approach. This is summarised in Figs. E.1 and E.2.

Table E.1 Diagram flow to construct a supergravity model.

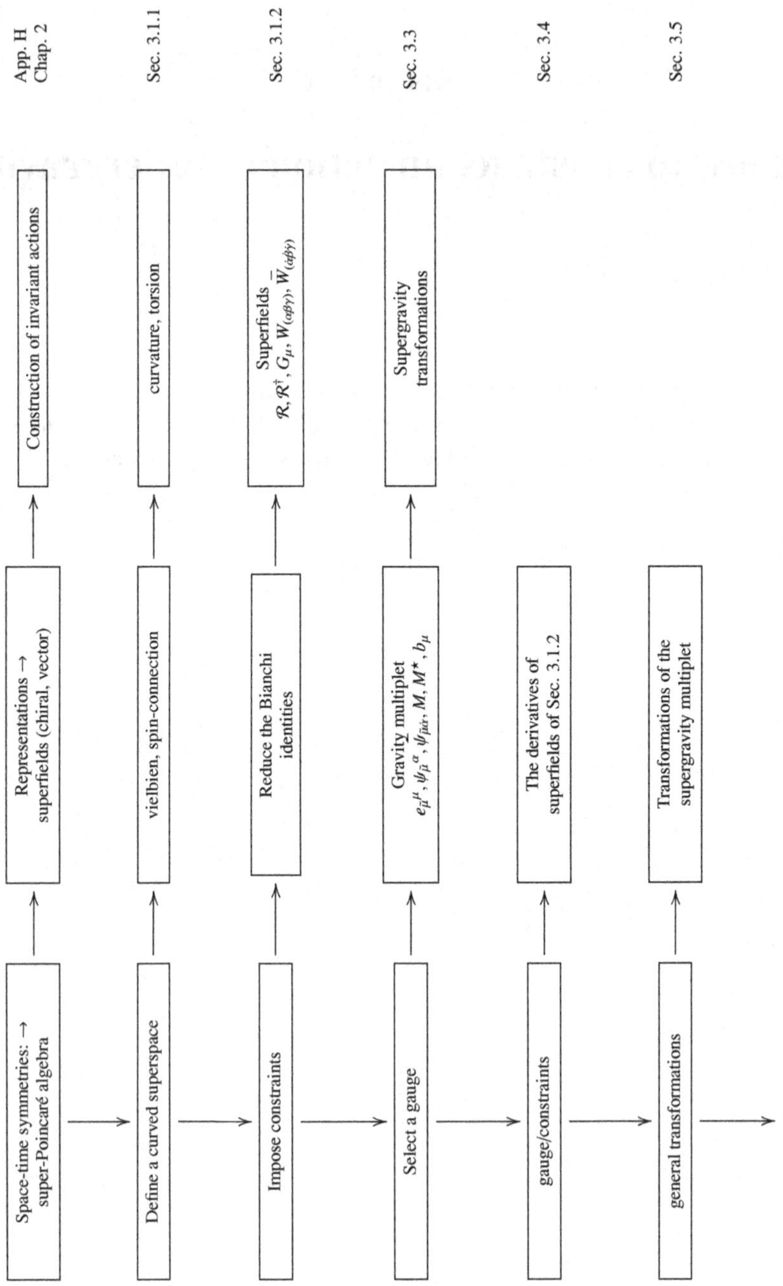

Step	→	→	Reference
Space-time symmetries: → super-Poincaré algebra	Representations → superfields (chiral, vector)	Construction of invariant actions	App. H Chap. 2
↓ Define a curved superspace	vielbien, spin-connection	curvature, torsion	Sec. 3.1.1
↓ Impose constraints	Reduce the Bianchi identities	Superfields $\mathcal{R}, \mathcal{R}^\dagger, G_\mu, W_{(\alpha\beta\gamma)}, \overline{W}_{(\dot\alpha\dot\beta\dot\gamma)}$	Sec. 3.1.2
↓ Select a gauge	Gravity multiplet $e_{\tilde\mu}{}^{\mu}, \psi_{\tilde\mu}{}^{\alpha}, \psi_{\tilde\mu\dot\alpha}, M, M^\star, b_\mu$	Supergravity transformations	Sec. 3.3
↓ gauge/constraints	The derivatives of superfields of Sec. 3.1.2		Sec. 3.4
↓ general transformations	Transformations of the supergravity multiplet		Sec. 3.5
↓			

Table E.2 Diagram flow to construct a supergravity model cont'd.

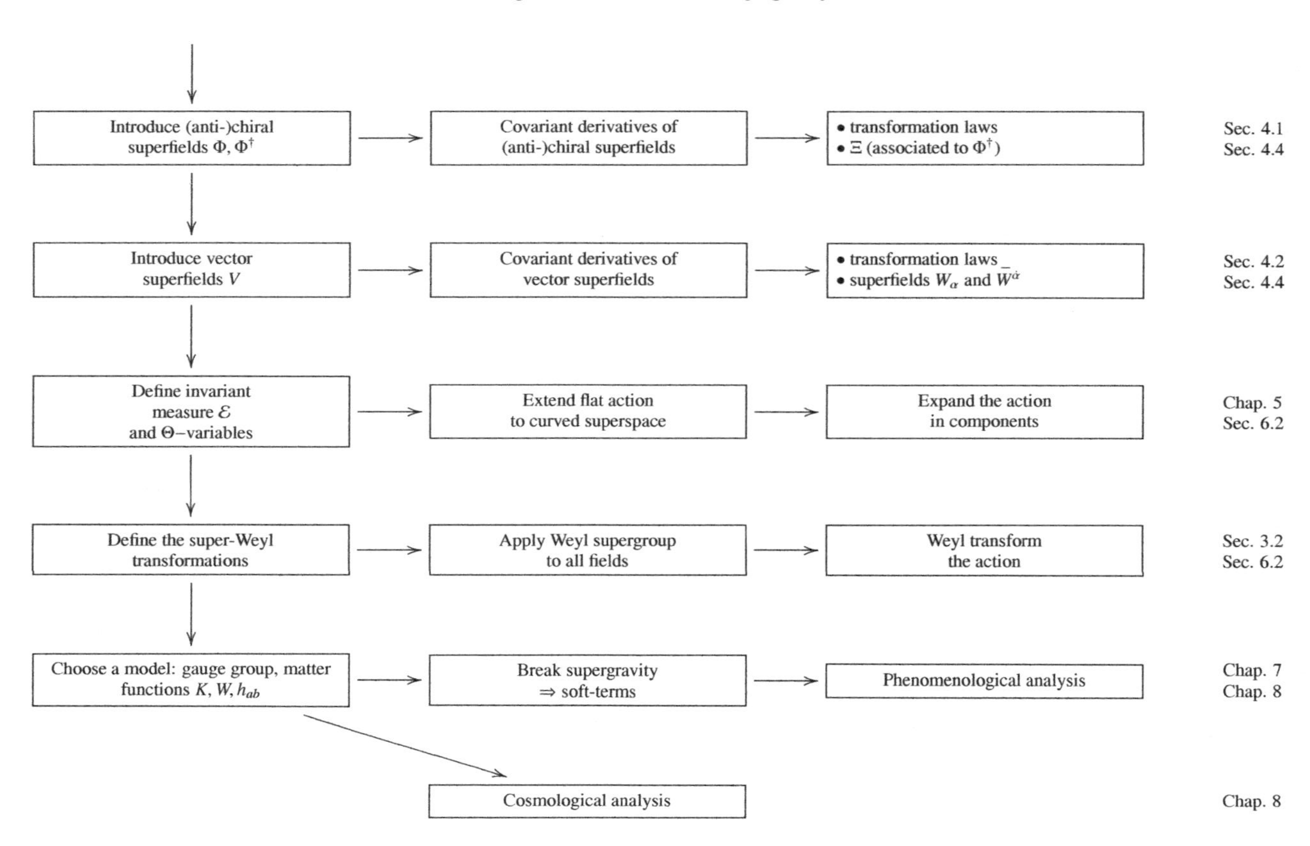

Appendix F

Kähler manifolds

F.1 General properties

In this Appendix we introduce some general properties of Kähler manifolds. One may consult *e.g.* [Kobayashi and Nomizu (1963); Nash and Sen (1983); Nakahara (2003)]. A real m-dimensional manifold $\mathcal{M}$ is a space that is locally isomorphic to the vector space $\mathbb{R}^m$. Basically we can cover $\mathcal{M}$ by a set of open sets U_i such that there exist local charts $f_i(P) = (X^1, \cdots, X^m)$ from U_i to $\mathbb{R}^m$ (f_i are homeomorphisms from $U_i \to U'_i \subset \mathbb{R}^m$). The variables $(X^1, \cdots, X^m)$ can be seen as the coordinates of the point P in U_i. There are also some conditions to be satisfied if we have two local charts such that $U_i \cap U_i \neq \emptyset$ in order to achieve a consistent construction. When m is even, *i.e.*, if we consider a $2n$-dimensional real manifold, we can define a complex manifold by setting

$$\begin{aligned} \phi^i &= X^i + iX^{i+n} \,, \\ \phi^\dagger_{i^*} &= X^i - iX^{i+n} \,. \end{aligned} \tag{F.1}$$

However, there is absolutely no guaranty that under a change of coordinates $\phi^i \to \phi'^i$ the splitting between the "holomorphic" and the "anti-holomorphic" parts for vectors of $T_P(\mathcal{M})$ (the tangent space of $\mathcal{M}$ at point P) is preserved. A complex manifold is a manifold such that this splitting is preserved, and is such that the change of coordinates $X^i \to X'^i$ corresponds to the set of (anti-) holomorphic transformations

$$\begin{aligned} \phi^i &\to \phi'^i = f^i(\phi) \,, \\ \phi^\dagger_{i^*} &\to \phi'^\dagger_{i^*} = f^*_{i^*}(\phi^\dagger) \,. \end{aligned}$$

A vector field then transforms as

$$\begin{aligned} V^i &\to V'^i = \frac{\partial \phi'^i}{\partial \phi^j} V^j \,, \\ V^\dagger_{i^*} &\to V'^\dagger_{i^*} = \frac{\partial \phi'^\dagger_{i^*}}{\partial \phi^\dagger_{j^*}} V^\dagger_{j^*} \,. \end{aligned}$$

A condition for a $2n-$dimensional real manifold to be an n-dimensional complex manifold if given in terms of the almost complex structure $J_i{}^j$ (which satisfies $J_i{}^j J_j{}^k = -\delta_i{}^k$). A real manifold is a complex manifold if the Nijenhuis tensor vanishes

$$N_{ij}{}^k = J_i{}^\ell\left(\partial_\ell J_j{}^k - \partial_j J_\ell{}^k\right) - J_j{}^\ell\left(\partial_\ell J_i{}^k - \partial_i J_\ell{}^k\right) = 0\,,$$

and the almost complex structure is called a complex structure. For a complex Riemannian manifold the metric takes the form[1]

$$\mathrm{d}s^2 = g_{ij}\mathrm{d}\phi^i\mathrm{d}\phi^j + g^{i^*j^*}\mathrm{d}\phi^\dagger_{i^*}\mathrm{d}\phi^\dagger_{j^*} + 2g^{i^*}{}_i\mathrm{d}\phi^i\mathrm{d}\phi^\dagger_{i^*}\,,$$

in the system of coordinates (F.1). The metric is said hermitian if $g_{ij} = 0$ and $g^{i^*j^*} = 0$ and is said Kähler if[2]

$$\partial_i g^{i^*}{}_j - \partial_j g^{i^*}{}_i = 0\,,$$

or equivalently if there exists (locally) a function K, the Kähler potential, such that

$$g^{i^*}{}_i = \frac{\partial^2 K}{\partial\phi^i\partial\phi^\dagger_{i^*}}\,.$$

Then, if we denote $I = (i, i^*)$ and define the Christoffel symbols

$$\Gamma_I{}^K{}_J = \frac{1}{2}g^{KL}\left(\partial_I g_{JL} + \partial_J g_{IL} - \partial_L g_{IJ}\right), \tag{F.2}$$

the only non-vanishing components are [Wess and Bagger (1992)]

$$\begin{aligned}
\Gamma_i{}^k{}_j &= g^k{}_{i^*}\frac{\partial g^{i^*}{}_j}{\partial\phi^i} \\
&= g^k{}_{i^*}\frac{\partial^3 K}{\partial\phi^i\partial\phi^j\partial\phi^\dagger_{i^*}}\,, \\
\Gamma^{i^*}{}_{k^*}{}^{j^*} &= g^i{}_{k^*}\frac{\partial g^{j^*}{}_i}{\partial\phi^\dagger_{i^*}} \\
&= g^i{}_{k^*}\frac{\partial^3 K}{\partial\phi^i\partial\phi^\dagger_{i^*}\partial\phi^\dagger_{j^*}}\,,
\end{aligned}$$

[1] In this system of coordinates, the almost complex structure takes the very simple form

$$J = \begin{pmatrix} i\delta^i{}_j & 0 \\ 0 & -i\delta_{i^*}{}^{j^*}\end{pmatrix}.$$

[2] Note that in this Appendix, and only here, the metric on the Kähler manifold is noted $g^{i^*}{}_i$ and not $K^{i^*}{}_i$.

with $g^i{}_{i^*}$ is the inverse metric tensor. If we define the curvature tensor by

$$R_{IJ}{}^L{}_K = \partial_I \Gamma_J{}^L{}_K - \partial_J \Gamma_I{}^K{}_L - \Gamma_I{}^M{}_K \Gamma_J{}^L{}_M + \Gamma_J{}^M{}_K \Gamma_I{}^L{}_M \,,$$

the non-vanishing part of $R_{IJKL} = g_{KM} R_{IJ}{}^M{}_L$ [Wess and Bagger (1992)] is given by[3]

$$\begin{aligned}
R_i{}^{i^*}{}_j{}^{j^*} &= g^{j^*}{}_k \frac{\partial \Gamma_i{}^k{}_j}{\partial \phi^\dagger_{i^*}} \\
&= \frac{\partial^4 K}{\partial \phi^i \partial \phi^j \partial \phi^\dagger_{i^*} \partial \phi^\dagger_{j^*}} - g^k{}_{k^*} \frac{\partial g^{k^*}{}_j}{\partial \phi^i} \frac{\partial g^{j^*}{}_k}{\partial \phi^\dagger_{i^*}} \\
&= \frac{\partial^4 K}{\partial \phi^i \partial \phi^j \partial \phi^\dagger_{i^*} \partial \phi^\dagger_{j^*}} - g^{k^*}{}_k \Gamma_i{}^k{}_j \Gamma^{i^*}{}_{k^*}{}^{j^*} \,,
\end{aligned}$$

and the other components following from the symmetries of the curvature tensor.[4]

Thus we see that the geometry of a Kähler manifold is specified by the Kähler potential, since any geometrical quantity is expressed in terms of the Kähler potential and its derivatives. Furthermore, the metric is obviously invariant under Kähler transformations of the form

$$K(\phi, \phi^\dagger) \to K(\phi, \phi^\dagger) + F(\phi) + F^*(\phi^\dagger) \,,$$

with F a holomorphic function.

F.2 Reparametrisation

Now, we consider a change of parametrisation of the Kähler manifold, that is a diffeomorphism,

$$\phi^i \to \phi'^i(\phi) \,,$$

and we want to obtain the corresponding transformation of the fields χ and F associated to ϕ in the superfield Φ. Indeed, the reparametrisation above induces the transformations $\chi^i \to \chi'^i$ and $F^i \to F'^i$. These transformations are not difficult to compute (for instance using the supergravity transformation (4.29), (4.30) and (4.31)) since now the transformed multiplet (ϕ'^i, χ'^i, F'^i) transforms in the same way under the supergravity transformations (4.29), (4.30) and (4.31) as the multiplet (ϕ^i, χ^i, F^i). This leads to (see *e.g.*, [Freedman and Van Proeyen (2012)])

---

[3]The components $R_i{}^{j^*}{}^k{}_\ell$ and their complex conjugate counterparts are also non-vanishing, but they are not relevant for us here.

[4]The are also some simplifications of the Bianchi identities, but we will not list them here.

$$\begin{aligned} \phi^i &\to \phi'^i(\phi) \,, \\ \chi^i &\to \chi'^i = \frac{\partial \phi'^i}{\partial \phi^j}\chi^j \,, \\ F^i &\to F'^i = \frac{\partial \phi'^i}{\partial \phi^j}F^j + \frac{1}{2}\frac{\partial^2 \phi'^i}{\partial \phi^j \partial \phi^k}\chi^j \cdot \chi^k \,. \end{aligned} \tag{F.3}$$

The transformation of χ^i shows that the fermionic fields transform like vectors of the Kähler manifold, legitimating *a posteriori* the introduction of the covariant derivatives (2.19). The auxiliary fields F^i are not vector fields of the Kähler manifold, in the same way, the Christoffel symbol is not a tensor. This gives some insight into the expression of F^i in (6.25) when we have eliminated it through its equation of motions.

F.3 Killing vectors and symmetries of the Kähler manifold

If we consider a Riemanian manifold $\mathcal{M}$ with metric g_{IJ} after an infinitesimal reparametrisation

$$\phi^I \to \phi'^I = \phi^I + k^I(\phi) \tag{F.4}$$

we have

$$g_{IJ}(\phi) \to g'_{IJ}(\phi') = \frac{\partial \phi^K}{\partial \phi'^I}\frac{\partial \phi^L}{\partial \phi'^J}g_{KL}(\phi) \,. \tag{F.5}$$

If we assume that the reparametrisation (F.4) is a symmetry of the metric, that is if $g'_{IJ}(\phi') = g_{IJ}(\phi)$, then (F.5) leads to the Killing equation

$$k^K \partial_K g_{IJ} + g_{JK}\partial_I k^K + g_{IK}\partial_J k^K = 0 \,.$$

Introducing $k_I = g_{IJ}k^J$, using (F.2) and the covariant derivative $\nabla_I k_J = \partial_I k_J - \Gamma_I{}^K{}_J k_K$, the Killing equation may be rewritten as

$$\nabla_I k_J + \nabla_J k_I = 0 \,. \tag{F.6}$$

If we now assume to have a set k_a^I of Killing vectors which generate a continuous symmetry (of parameters α^a), we can write

$$\delta\phi^I = i\alpha^a(-ik_a^J)\partial_J\phi^I = \alpha^a k_a^I \,,$$

and it is not difficult to see that the vector fields $T_a = -ik_a^I\partial_I$ generate the Lie algebra

$$[T_a, T_b] = (-k_a^I\partial_I k_b^J + k_b^I\partial_I h_a^J)\partial_J = i f_{ab}{}^c T_c \,.$$

In general, considering a given manifold it is difficult to find some continuous symmetries. In the case of a Kähler manifold the conditions (F.6) lead to two conditions: the preservation of the tensor metric $g^{i^*}{}_i$ and the preservation of the hermiticity condition $g_{ij} = 0$[5]

$$\begin{aligned} \nabla_i k^{\dagger i^*} + \nabla^{i^*} k_i &= 0 \,, \\ \nabla_i k_j + \nabla_j k_i &= 0 \,. \end{aligned} \tag{F.7}$$

Moreover, when the manifold is Kähler, we have holomorphic and anti-holomorphic Killing vectors $-ik^i_a(\phi)$ and $ik^\dagger_{ai^*}(\phi^\dagger)$:

$$\begin{aligned} \delta\phi^i &= i\alpha^a(-ik^i_a) = \alpha^a k^i_a \,, \\ \delta\phi^\dagger_{i^*} &= -i\alpha^a(ik^\dagger_{ai^*}) = \alpha^a k^\dagger_{ai^*} \,. \end{aligned} \tag{F.8}$$

To further specify the properties of the Killing vectors associated to a Kähler manifold, we follow the pedagogical approach of Freedman and Van Proeyen (2012), and in particular Chapter 13 where many details are given (see also [Wess and Bagger (1992)]). It can be shown that the Killing vectors can be obtained from the real functions $\mathcal{P}_a(\phi, \phi^\dagger)$, called the moment maps or the Killing potentials,

$$\begin{aligned} k_{ia} &= g^{i^*}{}_i k^\dagger_{ai^*} = i\frac{\partial \mathcal{P}_a}{\partial \phi^i} \,, \\ k^{i^*}_a &= g^{i^*}{}_i k^i_a = -i\frac{\partial \mathcal{P}_a}{\partial \phi^\dagger_{i^*}} \,, \end{aligned}$$

where $\mathcal{P}_a$ are solutions of the equations[6]

$$\nabla_i \partial_j \mathcal{P}_a = 0 \,, \tag{F.9}$$

and satisfy the algebra

$$(-ik^i_a\partial_i + ik_{ai^*}\partial^{i^*})\mathcal{P}_b = if_{ab}{}^c\mathcal{P}_c \,.$$

This means that for an abelian gauge group the solution of (F.9) is not uniquely defined, and in particular, in this case $\mathcal{P}$ can be shifted by an arbitrary constant

$$\mathcal{P} \to \mathcal{P} - \frac{1}{2}\xi \,. \tag{F.10}$$

This is the origin of the Fayet-Iliopoulos term in supergravity.

[5] In fact the Killing vectors have also to preserve the almost complex structure. It turns out that in the system of coordinates (F.1) this translates into a condition of holomorphicity, *i.e.*, the Killing vectors $k^i(\phi)$ ($k_{i^*}(\phi^\dagger)$) must be holomorphic (anti-holomorphic).

[6] In fact solving these equations leads to the moment maps and consequently to the Killing vectors.

Throughout this book we were supposing to know *a priori* the symmetries of the Kähler manifold since they are related to Yang-Mills invariance. In other words, we were starting from the very beginning with fields in a given representation of a gauge group G, and such that the Kähler manifold is invariant under the action of the gauge group G. In this case, the Killing equations automatically lead to the fact that T_a are the generators of the Lie algebra $\mathfrak{g}$ in the corresponding representation (complex conjugate representation). To make the connection with the general approach sketched, the Killing vectors simply read

$$\begin{aligned} k_a^i &= i(T_a\phi)^i \, , \\ k^\dagger_{ai^*} &= -i(\phi^\dagger T_a) \, , \end{aligned} \tag{F.11}$$

and the moment maps read (since the Kähler potential is gauge invariant)

$$\begin{aligned} \mathcal{P}_a &= -\frac{1}{2}\Big[K_i(T_a\phi)^i + K^{i^*}(\phi^\dagger T_a)_{i^*}\Big] \\ &= -K_i(T_a\phi)^i = -K^{i^*}(\phi^\dagger T_a)_{i^*} \, , \end{aligned} \tag{F.12}$$

and we readily have[7]

$$\begin{aligned} k_a^i &= -ig^i{}_{j^*}\partial^{j^*}\mathcal{P}_a \, , \\ k^\dagger_{ai^*} &= ig^j{}_{i^*}\partial_j\mathcal{P}_a \, . \end{aligned} \tag{F.13}$$

Of course in this approach, the symmetries are realised linearly. But in general, there are *a priori* no reasons that the transformations generated by the k's are linear, meaning that the Killing symmetries may be non-linear. In this case, it is more involved to have a local invariant theory. The first step is to define appropriate covariant derivatives. Following the standard procedure, from the transformations law of the scalars fields $(\phi^i, \phi^\dagger_{i^*})$ and spinor fields $(\chi^i, \bar{\chi}_{i^*})$

(F.3) we set [Wess and Bagger (1992); Freedman and Van Proeyen (2012)]

$$\begin{aligned} \mathcal{D}_\mu\phi^i &= \partial_\mu\phi^i + igv_\mu^a\Big(-ik_a^i(\phi)\Big) \, , \\ \mathcal{D}_\mu\phi^\dagger_{i^*} &= \partial_\mu\phi^\dagger_{i^*} - igv_\mu^a\Big(ik_{ai^*}(\phi)\Big) \, , \\ \mathcal{D}_\mu\chi^i &= \partial_\mu\phi^i + igv_\mu^a\left(-i\frac{\partial k_a^i(\phi)}{\partial\phi^j}\right)\chi^j \, , \\ \mathcal{D}_\mu\bar{\chi}_{i^*} &= \partial_\mu\bar{\chi}_{i^*} - igv_\mu^a\left(i\frac{\partial k_{ai^*}(\phi)}{\partial\phi^\dagger_{j^*}}\right)\bar{\chi}_{j^*} \, . \end{aligned} \tag{F.14}$$

[7] Note that the D–term of the scalar potential can now be written like $V_D = \frac{g^2}{2}h^{Rab}\mathcal{P}_a\mathcal{P}_b$.

Then the construction of invariant actions is more involved, but basically, the final result is similar to the covariant derivative with respect to the gauge transformations being defined by (F.14) instead of the usual ones. See [Wess and Bagger (1992)] for more details.

Appendix G

Two-loop renormalisation group equations

In Chapter 8 we have analysed how supergravity could be implemented in particles physics and a specific model was investigated. Further, supergravity must be broken in a hidden sector by an appropriate mechanism. In Chapter 7 gravity mediated supersymmetry breaking was presented. However, all the parameters, and in particular the soft-supersymmetric breaking terms, are obtained at the energy where supergravity is broken, say the Planck or the GUT scale. Consequently, it is necessary to compute their values at low energy. The content of this appendix is precisely to state without proof the so called two-loop Renormalisation-Group-Equations which relate the low energy parameters to their high energy values.

To fix the notations, consider a general renormalisable supersymmetric theory with Lie algebra $\mathfrak{g} = \times_I \mathfrak{g}_I$ where possibly one of the $\mathfrak{g}_I$ is an abelian $\mathfrak{u}(1)$. Assume further that the matter fields are in a representation $\mathcal{R} = \otimes_I \mathcal{R}_I$ (possibly reducible) of $\mathfrak{g}$. Denote $T^a(\mathcal{R}_I), a = 1, \cdots, \dim \mathfrak{g}_I$ the generators of $\mathfrak{g}_I$ in the representation $\mathcal{R}_I$ and let g_I be the coupling constant of $\mathfrak{g}_I$. Define the quadratic Casimir operator C_I and the Dynkin label τ_I in the representation $\mathcal{R}_I$ by

$$T^a(\mathcal{R}_I) T^b(\mathcal{R}_I) \delta_{ab} = C_I(\mathcal{R}_I) ,$$

$$\mathrm{Tr}[T^a(\mathcal{R}_I) T^b(\mathcal{R}_I)] = \tau_I(\mathcal{R}_I) \delta^{ab} .$$

A simple computation easily shows that

$$\tau_I(\mathcal{R}_I) \dim \mathfrak{g}_I = C_I(\mathcal{R}_I) \dim \mathcal{R}_I .$$

If $f(I)^{ab}{}_c$ are the structure constant of the Lie algebra $\mathfrak{g}_I$ and since $\mathrm{ad}(T^a(I))^b{}_c = -i f(I)^{ab}{}_c$, the quadratic Casimir operator and the Dynkin label in the adjoint representation are equal and take the form

$$C_I(\mathrm{adj}) \delta^c{}_e = -\delta_{ab} f(I)^{ac}{}_d f(I)^{bd}{}_e ,$$

$$\tau_I(\mathrm{adj}) \delta^{ab} = -f(I)^{ac}{}_d f(I)^{bd}{}_c .$$

The Casimir operator in the adjoint representation is noted $C_I(\mathfrak{g}_I)$ and $d_I = \dim(\mathfrak{g}_I)$. If $\mathfrak{g}_{I_0} = \mathfrak{u}(1)$ then $C_{I_0}(\mathcal{R}_{I_0}) = \sum_i q_i^2$ where q_i is the $\mathfrak{u}(1)$ charge of the chiral superfield Φ^i.

Let Φ^i be the chiral superfields of the model in the representation $\mathcal{R}$ of $\mathfrak{g}$. We further introduce

$$W(\Phi) = \frac{1}{6}\lambda_{ijk}\Phi^i\Phi^j\Phi^k + \frac{1}{2}\mu_{ij}\Phi^i\Phi^j + \zeta_i\Phi^i ,$$

a renormalisable gauge invariant superpotential. We also introduce ϕ^i the scalar counterpart of the superfield Φ^i and λ_I the gauginos associated to the Lie algebra $\mathfrak{g}_I$ (we implicitly assume to have a number of gauginos equal to the dimension of $\mathfrak{g}_I$). The soft-supersymmetric breaking terms are given by

$$\begin{aligned} V_{\text{soft}} &= \frac{1}{2}\phi_i^\dagger (m^2)^i{}_j\phi^j + \sum_I \Big[M_I\lambda_I\cdot\lambda_I + \text{h.c.}\Big] \\ &+\Big[\frac{1}{6}a_{ijk}\phi^i\phi^j\phi^k + \frac{1}{2}b_{ij}\phi^i\phi^j + c_i\phi^i + \text{h.c.}\Big] . \end{aligned}$$

We also have the complex conjugate parameters $(\lambda_{ijk})^* = \lambda^{ijk}, (a_{ijk})^* = a^{ijk}$, *etc.*

The supersymmetric renormalisation group equations are known at the two-loop level [Derendinger and Savoy (1984); Falck (1986); Martin and Vaughn (1994); Yamada (1993, 1994b,a); Jack et al. (1994)]. We do not give any detail about the way these equations have been obtained but only mention that some care must be taken in the process. Usually, quantum corrections, *i.e.*, loop-diagrams lead to divergent integrals which must be regularised. This is precisely the so-called renormalisation procedure. A standard method of regularisation, called dimensional regularisation ['t Hooft and Veltman (1972); Bollini and Giambiagi (1972)], consists in an analytic continuation which enables to perform integrations in $D = 4-2\varepsilon, \varepsilon \in \mathbb{C}$ spacetime dimensions and then identify the divergences as poles of the Euler Γ−function when $\varepsilon \to 0$. However, such a method breaks supersymmetry explicitly since the number of degrees of freedom of spinors and vectors depends on the spacetime dimension. Consequently, an alternative method has been developed in [Siegel (1979c)] where the spacetime dimension is also $D = 4 - 2\varepsilon$ but the number of components of spinors and vectors are unaffected. See [Capper et al. (1980)] for a pedagogical introduction. In this appendix we give the two-loop renormalisation group equations, following [Martin and Vaughn (1994)] and using the dimensional reduction scheme with a modified minimal subtraction ($\overline{\text{DR}}$). Note however that a comparison between the two schemes of regularisation is given in [Martin and Vaughn (1993)].

We only consider the case where there is at most one $\mathfrak{u}(1)$ factor. For more than one factor we refer to [Martin and Vaughn (1994); Fonseca et al. (2012)].

Renormalisation group equations of the coupling constants and of the gaugino masses.

The two equations for the coupling constants and the gaugino masses can be written in a unified way. If we let p_I stand for either g_I or M_I, *i.e.*, $p_I = g_I, M_I$ we have

$$\frac{\mathrm{d}p_I}{\mathrm{d}t} = \frac{1}{16\pi^2}\beta^{(1)}_{p_I} + \frac{1}{(16\pi^2)^2}\beta^{(2)}_{p_I} ,$$

with $t = \ln E$ the renormalisation parameter. The β–functions are

$$\begin{aligned}
\beta^{(1)}_{g_I} &= g_I^3[\tau_I(\mathcal{R}_I) - 3C_I(\mathfrak{g}_I)] ,\\
\beta^{(2)}_{g_I} &= g_I^3[- 6g_I^2C_I(\mathfrak{g}_I)^2 + 2g_I^2C_I(\mathfrak{g}_I)\tau_I(\mathcal{R}_I)\\
&\quad + 4\sum_J g_J^2\tau_I(\mathcal{R}_I)C_J(\mathcal{R}_J)] - g_I^3\lambda^{ijk}\lambda_{ijk}\frac{C_I(k)}{d_I} ,
\end{aligned}$$

for the coupling constants, and

$$\begin{aligned}
\beta^{(1)}_{M_I} &= g_I^2[2\tau_I(\mathcal{R}_I) - 6C_I(\mathfrak{g}_I)]M_I ,\\
\beta^{(2)}_{M_I} &= g_I^2[- 24g_I^2C_I(\mathfrak{g}_I)^2M_I + 8g_I^2C_I(\mathfrak{g}_I)\tau_I(\mathcal{R}_I)M_I\\
&\quad + 8\sum_J g_J^2\tau_I(\mathcal{R}_I)C_J(\mathcal{R}_J)(M_I + M_J)] + 2g_I^2[a^{ijk} - M_I\lambda^{ijk}]\lambda_{ijk}\frac{C_I(k)}{d_I} ,
\end{aligned}$$

for the gaugino masses. In the formula above:

- $C_I(i)$ is the Casimir operator associated to the representation of $\mathfrak{g}_I$ the superfield Φ^i belongs to;

- if the representation $\mathcal{R}_I$ is reducible $\tau_I(\mathcal{R}_I)$ is understood as a sum over all the irreducible components, *i.e.*, a sum over all chiral multiplets;

- $\tau_I(\mathcal{R}_I)C_J(\mathcal{R}_J)$ is the sum of all Dynkin labels weighted by the Casimir operator;

- the sum $\sum_J$ is understood as a summation over all the Lie algebras $\mathfrak{g}_J$.

Renormalisation group equations of the parameters of the superpotential.

The renormalisation group equations of the parameters of the superpotential read

$$\frac{\mathrm{d}\lambda^{ijk}}{\mathrm{d}t} = \lambda^{ij\ell}\Big[\frac{1}{16\pi^2}\gamma_\ell^{(1)k} + \frac{1}{(16\pi^2)^2}\gamma_\ell^{(2)k}\Big] + (k \leftrightarrow i) + (k \leftrightarrow j)\,,$$

$$\frac{\mathrm{d}\mu^{ij}}{\mathrm{d}t} = \mu^{ik}\Big[\frac{1}{16\pi^2}\gamma_k^{(1)j} + \frac{1}{(16\pi^2)^2}\gamma_k^{(2)j}\Big] + (i \leftrightarrow j)\,,$$

$$\frac{\mathrm{d}\zeta^{i}}{\mathrm{d}t} = \zeta^{j}\Big[\frac{1}{16\pi^2}\gamma_j^{(1)i} + \frac{1}{(16\pi^2)^2}\gamma_j^{(2)i}\Big]\,,$$

where

$$\gamma_i^{(1)j} = \frac{1}{2}\lambda_{ik\ell}\lambda^{jk\ell} - 2\delta_i^j\sum_I g_I^2 C_I(i)\,,$$

$$\begin{aligned}\gamma_i^{(2)j} &= -\frac{1}{2}\lambda_{imn}\lambda^{npq}\lambda_{pqr}\lambda^{mrj} + \sum_I g_I^2\lambda_{ipq}\lambda^{jpq}[2C_I(p) - C_I(i)]\\ &\quad +2\delta_i^i\Big[\sum_I g_I^4 C_I(i)\tau_I(\mathcal{R}_I) + 2\sum_I\sum_J g_I^2 g_J^2 C_I(i)C_J(i) - 3\sum_I g_I^4 C(\mathfrak{g}_I)C_I(i)\Big]\,.\end{aligned}$$

As before, $C_I(i)$ is the Casimir operator associated to the chiral superfield Φ^i and corresponds to the representation of $\mathfrak{g}_I$ it belongs to. In contrast $\tau_I(\mathcal{R}_I)$ is the Dynkin index summed over *all* chiral superfields. Finally, we observe that, as a consequence of the non-renormalisable theorem [Grisaru et al. (1979); Seiberg (1993); Weinberg (2000)], the superpotential is not renormalised. Thus, the evolution of the corresponding parameters is controlled by the anomalous dimensions γ of the fields and comes from the wave-function renormalisation.

Renormalisation group equations of the soft-supersymmetric breaking terms.

The renormalisation group equations of the trilinear, bilinear and linear terms are

$$\frac{\mathrm{d}a^{ijk}}{\mathrm{d}t} = \frac{1}{16\pi^2}[\beta_a^{(1)}]^{ijk} + \frac{1}{(16\pi^2)^2}[\beta_a^{(2)}]^{ijk}\,,$$

$$\frac{\mathrm{d}b^{ij}}{\mathrm{d}t} = \frac{1}{16\pi^2}[\beta_b^{(1)}]^{ij} + \frac{1}{(16\pi^2)^2}[\beta_b^{(2)}]^{ij}\,,$$

$$\frac{\mathrm{d}c^{i}}{\mathrm{d}t} = \frac{1}{16\pi^2}[\beta_c^{(1)}]^{i} + \frac{1}{(16\pi^2)^2}[\beta_c^{(2)}]^{i}\,,$$

with

$$[\beta_a^{(1)}]^{ijk} = \frac{1}{2}a^{ij\ell}\lambda_{\ell mn}\lambda^{mnk} + \lambda^{ij\ell}\lambda_{\ell mn}a^{mnk}$$

$$-2\sum_I \left(a^{ijk} - 2M_I\lambda^{ijk}\right)g_I^2 C_I(k) + (k \leftrightarrow i) + (k \leftrightarrow j)\,,$$

$$[\beta_a^{(2)}]^{ijk} = -\frac{1}{2}a^{ij\ell}\lambda_{\ell mn}\lambda^{npq}\lambda_{pqr}\lambda^{mrk} - \lambda^{ij\ell}\lambda_{\ell mn}\lambda^{npq}\lambda_{pqr}a^{mrk} - \lambda^{ij\ell}\lambda_{\ell mn}a^{npq}\lambda_{pqr}\lambda^{mrk}$$

$$+\sum_I g_I^2\left[a^{ij\ell}\lambda_{\ell pq}\lambda^{pqk} + 2\lambda^{ij\ell}\lambda_{\ell pq}a^{pqk} - 2M_I\lambda^{ij\ell}\lambda_{\ell pq}\lambda^{pqk}\right]$$

$$\times\left[2C_I(p) - C_I(k)\right]$$

$$+\sum_I g_I^4\left[2a^{ijk} - 8M_I\lambda^{ijk}\right]\left[C_I(k)\tau_I(\mathcal{R}_I) - 3C(\mathfrak{g}_I)C_I(k)\right]$$

$$+2\sum_I\sum_J g_I^2 g_J^2\left[2a^{ijk} - 8M_I\lambda^{ijk}\right]C_I(k)C_J(k) + (k \leftrightarrow i) + (k \leftrightarrow j)\,.$$

for the trilinear terms,

$$[\beta_b^{(1)}]^{ij} = \frac{1}{2}b^{i\ell}\lambda_{\ell mn}\lambda^{mnj} + \frac{1}{2}\lambda^{ij\ell}\lambda_{\ell mn}b^{mn} + \mu^{i\ell}\lambda_{\ell mn}a^{mnj}$$

$$-2\sum_I\left[b^{ij} - 2M_I\mu^{ij}\right]g_I^2 C_I(i) + (i \leftrightarrow j)\,,$$

$$[\beta_b^{(2)}]^{ij} = -\frac{1}{2}b^{i\ell}\lambda_{\ell mn}\lambda^{pqn}\lambda_{pqr}\lambda^{mrj} - \frac{1}{2}\lambda^{ij\ell}\lambda_{\ell mn}b^{mr}\lambda_{pqr}\lambda^{pqn}$$

$$-\frac{1}{2}\lambda^{ij\ell}\lambda_{\ell mn}\mu^{mr}\lambda_{pqr}a^{pqn} - \mu^{i\ell}\lambda_{\ell mn}a^{npq}\lambda_{pqr}\lambda^{mrj}$$

$$-\mu^{i\ell}\lambda_{\ell mn}\lambda^{npq}\lambda_{pqr}a^{mrj} + 2\sum_I\lambda^{ij\ell}\lambda_{\ell pq}(b^{pq} - \mu^{pq}M_I)g_I^2 C_I(p)$$

$$+\sum_I g_I^2\left[b^{i\ell}\lambda_{\ell pq}\lambda^{pqj} + 2\mu^{i\ell}\lambda_{\ell pq}a^{pqj} - 2\mu^{i\ell}\lambda_{\ell pq}\lambda^{pqj}M_I\right]\left[2C_I(p) - C_I(i)\right]$$

$$+\sum_I g_I^4\left[2b^{ij} - 8\mu^{ij}M_I\right]\left[C_I(i)\tau_I(\mathcal{R}_I) - 3C(\mathfrak{g}_I)C_I(i)\right]$$

$$+2\sum_I\sum_J g_I^2 g_J^2\left[2b^{ij} - 8\mu^{ij}M_I\right]C_I(i)C_J(i) + (i \leftrightarrow j)\,.$$

for the bilinear terms, and

$$[\beta_c^{(1)}]^i = \frac{1}{2}\lambda^{i\ell n}\lambda_{p\ell n}c^p + \zeta^p\lambda_{p\ell n}a^{i\ell n} + \mu^{ik}\lambda_{k\ell n}b^{\ell n} + 2\lambda^{ikp}(m^2)^\ell{}_p\mu_{k\ell} + a^{ik\ell}b_{k\ell}\,,$$

$$\begin{aligned}
[\beta_c^{(2)}]^i = {} & 2\sum_I g_I^2 C_I(\ell)\lambda^{ik\ell}\lambda_{pk\ell}c^p - \frac{1}{2}\lambda^{ikq}\lambda_{qst}\lambda^{\ell st}\lambda_{pk\ell}c^p \\
& -4\sum_I g_I^2 C_I(\ell)(\lambda^{ik\ell}M_I - a^{ik\ell})\lambda_{pk\ell}\zeta^p \\
& -\left[\lambda^{ikq}\lambda_{qst}a^{\ell st}\lambda_{pk\ell} + a^{ikq}\lambda_{qst}\lambda^{\ell st}\lambda_{pk\ell}\right]\zeta^p \\
& -4\sum_I g_I^2 C_I(\ell)\lambda_{jn\ell}\left[\mu^{n\ell}M_I - b^{n\ell}\right]\mu^{ij} \\
& -\left[\lambda_{jnq}a^{qst}\lambda_{\ell st}\mu^{n\ell} + \lambda_{jnq}\lambda^{qst}\lambda_{\ell st}b^{n\ell}\right]\mu^{ij} \\
& +4\sum_I g_I^2 C_I(\ell)\Big[2\lambda^{ik\ell}\mu_{k\ell}|M_I|^2 - \lambda^{ik\ell}b_{k\ell}M_I - a^{ik\ell}\mu_{k\ell}M_I^\dagger + a^{ik\ell}b_{k\ell} \\
& +\lambda^{ip\ell}(m^2)^k{}_p\mu_{k\ell} + \lambda^{ikp}(m^2)^\ell{}_p\mu_{k\ell}\Big] - \Big[\lambda^{ikq}\lambda_{qst}a^{\ell st}b_{k\ell} + a^{ikq}\lambda_{qst}\lambda^{\ell st}b_{k\ell} \\
& +a^{ikq}a_{qst}\lambda^{\ell st}\mu_{k\ell} + \lambda^{ipq}(m^2)^k{}_p\lambda_{qst}\lambda^{\ell st}\mu_{k\ell} + \lambda^{ikq}\lambda_{qst}\lambda^{pst}(m^2)^\ell{}_p\mu_{k\ell} \\
& +\lambda^{ikp}(m^2)^q{}_p\lambda_{qst}\lambda^{\ell st}\mu_{k\ell} + 2\lambda^{ikq}\lambda_{qsp}(m^2)^p{}_t\lambda^{\ell st}\mu_{k\ell} + \lambda^{ikq}a_{qst}a^{\ell st}\mu_{k\ell}\Big]\,,
\end{aligned}$$

and for the linear terms.

Finally, for the scalar mass^2 terms we have

$$\frac{\mathrm{d}(m^2)^i{}_i}{\mathrm{d}t} = \frac{1}{16\pi^2}[\beta_{m^2}^{(1)}]^i{}_j + \frac{1}{(16\pi^2)^2}[\beta_{m^2}^{(2)}]^i{}_j\,,$$

with

$$\begin{aligned}
[\beta_{m^2}^{(1)}]^j{}_i = {} & \frac{1}{2}\lambda_{ipq}\lambda^{pqn}(m^2)^j{}_n + \frac{1}{2}\lambda^{jpq}\lambda_{pqn}(m^2)^n{}_i + 2\lambda_{ipq}\lambda^{jpr}(m^2)^q{}_r + a_{ipq}a^{jpq} \\
& -8\delta_i^j\sum_I |M_I|^2 g_I^2 C_I(i) + 2\sum_I g_I^2 (T^a(I))^j{}_i \mathrm{Tr}[T^b(I)m^2]\delta_{ab}\,,
\end{aligned}$$

and

$$
\begin{aligned}
[\beta^{(2)}_{m^2}]^j{}_i = & -\frac{1}{2}(m^2)^\ell{}_i\lambda_{\ell mn}\lambda^{mrj}\lambda_{pqr}\lambda^{pqn} - \frac{1}{2}(m^2)^j{}_\ell\lambda^{\ell mn}\lambda_{mri}\lambda^{pqr}\lambda_{pqn} \\
& -\lambda_{i\ell m}\lambda^{jnm}(m^2)^\ell{}_r\lambda_{npq}\lambda^{rpq} - \lambda_{i\ell m}\lambda^{jnm}(m^2)^r{}_n\lambda_{rpq}\lambda^{\ell pq} \\
& -\lambda_{i\ell m}\lambda^{jnr}(m^2)^\ell{}_n\lambda_{pqr}\lambda^{pqm} - 2\lambda_{i\ell m}\lambda^{j\ell n}\lambda_{npq}\lambda^{mpr}(m^2)^q{}_r \\
& -\lambda_{i\ell m}\lambda^{j\ell n}a_{npq}a^{mpq} - a_{i\ell m}a^{j\ell n}\lambda_{npq}\lambda^{mpq} - a_{i\ell m}\lambda^{j\ell n}\lambda_{npq}a^{mpq} \\
& -\lambda_{i\ell m}a^{j\ell n}a_{npq}\lambda^{mpq} + \sum_I g_I^2\Big[(m^2)^\ell{}_i\lambda_{\ell pq}\lambda^{jpq} + \lambda_{ipq}\lambda^{\ell pq}(m^2)^j{}_\ell \\
& +4\lambda_{ipq}\lambda^{jp\ell}(m^2)^q{}_\ell + 2a_{ipq}a^{jpq} - 2a_{ipq}\lambda^{jpq}M_I \\
& \quad - 2\lambda_{ipq}a^{jpq}M_I^\dagger + 4\lambda_{ipq}\lambda^{jpq}|M_I|^2\Big]\Big[C_I(p) + C_I(q) - C_I(i)\Big] \\
& -2\sum_I g_I^2(T^a(I))^j{}_i(T^b(I)m^2)^\ell{}_r\delta_{ab}\lambda_{\ell pq}\lambda^{rpq} \\
& +8\sum_I\sum_J g_I^2g_J^2(T^a(I))^j{}_i\mathrm{Tr}[T^b(I)C_J(\mathcal{R}_J)m^2]\delta_{ab} \\
& +\delta^j_i\Big\{\sum_I\Big[24g_I^4|M_I|^2\{C_I(i)\tau_I(\mathcal{R}_I) - 3C(\mathfrak{g}_I)C_I(i)\} \\
& \quad + 8g_I^4C_I(i)\{\mathrm{Tr}(\tau_I(\mathcal{R}_I)m^2) - C(\mathfrak{g}_I)|M_I|^2\}\Big] \\
& \quad + \sum_J g_I^2g_J^2C_I(i)C_J(i)(32|M_I|^2 + 8M_IM_J^\dagger + 8M_JM_I^\dagger)\Big]\Big\}\,.
\end{aligned}
$$

To simplify the notations, we have denoted the generators of $\mathfrak{g}_I$ by $T^a(I)$. In the renormalisation group equations of the mass^2-terms, the traces are taken over all chiral superfields, with $C_I(i)$ the corresponding quadratic Casimir operator. Of course in the case of a non-abelian Lie algebra all the traces involving $T(I)^a$ vanish identically. The β_c-functions are taken from [Fonseca et al. (2012)]. Moreover, to obtain the renormalisation group equations with several $\mathfrak{u}(1)-$factors see [Martin and Vaughn (1994); Fonseca et al. (2012)] or for the renormalisation group equations for the vacuum expectation values see [Fonseca et al. (2012)].

For the application of the renormalisation group equations to the MSSM see [Martin and Vaughn (1994)] and for a pedagogical computation of the various coefficients see *e.g.* the Ph.D thesis of Alloul (2013).

Appendix H

Lie superalgebras associated to supergravity and supersymmetry

Supersymmetry and supergravity are based upon Lie superalgebras. Simple complex Lie superalgebras where classified in [Kac (1977a,b); Scheunert et al. (1976a,b); Freund and Kaplansky (1976); Parker (1980)]. A complex Lie superalgebra is a $\mathbb{Z}_2$−graded $\mathbb{C}$−vector space $\mathfrak{g} = \mathfrak{g}_0 \oplus \mathfrak{g}_1$. Let us suppose that $\mathfrak{g}$ is finite dimensional with basis $\mathfrak{g}_0 = \{B_i, i = 1, \cdots \dim(\mathfrak{g}_0)\}$ and $\mathfrak{g}_1 = \{F_a, a = 1, \cdots \dim(\mathfrak{g}_1)\}$. We denote generically $T_A = B_i, F_a$ the elements of $\mathfrak{g}$, and $|A|$ its grading. Then, $\mathfrak{g}$ satisfies the following properties,

(1) the zero-graded (or even) part, $\mathfrak{g}_0$ is a Lie algebra with brackets[1]

$$[B_i, B_j] = f_{ij}{}^k B_k \ ;$$

(2) the graded-one (or odd) part, $\mathfrak{g}_1$ is a representation of $\mathfrak{g}_0$ that is

$$[B_i, F_a] = R_{ia}{}^b F_b \ ;$$

(3) the symmetric product of two elements in $\mathfrak{g}_1$ closes in $\mathfrak{g}_0$

$$\{F_a, F_b\} = Q_{ab}{}^i B_i \ ;$$

(4) the following Jacobi identities are satisfied

$$(-)^{|A||C|}[T_A, [T_B, T_C]_{|B||C|}]_{|A|(|B|+|C|)} + \text{perm.} = 0 \ .$$

Two kinds of simple complex Lie superalgebras are relevant in supersymmetry/supergravity: the unitary superalgebras $A(m-1|n-1)$ and the orthosymplectic superalgebras $B(m|n)$ and $C(m|n)$.

The unitary (complex) superalgebras (for $m \neq n$) have the following defining representation: it is generated by the set of $(m+n) \times (m+n)$ complex matrices

$$M = \begin{pmatrix} A & B \\ C & D \end{pmatrix} ,$$

[1]Differently from the conventions adopted in this book, there is no i factor in the commutation relations $[B_i, B_j]$. Consequently for a unitary representation $B_i \to M_i$, the matrices M_i are anti-hermitian, *i.e.*, $M_i^\dagger = -M_i$.

with vanishing supertrace: $\mathrm{Str}(M) = \mathrm{Tr}(A) - \mathrm{Tr}(D) = 0$. The even part of $\mathfrak{g}$ is $\mathfrak{g}_0 = \mathfrak{sl}(m, \mathbb{C}) \oplus \mathfrak{sl}(n, \mathbb{C}) \oplus \mathfrak{u}(1, \mathbb{C})$ and the odd part is given by $\mathfrak{g}_1 = (\underline{\mathbf{m}}, \bar{\underline{\mathbf{n}}}) \oplus (\bar{\underline{\mathbf{m}}}, \underline{\mathbf{n}})$. This algebra is also denoted $\mathfrak{sl}(m|n)$

$$\begin{aligned} A(m-1|n-1) &= \mathfrak{sl}(m|n) \\ &= \Big(\mathfrak{sl}(m, \mathbb{C}) \oplus \mathfrak{sl}(n, \mathbb{C}) \oplus \mathfrak{u}(1, \mathbb{C})\Big) \oplus \Big((\underline{\mathbf{m}}, \bar{\underline{\mathbf{n}}}) \oplus (\bar{\underline{\mathbf{m}}}, \underline{\mathbf{n}})\Big) . \end{aligned}$$

We introduce the following index notations: $1 \le i, j \le m$ (indices related to $\mathfrak{sl}(m, \mathbb{C})$), $m+1 \le i', j' \le n+m$, (indices related to $\mathfrak{sl}(n, \mathbb{C})$), and $I = (i, i')$. Let $e^I{}_J$ be the canonical basis of the set of $(m+n) \times (m+n)$ matrices satisfying

$$e^I{}_J e^K{}_L = \delta^K{}_J e^I{}_L \,, \tag{H.1}$$

and let $I_m = e^i{}_i$, $I_n = e^{i'}{}_{i'}$. The generators of $\mathfrak{sl}(m|n)$ are defined as follows

$$\begin{aligned}
&\mathfrak{sl}(m, \mathbb{C}) \to J^i{}_j = e^i{}_j - \frac{1}{m}\delta^i{}_j I_m \,, && 1 \le i, j \le m \\
&\mathfrak{sl}(n, \mathbb{C}) \to J'^{i'}{}_{j'} = e^{i'}{}_{j'} - \frac{1}{n}\delta^{i'}{}_{j'} I_n \,, && m+1 \le i', j' \le m+n \\
&\mathfrak{u}(1, \mathbb{C}) \to H = \frac{1}{m-n}(n I_m + m I_n) \,, && \\
&(\underline{\mathbf{m}}, \bar{\underline{\mathbf{n}}}) \to F^i{}_{j'} = e^i{}_{j'} && 1 \le i \le m, 1+m \le j' \le m+n \\
&(\bar{\underline{\mathbf{m}}}, \underline{\mathbf{n}}) \to F^{i'}{}_j = e^{i'}{}_j && 1 \le j \le m, 1+m \le i' \le m+n \,.
\end{aligned}$$

Note however that the generators of $\mathfrak{sl}(m, \mathbb{C})$ and $\mathfrak{sl}(n, \mathbb{C})$ are not all independent since $J^i{}_i = 0$ and $J'^{i'}{}_{i'} = 0$[2]. The (anti-) commutation relations are then easily obtained and stem directly from (H.1). The commutations relations of $\mathfrak{g}_0$ and the action of $\mathfrak{sl}(m, \mathbb{C}) \oplus \mathfrak{sl}(n, \mathbb{C}) \oplus \mathfrak{u}(1, \mathbb{C})$ on $\mathfrak{g}_1$ are standard (*e.g.* $(\underline{\mathbf{m}}, \bar{\underline{\mathbf{n}}})$ transforms in the fundamental representation of $\mathfrak{sl}(m, \mathbb{C})$ and the anti-fundamental representation of $\mathfrak{sl}(n, \mathbb{C})$) and will not be given here. We thus have

$$\begin{aligned}
\big[\mathfrak{sl}(m, \mathbb{C}), \mathfrak{sl}(m, \mathbb{C})\big] &\subset \mathfrak{sl}(m, \mathbb{C}) \,, \\
\big[\mathfrak{sl}(n, \mathbb{C}), \mathfrak{sl}(n, \mathbb{C})\big] &\subset \mathfrak{sl}(n, \mathbb{C}) \,, \\
\big[\mathfrak{sl}(m, \mathbb{C}), (\underline{\mathbf{m}}, \bar{\underline{\mathbf{n}}})\big] &\subset (\underline{\mathbf{m}}, \bar{\underline{\mathbf{n}}}) \,, \\
\big[\mathfrak{sl}(n, \mathbb{C}), (\underline{\mathbf{m}}, \bar{\underline{\mathbf{n}}})\big] &\subset (\underline{\mathbf{m}}, \bar{\underline{\mathbf{n}}}) \,, \\
\big[\mathfrak{sl}(m, \mathbb{C}), (\bar{\underline{\mathbf{m}}}, \underline{\mathbf{n}})\big] &\subset (\bar{\underline{\mathbf{m}}}, \underline{\mathbf{n}}) \,, \\
\big[\mathfrak{sl}(n, \mathbb{C}), (\bar{\underline{\mathbf{m}}}, \underline{\mathbf{n}})\big] &\subset (\bar{\underline{\mathbf{m}}}, \underline{\mathbf{n}}) \,,
\end{aligned}$$

[2]The presentation of $\mathfrak{sl}(m|n)$ is complicated by the fact that it is obtained from the non-simple superalgebra $\mathfrak{gl}(m|n)$ generated by the $e^I{}_J$ (*i.e.* when the matrices are non-super-traceless).

and

$$\begin{aligned}
[H, F^i{}_{j'}] &= F^i{}_{j'} \,, \\
[H, F^{i'}{}_j] &= -F^{i'}{}_j \,, \\
\{F^i{}_{j'}, F^{k'}{}_\ell\} &= \delta^{k'}{}_{j'} e^i{}_\ell + \delta^i{}_\ell e^{k'}{}_{j'} \\
&= \delta^{k'}{}_{j'}(e^i{}_\ell - \frac{1}{m}\delta^i{}_\ell I_m) + \delta^i{}_\ell(e^{k'}{}_{j'} - \frac{1}{n}\delta^{k'}{}_{j'} I_n) + \delta^i{}_\ell \delta^k{}_j \frac{m-n}{nm} H \,.
\end{aligned}$$

Note that in the last anti-commutator the R.H.S. of the first line belongs to $\mathfrak{g}$ but each term does not automatically belong to either $\mathfrak{sl}(m,\mathbb{C})$ or $\mathfrak{sl}(n,\mathbb{C})$. In the second line the first/second term always belongs to $\mathfrak{sl}(m,\mathbb{C})$ and $\mathfrak{sl}(n,\mathbb{C})$ respectively. One observes that the two representations of $\mathfrak{g}_0$ have an opposite $\mathfrak{u}(1,\mathbb{C})$ charge. One can see *e.g.*, [Frappat et al. (2000)] to have explicit anti-commutation relations in the Cartan-Weyl basis.

There are two series of complex orthosymplectic algebras[3] that can be unified

$$\mathfrak{osp}(m|2n) = \Big(\mathfrak{so}(m,\mathbb{C}) \oplus \mathfrak{sp}(2n,\mathbb{C})\Big) \oplus (\underline{\mathbf{m}}, \underline{\mathbf{2n}}) \,, \tag{H.2}$$

that is $\mathfrak{g}_0 = \mathfrak{so}(m,\mathbb{C}) \oplus \mathfrak{sp}(2n,\mathbb{C})$ and $\mathfrak{g}_1 = (\underline{\mathbf{m}}, \underline{\mathbf{2n}})$. Denoting $R_{ij} = -R_{ji}, i, j = 1, \cdots, m$ the generators of $\mathfrak{so}(m,\mathbb{C})$, $S_{ab} = S_{ba}, a, b = 1, \cdots, 2n$ the generators of $\mathfrak{sp}(2n,\mathbb{C})$ and F_{ia} the generators of $(\underline{\mathbf{m}}, \underline{\mathbf{2n}})$, the orthosymplectic superalgebra complies with

$$\begin{aligned}
\big[\mathfrak{so}(m,\mathbb{C}), \mathfrak{so}(m,\mathbb{C})\big] &\subset \mathfrak{so}(m,\mathbb{C}) \,, \\
\big[\mathfrak{sp}(2n,\mathbb{C}), \mathfrak{sp}(2n,\mathbb{C})\big] &\subset \mathfrak{sp}(2n,\mathbb{C}) \,, \\
\big[\mathfrak{so}(m,\mathbb{C}), (\underline{\mathbf{m}}, \underline{\mathbf{2n}})\big] &\subset (\underline{\mathbf{m}}, \underline{\mathbf{2n}}) \,, \\
\big[\mathfrak{sp}(2n,\mathbb{C}), (\underline{\mathbf{m}}, \underline{\mathbf{2n}})\big] &\subset (\underline{\mathbf{m}}, \underline{\mathbf{2n}}) \,, \\
\{F_{ia}, F_{jb}\} &= \delta_{ij} S_{ab} + \Omega_{ab} R_{ij} \,,
\end{aligned} \tag{H.3}$$

with Ω_{ab} the symplectic form of $\mathfrak{sp}(2n,\mathbb{C})$. The four first brackets are not given here since they are standard. One can see *e.g.* [Frappat et al. (2000)] for explicit brackets in the Cartan-Weyl basis. These Lie superalgebras are complex Lie superalgebras, so $\mathfrak{g}_0$ is a complex Lie algebra. If one extends the spacetime symmetries in the context of Lie superalgebra, since the Poincaré algebra is a real Lie algebra, one must consider real Lie superalgebras.

[3] Corresponding respectively to $m = 2k$ or $m = 2k + 1$.

Now we are ready to apply these structures to extend the spacetime symmetries. The first extension is associated to the superalgebra $\mathfrak{sl}(4|1)$ [Ferrara et al. (1977b)] or more precisely its real form $\mathfrak{su}(2,2|1)$[4]

$$\mathfrak{su}(2,2|1) = \Big(\mathfrak{su}(2,2) \oplus \mathfrak{u}(1)\Big) \oplus \Big(\underset{\sim}{\mathbf{4}}^{-\frac{1}{2}} \oplus \underset{\sim}{\bar{\mathbf{4}}}^{\frac{1}{2}}\Big) .$$

The bosonic part is the real form of $\mathfrak{sl}(4,\mathbb{C})$ corresponding to the split real Lie algebra $\mathfrak{su}(2,2)$ and to the compact real form of $\mathfrak{u}(1,\mathbb{C})$ noted $\mathfrak{u}(1)$ as usual. Introducing $g = \text{diag}(1,1,-1,-1)$ the Lie group $SU(2,2)$ is defined by

$$SU(2,2) = \left\{ M \in \mathcal{M}_4(\mathbb{C}) \text{ s.t. } M^\dagger g M = g \,, \ \det(M) = 1 \right\} .$$

Considering an infinitesimal transformation $M = 1 + m$, m belongs to the Lie algebra $\mathfrak{su}(2,2)$ defined by

$$\mathfrak{su}(2,2) = \left\{ m \in \mathcal{M}_4(\mathbb{C}) \text{ s.t. } m^\dagger g + g m = 0 \ \text{tr}(m) = 0 \right\} .$$

The Lie algebra $\mathfrak{su}(2,2)$ is fifteen dimensional and a basis is constituted of the matrices T_a satisfying the relations $gT_a = -T_a^\dagger g$ and $\text{Tr}(T_a) = 0$. Moreover for the real form we consider, the generator associated to $\mathfrak{u}(1)$ is given by $T = iH$ (in our notations since there no i–factor in front of the structure constant, the $\mathfrak{u}(1)$ charge is anti-hermitian).

The brackets of $\mathfrak{su}(2,2|1)$ can be obtained from (H.2) with $m = 4, n = 1$. However the group isomorphism $SU(2,2) \cong \overline{SO(2,4)}$ (with $\overline{SO(2,4)}$ the universal covering group of the conformal group $SO(2,4)$) leads to have a more enlightening presentation. Recall that the conformal group is the set of transformations which preserve the metric up to a scale factor. Within this isomorphism, the odd part is simply constituted by two copies of Weyls spinors of $\mathfrak{so}(2,4)$: $\underset{\sim}{\mathbf{4}}^{-\frac{1}{2}}$ is a complex left-handed spinor and $\underset{\sim}{\bar{\mathbf{4}}}^{\frac{1}{2}}$ a complex right-handed spinor. It is further assumed that $\underset{\sim}{\mathbf{4}}^{-\frac{1}{2}} \oplus \underset{\sim}{\bar{\mathbf{4}}}^{\frac{1}{2}}$ is a Majorana spinor. Note also that the two Weyl spinors have opposite $U(1)$ charge[5].

To obtain the even-even part of the algebra we introduce $J_{MN} = -J_{NM}, -1 \le M, N \le 4$, the generators of $\mathfrak{so}(2,4)$. We start from the relations $(-1 \le M, N \le 4)$

$$[J_{MN}, J_{PQ}] = -\eta_{NP} J_{MQ} + \eta_{PM} J_{NQ} - \eta_{NQ} J_{PM} + \eta_{QM} J_{PN} \,, \tag{H.4}$$

[4]For a study of real forms (or complexification), but in the context of Lie algebras, one can see *e.g.* [Campoamor-Stursberg and Rausch de Traubenberg (2019)].

[5]Since the structure constant are real in our commutation relations, the charges are purely imaginary.

with $\eta_{MN} = \text{diag}(1, 1, -1, -1, -1, -1)$. Introducing

$$L_{MN} = \begin{cases} L_{\mu\nu} &= J_{\mu\nu} \\ L_{-1\mu} &= \frac{1}{2}(P_\mu + K_\mu) \\ L_{4\mu} &= \frac{1}{2}(P_\mu - K_\mu) \\ L_{-14} &= D\,, \end{cases}$$

we obtain

$$\begin{aligned}
&[J_{\mu\nu}, J_{\rho\sigma}] = -\eta_{\nu\rho}J_{\mu\sigma} + \eta_{\rho\mu}J_{\nu\sigma} - \eta_{\nu\sigma}J_{\rho\mu} + \eta_{\sigma\mu}J_{\rho\nu}\,, \\
&[J_{\mu\nu}, P_\rho] = -\eta_{\nu\rho}P_\mu + \eta_{\mu\rho}P_\mu\,, \\
&[J_{\mu\nu}, K_\rho] = -\eta_{\nu\rho}K_\mu + \eta_{\mu\rho}K_\mu\,, \\
&[J_{\mu\nu}, D] = 0\,, \qquad\qquad [D, P_\nu] = P_\mu\,, \\
&[P_\mu, P_\nu] = 0\,, \qquad\qquad [D, K_\nu] = -K_\mu\,, \\
&[K_\mu, K_\nu] = 0\,, \qquad\qquad [D, D] = 0\,. \\
&[P_\mu, K_\nu] = 2J_{\mu\nu} - 2\eta_{\mu\nu}D\,,
\end{aligned} \tag{H.5}$$

To obtain the even-odd part of the algebra, we build the $\mathfrak{so}(2, 4)$ Dirac-γ matrices from the $\mathfrak{so}(1, 3)$ Dirac-γ matrices. Recalling that

$$\gamma^\mu = \begin{pmatrix} 0 & \sigma^\mu \\ \bar{\sigma}^\mu & 0 \end{pmatrix}\,,$$

$$\gamma^5 = \begin{pmatrix} -\sigma^0 & 0 \\ 0 & \sigma^0 \end{pmatrix}\,,$$

we introduce

$$\Gamma^M = \begin{cases} \Gamma^\mu &= \gamma^\mu \otimes \sigma^0 \\ \Gamma^{-1} &= \gamma^5 \otimes (\sigma^1) \\ \Gamma^4 &= \gamma^5 \otimes (i\sigma^2) \\ \Gamma^6 &= \gamma^5 \otimes \sigma^3\,, \end{cases}$$

and therefore

$$\Gamma^{MN} = \frac{1}{4}[\Gamma^M, \Gamma^N] = \begin{cases} \Gamma^{\mu\nu} = & \gamma^{\mu\nu} \otimes \sigma^0 \\ P^\mu & \to \gamma^5\gamma^\mu \otimes \begin{pmatrix} 0 & 1 \\ 0 & 0 \end{pmatrix} \\ K^\mu & \to \gamma^5\gamma^\mu \otimes \begin{pmatrix} 0 & 0 \\ 1 & 0 \end{pmatrix} \\ D & \to \frac{1}{2}I \otimes \sigma^3\,. \end{cases} \tag{H.6}$$

The chirality matrix is given by Γ^6, the $\mathfrak{so}(2,4)$ charge conjugation matrix (defined as the product of symmetric $\Gamma-$matrices) by

$$C_6 = (-i\Gamma^0\Gamma^2)(\Gamma^1) = (\underbrace{-i\gamma^0\gamma^2}_{C_4=\begin{pmatrix}\varepsilon^{\alpha\beta} & 0\\ 0 & \varepsilon_{\dot\alpha\dot\beta}\end{pmatrix}})(\gamma^5\otimes\sigma^1)\,,$$

$$= \begin{pmatrix} 0 & 0 & -\varepsilon^{\alpha\beta} & 0 \\ 0 & 0 & 0 & \bar\varepsilon_{\dot\alpha\dot\beta} \\ -\varepsilon^{\alpha\beta} & 0 & 0 & 0 \\ 0 & \bar\varepsilon_{\dot\alpha\dot\beta} & 0 & 0 \end{pmatrix}$$

with C_4 the $\mathfrak{so}(1,3)$ charge conjugation matrix and the $\mathcal{B}-$ matrix (defined as the product of purely imaginary matrices) by

$$\mathcal{B} = -i\Gamma^2 = \mathcal{B}_4\otimes\sigma^0 = \begin{pmatrix} 0 & \varepsilon_{\dot\alpha\dot\beta} \\ \varepsilon^{\alpha\beta} & 0\end{pmatrix}\otimes\sigma^0\,,$$

which satisfy

$$\begin{aligned}(\Gamma^M)^t &= -C\Gamma^M C^{-1}\,,\\ (\Gamma^M)^* &= -\mathcal{B}\Gamma^M\mathcal{B}^{-1}\,,\end{aligned}$$

where A^t is the transpose of the matrix A and A^* the complex conjugated matrix.

A Majorana spinor is the defined by

$$\Psi^A = \begin{pmatrix} S_\alpha \\ \bar S^{\dot\alpha} \\ Q_\alpha \\ \bar Q^{\dot\alpha}\end{pmatrix}, \tag{H.7}$$

with Q_α, S_α two left-handed $\mathfrak{so}(1,3)-$ spinors and $\bar Q_{\dot\alpha} = (Q_\alpha)^\dagger, \bar S_{\dot\alpha} = S_\alpha$. The two Weyl spinors are given by

$$\begin{aligned}\mathbf{4}^{-\frac{1}{2}} &= \frac{1}{2}(1-\Gamma^6)\Psi \implies \begin{pmatrix} S_\alpha \\ \bar Q^{\dot\alpha}\end{pmatrix} \equiv \lambda_a\,,\\ \bar{\mathbf{4}}^{\frac{1}{2}} &= \frac{1}{2}(1+\Gamma^6)\Psi \implies \begin{pmatrix} \bar S^{\dot\alpha} \\ Q_\alpha\end{pmatrix} \equiv \bar\rho^{\bar a}\,,\end{aligned} \tag{H.8}$$

and

$$\Psi_A = (C_6)_{AB}\Psi^B = \begin{pmatrix} -Q^\beta & \bar Q_{\dot\beta} & -S_\beta & \bar S^{\dot\beta}\end{pmatrix}.$$

From

$$\left[L_{MN}, \Psi^A\right] = (\Gamma_{MN})^A{}_B\Psi^B\,,$$

we deduce the even-odd part of the algebra:

$$\begin{aligned}
&[J_{\mu\nu}, Q_\alpha] = (\sigma_{\mu\nu})_\alpha{}^\beta Q_\beta\,, && [J_{\mu\nu}, \bar{Q}^{\dot\alpha}] = (\bar\sigma_{\mu\nu})^{\dot\alpha}{}_{\dot\beta}\bar{Q}^{\dot\beta}\,,\\
&[J_{\mu\nu}, S_\alpha] = (\sigma_{\mu\nu})_\alpha{}^\beta S_\beta\,, && [J_{\mu\nu}, \bar{S}^{\dot\alpha}] = (\bar\sigma_{\mu\nu})^{\dot\alpha}{}_{\dot\beta}\bar{S}^{\dot\beta}\,,\\
&[P_\mu, Q_\alpha] = 0\,, && [P_\mu, \bar{Q}^{\dot\alpha}] = 0\,,\\
&[P_\mu, S_\alpha] = \sigma_{\mu\alpha\dot\alpha}\bar{Q}^{\dot\alpha}\,, && [P_\mu, \bar{S}^{\dot\alpha}] = -\bar\sigma_\mu{}^{\dot\alpha\alpha}S_\alpha\,,\\
&[K_\mu, Q_\alpha] = -\sigma_{\mu\alpha\dot\alpha}\bar{S}^{\dot\alpha}\,, && [K_\mu, \bar{Q}^{\dot\alpha}] = \bar\sigma_\mu{}^{\dot\alpha\alpha}S_\alpha\,,\\
&[K_\mu, S_\alpha] = 0\,, && [K_\mu, \bar{S}^{\dot\alpha}] = 0\,,\\
&[D, Q_\alpha] = -\tfrac{1}{2}Q_\alpha\,, && [D, \bar{Q}^{\dot\alpha}] = -\tfrac{1}{2}\bar{Q}^{\dot\alpha}\,,\\
&[D, S_\alpha] = \tfrac{1}{2}S_\alpha\,, && [D, \bar{S}^{\dot\alpha}] = \tfrac{1}{2}\bar{S}^{\dot\alpha}\,,\\
&[T, Q_\alpha] = -\tfrac{i}{2}Q_\alpha\,, && [T, \bar{Q}^{\dot\alpha}] = \tfrac{i}{2}\bar{Q}^{\dot\alpha}\,,\\
&[T, S_\alpha] = \tfrac{i}{2}S_\alpha\,, && [T, \bar{S}^{\dot\alpha}] = -\tfrac{i}{2}\bar{S}^{\dot\alpha}\,,
\end{aligned} \tag{H.9}$$

where have used that $\underset{\sim}{4}$ and $\underset{\sim}{\bar{4}}$ have opposite $U(1)$ charges (see (H.8)).

Finally, to obtain the odd-odd part of the algebra we use

$$\{\Psi_A, \Psi_B\} = 4i(C_6\Gamma^{MN})_{AB}J_{MN} + \lambda(C_6\Gamma^6)_{AB}T\,,$$

$$\{\Psi^A, \Psi_B\} = 4i(\Gamma^{MN})^A{}_B J_{MN} + \lambda(\Gamma^6)^A{}_B T\,,$$

where λ is a constant to be identified. In particular, studying the decomposition of the matrices $C_6\Gamma^{MN}$ and $C_6\Gamma^6$, we obtain,

$$\{Q_\alpha, S_\beta\} = -4i(\sigma^{\mu\nu})_{\alpha\beta}J_{\mu\nu} - 2i\varepsilon_{\alpha\beta}D + \lambda\varepsilon_{\alpha\beta}T\,.$$

The Jacobi identity with $(Q_\alpha, S_\beta, Q_\gamma)$ implies that $\lambda = -6$.[6]

Collecting all brackets we then obtain the odd-odd part of the algebra

$$\begin{aligned}
&\{Q_\alpha, \bar{Q}_{\dot\alpha}\} = -2i\sigma^\mu{}_{\alpha\dot\alpha}P_\mu\,,\\
&\{S_\alpha, \bar{S}_{\dot\alpha}\} = 2i\sigma^\mu{}_{\alpha\dot\alpha}K_\mu\,,\\
&\{Q_\alpha, S_\beta\} = -4i(\sigma^{\mu\nu})_{\alpha\beta}L_{\mu\nu} - 2i\varepsilon_{\alpha\beta}D - 6\varepsilon_{\alpha\beta}T\,,\\
&\{\bar{Q}_{\dot\alpha}, \bar{S}_{\dot\beta}\} = 4i(\bar\sigma^{\mu\nu})_{\dot\alpha\dot\beta}L_{\mu\nu} - 2i\varepsilon_{\dot\alpha\dot\beta}D + 6\varepsilon_{\dot\alpha\dot\beta}T\,,\\
&\{Q_\alpha, Q_\beta\} = 0\,,\\
&\{S_\alpha, S_\beta\} = 0\,,\\
&\{Q_\alpha, \bar{S}_{\dot\beta}\} = 0\,.
\end{aligned} \tag{H.10}$$

[6] We have used the relationship $(\sigma_{\mu\nu})_\alpha{}^\beta(\sigma^{\mu\nu})_\gamma{}^\delta = -(\varepsilon_{\alpha\gamma}\varepsilon^{\beta\delta} + \delta_\alpha{}^\gamma\delta_\gamma{}^\beta)$.

Note that the third line is the complex conjugate of the fourth since all operators are anti-hermitian and $[(\sigma_{\mu\nu})_{\alpha\beta}]^* = -(\bar{\sigma}_{\mu\nu})_{\dot{\alpha}\dot{\beta}}$. The superalgebra defined by (H.5), (H.9) and (H.10) is the superconformal algebra. It is very important in supergravity for several reasons. Firstly, it is observed that its bosonic part admits the subalgebra

$$\mathbb{R} \ltimes I\mathfrak{so}(1,3) \subset \mathfrak{so}(2,4) ,$$

of Poincaré transformations and dilatation (generated by D). This means that if dilatations act non-trivially, one has to specify, besides the usual representations of the Poincaré algebra, the behaviour of the fields w.r.t. to dilatation, *i.e.*, the conformal weight.

At the level of superfields, the transformations just become the Howe–Tucker transformations of Sec. 3.2.1 (see *e.g.*, the transformations (5.40)). More particularly if we set

$$\Sigma = (\sigma + i\alpha) + \sqrt{2}\Theta \cdot \omega - \Theta \cdot \Theta \mathcal{F} , \tag{H.11}$$

σ corresponds to a dilatation, α to a chiral rotation and ω to a conformal supersymmetric transformation. Indeed, comparing (5.37), (5.41) and (5.52) with Freedman and Van Proeyen (2012), we observe that the transformation associated to the parameter ω is a conformal supersymmetric transformation (generate by S_α). Next, the closure of the Howe–Tucker algebra imposes us to consider the special conformal transformation generated by K_μ to belong to the algebra. However, the conformal superalgebra and in particular the second equation in (H.10) makes it possible to obtain the action of the special conformal transformation on any field. In fact it is immediate to get this for all fields, except b_μ the action of K_μ is trivial. Computing

$$\Big(\delta_\omega \delta_{\omega'} - \delta_{\omega'} \delta_\omega\Big) b_\mu ,$$

with $\delta_\omega = \delta_{\Sigma = \Theta \cdot \omega}$ in (D.2) explicitly shows that the action of K_μ on b_μ is also trivial. Next, the transformation generated by $\mathcal{F}$ induces a shift on the fields M and $M^\star$. It is easy to see that this new transformation (which is not present in the superconformal algebra) is an ideal of the transformation generated by Howe–Tucker transformations. More precisely, if U denotes any infinitesimal Howe–Tucker transformation and if U_M generates the $M-$shift, then we have $[U, U_M] = \lambda_U U_M$ (with $\lambda_U \in \mathbb{R}$ but which could be equal to zero for some U's)[7]. The various transformations under a generic Howe–Tucker transformations are summarised in (D.2), (D.3) and (D.5). Finally, if we denote by U_σ (resp. U_ω) the generators

[7]To prove this is easy we just have to compute $(UU_M - U_M U)M$, *i.e.*, the infinitesimal action of $UU_M - U_M U$ on M using (D.2).

of the conformal transformations (resp. the superconformal transformations, *i.e.*, gravitino shift in the language of Howe–Tucker) looking to (D.2), (D.3) and (D.5) it is easy to show that

$$[\sigma U_\sigma, \omega \cdot U_\omega] = \sigma\omega \cdot U_\omega \,,$$

just by computing $\sigma U_\sigma \omega \cdot U_\omega - \omega \cdot U_\omega \sigma U_\sigma$ acting on the fields which transform non-trivially under the superconformal transformations, namely ψ_μ, b_μ and F^i. Now, if we assume that the parameters of the transformation are field dependent and given by (6.36), and if we set

$$\begin{aligned} M_\sigma &= \frac{1}{2}\ln\Big(\frac{-3}{\Omega}\Big)U_\sigma \,, \\ M_\omega &= -\frac{1}{2}\Omega^{-1}\Omega_i\chi^i \cdot U_\omega \,, \end{aligned}$$

it is immediate to see that

$$[M_\sigma, M_\omega] = 0$$

and thus, at the finite level, we have

$$e^{M_\omega} e^{M_\sigma} = e^{M_\omega + M_\sigma} \,. \tag{H.12}$$

So the field redefinition performed in Chapter 6 can be done either with the parameters given in the superfield Σ in (6.36), and corresponding to the R.H.S. of Eq.[H.12] or by performing a dilation of the graviton followed by a gravitino shift as in the L.H.S. of Eq.[H.12]. In Chapter 6 the field redefinition was done in two steps. All these considerations show that the Howe–Tucker transformations can be understood from an algebraic point of view.

This observation triggered a very efficient method to obtain Lagrangians in supergravity, namely by means of the construction of conformal supergravity (*i.e.* a theory invariant within the full superconformal algebra instead of the super-Poincaré algebra). This method is very powerful and then, degauging the superconformal invariant Lagrangian, one obtains Poincaré supergravity. One can turn to [Freedman and Van Proeyen (2012); Butter (2010)] for such an approach. It should be emphasised that the superconformal algebra is the more general algebraic structure compatible with the no-go theorem [Coleman and Mandula (1967); Haag et al. (1975)] that one can use in the context of Lie superalgebras. This gives a deeper understanding of the superconformal approach to supergravity[8]. Finally, one can observe that the Poincaré superalgebra is a subalgebra of the superconformal algebra.

---

[8]There are also extensions involving several supercharges or in higher dimensional spacetime (see [Nahm (1978)] for a general study).

The second type of superalgebra relevant in supergravity is obtained from the orthosymplectic algebra $\mathfrak{osp}(1|4) = \mathfrak{sp}(4,\mathbb{C}) \oplus \underset{\sim}{\mathbf{4}}$. It is known that we have the following isomorphism of Lie algebras $\mathfrak{sp}(4,\mathbb{C}) \cong \mathfrak{so}(5,\mathbb{C})$. An appropriate real form is obtained by considering a real form of $\mathfrak{so}(5,\mathbb{C})$ for which Majorana spinors exist. Among the real forms $\mathfrak{so}(p,q), p+q=5$ only the real form $\mathfrak{so}(2,3)$ admits Majorana spinors. We thus consider the real superalgebra (we keep the same name, since no confusion is possible)

$$\mathfrak{osp}(1|4) = \mathfrak{so}(2,3) \oplus \underset{\sim}{\mathbf{4}}\,, \tag{H.13}$$

with now $\underset{\sim}{\mathbf{4}}$ being understood as a Majorana spinor of $\mathfrak{so}(2,3)$. Introducing the generators L_{MN} of $\mathfrak{so}(2,3)$, $\eta_{MN} = \text{diag}(1,-1,-1,-1,1)$ $(M,N = 0,\cdots 4)$ and using the relation analogous to (H.4), we get the even-even part of the algebra

$$\begin{aligned}
[J_{\mu\nu}, J_{\rho\sigma}] &= -\eta_{\nu\rho}J_{\mu\sigma} + \eta_{\rho\mu}J_{\nu\sigma} - \eta_{\nu\sigma}J_{\rho\mu} + \eta_{\sigma\mu}J_{\rho\nu}\,,\\
[J_{\mu\nu}, J_{4\rho}] &= -\eta_{\nu\rho}J_{4\mu} + \eta_{\mu\rho}J_{4\nu}\,,\\
[J_{4\mu}, J_{4\nu}] &= J_{\mu\nu}\,.
\end{aligned} \tag{H.14}$$

We consider now, the $\mathfrak{so}(2,3)-$ generators in the spinor representation where

$$\Gamma^M = \begin{cases} \Gamma^\mu = \gamma^\mu\,,\\ \Gamma^4 = \gamma^5\,,\end{cases}$$

thus

$$\Gamma^{MN} = \frac{1}{4}[\Gamma^M, \Gamma^N] = \begin{cases} \Gamma^{\mu\nu} = \gamma^{\mu\nu}\,,\\ \Gamma^{4\mu} = \frac{1}{2}\gamma^5\gamma^\mu\,.\end{cases}$$

The charge conjugation matrix is given by

$$\begin{aligned}
C_5 &= \underbrace{(-i\Gamma^0\Gamma^2)}_{C_4}(-i\Gamma^4) = -iC_4\gamma^5\,,\\
&= \begin{pmatrix} -i\varepsilon^{\alpha\beta} & 0\\ 0 & i\varepsilon_{\dot\alpha\dot\beta}\end{pmatrix}.
\end{aligned}$$

with C_4, the usual charge conjugation matrix of $\mathfrak{so}(1,3)$. A Majorana spinor is defined by $((F_\alpha)^\dagger = \bar{F}_{\dot\alpha})$

$$\Psi^A = \begin{pmatrix} F_\alpha \\ \bar{F}^{\dot\alpha}\end{pmatrix},$$

so that

$$\Psi_A = C_{5AB}\Psi^B = \begin{pmatrix} -iF^\alpha & i\bar{F}_{\dot\alpha}\end{pmatrix}.$$

In particular this means that

$$\begin{aligned}
\Psi_\alpha &= F_\alpha , \\
\Psi^\alpha &= -iF^\alpha , \\
\bar{\Psi}^{\dot\alpha} &= \bar{F}^{\dot\alpha} , \\
\bar{\Psi}_{\dot\alpha} &= i\bar{F}_{\dot\alpha} .
\end{aligned}$$

The even-odd part of the algebra takes the form

$$[J_{MN}, \Psi^A] = (\Gamma_{MN})^A{}_B \Psi^B ,$$

and therefore

$$\begin{aligned}
[J_{\mu\nu}, F_\alpha] &= (\sigma^{\mu\nu})_\alpha{}^\beta F_\beta , \\
[J_{\mu\nu}, \bar{F}^{\dot\alpha}] &= (\bar{\sigma}^{\mu\nu})^{\dot\alpha}{}_{\dot\beta} \bar{F}^{\dot\beta} , \\
[J_{4\mu}, F_\alpha] &= -\frac{1}{2}\sigma_{\mu\alpha\dot\alpha}\bar{F}^{\dot\alpha} , \\
[J_{4\mu}, \bar{F}^{\dot\alpha}] &= \frac{1}{2}\bar{\sigma}_\mu{}^{\dot\alpha\alpha} F_\alpha .
\end{aligned} \tag{H.15}$$

Finally the last relation of (H.3) leads to the odd-odd part of the algebra

$$\begin{aligned}
\{F_A, F^B\} &= 4(\Gamma^{MN})_A{}^B J_{MN} , \\
\{F_A, F_B\} &= 4(C\Gamma^{MN})_{AB} J_{MN}
\end{aligned}$$

implies

$$\begin{aligned}
\{F_\alpha, F_\beta\} &= 4i(\sigma^{\mu\nu})_{\alpha\beta} J_{\mu\nu} , \\
\{\bar{F}_{\dot\alpha}, \bar{F}_{\dot\beta}\} &= -4i(\bar{\sigma}^{\mu\nu})_{\dot\alpha\dot\beta} J_{\mu\nu} , \\
\{F_\alpha, \bar{F}_{\dot\alpha}\} &= -2i\sigma^\mu{}_{\alpha\dot\alpha} J_{4\mu} .
\end{aligned}$$

The algebra (H.14), (H.15) and (H.16) is relevant for anti-de-Sitter supergravity (see *e.g.* [Freedman and Van Proeyen (2012)]). Two observations are in order. First, since there is no Majorana spinor for $\mathfrak{so}(1,4)$, it is not possible to construct de-Sitter supergravity. Second, the Poincaré superalgebra is not a subalgebra of the anti-de-Sitter superalgebra. The Poincaré superalgebra can indeed be obtained through a contraction of the anti-de-Sitter superalgebra. In fact if we perform an Inönü-Wigner contraction

$$\begin{aligned}
P_\mu &= \frac{1}{R} J_{4\mu} , \\
Q_A &= \frac{1}{\sqrt{R}} F_A ,
\end{aligned}$$

where we take the limit $R \to \infty$, we obtain the Poincaré superalgebra [Ferrara and Lledo (2002)]

$$\begin{aligned}
[J_{\mu\nu}, J_{\rho\sigma}] &= -\eta_{\nu\rho} J_{\mu\sigma} + \eta_{\rho\mu} J_{\nu\sigma} - \eta_{\nu\sigma} J_{\rho\mu} + \eta_{\sigma\mu} J_{\rho\nu} , \\
[J_{\mu\nu}, P_\rho] &= -\eta_{\nu\rho} P_\mu + \eta_{\mu\rho} P_\nu , \\
[P_\mu, P_\nu] &= 0 , \\
[J_{\mu\nu}, Q_\alpha] &= (\sigma_{\mu\nu})_\alpha{}^\beta Q_\beta , \\
[J_{\mu\nu}, \bar{Q}_{\dot\alpha}] &= (\bar\sigma_{\mu\nu})^{\dot\alpha}{}_{\dot\beta} \bar{Q}^{\dot\beta} , \\
[P_\mu, Q_\alpha] &= 0 , \\
[P_\mu, \bar{Q}_{\dot\alpha}] &= 0 , \\
\{Q_\alpha, \bar{Q}_{\dot\alpha}\} &= -2i\sigma^\mu{}_{\alpha\dot\alpha} P_\mu , \\
\{Q_\alpha, Q_\beta\} &= 0 , \\
\{\bar{Q}_{\dot\alpha}, \bar{Q}_{\dot\beta}\} &= 0 .
\end{aligned} \tag{H.16}$$

Finally, one observes that the Poincaré superalgebra can be extended by an $\mathfrak{u}(1)$–symmetry (the so-called R–symmetry) to

$$\begin{aligned}
[T, Q_\alpha] &= -\frac{i}{2} Q_\alpha , \\
[T, \bar{Q}^{\dot\alpha}] &= \frac{i}{2} \bar{Q}^{\dot\alpha} .
\end{aligned} \tag{H.17}$$

Note the algebra (H.16) and (H.17) is a subalgebra of the superconformal algebra (H.5), (H.9) and (H.10). However, for supersymmetric theories in particle physics, only a discrete subgroup of the R–symmetry is involved, namely the R–parity (see Chapter 8).

Bibliography

Akulov, V., Volkov, D. and Soroka, V. (1975). Gauge fields on superspaces with different holonomy groups, JETP Lett. **22**, pp. 187–188.

Akulov, V., Volkov, D. and Soroka, V. (1977). Generally covariant theories of gauge fields on superspace, Theor.Math.Phys. **31**, p. 285, doi:10.1007/BF01041233.

Alloul, A. (2013). Top-down and bottom-up excursions beyond the standard model: The example of left-right symmetries in supersymmetry, Ph.D. thesis, U. Strasbourg, `arXiv:1404.4564 [hep-ph]`, `http://inspirehep.net/record/1291146/files/arXiv:1404.4564.pdf`.

Alvarez-Gaumé, L., Claudson, M. and Wise, M. B. (1982). Low-energy supersymmetry, Nucl. Phys. **B207**, p. 96, doi:10.1016/0550-3213(82)90138-9.

Alvarez-Gaumé, L. and Freedman, D. Z. (1981). Geometrical structure and ultraviolet finiteness in the supersymmetric σ-model, Commun. Math. Phys. **80**, p. 443, doi:10.1007/BF01208280.

Alvarez-Gaumé, L., Polchinski, J. and Wise, M. B. (1983). Minimal low-energy supergravity, Nucl. Phys. **B221**, p. 495, doi:10.1016/0550-3213(83)90591-6.

Amaldi, U., de Boer, W. and Furstenau, H. (1991). Comparison of grand unified theories with electroweak and strong coupling constants measured at LEP, Phys.Lett. **B260**, pp. 447–455, doi:10.1016/0370-2693(91)91641-8.

Arkani-Hamed, N., Giudice, G. F., Luty, M. A. and Rattazzi, R. (1998). Supersymmetry-breaking loops from analytic continuation into superspace, Phys. Rev. **D58**, p. 115005, doi:10.1103/PhysRevD.58.115005, `arXiv:hep-ph/9803290`.

Arnowitt, R. L., Nath, P. and Zumino, B. (1975). Superfield densities and action principle in curved superspace, Phys. Lett. **B56**, p. 81, doi:10.1016/0370-2693(75)90504-3.

Avdeev, L., Tarasov, O. and Vladimirov, A. (1980). Vanishing of the three-loop charge renormalization function in a supersymmetric gauge theory, Phys.Lett. **B96**, pp. 94–96, doi:10.1016/0370-2693(80)90219-1.

Bagger, J. and Witten, E. (1982). The gauge invariant supersymmetric nonlinear sigma model, Phys.Lett. **B118**, pp. 103–106, doi:10.1016/0370-2693(82)90609-8.

Bagger, J. A. (1983). Coupling the gauge-invariant supersymmetric non-linear sigma model to supergravity, Nucl.Phys. **B211**, p. 302, doi:10.1016/0550-3213(83)90411-X.

Bagger, J. A., Moroi, T. and Poppitz, E. (2000). Anomaly mediation in supergravity theories, JHEP **0004**, p. 009, doi:10.1088/1126-6708/2000/04/009, `arXiv:hep-th/9911029 [hep-th]`.

Bailin, D. and Love, A. (1994). Supersymmetric Gauge Field Theory and String Theory (IOP Publishing).

Bailin, D. and Love, A. (2004). Cosmology in Gauge Field Theory and String Theory (IOP Publishing).

Barbieri, R., Ferrara, S., Nanopoulos, D. V. and Stelle, K. (1982a). Supergravity, R-invariance and spontaneous supersymmetry breaking, Phys.Lett. **B113**, p. 219, doi:10.1016/0370-2693(82)90825-5.

Barbieri, R., Ferrara, S. and Savoy, C. A. (1982b). Gauge models with spontaneously broken local supersymmetry, Phys. Lett. **B119**, p. 343, doi:10.1016/0370-2693(82)90685-2.

Bargmann, V. and Wigner, E. P. (1948). Group theoretical discussion of relativistic wave equations, Proc.Nat.Acad.Sci. **34**, p. 211.

Batalin, I. and Vilkovisky, G. (1983). Quantization of gauge theories with linearly dependent generators, Phys.Rev. **D28**, pp. 2567–2582, doi:10.1103/PhysRevD.28.2567.

Baumann, D. (2011). Inflation, Physics of the large and the small, TASI 09, proceedings of the Theoretical Advanced Study Institute in Elementary Particle Physics , pp. 523–686, doi:10.1142/9789814327183_0010, `arXiv:0907.5424 [hep-th]`.

Bechtle, P. et al. (2015). Killing the cMSSM softly, PoS **EPS-HEP2015**, p. 139.

Becker, K., Becker, M. and Schwarz, J. (2007). String Theory and M-Theory: A Modern Introduction (Cambridge: Cambridge University Press.), doi:10.1017/CBO9780511816086.

Berezin, F. A. (1987). Introduction to superanalysis. Ed. by A. A. Kirillov. transl. from the russian by J. Niederle and R. Kotecký, Mathematical Physics and Applied Mathematics 9 (Dordrecht, Netherlands: Reidel) .

Bergshoeff, E., de Roo, M. and de Wit, B. (1983). Conformal supergravity in ten dimensions, Nucl.Phys. **B217**, p. 489, doi:10.1016/0550-3213(83)90159-1.

Bergshoeff, E., de Roo, M., de Wit, B. and van Nieuwenhuizen, P. (1982). Ten-dimensional Maxwell-Einstein supergravity, its currents, and the issue of its auxiliary fields, Nucl.Phys. **B195**, pp. 97–136, doi:10.1016/0550-3213(82)90050-5.

Bezrukov, F., Gorbunov, D. and Shaposhnikov, M. (2009). On initial conditions for the hot big bang, JCAP **0906**, p. 029, doi:10.1088/1475-7516/2009/06/029, `arXiv:0812.3622 [hep-ph]`.

Bezrukov, F. L. and Shaposhnikov, M. (2008). The standard model Higgs boson as the inflaton, Phys. Lett. **B659**, pp. 703–706, doi:10.1016/j.physletb.2007.11.072, `arXiv:0710.3755 [hep-th]`.

Binetruy, P. (2006). Supersymmetry: Theory, Experiment and Cosmology (Oxford, UK: Oxford Univ. Pr.).

Binetruy, P., Dvali, G., Kallosh, R. and Van Proeyen, A. (2004). Fayet-Iliopoulos terms in supergravity and cosmology, Class.Quant.Grav. **21**, pp. 3137–3170, doi:10.1088/0264-9381/21/13/005, `arXiv:hep-th/0402046 [hep-th]`.

Binetruy, P., Girardi, G. and Grimm, R. (2001). Supergravity couplings: a geometric formulation, Phys.Rept. **343**, pp. 255–462, doi:10.1016/S0370-1573(00)00085-5, `arXiv:hep-th/0005225 [hep-th]`.

Bollini, C. and Giambiagi, J. (1972). Dimensional renormalization: The number of dimensions as a regularizing parameter, Nuovo Cimento Lett. **4**, pp. 20–26, doi:10.1007/BF02895558.

Brans, C. and Dicke, R. H. (1961). Mach's principle and a relativistic theory of gravitation, Phys.Rev. **124**, p. 925, doi:10.1103/PhysRev.124.925.

Breitenlohner, P. (1977). Some invariant Lagrangians for local supersymmetry, Nucl.Phys. **B124**, p. 500, doi:10.1016/0550-3213(77)90417-5.

Brignole, A., Ibanez, L. E. and Munoz, C. (1998). Soft supersymmetry breaking terms from supergravity and superstring models, Adv. Ser. Direct. High Energy Phys. **18**, pp. 125–148, doi:10.1142/9789812839657_0003, arXiv:hep-ph/9707209 [hep-ph].

Brink, L., Gell-Mann, M., Ramond, P. and Schwarz, J. H. (1978). Supergravity as geometry of superspace, Phys.Lett. **B74**, p. 336, doi:10.1016/0370-2693(78)90671-8.

Brink, L., Lindgren, O. and Nilsson, B. E. (1983). The ultra-violet finiteness of the N=4 Yang-Mills theory, Phys.Lett. **B123**, p. 323, doi:10.1016/0370-2693(83)91210-8.

Brink, L., Schwarz, J. H. and Scherk, J. (1977). Supersymmetric Yang-Mills theories, Nucl.Phys. **B121**, p. 77, doi:10.1016/0550-3213(77)90328-5.

Buchbinder, I. and Kuzenko, S. (1998). Ideas and Methods of Supersymmetry and Supergravity: Or a Walk Through Superspace (Studies in High Energy Physics, Cosmology and Gravitation. Bristol: Institute of Physics (IOP).).

Butter, D. (2010). N=1 conformal superspace in four dimensions, Annals Phys. **325**, pp. 1026–1080, doi:10.1016/j.aop.2009.09.010.

Campbell, I. and West, P. C. (1984). N = 2, D = 10 non-chiral supergravity and its spontaneous compactification, Nucl.Phys. **B243**, p. 112, doi:10.1016/0550-3213(84) 90388-2.

Campoamor-Stursberg, R. and Rausch de Traubenberg, M. (2019). Group Theory in Physics: A Practitioner's Guide (World Scientific), doi:10.1142/11081.

Capper, D. M., Jones, D. R. T. and van Nieuwenhuizen, P. (1980). Regularization by dimensional reduction of supersymmetric and nonsupersymmetric gauge theories, Nucl. Phys. **B167**, pp. 479–499, doi:10.1016/0550-3213(80)90244-8.

Carena, M. S., Pokorski, S. and Wagner, C. (1993). On the unification of couplings in the minimal supersymmetric standard model, Nucl.Phys. **B406**, pp. 59–89, doi:10.1016/ 0550-3213(93)90161-H, arXiv:hep-ph/9303202 [hep-ph].

Castano, D., Freedman, D. and Manuel, C. (1996). Consequences of supergravity with gauged U(1)-R symmetry, Nucl.Phys. **B461**, pp. 50–70, doi:10.1016/0550-3213(95) 00584-6, arXiv:hep-ph/9507397 [hep-ph].

Caswell, W. E. and Zanon, D. (1981). Zero three-loop beta function in the N = 4 supersymmetric Yang-Mills theory, Nucl.Phys. **B182**, p. 125, doi:10.1016/0550-3213(81) 90461-2.

Chamseddine, A. H. (1981). Interacting supergravity in ten dimensions: The role of the six-index gauge field, Phys.Rev. **D24**, p. 3065, doi:10.1103/PhysRevD.24.3065.

Chamseddine, A. H., Arnowitt, R. L. and Nath, P. (1982). Locally supersymmetric grand unification, Phys. Rev. Lett. **49**, p. 970, doi:10.1103/PhysRevLett.49.970.

Chamseddine, A. H. and Dreiner, H. K. (1995). Anomaly - free gauged U(1)-prime in local supersymmetry and baryon number violation, Nucl.Phys. **B447**, pp. 195–210, doi:10.1016/0550-3213(95)00253-O, arXiv:hep-ph/9503454 [hep-ph].

Chapline, G. and Manton, N. (1983). Unification of Yang-Mills theory and supergravity in ten dimensions, Phys.Lett. **B120**, pp. 105–109, doi:10.1016/0370-2693(83)90633-0.

Cheung, C., D'Eramo, F. and Thaler, J. (2011). Supergravity computations without gravity complications, Phys.Rev. **D84**, p. 085012, doi:10.1103/PhysRevD.84.085012, `arXiv:1104.2598 [hep-ph]`.

Coleman, S. R. and Mandula, J. (1967). All possible symmetries of the S matrix, Phys.Rev. **159**, pp. 1251–1256, doi:10.1103/PhysRev.159.1251.

Cremmer, E., Fayet, P. and Girardello, L. (1983a). Gravity induced supersymmetry breaking and low-energy mass spectrum, Phys. Lett. **B122**, p. 41, doi:10.1016/0370-2693(83)91165-6.

Cremmer, E., Ferrara, S., Girardello, L. and Van Proeyen, A. (1982). Coupling supersymmetric Yang-Mills theories to supergravity, Phys. Lett. **B116**, p. 231, doi:10.1016/0370-2693(82)90332-X.

Cremmer, E., Ferrara, S., Girardello, L. and Van Proeyen, A. (1983b). Yang-Mills theories with local supersymmetry: Lagrangian, transformation laws and super-Higgs effect, Nucl. Phys. **B212**, p. 413, doi:10.1016/0550-3213(83)90679-X.

Cremmer, E., Ferrara, S., Kounnas, C. and Nanopoulos, D. V. (1983c). Naturally vanishing cosmological constant in N=1 supergravity, Phys.Lett. **B133**, p. 61, doi:10.1016/0370-2693(83)90106-5.

Cremmer, E. and Julia, B. (1979). The SO(8) supergravity, Nucl.Phys. **B159**, p. 141, doi:10.1016/0550-3213(79)90331-6.

Cremmer, E., Julia, B. and Scherk, J. (1978a). Supergravity theory in 11 dimensions, Phys.Lett. **B76**, pp. 409–412, doi:10.1016/0370-2693(78)90894-8.

Cremmer, E., Julia, B., Scherk, J., Ferrara, S., Girardello, L. and van Nieuwenhuizen, P. (1978b). Super-Higgs effect in supergravity with general scalar interactions, Phys. Lett. **B79**, p. 231, doi:10.1016/0370-2693(78)90230-7.

Cremmer, E., Julia, B., Scherk, J., Ferrara, S., Girardello, L. and van Nieuwenhuizen, P. (1979). Spontaneous symmetry breaking and Higgs effect in supergravity without cosmological constant, Nucl. Phys. **B147**, p. 105, doi:10.1016/0550-3213(79)90417-6.

Cremmer, E. and Scherk, J. (1978). The supersymmetric nonlinear sigma model in four-dimensions and its coupling to supergravity, Phys.Lett. **B74**, p. 341, doi:10.1016/0370-2693(78)90672-X.

Cvetic, M. and Soleng, H. H. (1997). Supergravity domain walls, Phys.Rept. **282**, pp. 159–223, doi:10.1016/S0370-1573(96)00035-X, `arXiv:hep-th/9604090 [hep-th]`.

Das, A., Fischler, M. and Rocek, M. (1977). SuperHiggs effect in a new class of scalar models and a model of super QED, Phys.Rev. **D16**, pp. 3427–3436, doi:10.1103/PhysRevD.16.3427.

de Wit, B. and van Nieuwenhuizen, P. (1978). The auxiliary field structure in chirality extended supergravity, Nucl.Phys. **B139**, p. 216, doi:10.1016/0550-3213(78)90188-8.

Derendinger, J.-P. (1989). Lecture notes on globally supersymmetric theories in four-dimensions and two-dimensions, 3rd Hellenic School on Elementary Particle Physics, Corfu, Greece `https://www.worldscientific.com/worldscibooks/10.1142/1191`.

Derendinger, J. P. and Savoy, C. A. (1982). Gaugino masses and a new mechanism for proton decay in supersymmetric theories, Phys. Lett. **B118**, p. 347, doi:10.1016/0370-2693(82)90201-5.

Derendinger, J. P. and Savoy, C. A. (1984). Quantum effects and SU(2)×U(1) breaking in supergravity gauge theories, Nucl. Phys. **B237**, p. 307, doi:10.1016/0550-3213(84) 90162-7.

Deser, S. and Zumino, B. (1976). Consistent supergravity, Phys. Lett. **B62**, p. 335, doi: 10.1016/0370-2693(76)90089-7.

Deser, S. and Zumino, B. (1977). Broken supersymmetry and supergravity, Phys.Rev.Lett. **38**, p. 1433, doi:10.1103/PhysRevLett.38.1433.

D'Hoker, E. and Freedman, D. Z. (2002). Supersymmetric gauge theories and the AdS/CFT correspondence, Strings, Branes and Extra Dimensions: TASI 2001 Proceedings , pp. 3–158arXiv:hep-th/0201253 [hep-th].

Dimopoulos, S. and Georgi, H. (1981). Softly broken supersymmetry and SU(5), Nucl. Phys. **B193**, p. 150, doi:10.1016/0550-3213(81)90522-8.

Dimopoulos, S. and Raby, S. (1981). Supercolor, Nucl. Phys. **B192**, p. 353, doi:10.1016/ 0550-3213(81)90430-2.

Dimopoulos, S., Raby, S. and Wilczek, F. (1981). Supersymmetry and the scale of unification, Phys.Rev. **D24**, pp. 1681–1683, doi:10.1103/PhysRevD.24.1681.

Dine, M. (2007). Supersymmetry and String Theory. Beyond the Standard Model. (Cambridge: Cambridge University Press.), doi:10.2277/0521858410.

Dine, M. and Fischler, W. (1982). A phenomenological model of particle physics based on supersymmetry, Phys. Lett. **B110**, p. 227, doi:10.1016/0370-2693(82)91241-2.

Dine, M., Fischler, W. and Srednicki, M. (1981). Supersymmetric technicolor, Nucl. Phys. **B189**, p. 575, doi:10.1016/0550-3213(81)90582-4.

Dine, M. and Nelson, A. E. (1993). Dynamical supersymmetry breaking at low energies, Phys. Rev. **D48**, p. 1277, doi:10.1103/PhysRevD.48.1277, arXiv:hep-ph/9303230.

Dine, M., Nelson, A. E., Nir, Y. and Shirman, Y. (1996). New tools for low energy dynamical supersymmetry breaking, Phys. Rev. **D53**, p. 2658, doi:10.1103/PhysRevD.53. 2658, arXiv:hep-ph/9507378.

Dine, M., Nelson, A. E. and Shirman, Y. (1995). Low energy dynamical supersymmetry breaking simplified, Phys. Rev. **D51**, p. 1362, doi:10.1103/PhysRevD.51.1362, arXiv:hep-ph/9408384.

Djouadi, A., Kneur, J.-L. and Moultaka, G. (2007). Suspect: A Fortran code for the supersymmetric and Higgs particle spectrum in the MSSM, Comput. Phys. Commun. **176**, pp. 426–455, doi:10.1016/j.cpc.2006.11.009, arXiv:hep-ph/0211331 [hep-ph].

Djouadi, A., Rosier-Lees, S., Bezouh, M., Bizouard, M., Boehm, C., Borzumati, F., Briot, C., Carr, J., Causse, M., Charles, F., Chereau, X., Colas, P., Duflot, L., Dupperin, A., Ealet, A., El-Mamouni, H., Ghodbane, N., Gieres, F., Gonzalez-Pineiro, B., Gourmelen, S., Grenier, G. J., Gris, P., Grivaz, J., Hebrard, C., Ille, B., Kneur, J., Kostantinidis, N., Layssac, J., Lebrun, P., Ledu, R., Lemaire, M., LeMouel, C., Lugnier, L., Mambrini, Y., Martin, J., Montarou, G., Moultaka, G., Muanza, S., Nuss, E., Perez, E., Renard, F., Reynaud, D., Serin, L., Thevenet, C., Trabelsi, A., Zach, F. and Zerwas, D. (1998). The minimal supersymmetric standard model: Group summary report, GDR- Supersymétrie, Montpellier, France arXiv:hep-ph/9901246 [hep-ph], http://inspirehep.net/record/481987/files/arXiv:hep-ph_9901246.pdf.

Dragon, N. (1979). Torsion and curvature in extended supergravity, Z. Phys. **C2**, pp. 29–32, doi:10.1007/BF01546233.

Dragon, N., Ellwanger, U. and Schmidt, M. G. (1987). Supersymmetry and supergravity, Prog.Part.Nucl.Phys. **18**, p. 1, doi:10.1016/0146-6410(87)90008-1.

Duff, M., Nilsson, B. and Pope, C. (1986). Kaluza-Klein supergravity, Phys.Rept. **130**, pp. 1–142, doi:10.1016/0370-1573(86)90163-8.

Einhorn, M. B. and Jones, D. R. T. (2010). Inflation with non-minimal gravitational couplings in supergravity, JHEP **03**, p. 026, doi:10.1007/JHEP03(2010)026, arXiv:0912.2718 [hep-ph].

Ellis, J. R., Hagelin, J., Nanopoulos, D. V., Olive, K. A. and Srednicki, M. (1984a). Supersymmetric relics from the big bang, Nucl.Phys. **B238**, pp. 453–476, doi: 10.1016/0550-3213(84)90461-9.

Ellis, J. R., Kelley, S. and Nanopoulos, D. V. (1991). Probing the desert using gauge coupling unification, Phys.Lett. **B260**, pp. 131–137, doi:10.1016/0370-2693(91) 90980-5.

Ellis, J. R., Kounnas, C. and Nanopoulos, D. V. (1984b). No-scale supersymmetric GUTs, Nucl.Phys. **B247**, pp. 373–395, doi:10.1016/0550-3213(84)90555-8.

Ellis, J. R., Lahanas, A., Nanopoulos, D. V. and Tamvakis, K. (1984c). No-scale supersymmetric standard model, Phys.Lett. **B134**, p. 429, doi:10.1016/0370-2693(84) 91378-9.

Ellis, J. R., Nanopoulos, D. V. and Tamvakis, K. (1983). Grand unification in simple supergravity, Phys. Lett. **B121**, p. 123, doi:10.1016/0370-2693(83)90900-0.

Ellwanger, U., Hugonie, C. and Teixeira, A. M. (2010). The next-to-minimal supersymmetric standard model, Phys. Rept. **496**, pp. 1–77, doi:10.1016/j.physrep.2010.07.001, arXiv:0910.1785 [hep-ph].

Ellwanger, U., Rausch de Traubenberg, M. and Savoy, C. A. (1997). Phenomenology of supersymmetric models with a singlet, Nucl. Phys. **B492**, pp. 21–50, doi: 10.1016/S0550-3213(97)80026-0, 10.1016/S0550-3213(97)00128-4, arXiv:hep-ph/9611251 [hep-ph].

Falck, N. K. (1986). Renormalization group equations for softly broken supersymmetry: The most general case, Z. Phys. **C30**, p. 247, doi:10.1007/BF01575432.

Farrar, G. R. and Fayet, P. (1978). Phenomenology of the production, decay, and detection of new hadronic states associated with supersymmetry, Phys.Lett. **B76**, pp. 575–579, doi:10.1016/0370-2693(78)90858-4.

Fayet, P. (1975). Supergauge invariant extension of the Higgs mechanism and a model for the electron and its neutrino, Nucl. Phys. **B90**, p. 104, doi:10.1016/0550-3213(75) 90636-7.

Fayet, P. (1976). Supersymmetry and weak, electromagnetic and strong interactions, Phys.Lett. **B64**, p. 159, doi:10.1016/0370-2693(76)90319-1.

Fayet, P. (1977). Spontaneously broken supersymmetric theories of weak, electromagnetic and strong interactions, Phys.Lett. **B69**, p. 489, doi:10.1016/0370-2693(77) 90852-8.

Fayet, P. (1978). Massive gluinos, Phys. Lett. **B78**, p. 417, doi:10.1016/0370-2693(78) 90474-4.

Fayet, P. (1984). Supersymmetric theories of particles and interactions, Phys.Rept. **105**, p. 21, doi:10.1016/0370-1573(84)90113-3.

Fayet, P. and Iliopoulos, J. (1974). Spontaneously broken supergauge symmetries and goldstone spinors, Phys. Lett. **B51**, p. 461, doi:10.1016/0370-2693(74)90310-4.

Ferrara, S., Freedman, D., van Nieuwenhuizen, P., Breitenlohner, P., Gliozzi, F. and Scherk, J. (1977a). Scalar multiplet coupled to supergravity, Phys.Rev. **D15**, p. 1013, doi: 10.1103/PhysRevD.15.1013.

Ferrara, S., Girardello, L., Kugo, T. and Van Proeyen, A. (1983). Relation between different auxiliary field formulations of N=1 supergravity coupled to matter, Nucl.Phys. **B223**, p. 191, doi:10.1016/0550-3213(83)90101-3.

Ferrara, S., Gliozzi, F., Scherk, J. and Van Nieuwenhuizen, P. (1976a). Matter couplings in supergravity theory, Nucl.Phys. **B117**, p. 333, doi:10.1016/0550-3213(76)90401-6.

Ferrara, S., Grisaru, M. T. and van Nieuwenhuizen, P. (1978). Poincaré and conformal supergravity models with closed algebras, Nucl.Phys. **B138**, p. 430, doi:10.1016/0550-3213(78)90389-9.

Ferrara, S., Kaku, M., Townsend, P. and van Nieuwenhuizen, P. (1977b). Unified field theories with U(N) internal symmetries: Gauging the superconformal group, Nucl.Phys. **B129**, p. 125, doi:10.1016/0550-3213(77)90023-2.

Ferrara, S., Kallosh, R., Linde, A., Marrani, A. and Van Proeyen, A. (2010). Jordan frame supergravity and inflation in NMSSM, Phys. Rev. **D82**, p. 045003, doi:10.1103/PhysRevD.82.045003, `arXiv:1004.0712 [hep-th]`.

Ferrara, S., Kallosh, R., Linde, A., Marrani, A. and Van Proeyen, A. (2011). Superconformal symmetry, NMSSM, and inflation, Phys. Rev. **D83**, p. 025008, doi: 10.1103/PhysRevD.83.025008, `arXiv:1008.2942 [hep-th]`.

Ferrara, S. and Lledo, M. (2002). Considerations on superPoincaré algebras and their extensions to simple superalgebras, Rev.Math.Phys. **14**, pp. 519–530, doi:10.1142/S0129055X0200134X, `arXiv:hep-th/0112177 [hep-th]`.

Ferrara, S., Scherk, J. and van Nieuwenhuizen, P. (1976b). Locally supersymmetric Maxwell-Einstein theory, Phys.Rev.Lett. **37**, p. 1035, doi:10.1103/PhysRevLett.37.1035.

Ferrara, S. and van Nieuwenhuizen, P. (1978a). The auxiliary fields of supergravity, Phys.Lett. **B74**, p. 333, doi:10.1016/0370-2693(78)90670-6.

Ferrara, S. and van Nieuwenhuizen, P. (1978b). Tensor calculus for supergravity, Phys.Lett. **B76**, p. 404, doi:10.1016/0370-2693(78)90893-6.

Ferrara, S., Wess, J. and Zumino, B. (1974). Supergauge multiplets and superfields, Phys. Lett. **B51**, p. 239, doi:10.1016/0370-2693(74)90283-4.

Ferrara, S. and Zumino, B. (1974). Supergauge invariant Yang-Mills theories, Nucl. Phys. **B79**, p. 413, doi:10.1016/0550-3213(74)90559-8.

Ferrara, S. and Zumino, B. (1978). Structure of linearized supergravity and conformal supergravity, Nucl. Phys. **B134**, p. 301, doi:10.1016/0550-3213(78)90548-5.

Fonseca, R. M., Malinsky, M., Porod, W. and Staub, F. (2012). Running soft parameters in SUSY models with multiple U(1) gauge factors, Nucl. Phys. **B854**, pp. 28–53, doi:10.1016/j.nuclphysb.2011.08.017, `arXiv:1107.2670 [hep-ph]`.

Fradkin, E. and Tseytlin, A. A. (1985). Conformal supergravity, Phys.Rept. **119**, pp. 233–362, doi:10.1016/0370-1573(85)90138-3.

Frappat, L., Sciarrino, A. and Sorba, P. (2000). Dictionary on Lie Algebras and Superalgebras (San Diego, CA: Academic Press.).

Freedman, D. Z. (1977). Supergravity with axial gauge invariance, Phys.Rev. **D15**, p. 1173, doi:10.1103/PhysRevD.15.1173.

Freedman, D. Z. and Das, A. K. (1977). Gauge internal symmetry in extended supergravity, Nucl.Phys. **B120**, p. 221, doi:10.1016/0550-3213(77)90041-4.

Freedman, D. Z. and Schwarz, J. H. (1977). Unification of supergravity and Yang-Mills theory, Phys.Rev. **D15**, p. 1007, doi:10.1103/PhysRevD.15.1007.

Freedman, D. Z. and van Nieuwenhuizen, P. (1976). Properties of supergravity theory, Phys.Rev. **D14**, p. 912, doi:10.1103/PhysRevD.14.912.

Freedman, D. Z., van Nieuwenhuizen, P. and Ferrara, S. (1976). Progress toward a theory of supergravity, Phys. Rev. **D13**, pp. 3214–3218, doi:10.1103/PhysRevD.13.3214.

Freedman, D. Z. and Van Proeyen, A. (2012). Supergravity (Cambridge Univ. Press), doi: 10.1017/CBO9781139026833.

Freund, P. G. (1986). Introduction to Supersymmetry (Cambridge University Press).

Freund, P. G. O. and Kaplansky, I. (1976). Simple supersymmetries, J.Math.Phys. **17**, p. 228, doi:10.1063/1.522885.

Fuks, B. and Rausch de Traubenberg, M. (2011). Supersymétrie : exercices avec solutions (Ellipses), http://editions-ellipses.fr/supersymetrie-exercices-avec-solutions-p-7697.html.

Galperin, A., Ogievetsky, V. and Sokatchev, E. (1985). On matter couplings in N = 1 supergravities, Nucl.Phys. **B252**, p. 435, doi:10.1016/0550-3213(85)90456-0.

Gates, J., S. J., Stelle, K. and West, P. C. (1980). Algebraic origins of superspace constraints in supergravity, Nucl.Phys. **B169**, p. 347, doi:10.1016/0550-3213(80)90037-1.

Gates, J., S.J., Rocek, M. and Siegel, W. (1982). Solution to constraints for $n = 0$ supergravity, Nucl.Phys. **B198**, p. 113, doi:10.1016/0550-3213(82)90548-X.

Gates, S., Grisaru, M. T., Rocek, M. and Siegel, W. (1983). Superspace, or one thousand and one lessons in supersymmetry, Front.Phys. **58**, p. 1, `arXiv:hep-th/0108200 [hep-th]`.

Gates, S. J. and Siegel, W. (1980). Understanding constraints in superspace formulations of supergravity, Nucl.Phys. **B163**, p. 519, doi:10.1016/0550-3213(80)90414-9.

Giani, F. and Pernici, M. (1984). N=2 supergravity in ten dimensions, Phys.Rev. **D30**, pp. 325–333, doi:10.1103/PhysRevD.30.325.

Girardello, L. and Grisaru, M. T. (1982). Soft breaking of supersymmetry, Nucl.Phys. **B194**, p. 65, doi:10.1016/0550-3213(82)90512-0.

Girardi, G., Grimm, R., Muller, M. and Wess, J. (1984). Superspace geometry and the minimal, non minimal, and new minimal supergravity multiplets, Z.Phys. **C26**, p. 123, doi:10.1007/BF01572550.

Giudice, G. F. and Masiero, A. (1988). A natural solution to the $\mu-$problem in supergravity theories, Phys. Lett. **B206**, pp. 480–484, doi:10.1016/0370-2693(88)91613-9.

Giudice, G. F. and Rattazzi, R. (1999). Theories with gauge-mediated supersymmetry breaking, Phys. Rept. **322**, p. 419, doi:10.1016/S0370-1573(99)00042-3, `arXiv:hep-ph/9801271`.

Giunti, C., Kim, C. and Lee, U. (1991). Running coupling constants and grand unification models, Mod.Phys.Lett. **A6**, pp. 1745–1755, doi:10.1142/S0217732391001883.

Glashow, S. (1961). Partial symmetries of weak interactions, Nucl.Phys. **22**, pp. 579–588, doi:10.1016/0029-5582(61)90469-2.

Glashow, S., Iliopoulos, J. and Maiani, L. (1970). Weak interactions with lepton-hadron symmetry, Phys.Rev. **D2**, pp. 1285–1292, doi:10.1103/PhysRevD.2.1285.

Gliozzi, F., Scherk, J. and O., D. I. (1977). Supersymmetry, supergravity theories and the dual spinor model, Nucl.Phys. **B122**, pp. 253–290, doi:10.1016/0550-3213(77) 90206-1.

Goldberg, H. (1983). Constraint on the photino mass from cosmology, Phys.Rev.Lett. **50**, p. 1419, doi:10.1103/PhysRevLett.50.1419.

Golfand, Y. and Likhtman, E. (1971). Extension of the algebra of Poincaré group generators and violation of P invariance, JETP Lett. **13**, pp. 323–326.

Gomis, J., Paris, J. and Samuel, S. (1995). Antibracket, antifields and gauge-theory quantization, Phys.Rept. **259**, pp. 1–145, doi:10.1016/0370-1573(94)00112-G, `arXiv:hep-th/9412228 [hep-th]`.

Grimm, R. (2011). private communication.

Grimm, R., Muller, M. and Wess, J. (1984). Supersymmetric gauge theories coupled to nonminimal and new minimal supergravity, Z.Phys. **C26**, pp. 427–432, doi:10.1007/ BF01452570.

Grimm, R., Wess, J. and Zumino, B. (1978). Consistency checks on the superspace formulation of supergravity, Phys.Lett. **B73**, p. 415, doi:10.1016/0370-2693(78)90753-0.

Grimm, R., Wess, J. and Zumino, B. (1979). A complete solution of the Bianchi identities in superspace, Nucl.Phys. **B152**, p. 255, doi:10.1016/0550-3213(79)90102-0.

Grisaru, M. T., Rocek, M. and Siegel, W. (1980). Zero value for the three-loop β function in N=4 supersymmetric Yang-Mills theory, Phys.Rev.Lett. **45**, pp. 1063–1066, doi: 10.1103/PhysRevLett.45.1063.

Grisaru, T., M, Siegel, W. and Rocek, M. (1979). Improved methods for supergraphs, Nucl.Phys. **B159**, p. 429, doi:10.1016/0550-3213(79)90344-4.

Gross, D. and Wilczek, F. (1973). Asymptotically free gauge theories. 1, Phys.Rev. **D8**, pp. 3633–3652, doi:10.1103/PhysRevD.8.3633.

Gross, D. and Wilczek, F. (1974). Asymptotically free gauge theories. 2. Phys.Rev. **D9**, pp. 980–993, doi:10.1103/PhysRevD.9.980.

Gubser, S., Klebanov, I. R. and Polyakov, A. M. (1998). Gauge theory correlators from non-critical string theory, Phys.Lett. **B428**, pp. 105–114, doi:10.1016/S0370-2693(98) 00377-3, `arXiv:hep-th/9802109 [hep-th]`.

Haag, R., Lopuszanski, J. T. and Sohnius, M. (1975). All possible generators of supersymmetries of the S-matrix, Nucl.Phys. **B88**, p. 257, doi:10.1016/0550-3213(75) 90279-5.

Haber, H. E. and Kane, G. L. (1985). The search for supersymmetry: probing physics beyond the standard model, Phys.Rept. **117**, pp. 75–263, doi:10.1016/0370-1573(85) 90051-1.

Hall, L. J., Lykken, J. D. and Weinberg, S. (1983). Supergravity as the messenger of supersymmetry breaking, Phys. Rev. **D27**, pp. 2359–2378, doi:10.1103/PhysRevD.27. 2359.

Han, C., Hikasa, K.-i., Wu, L., Yang, J. M. and Zhang, Y. (2017). Status of cMSSM in light of current LHC Run-2 and LUX data, Phys. Lett. **B769**, pp. 470–476, doi: 10.1016/j.physletb.2017.04.026, `arXiv:1612.02296 [hep-ph]`.

Henneaux, M. (1990). Lectures on the antifield-BRST formalism for gauge theories, Nucl. Phys. Proc. Supl. **18A**, p. 47, doi:10.1016/0920-5632(90)90647-D.

Howe, P. S., Stelle, K. and Townsend, P. (1984). Miraculous ultraviolet cancellations in supersymmetry made manifest, Nucl.Phys. **B236**, p. 125, doi:10.1016/0550-3213(84)90528-5.

Howe, P. S. and Tucker, R. W. (1978). Scale invariance in superspace, Phys.Lett. **B80**, p. 138, doi:10.1016/0370-2693(78)90327-1.

Howe, P. S. and West, P. C. (1984). The complete N =2, D = 10 supergravity, Nucl.Phys. **B238**, p. 181, doi:10.1016/0550-3213(84)90472-3.

Hull, C., Karlhede, A., Lindstrom, U. and Rocek, M. (1986). Nonlinear sigma models and their gauging in and out of superspace, Nucl.Phys. **B266**, p. 1, doi:10.1016/0550-3213(86)90175-6.

Huq, M. and Namazie, M. (1985). Kaluza-Klein supergravity in ten dimensions, Class.Quant.Grav. **2**, p. 293, doi:10.1088/0264-9381/2/3/007.

Ibañez, L. E. (1982). Locally supersymmetric SU(5) grand unification, Phys. Lett. **B118**, p. 73, doi:10.1016/0370-2693(82)90604-9.

Ibanez, L. E. and Ross, G. G. (1981). Low-energy predictions in supersymmetric grand unified theories, Phys.Lett. **B105**, p. 439, doi:10.1016/0370-2693(81)91200-4.

Iliopoulos, J. and Zumino, B. (1974). Broken supergauge symmetry and renormalization, Nucl. Phys. **B76**, p. 310, doi:10.1016/0550-3213(74)90388-5.

Intriligator, K. A. and Seiberg, N. (2007). Lectures on supersymmetry breaking, Class.Quant.Grav. **24**, pp. S741–S772, doi:10.1088/0264-9381/24/21/S02, `arXiv:hep-ph/0702069 [hep-ph]`.

Jack, I., Jones, D. T., Martin, S. P., Vaughn, M. T. and Yamada, Y. (1994). Decoupling of the epsilon scalar mass in softly broken supersymmetry, Phys.Rev. **D50**, pp. 5481–5483, doi:10.1103/PhysRevD.50.R5481, `arXiv:hep-ph/9407291 [hep-ph]`.

Jones, D. (1977). Charge renormalization in a supersymmetric Yang-Mills theory, Phys.Lett. **B72**, p. 199, doi:10.1016/0370-2693(77)90701-8.

Kac, V. (1977a). Lie superalgebras, Adv.Math. **26**, pp. 8–96, doi:10.1016/0001-8708(77)90017-2.

Kac, V. (1977b). A sketch of Lie superalgebra theory, Commun.Math.Phys. **53**, pp. 31–64, doi:10.1007/BF01609166.

Kaku, M. and Townsend, P. (1978). Poincaré supergravity as broken superconformal gravity, Phys.Lett. **B76**, p. 54, doi:10.1016/0370-2693(78)90098-9.

Kaku, M., Townsend, P. and van Nieuwenhuizen, P. (1978). Properties of conformal supergravity, Phys.Rev. **D17**, p. 3179, doi:10.1103/PhysRevD.17.3179.

Kallosh, R., Kofman, L., Linde, A. D. and Van Proeyen, A. (2000). Superconformal symmetry, supergravity and cosmology, Class.Quant.Grav. **17**, pp. 4269–4338, doi:10.1088/0264-9381/17/20/308, `arXiv:hep-th/0006179 [hep-th]`.

Kane, G. L., Kolda, C. F., Roszkowski, L. and Wells, J. D. (1994). Study of constrained minimal supersymmetry, Phys. Rev. **D49**, pp. 6173–6210, doi:10.1103/PhysRevD.49.6173, `arXiv:hep-ph/9312272 [hep-ph]`.

Kaul, R. K. (1982). Gauge hierarchy in a supersymmetric model, Phys. Lett. **B109**, p. 19, doi:10.1016/0370-2693(82)90453-1.

Kaul, R. K. and Majumdar, P. (1982). Cancellation of quadratically divergent mass corrections in globally supersymmetric spontaneously broken gauge theories, Nucl. Phys. **B199**, p. 36, doi:10.1016/0550-3213(82)90565-X.

Kobayashi, S. and Nomizu, K. (1963). Foundations of Differential Geometry, vol II (John Wiley).

Kugo, T. and Uehara, S. (1983a). Conformal and Poincaré tensor calculi in N=1 supergravity, Nucl.Phys. **B226**, p. 49, doi:10.1016/0550-3213(83)90463-7.

Kugo, T. and Uehara, S. (1983b). Improved superconformal gauge conditions in the N = 1 supergravity Yang-Mills matter system, Nucl.Phys. **B222**, p. 125, doi:10.1016/0550-3213(83)90612-0.

Kugo, T. and Uehara, S. (1985). N= 1 superconformal tensor calculus multiplets with external Lorentz indices and derivative operations, Prog.Theor.Phys. **73**, p. 235, doi:10.1143/PTP.73.235.

Lahanas, A. and Nanopoulos, D. V. (1987). The road to no-scale supergravity, Phys.Rept. **145**, p. 1, doi:10.1016/0370-1573(87)90034-2.

Langacker, P. and Luo, M.-X. (1991). Implications of precision electroweak experiments for m_t, ρ_0, $\sin^2\theta_w$ and grand unification, Phys.Rev. **D44**, pp. 817–822, doi:10.1103/PhysRevD.44.817.

Lee, H. M. (2010). Chaotic inflation in jordan frame supergravity, JCAP **1008**, p. 003, doi:10.1088/1475-7516/2010/08/003, `arXiv:1005.2735 [hep-ph]`.

Maldacena, J. M. (1998). The large N limit of superconformal field theories and supergravity, Adv.Theor.Math.Phys. **2**, pp. 231–252, `arXiv:hep-th/9711200 [hep-th]`.

Maldacena, J. M. (2003). Tasi 2003 lectures on AdS/CFT, Progress in string theory. Proceedings, Summer School, TASI 2003 , pp. 155–203`arXiv:hep-th/0309246 [hep-th]`.

Mandelstam, S. (1983). Light-cone superspace and the ultraviolet finiteness of the N=4 model, Nucl.Phys. **B213**, pp. 149–168, doi:10.1016/0550-3213(83)90179-7.

Martin, S. and Vaughn, M. (1994). Two loop renormalization group equations for soft supersymmetry breaking couplings, Phys.Rev. **D50**, p. 2282, doi:10.1103/PhysRevD.50.2282,10.1103/PhysRevD.78.039903, `arXiv:hep-ph/9311340 [hep-ph]`.

Martin, S. P. (1997). A supersymmetry primer, [Adv. Ser. Direct. High Energy Phys.18,1 (1998)] , pp. 1–98doi:10.1142/9789812839657_0001,10.1142/9789814307505_0001, `arXiv:hep-ph/9709356 [hep-ph]`.

Martin, S. P. and Vaughn, M. T. (1993). Regularization dependence of running couplings in softly broken supersymmetry, Phys. Lett. **B318**, pp. 331–337, doi:10.1016/0370-2693(93)90136-6, `arXiv:hep-ph/9308222 [hep-ph]`.

Moultaka, G., Rausch de Traubenberg, M. and Tant, D. (2019a). Low energy supergravity revisited (i), Int. J. Mod. Phys. **A34**, 01, p. 1950004, doi:10.1142/S0217751X19500040, `arXiv:1611.10327 [hep-th]`.

Moultaka, G., Rausch de Traubenberg, M. and Venin, V. (2019b). In preparation.

Muller (1989). Consitent Supergravity Theories (Springer-Verlag).

Muller, M. (1982). The density multiplet in superspace, Z. Phys. **C16**, p. 41, doi:10.1007/BF01573745.

Nahm, W. (1978). Supersymmetries and their representations, Nucl.Phys. **B135**, p. 149, doi:10.1016/0550-3213(78)90218-3.

Nakahara, M. (2003). Geometry, Topology and Pysics (Taylor & Francis group).

Nappi, C. R. and Ovrut, B. A. (1982). Supersymmetric extension of the SU(3)×SU(2)×U(1) model, Phys. Lett. **B113**, p. 175, doi:10.1016/0370-2693(82)90418-X.

Nash, C. and Sen, S. (1983). Topology and Geometry for Physicists (Academic Press).

Nath, P. (2016). Supersymmetry, Supergravity, and Unification, Cambridge Monographs on Mathematical Physics (Cambridge University Press), doi:10.1017/9781139048118, http://www.cambridge.org/academic/subjects/physics/theoretical-physics-and-mathematical-physics/supersymmetry-supergravity-and-unification?format=HB&isbn=9780521197021.

Nath, P. and Arnowitt, R. L. (1975). Generalized super-gauge symmetry as a new framework for unified gauge theories, Phys.Lett. **B56**, p. 177, doi:10.1016/0370-2693(75)90297-X.

Nath, P. and Arnowitt, R. L. (1976). Supergravity and gauge supersymmetry, Phys.Lett. **B65**, p. 73, doi:10.1016/0370-2693(76)90537-2.

Ne'eman, Y. and Regge, T. (1978). Gravity and supergravity as gauge theories on a group manifold, Phys.Lett. **B74**, p. 54, doi:10.1016/0370-2693(78)90058-8.

Nilles, H. P. (1983). Supergravity generates hierarchies, Nucl. Phys. **B217**, p. 366, doi:10.1016/0550-3213(83)90152-9.

Nilles, H. P. (1984). Supersymmetry, supergravity and particle physics, Phys. Rept. **110**, p. 1, doi:10.1016/0370-1573(84)90008-5.

Nilles, H. P., Srednicki, M. and Wyler, D. (1983). Weak interaction breakdown induced by supergravity, Phys. Lett. **B120**, p. 346, doi:10.1016/0370-2693(83)90460-4.

Ogievetsky, V. and Sokatchev, E. (1978). Structure of supergravity group, Phys.Lett. **B79**, p. 222, doi:10.1016/0370-2693(78)90228-9.

Ogievetsky, V. and Sokatchev, E. (1980a). Equation of motion for the axial gravitational superfield, Sov.J.Nucl.Phys. **32**, p. 589.

Ogievetsky, V. and Sokatchev, E. (1980b). The gravitational axial superfield and the formalism of differential geometry, Sov.J.Nucl.Phys. **31**, p. 424.

Ogievetsky, V. and Sokatchev, E. (1980c). The simplest group of Einstein supergravity, Sov.J.Nucl.Phys. **31**, p. 140.

Ogievetsky, V. and Sokatchev, E. (1980d). Torsion and curvature in terms of the axial gravitational superfield, Sov.J.Nucl.Phys. **32**, p. 447.

Ohta, N. (1983). Grand unified theories based on local supersymmetry, Prog. Theor. Phys. **70**, p. 542, doi:10.1143/PTP.70.542.

O'Raifeartaigh, L. (1975). Spontaneous symmetry breaking for chirals scalar superfields, Nucl. Phys. **B96**, p. 331, doi:10.1016/0550-3213(75)90585-4.

Parker, M. (1980). Classification of real simple Lie superalgebras of classical type, J.Math.Phys. **21**, pp. 689–697, doi:10.1063/1.524487.

Penrose, R. (1960). A spinor approach to general relativity, Annals Phys. **10**, pp. 171–201, doi:10.1016/0003-4916(60)90021-X.

Poggio, E. C. and Pendleton, H. (1977). Vanishing of charge renormalization and anomalies in a supersymmetric gauge theory, Phys.Lett. **B72**, p. 200, doi:10.1016/0370-2693(77)90702-X.

Politzer, H. D. (1974). Asymptotic freedom: an approach to strong interactions, Phys.Rept. **14**, pp. 129–180, doi:10.1016/0370-1573(74)90014-3.

Polonyi, J. (1977). Generalization of the massive scalar multiplet coupling to the supergravity, Hungary Central Inst Res - KFKI-77-93 (unpublished) .

Randall, L. and Sundrum, R. (1999). Out of this world supersymmetry breaking, Nucl.Phys. **B557**, pp. 79–118, doi:10.1016/S0550-3213(99)00359-4.

Rarita, W. and Schwinger, J. (1941). On a theory of particles with half-integral spin, Phys.Rev. **60**, p. 61, doi:10.1103/PhysRev.60.61.

Sakai, N. (1981). Naturalnes in supersymmetric GUTS, Zeit. Phys. **C11**, p. 153, doi:10.1007/BF01573998.

Salam, A. and Strathdee, J. (1974a). On goldstone fermions, Phys.Lett. **B49**, pp. 465–467, doi:10.1016/0370-2693(74)90637-6.

Salam, A. and Strathdee, J. A. (1974b). Super-gauge transformations, Nucl. Phys. **B76**, p. 477, doi:10.1016/0550-3213(74)90537-9.

Salam, A. and Strathdee, J. A. (1975). Superfields and Fermi-Bose symmetry, Phys. Rev. **D11**, p. 1521, doi:10.1103/PhysRevD.11.1521.

Salam, A. and Strathdee, J. A. (1978). Supersymmetry and superfields, Fortschr. Phys. **26**, p. 57, doi:10.1002/prop.19780260202.

Salam, A. and Ward, J. C. (1964). Electromagnetic and weak interactions, Phys.Lett. **13**, pp. 168–171, doi:10.1016/0031-9163(64)90711-5.

Scheunert, M., Nahm, W. and Rittenberg, V. (1976a). Classification of all simple graded Lie algebras whose Lie algebra is reductive. i, J.Math.Phys. **17**, p. 1626, doi:10.1063/1.523108.

Scheunert, M., Nahm, W. and Rittenberg, V. (1976b). Classification of all simple graded Lie algebras whose Lie algebra is reductive. II. construction of the exceptional algebras, J.Math.Phys. **17**, p. 1640, doi:10.1063/1.523109.

Schwarz, J. H. (1983). Covariant field equations of chiral N = 2, D = 10 supergravity, Nucl.Phys. **B226**, p. 269, doi:10.1016/0550-3213(83)90192-X.

Schwarz, J. H. and West, P. C. (1983). Symmetries and transformations of chiral N = 2, D = 10 supergravity, Phys.Lett. **B126**, p. 301, doi:10.1016/0370-2693(83)90168-5.

Seiberg, N. (1993). Naturalness versus supersymmetric non-renormalization theorems, Phys.Lett. **B318**, pp. 469–475, doi:10.1016/0370-2693(93)91541-T, `arXiv:hep-ph/9309335 [hep-ph]`.

Siegel, W. (1978). Solution to constraints in Wess-Zumino supergravity formalism, Nucl.Phys. **B142**, p. 301, doi:10.1016/0550-3213(78)90205-5.

Siegel, W. (1979a). Superconformal invariance of superspace with nonminimal auxiliary fields, Phys.Lett. **B80**, p. 224, doi:10.1016/0370-2693(79)90203-X.

Siegel, W. (1979b). Supergravity supergraphs, Phys.Lett. **B84**, p. 197, doi:10.1016/0370-2693(79)90283-1.

Siegel, W. (1979c). Supersymmetric dimensional regularization via dimensional reduction, Phys. Lett. **84B**, pp. 193–196, doi:10.1016/0370-2693(79)90282-X.

Siegel, W. and Gates, J., S. J. (1979). Superfield supergravity, Nucl.Phys. **B147**, p. 77, doi:10.1016/0550-3213(79)90416-4.

Sohnius, M. F. (1983). Soft gauge algebras, Z.Phys. **C18**, p. 229, doi:10.1007/BF01571364.

Sohnius, M. F. and West, P. C. (1981a). An alternative minimal off-shell version of N = 1 supergravity, Phys.Lett. **B105**, p. 353, doi:10.1016/0370-2693(81)90778-4.

Sohnius, M. F. and West, P. C. (1981b). Conformal invariance in N = 4 supersymmetric Yang-Mills theory, Phys.Lett. **B100**, p. 245, doi:10.1016/0370-2693(81)90326-9.

Soni, S. K. and Weldon, H. A. (1983). Analysis of the supersymmetry breaking induced by N=1 supergravity theories, Phys. Lett. **126B**, pp. 215–219, doi:10.1016/0370-2693(83)90593-2.

Stelle, K. and West, P. C. (1978a). Relation between vector and scalar multiplets and gauge invariance in supergravity, Nucl.Phys. **B145**, p. 175, doi:10.1016/0550-3213(78) 90420-0.

Stelle, K. and West, P. C. (1978b). Tensor calculus for the vector multiplet coupled to supergravity, Phys.Lett. **B77**, p. 376, doi:10.1016/0370-2693(78)90581-6.

Stelle, K. S. and West, P. C. (1978c). Minimal auxiliary fields for supergravity, Phys. Lett. **B74**, p. 330, doi:10.1016/0370-2693(78)90669-X.

Strathdee, J. (1987). Extended Poincaré supersymmetry, Int.J.Mod.Phys. **A2**, p. 273, doi: 10.1142/S0217751X87000120.

Stueckelberg, E. (1938a). Interaction forces in electrodynamics and in the field theory of nuclear forces. i, Helv. Phys. Acta **11**, p. 225.

Stueckelberg, E. (1938b). Interaction forces in electrodynamics and in the field theory of nuclear forces. ii, Helv. Phys. Acta **11**, p. 299.

Stueckelberg, E. (1938c). Interaction forces in electrodynamics and in the field theory of nuclear forces. iii, Helv. Phys. Acta **11**, p. 312.

't Hooft, G. and Veltman, M. J. G. (1972). Regularization and renormalization of gauge fields, Nucl. Phys. **B44**, pp. 189–213, doi:10.1016/0550-3213(72)90279-9.

Tanabashi, M., Hagiwara, K., Hikasa, K., Nakamura, K., Sumino, Y., Takahashi, F., Tanaka, J., Agashe, K., Aielli, G., Amsler, C., Antonelli, M., Asner, D. M., Baer, H., Banerjee, S., Barnett, R. M., Basaglia, T., Bauer, C. W., Beatty, J. J., Belousov, V. I., Beringer, J., Bethke, S., Bettini, A., Bichsel, H., Biebel, O., Black, K. M., Blucher, E., Buchmuller, O., Burkert, V., Bychkov, M. A., Cahn, R. N., Carena, M., Ceccucci, A., Cerri, A., Chakraborty, D., Chen, M.-C., Chivukula, R. S., Cowan, G., Dahl, O., D'Ambrosio, G., Damour, T., de Florian, D., de Gouvêa, A., DeGrand, T., de Jong, P., Dissertori, G., Dobrescu, B. A., D'Onofrio, M., Doser, M., Drees, M., Dreiner, H. K., Dwyer, D. A., Eerola, P., Eidelman, S., Ellis, J., Erler, J., Ezhela, V. V., Fetscher, W., Fields, B. D., Firestone, R., Foster, B., Freitas, A., Gallagher, H., Garren, L., Gerber, H.-J., Gerbier, G., Gershon, T., Gershtein, Y., Gherghetta, T., Godizov, A. A., Goodman, M., Grab, C., Gritsan, A. V., Grojean, C., Groom, D. E., Grünewald, M., Gurtu, A., Gutsche, T., Haber, H. E., Hanhart, C., Hashimoto, S., Hayato, Y., Hayes, K. G., Hebecker, A., Heinemeyer, S., Heltsley, B., Hernández-Rey, J. J., Hisano, J., Höcker, A., Holder, J., Holtkamp, A., Hyodo, T., Irwin, K. D., Johnson, K. F., Kado, M., Karliner, M., Katz, U. F., Klein, S. R., Klempt, E., Kowalewski, R. V., Krauss, F., Kreps, M., Krusche, B., Kuyanov, Y. V., Kwon, Y., Lahav, O., Laiho, J., Lesgourgues, J., Liddle, A., Ligeti, Z., Lin, C.-J., Lippmann, C., Liss, T. M., Littenberg, L., Lugovsky, K. S., Lugovsky, S. B., Lusiani, A., Makida, Y., Maltoni, F., Mannel, T., Manohar, A. V., Marciano, W. J., Martin, A. D., Masoni, A., Matthews, J., Meißner, U.-G., Milstead, D., Mitchell, R. E., Mönig, K., Molaro, P., Moortgat, F., Moskovic, M., Murayama, H., Narain, M., Nason, P., Navas, S., Neubert, M., Nevski, P., Nir, Y., Olive, K. A., Pagan Griso, S., Parsons, J., Patrignani, C., Peacock, J. A., Pennington, M., Petcov, S. T., Petrov, V. A., Pianori, E., Piepke, A., Pomarol, A., Quadt, A., Rademacker, J., Raffelt, G., Ratcliff, B. N., Richardson, P., Ringwald, A., Roesler, S., Rolli, S., Romaniouk, A., Rosenberg, L. J., Rosner, J. L., Rybka, G., Ryutin, R. A., Sachrajda, C. T., Sakai, Y., Salam, G. P., Sarkar, S., Sauli, F., Schneider, O., Scholberg, K., Schwartz, A. J., Scott, D., Sharma, V., Sharpe, S. R., Shutt, T., Silari, M., Sjöstrand, T., Skands, P., Skwarnicki, T., Smith, J. G., Smoot, G. F., Spanier, S., Spieler, H., Spiering,

C., Stahl, A., Stone, S. L., Sumiyoshi, T., Syphers, M. J., Terashi, K., Terning, J., Thoma, U., Thorne, R. S., Tiator, L., Titov, M., Tkachenko, N. P., Törnqvist, N. A., Tovey, D. R., Valencia, G., Van de Water, R., Varelas, N., Venanzoni, G., Verde, L., Vincter, M. G., Vogel, P., Vogt, A., Wakely, S. P., Walkowiak, W., Walter, C. W., Wands, D., Ward, D. R., Wascko, M. O., Weiglein, G., Weinberg, D. H., Weinberg, E. J., White, M., Wiencke, L. R., Willocq, S., Wohl, C. G., Womersley, J., Woody, C. L., Workman, R. L., Yao, W.-M., Zeller, G. P., Zenin, O. V., Zhu, R.-Y., Zhu, S.-L., Zimmermann, F., Zyla, P. A., Anderson, J., Fuller, L., Lugovsky, V. S. and Schaffner, P. (2018). Review of particle physics, Phys. Rev. D **98**, p. 030001, doi:10.1103/PhysRevD.98.030001, `https://link.aps.org/doi/10.1103/PhysRevD.98.030001`.

Terning, J. (2006). Modern Supersymmetry: Dynamics and Duality (International Series of Monographs on Physics 132. Oxford: Oxford University Press.).

Townsend, P. (1977). Cosmological constant in supergravity, Phys.Rev. **D15**, pp. 2802–2804, doi:10.1103/PhysRevD.15.2802.

Townsend, P. and van Nieuwenhuizen, P. (1979). Simplifications of conformal supergravity, Phys.Rev. **D19**, p. 3166, doi:10.1103/PhysRevD.19.3166.

van Nieuwenhuizen, P. (1981). Supergravity, Phys. Rept. **68**, p. 189, doi:10.1016/0370-1573(81)90157-5.

Volkov, D. and Akulov, V. (1973). Is the neutrino a goldstone particle? Phys.Lett. **B46**, pp. 109–110, doi:10.1016/0370-2693(73)90490-5.

Weinberg, S. (1967). A model of leptons, Phys.Rev.Lett. **19**, pp. 1264–1266, doi:10.1103/PhysRevLett.19.1264.

Weinberg, S. (1972a). Gravitation and Cosmology (John Wiley & Sons, New-York).

Weinberg, S. (1972b). Mixing angle in renormalizable theories of weak and electromagnetic interactions, Phys.Rev. **D5**, pp. 1962–1967, doi:10.1103/PhysRevD.5.1962.

Weinberg, S. (1995). The Quantum theory of fields. Vol. 1: Foundations (Cambridge University Press).

Weinberg, S. (2000). The Quantum Theory of Fields, Vol 3 (Cambridge University Press), ISBN 0-521-55001-7.

Wess, J. and Bagger, J. (1992). Supersymmetry and Supergravity, 2nd edn. (Princeton University Press), ISBN 0-691-08556-0; 0-691-02530-4.

Wess, J. and Zumino, B. (1974a). A Lagrangian model invariant under supergauge transformations, Phys. Lett. **B49**, p. 52, doi:10.1016/0370-2693(74)90578-4.

Wess, J. and Zumino, B. (1974b). Supergauge invariant extension of quantum electrodynamics, Nucl. Phys. **B78**, p. 1, doi:10.1016/0550-3213(74)90112-6.

Wess, J. and Zumino, B. (1974c). Supergauge transformations in four dimensions, Nucl. Phys. **B70**, p. 39, doi:10.1016/0550-3213(74)90355-1.

Wess, J. and Zumino, B. (1977). Superspace formulation of supergravity, Phys. Lett. **B66**, p. 361, doi:10.1016/0370-2693(77)90015-6.

Wess, J. and Zumino, B. (1978a). The component formalism follows from the superspace formulation of supergravity, Phys.Lett. **B79**, p. 394, doi:10.1016/0370-2693(78)90390-8.

Wess, J. and Zumino, B. (1978b). Superfield Lagrangian for supergravity, Phys.Lett. **B74**, p. 51, doi:10.1016/0370-2693(78)90057-6.

West, P. C. (1986). Introduction to Supersymmetry and Supergravity (Singapore, Singapore: World Scientific).

White, P. (1992). Analysis of the superconformal cohomology structure of N=4 super Yang-Mills, Class.Quant.Grav. **9**, pp. 413–444, doi:10.1088/0264-9381/9/2/009.

Wigner, E. P. (1939). On unitary representations of the inhomogeneous Lorentz group, Annals Math. **40**, pp. 149–204.

Witten, E. (1981a). Dynamical breaking of supersymmetry, Nucl. Phys. **B188**, p. 513, doi: 10.1016/0550-3213(81)90006-7.

Witten, E. (1981b). Lecture notes on supersymmetry, Lecture given at ICTP, Trieste .

Witten, E. (1998). Anti de Sitter space and holography, Adv.Theor.Math.Phys. **2**, pp. 253–291, `arXiv:hep-th/9802150 [hep-th]`.

Witten, E. and Bagger, J. (1982). Quantization of Newton's constant in certain supergravity theories, Phys.Lett. **B115**, p. 202, doi:10.1016/0370-2693(82)90644-X.

Yamada, Y. (1993). Two loop renormalization of gaugino mass in supersymmetric gauge model, Phys.Lett. **B316**, pp. 109–111, doi:10.1016/0370-2693(93)90665-5, `arXiv:hep-ph/9307217 [hep-ph]`.

Yamada, Y. (1994a). Two loop renormalization group equations for soft SUSY breaking scalar interactions: Supergraph method, Phys.Rev. **D50**, pp. 3537–3545, doi: 10.1103/PhysRevD.50.3537, `arXiv:hep-ph/9401241 [hep-ph]`.

Yamada, Y. (1994b). Two loop renormalization of gaugino masses in general supersymmetric gauge models, Phys.Rev.Lett. **72**, pp. 25–27, doi:10.1103/PhysRevLett.72.25, `arXiv:hep-ph/9308304 [hep-ph]`.

Yang, C.-N. and Mills, R. L. (1954). Conservation of isotopic spin and isotopic gauge invariance, Phys.Rev. **96**, pp. 191–195, doi:10.1103/PhysRev.96.191.

Zumino, B. (1978). Supergravity and superspace, In Cargese 1978, Proceedings, Recent Developments In Gravitation, 405-459 .

Zumino, B. (1979). Supersymmetry and Kähler manifolds, Phys. Lett. **B87**, p. 203, doi: 10.1016/0370-2693(79)90964-X.

Index

Printed in the United States
By Bookmasters